COMPUTER AND COMPUTING TECHNOLOGIES IN AGRICULTURE, VOLUME I

IFIP – The International Federation for Information Processing

IFIP was founded in 1960 under the auspices of UNESCO, following the First World Computer Congress held in Paris the previous year. An umbrella organization for societies working in information processing, IFIP's aim is two-fold: to support information processing within its member countries and to encourage technology transfer to developing nations. As its mission statement clearly states,

> *IFIP's mission is to be the leading, truly international, apolitical organization which encourages and assists in the development, exploitation and application of information technology for the benefit of all people.*

IFIP is a non-profitmaking organization, run almost solely by 2500 volunteers. It operates through a number of technical committees, which organize events and publications. IFIP's events range from an international congress to local seminars, but the most important are:

- The IFIP World Computer Congress, held every second year;
- Open conferences;
- Working conferences.

The flagship event is the IFIP World Computer Congress, at which both invited and contributed papers are presented. Contributed papers are rigorously refereed and the rejection rate is high.

As with the Congress, participation in the open conferences is open to all and papers may be invited or submitted. Again, submitted papers are stringently refereed.

The working conferences are structured differently. They are usually run by a working group and attendance is small and by invitation only. Their purpose is to create an atmosphere conducive to innovation and development. Refereeing is less rigorous and papers are subjected to extensive group discussion.

Publications arising from IFIP events vary. The papers presented at the IFIP World Computer Congress and at open conferences are published as conference proceedings, while the results of the working conferences are often published as collections of selected and edited papers.

Any national society whose primary activity is in information may apply to become a full member of IFIP, although full membership is restricted to one society per country. Full members are entitled to vote at the annual General Assembly, National societies preferring a less committed involvement may apply for associate or corresponding membership. Associate members enjoy the same benefits as full members, but without voting rights. Corresponding members are not represented in IFIP bodies. Affiliated membership is open to non-national societies, and individual and honorary membership schemes are also offered.

COMPUTER AND COMPUTING TECHNOLOGIES IN AGRICULTURE, VOLUME I

First IFIP TC 12 International Conference on Computer and Computing Technologies in Agriculture (CCTA 2007), Wuyishan, China, August 18-20, 2007

Edited by

DAOLIANG LI
China Agricultural University

Computer and Computing Technologies in Agriculture, Vol. 1

Edited by Daoliang Li

p. cm. (IFIP International Federation for Information Processing, a Springer Series in Computer Science)

ISSN: 1571-5736 / 1861-2288 (Internet)

ISBN: 978-1-4419-4583-9 e-ISBN: 978-0-387-77251-6

Printed on acid-free paper

9 8 7 6 5 4 3 2 1

springer.com

CONTENTS

FOREWORD

The papers in this volume comprise the refereed proceedings of the the First International Conference on Computer and Computing Technologies in Agriculture (CCTA 2007), in Wuyishan, China, 2007.

This conference is organized by China Agricultural University, Chinese Society of Agricultural Engineering and the Beijing Society for Information Technology in Agriculture. The purpose of this conference is to facilitate the communication and cooperation between institutions and researchers on theories, methods and implementation of computer science and information technology. By researching information technology development and the resources integration in rural areas in China, an innovative and effective approach is expected to be explored to promote the technology application to the development of modern agriculture and contribute to the construction of new countryside.

The rapid development of information technology has induced substantial changes and impact on the development of China's rural areas. Western thoughts have exerted great impact on studies of Chinese information technology development and it helps more Chinese and western scholars to expand their studies in this academic and application area. Thus, this conference, with works by many prominent scholars, has covered computer science and technology and information development in China's rural areas; and probed into all the important issues and the newest research topics, such as Agricultural Decision Support System and Expert System, GIS, GPS, RS and Precision Farming, CT applications in Rural Area, Agricultural System Simulation, Evolutionary Computing, etc. In the following sessions, this conference could hopefully set up several meeting rooms to provide an opportunity and arena for discussing these issues and exchanging ideas more effectively. We are also expecting to communicate and have dialogues on certain hot topics with some foreign scholars.

With the support of participants and hard working of preparatory committee, the conference achieved great success on the participation. We received around 427 submitted papers and 180 accepted papers will be published in the Springer Press. It has evidenced our remarkable achievements made in our studies of the New Period, forming a necessary step in the development of the research theory in China and a worthy legacy for information technology studies in the new century. The conference is planned to be organized annually. We believe that it can provide a platform for exchanging ideas and sharing outcomes and also contribute to China's agricultural development.

Finally, I would like to extend the most earnest gratitude to our co-sponsors, Chinese Society of Agricultural Engineering and the Beijing Society for Information Technology in Agriculture, also to Nongdaxingtong Technology Ltd., all members and colleagues of our preparatory committee, for their generous efforts, hard work and precious time! On behalf of all conference committee members and participants, I also would like to express our genuine appreciation to Fujian Provincial Agriculture Department, Nanping City Bureau of Agriculture and Wuyishan City government. Without their support, we can not meet in such a beautiful city.

This is the first in a new series of conferences dedicated to real-world applications of computer and computing technologies in agriculture around the world. The wide range and importance of these applications are clearly indicated by the papers in this volume. Both are likely to increase still further as time goes by and we intend to reflect these developments in our future conferences.

Daoliang Li

Chair of programme committee, organizing committee

ORGANIZING COMMITTEE

Chair

Prof. Daoliang Li, China Agricultural University, China
Director of EU-China Center for Information & Communication technologies in Agriculture

Members [in alpha order]

Mr. Chunjiang Zhao, Director, Beijing Agricultural Informatization Academy, China

Mr. Ju Ming, Vice Section Chief, Foundation Section, Science & Technology Department, Ministry of Education of the People's Republic of China

Prof. Haijian Ye, China Agricultural University, China

Prof. Jinguang Qin, Chinese Society of Agricultural Engineering, China

Prof. Qingshui Liu, China Agricultural University, China

Prof. Rengang Yang, China Agricultural University, China

Prof. Renjie Dong, China Agricultural University, China

Prof. Songhuai Du, China Agricultural University, China

Prof. Wanlin Gao, China Agricultural University, China

Prof. Weizhe Feng, China Agricultural University, China

PROGRAM COMMITTEE

Chair

Prof. Daoliang Li, China Agricultural University, China
Director of EU-China Center for Information & Communication technologies in Agriculture

Members [in alpha order]

Dr. Alex Abramovich, Maverick Defense Technologies Ltd., Israel
Dr. Boonyong Lohwongwatana, Asian Society for Environmental Protection (ASEP), Thailand
Dr. Feng Liu, Mercer University, GA, USA
Dr. Haresh A. Suthar, Industry (Masibus Automation & Instrumentation (p)) Ltd., India
Dr. Javad Khazaei, University of Tehran, Iranian
Dr. Jinsheng Ni, Beijing Oriental TITAN Technology. Co. Ltd.
Dr. Joanna Kulczycka, Polish Academy of Sciences Mineral and Energy Economy Research Institute, Poland
Dr. John Martin, University of Plymouth, Plymouth, UK
Dr. Kostas Komnitsas, Technical University of Crete, Greece
Dr. M. Anjaneya Prasad, College of Engg. Osmania University, India
Dr. Pralay Pal, Engineering Automation Deputy General Manager (DGM), India
Dr. Shi Zhou, Zhejiang University, China
Dr. Sijal Aziz, Executive Director WELDO, Pakistan
Dr. Soo Kar Leow, Monash University, Malaysia
Dr. Wenjiang Huang, National Engineering Research Center for Information Technology in Agriculture, China
Dr. Yong Yue, University of Bedfordshire, UK
Dr. Yuanzhu Zhang, Suzhou Center of Aquatic Animals Diseases Control, China
Mr. Weiping Song, DABEINONG Group
Prof. A.B. Sideridis, Informatics Laboratory of the Agricultural University of Athens, Greece
Prof. Andrew Hursthouse, University of Paisley, UK

Prof. Apostolos Sarris, Institute for Mediterranean Studies, Greece
Prof. Chunjiang Zhao, China National Engineering Center for Information Technology in Agriculture, China
Prof. Dehai Zhu, China Agricultural University, China
Prof. Fangquan Mei, Institute of Information, China Agricultural Science, China
Prof. Gang Liu, China Agricultural University, China
Prof. Guohui Gan, Institute of Geographic Sciences and Natural Resources
Prof. Guomin Zhou, Institute of Information, China Agricultural Science, China
Prof. Iain Muse, Development into Community Cooperation Policies and International Research Areas, Belgium
Prof. Jacques Ajenstat, University of Quebec at Montreal, Canada
Prof. K.C. Ting, Department of Agricultural and Biological Engineering, University of Illinois at Urbana-Champaign
Prof. Kostas Fytas, Laval University, Canada
Prof. Liangyu Chen, Countryside Center, Ministry of Science & Technology, China
Prof. Linnan Yang, Yunnan Agricultural University, China
Prof. Liyuan He, Huazhong Agricultural University (HZAU), China
Prof. Maohua Wang, member of Chinese Academy of Engineering, China Agricultural University, China
Prof. Maria-Ioanna Salaha, Wine Institute-National Agricultural Research Foundation, Greece
Prof. Michael Petrakis, National Observatory of Athens, Greece
Prof. Michele Genovese, Unit Specific International Cooperation Activities, International Cooperation Directorate, DG Research, UK
Prof. Minzan Li, China Agricultural University, China
Prof. Nigel Hall, Harper Adams University College, England
Prof. Raphael Linker, Civil and Environmental Engineering Dept., Technion
Prof. Rohani J. Widodo, Maranatha Christian University, Indonesia
Prof. Xiwen Luo, South China Agricultural University, China
Prof. Yanqing Duan, University of Bedfordshire, UK
Prof. Yeping Zhu, Institute of Information, China Agricultural Science, China
Prof. Yiming Wang, China Agricultural University, China
Prof. Yu Fang, Information Center, Ministry of Agriculture, China
Prof. Yuguo Kang, Chinese cotton association, China
Prof. Zetian Fu, China Agricultural University, China
Prof. Zuoyu Guo, Information Center, Ministry of Agriculture, China

SECRETARIAT

Secretary-general

Baoji Wang (China Agricultural University, China)
Liwei Zhang (China Agricultural University, China)

Secretaries

Xiuna Zhu (China Agricultural University, China)
Yanjun Zhang (China Agricultural University, China)
Xiang Zhu (China Agricultural University, China)
Liying Xu (China Agricultural University, China)
Bin Xing (China Agricultural University, China)
Xin Qiang (China Agricultural University, China)
Yingyi Chen (China Agricultural University, China)
Chengxian Yu (China Agricultural University, China)
Jie Yang (China Agricultural University, China)
Jing Du (China Agricultural University, China)

MEASUREMENT AND PREDICTION OF STRESS-STRAIN FOR EXTRUDED OILSEED USING NEURAL NETWORKS UNDER UNIAXIAL COLD PRESSING

Xiao Zheng [1,*], Guoxiang Lin [1], Dongping He [2], Jingzhou Wang [1]

[1] *Department of Mechanical Engineering, Wuhan Polytechnic University, Wuhan, Hubei Province, P. R. China, 430023*

[2] *Department of Food Science and Engineering, Wuhan Polytechnic University, Wuhan, Hubei Province, P. R. China, 430023*

* *Corresponding author, Address: Department of Mechanical Engineering, Wuhan Polytechnic University, Wuhan 430023, Hubei Province, P. R. China, Tel: +86-27-83956425, Fax: +86-27-83956425, Email: zhengxiao580405@163.com*

Abstract: A visualization of testing apparatus was developed to measure property of oilseeds relevant to physical mechanics during mechanical pressing for oil extraction. Stress-strain relationships were measured for extruded peanut, soybean, sesame and linseed compressed at thirteen pressures under uniaxial cold pressing. The prediction model of the stress-strain relationship was developed based on BP neural network. Results indicated that the stress-strain relationships were nonlinear. Over 50% strains for extruded soybean, sesame and linseed occurred at stress below 20MPa. Over 60% strain for extruded peanut occurred at stress below 10MPa. No more than 13% strain occurred at stress over 20MPa for extruded soybean sesame and linseed, and no more than 13% strain occurred at stress over 10MPa for extruded peanut. The maximum error between prediction and measurement for the stress-strain relationship was less than 0.0084 and the maximum training times was less than 88.

Keywords: measurement, prediction, stress-strain, neural networks, oilseed, cold pressing

1. INTRODUCTION

Peanut, soybean and sesame oil are important edible oil in the world. The mechanical pressing is the most common method for oil extraction in the world.

Zheng, X., Lin, G., He, D. and Wang, J., 2008, in IFIP International Federation for Information Processing, Volume 258; Computer and Computing Technologies in Agriculture, Vol. 1; Daoliang Li; (Boston: Springer), pp. 1–10.

Vegetable oilseed expresses complex mechanics behavior during pressing (Mrema et al., 1985). The conventional method of oil extraction suggests that oilseeds must be thermal (cooking) pretreatments before pressing, which is called the hot pressing (Rasehom et al., 2000; Bargale et al., 1999, 2000). More recently, the cold pressing for oil extraction, which needn't be cooked prior to pressing, is very popular in China as well as in other many countries. The main reason for popularity of the cold pressing is that the cold pressings yields limpid color and fruity oilseed oil with lower phosphorus and fatty acid (Rasehom et al., 2000; Zheng Xiao et al., 2004). However, compared to the hot pressing, the cold pressing is inefficient with lower throughputs and higher residual oil contents in the defiled cake. It indicates that the oil press used to the hot pressing needs further improve for the purpose of the cold pressing (Rasehom et al., 2000; Zheng Xiao et al., 2004).

The stress-strain relationship is the most important performance of physical mechanics for extruded oilseeds. The stress-strain model for extruded oilseeds by cold pressing is essential to rigorous theoretical analysis of mechanisms and physical processes. It lays a foundation for problems of permeability, differential equation for seepage (Zheng Xiao et al., 2004). Davison et al. (1975, 1979) have studied mechanical properties of single rapeseeds. Sukumaran et al. (1989) have studied bulk properties of rapeseeds under compression. However, the research relating to stress-strain relationship for extruding oilseed has not yet been reported in the world up till now except research for rapeseed and dehulled rapeseed (Zheng Xiao et al., 2004).

It is found very difficult to develop the theoretical model for stress-strain relationship for oilseeds due to the complexity of physical mechanics performance during pressing (Zheng Xiao et al., 2004). At present, multivariable nonlinear regression analysis is most common method to develop empirical formula to predict stress-strain relationship for complex material. However, the difference in the variable used in the analytical model and the details of the experiment will lead to significant diversity in the calculation formulas, and furthermore there is usual a difficulty to determine suitable regression equation used in multiple regression analysis, which requires considerable technique and experience due to understanding of the data characteristic of stress-strain experiment. The objectives of this study were to measure stress-strain for extruding oilseeds by uniaxial cold pressing, and develop neural network modeling to predict the stress-strain relationship.

2. STRESS-STRAIN EXPERIMENT

2.1 Design of visualized compression cell

A visualization of test apparatus used for the experiment was specially designed. Its schematic diagram is shown in Fig. 1. It mainly consists of a loading piston, an outer cylinder, an inner cylinder, a sealing ring, a support plate, a porous stone and a base plate. The test apparatus is mounted in a universal hydraulic test machine capable of applying compressive loads of 300KN. The pressing chamber is provided with a 44mm diameter×95mm deep bore through which the loading piston compresses sample. The visual cylinder is made of plexiglas. An outer cylinder made of mild steel is essential to visual cylinder in order to increase its strength and rigidity. The outer cylinder is provided with two observed windows with a 20mm width × 25mm height. The performance and phenomenon of compressive process of oilseed samples can be observed through the visual inner cylinder. Support plate made of stainless-steel with several 3 mm diameter traverse holes distributed uniformly is designed to prevent porous stone from breaking. In order to ensure uniform fluid pressure within oilseed cakes, both the bottom of loading piston and the top of base plate are provided with radial and circular grooves 5 mm width × 5 mm depth. The top and bottom of oilseed sample are respectively provided with a porous stone in order to expel liquid (including oil and water) and air from oilseed during compression.

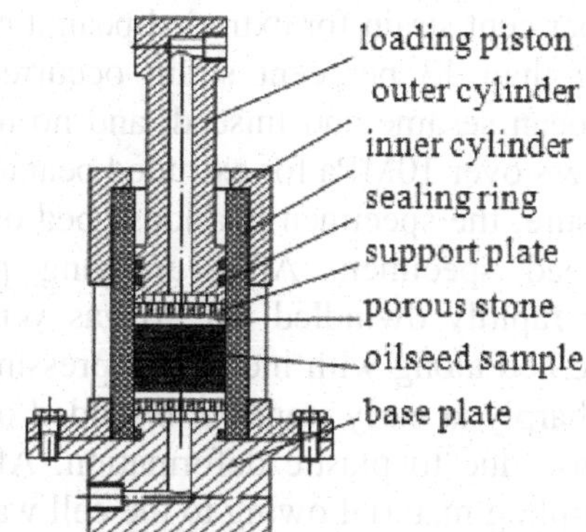

Fig. 1. Schematic diagram of visualized compression cell

2.2 Measurement of stress-strain

A 30g sample was chosen as testing specimen for the experiment of peanut, soybean, sesame and linseed. On the top and the bottom of oilseed specimen two fast speed filter papers were respectively inserted for the purpose of preventing porous stones from blocking up with bits of broken oilseed. After the specimen was poured into the compression cell, the cell was mounted in a computer-controlled precision universal test machine. Initial thickness of specimen of peanut, soybean, sesame and linseed were measured, which were 33.3mm, 28.2mm, 29.8mm and 27.7mm respectively. Equal rate of applied pressure (0.1MPa.s-1) was used in the experiment. Four series of experiments were carried out under 18°*C* of room temperature and under double surface for flow of fluids through a porous stone. Each desired stress was 60MPa.

2.3 Measured results and discussion

Defined applied stress σ and axial strain ε are as follows

$$\sigma = \frac{F}{A} \tag{1}$$

$$\varepsilon = \frac{\Delta H}{H_0} \tag{2}$$

Where: F is the applied force acting on the specimen surface (N), A is the area of section of the specimen (mm^2), H_0 is the initial height of the specimen (mm), and ΔH is the displacement of the specimen (mm).

Table 1 shows the measured results of strain with stress for extruded peanut, soybean, sesame and linseed compressed at thirteen stresses. Over 50 per cent strain for extruded soybean, sesame and linseed occurred at stress below 20MPa. Over 60 per cent strain for extruded peanut occurred at stress below 10MPa. No more than 13 per cent strain occurred at stress over 20MPa for extruded soybean sesame and linseed, and no more than 13 per cent strain occurred at stress over 10MPa for extruded peanut.

Prior to applying pressure, the specimen is a loose bed owing to a lots of pore space within oilseed specimen. After applying pressure on the specimen, pore space is rapidly dwindled due to gas vented rapidly and elastic deformation in the bed along with increasing pressing pressure. That is why the strains vary sharply at early stage for extruded oilseeds. The bed of oilseeds becomes dense due to plastic deformation. After that the bed becomes a fluid-solid coupling material owing to the cell wall of oilseed and granule broken. Last, the bed becomes oilseed cake as result of bond between broken oilseeds granule. The cake becomes denser and denser as oil is expelled. It explains the reason that no more than 13 per cent strain occurred at later stage for extruded oilseeds.

Table 1. Variation of axial strain ε (%) with stress σ (MPa) for the extruded peanut, soybean, sesame and linseed

oilseed	Stress (MPa)												
	0	5	10	15	20	25	30	35	40	45	50	55	60
peanut	0	53.05	61.24	64.14	66.22	68.02	69.42	70.39	71.26	72.04	72.97	73.63	74.49
soybean	0	27.54	40.83	46.74	50.50	52.98	55.16	56.10	56.71	57.38	57.98	58.44	59.02
sesame	0	39.39	54.34	60.54	63.90	67.65	70.06	72.07	73.68	74.63	75.31	75.95	76.69
linseed	0	29.08	44.75	52.71	58.51	62.74	65.40	66.71	67.95	69.06	70.15	70.72	71.26

3. NEURAL NETWORKS IDENTIFICATION ALGORITHM

It had been proved in theory that feed-forward neural networks trained with the back propagation (BP) can approximate continuous function and curve with arbitrary precision. The BP algorithm is a training learning process, which is divided into two processes, called forward-propagation and back-propagation respectively. Forward propagation is that input data from input layers are transmitted into hidden layer and into output layers after treated by hidden layers and output layers. If the practical output of neural networks is not expected output, the error between practical output and expected output will return through original path to change weights between layers, that is back-propagation. Forward propagation and back propagation are repeated until the prescribed error is met. The training learning process of artificial neural networks is actually one process of identification. So, BP neural network have been widely used in system identification to identify complex nonlinear system (Yang Jian et al., 2006; Sun Tao et al., 2005). The experiment indicated that stress-strain relationship for oilseeds during pressing was nonlinear. In this study, neural networks modeling techniques with BP network was used to predict the stress-strain relationship. Fig. 2 is the network model, which have *r*-inputs and one hidden layer.

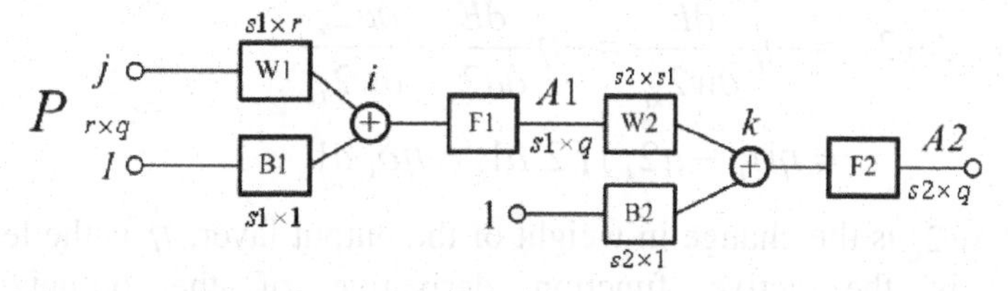

Fig. 2. Neural network model

P is input matrix, *W*1 is the weight matrix of the input layer, *B*1 is the deviation matrix of the input layer, *F*1 is the active function of the hidden layer, *A*1is the output matrix of the hidden layer, *W*2 is the weight matrix of the output layer, $B2$ is the deviation matrix of the output layer, $F2$ is the active function of the output layer, and $A2$ is the output matrix of the output layer.

3.1 Forward transfer of information

The ith node output for hidden layer is

$$a1_i = f1(\sum_{j=1}^{r} w1_{ij} P_j + b1_i), i = 1,2,\cdots,s1 \tag{3}$$

Where: $a1_i$ is the ith node output of the hidden layer, $f1(.)$is the active function of the hidden layer, $w1_{ij}$ is the connection weight from the jth input node to the ith hidden node, P_j is the jth input, and $b1_i$ is the ith node bias value of the hidden layer.

The kth node output for output layer is

$$a2_k = f2(\sum_{i=1}^{s1} w2_{ki} a1_i + b2_k), k = 1,2,\cdots,s2 \tag{4}$$

Where: $a2_k$ is the kth node output of the output layer, $f2(.)$ is the active function of the out layer, $w2_{ki}$ is the connection weight from the ith output node of the hidden layer to the kth output node of the output layer, and $b2_k$ is the kth node bias value of the output layer.

Adopting the error function as follows

$$E(W,B) = \frac{1}{2}\sum_{k=1}^{s2}(t_k - a2_k)^2 \tag{5}$$

Where: $E(W,B)$ is the error function of the output, t_k is the kth node objective value of the output layer, and $a2_k$ is the kth node output of the output layer.

3.2 Change weight using gradient descent algorithm

The weight from ith input to kth output is

$$\Delta w2_{ki} = -\eta \frac{\partial E}{\partial w2_{ki}} = -\eta \frac{\partial E}{\partial a2_k} \cdot \frac{\partial a2_k}{\partial w2_{ki}}$$
$$= \eta(t_k - a2_k) f2' a1_i = \eta \delta_{ki} a1_i \tag{6}$$

Where: $\Delta w2_{ki}$ is the change in weight of the output layer, η is the learning rate, $f2'$ is the active function derivative of the output layer, $\delta_{ki} = (t_k - a2_k) f2' = e_k f2'$, $e_k = t_k - a2_k$, where δ_{ki} is the error from the ith output node of the hidden layer to the kth output node of the output layer, and e_k is the kth output error of the output layer. In the same way

$$\Delta b2_{ki} = -\eta \frac{\partial E}{\partial b2_{ki}} = -\eta \frac{\partial E}{\partial a2_k} \cdot \frac{\partial a2_k}{\partial b2_{ki}}$$

$$= \eta(t_k - a2_k) f2' = \eta \delta_{ki} \tag{7}$$

Where: $\Delta b2_k$ is the change of the *k*th node bias value of the output layer. The weight from *j*th input to *i*th output is

$$\Delta w1_{ij} = -\eta \frac{\partial E}{\partial w1_{ij}} = -\eta \frac{\partial E}{\partial a2_k} \cdot \frac{\partial a2_k}{\partial a1_i} \cdot \frac{\partial a1_i}{\partial w1_{ij}}$$

$$= \eta \sum_{k=1}^{s2} (t_k - a2_k) f2' w2_{k1} f1' p_j = \eta \delta_{ij} p_j \tag{8}$$

Where: $\Delta w1_{ij}$ is the weight change of the hidden layer, $f1'$ is the active function derivative of the hidden layer, $\delta_{ij} = e_i f1', e_i = \sum_{k=1}^{s2} \delta_{ki} w2_{ki}$, $\delta_{ki} = e_k f2', e_k = t_k - a2_k$, where: δ_{ij} is the error from the *j*th input node of the input layer to the *i*th output node of the hidden layer, and e_i is the *i*th node output error of the hidden layer. In the same way

$$\Delta b1_i = \eta \delta_{ij} \tag{9}$$

Where: $\Delta b1_i$ is the bias value change of the hidden layer.

A three layer feed-forward neural networks trained with the back propagation (BP) algorithm was adopted in this paper. Both input layer and output layer had one node, which represented applied pressures sequence and measured strains sequence respectively. Hidden layer had five nodes. 0.01 and 1000 were used as the error tolerance and the maximum number of training cycle respectively. Sigmoid function $f1(s) = (1 + e^{-s})^{-1}$ was selected as active function $f1(s)$. Linear function was selected as active function $f2(s)$. The measured results had been taken as samples. 11 and 2 data were chosen randomly as training and testing sample respectively. The error function is

$$E = \frac{1}{2} \sum_{k=1}^{11} (t_k - a2_k)^2 \tag{10}$$

3.3 Results and discussion

Fig. 3 shows the curves of stress-strain predicted and measured, and Fig. 4 shows the curves of relationship between training times and error for peanut, soybean, sesame and linseed during the training process. The values of sum errors of the prediction for peanut, soybean, sesame and linseed were 0.00282, 0.0083, 0.0084 and 0.0047 respectively. The training times were 11,

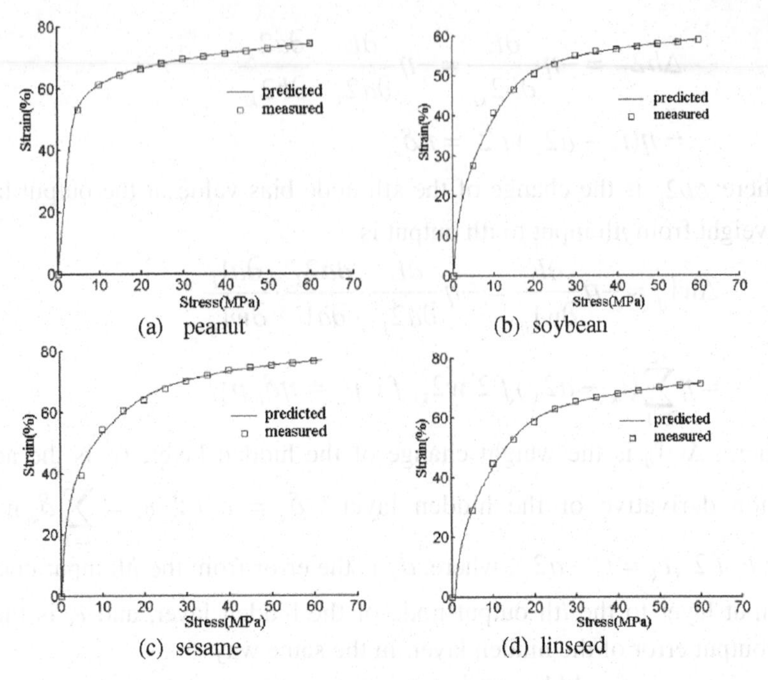

Fig. 3. Comparison of prediction with measurement

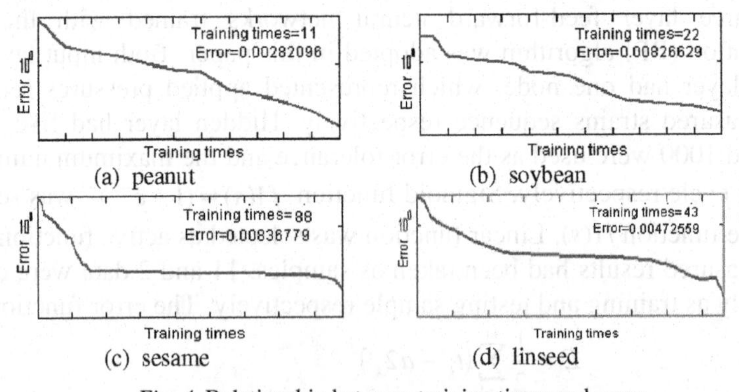

Fig. 4. Relationship between training times and error

22, 88, and 43 respectively. It was found that there was a lack smooth for prediction curves for oilseeds due to over-fitting when the error tolerance is less than 0.001.

4. CONCLUSIONS

Apparatus and procedures were developed to measure the stress-strain relationships for extruded peanut, soybean, sesame and linseed. Stress-strain relationships were measured compressed at twelve pressures (5, 10, 15, 20, 25, 30, 35, 40, 45, 50, 55, 60MPa) under uniaxial cold pressing. The model was developed to predict the stress-strain relationship for extruded oilseeds based on BP neural network.

Results indicated that the stress-strain relationships were nonlinear. Over 50 per cent strains for extruded soybean, sesame and linseed occurred at stress below 20MPa. Over 60 per cent strain for extruded peanut occurred at stress below 10MPa. No more than 13 per cent strain occurred at stress over 20MPa for extruded soybean sesame and linseed, and no more than 13 per cent strain occurred at stress over 10MPa for extruded peanut. There were significant increases in the values for the strains at early stage for extruded oilseeds, and there were no significant increases in the values for the strains at later stage for extruded oilseeds.

BP neural network can be used to predict the stress-strain relationship for oilseeds, which not only overcomes the difficulty for theoretical model development, but also avoids requiring considerable technique and experience for nonlinear regression analysis. No more than 0.0084 maximum error showed that the model predicted the stress-strain relationships with highly accuracy. In view of the predicted results and the simple model consisting of input and output layer with one node, and hidden layer with five nodes, the method of stress-strain prediction for oilseeds by using artificial neural networks is both feasible and effective.

ACKNOWLEDGEMENTS

Funding for this research was provided by Hubei Provincial Department of Education (P. R. China). The first author is grateful to the Wuhan Polytechnic University for providing him with pursuing a PhD degree at the Wuhan University of Technology.

REFERENCES

C. R. Sukumaran, B. P. N. Singh. Compression of bed of rapeseeds: the oil-point, Journal of Agricultural Engineering Research, 1989, 42:77-84

E. Davion, A. G. Meiering, F. J. Middendof. A theoretical stress model of rapeseed, Canadian Agricultural Engineering, 1979, 21(1):45-46

E. Davion, F. J. Middendof, W. K. Bilanski. Mechanical properties of rapeseed, Canadian Agricultural Engineering, 1975, 17(1):50-53

G. C. Mrema, P. B. Mcnulty. Mathematical Model of Mechanical Oil Expression from Oilseeds. Journal of Agricultural Engineering Research, 1985, 31:361-370

H. J. Rasehom, H. D. Deicke, Xin Yaonian. Theory and praxis of decortication and cold pressing of rape seed, China oils and fats, 2000, 25(6): 50-54 (in Chinese)

P. C. Bargale, R. J. Ford, F. W. Sosulski, et al. J Irudayaj. Mechanical Oil Expression from Extruded Soybean Samples. Journal of the American oil chemists society, 1999, 76(2):223-229

P. C. Bargale, Jaswant Singh. Oil expression characteristics of rapeseed for a small capacity screw press, Journal of food Science Technolage, 2000, 37(2):130-134

P. C. Bargale, R. Ford, D. Jwulfsohn, et al. Measurement of consolidation and permeability properties of extruded soy under mechanical pressing, Journal of Agricultural Engineering Research, 1999, 74:155-165

Sun Tao, Cao Guangyi, Zhu Xinjian. Nonlinear modeling of PEMFC based on neural networks identification, Journal of Zhejiang University Science, Vol. 64, No. 5, 2005, 64(5):365-370

Yang Jian, Xu Bing, Yang Huayong. Noise identification for hydraulic axial piston pump based on artificial neural networks, Chinese Journal of mechanical engineering, 2006, 19(1):120-123

Zheng Xiao, Wan nong, Lin Guoxiang, et al. Research on microstructure of cold pressed cakes from decorticated rapeseed based on porosity, China Oils and Fats, 2004, 29(12):14-17 (in Chinese)

Zheng Xiao, Zeng Shan, Lin Guoxiang, et al. Research on stress-strain of rapeseed and decorticated rapeseed by uniaxial cold pressing under single surface for flow of fluids through a porous medium, China oils and fats, 2004, 29(7):11-14 (in Chinese)

PARAMETERIZED COMPUTER AIDED DESIGN OF STUBBLE CLEANER

Lige Wen [1,2], Jianqiao Li [1,*], Xiuzhi Zhang [2], Benard Chirende [1]
[1] *The Key Laboratory for Terrain-Machine Bionics Engineering, Ministry of Education, Jilin University,Changchun, China, 130025*
[2] *College of Mechanical Science and Engineering, Jilin University, Changchun, China, 130025*
* *Corresponding author, Address: No. 5988, Renmin Street, Changchun, P. R. China, 130025, Tel: +86-431-85095760-8407, Fax: +86-431-85095575, Email: jqli@jlu.edu.cn*

Abstract: The traditional agricultural machine design were represented as 2D drawing, which is difficult to modify, and also not intuitive in the solid way for the form and structure. Therefore, software, UG-NX3, was used to conduct parameterized design of stubble cleaner parts. The parts were designed associatively and assembled virtually. The structure of the whole machine and the spatial distribution of parts can be seen and analyzed intuitively. Under UG-NX3 circumstance, the designed 3D model can be transformed automatically to 2D drafting which is used in fabrication and production. The result proved that computer aided parameterized design can allow dynamical operation, preview and repeated modification of the design, reliably and quickly. Therefore, the optimum design efficiency of stubble cleaner is improved; 3D modeling time and 2D drafting time is greatly decreased.

Keywords: Stubble cleaner, computer aided design, parameterized design

1. INTRODUCTION

In recent years, with the rapid development of computer technology, 3D design and virtual assembling technology were introduced into the mechanical design field, and as a result, product update frequency and design efficiency were greatly increased (Wen, 2003). These machines are also gradually incorporated into the agricultural machine design field (Yang

Wen, L., Li, J., Zhang, X. and Chirende, B., 2008, in IFIP International Federation for Information Processing, Volume 258; Computer and Computing Technologies in Agriculture, Vol. 1; Daoliang Li; (Boston: Springer), pp. 11–18.

et al., 2004; Yang et al., 2002; Yuan et al., 2006). Computer aided design will be the necessary trend of agriculture machine design field (Yan et al., 2004). Prior to using the virtual assembling technology, design for stubble cleaner was mainly 2D design in which the structure of parts and the whole machine could not be easily visualized, hence associative relationship among parts could be hardily established. In addition, parts assembly and interference checkup could not be conducted. In order to resolve the above problems, parameterized design and virtual assembly were used to design stubble machines based on UG-NX3 (Fu, 2005; Zhao et al., 2005), and this is in line with the trend of stubble cleaner development (Wu et al., 2000).

2. PARAMETERIZED DESIGN AND ASSOCIATIVE DESIGN

2.1 Parameterized design

Parameterized design method is a new kind of 3D design method, and UG- NX3 is one of the representational 3D parameterized design software. There is driving parameter and calculation parameter under the environment of UG-NX3. Driving parameter means that a variable can be evaluated and its value can be changed at will, whilst calculation parameter is the parameter obtained through calculation based on driving parameter. UG-NX3 has strong sketch functions, utilizing dimension constraint and geometry constraint to drive the sketch, dimensions and shapes of the sketch alter with respect to the change of constraints. Design modification can be done repeatedly and quickly, and the time spent on 3D modeling is markedly reduced.

2.2 Associative design

Associative design means setting up associative relationship among parts such that the associative parts change their sizes and structures simultaneously. Associative design method avoids interference among parts, and makes modifying work convenient and accurate. UG-NX3 provides two assembly methods, one is Bottom-Up design method and the other is Top-Down design method. The former first establishes models of all parts, and then assembles them. Associative constraint must be set up. In contrast, the latter sets up main parts first and other parts come into being according to their associative relationship with the main parts and the dimensions of the main parts. The latter suits people's design habit very well and avoids assembling interposition, therefore improving design efficiency. WAVE Geometry Linker in UG-NX3 serves the Top-Down design method. Parameter modeling establishes interrelated relationships within a part, while WAVE Geometry Linker extends this notion to set up associative relationship among different parts.

3. PARAMETERIZED DESIGN FOR PARTS OF STUBBLE CLEANER

3.1 Parts of stubble cleaner

The main function of stubble cleaner is cutting the crop stubble into small pieces and mixing them with earth uniformly and rotarily ridging the earth. Stubble cleaner mainly consists of four components.

✧ Frame, which supports the whole machine and joins with power framework, and other parts assembled on it.

✧ Gear-box, which changes the power transmission direction and speed of motion, it is composed of gears with straight tooth, transmission shaft and axletree seat.

✧ Stubble roller, which cuts up the stubble and ploughs up the earth, it consists of blade tray, stubble-cutting blade, square shaft and axletree seat.

✧ Shield, whose function is to break up the earth block, ridge the earth, and mix the earth and the stubble.

Next step is to determine main parts of every component, and to make clear the associative relationship among parts.

3.2 3D design of parts

Part modeling process is as shown in Figure 1. Firstly, the main part of every component is chosen; next, structure is analyzed, sketch is constructed, driving dimension and calculating dimension are given; then associative design to other dimensions is conducted and then feature instantiation is ensured; finally, feature-based 3D parameterized model of parts of stubble cleaner can be obtained.

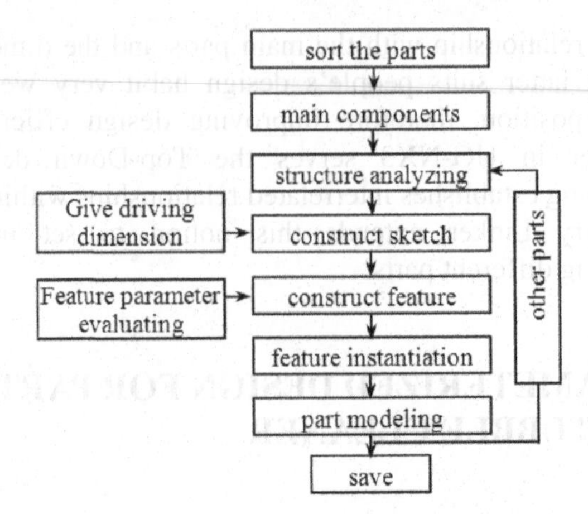

Figure 1. Modeling process for parts

3.2.1 Parameterized design of spur gear in gear-box

There is one pair of bevel gears and two pairs of spur gears in the cleaner. One of their 3D models is shown in Figure 2. The two bevel gears which are used to change the direction of rotating speed have the same tooth number, and their sizes remain constant. Spur gears are used to change rotating speed, because stubble cleaners are usually designed to operate at different rotating speeds.

The course of parameterized design of spur gears is as follows. Driving parameter of the gear is composed of pressure angle, modulus, tooth number, while calculation parameter consists of addendum circle diameter, gear root diameter, graduated circle diameter and base circle diameter etc.. When modifying driving parameter, the structure and size of the gear change correspondingly. Figure 3 shows that when gear tooth number changes from 19 to 24, the structure of the gear is also altered. Associative relationship is set up between assembly hole diameter of the gear and shaft diameter.

Figure 2. Straight bevel gear of stubble cleaner

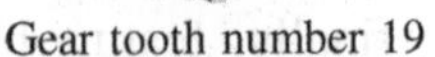

Gear tooth number 19

Gear tooth number 24

Figure 3. Changing of gear structure

3.2.2 Design of stubble blade and blade tray

The main components in stubble roller are left blade & right blade and four blade trays assembled at different angles. 3D model of stubble blade and one of the blade trays are as shown in Figure 4. Respective assembling holes in blade tray and stubble blade are associative, when the dimension of the assembling hole in blade changes, that of blade tray changes accordingly. The dimensions of the center square holes of the stubble trays are related to the dimensions of the square shaft.

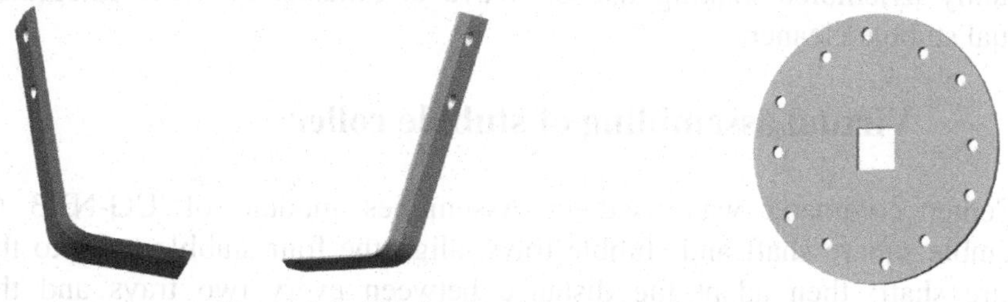

Figure 4. Stubble blades and one of the blade trays

3.2.3 Design of cushion Using WAVE Geometry Linker

Modeling method of axletree seat cushion is utilized as an example in order to introduce WAVE Geometry Linker design method, which is a Top-Down assembly method. The process of modeling axletree seat cushion using WAVE Geometry Linker is as follows:

✧ Open "assemblies" module under UG-NX3 environment, then set up a new file; choose Non-Master Part in "New Part File" dialog box. UG-NX3 provides a function of Master Part, the formerly established model such as axletree seat is set as a Master Part, the dimensions of which can be modified, resulting in that of Non-Master Part changing accordingly. Dimensions cannot be directly changed on the Non-Master Part.

✧ Use "Add Existing" command to recall the existing axletree seat, then making use of the command "WAVE Geometry Linker", choosing assembling surface of the axletree seat where the cushion will be assembled. Now, the surface feature can be obtained, and then using Pad command, giving the thickness, the cushion is obtained as in Figure 5.

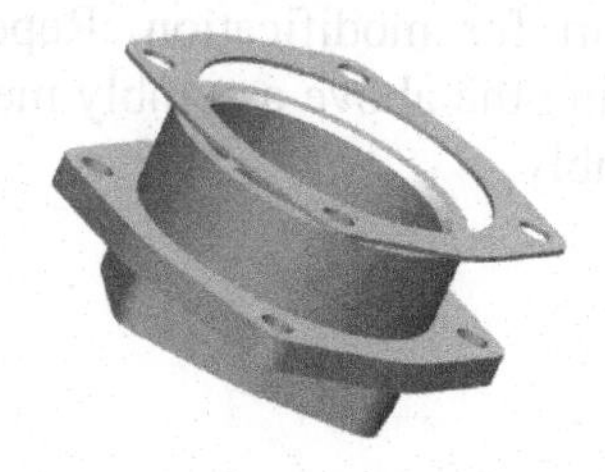

Figure 5. Create cushion using WAVE

Finally, the dimension associative relationship between cushion and axletree seat is established.

4. VIRTUAL ASSEMBLY

The stubble roller was virtually assembled by 3D models designed above firstly. Then it together with other main components and minor parts was virtually assembled making use of Wave commands to make integrated virtual stubble cleaner.

4.1 Virtual assembling of stubble roller

Center command was used in Assemblies module of UG-NX3 to assemble square shaft and stubble trays, align the four stubble trays to the square shaft; then adapt the distance between every two trays and the distances between the outboard tray and the square shaft end should be same. Next, assemble one blade on a stubble tray with Mate command and Align command and assemble other 5 blades on the stubble tray using Component Arrays command. Utilize the same method to assemble 3, 3 and 6 blades respectively on the other 3 stubble trays and they make a total of 18 blades, 6 blades on the outboard two trays respectively and 3 ones on the inboard two trays respectively. Then we get a stubble roller set as Figure 6 shows. The 18 blades were arranged helically and have a 20º gap when cutting into the earth. It is seen clearly at different direction in Figure 6 that there is no interference in the assemble body.

4.2 Virtual assembling of the whole stubble cleaner

Figure 7 is 3D assembly model of the whole stubble cleaner in which the shield board of the left roller, one of the shield boards of the right roller and the side shield boards are hidden. The structure of the whole machine and the relationship among parts can be seen intuitively. Under the assembly module, interference among parts can be checked up, and the gap between parts can be examined. If unreasonable condition exists, then it is necessary to return the Master Part for modification. Repeat the process until the assembly is dead-on. Using the above assembly method can minimize errors from unreasonable assembly.

Figure 6. Virtually assemble the stubble roller

Figure 7. 3D modeling of stubble cleaner

5. DRAFT CREATION

UG-NX3 has the function to transform 3D module to 2D drafting automatically, which is required in producing and processing. Different kinds of views can be obtained under UG-NX3 circumstance, such as sections, axonometric drawing, partial enlarged view, and so on. Dimensions and shape, location tolerance and roughness can be given to drafting here. Drafting can be output as a file in kinds of formats, in which *.dxf file and *.dwg file can be read by AutoCAD. 3D model and 2D drafting are structure and size interrelated, when the 3D model is modified 2D drafting changes accordingly. Therefore the data consistency of 3D entity and 2D drafting is ensured.

6. CONCLUSIONS

✧ Parts and components of stubble cleaner were parameterized designed using UG-NX3, relationship between parts and a part lib was built up. When the size of a part is changed, that of the relational parts will be changed correspondingly, which makes it convenient to change the measurement and the profile when needed later.

✧ Conducting parameterized design, associative design and virtual assembly, and gap and interference checkup to stubble cleaner parts under UG-NX3 environment makes product design more intuitional. It improves the efficiency of remodel design and serial design of stubble cleaner, and ensures high design precision.

✧ Using this method, 2D drafting of stubble cleaner and parts can be gained conveniently and speedily. And the presented method realizes associative modification and ensures data consistency of 3D entity and 2D drawing.

✧ The method used here would provide a reference and template for other agricultural framework design using UG-NX3 software.

ACKNOWLEDGEMENTS

The authors would like to acknowledge the referees for their helpful comments to improve the presentation of this paper. The item is supported by the National Natural Science Foundation of China (Grant No.50635030) and Projection of Science Committee of Jilin Province, China (Item No. 20050539).

REFERENCES

Fu Benguo, editor in chief, UG NX 3.0 3D Mechanical Design, Publishing House of Mechanical Industry, Beijing, China: 2005

Translate and edit: Zhao Bo, Zhang Qin, UG-NX2 Correlation Parameterized Design Training Tutorial, Press of Tsing Hua University, Beijing, China: 2005

Wen Bangchun, Zhou Zhicheng, Han Qingkai, et al., Important Role of Products Design of Modern Machinery in the Research and Development of New Products—Study on the Threefold Method "Dynamic Design, Intelligent Design and Partial Virtual Design" Face to Products Generalization Quality, Chinese Journal of Mechanical Engineering, 2003, 39(10): 43–52

Wu Zi-yue, Gao Huan-wen. Present state and development on technology of stubble chopping, Journal of China Agricultural University, 2000, 5(4): 46–49

Yan Chu-liang, Yang Fang-fei, Zhang Shu-ming, Digitized Design Technology and its Application in Agricultural Machinery Design, Transactions of Chinese Society of Agricultural Machinery, 2004, 35(6): 211–214

Yang Fu-zeng, Fu Xiang-hua, Yang Fang, et al., Research on Parametric Modeling of Parts of Seeding Machine Based on Pro/E, Transactions of Chinese Society of Agricultural Machinery, 2002, 33(4): 66–68

Yang Xin, liu Jun-feng, Feng Xiao-jing, Feature Modeling and Assembling Conjunction Design on Wheat Precision Seedmeter, Transactions of The Chinese Society of Agricultural Enginnering, 2004, 20(3): 89–92

Yuan Rui, Ma Xu, Ma Cheng-lin, et al., Virtual manufacturing and motion simulation of precision planter unit, Journal of Jilin University (Engineering and Technology Edition), 2006, 36(4): 523–528

REGIONAL COUNTRY INFORMATION SERVICE PLATFORM BASED ON HYBRID NETWORK

Songbin Zhou[1,2,*], Guixiong Liu[1], Taobo Cheng[1,2]
[1] *School of Mechanical Engineering, South China University of Technology, Guangzhou 510070, P. R. China, zsb@autocenter.gd.cn, megxliu@scut.edu.cn, ctb@autocenter.gd.cn*
[2] *Automation Engineering R&M Center, Guangdong Academy of Sciences Guangzhou 510070, P. R. China*
[*] *Corresponding author, Address: School of Mechanical Engineering, South China University of Technology, Guangzhou 510070, P. R. China, Tel: +86-20-8710568, Email: zhousongbin@126.com*

Abstract: In view of the current situation of the country where basic facilities are lagging and inhabitation is dispersed and the weaknesses of the various rural information technologies, a multilayered, distributed and regional country information service platform based on wireless field-bus hybrid network is proposed. Its structure, hardware frame and communication mechanism based on its application in a village are analyzed. The research shows that the platform can solve the 'the last one-kilometer' problem in China's rural informatization and thus has significant popularization value.

Keywords: hybrid network, information service platform, rural informatization

1. INTRODUCTION

In recent years, China has made great progress in rural informatization (RI) construction, however it hasn't found a good solution to the issue of 'the last one-kilometer', and as a result, agricultural information cannot reach farmers quickly and effectively (Chen, 2006). There are various existing in the service platforms of our present rural information technologies like Short Message Service (SMS) (He, 2005), accessing internet via dial-up, Asymmetrical Digital Subscriber Loop (ADSL) and Cable Antenna Television

Zhou, S., Liu, G. and Cheng, T., 2008, in IFIP International Federation for Information Processing, Volume 258; Computer and Computing Technologies in Agriculture, Vol. 1; Daoliang Li; (Boston: Springer), pp. 19–26.

(CATV), especially high running expenses, complicated operation and poor interaction. Therefore, it's highly necessary to develop a technology that is suitable for RI.

Field-bus has the advantages of good interoperability, decentralized control function, easy maintenance; and wireless communication shows the advantages of low cost, short construction period and expandability. Hybrid network can bring out the advantages of each network and expand application range of the system and get high performance index (Vitturi, 2005). The multilayered, distributed and regional country information service platform based on wireless field-bus hybrid network that is to be discussed in the paper can adapt to the rural areas where basic facilities are lagging and inhabitation is dispersed. The platform gives farmers access to all kinds of agricultural information like dated agricultural condition and has functions of browsing, query and ordering, besides, it provides interaction functions between farmers and agricultural information system as well as among farmers.

1. REGIONAL COUNTRY INFORMATION SERVICE PLATFORM BASED ON HYBRID NETWORK

The platform adopts the hierarchical hybrid structure of wireless and field-bus technology (see figure 1), with the upper level adopting wireless communication technology network and the lower level adopting field-bus network. The platform is consisted of information center, information host and terminal.

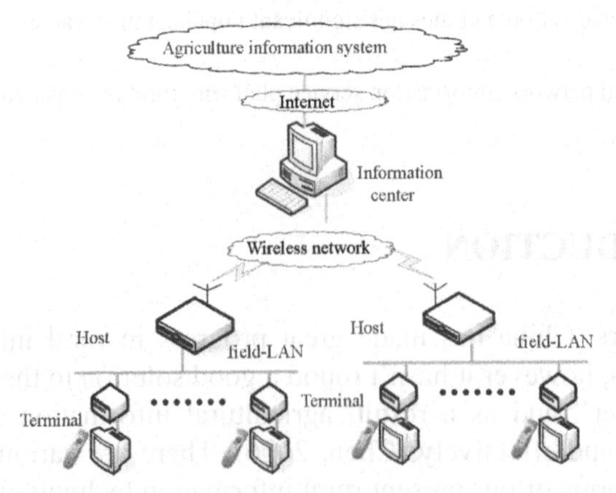

Figure 1. System structure graph

Information center communicates with all agricultural systems through internet. Having got the agricultural information, the center stores it in local database and then sends it to information host through wireless network. After receiving the information, the host stores it in its memory and forms field-LAN together with terminals through field-bus. Placed in farmers' houses and operated through infrared controller, the terminals acquire information from the host and send it to TV set in the means of video signals.

The platform is applied in a village of Qingyuan City, Guangdong Province. Placed in the office of the Village Committee, the information center gets agricultural information through internet and manages 10 field-LANs. Each field-LAN is formed by one host and 60 terminals. In its application, the distance between host and terminal can reach 800m, and the distance between field-LAN locations can reach 1000m. The entire platform can cover a village with a 3600m diameter, providing dated agricultural information for 600 farmers.

3. SYSTEM DESIGN

3.1 Hardware Design

At present, there are different types of long-distance wireless communication technologies such as GPRS/CDMA, Wireless Digital Transmission, WLAN Bridge, while the low-cost field-bus has RS485 and CAN, *etc.* In consideration of cost and distance, the long-distance wireless network is constructed by using CC1000 digital transmission module with 12dB outdoor omni-directional antenna. CC1000 is an ultra low-power consumption transceiver designed by CHIPCON. It is intended mainly for the Industrial, Scientific and Medical (ISM) and Short Range Device (SRD) frequency bands at 315,433,868 and 915MHz. It can realize wireless digital transmission between two facilities through RS-232 interface by relying on a few external components. Field-bus network adopts two-wire RS-485, which shows the characteristics of long-distance (the length of cable can reach 1200m at 100Kbps rate), multiple-node, low-cost transmission, and is also good at preventing common-mode disturbance (Dong, 2002).

Host and terminal adopt LPC2131 processor based on ARM7TDMI, which has small encapsulation and low power with high-speed Flash memory and several 32-bit timers, 47 GPIO and 9 edge- or level- triggered external interrupts. As illustrated in figure 2, information host is mainly consisted of processor, power circuit, memory, field-bus circuit, wireless communication circuit and LED display circuit. Processor receives information from wireless module and stores in RAM, at the same time it communicates with each terminal through field-bus. As illustrated in figure 3, terminal hardware includes the same components as host except that it has output circuit, buzzer and infrared receiver module rather than external memory and wireless module. Video output interface connects with AV interface of TV set, buzzer signals new information or alarm, infrared receiver module receives controller signals and LED signals operation and communication. The terminal is a thin client who can decrease the cost.

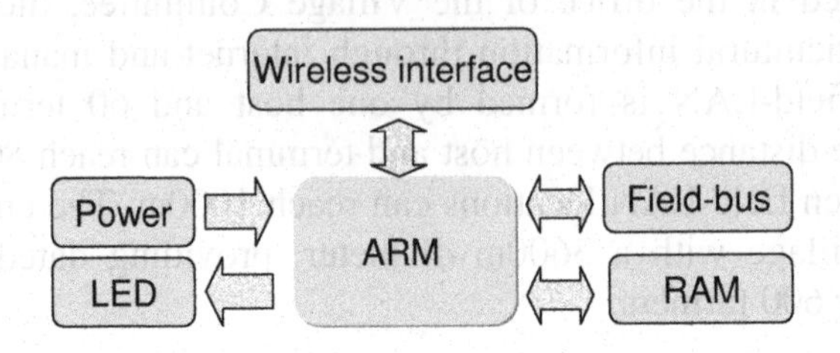

Figure 2. Host hardware frame

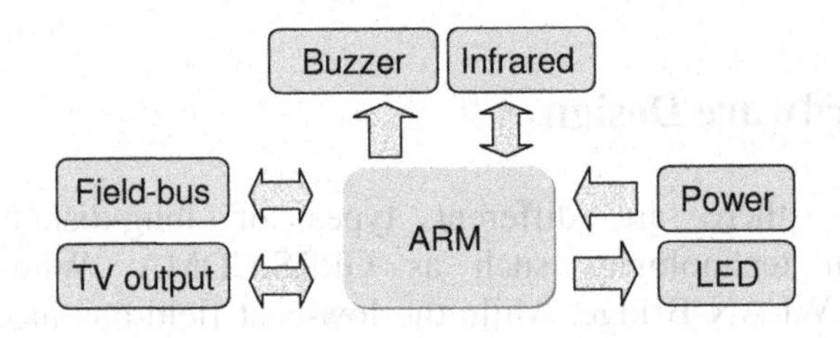

Figure 3. Terminal hardware frame

3.2 Software Design and Information Communication Mechanism

Host and terminal adopts uC/OS-II operating system, which is an open-coded, solidifiable real-time multiple-task operating system (Jing, 2006). For most distributed systems, their communication task is sending orders and collecting data by upper-level information center, but for this platform the main task is receiving orders from terminals by information center and sending out huge number of data completely and correctly. So the key of the software design is to ensure integrality and accurate transmission of the huge amount of information in the hybrid network.

Time synchronization is the basis in realizing each communication task, and it is realized through the following steps in this platform: each host renews its time after reading the time record in the information center when it is charged with power; then the terminal reads the time in host and renews its time record. Finally the communication mechanism between information center and host as well as between host and terminal is design to ensure integrality and accuracy of the information in transmission.

1) Information Center-Host Communication Mechanism

Because there is the possibility of drop net and power-off, there exists the problem of missing information from off-line host if the information center adopts broadcasting mode. Therefore point-to-point polling mode is adopted. Every piece of information includes issue time, save time, type, content, grade and verification. Issue time means the time when the information is put in; save time refers to the time when the information is saved. Time is saved in double-precision floating point, according to which data is sorted, searched and queried. Type is divided into common type and urgent type.

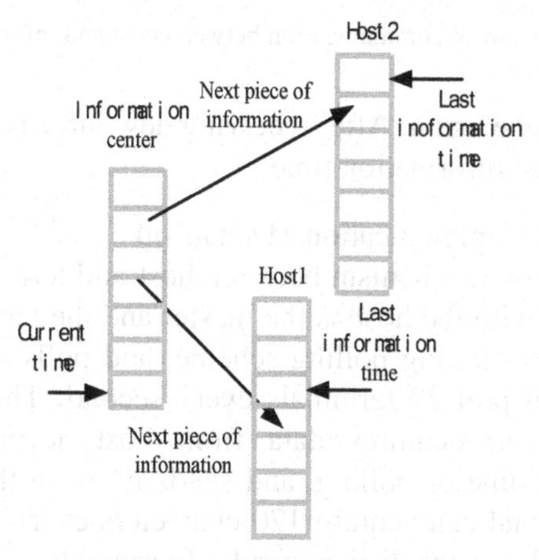

Figure 4. Information mapping method

Last information time refers to the save time of the last piece of information received by the host. As illustrated in figure 4, the information center stores the information in local-time sequence database after receiving it from each agricultural information system, and searches for the next piece of information and sends it to the polling host according to the last information time. When the memory is full, new pieces of information replace the earlier ones. Figure 5 demonstrates the flow chart between information center and host. First, information center inquires host and if there is no answer it then inquires the next one; then host uploads information packet, otherwise it sends out local information time; after information center gets response, it sends urgent information if there were any, otherwise it searches the piece of information next to the local last information time, then it inquires the next host. After polling all hosts, it

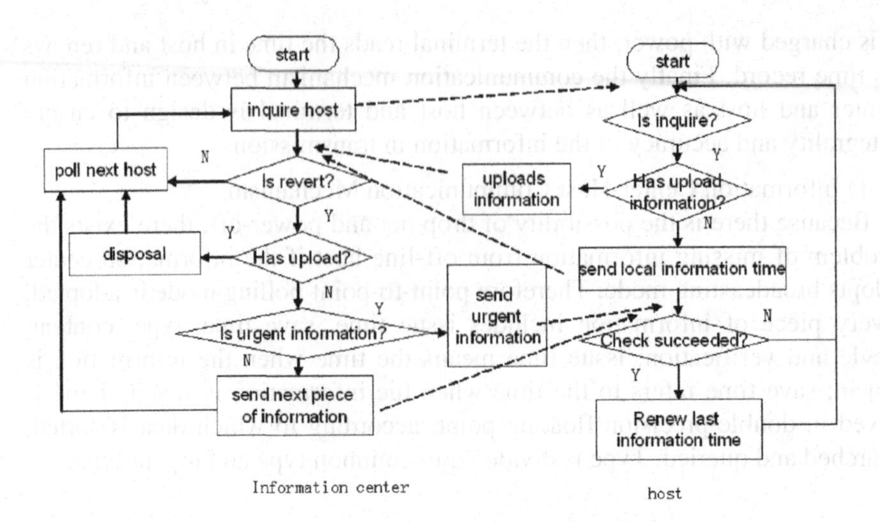

Figure 5. Flow chart of communication between host and information center

starts from the first one again. After receiving new information, host verifies, stores and renews last information time.

2) Host-Terminal Communication Mechanism

The communication mechanism between host and terminal mainly adopts master-slave mode, with the host as the master and the terminal as the slave. The host adopts time-sharing polling scheme, and polls one terminal every 50ms, that is, it can poll 20 terminals every second. The terminal doesn't store information and acquires data from host according to farmers' requirements in the time of polling, and sends it out in the means of video signals. Every terminal can acquire 120 characters every time at the rate of 19.2kb/s. Figure 6 shows the display result of terminals.

Figure 6. Terminal application

4. CONCLUSIONS

The current agricultural information technologies show respective weaknesses in their application in the rural areas. Mobile phone short message service, which uses mobile communication network as the communication platform and receives message through mobile phone, has the advantage of wide coverage and easy operation, yet on the other hand it has high cost (every family needs to pay (50-100) Yuan per month if it receives 1000 messages per month), limited information volume and monotonous information presentation. Publishing information through CATV network, broadcast television service shows the characteristics of diversified presentations, but it has to occupy TV channels, has poor interaction between information and users and cannot store the information. Besides, every family needs to pay (10-20) Yuan per month. Accessing internet via dial-up or ADSL has powerful functions but has high cost and complicated system operation, every family needs to pay (80-100) Yuan per month. The platform proposed in this paper adopts hybrid network and shared network access, and each family only needs to pay (2-5) Yuan per month. Therefore it has great advantage comparing with the methods above.

Regional country information service platform based on hybrid network is suitable for the application in the rural areas in China and other developing countries with low-cost, convenient operation and flexible network.

ACKNOWLEDGEMENTS

This work was supported by the National Agriculture Scientific and Technological Achievements Transformation Foundation of China (No. 05EFN214400213).

REFERENCES

Chen Yunhui. Analysis of the way of Span the ‘Last One Kilometer’ of Agricultural Informatization. Chinese Agricultural Science Bulletin, 2006, 22(9): 462–465

Dong Qian, Xie Jianying. Fieldbus network implementation based on RS-485. Proceedings of the 4th World Congress on Intelligent Control and Automation, 2002, Vol. 4: 2790–2793

He Yong, Yu Haihong, Qiu Zhengjun. Study on farm information acquisition by using wireless remote methods and treatment systems. Proceedings of the Third International Conference on Information Technology and Applications (ICITA’05), 2005, Vol. 1: 526–530

Jing Bo, Zhang Jie, Qin Zheng, et al. Design of Micro Wireless Network Measurement and Control Server Based on ARM and μC/OS-II. Proceedings of the 6th World Congress on Intelligent Control and Automation (WCICA2006), 2006, Vol. 1: 4453–4457

Vitturi S, Miorandi D. Hybrid Ethernet/IEEE 802.11 networks for real-time industrial communications. Proceedings of the 10th IEEE International Conference on Emerging Technologies and Factory Automation (ETFA 2005), 2005, Vol. 2: 19–22

THE SHORT MESSAGE MANAGEMENT SYSTEM BASED ON J2ME

Shumin Zhou*, Guoyun Zhong, Tiantai Zhang
School of Information Engineering, East China Institute of Technology, Fuzhou, China, 344000
* *Corresponding author, Address: School of Information Engineering, East China Institute of Technology, 56 Xufu Road, Fuzhou, Jiangxi Province, China, 344000, Tel: +86-794-8258390, Fax: +86-794-8258390, Email: smzhou@ecit.edu.cn*

Abstract: In order to apply Java to mobile communication device and embedded device, the J2ME is issued by SUN. It is also be used on 3G mobile multiplication service by telecommunication company. The paper introduces the methods and processes of developing short message management system adopt J2ME and Siemens Mobility Toolkit 6688i. The system applies 12 classes to realize message grouping, message list and message editing. By this system, the efficiency of the short message management in the mobile phone is increased.

Keywords: J2ME, Short Message, MIDlet

1. INTRODUCTION

Short Message Service (SMS) (ETSI/TC, 1997) has become a mature wireless communication service. SMS provides a connectionless transfer of messages with low-capacity and low-time performance (Gerd et al., 2003). In order to improve the application of mobile phone, the related programming languages were issued. The development of the mobile phone software has mainly experienced three stages. At first stage, it just focused on the basic voice service, afterwards the simple value-added service applied, then the function which access Internet by the WAP was developed. J2ME is issued by SUN in order to apply Java to mobile communication device and embedded device and consumed electrical equipment (Microsystems, 1999). Two problems can be solved by J2ME: (1) Java language runs across different platform, software development businessman develops related

Zhou, S., Zhong, G. and Zhang, T., 2008, in IFIP International Federation for Information Processing, Volume 258; Computer and Computing Technologies in Agriculture, Vol. 1; Daoliang Li; (Boston: Springer), pp. 27–34.

application easily, so it can also be installed conveniently to mobile phone; (2) J2ME has offered high-grade programming of the internet protocol such as HTTP, TCP and UDP, so JAVA application can be accessed on Internet freely. In the mobile communication toady, global operators are arranging 3G strategy, they have met the problem how to consume these bandwidth increased. Now J2ME is seemed as a best solving scheme, offer the multiplication service of mobile data business based on J2ME for user. Mobile multiplication service, such as mobile business, mobile information service, mobile internet business and Virtual special net business, mobile office, mobile medical treatment, will become true by using J2ME.

Short message is now mobile service used the most broadly exclude the conversation of speech sound outside, therefore this paper adopts J2ME to realize Short message management system, makes the JAVA user of mobile telephone carry out various managements for short message.

2. THE STRUCTURE OF CLDC/MIDP

J2ME is composed of two major components: configuration and profile, which together establish an implementation specification for consumer electronics and embedded device manufactures (Sun Microsystems, 2002). J2ME is built up on a three-layered architecture (from bottom to top): JVM Layer, Configuration Layer, and Profile Layer. The JVM layer is an implementation of a specific JVM that is customized for an operating system on small devices. For example, the K virtual machine (KVM) is a highly portable virtual machine, designed for small memory, limited-resource, and network-connected devices such as cellular telephones, pagers, and small point of sale (POS) systems. The configuration layer defines a minimum set of Java virtual machine features and Java class libraries available on a particular category of devices. At present there are two J2ME configurations: the Connected Limited Device Configuration (CLDC) and the Connected Device Configuration (CDC). CLDC is designed for developing applications on resource-constrained devices with parsimonious memory, say, less than 512KB. On the contrary, CDC is developed for embedded devices with relatively larger amounts of memory and more robust resources like set-top box, screen phone or refrigerator. The configuration layer also specifies core class libraries of device-dependence so that the applications can be developed regardless of the devices they will be deployed to.

The top layer is the so-called profile layer, which is independent on the connection layer. A profile is an industry-defined specification of the Java APIs which fits for a specific industry or class of devices. For example, the Mobile Information Device Profile (MIDP) coupled with CLDC specifies the issues of user interface, persistent storage, and networking for mobile information devices such as cellular phones. The so-called K-Java is a complete J2ME application run-time environment fully adhering to the CLDC specification whereas PersonalJava (the so-called pJava) is partly compatible to the CDC specification since the review of CDC specification is still underway. The former one is suitable for consumer electronics with colorful terminals like Windows CE devices while the latter is good for resource-constrained devices such as PalmPilot. J2ME is designed on the basis of philosophy that applications developed on a particular profile can be deployed straightforward onto different configurations without having to modify the code. However, such ideal has not been realized currently. In this paper, we adopt the MIDP/CLDC approach to develop the surveillance system on the K-Java environment. It is expected the hurdle of platform heterogeneity can be removed when CDC is finalized.

Table layer also contain a group of API, defined aiming at mainly the specific equipment of certain a clan department. Table layer realizes on specific disposition layer, programmer takes the responsibility for compiling application program on specific Table layer. J2ME has defined many Tables, but only MIDP has realization reference, MIDP (Mobile Information Device Pro2file) is designed for mobile telephone and two-way paging equipment. It is very similar that MIDP application is called as MIDlet with Java Applet, they have similar life cycle. The life cycle of MIDlet can 3 stages: (1) start, get resource and run; (2) suspend, release resource and change to wait state; (3) die out, release the resource owned, finish thread and all activities. MIDP and CLDC have formed a complete configuration together. Fig. 1 has shown the structure of CLDC/MIDP of development of mobile telephone program.

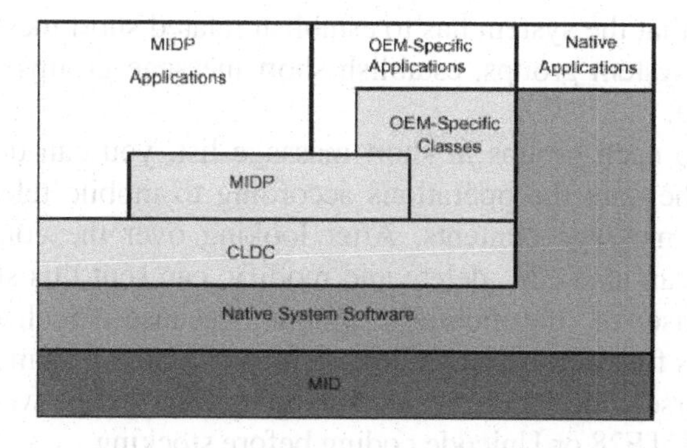

Figure 1. The structure of CLDC/MID

3. THE DEVELOPMENT OF SHORT MESSAGE MANAGEMENT SYSTEM

Developing a MIDlet program or other wireless Java application software following three steps: (1) compile application software; (2) test application software in simulator; (3) download the compiled software to the mobile phone, then operate and test.

Short message management system belongs to MIDlet (Ganhua et al., 2003), which is a short message management system developed with API of Siemens. Purpose of compiling the system is that user can manage the short message on mobile telephone effectively, its major functions include dividing short messages into groups, the editor, insert, delete operation of short message list and message content etc., additionally realize short message to dispatch operation.

3.1 Development tool and development environment

The popular J2ME development tools include: (1) SUN Wireless Toolkit + UltraEdit + various simulators; (2) Builder + Mobile Set + Nokia SDK; (3) JCreator + MotoSDK. The JDK + UltraEdit + SMTK6688i was used to develop J2ME which is provide by Simens in our system.

3.2 System profile

The system is composed of 12 operating class, can realize management of short message and some related operations based on this function. User can use the groups that the system has to establish related short message list, also can delete the system groups, establish short message groups, and also can change its name.

According to each groups in short message list, you can delete and add and modify. They are the operations according to mobile telephone of the user and short message contents. After looking over the content of short message, user can also edit, delete and modify, can sent this short message to the current user of the mobile telephone. Because it realizes by a class which improves function based on com.siemens.mp.io. of Siemens class File, we must converse code before we do file operation, that is, we must change ASCII yard to UTF28 or Unicode coding before stocking.

Above analyzed, we can get the structure of systematic function as Fig. 2 shows.

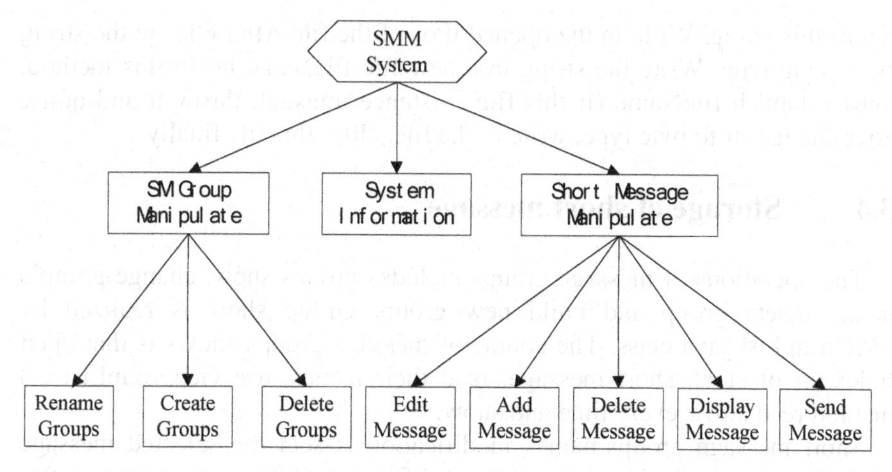

Figure 2. SMM system structure

3.3 The file operation of short message group

Because the J2ME language of standard do not offer file operating class, this system used a ManipFile class that compiled on com.siemens.mp.io.File of Siemens to realize related file operation. SMM.java class is the frame of whole system, its major function is use SMM structure function to example the example of current program, its value is this indicator of current program, this value will be used for getting to show equipment and system the sentence during the whole process. In addition, the SMM.java class realizes the interface between the systemc and application program management equipment, application program management equipment is the software that installed on MIDP equipment in advance and acts the role of operation environment. It can use the life period method of MIDlet in order to control the state of MIDlet. SMMInfo.java class shows copyright of the system, the informations and helps by inheriting Form. Add two buttons on Form, it will respond to the operation that user press one button by realizing CommandListener interface, for example, it will enter to show short message list after pressing cmdEnter, and exit cmdExit.

ManipFile. java class is the key class concerned with file operating during the system operation course, during the system operation course, all operations concerned with file operating have corresponding definitions in this class. File the structure function of operating class, open fileName file, if file does not exist, build a new fileName file f isNew is true, otherwise, it throw unusual and quit program. After opening file, structure function with initialize operation such as the descriptions of the file, the length and the name of this file, and the current file indicator etc. According to the file indicator location that offers, read the content as (iStart2iEnd) from file and return this string. In current file, read one row of character from nStart, and return this string. Write in the opened iPos of the file After change the string str to byte type. Write the string in a new file fileName in. In this method, must establish fileName (if this file existence unusual, throw it and quit), after change str to byte type, write in the file, close this file finally.

3.4 Storage of short message

The operations of message groups includes groups show, change group's name, delete group and build new group. Group show is realized by SMGroupList.java class. The course of message groups shows is that open Index.txt of stock short message, read their names, use GetGroupList ( ) method read a row every time and show.

Short message groups'names modification is sent the selected message group's name to the groups' name modification class as parameter. After modification, must write the result of modification to stock groups Index. txt file.During the course of stocking, must use the method InsertStrToFile of ManipFile. java class. Adding short message groups is also establishment the method of groups requirement completion after need should groups name join. txt file end, use ManipFile. FileAppend of java class ( ) method.

In the course of deleting message groups, must conside the possibility of short message groups, because before deleting, should warn; If confirme to deleted, besides delete the current groups' names from Index. txt file, and delete the file stocking current message. Because short message groups file named with sequence number, so just according to current groups sequence number, we can find very easily short message groups file and delete it, then, rename short message files after this file again according to digital order.

If some errors such as transfer failure, invalidate retry occurred in SM sending module, SM receipting module and link maintain module, it will be terminated in the corresponding module, The manage module will take some measures such as resume the link, restart the module and so on according to the terminate sate.

3.5 The list of short message

Message list management have realized the operations – show, delete, modification, add etc in divided group messages. Short message list show by SMSList.java class. Construct function shows current message list and open corresponding file to read short message list of this groups and show in Form by recepting the parameter and sequence number of the title. In the course of adding short message content, we must input short message content and mobile telephone which is made up with 11 numbers, which is controlled by the sentence "numTxtF ld = newTextField ("the number of mobile telephone" ",", 11, TextField. PHONENUMBER)". Because generally length of the message content is also limited, which can be control according to specific condition by the sentence "smsTxtFld = new TextField ("short letter content" ",", 200, TextField. ANY) (Maximum will be 200 characters short the length definition of news content)". The editor of short message list is similar with short message groups. During the deleting of short message list, we must consider the length of short message and find out the location of this message, then carry out related insert operation, replace the short message content to delete with space. List show and content input interface as Fig. 3 show.

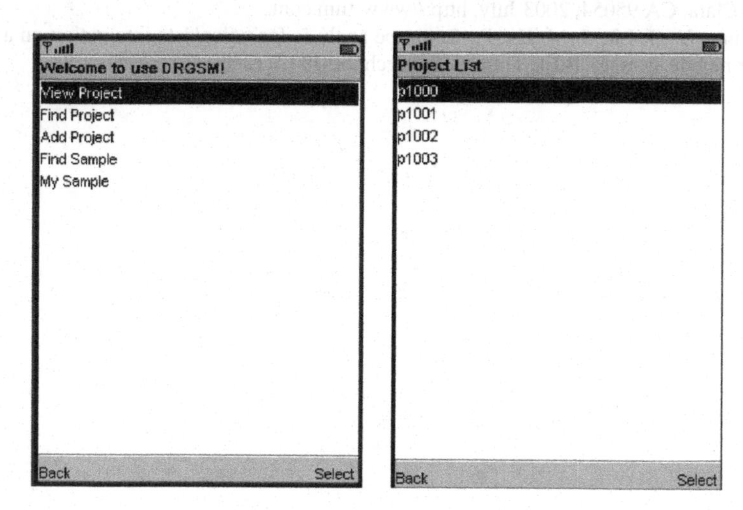

Figure 3. Short Message List

4. CONCLUSIONS AND FUTURE WORKS

This system is used Siemens API for special purpose to develop , in the process of development, it uses Siemens6688i simulator to carry out debug and operate, and test successfully on the mobile telephone of Siemens, if it to be used in other simulator or the mobile telephone excepted Siemens, need to carry out further modification and test.

REFERENCES

ETSI/TC, "Use of Data Terminal Equipment-Data Circuit Terminating; Equipment (DTE-DCE) Interface for Short Message Service (SMS) and Cell Broadcast Service (CBS)," v. 5.3.0, Tech. rep., Rec. GSM 07.05, 1997.

Gerd Beuster, Thomas Kleemann and Bernd Thomas, "Multi-Agent Location Based Information Systems for Mobile Users in 3G Networks" Submitted to WI2003-6, International Tagung Wirts chafts informatik 2003.

Ganhua Li and Yunhui Liu, Robotics. A distributed and adaptive data flow system for SMS[C]. Intelligent Systems and Signal Processing, vol. 2 1350, 2003

Sun Microsystems, J2ME CLDCAPI 1.1, Inc. 2003

Mobile Information Device Profile (MIDP). http://Java.sun.com/products/midp/, 2002-09. [EB/OL]

J. Desbonnet, P.M. Corcoran, System architecture and implementation of a CE Bus/Internet gateway, IEEE Transactions on Consumer Electronics 43 (1997) 1057–1062.

J. Ellis, J2ME Web Services Specification, Sun Microsystems, Inc., 4150 Network Circle Santa Clara, CA 95054, 2003 July, http://www.sum.com.

Y.-R. Haung, Y.-B. Lin, J.-M. Ho. Performance analysis for voice/data integration on a finite buffer mobile system, IEEE Trans. Veh. Technol. 49 (2) (2000) 367–378.

THE ENLIGHTENMENT TO CHINA OF E-GOVERNMENT APPLICATION IN RURAL AREAS FROM JAPAN, SOUTH KOREA AND INDIA

Wenyun Liu[1], Jinglei Wang[1,*]

[1] *Institute of Scientific & Technical Information, Shan Dong University of Technology, Zibo, China, 255049*

[*] *Corresponding author, Address: P.O. Box 342, Institute of Scientific & Technical Information, Shan Dong University of Technology, 12 Zhangzhou Road, Zibo, 255049, P. R. China, Tel: +86-0533-2781428, Email: reeeee@126.com*

Abstract: E-government is one of the major forms of government informatization. It has played an important role in the rural areas of some Asian countries. This article gives an analysis of e-government application in some rural areas in Asia and the status of e-government application in rural China, and gets some enlightenment to solve our problems.

Keywords: e-government, rural, enlightenment, Asia

1. THE APPLICATION OF E-GOVERNMENT IN RURAL ASIA

1.1 Japan: Pays attention to agricultural information system. And its computer network develops rapidly

First, Japan pays attention to the establishment of market rules and development policy on rural informationization. According to agricultural production and market operation rules, the government established a number of specialized advisory committee and formulated a series of system rules and operational rules to restrict the code of conduct of all aspects of the market. Moreover, the government established development policies

Liu, W. and Wang, J., 2008, in IFIP International Federation for Information Processing, Volume 258; Computer and Computing Technologies in Agriculture, Vol. 1; Daoliang Li; (Boston: Springer), pp. 35–42.

according to the actual needs to promote the market operated orderly. Second, the government pays attention to agricultural infrastructure construction. Successive Japanese governments have attached great importance to rural telecommunication, broadcasting, and TV development. Currently, Japanese Ministry of Agriculture, Forestry and Fisheries are in the process of formulating a "21st Century Agriculture, Forestry and Fisheries fields of information strategy", and the basic idea is greatly augment the rural information and communication infrastructure, including the laying of fiber optic cable in order to build advanced communications network.

Japanese agricultural market information service is composed of two systems (Yang, 2005). One is the "Federation of Agricultural Central Market," the other is various agricultural products production volume and price forecasting system composed of the national 1800 "comprehensive agricultural group". With the accurate market information offered by the two systems, each farmer can know well of what is good market, the price, and each product's production quantity on the domestic market and even on the international market. They can identify and adjust their product and production according to their actual ability to make the production in a clear, highly orderly state.

Japan had developed more than 400 agricultural networks by the end of 1994. The computer prevalence rate had reached 93% in agricultural production sector. The agricultural and technological information network system developed in recently two years, farmers can query and use the network data at any time. Meanwhile, the Japanese government has attached great importance to the popularization and application of computers in rural areas, Japanese farmers will receive subsidies to purchase computer.

1.2 South Korea: "information village" project

The "information village" program started in 2001. The same year in March, the "information village planning groups" made up of the Ministry of Agriculture and Forestry and the Information and Communication Ministry, the Ministry and other government departments, administrative autonomy started planning and the experimental work, and was specifically responsible for the operation of the Ministry of administrative autonomy.

The ministry of administrative autonomy fixed on the first batch of more than 20 identified "information village" model as of May 2002, and gradually generalized when experience was gained. The project was welcomed by farmers quickly in the country and expanded rapidly. There were 78 and 88 "information model village" set up in central city in 2003 and 2004. There were newer 89 in 2005. "Information model villages" were in a total of 280 and 79,300 farmers participated as of May 2005.

Every "information model village" is a system engineering and essential construction project include: First, the construction of high-speed Internet infrastructure, including laying specific optical cable, constructing equipment installed mainframe room, achieving broadband Internet access to farmers; Second, it is necessary to establish rural information centers, and provide computer word processors and other hardware, and make the achievement of the information network connected with the local administration; Third, it establish farmers computer using environment, the existing model village has been able to equipped more than 73% of the rural households with computers; The fourth is to establish management operation system, the business committee is formed by the model village's villages, information-based instruction personnel and information center management personnel participate and run together. The fifth is the education and training of personnel. South Korea had trained 233,500 villagers to use computer and Internet, and cultivated a group of information technology and key management personnel in rural areas as of May 2004 (Cao, 2006).

The "information village" program has taken obvious effect. It behaved outstandingly that information network greatly facilitates the flow and transactions of produce and enliven the experience of rural tourism and cultivated the rural and native products brands. It made agriculture and rural areas got benefits and farmers' income increased. It contributed to increasing region's economic competitive promoted balanced regional development.

1.3 India: GYANDOOR program and CSC program

GYANDOOR program is a unique e-government project. As shown in Figure 1:

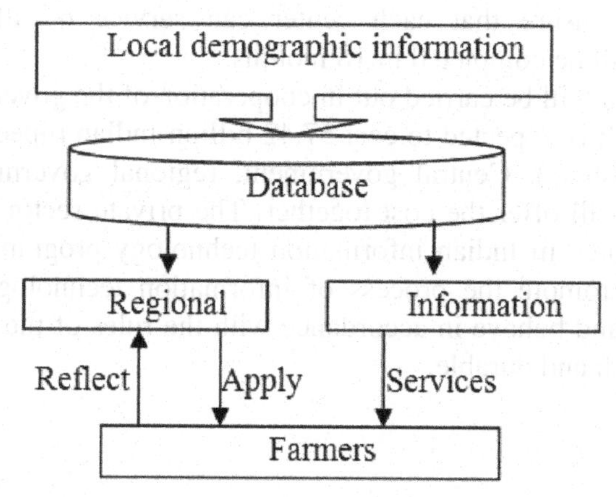

Figure 1. GYANDOOR program

The project started in November 1999, mainly serviced rural areas and the tribal Dhar. A lot of important local demographic information such as income, estate, native place, ownership of land, creditor's rights are stored in computer. The network runs through regional capitals and 21 independent operate information centers. Most of these information centers locate in the town center street, which people often go across. There, each villager can enjoy a range of services such as a certificate of origin supply, housing ownership certificates, local welfare and the latest information on agricultural prices. Even if some farmers are illiteracy, they can also understand the price figures appeared on the screen with the operator's help.

In addition to the services above, the villagers could also reflect problems to the senior manager of the region through network, such as lack of agricultural experts, pump failure, official misconduct and dealers to sell fake seeds, inferior fertilizers and so on. There must be a reply in 7 days under the supervision. India is a big agricultural country, 70% of the population scattered in 600,000 villages, although the teledensity (fixed telecommunication and mobile communications) per capita share rate is only 7% in India, rural telecommunication services coverage rate reached 83%. Moreover, they also set up a new target: any village within a radius of five kilometers can have the wireless Internet access.

Recently Indian cabinet approved the establishment of 100,000 broadband common service centers (CSC program). The aim is to popularize the digital services to every citizen, and increase the opportunities to the economic development of remote areas, and reduce the digital divide. Indian government also said that broadband common service center is one of the three major infrastructures of e-government. According to the program, the 100,000 broadband common service centers will distribute in rural areas reasonably to ensure that each center can service 6 villages. All the construction will be completed in 18 months.

This program will be carried out in cooperation of the government and the private sector. It is expected to cost 57.42 billion Indian rupees (about 1.254 billion U.S. dollars). Central government, regional governments and the private sector will offer the cost together. The private sector's participation is a positive force in Indian information technology program. Its existence will actively promote the process of information technology in the rural areas to work and behave in accordance with the rules of the market. It will be more flourish and durable.

2. THE STATUS OF E-GOVERNMENT APPLICATION IN RURAL CHINA

2.1 The deficiencies of e-government application in rural China

The deficiencies of e-government application in rural China are shown in table 1.

Table 1. Deficiencies

Deficiencies	Details
Low Internet popularity rate in rural areas (CNNIC, 2006).	As shown in Figure 2, the 17th Chinese Internet Development status Statistic shows that there is a huge difference between urban and rural areas in Internet popularity rate. It shows that the number of Internet users in rural China is only one-fifth of urban Internet users, and rural Internet popularity rate is only one-sixth of urban Internet penetration rate.
There is no uniform standard in website construction or inability to share information resources.	On one hand, the construction is low-level repetitive. On the other hand, there is no unified standard data. Valuable information resources belong to the agricultural sector, agricultural enterprises and related research institutions respectively even in the same agricultural information service website, and there is no cross between them.
The agricultural websites are inadequate to farmers' needs (Fan, 2007).	Moreover, the content of the website is identical; its update speed is slow. Generally speaking, the financial investment in the agricultural gives more convenience than its benefits to farmers or agricultural enterprises. The information that reflects phenomenon is too much, but the information analyzed or assisting leaders to make macro decision and operators to make micro decision is too little.
Information form is single and not timely (Wang, 2006).	Farmers' demand to information is various, for example, many agricultural crop farming skills need a large number of pictures and video to explain and demonstrate. But now, agricultural web pages are more static than dynamic and lack of website navigation, The site isn't vivid enough. It goes against farmers' learning and using. The sites can't update in time. The information on the website is incomplete and inaccurate, the content is dull and the value is low in use and practicality.

NOTES

1. The following is the 17th Chinese Internet Development status Statistic, as shown in Figure 2:

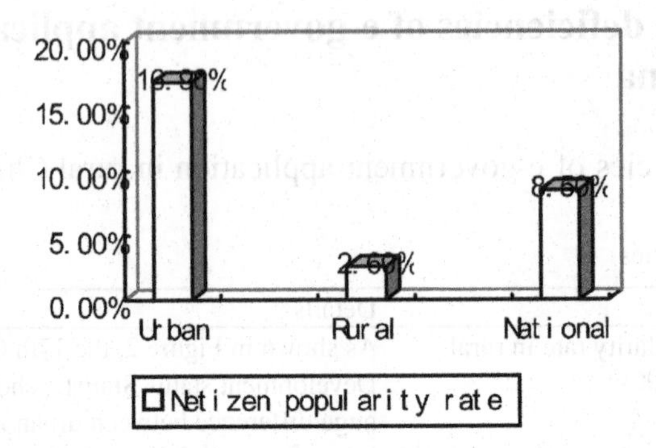

Figure 2. The 17th Chinese Internet Development Status Statistic issued by CNNIC

2.2 E-government achievements in rural China

E-government achievements in rural China are shown in table 2.

Table 2. Achievements

Achievements	Details
Golden Agriculture Project (Liu, 2005).	This project will further promote the standardization of Chinese agricultural construction. It can improve the government's management and service levels
We have built up an agricultural information network system elementarily (Liu, 2003).	The Ministry of Agriculture has established a national agricultural portal website with Chinese agricultural information network as its core including more than 20 professional networks. We have built national agricultural management satellite communications system and national fishery command center, and achieved national industry-wide command and control network.
The rural information service is gradually strengthened (Wang, 2007).	Agricultural departments at various levels actively change their functions and responsibilities to fulfill the government information service, make a comprehensive exploration of information services, and gradually become institutionalized.

3. THE ENLIGHTENMENT TO CHINA

We can get some enlightenment from the three counties above and we should take measures to solve our problems.

Table 3. Enlightenment & Measures

No.	Enlightenment & Measures
1	We must strengthen the building of rural infrastructure and improve the level of information technology in the rural areas (Wang, 2006).
2	The government should provide analyzing and forecasting services and strengthen the role of guidance.
3	We should strengthen the systemic training of farmers. We should improve farmers' quality, and promote the effective use of information.
4	We should formulate unified standards to strengthen rural information system.
5	We should adopt various methods to disseminate information to farmers according to local conditions.
6	We should create the information needs of farmers, and strengthen the farmers' sense of information.
7	We should establish and improve the rural information service system.

We conceive an agricultural information system, as shown in Figure 3:

The core of the system is local agricultural Information center. It is responsible for carrying transferring and controlling messages; Portal is

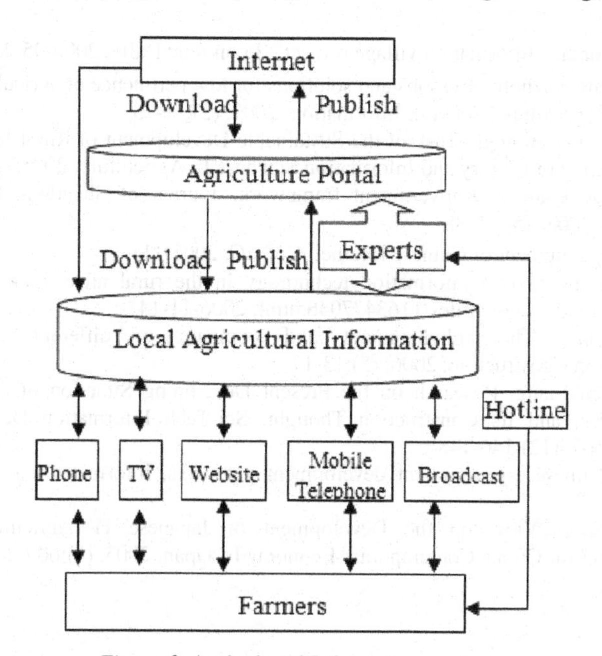

Figure 3. Agricultural Information system

responsible for information collection, processing and dissemination. Experts are responsible for the technical support, the provision of information and management. Farmers can obtain the useful information from TV, website, Mobile telephone, Broadcast and hotline.

This system has the following advantages:

(1) It is easy for farmers to obtain the information; they can obtain the information in many ways, such as TV, website, Mobile telephone, Broadcast and hotline.

(2) Farmers can obtain the information immediately. They can obtain the information through SMS platform by mobile telephone.

(3) Farmers can obtain authoritative information form the experts; the information they obtained is very professional.

(4) The system is very integrated. Portal, information centers, experts, a fixed telephone, television, websites, mobile phones and radio form an integrated information system in rural areas. It overcomes the problem that farmers don't have computer, they can obtain information by phone or TV etc.

May this system play an important role in rural informatization.

REFERENCES

Cao Shigong, Korea: "Information village project", Economic Daily, 2006-05-23

Fan Jizheng; Yang Jinzhong, Reasons and solutions for low pertinence of agriculture websites user groups, Agriculture Network Information, 2007, (2):71-73

Liu Shihong, On the Strategic Task of the Sustainable Development of Rural IT Application in China, Journal of Library and Information Sciences In Agriculture, 2005, (2):5-9

Liu Wenyun, On Chinese e-government framework. Journal of Shandong University of Technology, 2003, (5):35-38

Report on the development of Chinese Internet, CNNIC, 2006, (1)

The way Indian popularize information technology in the rural areas is worth learning, http://www.ciia.org.cn/genfiles/1163477046.html, 2006-11-14

Wang Xiangdong, The enlightenment and compare of different e-government, Informatization Construction, 2006, (5):13-17

Wang Yan, Chen Liang, Research on the Present Developing Situation of Chinese Rural Informatization and Its Construction Thought, Sci-Tech Information Development & Economy, 2007, (12):146-148

Wang Yukai, Chinese e-government development prospects, Informatization Construction, 2006, (1):10-15

Yang Yi, A Brief View on the Development of Japanese IT Agriculture and the Enlightenment for China, Contemporary Economy In Japan, 2005, (6):60-62

DESIGN AND IMPLEMENTATION OF A MOBILE MANAGEMENT SYSTEM FOR CAMPUS SERVER

Shijue Zheng, Zhenhua Zheng [*]

Department of Computer Science, Huazhong Normal University, Wuhan, China, 430079

[*] *Corresponding author, Address: Department of Computer Science, Huazhong Normal University, Wuhan, China, 430079, Email: wjzzh2002@yahoo.com.cn*

Abstract: Server Management plays an important role in Campus Network Management which is crucial to build an effective campus network. However, if the server makes some small mistakes while he doesn't stay in the server room, he should come back quickly from another place which is far from the server room to solve just a small problem. The authors design a mobile management system for campus servers, and implement it with Microsoft .NET Mobile Web SDK and SQL Server 2000. The system is based on B/S structure and WAP (Wireless Application Protocol), and it has been proved to be effective and reliable through the software testing.

Keywords: Microsoft .NET; mobile management system; WAP; campus server

1. INTRODUCTION

In our department, someone is put in the charge of Server Management. He should stay in the narrow server room to monitor the server all day. The job is dull but not effective, because it is not necessary to keep your eye on a server which makes few mistakes. However, if the server makes some small mistakes while he doesn't stay in the server room, he should come back quickly from another place which is far from the server room to solve just a small problem, e.g. restart the server. So it is a waste of human resources. Somebody else has developed a system which can be used to manage the

Zheng, S. and Zheng, Z., 2008, in IFIP International Federation for Information Processing, Volume 258; Computer and Computing Technologies in Agriculture, Vol. 1; Daoliang Li; (Boston: Springer), pp. 43–50.

server in a distance place through Web. However the network manager still needs a PC which connects to the Internet to use the system, so it is lack of mobility. To solve the problem we design a mobile management system for campus servers, and implement it with Microsoft.NET Mobile Web SDK and Microsoft SQL Server 2000. The system is based on B/S structure and WAP (Wireless Application Protocol), and it has been proved to be effective and reliable through the software testing (Microsoft MSDN, 2001). If the server has installed the mobile management system, the network manager can manage it in anywhere with a mobile phone or PDA which supports WAP. After signed in, the network manager can submit any command lines to manage the server, even shut down it (Jiang. A, 2002).

2. SYSTEM DESIGN

2.1 Preliminary Design

The mobile management system is based on B/S structure. The client can be any mobile devices which has a browser that supports WAP. On the server-side there is the mobile web application built by the authors and the Microsoft SQL Server. In normal condition, the mobile web application runs quietly on the server-side. When the network manager wants to manage the server, he just uses his mobile device to download the server's pages which are produced by the mobile web application (3Com, 2002). When he has signed in, he can see the management center page. To complete a certain mission he just inputs the corresponding command lines and then submits them to the server. Then the mobile web application receives the command lines and transmits them to the SQL Server. The SQL Server executes these command lines with an extended stored procedure called xp_cmdshell which has the ability to execute operating-system commands and then returns the result as a dataset. After that the mobile web application captures the result dataset, and then transforms it to a certain format (e.g. WML) which is decided by the client-side device, and finally sends it to the browser. Now, the manager has completed his mission and seen the results of his operating to the server. As figure 1 shows, is the infrastructure of the mobile management system (Antinisca, 2007).

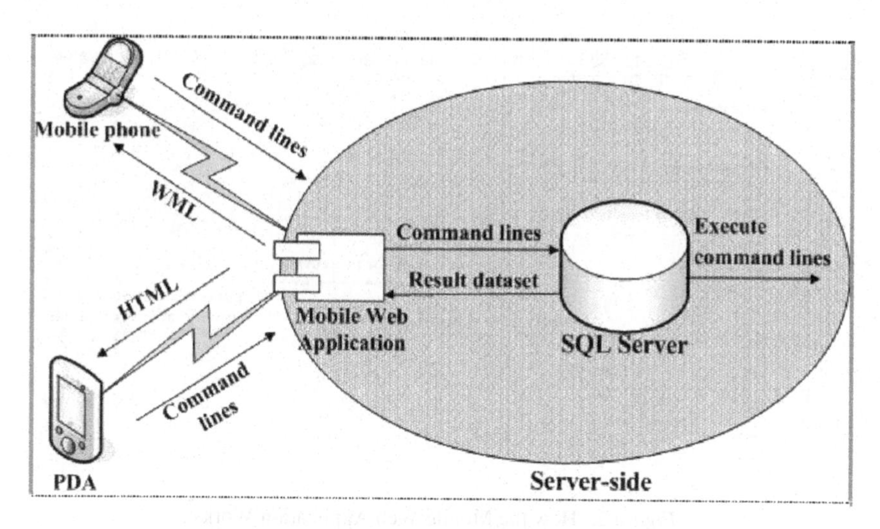

Figure 1. Infrastructure of the Mobile Management System

2.2 Detailed Design

One of the advantages of B/S structure is that there is no need to develop software for client-side, so the user has only to own a mobile device which supports WAP. We put more emphasis on the design and implementation of the mobile web application because it's the interface between the mobile device and the server (Gianluig, 2004). The mobile web application consists of three functional modules, which are the sign-in module, the execute-cmd module and the show-result module. The sign-in module can be used to provide safety for the server by refusing disabled users' accessing. The execute-cmd module is responsible for the executing of the command lines. However he can't do it separately, he must call the SQL Server's stored procedure (Volker, 2004). And then the show-result module receives the result as a dataset from the SQL Server. Finally the show-result module uses the dataset to produce a page with certain format and sends it to the client-side browser (Amar, 2004). The whole mobile web application works as Figure 2 illustrates.

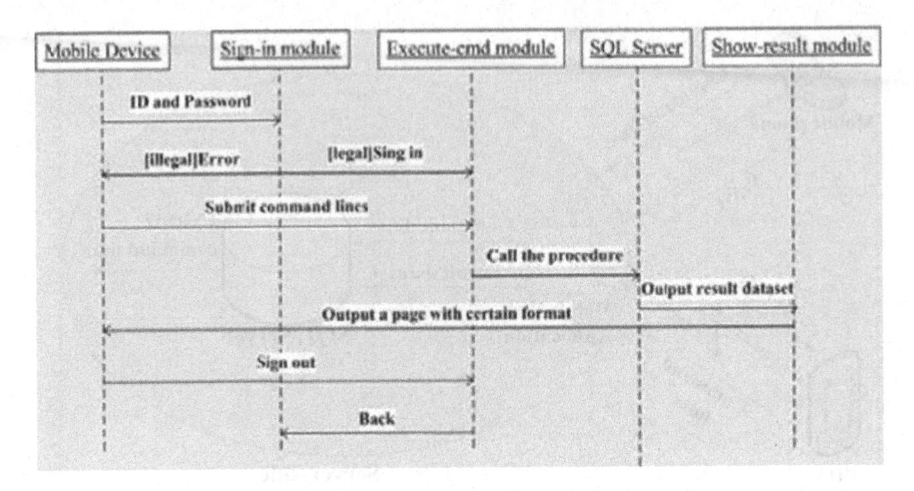

Figure 2. How the Mobile Web Application Works

3. SYSTEM IMPLEMENTATION

3.1 System Environment Configuration

(1) Hardware requirements for the server (see table 1).

(2) Software requirements for the server. The operation system should be Windows 2000/2003 Server with IIS5.0 (or higher) and with Microsoft .Net Framework 1.1 installed (Fredrik, 2000).

(3) Requirements for the client. Any mobile device with the browser which supports WAP.

3.2 Configuration of SQL Server

The system uses the extended stored procedure named xp_cmdshell in SQL Server, which is a command shell of windows operation system. It has the ability to execute any command of operation systems, such as "copy", "del", "md" and even "format" or "tsshutdn" (the command to shutdown the computer in Windows Server 2003). So only the role that belongs to the System Administrators Group can use this procedure. We

Table 1. Hardware requirements for the server

Level	Hardware configuration
Lower	P4 2.8G/512M RAM 80G Hard disk
Middle	2*P4 2.8G/1G RAM /72G SCSI Hard disk
Higher	4*P4 2.8G/2G RAM /2*72G SCSI Hard disk

have created a role called MobAdmin which belongs to such a group. We use it to login the SQL Server (Openwave Systems Inc, 2003).

3.3 Development of the Mobile Web Application

We use Visual Studio .NET 2003 IDE to create a mobile web application project named MobSys. It consists of three forms. (See Figure 3).

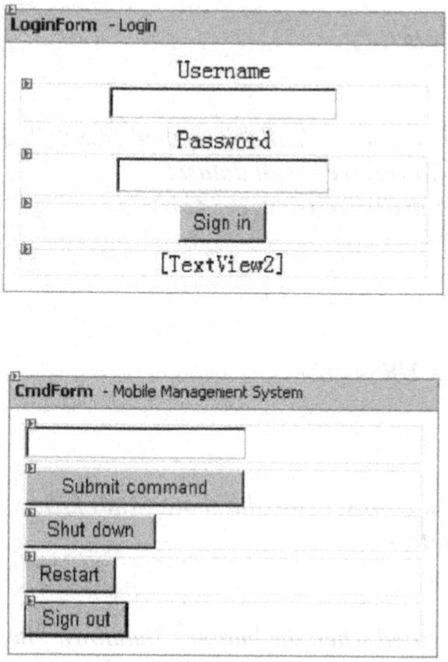

Figure 3. Forms of the Mobile Web Application

Before programming we should store some configuration information in a file called Web.Config. In order to make the system own good portability we store database connection string, username and password in the file Web.Config. The key codes are as follows.

```
<appSettings>
<!—Database connection string -->
<add key="ConnectionString"
value="Server=(local);uid=MobAdmin;pwd=74138;Database=master;" />
<!—username and password -->
<add key="username" value="admin" />
<add key="password" value="123" />
</appSettings>
```

Then we write C# codes to implement three functional modules designed above. Some of the codes and notes are as follows.

```
//the method to receive command lines and output result as dataset
private SqlDataReader EXEC(string CmdStr)
{
//connect to SQL Server
```

```
SqlConnection conn=new
SqlConnection(ConfigurationSettings.AppSettings["ConnectionString"]);
   conn.Open();
   //call the stored procedure xp_cmdshell
   SqlCommand SqlCmd=new SqlCommand ("exec xp_cmdshell '"+CmdStr+"'",conn);
   return (SqlDataReader)SqlCmd.ExecuteReader();
   }
   // following codes are executed when the button "submit command" has been pressed
    private void Command1_Click(object sender, System.EventArgs e)
    {
     SqlDataReader rs;
     string ResultStr="Done.<br/>";
     try
     {
     rs=EXEC(TextBox1.Text.Trim()); //call the method "EXEC" defined before
      while(rs.Read()) //loop to receive result dataset
      { ResultStr+=rs.GetValue(0).ToString()+"<br/>";
      }
     }
     catch
     {
      ResultStr="ERROR OCCURS<br/>";
     }
     finally
     {
      ActiveForm=ResultForm;    //Activate the show-result form
      TextView1.Text=ResultStr;
     }
    }
   //following codes are executed when the button "Shutdown" has been pressed
   private void Command2_Click(object sender, System.EventArgs e)
   {
    EXEC("tsshutdn "); //tsshutdn is the command to shutdown the computer in Win2003
   }
```

4. INTEGRATED TESTING

In order to guarantee the stability and reliability of the system we have tested it in many mobile devices and simulators. In the following devices and simulators, the mobile management system runs well (Wireless Application Protocol Forum, 2004).

- Pocket PC
- Sony CMD-z5 with Microsoft Mobile Explorer Mitsubishi T250
- Nokia 7110
- Sprint Touchpoint
- Samsung Touchpoint
- Simulator for Microsoft Mobile Explorer version 2.01
- Simulator for Phone.com UP 3.2
- Simulator for Nokia 7110

- Simulator for Phone.com UP 4.0
- Personal computer with Microsoft Internet Explorer 5.5
- Openwave V7 Simulator

Figure 4 shows the interface of the system. Figure 5 shows the show-result form when the command to backup database files has been executed. Figure 6 shows what the server does when the button "Shut down" has been pressed.

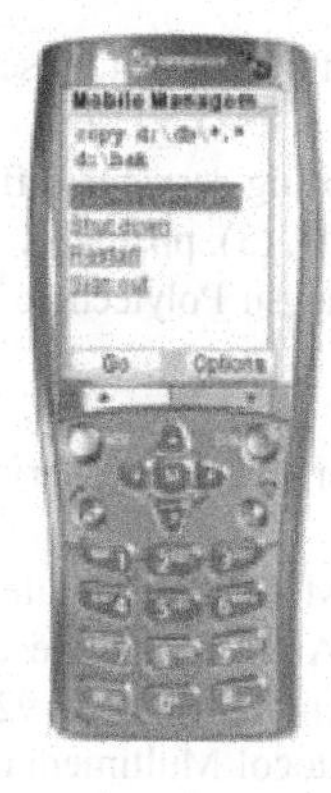

Figure 4. The Interface of the System

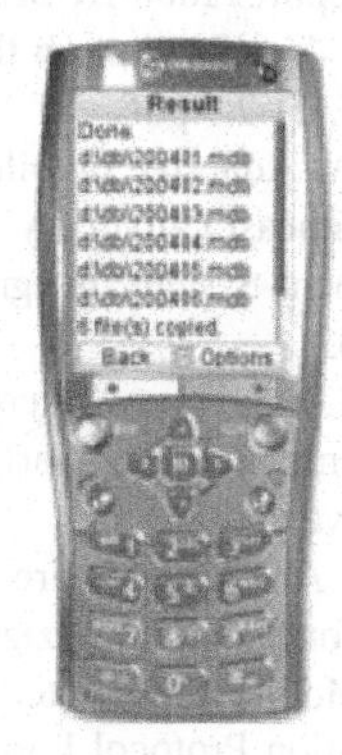

Figure 5. The Result When the Command to Backup Files Has Been Executed

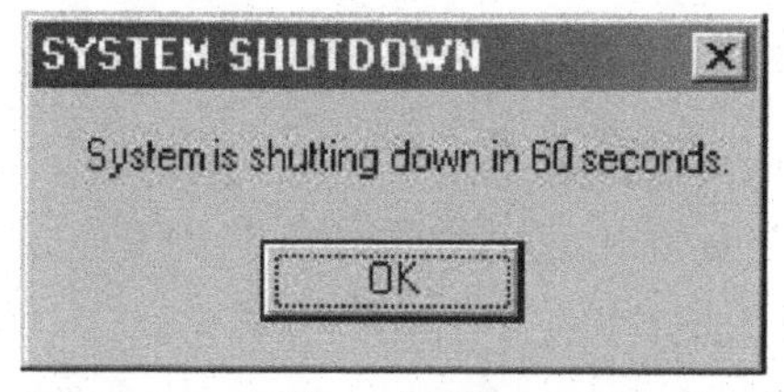

Figure 6. The Server's Response When the Button "Shut down" Has Been Pressed

5. CONLUSION

In order to manage the campus server in anywhere we have designed and implemented a mobile management system. It has been proved to be effective and reliable through the software testing. If the server has installed the mobile management system, the network manager can manage it in anywhere with a mobile phone or PDA which supports WAP (Library Management System, 2006). After signed in, the network manager can submit any command lines to manage the server, even shut down it.

REFERENCES

3Com White Papers. http://www.3com.com.cn/solutions/whitepapers/

Amar. R, B. WAP Management with user defined content reduction. http://www.cs.utexas.edu/~cdj/wia_files/submissions/001Final.pdf

Antinisca. D and Cecilia. M. Performance Analysis and Prediction of Physically Mobile Systems.proceedings of WOSP'07, February 5–8, 2007.

Fredrik. B. A Management System on a WAP Platform. http://www.ee.kth.se/php/modules/publications/reports/2000/IR-SB-EX0015.pdf

Gianluigi. F et al. Verification on the web of Mobile Systems. http://www.it.uu.se/profundis/Year2/Deliv2/A.1.2.9.pdf

Jiang. A, Zhao. A. Adaptive mobile management scheme supporting services differentiation. Journal of Southeast University (Natural Science Edition), 2004, (3), pp.22-26.

Library Management System. Integrated with WAP.Journal of Jiangsu Polytechnic University. 2006, (3) 49-52

Microsoft MSDN. http://www.microsoft.com/china/MSDN/

Openwave Systems Inc. Comparison of WAP Push and Short Message Service (SMS). www.openwave.com.

Volker G et al. An Architecture Description Language for Mobile Distributed Systems. http://ebus.informatik.unileipzig.de/papers/paperuploads/An_Architecture_Description_Language_for_Mobile_Distributed_SystemsVolker_Gruhn__Clemens_Schaefer9285.pdf.

Wireless Application Protocol Forum. Wireless Application Protocol Multimedia Messaging Service Architecture Overview Specification.http://www.wapforum.org

HYPERSPECTRAL LASER-INDUCED FLUORESCENCE IMAGING FOR NONDESTRUCTIVE ASSESSING SOLUBLE SOLIDS CONTENT OF ORANGE

Muhua Liu [1,*], Luring Zhang [1], Enyou Guo [1]
[1] *Engineering College, Jiangxi Agricultural University, Nanchang, China*
[*] *Corresponding author, address: Engineering College, Jiangxi Agricultural University, Nanchang, 330045, P. R. China, Tel: 086-0791-3813260; Fax: 086-0791-3813260; Email: suikelmh@sohu.com*

Abstract: Laser-induced fluorescence imaging is a promising technique for assessing quality of fruit. This paper reports on using a hyperspectral laser-induced fluorescence imaging technique for measurement of laser-induced fluorescence from orange for predicting soluble solids content (SSC) of fruit. A laser (632 nm) was used as an excitation source for inducing fluorescence in oranges. Fluorescence scattering images were acquired from 'Nanfeng' orange and navel orange by a hyperspectral imaging system at the instance of laser illumination. Subsequent analysis of Fluorescence scattering images consisted in selecting regions of interest (ROIs) of 100×50 pixels, and ROIs were segment around the laser illumination point from Fluorescence scattering images. The hyperspectral fluorescence image data in the wavelength range of 700-1100 nm were represented by mean grey value of the ROIs. The fruit soluble solids content were measured using hand-held refractometer. A line regressing method was used for developing prediction models to predict fruit soluble solids content. Excellent predictions were obtained for soluble solids content with the correlation coefficient of prediction of 0.998 ('Nanfeng' orange) and 0.96 (navel orange). The results show that hyperspectral laser-induced fluorescence imaging is a very good method for nondestructive assessing soluble solids content of orange.

Keywords: hyperspectral imaging, laser-induced fluorescence, orange, soluble solids content

Liu, M., Zhang, L. and Guo, E., 2008, in IFIP International Federation for Information Processing, Volume 258; Computer and Computing Technologies in Agriculture, Vol. 1; Daoliang Li; (Boston: Springer), pp. 51–59.

1. INTRODUCTION

Soluble solids content (or sugar index) is one of parameters for determining orange quality and maturity. Soluble solids content (*SSC*) may be determined from the juice extracted from fruit flesh using the refract metric method. This measurement method is destructive, inefficient and time consuming. A nondestructive sensing technique that is capable of measuring fruit quality parameters will be of great value in ensuring consistent quality fruit for the consumer. Many researches had been reported on the development of nondestructive sensing techniques for assessing soluble solids content of fruit. The technology and techniques include surface reflectance and transmittance of varying light energies. One good method is visible/near-infrared spectroscopy (VIS/NIRS). VIS/NIRS have became a non-destructive estimation of soluble solids content, oil contents, water content, dry matter content, acidity, firmness and others physiological properties of a number of fruit products indistinctly including citrus (Steuer et al., 2001); mandarin (Kawano et al., 1993; Miyamoto et al., 1997; McGlone et al., 2003; Antihus H. G. et al., 2006); and apple (Lu et al., 2000; McGlone et al., 2002; Park et al., 2003; Lu et al., 2005). However, spectroscopic assessment with relatively small point-source measurement have disadvantage compared to an imaging approach that characterizes the spatial variability of a sample material (Kim et al., 2001). Today, two recent techniques to access the quality of agricultural products are hyperspectral and mutispectral imaging. Hyperspectral and multispectral imaging is a relatively new technique for measuring the quality of food and agricultural products. The technique allows us to acquire both spectral and spatial information from a sample, thus offering some unique advantages over conventional imaging and spectroscopy techniques in detecting quality and safety of food and agricultural products (Lu et al., 1998). Latest research (Lu, 2003, 2004; Peng et al., 2005; Lu et al., 2006) showed that hyper- or multi-spectral scattering imaging, which measures light scattering and absorption, provides good assessment of fruit firmness. Several recent studies (Kim et al., 2003; Vargas et al., 2004; Ariana et al., 2006) reported that hyper- and/or multi-spectral fluorescence imaging is useful for detecting defects and safety of agricultural and food products.

In this research, a hyperspectral fluorescence imaging technique was used to measure laser-induced fluorescence scattering for assessing *SSC* quality of orange fruit. Specific objectives of this research were to:

- use a hyperspectral fluorescence imaging technique for acquiring fluorescence scattering images from orange fruit with the illumination of a laser of 632 nm.
- develop line regressing method relating fluorescence spectral features to *SSC* quality parameters of orange fruit.

2. MATERIALS AND METHODS

2.1 Orange samples

Two hundred and sixty five 'Nanfen' oranges harvested from the orchard of Nanfeng country of Jiangxi province, P. R. China in 2006. Two hundred and sixty five navel oranges harvested from the orchard of Xinfeng country of Jiangxi province, P. R. China in 2006. The oranges were kept in dark at room temperature (22°C) for at least 24 h before fluorescence and standard quality measurements were performed.

2.2 Hyperspectral imaging system

A hyperspectral fluorescence imaging system, schematically shown in Fig. 1, was assembled to acquire laser-induced fluorescence scattering images from orange fruit. The system consisted of a hyperspectral imaging unit, a 632 nm laser unit, a fruit holder, and the imaging chamber. The hyperspectral imaging unit was composed of a high sensitivity back-thinned charge coupled device (CCD) camera, an imaging spectrograph (ImSpector V10E, Spectral Imaging Ltd., Oulu, Finland), and a computer used to control the CCD camera and acquire images. The laser unit was equipped with a laser drive and temperature control, and its 632 nm laser was to generate a circular beam of 1.5mm diameter at the fruit. As the laser beam hit the fruit, a portion of the monochromatic light was absorbed by the fruit tissue with subsequent release of fluorescence (i.e., light of longer wavelengths). This resulted in a fluorescence emission distribution image around the illumination point at the surface of the fruit. The imaging spectrograph line scanned the fruit to collect the fluorescence scattering image at the surface of the fruit. As the light passed through the imaging spectrograph, it was dispersed into different wavelengths while its spatial information was preserved. The dispersed light signals were then projected onto the CCD detector, creating a two-dimensional image: one dimension represents the spatial and the other dimension spectral. The line scanning position of the imaging spectrograph was 2mm off from the beam incident center to avoid saturation caused by high intensity signals. The hyperspectral imaging system had an effective spectral region of 408-1117 nm with a nominal spectral resolution of 2.8 nm and a spatial resolution of 0.2 mm/pixel (Fig. 2).

2.3 Fluorescence imaging acquiring and fruit quality measurement

2.3.1 Fluorescence imaging acquiring

Images were collected in a dark room with only the laser light source. Fluorescence measurements were first performed on individual orange fruit, followed by standard measurements for *SSC*. Each orange with the stem-calyx end horizontal was placed on the plate which was drive by motor in the imaging chamber, which was completely shielded from ambient light during the measurement (Fig. 1). The hyperspectral imaging system captured the fluorescence scattering image of orange, and these images were saved in a computer for further analysis.

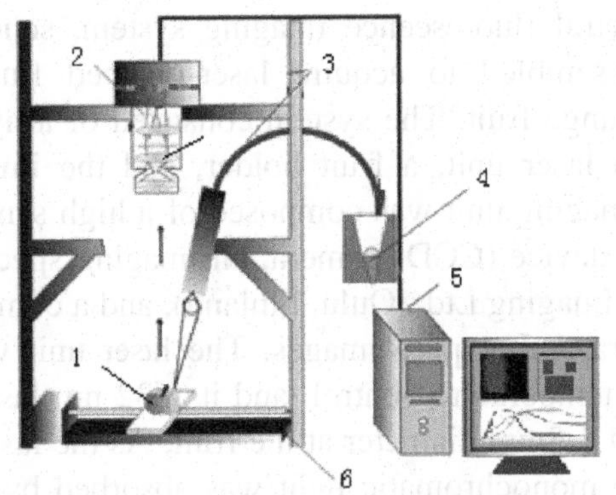

1.orange, 2.CCD camera, 3.fiber, 4.laser, 5.computer

Fig. 1. Hyperspectral imaging system

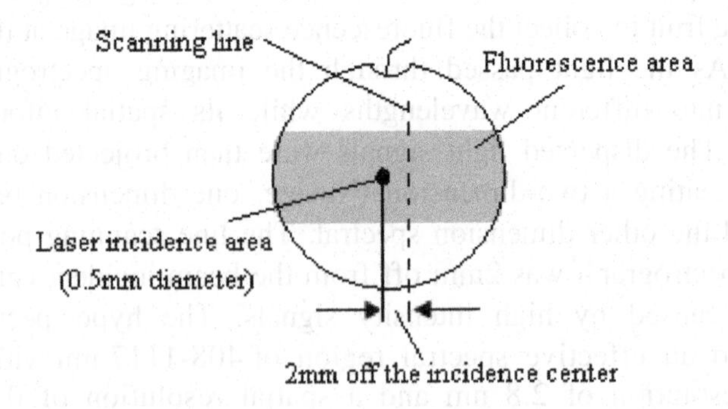

Fig. 2. The line scanning position of the hyperspetral imaging system, the laser incidence center and fluorescence area

2.3.2 Orange SSC measurement

Soluble solids content expressed in Brix was measured from the juice released using a hand-held refractometer (Model WZ—103, Zhongyou Optical Instrument Corp., China).

2.4 Fruit quality prediction models

A typical hyperspectral fluorescence image is displayed in Fig. 3. Each scattering image is composed of hundreds of spectra with each coming from a different position at the fruit surface. Subsequent analysis of Fluorescence scattering images consisted in selecting regions of interest (ROIs) of 100×50 pixels, and ROIs were segment around the laser illumination point track from Fluorescence scattering images. To properly characterize each scattering image, mean of CCD count was calculated from ROIs of the spatial scattering profiles (100×50 pixels) of the hyperspectral image for each wavelength from 700 to 1100 nm. This mean value calculated from ROIs was then assumed to represent the fluorescence intensity of fruit images.

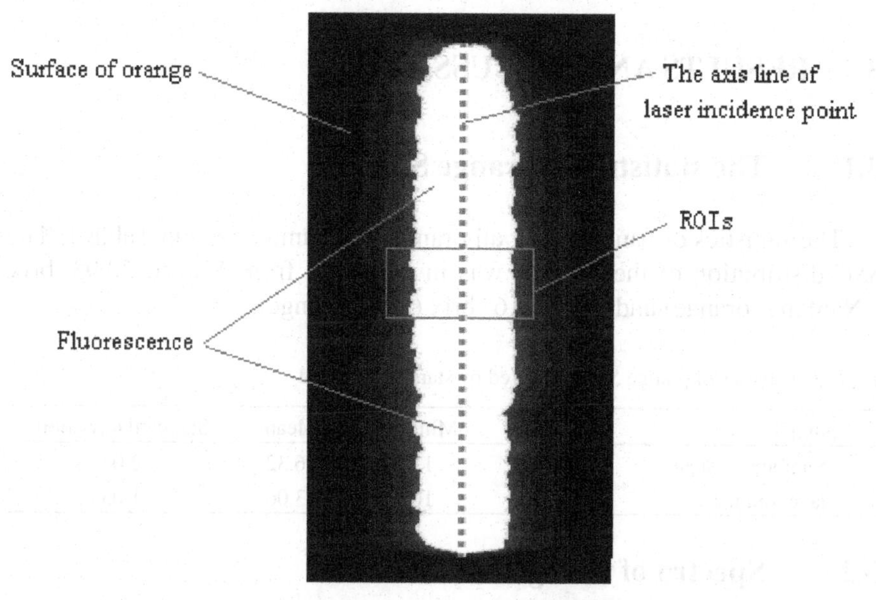

Fig. 3. Hyperspectral fluorescence image from a "Nanfeng" orange

A line regressing method and combining principal component (PC) analysis was proposed for developing orange quality prediction models. Principal component analysis provides an effective means for analysis

essential spectral region. As such, PC analysis was performed on mean spectra of hyperspectral fluorescence image. PC scores and coefficients of mean spectra were used to select spectral region. Then line regressing method to develop a calibration model for predicting orange *SSC*. The procedure of developing and validating the prediction model for each fruit quality parameter is demonstrated through the algorithm for orange fruit *SSC* prediction.

First, all samples were arranged in ascending order for their *SSC* values. Then samples were then divided into the training groups (200 samples) and validation groups (60 samples). The training samples were analyses using principal component (PC) for look for effective spectral region. 920–992 nm and 949-1000 nm were respectively thought as best spectral region for 'Nanfeng' orange and navel orange. Then the *SSC* prediction model was developed using line regressing method. Finally, the obtained model was used to predict orange *SSC* for the validation samples. This procedure was performed using programmer file created in Mathlab (The Math Works Inc., Natick, MA, USA). In this paper, we only present the results from the best prediction model, as measured by R and the standard error of prediction (SEP).

3. RESULTS AND DISCUSSION

3.1 The statistics of orange SSC

The statistics of fruit *SSC* for all oranges are summarized in Table 1. The *SSC* distribution of the oranges was in the range from 12.6 to 20.93° brix ('Nanfeng' orange) and 10.8 to 16° brix (navel orange).

Table 1. statistics of orange *SSC* measured by standard method

Sample	Maximum	Minimum	Mean	Standard deviation
'Nanfeng' orange	20.93	12.6	16.32	2.04
Navel orange	16	10.8	13.06	1.05

3.2 Spectra of orange

Spectra of orange scattering image was characterized by the mean of intensity CCD count that calculated from The ROIs of the spatial scattering profiles (100×50 pixels) of the hyperspectral image for each wavelength from 700 to 1100 nm. Fig. 4 shows the spectra of one 'Nanfeng' orange and the result was same for navel orange. It shows that laser-induced fluorescence was characterized by period of stranger and weak intensity, and the wavelength interval was about 20 nm.

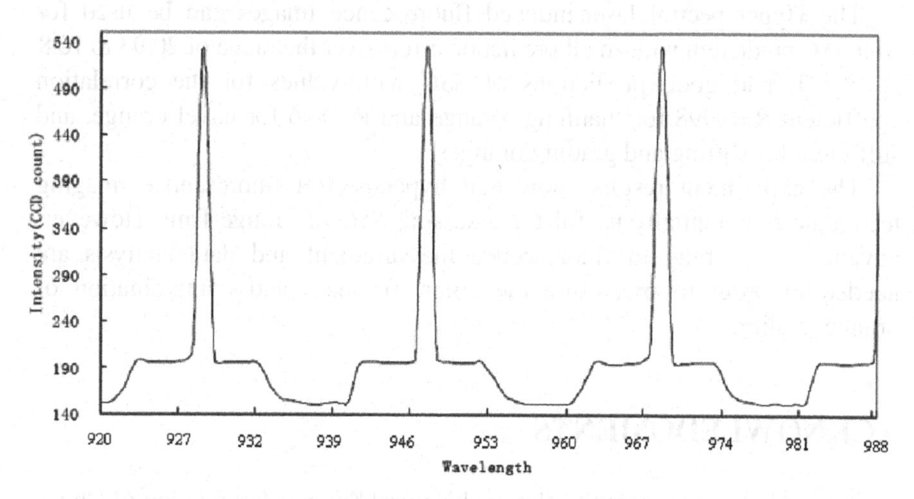

Fig. 4. The mean of CCD intensity for one 'Nanfeng' orange in 920–992 nm

3.3 Hyperspectral laser-induced fluorescence prediction models

After analyzing, spectra data of 920–992 nm ('Nanfeng' orange) and 949-1000 nm (navel orange) were respectively selected for building models. The fruit *SSC* prediction results from fluorescence imaging are shown in Table 2. The fluorescence prediction of fruit *SSC* found in this research is sufficient for sorting and grading oranges. An excellent correlation was obtained between fluorescence measurement and fruit *SSC*, The *R* and *SEP* were 0.998, 0.029 for 'Nanfeng' orange and 0.96, 0.28 for navel orange. The *SSC* prediction results by the fluorescence prediction model are considerably good than those obtained using visible/shortwave NIR spectroscopy in the wavelengths of 500–1100 nm.

Table 2. prediction model results for SSC

Sample	The correlation coefficient (R)	The standard error of prediction (SEP)
'Nanfeng' orange	0.998	0.029
Navel orange	0.96	0.28

4. CONCLUSIONS

Past research show that VIS/NIR spectroscopy is a suitable tool for measuring of fruit SSC. A new method based on Hyperspectral laser-induced fluorescence imaging was used to nondestructive measuring fruit SSC in this study.

The Hyperspectral laser-induced fluorescence images can be used for fruit *SSC* prediction with small prediction errors over the range of 20.93 to 10.8 of SSC. It had good predictions of *SSC* with values for the correlation coefficient R=0.998 for 'nanfeng' orange and R=0.96 for navel orange, and sufficient for sorting and grading oranges.

The experiment results show that hyperspectral fluorescence imaging technique is potentially useful for assessing *SSC* of orange fruit. However, advance researching in fluorescence measurement and data analysis are needed in order to predicting the color, firmness and contamination of orange quality.

ACKNOWLEDGMENTS

The authors appreciate the National Natural Science Foundation of China for assistance.

REFERENCES

Antihus, H.G., He, Y., & Pereira, G.A. Non-destructive measurement of acidity, soluble solids and firmness of Satsuma mandarin using Vis/NIR-spectroscopy techniques. Journal of Food Engineering, 2006, 77, 313–319.

Ariana, D.P., Shrestha, B.P., & Guyer, D.E. Integrating reflectance and fluorescence imaging for apple disorder classification. Comput. Electron. Agric. 2006, 50, 148–161.

DeEll, J.R., Kooten, O., Prange, R.K., & Murr, D.P. Applications of chlorophyll fluorescence techniques in postharvest physiology. Horticultural Reviews, 1999, 23, 69–107.

Kawano, S., Fujiwara, T., & Iwamoto, M.J. Nondestructive determination of sugar content in Satsuma mandarin using near infrared (NIR) transmittance. Journal Japanese Society Horticultural Science, 1993, 62 (2), 465–470.

Kim, M.S., Chen, Y.R., & Mehl, P.M. Hyperspectral reflectance and fluorescence imaging system for food quality and safety. Transactions of the ASAE. 2001, 44(3), 721–729.

Kim, M.S., Lefcourt, A.M., & Chen, Y.R. Multispectral laser-induced fluorescence imaging system for large biological samples. Appl. Opt. 2003, 42, 3927–3934.

Lu, R. Imaging spectroscopy for assessing internal quality of apple fruit. ASAE Paper No. 036012. 2003 ASAE, St. Joseph, MI.

Lu, R. Multispectral imaging for predicting firmness and soluble solids content of apple fruit. Postharvest Biology and Technology, 2004, 31, 147–157.

Lu, R., Bailey, B.B. NIR measurement of apple fruit soluble solids content and firmness as affected by postharvest storage. ASAE Paper No. 056070. 2005, ASAE, St. Joseph, MI.

Lu, R., Chen, Y.R. Hyperspectral imaging for safety inspection of food and agricultural products. SPIE Proc. 1998, 3544, 121–133.

Lu, R., Guyer, D.E., & Beaudry, R.M. Determination of firmness and sugar content of apples using near-infrared diffuse reflectance. J. Texture Stud. 2000, 31, 615–630.

McGlone, V.A., Fraser, D.G., Jordan, R.B., & Kunnemeyer, R. Internal quality assessment of mandarin fruit by Vis/NIR spectroscopy. Journal of Near Infrared Spectroscopy, 2003, 11, 323–332.

McGlone, V.A., Jordan, R.B., & Martinen, P.J. Vis/NIR estimation at harvest of pre- and post-storage quality indices for 'Royal Gala' apple. Postharvest Biology and Technology, 2002, 25, 135–144.

Miyamoto, K., Kawauchi, M., & Fukuda, T. Classification of high acid fruits by PLS using the near infrared transmittance (NIT) spectra of intact Satsuma mandarins. Journal of Near Infrared Spectroscopy, 1997, 6 (1–4), 267–271.

Park, B., Abbott, J.A., Lee, K.J., Choi, C.H., & Choi K.H. Near-infrared diffuse reflectance for quantitative and qualitative measurement of soluble solids and firmness of delicious and gala apples. Trans. ASAE, 2003, 46, 1721–1731.

Peng, Y., Lu, R. Modeling multispectral scattering profiles for prediction of apple fruit firmness. Trans. ASAE, 2005, 48, 235–242.

Steuer, B., Schulz, H., & Lager, E. Classification and analysis of citrus oils by NIR spectroscopy. Food Chemistry, 2001, 72, 113–117.

Vargas, A.M., Kim, M.S., Tao, Y., Lefcourt, A., & Chen, Y.R. Safety inspection of cantaloupes and strawberries using multispectral fluorescence imaging techniques. ASAE Paper No. 043056. 2004, ASAE, St. Joseph, MI.

A FAST LONGEST COMMON SUBSEQUENCE ALGORITHM FOR BIOSEQUENCES ALIGNMENT

Wei Liu [1,*], Lin Chen [2,3]

[1] *Institute of Information Science and Technology, Nanjing University of Aeronautics and Astronautics, Nanjing 210093, China*

[2] *Department of Computer Science, Yangzhou University, Yangzhou 225009, China*

[3] *State Key Lab of Novel Software Technology, Nanjing University, Nanjing 210093, China*

[*] *Corresponding author, Address: P.O. Box 274, Institute of Information Science and Technology, Nanjing University of Aeronautics and Astronautics, 29 Yudao ST., Nanjing, 210093, P. R. China, Email: yzliuwei@126.com*

Abstract: Searching for the longest common substring (LCS) of biosequences is one of the most important tasks in Bioinformatics. A fast algorithm for LCS problem named FAST_LCS is presented. The algorithm first seeks the successors of the initial identical character pairs according to a successor table to obtain all the identical pairs and their levels. Then by tracing back from the identical character pair at the largest level, the result of LCS can be obtained. For two sequences X and Y with lengths n and m, the memory required for FAST_LCS is $\max\{8*(n+1)+8*(m+1),L\}$, here L is the number of identical character pairs and time complexity of parallel implementation is $O(|LCS(X,Y)|)$, here, $|LCS(X,Y)|$ is the length of the LCS of X,Y. Experimental result on the gene sequences of *tigr* database using MPP parallel computer Shenteng 1800 shows that our algorithm can get exact correct result and is faster and more efficient than other LCS algorithms.

Keywords: bioinformatics; longest common subsequence; identical character pair

Liu, W. and Chen, L., 2008, in IFIP International Federation for Information Processing, Volume 258; Computer and Computing Technologies in Agriculture, Vol. 1; Daoliang Li; (Boston: Springer), pp. 61–69.

1. INTRODUCTION

Biological sequence (Bailin Hao et al., 2000) can be represented as a sequence of symbols. When biologists find a new sequence, they want to know what other sequences it is most similar to. Sequence comparison (Edmiston E.W. et al., 1988)has been used successfully to establish the link between cancer-causing genes and a gene evolved in normal growth and development. One way of detecting the similarity of two or more sequences is to find their longest common subsequence (LCS).

The longest common subsequence problem is to find a substring that is common to two or more given strings and is the longest one of such strings. Presented in 1981, Smith-Waterman algorithm (Smith T.F. et al., 1990) was a well known LCS algorithm which was evolved by the Needleman-Wunsch (Needleman, S.B. et al., 1970) algorithm, and can guarantee the correct result. Aho et al. (A. Aho et al., 1976) gave a lower bound of $O(mn)$ on time for the LCS problem using a decision tree model. It is shown in (O. Gotoh, 1982) that the problem can be solved in $O(mn)$ time using $O(mn)$ space by dynamic programming. Mayers and Miller (E.W. Mayers et al., 1998) use the skill proposed by Hirschberg (D.S. Hirschberg, 1975) to reduce the space complexity to $O(m+n)$ on the premise of the same time complexity. To further reduce the computation time, some parallel algorithms (Y. Pan et al., 1998, Jean Frédéric Myoupo et al., 1999, L. Bergroth et al., 2000, A. Aggarwal et al., 1988) have been proposed for the LCS problem on different computational models. On CREW-PRAM model, Aggarwal (A. Aggarwal et al.,1988) and Apostolico et al. (A. Apostolico et al., 1990) independently proposed an $O(\log m \log n)$ time algorithm using $mn/\log m$ processors. Many parallel LCS algorithms have also been proposed on systolic arrays. Robert et al. (K. Nandan Babu et al., 1997) proposed a parallel algorithm with $n+5m$ steps using $m(m+1)$ processing elements. Freschi and Bogliolo (V. Freschi et al., 2000) addressed the problem of computing the LCS between run-length-encoded (RLE) strings. Their algorithm requirs $O(m+n)$ steps on a systolic array of $M+N$ processing elements, where M and N are the lengths of the original strings and m and n are the number of runs in their RLE representation.

In this paper, we present a fast algorithm named FAST_LCS for LCS problem. The algorithm first seeks the successors of the initial identical character pairs according to a successor table to obtain all the identical pairs and their levels. Then by tracing back from the identical character pair at the largest level, the result of LCS can be obtained. For two sequences X and Y with lengths n and m, the memory required for FAST_LCS is $\max\{8*(n+1)+8*(m+1), L\}$, here L is the number of identical character pairs and time complexity of parallel implementation is $O(|LCS(X,Y)|)$, here, $|LCS(X,Y)|$ is the length of the LCS of X,Y. Experimental result on the gene sequences of *tigr* database using MPP parallel computer Shenteng 1800 shows that our algorithm can get exact correct result and is faster and more efficient than other LCS algorithms.

2. THE IDENTICAL CHARACTER PAIR AND ITS SUCCESSOR TABLE

Let X($x_1, x_2, \ldots, x_n$), $Y = (y_1, y_2, \ldots, y_m)$ be two biosequences, where x_i, y_i□{A,C,G,T}. We can define an array CH of the four characters so that $CH(1)$="A", $CH(2)$="C", $CH(3)$="G" and $CH(4)$="T". To find their longest common subsequence, we first build the successor tables of the identical characters for the two strings. The successor tables of X and Y are denoted as TX and TY which are $4*(n+1)$ and $4*(m+1)$ two dimensional arrays. $TX(i, j)$ is defined as follows.

Definition 1. For the sequence $X = (x_1, x_2, \ldots, x_n)$, its successor table TX of identical character is defined as:

$$TX(i,j) = \begin{cases} \min\{k \mid k \in SX(i,j)\} & SX(i,j) \neq \phi \\ - & \text{otherwise} \end{cases} \quad (1)$$

Here, $SX(i, j)=\{k| x_k=CH(i), k>j\}$, $i = 1,2,3,4$, $j = 0,1,\ldots n$. It can be seen from the definition that if $TX(i, j)$ is not "-", it indicates the position of the next character identical to $CH(i)$ after the jth position in sequence X. If $TX(i, j)$ is equal to "-", it means there is no character $CH(i)$ after the jth position.

Example 1. Let X ="$T\ G\ C\ A\ T\ A$", Y="$A\ T\ C\ T\ G\ A\ T$". Their successor tables TX and TY are:

TX:

i	*CH(i)*	0 1 2 3 4 5 6
1	A	4 4 4 4 6 6 -
2	C	3 3 3 - - - -
3	G	2 2 - - - - -
4	T	1 5 5 5 5 - -

TY:

i	*CH(i)*	0 1 2 3 4 5 6
1	A	4 4 4 4 6 6 -
2	C	3 3 3 - - - -
3	G	2 2 - - - - -
4	T	1 5 5 5 5 - -

Definition 2. For the sequences X and Y, if $x_i = y_j$, we call them an identical character pair of X and Y, and denote it as (i, j). The set of all the identical character pairs of X and Y is denoted as $S(X, Y)$.

Definition 3. Let (i, j) and (k, l) be two identical character pairs of X and Y. If $i<k$ and $j<l$, we call (i, j) a predecessor of (k, l), or (k, l) a successor of (i, j), and denote them as $(i, j)<(k, l)$.

Definition 4. Let $P(i, j) = \{(r, s)|(i, j) < (r, s), (r, s)$□$S(X,Y)\}$ be the set of all the successors of identical pair (i, j), if (k, l)□$P(i, j)$ and there is no

(k', l')□$P(i, j)$ satisfying the condition: $(k', l') < (k, l)$, we call (k, l) the direct successor of (i, j), and denoted it as $(i, j) \prec (k, l)$.

Definition 5. If an identical pair (i, j)□$S(X,Y)$ and there is no (k, l)□$S(X,Y)$ so that $(k, l) < (i, j)$, we call (i, j) an initial identical pair.

Definition 6. For an identical pair (i, j)□$S(X,Y)$, its level is defined as follows:

$$level(i,j) = \begin{cases} 1 & \text{if } (i,j) \text{ is an initial identical character pair} \\ \max\{level(k,l)+1 \mid (k,l) < (i,j)\} & \text{otherwise} \end{cases} \quad (2)$$

From the definitions above, the following lemma can be easily deduced:

Lemma 1. Denote the length of the longest common subsequence of X, Y as |LCS(X, Y)|, then |LCS(X, Y)|=max{level (i, j) | $(i, j) \in S(X, Y)$}.

Proof of Lemma 1 is omitted due to space limitation.

3. THE OPERATIONS OF PRODUCING SUCCESSORS AND PRUNING

For an identical character pair (i, j)□$S(X, Y)$, the operation of producing all its direct successors is as follows:

$$(i,j) \rightarrow \{(TX(k,i), TY(k,j)) \mid k = 1,2,3,4, TX(k,i) \neq \text{'-' and } TY(k,j) \neq \text{'-'}\} \quad (3)$$

From (3) we can see that this operation is to couple the elements of the ith column of *TX* and the jth column of *TY* to get the pairs.

Lemma 2. For an identical character pair (i, j), the method illustrated above can produce all its successors.

Proof of Lemma 2 is omitted due to space limitation.

In such process of generating the successors, pruning technique can be implemented to remove the identical pairs so as to reduce the searching space and accelerate the speed of process. These prune operations are based on the following theorems.

Theorem 1. If two identical character pairs (i, j) and (k, l) generated at the same time step satisfy $(k, l)>(i, j)$, then (k, l) can be pruned without affecting the algorithm to get the longest common subsequence of X and Y.

Theorem 2. If on the same level, there are two identical character pairs (i_1, j) and (i_2, j) satisfying $i_1 < i_2$, then (i_2, j) can be pruned without affecting the algorithm to get the longest common subsequence of X and Y.

Proof of Theorem 1 and 2 is omitted due to space limitation.

4. FRAMEWORK OF THE ALGORITHM AND COMPLEXITY ANALYSIS

Based on the operations mentioned above, we present a fast parallel longest common subsequence algorithm FAST_LCS. The algorithm first begins with the initial identical character pairs, then continuously searches for their successors using the successor tables. In this phase, the pruning technology is implemented to discard those search branches that obviously can't obtain the optimum solution so as to reduce the search space and speed up the process of searching. In the algorithm, a table called *pairs* is used to store the identical character pairs obtained in the algorithm. In the table *pairs*, each record takes the form of (*k*, *i*, *j*, *level*, *pre*, *state*) where the data items denote the index of the record, the identical character pair (*i*, *j*), its level, index of its direct predecessor and its current state. Each record in *pairs* has two states. For the identical pairs whose successors have not been searched, it is in *active* state, otherwise it is in *inactive* state. In every step of search process, the algorithm searches for the successors of all the identical pairs in *active* state in parallel. Repeat this search process until there is no identical pair in *active* state in the table. The phase of tracing back starts from the identical pairs with the maximal level in the table, and traces back according to the *pred* of each identical pair. This tracing back process ends when it reaches an initial identical pair, and the trail indicates the longest common subsequence. If there are more than one identical pair with the maximal level in the table, the tracing back procedure for those identical pairs can be carried out in parallel and several longest common subsequences can be obtained concurrently. The framework of the algorithm FAST_LCS is as follows:

Algorithm-FAST_LCS (*X*,*Y*)
Input *X* and *Y*: Sequences with lengths of *m* and *n* respectively;
Output LCS: The longest common subsequence of *X*,*Y*;
Begin

1. Build tables *TX* and *TY*;
2. Find all the initial identical character pairs: (*TX*(*k*, 0), *TY*(*k*, 0)), *k*=1,2,3,4;
3. Add the records of the initial identical pairs (*k*, *TX*(*k*, 0), *TY*(*k*, 0), 0, ϕ, *active*), *k*=1,2,3,4 to the table *pairs*.
   /* For all the initial identical pairs, their *level*=1, *pre*=ϕ and *state*=*active**/
4. Repeat

For all *active* identical pairs (*k, i, j, level, pre, active*) in *pairs* parallel-do

Produce all the successors of (*k, i, j, level, pre, active*).

For each identical character pair (*g, h*)in the successor set of (*k, i, j, level, pre, active*), a new record (*k', g, h, level*+1, *k, active*) is generated and inserted into the table *pairs*.

Change the state of (*k, i, j, level, pre, active*) into *inactive*.

End for

Use prune operation on all the successors produced in this level to remove all the redundant identical pairs from table *pairs*.

5. Until there is no record in *active* state in table *pairs*.
6. Compute r = the maximal level in the table pairs.
7. For all identical pairs (*k, i, j, r, l, inactive*) with *level* = r in *pairs* parallel-do

$Pred = l$; $LCS(r) = x_i$.

While $pred \neq \phi$ do

7.1.1 get the *Pred*-th record (*prel, g, h, r', l', inactive*) from table *pairs*.

7.1.2 $Pred = l'$; $LCS(r') = x_g$.

7.3 end while

End.

Assume that the number of the identical character pairs of *X*, *Y* is *L*. In our algorithm, the time complexity for sequentially executing of the algorithm FAST_LCS (*X*, *Y*) is $O(L)$ and the storage complexity of our algorithm is $\max\{8*(n+1)+8*(m+1),L\}$. In parallel implementation of the algorithm, the time complexity of parallel computing is $O(|LCS(X,Y)|)$, here, $|LCS(X,Y)|$ is the length of the longest common subsequence of *X*, *Y*.

5. EXPERIMENTAL RESULTS

5.1 The results of sequential computing-two sequences

We test our algorithm FAST_LCS on the rice gene sequences of *tigr* database and compare the performance of FAST_LCS with that of Smith-Waterman algorithm and FASTA algorithm which are currently the most widely used LCS algorithms.

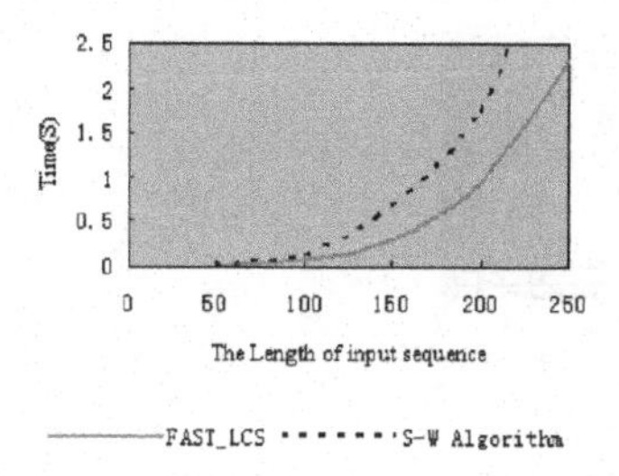

Fig. 1. Comparison of the computation time of FAST_LCS with that of Smith-Waterman algorithm

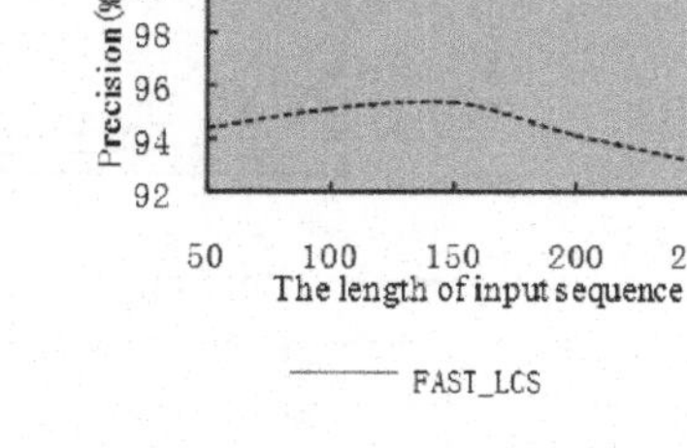

Fig. 2. Comparison of the precision of FAST_LCS with that of FASTA using the same computation time

Fig. 1 shows the comparison of the computation time of our algorithm and that of Smith-Waterman algorithm. From the figure, we see that our algorithm is obviously faster than Smith-Waterman algorithm for sequences sets of all different lengths, especially when the length of sequences become greater than 150. This means our algorithm is much faster and more efficient than Smith-Waterman's for LCS problem of long sequences.

We also compare the precision of our algorithm with that of FASTA on the premise of the same computing time. Here precision is defined as:

$$\text{Precision} = \frac{\text{The length of the common subsequence computed by the algorithm}}{\text{The length of the longest common subsequence in correct match}}$$

From Fig. 2, we can see that our algorithm can obtain exactly correct result no matter how long the sequence could be, while the precision of FASTA declines when the length of the sequences is increased. Therefore the precision of our algorithm is higher than FASTA algorithm.

5.2 The results of parallel computing

We also test our algorithm on the rice gene sequence from *tigr* database on the massive parallel processors Shenteng 1800 using MPI (C bounding). In the parallel implementation of our algorithm FAST_LCS, the identical character pairs in *active* state are assigned and processed in different processors. The experimental results by using different numbers of processors are shown in Fig. 3. From Fig. 3, we can see that the computation speed will become faster as the number of processors increases. But the speedup will slow down when the number of processors is larger than 6. Because of the overhead of communication between processors which increases the total time of the algorithm, the speedup of our algorithm can not increase linearly with the increasing of processors exactly. This is in conformity with the Amdahl's Law.

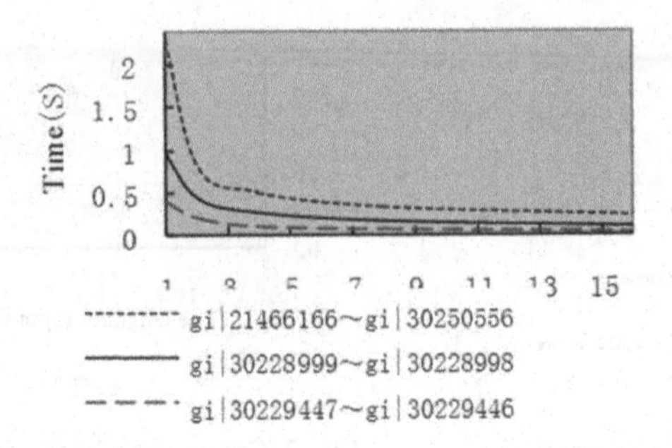

Fig. 3. Parallel computational time of FAST_LCS using different processor numbers

6. CONCLUSION

On the premise of guaranteeing precision, this paper present a parallel longest common subsequence algorithm FAST_LCS based on the identical character pair to improve the speed of LCS problem. Our algorithm first seeks the successors of the initial identical character pairs according to a successor table to obtain all the identical pairs and their levels. Then by tracing back from the identical character pair with the largest level, the result of LCS can be obtained. For two sequences X and Y with length n and m, the memory required for FAST_LCS is $\max\{4*(n+1)+4*(m+1),L\}$, here L is the number of identical character pair and time complexity of parallel computing is $O(|LCS(X,Y)|)$, here $|LCS(X,Y)|$ is the length of the LCS of X,Y. Experimental result on the gene sequences of *tigr* database using MPP parallel computer Shenteng 1800 shows that our algorithm can get exact correct result and is faster and more efficient than other LCS algorithms.

ACKNOWLEDGEMENTS

This research was supported in part by Chinese National Natural Science Foundation under grant No. 60673060, and Natural Science Foundation of Jiangsu Province under contract BK2005047.

REFERENCES

A. Aggarwal and J. Park, 1988, Notes on Searching in Multidimensional Monotone Arrays, Proc. 29th Ann. IEEE Symp. Foundations of Comput. Sci. pp. 497-512.

A. Aho, D. Hirschberg, and J. Ullman, 1976, Bounds on the Complexity of the Longest Common Subsequence Problem, J. Assoc. Comput. Mach., Vol. 23, No. 1, 1976, pp. 1-12.

A. Apostolico, M. Atallah, L. Larmore, and S. Mcfaddin, 1990, Efficient Parallel Algorithms for String Editing and Related Problems, SIAM J. Computing, Vol. 19, pp. 968-988.

Bailin Hao, Shuyu Zhang, 2000, The manual of Bioinformatics, Shanghai science and technology publishing company.

D.S. Hirschberg, 1975, A Linear Space Algorithm for Computing Maximal Common Subsequences, Commun. ACM, Vol. 18, No. 6, pp. 341-343.

E.W. Mayers, W. Miller, 1998, Optimal Alignment in Linear Space, Comput. Appl. Biosci. Vol. 4, No. 1, pp. 11-17.

Edmiston E.W., Core N.G., Saltz J.H, et al., 1988, Parallel processing of biological sequence comparison algorithms. International Journal of Parallel Programming, Vol. 17, No. 3, pp. 259-275.

Jean Frédéric Myoupo, David Seme, 1999, Time-Efficient Parallel Algorithms for the Longest Common Subsequence and Related Problems, Journal of Parallel and Distributed Computing, Vol. 57, No. 2, pp. 212-223.

K. Nandan Babu, Wipro Systems, and Sanjeev Saxena, 1997, Parallel Algorithms for the Longest Common Subsequence Problem, 4th International Conference on High Performance Computing, December 18-21, 1997 - Bangalore, India.

L. Bergroth, H. Hakonen, and T. Raita, 2000, A survey of longest common subsequence algorithms, Seventh International Symposium on String Processing Information Retrieval, pp. 39-48.

Needleman, S.B. and Wunsch, C.D., 1970, A general method applicable to the search for similarities in the amino acid sequence of two proteins, J. Mol. Biol., Vol. 48, No. 3, pp. 443-453.

O. Gotoh, 1982, An improved algorithm for matching biological sequences, J. Molec. Biol. Vol. 162, pp. 705-708.

Smith T.F., Waterman M.S. 1990, Identification of common molecular subsequence. Journal of Molecular Biology, Vol. 215, pp. 403-410.

V. Freschi and A. Bogliolo, 2004, Longest common subsequence between run-length-encoded strings: a new algorithm with improved parallelism, Information Processing Letters, Vol. 90, No. 4, pp. 167-173.

Y. Pan, K. Li, 1998, Linear Array with a Reconfigurable Pipelined Bus System - Concepts and Applications, Journal of Information Science, Vol. 106, pp. 237-258.

CHINESE AGRICULTURAL STATUS, ISSUES AND STRATEGIES OF THE DEVELOPMENT OF ELECTRONIC COMMERCE

Hua Jiang [1,*], Jing Yang [2]

[1] *School of Economy and Management, Hebei university of engineering, Handan, China, 056038*

[2] *School of Kexin, Hebei university of engineering, Handan, China, 056038*

[*] *Corresponding author, Address: Information Management Department, School of Economy and Management, Hebei university of engineering, 199 Guangming South street, Handan, 056038, P. R. China, Tel: +86-0310-6146083, Email: hdjianghua@126.com*

Abstract: Since the 1990s, the introduction of the household contract responsibility system in China's rural areas caused by China's agricultural production and small-business contradiction between socialized market has become increasingly prominent. Establishing a responsive to both our agricultural production mode of operation at this stage to have camp features can adapt to the global market information dissemination and exchange needs of modern agricultural market service system is imminent. Agricultural e-commerce is a useful way to build a modern agricultural market service system. The status quo of China's agricultural products e-commerce applications analyzed on the basis of analysis of the prospects for the development of China's agricultural e-commerce, e-commerce and agricultural problems facing the implementation of its response.

Keywords: Agricultural products; E-commerce; Informationization; Agricultural e-commerce, China

1. INTRODUCTION

In the 20th century Since the 1990s, our main agricultural products such as grain, cotton, vegetables, fruit and fish products have a certain flow are sluggish, prices. Ostensibly the reasons for such a situation is: My irrational

Jiang, H. and Yang, J., 2008, in IFIP International Federation for Information Processing, Volume 258; Computer and Computing Technologies in Agriculture, Vol. 1; Daoliang Li; (Boston: Springer), pp. 71–84.

agricultural product mix, strong homogeneous products, the quality is poor and can not meet the needs of market diversification; Agricultural circulation links too long, the transaction costs are too high, intermediate links were retained interests. But fundamentally speaking, this phenomenon is the introduction of the household contract responsibility system in China's rural areas caused by China's agricultural production and small-business contradiction between socialized market has become increasingly prominent (Hu, 2005).

China's agricultural production process used in household production, small-scale farmers produce, capture individual farmers, the poor capacity for information analysis, reacting to market signals, production and operation of the larger farmers blindness and prudent. "Economic Daily News," the survey showed that more than 70% of our products to the farmers what crops they produce outlets and numerous, many farmers in the production of visual and experience decision-making alone, or to "represent" a sense of decision-making, production fall, often leading to enormous waste of social resources. How will the domestic market and unified global market demand timely and accurate transmission of information to the thousands upon thousands of farmers, thereby guiding farmers operating productive activities, the rate of commercialization of agricultural products and enhance the overall social benefits of further development of China's agriculture has become a major practical problems which need urgent solutions (Chen, 2003).

Practice has proved that the solution to these problems, rely on the establishment of traditional agricultural socialized service system is far from enough. My current existing rural service system are mostly based on the regional administrative building, these agricultural services organizations to rural collective economic, national economic and technological sector and professional associations and narrow the scope of services and technical means backward, not a scale, only to provide some farmers produce the most basic services, and farmers can not provide for prenatal, births and post-natal services in all directions. But in some places the collective economy has phantom actually are unable to assume the functions of providing services to farmers. But in some places the collective economy has phantom actually are unable to assume the functions of providing services to farmers. China's agricultural development, the need to establish a responsive to both our agricultural production mode of operation at this stage to have camp features can adapt to the global market information dissemination and exchange needs of modern agricultural market service system. Agricultural e-commerce is a useful way to build a modern agricultural market serverce system (Gu, 2005).

2. AGRICULTURAL E-COMMERCE DEVELOPMENT MODEL PROPOSED

Understanding of the internal structure of the agricultural sector itself is a prerequisite for modern agricultural market service system. In systems theory view, the entire agricultural sector could be seen as a huge system, the system can be used "agricultural industrial chain" to describe. "agricultural industry chain" refers to agricultural commodity production closely related industries, which include agricultural production to prepare research, agricultural supplies, and other industries that farmers former sector and agricultural raw materials for the processing, storage, transport, sale industries, agricultural sector, and agricultural production is the industry chain core or base, including the cultivation of crops and rearing livestock, farmers and poultry industries in the agricultural sector, the three major systems of agriculture sector constitutes subsystems. Agricultural industry chain, chain and chain-link between a couple or the whole agricultural market system (Yang, 2001).

Detailed analysis of the agricultural market system linking every couple or process, we may discover that both agricultural technical support, the purchase of agricultural goods or the purchase of agricultural products, marketing, transport, processing, marketing and advertising payments, etc., there were not accompanied by information flows, the information is the agricultural market system link.

The rapid development of modern information technology and applications, particularly in the emergence of e-commerce technology to enable the effective functioning of modern agriculture largest market possible. According to the definition of e-commerce presentations, e-commerce refers to the telecommunications network through the production, marketing and circulation activities, which not only means for Internet transactions, and that all use of electronic information technology to solve problems, reduce costs and increase value and create business opportunities, including through the network of information from raw materials, procurement, product display, ordered to export, storage and trade of electronic payment and a series of activities. E-commerce applications, the traditional agricultural production and circulation process will have a profound impact (Yi, 2006).

As e-commerce technology to eliminate the traditional business activities of the space-time exchange of information transmission and obstacles, so that direct supply and demand sides met on the Internet, reducing the flow of agricultural links, thus greatly reducing the farmers for advertising, information search, trade negotiations and other business activities cost (Xie, 2006).

Dependence on the national or international Internet e-commerce trading network, enabling enterprises to break through the barriers of market structure, the regional market from the constraints of e-commerce technology can create various forms of "virtual company." This company will disperse the thousands upon thousands of peasant organizations to scale up the production and operation, or a specific point of production. This new form of enterprise organization if all of the rights of farmers to their fullest assurances, and because it seems to assume a certain functional information with some entities (CNNIC, 2007). This "virtual company" in the form of agriculture can better adapt to our current mode of production characteristics.

In our basic problems facing the development of agriculture, we can find that the use of electronic commerce technologies for the development of China's agriculture provides an effective solution.

In e-commerce applications on the basis of our further agricultural development of e-business models. We believe that China's future agricultural development will be a new type of agricultural e-commerce development model, the new model of agriculture to modern information technology to support computer networks to domestic or international market demand for the reunification goal of the agricultural sector to be seen as a major system, from production to final consumption of agricultural and sideline products throughout most of the allocation of resources can be optimized, so that the agricultural sector in general and integrated by significantly improving efficiency and effectiveness (Daugherty, 1998).

Directory model, information intermediaries, virtual communities, on-line shops, e-procurement, value chain integration, third-party market is the main agricultural e-commerce model, in which third-party market model has extensive practical significance by used in the electronic market of agricultural products in China. This is because China's agriculture SMEs and farmers accounted for 99% of the total, which compared with large enterprises, has its own weaknesses, such as insufficient funds, small production scale, lack of personnel, marketing network and other narrow. And e-commerce is the one of mainstream survive ways of future enterprises, as more and more SMEs and farmers began to set foot in the area of electronic commerce (Gimnez, 2003). However, a complete e-commerce system is very complex, business needs to have considerable input, this is a major problem to the under-strength. Under these circumstances, third-party transactions market patterns emerged on the edge of staple agricultural products to provide a platform for transactions, and match with higher capacity. Usually the buyer and seller are very scattered cases have been successful. In short, different agricultural e-commerce model solve or alleviate the current agricultural trade existence of different issues, it is different network adaptability: Value chain integration and third-party market can effectively resolve agricultural trade links excessive; The flow of information, transparent can regulate the conduct of the parties to transactions, in four model: on-line shops, electronic procurement, value chain integration, third-party market and so on, standardized transaction process, scientific methods can reduce irregular transactions ills of traditional transactions; Agricultural e-business seven major models will have the information gathering, publishing functions, and the enterprises which use these models in order to gather losing and to provide better services reinforce the capacity of information services, so that the participants can obtain more comprehensive information related transactions, to a certain extent, to eliminate information asymmetry; Information intermediary model can effectively reduce agricultural trade information gathering costs; E-shops, e-procurement, value chain integration, third-party market transactions were in varying degrees to reduce transaction costs; third-party market model, through an effective means of online transactions and some transactions, to reduce transaction rate fluctuations; the same time, the transaction volume of agricultural products, production of the seasonal and regional features, agricultural e-commerce has different patterns adaptability. Given China's current trade in agricultural products with the characteristics and the adaptability of e-commerce, table 1 for matching analysis (Hu, 2006).

Table 1. Table of agricultural and trade characteristics and adapt to e-commerce model, matching analysis

	Directory model	Information intermediaries	Virtual communities	On-line shops	E-procurement	Value chain integration	Third-party market
Many links in the value chain						V	V
Irregular transactions		V		V	V	V	V
Information asymmetry	V	V	V	V	V	V	V
Large transaction costs		V		V	V	V	V
Large trading volume					V	V	V
Price volatility		V					V
The seasonal production	V	V	V	V	V	V	V
Regional of production	V	V	V	V			V

3. ISSUES AND STRATEGIESION THE IMPLEMENTATION OF AGRICULTURAL E-COMMERCE

3.1 Problems

3.1.1 The construction of agricultural information leading role played enough

From abroad shows that agriculture is protected by State weak industry, and government in building information technology to play a leading role in planning and policy formulation is, the strengthening of legislation and increasing investment. However, the domestic situation, the role of the State in these areas plays enough. 1996, the Ministry of Agriculture has drawn up "," 95 "planning period rural economic information system", but this paper is planning to focus on information, networking and information sharing of information taken seriously enough that some content has changed and the urgent need to revise the plan. In terms of investment, the Ministry of Agriculture a few years ago on "Don works," but the country has not been established, although the 2000 national produce 20 million Yuan of special funds for agricultural information system, but remains low, which to some extent affected the agricultural informationization construction. Although the Ministry of Agriculture Agricultural Information Network has begun operation, and increasingly rich content, some of the provincial agricultural information network construction has started, but in the municipal and county level is more backward, mostly single aircraft operations, some computer models is still very backward, especially at the mouth of some farmers even salaries are difficult, but have no time to take into account agricultural informationization construction (Zhu, 2005).

3.1.2 Agricultural information system is perfect and imperfect information services

Overall, the provinces, cities, counties, towns, and have reached, Gouji, access and administrative barriers to the development of a low-level redundant construction of the network, agricultural information systems harmonization and standardization of low level. At the same time, agricultural information collection, dissemination, although the pattern has taken shape, but agricultural information processing, analysis, use and agricultural information channels open, the agricultural information market by developing slowly, especially agricultural information services market, agricultural design (agricultural biological engineering technology) market, agricultural funds (mobilization, mobilization, input) market, processing of agricultural products market, transport and storage and packaging of agricultural products market, or have not yet formed development, agriculture information system is not yet perfect. Agricultural information service not comprehensive and perfect, the lack of specific information service is limited to the current agricultural new varieties and new technologies of communication and information dissemination on agricultural market supply and demand inadequate; Development and use of information resources insufficient depth, surface, the small proportion of direct information, forward-looking, anticipating information than major; General information are many, complex and authoritative, lack of availability of information, particularly long-term market analysis and projections, in conjunction with local development and utilization of information resources are very scarce. The total lack of agricultural information services, structural imbalance led to the production of new things. This should lead to information agencies attention. In addition, agricultural information system bodies were still under the traditional system or semi-conversion. Traditional working patterns and the rapid development of the new system can not meet the needs of rural economic work, its goals and interests and the interests of farmers in the production and the lack of close contact, resulting in increased efficiency in the use of agricultural information unpleasant.

3.1.3 Agricultural information dissemination channels are sluggish, backward way to receive information, one-way, information and the information lag awareness weak

Grassroots rely mainly on meetings, classes, made information, and cable broadcasting, cable television and other means to disseminate agricultural information; it is clear that behind changes in the market demands. At the same time, farmers passive, one-way access to the rural grassroots cadres and agricultural information dissemination of information agencies, the use of information resources and the lack of enthusiasm risks, resulting in inefficient agricultural information dissemination. Dissemination of agricultural information backward, passive, one-way nature of the information, it will inevitably result in agriculture is lagging behind and it is very easy to fall in agricultural production, leading to excess or shortage of agricultural products, causing price fluctuations that can not truly reflect the value of agricultural products, the impact of agricultural restructuring process. For information awareness weak farmers, small-scale production so that farmers have a habit of what, what helped the operational experience, their lack of awareness of the importance of information; At the same time, due to economic constraints and cultural foundation, and they have no means of timely, direct access to information from the Internet, nor the ability to access information for analysis screening. No information may be issued more online. Therefore, at present, to serve farmers on the dissemination of information, lack of information networks linking farmers with an effective carrier and affected agricultural information and proper role to play, and also makes it difficult to agricultural information from the Internet into the homes of families (chen, 2003).

3.1.4 The lack of agricultural information network professionals

Agricultural information network building needs a large number of not only good network technology, and they are familiar with the agricultural economy of professionals, agricultural dealers to provide timely and accurate agricultural information network for information collection, collation, analysis of market situation, the e-mail to network users, answering questions. And because insufficient attention to agriculture information network talent into less funding, coupled with inadequate training mechanisms, the current agricultural information network talent rather lack of information makes professional agricultural bank building, updating slowly (Yu, 2003; Yi, 2006).

3.2 Study countermeasures

3.2.1 Through the promotion, integration of existing information infrastructure resources, a multi-level cover rural agricultural information network

Agricultural information is a classic public products, a strong external nature, it is difficult to provide through the market. Meanwhile, in the network economy era, the rural infrastructure concept has become broad-based and should not be confined to the traditional sense water conservation, and transformation of low-yielding fields, should be added agricultural information projects. Therefore, agricultural information network infrastructure should be as an integral part of rural infrastructure, primarily by government at all levels to the building of joint ventures by the government to provide such information public agricultural products, and this is the new situation of reform and financial modalities of a new attempt. Governments at all levels can be used for agricultural informationization construction funds is obviously insufficient, the larger the gap, the key is to seize the following five areas: First, it should address the current agricultural and rural economic development information are sluggish, sales from this outstanding problems, the grass-roots farmers and the actual demand for the provision of agricultural information services, as recent agriculture information center; Second, the full utilization of existing information infrastructure resources, savings investment in the fund shortage restrictions, and fundamentally ease the insufficient supply of agricultural information technology facilities contradictions; Third is to provide free information service for farmers, the fundamental solution to farmers on agricultural information services to the enormous demand and the ability of contradictions between the actual demand seriously inadequate; Fourth, governments at all levels to take graded investment approach to new investment mechanism and modalities to mobilize local governments in agriculture informationization enthusiasm and initiative; Fifth agriculture information network works with the "online government" simultaneous construction projects, the realization of "two network integration" is a practical way (Du, 2002).

3.2.2 To the development and utilization of agricultural information resources, and strengthen agricultural information services organizations, and the building of agricultural information intermediary transmission mechanism

First, there must be selective in the development and use of agricultural information resources. Standardize agricultural information collection standards to agriculture, rural areas and farmers need market supply and demand information, technology information, management expertise as the main content, and strengthen the construction of agricultural information databases, information collection expanded coverage and improve the timeliness of information, openness and sharing. Secondly, the transmission mechanism should be established agricultural information intermediary, the information going into the households to accelerate agricultural enterprises. And agricultural information service stations are actually connected to the Internet and a vast number of peasants fundamental link farmers what information needs, what products to sell, through the township agricultural information services from Internet and, therefore, should be to the township agricultural information service stations for carriers to establish agricultural information intermediary transmission mechanism. And agricultural information kiosks functions and mode of operation through the following means to further improve: agricultural information website - to radio, television, newspapers and advocacy column as aids will be carefully screened useful information dissemination to farmers and to the agricultural information services directly extended to administrative villages and farmers, effectively convey information to farmers "last kilometre" problem, also through qualified local cable television network, telephone advice issued to agricultural information. In addition, the township agricultural information service stations should also strengthen coordination and professional farmers, agricultural extension services, agricultural materials supply enterprises, township enterprises horizontal exchange of information, through their production of timely transmission of information to farmers, and guide their production and operational conduct for these organizations in the exercise of their functions, it is also a fact of agricultural information services organizations.

3.2.3 Earnestly agricultural e-commerce pilot, and make efforts to enhance the flow of agricultural efficiency

Currently, agricultural e-commerce in general is still in the "online-information network, deal," the initial stage, the necessary supporting e-commerce for agricultural conditions and market mechanisms have not yet formed, a real sense of agricultural e-commerce has not yet happened. Agricultural e-commerce represents the general trend of global agricultural trade. To meet the new situation after China's WTO entry, with the network economy era agricultural trade Delivery trend must vigorously develop agricultural e-commerce suited to China's national conditions. E-commerce through the development of agricultural products, supply and marketing network to promote agricultural production, improve the flow of agricultural efficiency is the advance of information technology in agriculture an extremely important goal. Agricultural development is the key to advancing e-commerce agricultural standardized electronic authentication, sound laws and regulations, improve credit system to promote e-commerce of agricultural products create a good external environment. At the same time, we should reform the traditional logistics, conducting large-scale time-bound delivery business professionals located on-line via e-mail to communicate with consumers, the way to achieve fundamental changes in the circulation of agricultural products, and fundamentally improve the market competitiveness of our agricultural products (Wang, 2005).

3.2.4 Based on national conditions, the selective application of agriculture in agriculture it

Agriculture information technology is agricultural production, business management, strategic decision-making processes of natural, economic and social information collection, storage, transmission, processing, analysis and use of technology. In the long term, in the agricultural information network at the same time, consistent with our national conditions agriculture it systems such as agricultural experts, agricultural database, agricultural management information systems also need to develop to the complete elimination of agricultural and rural economic development exists the "digital divide" so that the rural community with the information society pace of advance. Currently, the focus is to help farmers apply the use of information technology to transform traditional agricultural production process, the promotion of agricultural efficiency, peasant incomes. At present, China's relative success in practice the application of information technology such as agriculture, "Agriculture expert system" has developed dozens of sets, covering agriculture, forestry, livestock, fisheries sectors, the application has achieved significant economic and social benefits should be the development of China's agriculture it a priority. Other agricultural information technology such as agriculture database, agricultural management information systems, decision support systems, simulation modeling system, and should be based on China's rural economic development, and timely, to be applied selectively. It should focus on agricultural applications of economic efficiency and social efficiency, advanced technology should be subordinated to the economic and social viability. Precision agriculture development should be targeted at the rural labour surplus, 58,000 farmland size, inadequate energy supplies, agricultural pesticide residues in high status to the protection of the environment, food safety, energy conservation, conservation of land resources, reduction of the agricultural development goals and promote sustainable agricultural development.

4. DEVELOPMENT PROSPECTS

4.1 A favorable macroeconomic environment

Countries for the "three agricultural" to the state ministries of information technology and e-commerce attention and support: National Policy agricultural wholesale market information, the Ministry of Agriculture for agricultural information attention to and actively support policies; Economic globalization externally driven.

4.2 The flow of technology to China's agricultural injected new vigor and vitality

From the traditional model of agricultural hand opponent transactions to the resources through the integration, and the use of advanced and convenient application of information technology platforms agricultural structures in the implementation of agricultural trade network to improve our agricultural value chain and enhance agricultural competitiveness has great catalyst. But agricultural e-commerce is not a simple substitution of traditional circulation method, which is a revolutionary change to the traditional agricultural economy. First, the ultimate agricultural products from production to market, its difficult standardized features constraining the flow speed Internet market by establishing a need for standardized agricultural requirements, and this is bound to lead to the upgrading of agricultural brands and core competitiveness enhancement; Second, online transactions more open, equitable, transparent, real prices of agricultural products to reflect market supply and demand, for the government at all levels to guide farmers and the general scientific organization of production to sales set off; Third, the online trading platform is the establishment of the original extension of the traditional agricultural trading market for trading diversification and for businesses to provide broader business opportunities.

4.3 Self-innovation-driven demand

Industrial development is based on production, but the market is the flow of decisions and the development of key industries. Impede the flow of agricultural products are sluggish agriculture and the rural economy has become a healthy development, the impact of incomes of peasants and rural stability an important factor. Even slower sales of agricultural products and agricultural structural, seasonal, regional surplus from the circulation links view, the existence of two main issues: First, the information flow, blindly follow the trend. Market information and information dissemination mechanism for the formation of backward means that farmers lack market information guide. Second, the agricultural means of a single transaction, market management irregularities. Now the traditional way - mainly the cash transactions, the bulk of modern agricultural market is not universal, futures, forwards, transaction forms less.

5. CONCLUSIONS

Years of constant practice and pragmatic domestic and foreign enterprises to explore, in an agricultural e-commerce platform for transactions and business processes, effectively solved a series of practical problems in agricultural circulation to the implementation of China's agricultural e-commerce accumulated some experience: First, guide enterprises to integrate business processes and resources to change the traditional way of doing business, a highly efficient industrial chain. At present, China's rapid development and agricultural circulation environment, enterprises are faced with competition will no longer be simple products and services competition, but business models and industrial chain competition. There is therefore a need to re-evaluate, inherit the original business processes and integration industry chain. Second, the use of e-commerce enable enterprises to the value-added part of the value chain, thereby showing its real value due; e-commerce applications for the formation of a new business model for enterprises play a strong role in promoting. Not only is it the original business model electronic. From a development perspective, the new model should be to bring businesses and new values, including the concept of updating, management improvement, information flow, efficiency, cost reduction, efficiency growth, channel development, cooperation coordination, brand promotion, standards harmonization, improvement of services, go to the integrated enterprise competitiveness. This is also the goal of implementing e-commerce strategies. Third is stable mentality, and gradually form a profitable model. Whether e-commerce operator, or traditional agricultural enterprises in e-commerce business is business, they must make a profit, only profit can guarantee the survival and development of enterprises.

In short, China's e-commerce development is a gradual process, it is impossible to replace all the traditional mode of operation, the exploration and development of agricultural e-commerce is no exception. The true value is reflected in the integration of e-commerce and supply chain optimization, improves service quality, lower operating costs in the process. Therefore, e-commerce Business enterprises need to take a step by step development strategy.

REFERENCES

Chen Tianbao, my agricultural development of e-commerce opportunities and challenges, the Beijing Vocational College of Agriculture Journal, 2003 (3): pp38-41.

CHEN Tianbao, Opportunity and Challenge of Developing Electronic Business Affair on Farm Produce in China, Journal of Beijing Agricultural Vocation College, sep. 2003 Vol. 17 No. 3: pp38-41.

CNNIC. China Internet Development Report [EB/OL], Http://www.ennic.net.cn, 2007-05

Daugherty P J, Ellinger A E, Guestin C M. Integrated logistics: the performance connection [M]. Council of Logistics Management Annual Conference Proceedings, 1998. Anaheim, California: pp383-388.

Du Hongmei. Deal with china's WTO and the circulation of agricultural products and the organization [J]. Agricultural modernization, 2002, 23 (5): pp395-397.
Gimnez C, Ventura E. Supply chain management as a competitive advantage in the Spanish grocery sector [J]. The International Journal of Logistic Management, 2003. 14 (1): pp77-88.
Gu Wen, Huang Liping, agricultural e-commerce waiting to happen, e-commerce, 2005 (4): pp69-71.
Hu Tianhua, Fu Tiexin, the Chinese agricultural development of e-commerce analysis, the agricultural economy, 2005 (5): pp23-27.
Hu Tianshi, Chinese agricultural e-commerce model analysis, http://www.e-gov.org.cn/dianzishangwu/hangyedianzishangwu/200612/45309.html.
Kaplan S, Sawhney M.E-hubs:the new B2B market places [J]. Harvard Business Review, 2000 (May/June).
Lin Hua, Channels and measures of agricultural e-commerce development, Agriculture and Technology, 2005, 6 Vol. 25, No 3. pp40-41.
Salin V. Information technology in agri-food supply chains [J]. International Food and Agribusiness Management Review, 1998(11).
Wang Ning, Huang Liping, Agricultural products logistics supply chain management research Based on the information network. Research of Agricultural Modernization. 2005 Vol. 26, No. 2: pp126-129, 144.
Xie Xingang, Shi Lijuan, The Applying Situation of Electronic business in Material Flow of Agricultural Products and Investigating its Counter measures, Agricultural Machinery Research, 2006 (1): pp53-54.
Xin Lixian, Agricultural market development trends of e-commerce. Commercial Time. 2006 (5): pp71, 73.
Yang Lin. Introduction to e-commerce [M]. Beijing: Machinery Industry Press, 2001.
Yi Fa-min, EC Platform and Electronic Integration of Agri-Production Supply Chain, Finance and Trade Research, 2006 (6): pp13-18.
Yi Famin. The circulation of agricultural products online trading platform for the construction of [J]. E-commerce, 2006, (2)
Yu Xiaoyan, Huang Liping, Build a new agricultural logistics operation model, Shanghai Business, 2004 (5): pp34-36.
Zhu Lina, Domestic agricultural e-commerce transactions main research, Journal of Yurman Finance & Economics University, 2005, Vol. 20, No. 5: pp84-85.

RESEARCH ON MONITORING TECHNOLOGY OF DIGITAL RESERVOIR

Chengming Zhang[1,2,*], Jixian Zhang[3], Yong Liang[2], YongXiang Sun[2], Zhixin Xie[2]

[1] *School of Geoinformation Science and Engineering, Shandong University of Science and Technology, Qingdao, China, 266510*

[2] *School of Information Science and Engineering, Shandong Agricultural University, Taian, China, 271018*

[3] *Chinese Academy of Surveying and Mapping, Beijing, China, 100039*

** Corresponding author: Address: School of Information Science and Engineering, Shandong Agricultural University, 61 Daizong Road, Taian, Shandong, 271018, P. R. China, Tel: +86-0538-8242497, Fax: +86-0538-8249275, Email: chming@sdau.edu.cn*

Abstract: Firstly, the concept of digital reservoir is proposed and discussed in the paper, then the method how to monitor dam's state by digital means and how to transmit the monitoring data by GPRS technology is presented, and the method can monitor the dam's state in real time and its advantages such as real-time, visibility, visualization are analyzed later. The indicators reflecting the safety of dam are selected, and the method how to build mathematical model to real time monitor the safety and the procedure of modeling in engineering are given. At last, the above proposed method is demonstrated in Xueye reservoir, Lai Wu, Shandong Province. As a result, it is prove that the monitoring technology of the dam is effective.

Keywords: digital Reservoir, monitoring technology of the dam, data collection

1. INTRODUCTION

Reservoirs are important water conservancy facilities and water resources protection bases. They are key facilities which ensure industrial and agricultural production and urban people's life. Also, they are foreland of the rapid response to flood prevention, drought control and flood warning. The management level of reservoir is directly related to the normal design efficiency and people's life and property's safety. As the main contents of the reservoir modernization, reservoir automatic monitoring system can

Zhang, C., Zhang, J., Liang, Y., Sun, Y. and Xie, Z., 2008, in IFIP International Federation for Information Processing, Volume 258; Computer and Computing Technologies in Agriculture, Vol. 1; Daoliang Li; (Boston: Springer), pp. 85–94.

realize the automatic collection and delivery of hydrological factors such as rainfall, water level and water scheduling, directly serve the flood forecasting and scheduling and the water resources management, achieve the optimal allocation of water resources, provide the scientific basis for decision-making on the efficient use of water resources, comprehensively upgrade the management level of the reservoir, and is an important means to realize reservoir management modernization (Liang et al., 2005).

The digital reservoir is usually understood as describing the whole reservoir with digital information technology to make it serve the human existence and development furthest (Liu, 2004). Strictly speaking, the digital reservoir is referred to describing vast information of the reservoir in different dimension and space-time by RS, GPS, GIS, telemetry, remote-control and virtual reality technology based on computer, multi-media, large-scale memory and wide-band networks technology for the human existence, development and daily work, life and entertainment. The core of digital reservoir is to realize the intelligence and visibility of vast information of the reservoir through computers and networks. The dam is main building of reservoir, whose safety concerns reservoir and people's safety. Safety monitoring is important way guaranteeing the dam's safety, which controls the dam's running through collecting the dam's information concerned and developing trend. Safety monitoring of the dam is the process from collection and processing of initial safety information to forming safety concept in the brain. The paper mainly researches information collection and processing of the dam by digital means.

2. COLLECTION OF THE DAM INFORMATION

Affected by complicated factors, the dam's press condition will real time changes and the flaw hiding the dam's structure and harm brought by the flaw both have the characteristic of gradation. The particularity determines that, on one hand, safety monitoring system of the dam has long-term stability and high precision, on the other hand, the primitive data collected and transmitted must be exact, reliable and timely, especially under the bad condition such as rainstorm and earthquake etc (Su et al., 2004; Xin, 2004). Digital technology changes traditional way of information collection and processing and provides technical support for raising the safe monitoring level of the dam (Zhang et al., 2004; Xu et al., 2003).

2.1 Indicators of data collection

It can be known from statistical data, that the cause of the dam accidents can be classified into three kinds: One is caused by design, construction and natural factors, such as selecting too high water level, too low concrete grate, not considering earthquake load etc. The factors are determined since the dam is built, this do not exists the course from quantitative change to qualitative change. The second is formed in the course of running and management of the dam, this exists the course from quantitative change to qualitative change such as washing, eroding, aging of concrete, rusting of metal etc. The third is blend of the first and second case, that is, flaw in the design and construction is not corrected in the running of the dam, with time goes and imperfect management the flaw will be developed into destruction. At present, safety monitoring of the dam mainly aims at the last two cases.

Information collection indicators should be selected on the base of analyzing primitive observed data and running data. Observe the dam from both time and space, collect the information and find sensitive part of the dam and corresponding observing items to grasp safety information rapidly and exactly.

In spatial, the information collection indicators emphasize outer character of the dam such as transmogrification and seepage. An outer behavior of the dam is general reflection of the dam's character. For concrete hidden trouble and danger degree, inner trouble detecting apparatus is needed. Monitoring data collected can be analyzed through building monitoring model to know the dam's safety situation and developing trend. In addition, the conditions of the reservoirs upstream and downstream, rainfall, floodwater regulation, dam, base, counterfeit, drainage structure, power station and the flood discharge construction should be included in the indicators of information collection. So the indicators can be classified into three kinds: transmogrification information, seepage information, hydrology information and meteorological information. Transmogrification information included dam surface transmogrification, inner transmogrification, crack and joint etc. Seepage information includes dam-body seepage, dam-radix seepage and the amount of seepage. Hydrology information and meteorological information mainly includes the water level of the reservoir, the water level of the drainage building, the amount of the precipitation, the temperature of the water, the temperature of the atmosphere etc.

The dam safety state change is a gradual course. Discontinuous observing data can't provide effective information for safety judgment, even brings completely wrong result. So information collection should follow the principle of continuity.

2.2 Technology of data collection

It is the purpose of collecting the big dam information that understanding the safe condition in the whole and the each part of safe levels, controlling the change of the condition of dam, eliminating the hidden trouble of the big dam, servicing safe running of reservoir. In the recent years, with the development of the RS technology, the GPS technology, the sensor technology and the geographic information system, the technology is developing along all-in-one, automatization, numeralization and intelligentization.

The RS is the detecting technology which can collect information without touching the target directly. The RS obtains and deals with the information of the earth's surface, shows the datum on the photos and digital images. The digital images can be used by farther processed which is called Image processing. The Image processing includes the operations such as image compression, image storage, image enhancement, image quantification and so on. Recently, the RS has been the valuable tools. It not only can obtain the visible information, such as sinking size, but also can process the invisible information, the temperature of water, gas and so on.

The sensor technology is the base of the collection of dam information. By reasonable disposal, the information such as the condition of rain, water, working, disaster, water quality can be converted into electronic signal through the sensor and carried to the controlling center.

The location technology of GPS is adopted in the collection of the dam information. The GPS receiver has small volume, higher measurement precision. The receiver can work at open country and carry on normally at the abominable condition filled with lightning storm, mill dust, hot or chilliness. The GPS technology which has the advantage of round-the clock, automatization and real time can be used to collect the displacement at real time.

The distortion questions of the dam such as the coast and sinking of upstream face and downstream face, the glide of the big dam base are concerned with the water pressure among the gaps closely. The distortion questions are caused due to the seepage distortion. So the collection of seepage information is the emphases of the whole collection. At present, the well-rounded technology is the sensor and has been used in the many special engineering fields.

The chromatography imaging technology has been used widely in the seepage information collection. It is the chromatography imaging technology that using the mathematical method to compute and rebuild the two or three dimensional images on the special lever of the target according to the one dimensional image datum which are caught from the object circumference without destroy. Because the distributing condition and rejected region of the lining material can be reflected quantificational, the complexity of devices is reduced and the safety of big dam is increased, at the same time the important reference is provided to the internal state detection of big dam, the bug searching and the aging evaluation.

Seeing from the current actual circumstance, although monitoring facilities and technology has been modified significantly, the operation, analysis and use of observational data is still one kind of passive processing mode. The experts think that the key is that explaining these data by numerical method and not by only experience. According to the theoretical model and the foundation behavior, the working condition of the big dam should be valuated. At first, the theoretical model should be built for the key of monitoring data of the object, secondly, prognosticating the safe condition of the big dam, during the period if the difference between influencing quantity and the predicted value is in the allowed range, the big dam are considered under the safe condition, otherwise, the big dam are considered under the dangerous condition, certainly all the operations are according to the influencing quantity (sinking, displacement) of the given time, environment (water level, temperature, pressure). The method introduced is called real time and quantificational detecting method.

3. DATA TRANSMISSION NETWORK

According to the availability of current domestic communication resources and the actual situation and also taking the operational management facilities of the system into account, a mixed-mode network is designed with the use of wireless mobile communications as the main channel and the cable communications (PSTN) as a backup channel. Equipped with the "double channel", the communication can be automatically switched to the backup channel in case of the main channel failure, and return to the main channel to transmit data after the main channel is normal again (Guo et al., 2007; Liu et al., 2006). Because all the stations are situated in the coverage areas of China Mobile's signal, the wireless mobile communications are taken as the main channel. Data are packaged into TCP/IP data packets in the GPRS modules of data collection terminals and are sent to the data-processing and control center through the GPRS wireless network (Li et al., 2004; Cui, 2004). Transmission network is shown in figure 1. Transmission network is made up of automatic sending end, data communication net, central workstation of data and computer network. These parts are used to gather dam information data sent by several terminals automatically and join them to the centre process system.

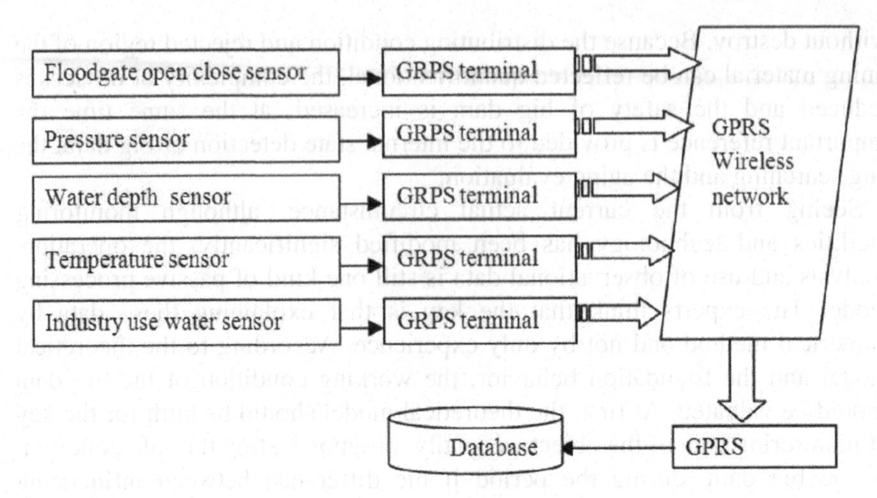

Figure 1. Transmission Network

Automatic sending end that is formed with units of sensor and single-chip system is installed at each observation station as terminals of the system. The single-chip in wired sending station transmits dam data got by sensors to computer through serial communication. Then, the computer sends processed data to the central business station which has fixed IP through internet. The central business station should have corresponding server data and software to receive and save data.

The wireless automatic sending end is mainly formed with units of sensor, single-chip system and units of GPRS data terminal. GPRS wireless data terminal (DTU) adopts GPRS wireless DDN data terminal in the system. Each module needs to install SIM card while using and each SIM card has unique ID in moving networks. GPRS DTU module also has user data interface, in order to supply power and carry on data interchange to the module. After all connections of terminal station are finished, GPRS data terminals can be managed through the setting of built-in establishment, management and debugging tools. GPRS modules adopted H7000 produced by Hongdian Company in Shenzhen. The unified mobile SIM cards are required to install in each module of collection points, and has the only ID in the mobile network just as the mobile phone. Specialized APN distributed by China Mobile is adopted by GPRS wireless data terminals and control center to access the wireless network. There are four modes for H7000 GPRS wireless DDN System, and different stations can choose arbitrary one mode according to the actual situation (Wu et al., 2007).

The wireless data transmission of automatic dam monitor network adopts the way that one central point to several points. Many types of equipment are allocated in the automatic data collection system, and they pack data into IP packets with their own GPRS data terminals. These data are connected into wireless GPRS network by the interface of aerial GPRS and then connected into Internet by ISP subsequently. The data at last reach unified data processing unit in the central work station through various kinds of gateways and routers.

The central management part of the network system of data processing is made up of one main server and several data processing servers. All data came from automatic hydrometric station of the terminal station enter the main server with fixed IP address at first through the network, the data processing server comes to finish the data processing task that the main server distribute.

4. TECHNOLOGY OF PROCESS DATA

4.1 Seepage monitoring model based on regression analysis

The regression analysis is to make the statistical analysis to fits questions. It measures the quantity changes between two or more variables that have general relations. After establishing a corresponding mathematical expression formula, it calculates another unknown quantity from a known quantity and provides the basis for estimation. Because of the complexity interrelation among diversified factors of the reservoir, the relation among the factors is difficult to show in function form. To find out the relation between them by statistical method needs a large number of data got through experiments or observations. The regression analysis confirms expression formula between y and x1, x2, xp mainly according to the statistical data.

The regression analysis can be divided into four styles, namely unitary linear regression, multiple-linear regression, unitary nonlinear regression and multiple-nonlinear regression. The relations between the variables are not always linearity; it is presented nonlinear relation sometimes. To confirm the curve regression equation, observation materials should be compared and be analyzed, especially the curve presented by the picture should be observed through pursuing a diffused dot diagram. The function of known figure should also be considered to choose the proper mathematics expression formula. While confirming the equation, values of unknown parameters in it need to be calculated. It is the least squares method that be commonly used in the calculating of parameters. The following step is proposed in this paper, combing the actual conditions in practical application:

(1) Gather and store the data surveyed on the spot or utilize history data, analyze these data synchronously; (2) doing calculation about curve of related quantity to calculate and set up regression model; Calculate the

correlative variables (3) Adjust the models through comparing the calculating value with the monitoring value; (4) Revise the model when necessary; (5) Store models after revising; (6) Determine the sequential movements according to whether the dispersion between the calculating value and the surveying value is within the range when manipulating in practice.

4.2 Experiment result

Experiment is explored in Xueye reservoir, Lai Wu, Shandong Province. Xueye reservoir lies on Yingwen River, a branch of Dawen River. It is a large reservoir and constructed in 1959. The main project consists of dams, off lets, a spillway and a water power plant. The dam is composed of a primary dam and a secondary dam. The primary dam is a composite structure with grit hull and clay core below 230.5m and a homogenous clay dam above 230.5m. The length and height of it are 1200m and 239.5m respectively. The secondary dam is a homogenous dam with a length of 552m. The spillway lies on the eastern side of the secondary dam, on which there is a steel flashboard with a dimension of 10m times by 6m. There are two off lets, that is, the eastern off let and the western one. The power plant is at the back of the primary dam, whose main functions are irrigation and floodwater regulation.

The water level data got from April to August, 2006, shown in table 1. The results of regression analysis are presented in figure 2. The comparison between these results and the subsequent data showed that the distribution of error is quite even.

5. CONCLUSIONS

Comparing with the traditional technology, this method of displacement information collecting and processing can combine the safety evaluation and the indexes of design standard and parameter such as safety factor and reliability. At the same time, this approach can also make full use of the successful experiences and methods of safety inspection so that it can be comprehended, mastered and applied into practice more easily. Transmitting the observation data by means of GPRS technology has obvious advantages in several aspects such as real-time and reliability.

Table 1. Water level data

No.	1	2	3	4	5	6	7
Value	216.783	216.843	216.943	216.733	216.123	216.933	217.083
No.	8	9	10	11	12	13	14
Value	217.213	217.033	217.013	217.113	217.203	216.993	216.833

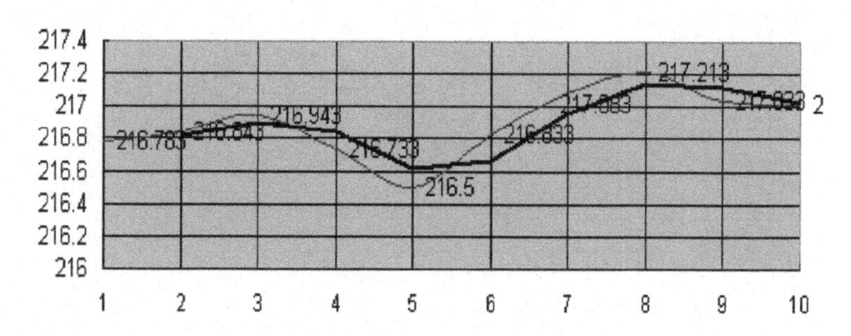

Figure 2. Regression analysis result

ACKNOWLEDGEMENTS

This study has been funded by Special Funds of Water Science and Technology of Shandong (Contract Number: 200357). Sincerely thanks are also due to the Xueye Reservoir for providing the data for this study.

REFERENCES

Cui Ziqian 2004, Data Acquisition System for Industrial Based on GPRS. Mechanization Research. 11:108-109 (in Chinese).

Guo Chunyan, Yu Junqi, Song Xianwen 2007, Design of Wireless Remote System for Reservoir-bedload. Control & Automation. 2:85-88 (in Chinese).

Li Tao, Fu Yongsheng 2004, Data Transmission Based on GPRS. Mobile Communications. 7:76-79 (in Chinese).

Liang Yong, Zhao Ao, Zhang Maoguo 2005, Studies on Digital Reservoir. Journal of Shandong Agricultural University (Natural Science). 2:313–316 (in Chinese).

Liu Chunling, Wang Yanfen, Zhang Shihu 2006, Realization of Embedded Telemetering System of Hydrological Signal Based on ARM for Irrigation District. Water Resources and Hydropower Engineering. 8:74-77 (in Chinese).

Liu Jianbiao 2004, Research of Digital Reservoir. Water Development Studies. 3:44-46 (in Chinese).

Su Huaizhi, Wen Zhiping 2004, New Intelligent Technologies Monitoring Dam Safety. J of China Three Gorges University (Natural Sciences). 3:202-205 (in Chinese).

Wu Qiulan, Liang Yong, Zhang Chengming, Ge Pingju 2007, Design of Wireless Hydrological Automatic Measurement System Based on GPRS. Computer Engineering. 2:280-282 (in Chinese).

Xin Sijin 2004, Study on Application of FBG Sensing Technology for Dam Safety Monitoring. Journal of Wuhan University of Technology. 4:31-33 (in Chinese).

Xu Weichao, Tian Bin 2003, Study of Safety Monitoring System of Dam Based on GIS. Journal of China Three Gorges University (Natural Sciences). 4:298-300 (in Chinese).

Zhang Qianfei, Wang Jinguo, Li Xuehong 2002, Study on Indefinite Information in Dam Safety Monitoring. Journal of Hohai University (Natural Sciences). 5:113-117 (in Chinese).

RESEARCH ON LOW ALTITUDE IMAGE ACQUISITION SYSTEM

Hongxia Cui [1,2], Zongjian Lin [2], Jinsong Zhang [3,*]
[1] *Department of Information Science and Engineering, University of Bohai, Jinzhou, Liaoning Province, 121000, China*
[2] *Chinese Academy of Surveying and Mapping, Beijing, 100039, China*
[3] *Department of public information network security supervise, Jinzhou police station, Jinzhou, Liaoning Province, 121000, China*
** Corresponding author, Address: Department of public information network security supervise, Jinzhou police station, Jinzhou, Liaoning Province, 121000, China, Tel: +86-0416-8919456; Email: lnchx316@sohu.com*

Abstract: In order to acquire quick, high resolution, affordable aerial photographs, an unmanned helium filled airship system equipped with a non-metric digital camera was developed in Chinese Academy of Surveying and Mapping. To meet the potential demand of large scale photogrammetric applications, the non-metric digital camera was calibrated rigourously based on the bundle adjustment with additional parameters and the workflow for generating digital orthophotos was proposed. At the paper end, the promising results obtained from an application with the system are shown and evaluated.

Keywords: UAV (unmanned aerial vehicle), photogrammetry, airship, camera, computer vision

1. INTRODUCTION

For the applications of aerial photogrammetry and remote sensing, the lack of appropriate image acquisition methods is usually one of the most important points of interest. Compared with classical flight campaigns, other UAV-based platform such as kite, balloon, airship and remote-controlled model helicopter or unmanned aircraft seems to provide a low-cost solution for quick, affordable image acquisition of area with small size (especially in urban areas).

Cui, H., Lin, Z. and Zhang, J., 2008, in IFIP International Federation for Information Processing, Volume 258; Computer and Computing Technologies in Agriculture, Vol. 1; Daoliang Li; (Boston: Springer), pp. 95–102.

During past decades, to meet the increasing demand for low-cost, high-resolution and large-scale imagery, UAV-based systems equipped with digital cameras have been used in many fields including 3D building reconstruction (Eisenbeiss et al., 2004), precision agriculture (Eisenbeiss et al., 2004), rapid emergency response operations (Davison et al., 2003), 3D vector map production (Lacroix et al., 2004), etc. Researchers in Chinese Academy of Surveying and Mapping have developed an unmanned aircraft (Cui et al., 2004) and an unmanned airship to provide low altitude platforms for diverse applications.

2. SYSTEM OVERVIEW

The UAVF-RS (unmanned aerial vehicle for remote sensing) system developed in Chinese Academy of Surveying and Mapping consists of the remote controlled helium-airship, the photographic system and the ground station.

2.1 Helium airship

The unmanned helium-airship (Fig. 1) consists of an envelope, a gondola, two fins, an electrical system.

It is designed to be able to carry photographic system. It is designed to fly in areas near to ground between 20 m and 1,000 m flying height. The helium-airship is controlled by the model uav pilot.

Two fins are attached to the envelope. There is the airship's control system with its electrical supply, fuel and propulsion in the gondola. Two petrol engines provide the propulsion. The electrical system is comprised of a radio receiver, a radio transmitter, some actuators, an onborad GPS system and power supply (two sealed lead-acid batteries).

The technical characteristics of the airship in this paper are the following:

Volume: 66cu.mts
Length: 12 meter
Maximum payload weight (Camera): 5 kg
Maximum take-off weight: 72kg
Range of control: 30km
Endurance: 2.5hrs
Maximum altitude: 1.0km
Maximum velocity: 20m/sec

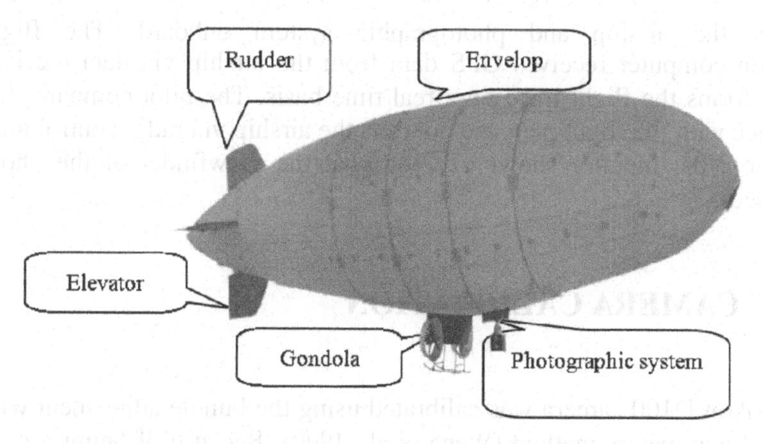

Fig. 1. Unmanned helium-airship

2.2 Photographic system

Fig. 2. Nikon D100 camera

Fig. 3. Ground station overview

The Nikon D100 camera with lens of 20 mm focal length (Fig. 2) is mounted on a two axis pan & tilt platform under the airship which is completely self-leveling. An automatic control system is developed to ensure the camera exposuring normally. Since the airship is always moving in the breeze, fast shutter speeds are needed.

A small video camera is placed up to the viewfinder of the Nikon D100 camera and is wired to a video transmitting unit. Therefore the on-board color-micro video camera provides the operator on the ground with an image of the area to be photographed.

2.3 Ground station

The Ground station (Fig. 3) includes a video monitor, a video receiver, a data receiver, a flight navigation computer and various switches for operating the airship and photographic system onboard. The flight navigation computer receives GPS data from the airship via data receiver unit and forms the flight trace on a real time basis. The pilot compares the flight trace with the flight plan and controls the airship via radio control unit. In addition, the monitor shows exactly what the viewfinder of the photo camera sees.

3. CAMERA CALIBRATION

The Nikon D100 camera was calibrated using the bundle adjustment with additional parameters method (Weng et al., 1990). Because of being a non-metric camera, its distortion calibration accuracy is evaluated in detail.

3.1 Bundle adjustment with additional parameters

The Bundle adjustment with additional parameters method is normally divided into four steps: First, Object space coordinates of control points on some calibration bar in a laboratory were measured by use of total station. Second, 8 images of the control points from four different positions were taken by the camera with the lens fixed at infinity focus. Third, coordinates of control points on each image are (semi-) automatically extracted. Fourth, calibration is implemented by use of bundle adjustment method.

According to the two image datasets captured on Aug. 15th and Oct. 11th in 2005 separately, the camera was calibrated two times by the software developed in this paper (Fig. 4).The calibration results are shown in Table 1.

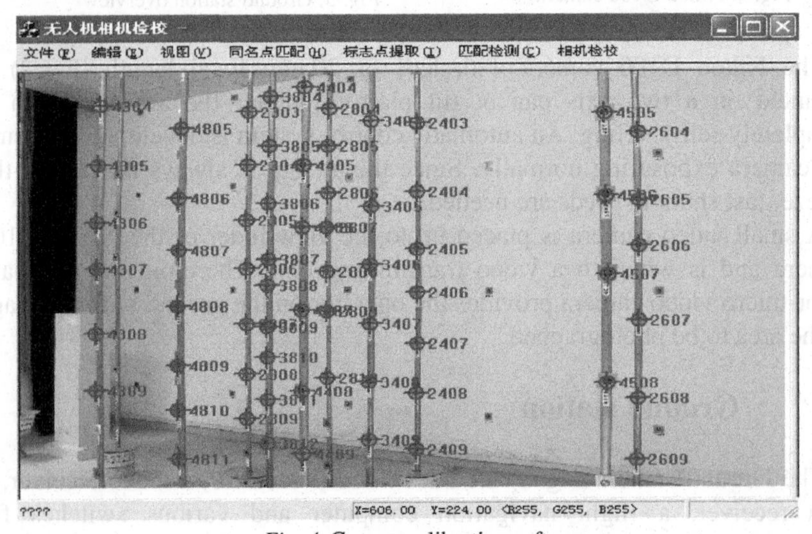

Fig. 4. Camera calibration software

Table 1. Parameters of interior orientation and distortion

	Aug. 15th, 2005		Oct. 11th, 2005	
	value	□σ_0	value	□σ_0
f(pixel)	2617.12	0.42	2617.31	0.76
x0(pixel)	1505.27	0.21	1505.8	0.13
y0 (pixel)	1000.45	0.2	1000.4	0.14
k1(pixel^{-2})	1.58 x 10^{-8}	3.69 x 10^{-10}	1.60 x 10^{-8}	1.67 x 10^{-10}
k2(pixel^{-4})	-1.26 x 10^{-15}	6.41 x 10^{-16}	-1.42 x 10^{-15}	4.29 x 10^{-16}
p1(pixel^{-2})	-3.09 x 10^{-7}	7.10 x 10^{-8}	-2.68 x 10^{-7}	9.08 x 10^{-9}
p2(pixel^{-2})	8.25 x 10^{-8}	3.02 x 10^{-8}	8.30 x 10^{-8}	3.49 x 10^{-8}
b1 (pixel^{-1})	4.87 x 10^{-4}	1.29 x 10^{-6}	3.64 x 10^{-4}	3.29 x 10^{-6}
b2(pixel^{-1})	9.88 x 10^{-5}	4.98 x 10^{-6}	7.83 x 10^{-5}	7.01 x 10^{-6}

Where the terms of f represents focal length, (x0, y0) the location of the principal point, Ki the coefficients of radial distortion and r the radial distance. Pi are the coefficients of the decanting distortion. The scale parameter b1 models a no-square pixel size and shear parameter b2 compensates for a nonorthogonality in the pixel array.

3.2 Distortion evaluation

During adjustment, the coordinates of check points are deemed unknown. Finally, adjustment results are compared with observation of total station to check accuracy. The relative accuracy is shown in Table 2, it is obvious that although the accuracy in the Z (height) direction is not as high as that in the planar direction.

Table 2. Relative accuracy

	Aug. 15th, 2005	Oct. 11th, 2005
Z	1/6230	1/6289
Planar	1/8474	1/8264

There are four principal sources of departure from collinearity, which are "physical" in nature (Fraser et al., 1999). These are the radial lens distortion, de-centering lens distortion, image plane un-flatness, and in-plane image distortion. Radial lens distortion is calculated by polynomial series, equation (1) (Kraus et al., 2000). The term K_1, K_2 are considered in this paper.

$$\begin{aligned}\Delta x_{RD} &= (x - x_0)r^2k_1 + (x - x_0)r^4k_2 \\ \Delta y_{RD} &= (y - y_0)r^2k_1 + (y - y_0)r^4k_2\end{aligned} \tag{1}$$

where $$r = \sqrt{(x - x_0)^2 + (y - y_0)^2}$$

Decentering distortion which can be acquired by equation (2) is the second category of lens distortion because of lack of centering of the lens elements along the optical axis.

$$\begin{aligned} \Delta x_{DD} &= (2(x-x_0)^2 + r^2)p_1 + 2(x-x_0)(y-y_0)p_2 \\ \Delta y_{DD} &= 2(x-x_0)(y-y_0)p_1 + (2(y-y_0)^2 + r^2)p_2 \end{aligned} \tag{2}$$

Out-of-plane image deformation is usually ignored. In-plane distortions include distortion caused by differential scaling between x and y image coordinates and non-orthogonality between image axes. These distortions are usually denoted "affine deformations" and can be mathematically obtained by equation (3).

$$\Delta x_{CD} = (x-x_0)b_1 + (y-y_0)b_2 \tag{3}$$

Thus, the total systematic distortion can be described by equation (4).

$$\begin{aligned} \Delta x &= \Delta x_{RD} + \Delta x_{DD} + \Delta x_{CD} \\ \Delta y &= \Delta y_{RD} + \Delta y_{DD} + \Delta y_{CD} \end{aligned} \tag{4}$$

Conclusions can be drawn by computing different distortion of every image point. First, the total distortion is up to 69.67 pixel and radial distortion is always over 97%. Second, the decentering distortion can not be ignored for the non-metric camera used in the paper. Third, the affine distortion is less than 0.1 pixel which will be not considered during the following image processing.

4. PHOTOGRAMMETRIC APPLICATION

With the developed airship and the camera calibrated in this paper, the low altitude images of a test project area have been taken. The flight plan and the flight trace of the airship is illustrated in Fig. 5.

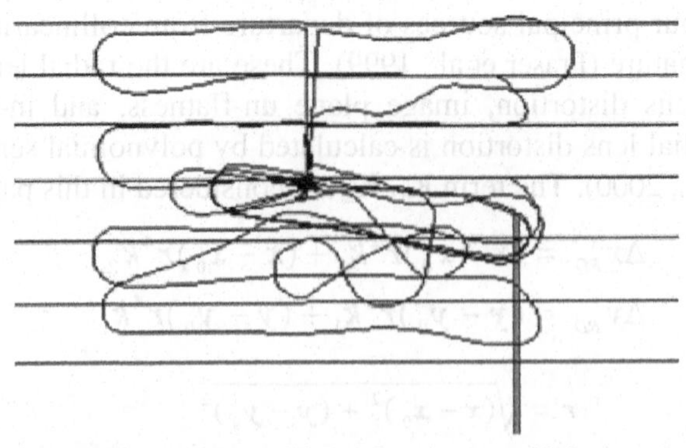

(straight lines: flight plan; curves: flight trace)

Fig. 5. Test flight

The airship-based system enables to acquire low altitude digital images. To deal with these images for photogrammetric application, the specific software has been developed. The workflow of the software to produce photogrammetric products is aerial triangulation, block adjustment, image matching, DEM production and 3D stereo models (Cramer et al., 2003). Currently, the software enables delivery of seamless high resolution orthophotos and 3D vector maps. Especially, one point on the ground can be imaged up to 15 images because the strips of images have at least 80% forward overlap and at least 20% side overlap. Based on such images, we propose to detect and eliminate the blunder and increase the number of multi-ray points robustly by means of relative orientaton, trifocal tensor (Hartley et al., 2003) as well as free net bundle adjustment step by step. Further, over 100 points at 1:500 maps have been measured three dimensionally in CAD-Software. Through comparing the coordinates of these points with those derived from the orthophoto, conclusions can be drawn that the accuracy meets that required for national map accuracy standards at scale of 1:2000. A further control was made by placing the raster map at scale of 1:500 over the orthophoto (Fig. 6).

Fig. 6. 1:2000 Orthophoto with 1:500 raster map

5. CONLUSIONS

From what has been discussed above, we may safely draw conclusions as follows:

(1) The advantages of this system over traditional aerial photogrammetric system with large format metric cameras are its quick, affordable image acquisition of area with small size (especially in urban areas).

(2) Non-metric camera is able to be applied in photogrammetric application at large scale after rigorous calibration. However, the decentering distortion can not be ignored and the affine distortion is not considered during image processing.

(3) The high redundancy of digital images is helpful to a fully automated workflow for photogrammetric production.

REFERENCES

Cramer M. GPS/inertial and digital aerial triangulation – recent test results. Photogrammetric Week , 2003:161-172.

Cui HX, Lin ZJ, Sun J. UAVRS-II unmanned air vehicle remote sensing system. Proceedings of the third international symposium on instrumentation science and technology, 2004, 2:355-359.

Davison AJ. Real-time simultaneous localization and mapping with a single camera. IEEE Int. Conf. on Computer Vision, 2003:1403-1410.

Eisenbeiss H. A mini unmanned aerial vehicle (UAV): system overview and image acquisition. International Workshop on Processing and visualization using high-resolution imagery, 2004.

Fraser C. Digital camera self-calibration. ISPRS Journal of Photogrammetry and Remote Sensing. 1997, 52:149-159.

Hartley R, Zisserman A. Multiple View Geometry in Computer Vision. Cambridge University Press, 2003.

Herwitz SR, Johnson LF, Dunagan SE. Demonstration of UAV-based imaging for agricultural surveillance and decision support. Computers and Electronics in Agriculture, 2004, 44:49-61.

Kraus K. Choice of Additional Parameters. Photogrammetry Advanced Methods and Applications. 2000, 2:133-136.

Lacroix S, Kung IK. High resolution 3D terrain mapping with low altitude imagery. 8th ESA Workshop on Advanced Space Technologies for Robotics and Automation, 2004.

Weng J, Cohen M, Claibration of stereo cameras Using a Non linear distortion model (CCD sensory). Proc. 10th. conf. Pattern recognition, 1990:246-253.

A THRESHOLD-BASED SIMILARITY RELATION UNDER INCOMPLETE INFORMATION

Xuri Yin
Simulation Laboratory of Military Traffic, Institute of Automobile, Management of PLA, Bengbu, 233011, China, yinxuri@163.com

Abstract: The conventional rough set theory based on complete information systems stems from the observation that objects with the same characteristics are indiscernible according to available information. Although rough sets theory has been applied in many fields, the use of the indiscernibility relation may be too rigid in some real situations. Therefore, several generalizations of the rough set theory have been proposed some of which extend the indiscernibility relation using more general similarity or tolerance relations. In this paper, after discussing several extension models based on rough sets for incomplete information, a novel relation based on thresholds is introduced as a new extension of the rough set theory, the upper-approximation and the lower approximation defined on this relation are proposed as well. Furthermore, we present the properties of this extended relation. The experiments show that this relation works effectively in incomplete information and generates rational object classification.

Keywords: rough sets, incomplete information, tolerance relation, similarity relation, constrained dissymmetrical similarity relation

1. INTRODUCTION

Rough set theory (Pawlak, 1982), which has been developed by Z. Pawlak and his co-workers since the early 1980s, has recently received more and more attention as a means of knowledge discovery. Rough set is a kind of mathematical tool, which is used to depict incompletion and uncertainty of

Yin, X., 2008, in IFIP International Federation for Information Processing, Volume 258; Computer and Computing Technologies in Agriculture, Vol. 1; Daoliang Li; (Boston: Springer), pp. 103–110.

the information. We can discover the connotative knowledge and the underlying rules through its analyzing and reasoning for the data (Yin et al., 2001).

In the classical rough set theory the information system must be complete. However, in the real world some attribute values may be missing due to errors in the data measure, the limitation of data comprehension as well as neglects during the data registering process. Therefore, several generalizations of the rough sets theory have been proposed to deal with the incomplete information systems. The LERS system first transforms an incomplete information system into complete information system, then generate rules (Chmielewski et al., 1998). Kryszkiewicz proposed a new method which produces rules from incomplete information system directly, he extended some concepts of the rough set theory in the incomplete information system, studied the tolerance relation in his papers (Kryszkiewicz, 1998; Kryszkiewicz, 1999). Stefanowski presented an extended rough set theory model based on similarity relations and tolerance relations (Stefanowski et al., 1999). In the paper (Yin et al., 2006), the constrained dissymmetrical similarity relation is introduced and showed that it is between the tolerance relation and the similarity relation. Another model based on constrained similarity relations was defined in the paper (Wang, 2002).

In this paper, several present extension models of rough set under incomplete information systems are discussed. Then the concept of threshold based similarity relation as a new extension of rough sets theory is introduced, and the upper-approximation and lower-approximation are redefined. Furthermore, the properties of this relation are discussed also. The experiments show that the proposed threshold based similarity relation can effectively process incomplete information and generate rational object classes. By threshold man can easily control the partition of universe in some way.

The rest of this paper is organized as follow: Section 2 introduces several extension models based on rough sets theory under incomplete information systems. Section 3 presents a concept of threshold based similarity relation as a new extension of rough sets theory and discusses its properties. Section 4 shows some examples. Finally, we conclude the paper with a summary in Section 5.

2. SEVERAL EXTENSION MODELS

Knowledge representation in rough set theory is done via information systems, which are a form of data table.

Definition 1. *Structure* $I = <U, \Omega, V_q, f_q>_{q \in \Omega}$ *is called an information system where:*

1. U is a nonempty and finite set of a group objects (or instances), called the universe of discourse, Assume the number of the objects is n, then U can be denoted to : $U = \{x_1, x_2, ..., x_n\}$,

2. Ω *is a nonempty and finite set which contains finite attributes, Assume the number of the attributes is m, then it can be denoted to:* $\Omega = \{q_1, q_2, ..., q_m\}$,

3. For each $q \in \Omega$, V_q *is enumerated domain of the attribute q,*

4. For each $q \in \Omega$, f_q *is a information function,* $f_q: U \rightarrow V_q$, *such that* $\forall x \in U, \exists y \in V_q, f_q(x) = y$.

The information system with such a domain V_q that contains missing values represented by "*" is an incomplete information system.

To process and analyze the incomplete information system, Kryszkiewicz proposed the tolerance relation T as follows (Kryszkiewicz, 1999):

$$\forall_{x,y \in U}(T_B(x,y) \Leftrightarrow \forall_{b \in B}((f_b(x) = *) \vee (f_b(y) = *) \vee (f_b(x) = f_b(y))$$

The tolerance relation T satisfies the reflexivity and symmetry, but not transitivity. The lower-approximation $\underline{X}_B^T$ and upper-approximation $\overline{X}_B^T$ can be defined as:

$$\underline{X}_B^T = \{x \mid x \in U \wedge I_B(x) \subseteq X\} \qquad \overline{X}_B^T = \{x \mid x \in U \wedge (I_B(x) \cap X \neq \phi)\} \quad (1)$$

where

$$I_B(x) = \{y \mid y \in U \wedge T_B(x,y)\} \quad (2)$$

Stefanowski and others proposed a dissymmetrical similarity relation S.

$$\forall_{x,y \in U}(S_B(x,y) \Leftrightarrow \forall_{b \in B}((f_b(x) = *) \vee (f_b(x) = f_b(y))$$

Obviously, the relation S is dissymmetrical, but transferable and reflexive. Also, Stefanowski defined the lower-approximation $\underline{X}_B^S$ and the upper-approximation $\overline{X}_B^S$ of the set $X \subseteq U$ based on the dissymmetrical similarity relation S (Stefanowski et al., 1998):

$$\underline{X}_B^S = \{x \mid x \in U \wedge \underline{R}_B^S(x) \subseteq X\} \qquad \overline{X}_B^S = \bigcup_{x \in B} \overline{R}_B^S(x) \quad (3)$$

where

$$\underline{R}_B^S(x) = \{y \mid y \in U \wedge S_B(x,y)\} \qquad \overline{R}_B^S(x) = \{y \mid y \in U \wedge S_B(y,x)\} \quad (4)$$

It can be proved that the lower-approximation and upper-approximation of the object set *X* which is based upon the dissymmetrical similarity relation S is an extension to which is based upon the tolerance relation T.

In the paper (Yin et al., 2006), we presented a constrained dissymmetrical similarity relation C, which is defined as:

$$\forall_{x,y\in U}(C_B(x,y)\Leftrightarrow\forall_{b\in B}(f_b(x)=*)\vee((P_B(x,y)\neq\varnothing)\wedge\forall_{b\in B}((b\in P_B(x,y))\rightarrow(f_b(x)=f_b(y))))$$

where

$$P_B(x,y)=\{b\mid b\in B\wedge(f_b(x)\neq *)\wedge(f_b(y)\neq *)\}$$

Obviously, relation C is reflexive, but not symmetric and transferable. The lower-approximation $\underline{X}_B^C$ and upper-approximation $\overline{X}_B^C$ based on the constrained dissymmetrical similarity relation C is defined:

$$\underline{X}_B^C=\{x\mid x\in U\wedge \underline{R}_B^C(x)\subseteq X\} \qquad \overline{X}_B^C=\bigcup_{x\in B}\overline{R}_B^C(x) \quad (5)$$

Here,

$$\underline{R}_B^C(x)=\{y\mid y\in U\wedge C_B(x,y)\} \qquad \overline{R}_B^C(x)=\{y\mid y\in U\wedge C_B(y,x)\} \quad (6)$$

Theorem 1. *Information system I=<U , Ω , V_q, f_q>$_{q\in\Omega}$, X⊆U, B⊆Ω,*

(1) $\underline{X}_B^T\subseteq\underline{X}_B^C$; $\overline{X}_B^C\subseteq\overline{X}_B^T$

(2) $\underline{X}_B^C\subseteq\underline{X}_B^S$; $\overline{X}_B^S\subseteq\overline{X}_B^C$

Theorem 1 shows that the constrained dissymmetrical similarity relation is just between tolerance relation and similarity relation (Yin et al., 2006).

3. THRESHOLD-BASED SIMILARITY RELATION

In the paper (Yin et al., 2006), the second kind of situation of relation C is one in which the object x and y must have the same definite attribute value in at least one attribute. But with the incensement of the attribute number in data sets, this kind of condition still appeared quite loosely. Therefore, we have the necessity to introduce a threshold value, ratio of the number of attributes that has the same definite attribute value for object x and y to the number of all attributes. By adjusting the threshold values, to a certain extent, man can flexibly determine the class of the objects to meet the needs for practical applications in the areas of data mining.

Definition 2. *Assume that information system I=<U, Ω, V_q, f_q>$_{q\in\Omega}$, B⊆Ω and is a nonempty, the threshold based similarity relation A can be defined as:*

$$\forall_{x,y\in U}(A_B(x,y)\Leftrightarrow\forall_{b\in B}(f_b(x)=*)\vee((\frac{P_B(x,y)}{|B|}\geq\alpha)\wedge\forall_{b\in B}((b\in P_B(x,y))\rightarrow(f_b(x)=f_b(y))))$$

where

$$P_B(x,y)=\{b\mid b\in B\wedge(f_b(x)\neq *)\wedge(f_b(y)\neq *)\}$$

$$0\leq\alpha\leq 1$$

The lower-approximation and upper-approximation based on the threshold based similarity relation A can be defined in the following.

Definition 3. *Assume that information system $I = <U, \Omega, V_q, f_q>_{q\in\Omega}$, $X\subseteq U$, $B\subseteq\Omega$ and is a nonempty, the lower-approximation $\underline{X}_B^A$ and upper-approximation $\overline{X}_B^A$ based on the threshold based similarity relation A can be defined as:*

$$\underline{X}_B^A = \{x \mid x \in U \wedge \underline{R}_B^A(x) \subseteq X\} \qquad \overline{X}_B^A = \bigcup_{x\in B} \overline{R}_B^A(x) \tag{7}$$

Here,

$$\underline{R}_B^A(x) = \{y \mid y \in U \wedge A_B(x,y)\} \qquad \overline{R}_B^A(x) = \{y \mid y \in U \wedge A_B(y,x)\} \tag{8}$$

Theorem 2. *Information system $I=<U, \Omega, V_q, f_q>_{q\in\Omega}$, $X\subseteq U$, $B\subseteq\Omega$ and is a nonempty, $0 \le \alpha \le 1$. Then*

(1) If $\alpha = 0$, then $A_B(x,y) = T_B(x,y)$;

(2) If $0 < \alpha \le 1/|B|$, then $A_B(x,y) = C_B(x,y)$;

(3) If $1/|B| < \alpha \le 1$, then $\underline{X}_B^A \supseteq \underline{X}_B^C, \overline{X}_B^A \subseteq \overline{X}_B^C$;

(4) If $1/|B| \le \alpha_1 \le \alpha_2 \le 1$, then $\underline{X}_B^{A_2} \supseteq \underline{X}_B^{A_1}, \overline{X}_B^{A_2} \subseteq \overline{X}_B^{A_1}$

Proof. (1) According to the definitions of relation T and relation A, it is obvious that If $\alpha = 0$, then $A_B(x,y) = T_B(x,y)$

(2) For any object x and y of U, if $0 < \alpha \le 1/|B|$, then $\frac{P_B(x,y)}{|B|} \ge \alpha \Leftrightarrow P_B(x,y) \ne \varnothing$. By the definition 2 and the definition of constrained dissymmetrical similarity relation we have

$$A_B(x,y) = C_B(x,y)$$

(3) Let $1/|B| < \alpha \le 1$, for any object x and y of U, if $\frac{P_B(x,y)}{|B|} \ge \alpha \Rightarrow P_B(x,y) \ne \varnothing$, so

$$\forall_{x,y\in U} (A_B(x,y) \Rightarrow C_B(x,y)) \qquad \forall_{x,y\in U} (A_B(y,x) \Rightarrow C_B(y,x))$$

$$\underline{R}_B^A(x) \subseteq \underline{R}_B^C(x) \qquad \overline{R}_B^A(x) \subseteq \overline{R}_B^C(x)$$

By the definitions above we can conclude:

$$\underline{X}_B^A \supseteq \underline{X}_B^C, \overline{X}_B^A \subseteq \overline{X}_B^C$$

(4) When $1/|B| < \alpha_1 \le \alpha_2 \le 1$, it is evident that

$$\forall_{x,y\in U} \left(\frac{P_B(x,y)}{|B|} \ge \alpha_2 \Rightarrow \frac{P_B(x,y)}{|B|} \ge \alpha_1\right)$$

So,

$$\forall_{x,y\in U}(A_{2B}(x,y)\Rightarrow A_{1B}(x,y)) \qquad \forall_{x,y\in U}(A_{2B}(y,x)\Rightarrow A_{1B}(y,x))$$

$$\underline{R}_B^{A_2}(x)\subseteq \underline{R}_B^{A_1}(x) \qquad \overline{R}_B^{A_2}(x)\subseteq \overline{R}_B^{A_1}(x)$$

According to the definitions above we have the following conclusion:

$$\underline{X}_B^{A_2}\supseteq \underline{X}_B^{A_1}, \overline{X}_B^{A_2}\subseteq \overline{X}_B^{A_1}$$

In the classical rough set theory, the lower-approximation and the upper approximation of the object set are defined by equivalence relation. In incomplete information systems, the tolerance relation and the similarity relation can be seen as an extension of equivalence relation. From theorem 2, we can see that the threshold based similarity relation proposed in this paper is an extension of the tolerance relation and the constrained dissymmetrical similarity relation.

4. EXAMPLES

We use two examples to analyze the threshold based relation proposed above, one of which is an incomplete information system from the paper (Stefanowski et al., 1999) and the other is a data set from the *UCI Machine Learning Repository*.

Table 1. An example of the incomplete information system

A	a_1	a_2	a_3	a_4	a_5	a_6	a_7	a_8	a_9	a_{10}	a_{11}	a_{12}
C_1	3	2	2	*	*	2	3	*	3	1	*	3
C_2	2	3	3	2	2	3	*	0	2	*	2	2
C_3	1	2	2	*	*	2	*	0	1	*	*	1
C_4	0	0	0	1	1	1	3	*	3	*	*	*
D	Φ	Φ	Ψ	Φ	Ψ	Ψ	Φ	Ψ	Ψ	Φ	Ψ	Φ

Firstly, the incomplete information system is given in Table 1, where U is the set of objects denoted as U = $\{a_1, a_2 . . . , a_{12}\}$ and B is the set of condition attributes denoted as $\{C_1, C_2, C_3, C_4\}$, D is the decision attribute, "*" denotes the missing value. Assume that X = $\{a_1, a_2, a_4, a_7, a_{10}, a_{12}\}$ (Stefanowski et al., 1999).

(1) If $\alpha = 0$, then we can conclude:

$$\underline{X}_B^{A}=\underline{X}_B^{T}=\phi \quad \overline{X}_B^{A}=\overline{X}_B^{T}=\{a_1,a_2,a_3,a_4,a_5,a_7,a_8,a_9,a_{10},a_{11},a_{12}\}$$

(2) If $0<\alpha\le 0.25$, then we can conclude:

$$\underline{X}_B^{A}=\underline{X}_B^{C}=\{a_{10}\} \quad \overline{X}_B^{A}=\overline{X}_B^{C}=\{a_1,a_2,a_3,a_4,a_5,a_7,a_9,a_{10},a_{11},a_{12}\}$$

(3) If $0.25<\alpha\le 0.5$, then we can conclude:

$$\underline{X}_B^{A}=\{a_1,a_{10},a_{11}\} \quad \overline{X}_B^{A}=\{a_1,a_2,a_3,a_4,a_5,a_7,a_9,a_{12}\}$$

(4) If $0.5 < \alpha \leq 0.75$, then we can conclude:

$$\underline{X}_B^A = \{a_1, a_4, a_5, a_7, a_8, a_{10}, a_{11}\} \quad \overline{X}_B^A = \{a_1, a_2, a_3, a_9, a_{12}\}$$

(5) If $0.75 < \alpha \leq 1$, then we can conclude:

$$\underline{X}_B^A = \{a_1, a_4, a_5, a_7, a_8, a_{10}, a_{11}, a_{12}\} \quad \overline{X}_B^A = \{a_1, a_2, a_3\}$$

Obviously, with $1/|B| < \alpha \leq 1$, we have

$$\underline{X}_B^A \supseteq \underline{X}_B^C, \overline{X}_B^A \subseteq \overline{X}_B^C.$$

Moreover, if $1/|B| \leq \alpha_1 \leq \alpha_2 \leq 1$, then $\underline{X}_B^{A_2} \supseteq \underline{X}_B^{A_1}, \overline{X}_B^{A_2} \subseteq \overline{X}_B^{A_1}$

Secondly, we choose a data set named shuttle-landing-control which is concerned about *Space Shuttle Autolanding Domain* from the *UCI Machine Learning Repository*. In order to validate its ability in dealing with practiced problems, we made some appropriate modification in it: replacing some real values with missing values randomly at the ratio of less than 15%. Just as the following Table 2.

Table 2. Modified shuttle-landing-control data set

A	a_1	a_2	a_3	a_4	a_5	a_6	a_7	a_8	a_9	a_{10}	a_{11}	a_{12}	a_{13}	a_{14}	a_{15}
C_1	*	2	1	1	1	*	1	1	1	*	1	1	*	1	1
C_2	*	*	2	1	3	*	4	4	4	3	*	3	3	3	3
C_3	*	*	*	*	2	*	*	*	*	*	1	1	1	1	1
C_4	*	*	*	*	2	*	*	*	*	1	*	2	*	1	2
C_5	*	*	*	*	*	*	1	2	*	1	*	1	*	3	3
C_6	2	1	1	1	1	1	1	1	1	*	1	*	1	1	1
D	Φ	Ψ	Ψ	Ψ	Ψ	Ψ	Φ	Φ	Φ	Φ	Φ	Φ	Φ	Ψ	Φ

For this data set, let $X \subseteq U$, $B \subseteq \Omega$ and is a nonempty, by the calculation we can also draw all conclusions of theorem 2.

The experiment results show that the threshold based similarity relation proposed in this paper is an extension of the tolerance relation and the constrained dissymmetrical similarity relation, that is, the tolerance relation and the constrained dissymmetrical similarity relation are a special case of the threshold based similarity relation. This relation makes objects' classification more reasonable, and it is more practicable and flexible than the present.

5. CONCLUSIONS

Using standard rough set theory we may describe complete information systems. However many real-life data sets for data analysis usually contain a mass of missing values. So, the research how to acquire knowledge from such an incomplete information system has become a hotspot.

In this paper, after analyzing several present models based on rough sets for incomplete information systems, we propose an extended model under the threshold based similarity relation. From both the theoretically proof and experiments it can be seen that the rough set model based on the threshold based similarity relation is classifies more reasonable and flexible than that based on tolerance relation or constrained dissymmetrical similarity relation. By threshold man can easily control the partition of universe in some way.

REFERENCES

Chmielewski, M.R., Grzymala-Busse, et al. (1998). The rule induction system LERS-A version for personal computers. *Found Compute Decision Sciences*, 3/4: 181–212.

Kryszkiewicz, M. Rough set approach to incomplete information systems. (1998). *Information Sciences*, 1–4: 39–49.

Kryszkiewicz, M. Rules in incomplete information system. (1999). *Information Sciences*, 4: 271–292.

Pawlak Z. Rough sets. (1982). International Journal of Information and Computer Science, 5: 341–356.

Stefanowski, J., Tsoukias, A. (1999). On the extension of rough sets under Incomplete Information. In: Proceedings of the 7th International Workshop on New Directions in Rough Sets, Data Mining, and Granular-Soft Computing, Yamaguchi: Physica-Verlag, 73–81.

Wang G.Y. (2002). Extension of rough set under incomplete information systems. *Journal of Computer Research and Development*, 10: 1238–1243 (in Chinese).

Yin, X.R, Jia, X.Y., Shang, L. (2006). A New Extension Model of Rough Set Under Incomplete Information. In: Proceedings of the First International Conference on Rough Sets and Knowledge Technology, LNAI 4062, Springer-Verlag Berlin Heidelberg, 141–146.

Yin, X.R, Zhou, Z.H., Li, N., Chen, S.F. (2001). An approach for data filtering based on rough set theory. In: Proceedings of the Second International Conference on Web-age Information Management. LNCS 2118, Springer-Verlag Berlin Heidelberg, 367–374.

WEB-BASED EXPERT SYSTEM OF WHEAT AND CORN GROWTH MANAGEMENT

Xuesong Suo [1,*], Nan Shi [2]
[1] *School of Mechanical and Electrical Engineering, Hebei Agricultural University, Baoding 071001, China*
[2] *College of Life Sciences, Hebei University, Baoding 071002, China*
[*] *Corresponding author, Address: School of Mechanical and Electrical Engineering, Hebei Agricultural University, Baoding 071001, China; Tel: +86-312-7526475, Fax: +86-312-7526579, Email: suocedar@163.com*

Abstract: Web-based expert system of wheat and corn growth management mainly to put the information of wheat and corn planting on web. It integrated much new knowledge of agriculture including weather, plant protection, expertise, GIS, Internet and computer technology. It provided the information of breed, soil, wheat and crop characteristic for the farmer of different area in Hebei Province on web. Furthermore, it provided all kinds of wheat and corn planting expert systems for different period (insemination, fertilization, remedy, irrigation) of wheat and corn growth. The real time forecasting and management of plant was realized in agricultural production by this system.

Keywords: ASP; Precision Agriculture; Expert System; WEB

1. INTRODUCTION

As the technical progress of the computer, particularly the continuous development of the network, it makes the exchanges of the information resources to come to an unprecedented fast and convenient. In the agriculture aspect, the large agriculture technologist has already succeeded of carried on the agriculture expert system. Moreover, in recent years a kind of agriculture expert system based on network arose, and because of the large capacity, whole and new of its information has strongly favored of the large farmer.

Suo, X. and Shi, N., 2008, in IFIP International Federation for Information Processing, Volume 258; Computer and Computing Technologies in Agriculture, Vol. 1; Daoliang Li; (Boston: Springer), pp. 111–119.

2. THE PURPOSE AND THE MEANING OF THE SYSTEM DEVELOPMENT

There is very big margin in the level of agriculture infomationization between the developed nation and our country. They mainly performance: the agriculture information equipment is weak; the shared mechanism of the agriculture information standard and resources are lack; especially the valid resources that can be provided to use for farmer; the technologists of the agriculture are lack, and the strength of studies disperses, the level is low; the agriculture information technique application degree is low, far and far, it can't satisfy the request of the new century, the new stage of our country agriculture and village economic development.

To develop the network agriculture expert system, and make use of the Internet to provide the information service and the decision opinion for the farmer is the directions of agriculture infomationization in our country. The Internet has great capacity knowledge, delivers quickly; there are no time and region limiting, so it is the important form of the future information service. It makes use of the information and knowledge obtain, handle, spread, inform to farmer's hand in time and accurately, carry out the agricultural infomationization of production, management and marketing, accelerate the alteration of traditional agriculture and significantly upgrade the agriculture efficiency and management level (Zixing Cai, 2003, Plant R.E., 2001).

3. TOTAL DESIGN

3.1 The development target of the system

This system has a foundation that many other technique being researched chronically in the agrology, cultivation, meteorology and the agriculture resources efficiently using, applies artificial intelligence technique, and asks for help of the local agriculture expert and plant expert's knowledge, builds up the foundation information database, technique database, model database of the wheat and corn crop control. Studies the intelligence decision of wheat and corn in the each stage of growth develops the wheat and corn crop expert system based on the GIS and WEB, to expand the application of intelligence agriculture, to carry out the service of agriculture information network, to accelerate the infomationization progress of the Chinese agriculture.

3.2 The development route of the system

Collect the space data and attribute data concerning wheat and corn in Hebei;

Establish the data input project that collected from the real growth of wheat, corn and the experience of agriculture technologist;

To establish the wheat and corn crop control strategy (seed, fertilizer, irrigate, medicine);

Make use of the GIS controls to develop program, carry out the issuance of digital map and the crop control decision on the net;

Realize the On-line search of control information concerning wheat and corn.

3.3 The network project of the system

That system takes the MapXtreme of the MapInfo company as the technique core, the MapXtreme is the map application server that circulates on the intranet or the internet. It makes the crop control information and resources of wheat, corn can be see on computer map with a variety of form using the MapXtreme technique, carrying out the expert system decision information to outward release, helping the farmer and manager to assign the agriculture resources reasonably (Dongjun, 2000).

4. FUNCTION OF THE SYSTEM

4.1 Design of the function

The main function of the system is to put the crop and control information of wheat and corn in whole growth course on the web. The system combines weather factor, the knowledge of plant, the expert's experience, geography information system, the Internet technique and computer technique together, provides the breeds condition, growth characteristics, soil type and different weather data in different district of Hebei on the web. In the meantime, it includes the crop management expert system (seeding, fertilization, pesticide and irrigation expert systems) of wheat and corn in each growth stage. It realizes the real-time management and the estimate of crop in whole growth time. It mainly includes four functions: the macroscopical grow information system of wheat in Hebei;

the macroscopical grow information system of corn in Hebei; the microcosmic grow information system of wheat in Hebei; the microcosmic grow information system of corn in Hebei.

4.2 The macroscopically grow information system of wheat in Hebei

4.2.1 The forecast and decision of wheat's insemination period

The temperature and sunshine are the main factors that decides the wheat insemination period, so the system applies the function of search for the temperature and sunshine in everyplace in Hebei firstly, then the system can make the diagram of statistical data, it means the different result value with the different color, for example the scarlet means the heat area, the pink color means the low temperature area, the result is very intuitionistic, shown in figure 1. In the decision function of wheat's insemination, the system can make the curve of insemination data based different year and different area, also can offer the text advice of insemination (Ramon, 2004).

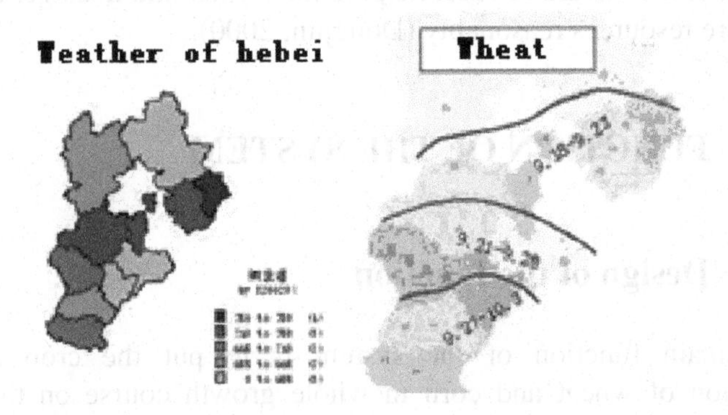

Figure 1. Insemination Pre-decision of Wheat

4.2.2 The distribute of wheat breed

System's digital map shows the distributes of different breed with the different icon in each region of Hebei, when the mouse clicks the region, system will show text for each region condition; when the mouse clicks the breed icon, system will show text for each breed, shown in figure 2.

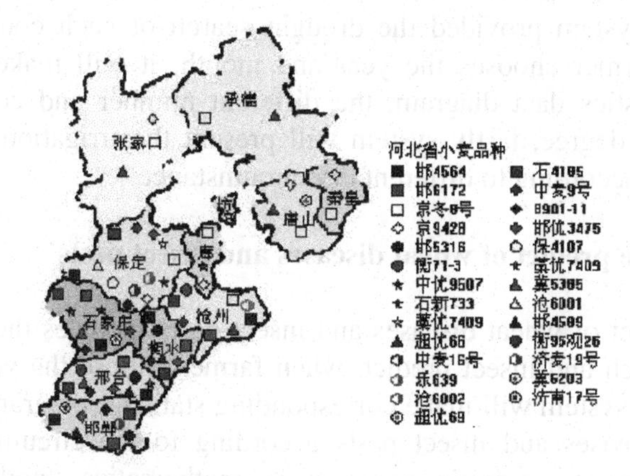

Figure 2. Breed Distribution of Wheat

4.2.3 The distribute of soil type

Choose the N, P, the K the system will make the homologous diagram of statistical data of soil chemical element, in the same methods, it means the different result value with the different color. When you search each area soil type, it will show the corresponding soil chemical element and the detailed text to each type, shown in figure 3.

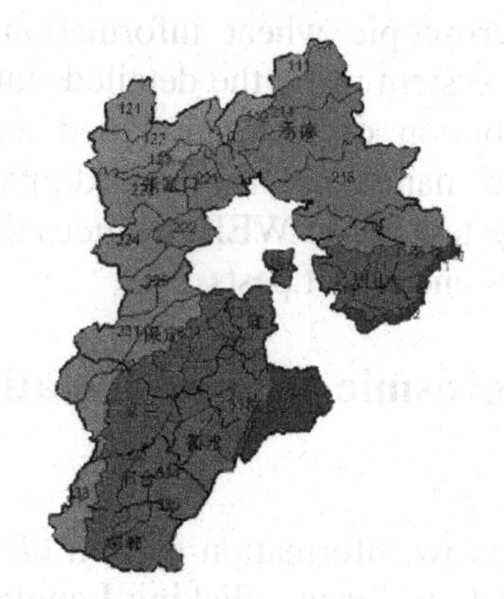

Figure 3. Distribution of Soil Type

4.2.4 The decision of wheat irrigation

The wheat irrigation decision according to the dry degree of the earth's surface and precipitation within the same period anciently. Make right irrigation decision according to the change of weather is an important path to carry out high yield. For making the farmer had a clear know to the dry degree, the system provided the drought search of each county in Hebei. When the former chooses the year and month, it will make homologous drought statistics data diagram, the different number and color mean the different dry degree, lastly system will present the irrigation opinion with detailed text according to different dry circumstance.

4.2.5 The predict of wheat diseases and insect pests

The predict of wheat diseases and insect pests includes the wheat insect diseases search and insect predict, when farmer choose the year and insect category, the system will make corresponding statistics diagram. The predict of wheat diseases and insect pests according to the circumstance of the weather forecast, expert's experience, mathematics model and crop condition, estimate the possible plant diseases and insect pests a month later, result with three kinds of data, index number, popular degree and loss rate, and shows the popular degree (heavy, medium, light, very light, out of fashion) of the plant diseases and insect pests on the digital map with different color conveniently (Zixing Cai, 2003).

4.3 The macroscopically grow information system of corn in Hebei

The same to macroscopic wheat information system based WEB in above-mentioned, the system made the detailed statistic to the different corn plant area, the distribution of different breed and output, lastly use the diagram make an auto manifestation on the digital map. At the same time builds up the expert system based WEB, includes the corn irrigation decision and corn plant diseases and insect pests.

4.4 The microcosmic grow information system of wheat in Hebei

The microcosmic grow information system of wheat in Hebei takes the Langfang as an example to set up, clicking Langfang on the map of Hebei, then enters that region, then can enter The microcosmic grow information system of wheat in Langfang.

4.4.1 The wheat insemination predict in Langfang

In the wheat insemination predict function, the system appears some consultation interfaces of local circumstances, after the farmer responds the corresponding choice, click reload the map, the system will release the digital map of wheat insemination predict in Langfang, its form is similar to the macroscopic wheat insemination period estimate, the system will also offer a advise according to the local circumstance. Its result interface such as figure 4 shows.

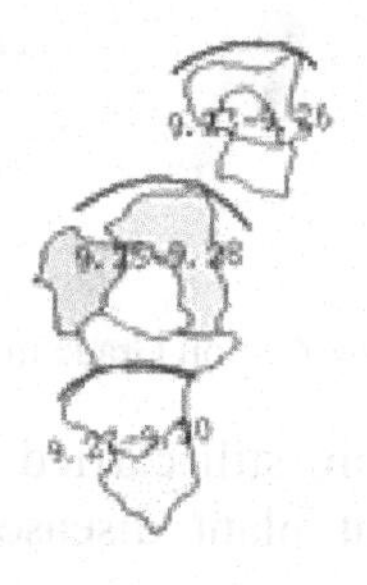

Figure 4. Wheat Insemination in Langfang

4.4.2 The wheat breed distribution in Langfang

After the enters this function the system's digital map shows the distributes of different breed with the different icon in each region of Langfang, when the mouse clicks the region, system will show text for each region condition; when the mouse clicks the breed icon, system will show text for each breed. The result interface such as figure 5 shows.

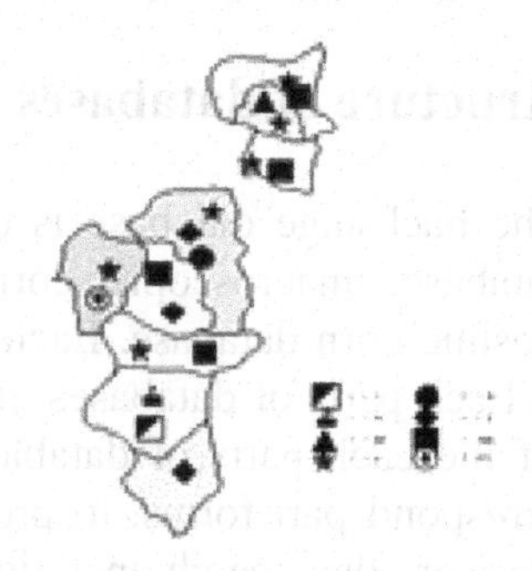

Figure 5. Wheat Breed in Langfang

4.4.3 The soil productivity distribution in Langfang

It can statistic the soil productivity distribution in every area of Langfang in the system, and makes caky diagram to release on the digital map, keeping the understanding of view very much to the customer. The result interface such as figure 6 shows.

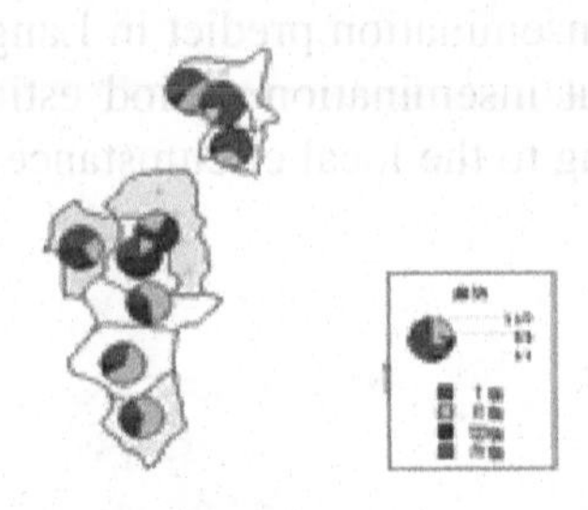

Figure 6. Soil Grade in Langfang

In the meantime the system still carried out the fertilizer decision, the irrigation decision and wheat plant diseases and insect pests predict in Langfang.

4.5 The microcosmic grow information system of corn in Hebei

The functions of microcosmic grow information system of corn in Hebei is similar to above, also takes the Langfang as an example, includes the different area of corn plant, the corn species distribution and soil land productivity grade. On the intelligent policy aspect, includes the fertilizer decision, the irrigation decision and wheat plant diseases and insect pests.

4.6 The total structure of databases

The total work of the backstage database is divided into four parts, is macroscopic wheat database, macroscopic corn database, microcosmic wheat database, microcosmic corn database. Each part of databases all uses the Access2000 format. Each parts of databases includes a total information form, the elucidation of the each parts of database is saving in this form, other data save into correspond part forms, to provide the system connects with the Mapinfo map data or other search operation.

5. THE CONCLUSION AND OUTLOOK

The network agriculture expert system can be convenient to diffuse agriculture science and technology fastly, having great capacity of information, the instruction is strong, the information is in time and accurate that wins the favor of the large farmer. As the same time, because of the booming of agriculture science and technology in our country will necessarily push the network agriculture expert system to a superior level. Again it, the fierce development of the computer hardware also provided the assurance for the healthy development of the expert system. So, the network agriculture expert system will necessarily become the importance part in the agriculture science and technology diffuseness in the future, and continuously develop accompany with the development of the agriculture science and technology.

REFERENCES

Dongjun Huang, Rui Liu, Songqiao Cheng. The development of expert system based on web [J]. The project and applications of computer, 2000, (9): 124-126.

Plant R.E. An Artificial Intelligence Based Method for Scheduling Crop Management Actions Agriculture System [J]. 2001, 31:127-155.

Ramon M. CU, Roland F. Line. An Expert Advisory System for Wheat Disease Management [J]. Plant disease. 2004, (2): 209-215.

Zixing Cai, Guangyou Xu, Artificial Intelligence: Principles and Applications [M], Qinghua University Press, Beijing, 137-143, 2003.

STUDY OF CORN OPTIMIZATION IRRIGATION MODEL BY GENETIC ALGORITHMS

Bing Zhang [1,*], ShouQi Yuan [2], JianSheng Zhang [1], Hong Li [2]
[1] *Changzhou institute of technology, Chang Zhou, China*
[2] *Research Center of Fluid Machinery Engineering and Technology, JiangSu University, Zhenjiang, China*
* *Corresponding author, Email: jiangsudaxuezb@163.com, Mobilephone: +86-13616110031*

Abstract: Many factors affecting irrigation model, including irrigation water volume, crop water requirement, production function of irrigation water, rainfall, soil water balance, water sensitive index in different stages of crop growth, the grain market price, irrigation water price, minimum yield, irrigation cost etc are considered. Then a multi-constraints and non-linear optimization irrigation model based on the maximal profit of irrigation water volume is set up, which is adaptive to our national conditions, and the real number encoding space of the model is searched by the powerful searching ability of genetic algorithm. The results show that this model can solve the optimization irrigation problem of summer corn, and genetic algorithm has very perfect searching function, and the optimal solution of the model can be found in very short time.

Keywords: genetic algorithms, real number encoding, Jensen model, objective function, optimization irrigation

1. INTRODUCTION

Nowadays, water resource becomes more and more scarce in many regions due to increasing demand, and the water needed by agricultural irrigation can't be increased any more, so how to allocate the limited water supply effectively over different stages of crop growth, and how to control

Zhang, B., Yuan, S., Zhang, J. and Li, H., 2008, in IFIP International Federation for Information Processing, Volume 258; Computer and Computing Technologies in Agriculture, Vol. 1; Daoliang Li; (Boston: Springer), pp. 121–132.

the quantity precisely for the maximum return from crop become the focus of agricultural irrigation engineering. With the sustainable agriculture developing, the key research on agricultural irrigation diverts from maximum yield to maximum return from crop with the same irrigation water volume.

It has always been a difficult problem to find optimal or near optimal solutions of irrigation model with traditionally dynamic optimization technology because of multi-factors, uncertainty and non-linearity in the model. To gain the optimal solutions, scientists adopt the dynamic programming stepwise approach (DPSA) and nonlinear programming (NLP), namely, a multi-stage decision-making process How to allocate the limited water supply over different stages of growth. But DPSA carries no guarantee that an optimum solution will be converged under any circumstance; and NLP is seldom used because of its nonlinearity.

Genetic Algorithms (GAs), as a kind of new global optimization search method, have many remarkable characteristics. They are computationally simple, adaptive and robust optimization techniques. GAs are based on the principles of population genetics and, constitute a special class among adaptive algorithms. They combine the adaptive process in nature with functional optimizations by simulating the selection of the best performing individuals in the populations. GAs are the iterative search method of "survival and detection" and start with an initial random population, then allocate initial solution to regions of the encoding search space. Among them, three main operators of GAs are selection, crossover, mutation; key content of GAs includes parameter encoding, the initial population settings, design of fitness function, penalty function, design of genetic operating and control parameters settings.

Previous studies on irrigation model have used binary encoding genetic algorithms. Necati Canpolat (1997) solves the seasonal irrigation scheduling using GAs. The goal of his irrigation model is finding the maximum net return. An irrigation schedule consists of a sequence dates on which water is to be applied to the crop, and for each date an amount of water to apply. The crop growth season of winter wheat is about 250 days, so the scheduling presents 2^{250} different possible irrigation decisions. Thus, the solution space for this problem is 2^{250}. At the same time, different genetic parameters are established for result comparison, and then get a group of best genetic parameters. But this model has not considered the other essential constraints, just considers crop yield and water supply. Besides, because the water irrigated each time is certain, it is impractical to be used to irrigation practice. Fu Qiang et al. (2003) set up the irrigation system of rice irrigated by well under deficit irrigation in SanJiang Plain using the real code acceleration genetic algorithm (RAGA) with the multidimensional dynamic planning (DP), and have made satisfactory results. Jose Fernando Ortega Alvarez et al. (2004) set up irrigation scheduling model which gaining the optimum irrigation water volume of different crops using binary encoding genetic algorithms. K. Srinivasa Raju and D. Nagesh Kumar (2004) study an irrigation model on the basis of maximum net return of many crops using GAs, and find that genetic algorithm is an effective tool to solve the optimization irrigation model, and get a group of optimum genetic algorithm parameters.

This article presents an irrigation model considered multi-factors and multi-constraints, and the model can be well solved by Genetic Algorithms.

2. SEARCH PROCEDURE OF GAS

In general, the objective function can be described as:

$$\max F_i = f(x) + \in \sum_{j=1}^{k} \delta_j (\phi_j)^2 \tag{1}$$

where F_i is fitness value, $f(x)$ is objective function value, k is total number of constraints, $\in$ is -1 for maximization and +1 for minimization, δ_j is penalty coefficient and ϕ_j is amount of violation. Once the problem is converted into an unconstrained problem, rest of the procedure remains the same. A detailed description of genetic algorithms is given by Deb (1999).

Search procedure of GAs is as follows:

Step 1, initialization: Setting initial evolution step $t = 0$; setting the maximal evolution step T; randomly creating an initial population $P(0)$ including M individuals.

Step 2, Individual evaluation: Using an evaluation function (eval (V) is employed in this study) to assess the fitness of each individual.

Step 3, Selection: Using a probabilistic selection process, the population for the next generation is formed. The parent individuals are selected based on their fitness. Individuals with higher fitness have a greater chance of contributing offspring. The selection mechanism plays an important role for searching towards better solutions.

Step 4, Crossover: Crossover involves the exchange of genetic material from two patents by randomly swapping parts of their chromosomes. Crossover provides a powerful exploration capability new individual for further evaluation within the hyper planes already represented in the population.

Step 5, Mutation: A mutation operator p_m is applied to the population. By modifying one or more of the gene values of an existing individual, mutation crates new individuals, generally resulting in increased variability of the population. It insures that the probability of reaching any point in the search space is never zero. A new population P(t+1) is produced from previous population P(t) by the operation of selection, crossover and mutation.

Step 6, Judgments of ending: if $t \leq T$, go to step 2; If t>T, individual of the highest fitness is determined as the optimal solution, and operations of GAs stop.

3. IRRIGATION MODEL

The net return per ha from crop is determined by input-output relationship. The crop net return to irrigation water is calculated as the crop yield multiplied by the selling price of the crop (some scientists have also considered byproduct's income, government's subsidy etc.). Input includes irrigating cost, fertilizer, agriculture chemicals, workforce's expenses and agricultural tax, etc. In this study, irrigation costs are the only variation costs of production to be considered, other inputs are assumed to be constant. The goal is to find a long-term irrigation schedule that will provide the maximum return for crop growth. The objective function, $f(Y_a, W)$, is the net return from crop:

$$f(Y_a, W) = Y_a * P_Y - W * P_w - \sum_{j=i}^{m} C_j \tag{2}$$

where f = net margin (yuan ha^{-1}); Y_a = real harvested yield (kg ha^{-1}); P_Y = selling price of the product (yuan kg^{-1}); W = total irrigation volume (m3 ha^{-1}); P_W = price of irrigation water (yuan m^{-3}); $\sum_{j=1}^{m} C_j$ = sum of other crop investments (yuan ha^{-1}); j represents number of other crop investments.

The actual crop yield Y_a and total irrigation volume W in the non-linear irrigation model mentioned above have strong coupling relation, and traditional irrigation is based on theory of soil water balance. Under deficit condition, some factors, such as deep leakage, surface runoff, can be neglected. Therefore, soil water balance can be simplified as:

$$\text{soil water balance: } W_i = ET_i - P_{ei} \tag{3}$$

where W_i = irrigation volume in the growth stage $i(mm)$; P_{ei} = rainfall in the growth stage $i(mm)$; ET_i = actual evapotranspiration in the growth stage $i(mm)$;

Different irrigation volume in each stage of crop growth has a complicated effect on crop yield, In this study, a numerical model described below, crop water production function (1968) is employed, which is popularly used, to predict crop yield corresponding to given irrigation schedule.

Jensen model: $$\frac{Y_a}{Y_m} = \prod_{i=1}^{n}\left(\frac{ET_i}{ET_{mi}}\right)^{\lambda_i} \tag{4}$$

where ET_i = actual evatransporation in stage $i\,(mm)$; ET_{mi} = maximal evatransporation in stage $i\,(mm)$; n = the stage number of crop growth divided; λ_i = water sensitive index in stage i; Y_m = maximum crop yield under abundant irrigation (kg ha^{-1})

A multi-factors crop irrigation model can be gained by considering Eqns (2), Eqns (3), and Eqns (4), which involves irrigation water volume, rainfall, water sensitive index, grain price and price of agricultural irrigation water influenced by market. What's more, the model Subjects to many constraints.

Irrigation model:

$$f(Y_a, W) = Y_m \prod_{i=1}^{n}\left(\frac{W_i + P_{ei}}{ET_{mi}}\right)^{\lambda_i} P_Y - P_W \sum_{i=1}^{n} W_i - \sum_{j=1}^{m} C_j \tag{5}$$

Subject to:

Constraint of total irrigation volume: $W_{\min} \le \sum_{i=1}^{n} W_i \le W_{\max}$

Constraint of irrigation volume in each stage: $W_{i\min} \le W_i \le W_{i\max}$

Constraint of minimal crop yield (basic grain consumption): $Y_a \ge Y_{\min}$

4. RESULTS AND DISCUSSION

Using the experimental data from the experiment region plot in Libao, Shanxi, to solve and verify the irrigation-margin model mentioned above, to find out the optimum water distribution and maximal return from irrigation of summer corn under deficit and multi-constraints condition,

4.1 Model Parameter Settings

Objective function:

$$f(Y_a, W) = Y_m \prod_{i=1}^{n} \left(\frac{W_i + P_{ei}}{ET_{mi}} \right)^{\lambda_i} P_Y - P_W \sum_{i=1}^{n} W_i - \sum_{j=1}^{m} C_j$$

max

where Y_m = maximum crop yield under abundant irrigation condition (6500kg ha^{-1}); P_Y = selling price of corn (0.9yuan kg^{-1}); P_W = price of irrigation water (6yuan mm^{-1}); $\sum_{j=1}^{m} C_j$ = sum of other crop investments (1500yuan ha^{-1}).

Table 1. Water sensitive index, maximal water requirement and rainfall in each stage of summer corn in Huoquan, Shanxi in 1998

Growth stage	Sowing-jointing	Jointing-tasseling	Tasseling-grouting	Grouting-ripening
Water sensitive index	0.0992	0.2368	0.3926	0.2014
Maximal evatransporation in each stage (*mm*)	94.56	112.25	124.04	90.57
Rainfall (*mm*)	70.2	52.1	47.5	35

Constraint matrix of irrigation volume in each stage:

$$\text{Constraints} = \begin{bmatrix} 10 & 30 \\ 15 & 80 \\ 25 & 100 \\ 20 & 80 \end{bmatrix} \quad (mm)$$

Constraint of minimal crop yield: $Y_a \geq 3500kg$

4.2 GAs Parameters

The search course of GAs is close related to settings of GAs parameters. Before searching, solution space should be encoded. In this study, real number encoding is chosen for the purpose of precision. Then find the fitness value for each individual. The parent individuals are selected based on their finesses. Individuals with higher fitness have a greater probability of contributing offspring. In the selection of solution, proportional selection (also called Roulette Wheel) is used; the way of crossover adopts one-point selection, involves the exchange of genetic material from two parents by randomly swapping parts of their chromosomes and formation of individuals. In order to improve the partial search capacity of genetic algorithms, maintain the diversity of the population and prevent the premature phenomenon, operator of mutation is utilized in the course of search. Because of the randomness of genetic operation such as selection, crossover and mutation, the individual with best fitness in the present population might be deleted. GAs adopts the tactic of saving best solution in the search course. Meanwhile, there are a lot of constraints in the model; GAs punish the solution dissatisfying constraints with penalty function.

GAs parameter settings: The way of encoding adopts real number encoding; number of individuals in the population is $M = 20$; the biggest step is T = 100; the way of selection is proportional selection; the way of crossover is one-point crossover; crossover probability is $p_c = 0.9$; mutation probability is $p_m = 0.01$, penalty function, which is used to punish the solutions dissatisfying constraints, is $F'(x) = 0.3F(x)$.

4.3 GAs Solution and Analysis

The experimental data of summer corn from experiment region in Libao, Shanxi under deficit irrigation is used to find out the optimal solution by GAs. The solution should promise the maximal return from irrigation and satisfy the constraints.

Figure 1, figure 2, and figure 3 show that at the beginning of the run, the optimal solution of the population is fairly low. As the number of generations pass, the optimal solution has a definite upward trend which

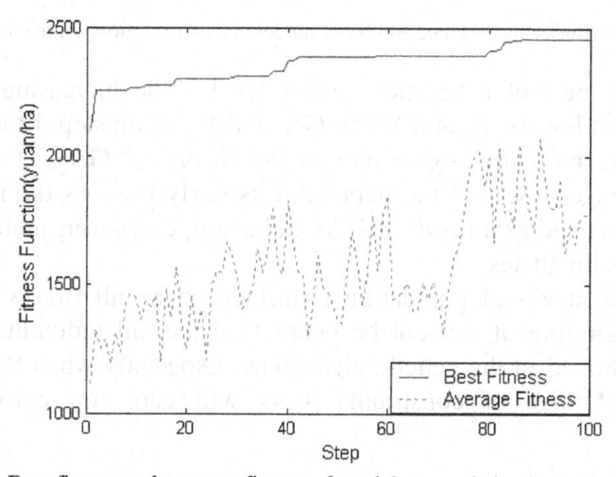

Figure 1. Best fitness and average fitness of model versus irrigation volume of 130mm

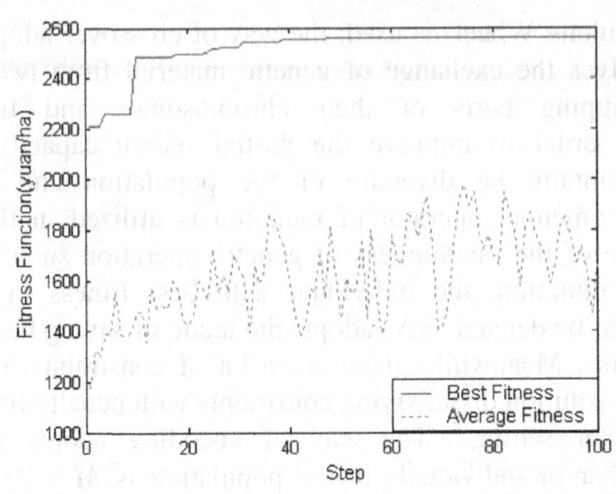

Figure 2. Best fitness and average fitness of model versus irrigation volume of 140mm

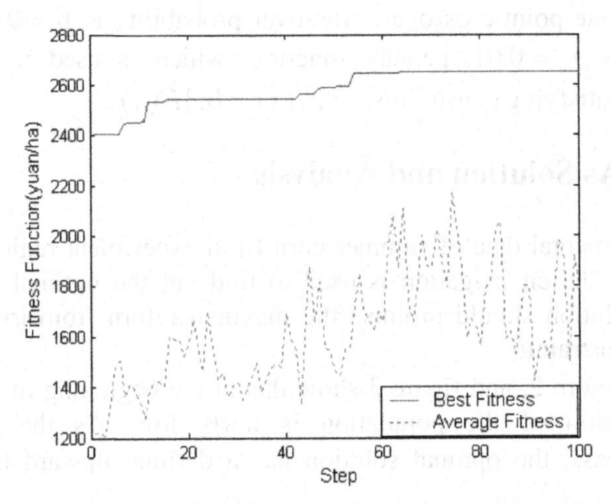

Figure 3. Best fitness and average fitness of model versus irrigation volume of 150mm

approaches the best of generation fitness levels. The increasing speed of optimal solution lowers. Again, When GAs reach certain step, fitness values keep invariable basically. According to the theory of GAs, at the initial stage, the average fitness of the population is fairly low. As the number of generations pass and as a result of GAs (selection, crossover, mutation), the average fitness fluctuates.

The average fitness of population is influenced by all fitness values of individuals; therefore, it can not be guaranteed that all individuals can be obviously improved in the genetic algorithms. Especially when the solution is out of the range of constraints, GAs will carry on corresponding punishment to fitness value according to penalty function; in other words, the fitness value change into a smaller one. So the average fitness of population fluctuates randomly.

In order to understand better the change of optimal solution and the influence of irrigation to return under different irrigation condition, the optimum solutions are compared in figure 4.

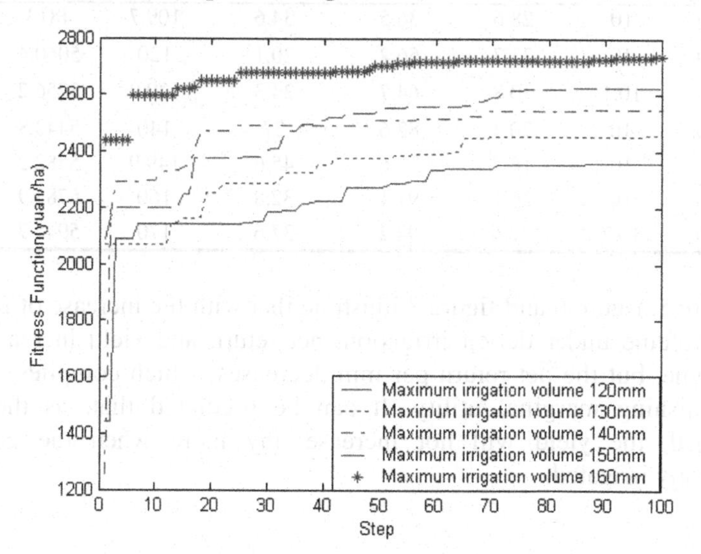

Figure 4. Comparison of optimal solution

From the comparison of optimal solution of irrigation volume under different irrigation level in figure 4, return from irrigation increases along with the increase of irrigation volume when under deficit irrigation condition. It obtains the maximum value when irrigation volume is 160 millimeters. When irrigation volume is 120 millimeters, it is the shallest.

From table 2, when the net return from corn is maximal, the optimum distribution of irrigation volume and corn yield response to irrigation schedule. Within certain irrigation range, under deficit irrigation condition, GAs choose upper limit of supplied irrigation water first. Since the water sensitive index in tasseling-grouting stage is the biggest, the proportion of irrigation volume is correspondingly the biggest, while the proportion in sowing - jointing stage is the smallest. GAs consider the influence of water sensitive index within certain irrigation range, which can be proved in table 2. With the increase of irrigate volume, net return and yield increase at the same time.

Table 2. Optimal distribution of irrigation volume, corn yield and net return under different irrigating condition

Irrigation volume (*mm*)	Optimal distribution of limited irrigation water supply in different stage of growth (*mm*)				Total irrigation volume (mm)	Crop yield kg/ha	Net return Yuan/ha
	Sowing-jointing	Jointing-tasseling	Tasseling-grouting	Grouting-ripening			
100–110	10	28.6	36.5	34.6	109.7	4813	2173.5
110–120	10	33.7	56.2	20.1	120	5060.4	2334.3
120–130	10.1	30.8	64.7	24.3	129.9	5256.2	2450.6
130–140	10	20.4	82.6	27	140	5442.8	2558.5
140–150	10	16.7	77.6	45.6	149.9	5582.2	2624.2
150–160	10	26.1	91.1	32.8	160	5783.9	2745.5
160–170	10.7	27.5	94.2	37.6	170	5945.7	2831.1

Figure 5, figure 6 and figure 7 illustrate that with the increase of irrigation water volume under deficit irrigation, net return and yield increase at the same time, but the net return per mm decreases, which complies with law ofdiminishing marginal utility. It can be predicted that, as the search continued, the yield will not increase any more when the corns are abundantly irrigated.

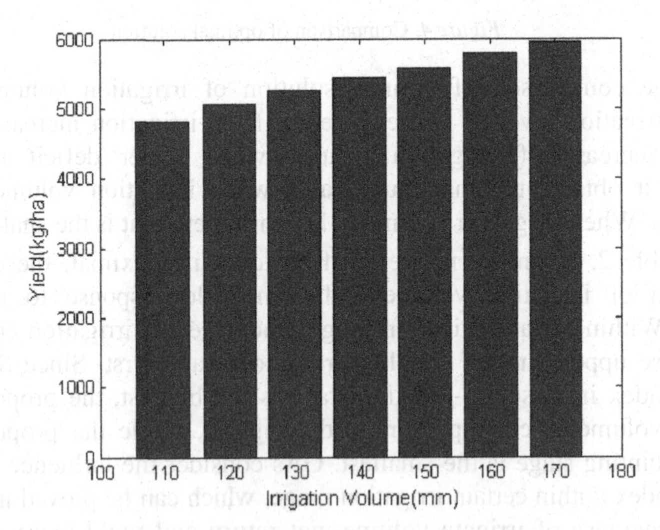

Figure 5. Relation between irrigation and yield

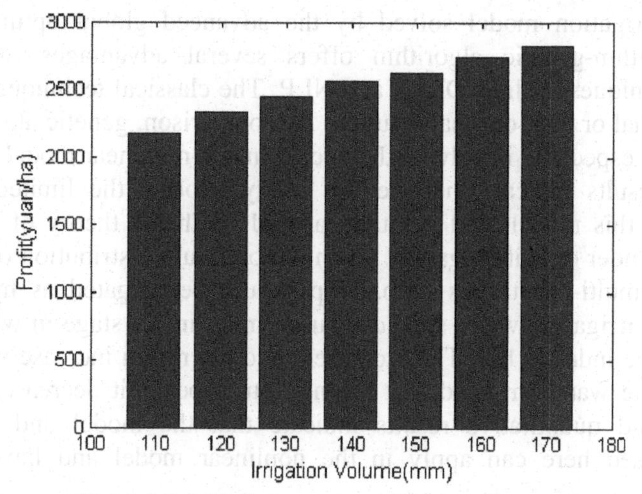

Figure 6. Relation between irrigation volume and profit

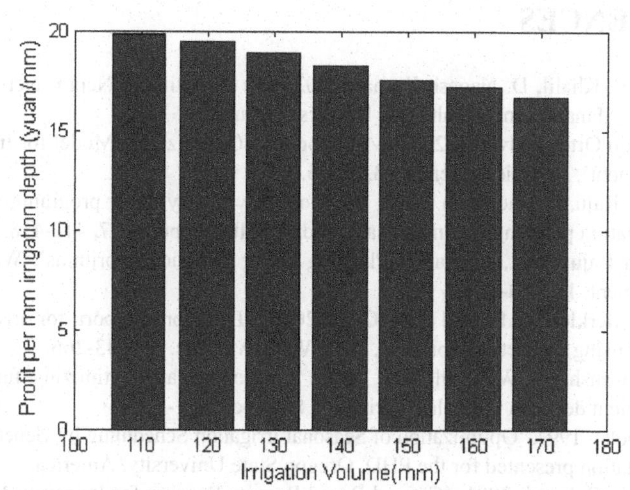

Figure 7. Relation between irrigation volume and profit per mm irrigation volume

5. CONCLUSIONS

(1) A multi-constraints and non-linear irrigation-return model is presented in this article, involving many factors, such as rainfall, water sensitive index, selling price of grain and irrigation water price, and combines soil water balance, production function of irrigation water. This irrigation model can meet the need of present agricultural irrigation practice of our country.

(2) The irrigation model solved by the advanced global optimization search algorithm-genetic algorithm offers several advantages over the classical techniques such as DPSA and NLP. The classical techniques can't find the optimal or near optimal solution. By comparison, genetic algorithms perform well, especially in solving the uncertain and nonlinear model.

(3) The results indicate that we can easily allocate the limited water supply using this model, and it totally accords with the theory of genetic algorithms. Under deficit irrigation, when the optimum distribution of water satisfies the multi-constraints term, crop should be irrigated as much as possible, and irrigation water should be used more in the stage in which its water sensitive index is big. The crop yield and net return increase with the increase of the water irrigated, but the net return per unit decrease. These qualitative and quantitative results indicate that the model and genetic algorithms used here can apply in the nonlinear model and have good prospect.

REFERENCES

Abedalrazq F. Khalil, D. Nagesh Kumar. 2002 'Use of Artificial Neural Networks in Canal Irrigation Management', Utah State University, America.

Jose Fernando Ortega Alvarez. 2004, 'An Economic Optimization Model for Irrigation Water Management', Irrigation science. 23, 61-75.

K. Srinivasa Raju, L. Duckstein. 2003, 'Multiobjective fuzzy linear programming for sustainable irrigation planning: an Indian case study', Soft Computing. 7, 412-418.

K. Srinivasa Raju. 2004, 'Irrigation Planning Using Genetic Algorithms', Water Resources Management. 18, 163-176.

Kuo, S.F., Merkley, G.P. and Liu, C.W. 2000, 'Decision support for irrigation project planning using a genetic algorithm', Agri. Water Manage. 45, 243-266.

N.S. Raghuwanshi, W.W. Wallender. 1999, 'Forecasting and optimizing furrow irrigation management decision variables'. Irrigation Science. 19, 1-6.

Necati Canpolat. 1997, 'Optimization of Seasonal Irrigation Scheduling by Genetic Algorithms' A dissertation presented for the PHD. Oregon State University. America.

Paul Dominic Colaizzi. 2001, 'Ground Based Remote Sensing for Irrigation Management in Precision Agriculture ', A dissertation presented for the PHD. The university of Arizona. America.

Radwan A. Al-Weshah. 2000, 'Optimal use of irrigation water in the Jordan valley: A case study', Water Resources Management. 14, 327-338.

Renato Silvio da Frota Ribeiro. 1998, 'Fuzzy logic based automated irrigation control system optimized via neural networks', A dissertation presented for the PHD. The university of Tennessee. America.

Zhang Bing, Yuan Shouqi, Cheng Li. 2004 'Study on the Optimized Distribution Model of the Limited Irrigation Volume on the Basis of MATLAB'. 2004 CIGR International Conference. Beijing.

ANALYSIS OF VIRTUAL REALITY TECHNOLOGY APPLICATIONS IN AGRICULTURE

Hailin Li
Institute of Electronic Technology, Information Engineering University, Zhengzhou, China, 450004
* *Corresponding author, Address: Institute of Electronic Technology, Information Engineering University, 12 Shangcheng East Road, Zhengzhou, 450004, P. R. China, Tel: +86-371-63538447, Email: lihl_c@yahoo.com.cn*

Abstract: Agricultural information technology, especially virtual reality (VR) technology, will act the important roles in agricultural modernization and realm. Computer science and IT are both playing important roles in the development of agriculture and rural areas of the world. Combining agriculture science with IT and VR, the virtual agriculture technology explored new ways of studying and applying agriculture information technology. On the basis of concept of virtual agriculture given, the composition, application range and development direction of virtual agriculture were analyzed. On basis of above mentioned, the architecture of virtual crops which is a typical application of virtual agriculture was analyzed and studied, and the model of virtual crops was modeled using plant three-dimensional rebuild technique. It can improve the applications of VR in agriculture fields and advance the process of agricultural modernization.

Keywords: VR; virtual agriculture; virtual crops; architecture; model

1. INTRODUCTION

Nowadays, the computer science and information technology are making the transition of Chinese agriculture from traditional agriculture to modernized agriculture, they have become the most efficient methods and tools for improving the agriculture productivity and utilization ratio of agricultural resources, and they are an important means for achieving agricultural information and modernization (Jia S G, 1999; Yang Y X et al., 1999). Agriculture virtual reality, or named virtual agriculture, which is a

Li, H., 2008, in IFIP International Federation for Information Processing, Volume 258; Computer and Computing Technologies in Agriculture, Vol. 1; Daoliang Li; (Boston: Springer), pp. 133–139.

major research field, is different from traditional research methods. It can integrate the agriculture science with information technology and explore new approaches to study and apply agricultural information technology to the condition of computer auxiliary design and simulation (Jin R Z et al., 2001; Zhu Y, 2004).

In the wake of development in computer science and virtual reality technology, the research and application of virtual agriculture, which will take great effect on agricultural production, scientific research, education, research of plant diseases and insect pests and development of new agricultural machines, are far-reaching significance strategic measures to China which is a great agricultural nation (Wang C W et al., 2000; Yang GC, 2005). The research of virtual agriculture can promote the basic agricultural theoretical research, and in turn it can promote the development of virtual agriculture and virtual reality, then advance agricultural scientific research (Su Y, 2006; Zhang Y L et al., 2001).

2. VIRTUAL AGRICULTURE TECHNOLOGY

2.1 Rationale of virtual agriculture

(1) Computable theory of crops-environment

The rationale of virtual agriculture is the truth that the relationship between crops and environment is computable. The researchers in Nanjing Agricultural University have set up a key Chinese ministry of agriculture laboratory that can regulate and control crops' growth, they made full research and simulation on the relationship between crops and environment. The results showed that the relationship between crops and environment is computable (Wang D B et al., 2006).

(2) Virtual reality technology

VR is a high-level human-computer interface featuring immersive sense, interactivity and proposition (Wang C W et al., 2000). It makes integrated use of computer graphic science, simulation technology, multi-media technology, artificial intelligence, computer network and multi-sensors technology to simulate the human sense organs such as vision, hearing and touch, and can get people immersed in the virtual states which can interact with it by the ways of language and gestures. VR has been acknowledged as one of the most important technologies that will produce great effects on human life in the 21st century (Yang G C, 2005; Holt D A et al., 2003).

The appearance and development of VR provide many new methods of solving problems for us. At present, the international societies have got ripe results of VR research on many applications, but VR technology of agriculture is still a future technology which is still in an exploring stage. Happily, the appearance of virtual agriculture and the tentative ideas and explorations on virtual plants, virtual farms and virtual agricultural experiments make us notice that the VR technology certainly will be used widely and will produce renovate train of thought and new means for agriculture science study and agricultural production (Yang G C., 2005; Guo Y et al., 2001; Li Z W et al., 2005).

2.2 Fundamental conception of virtual agriculture

Virtual agriculture extended from virtual reality which came out in the mid-1980s.Virtual agriculture is a renovation of traditional agriculture, and a new agricultural research method which gets agriculture domain as the object of study and gets advanced technology as means. It can reduce experiment costs, shorten the research time, and raise the research efficiency that agronomists carry out agricultural research in virtual environments.

At present, there are many different definitions of virtual agriculture all over the world, such as D. A. Holt and S. T. Sonka's (Holt D A et al., 2003), Academician Sun's (Chen S B et al., 2003; Sun J L, 2000) and Professor Yang's (Yang G C, 2005) definition. Synthesizing various opinions, the author gives a definition of virtual agriculture as agricultural simulation agriculture that on the basis of information technology and VR technology, and a significant subfield of VR. It views the objects of agriculture domain as core, takes advanced information technology and VR applying to complicated agricultural production, management, teaching, scientific research, planning, resource collocation and circulation of goods with the support of high-efficient and reliable communication network. And it takes networks and computers as a platform to simulate and reappear the studied objects of each link in the agriculture and achieve the aims of interaction and visualization of studied objects and environments.

2.3 Composition of virtual agriculture

The virtual agriculture is one of the greatest key technologies of digital agriculture. It has vast vistas on the fields of agricultural management of production, planning and resource configuration (Wang Y M, 2003).

In a broad sense, virtual agriculture includes: virtual crops, which is used to breed new varieties of rice, corn, wheat, soybean, cotton, and so on; virtual animals, which is used to cultivate new varieties of animal by-products and aquatic product such as pig, cattle, sheep, chicken, fish; virtual agricultural machinery manufacture, which is used to design and manufacture new energy conservational and high-efficient agricultural machinery to raise working efficiency and utilization ratio of agricultural devices and facilities, and to raise comprehensive utilization ratio of agricultural resources; virtual farm, which can simulate the market of agricultural products and management of production (Chen S B et al., 2003).

3. THE FUTURE OF VIRTUAL AGRICULTURE TECHNOLOGY

The virtual agriculture is an effective and practical technology, so it possesses good prospect of application and dissemination (Zhang W X et al., 2006). With the uninterrupted growth and ripeness, the virtual agriculture technology certainly will get greater progress on the under aspects (Wang D B et al., 2006; Li Z W et al., 2005; Zhang W X et al., 2006; Song Y H et al., 2000; frwork.htm, 2005; Zheng Y Y et al., 2004):

(1) The development and application of intelligent agricultural expert system

On the existing basis, to study the growth mechanism models of crops, fruit trees, live stocks and fowls, and depending on the knowledge of experts in agriculture, to achieve the integration with other information technologies, and then to develop all kinds of high-efficient, practical and intelligent agricultural expert systems.

(2) The application of virtual agriculture in the digital earth

Virtual reality and visualization are the technological basis of digital earth, its basic components are digital agriculture and digital city. Virtual agriculture uses visualization and virtual reality to simulate agriculture, and to achieve digital agriculture. That is, virtual agriculture is a main form of digital agriculture, and an indispensable part to digital earth.

(3) The application of virtual agriculture in the tour of agriculture

The tour of agriculture can make rational use of rural resources, simulate the activities of picking, marketing, fishing and playing to get tourists participate, experience and enjoy agricultural production, and arouse tourists' interests of loving labor, lives and nature.

4. AN APPLICATION CASE OF VIRTUAL AGRICULTURE-VIRTUAL CROPS

4.1 Virtual crops

Virtual reality can take strong sense of reality and true experience to users, and take a new interactive concept-immersed interactive environment. Applying this technology, the many years' data of crops growth can be simulated as a few minutes procedure of growth. It is possible to operate, observe, test and achieve crops data in short time (Su Z B et al., 2005).

Virtual crops technology applies virtual reality to simulate the structure, procedure of growth and environment of crops in three-dimensional space. It uses data collecting system to monitor environment factors changes and crops growth trends, and to study regular patterns of crops and environment. Virtual crops system has significant value for exploring crops ideal models, optimizing growth measures of crops, constructing shapes of crops, designing gardens and teaching (Sun J L, 2000; Song Y H et al., 2000; frwork.htm, 2005).

Virtual agriculture has different architectures of different targets. The architecture of virtual crops system is showed in Fig. 1.

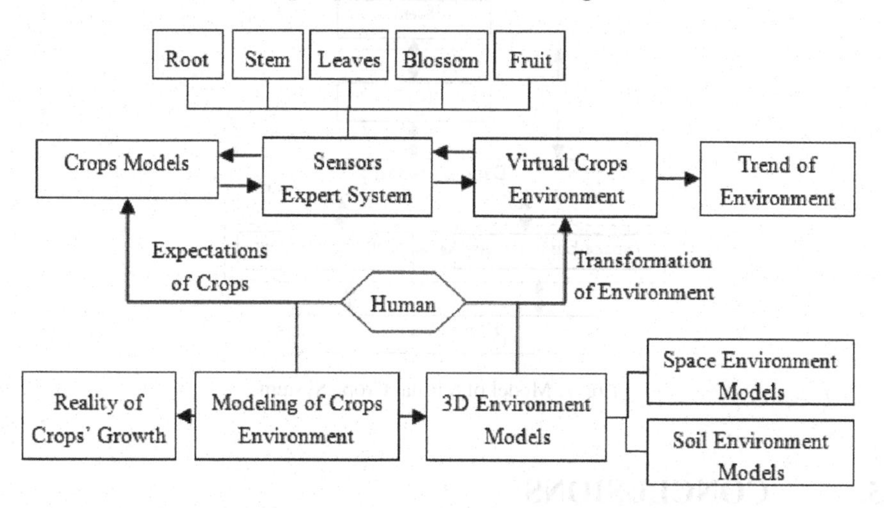

Fig. 1. The general achitecture of virtual crops system

4.2 Research on models of virtual crops

The models of virtual crops are built based on crops research of real world. Firstly, it should take accurate quantitative research on shapes and constructions of real crops, sum up growth law crops, then express them with proper methods (Su Y, 2006).

In this paper, the three-dimensional rebuild method of plants was used to build the model. It used instruments to collect spatial data and environment data of crops, to program in computer to call the achieved data, then, to achieve the three-dimensional simulation of crops. This is a simulated method of real crops. With the improvement of instrument precision, the realness of crops simulated will be higher (Shen W J et al., 2002).

The typical component parts of crops are root, stem, seeds, leaves, blossom and fruit (frwork.htm, 2005; Su Z B et al., 2005). The root is called underground part, it interacts with soil environment. The stem, leaves, blossom, fruit and seeds are called ground part, it interacts with spatial environment. There are interaction and cooperation among organs of crops. The human being as an environmental factor is a intelligent agent. The human being can change parameters of related environmental factors to observe the procedure of growth and evolution of virtual crops and get the purpose of study and teaching. The fig. 2 shows the model structure of virtual crops system.

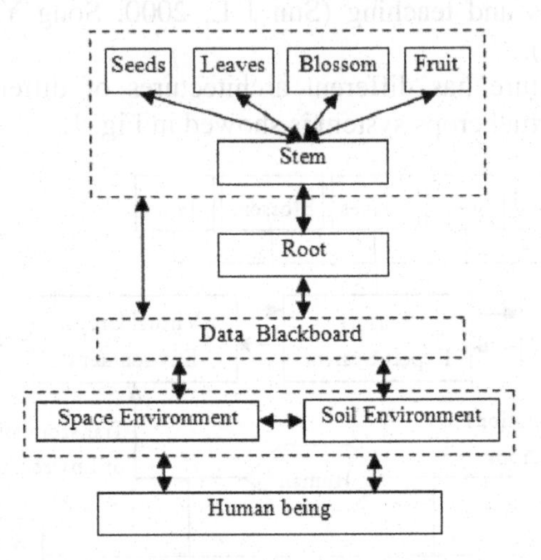

Fig. 2. Model of Virtual Crops System

5. CONCLUSIONS

Virtual agriculture is the result of combining agriculture science with information and virtual reality technology. It makes agronomists develop agricultural research in the virtual environment, this can reduce experiment costs, shorten the research time, get visualized process and experiment results directly, and improve the research efficiency of agricultural domain.

The applications of virtual crops have two problems to be improved and perfected: the interaction between virtual crops and environment and the virtual root system of crops. With the advances of the above problems, agricultural production will certainly produce earthshaking change in China. At the same time, the research of virtual agriculture technology has been risen as an international forward agriculture technology.

REFERENCES

Chen Shen-bin, Sun Jiu-lin. Virtual Agriculture and Virtual Reality-Potential Application of Scientific Database. http://www.pcvr.tom.cn, 2003.

Guo Yan, Li Bao-guo. The Research Progress of Virtual Plants [J]. Science Bulletin, 2001, 4: 273-280. (in Chinese)

Holt D A, Sonka S T. Virtual Agriculture: Developing and Transferring Agricultural Technology in the 21st Century. http://www.agr.uiuc.edu/virtagl.html, 2003.

http://www.cau.edu.cn/viscs/vplant/frwork.htm, 2005 (in Chinese)

Jia Sha-gang. Agriculture Informatization and Agriculture Scientific Revolution [J]. Computer and Agriculture, 1999, 2: 3-8 (in Chinese)

Jin Run-zhao, Wang Zhao-yi. Introduction to Virtual Reality and Applications in Agriculture. Journal of Tianjin Agricultural College, 2001, 2: 27-32 (in Chinese)

Li Zhi-wen, Han Xiao-ling. Present Situation and Development Trend of VR [J]. Information Technology, 2005, 3: 94-96 (in Chinese)

Shen Wen-jun, Zhao Chun-jiang, Shen Zuo-rui, et al. Virtual Reality Technology and Its Application in Agriculture. Research of Agricultural Modernization, 2002, 5: 378-381 (in Chinese)

Song You-hong, Jia Wen-tao, Guo Yan, et al. Advances in Virtual Crops Research [J]. Computer and Agriculture, 2000, 6: 6-8 (in Chinese)

Su Yu. Study on Progress Problems of Virtual Agriculture [J]. Shanxi Agriculture Sciences, 2006, 1: 101-103 (in Chinese)

Su Zhongbin, Meng Fanjiang, Kang Li, et al. Virtual plant modeling based on Agent technology. Transactions of the CSAE, 2005, 8: 114-117 (in Chinese)

Sun Jiu-lin. Agriculture Information Engineering: Theory, Method and Application [J]. Engineering Science of China, 2000, 3: 89-91 (in Chinese)

Wang Cheng-wei, Gao Wen, Wang Xing-ren. The Theory, Realization and Application of Virtual Reality Technology, Qsinghua University press, 2000 (in Chinese)

Wang Dao-bo, Zhou Xiao-guo, Zhang Guang-lu. The Technology and Application of Virtual Agriculture [J]. Agriculture Network Information, 2006, 3: 16-18 (in Chinese)

Wang Yi-ming. Situation and Development of Digital Agriculture [J]. Transactions of the CSAE, 2003, supplement: 9-10 (in Chinese)

Yang Guo-Cai. Research on Architecture of Virtual Agriculture [J]. Computer Science, 2005, 3: 125-126 (in Chinese)

Yang Yong-xia, Zhu De-hai, Tan Tai-lai, et al. The Development of IT and the Application Trend of Which in Agriculture Field [J]. Computer and Agriculture, 1999, 4: 1-6 (in Chinese)

Zhang Wei-xing, Zhu De-feng, Zhao Zhi, et al. A Brief Review of Virtual Reality Technology and Virtual Agriculture [J]. Guizhou Agriculture Sciences, 2006, 2: 115-118 (in Chinese)

Zhang Yun-lan, Zheng Jiang-ping, Chen Zheng-yu. Analysis on Agricultural Information Technology Status quo and Suggestion for Development in China [J]. Journal of Zhejiang University (Agric. & Life So1.), 2001, 2: 229-232 (in Chinese)

Zheng Yan-yan, Zhou Guo-min. VR and fruit pruning [J]. Agriculture Network Information, 2004, 8: 10-12 (in Chinese)

Zhu Yong. Effects of Information Technology in Agriculture Modernization [J]. Agriculture Modernization, 2004, 5: 43-44 (in Chinese)

RESEARCH AND DEVELOPMENT OF THE INFORMATION MANAGEMENT SYSTEM OF AGRICULTURAL SCIENCE AND TECHNOLOGY TO FARMER BASED ON GIS

Hao Zhang[1], Lei Xi[1], Xinming Ma[1,2,*], Zhongmin Lu[3], Yali Ji[1], Yanna Ren[1]

[1] *College of Information and Management Science, Henan Agricultural University, Zhengzhou, Henan, He Nan, China, 450002*

[2] *College of Agriculture, Henan Agricultural University, Zhengzhou, Henan, China, 450002*

[3] *Hua county Center of Agricultural Technology Popularization, Henan, China, 456400*

* *Corresponding author, Address: College of Information and Management Science, Henan Agricultural University, 63 Agricultural Road, Zhengzhou, 450002, P. R. China, Tel: +86-371-63558388, Fax: +86-371-63558090, Email: xinmingma@371.net*

Abstract: With the rapid progress of information technology, more and more people utilize high technology to boost the fast development of the national economy. Since the project of agricultural science and technology to farmer was put into practice by Ministry of Agriculture in 2005, 212 representative counties have been set, 200,000 typical households have been added, and 4 million peasants have enhanced production and increased income. According to the criterion of Software Engineering, the article collected the information of Agricultural science and technology to farmer and the geography information of all villages and towns in HUA county, designed the system structure of Agricultural science and technology to farmer, implemented the information query and management, and realized the special topic analysis to the information distribution by tools of OOP, GIS components and network database, integrating GIS and MIS smoothly. The system has been applied in HUA county, and facilitated information management, analysis and decision-making to the agricultural science and technology to farmer.

Keywords: agricultural science and technology to farmer, GIS, C/S, smooth integration

Zhang, H., Xi, L., Ma, X., Lu, Z., Ji, Y. and Ren, Y., 2008, in IFIP International Federation for Information Processing, Volume 258; Computer and Computing Technologies in Agriculture, Vol. 1; Daoliang Li; (Boston: Springer), pp. 141–150.

1. INTRODUCTION

With the rapid progress of information technology, more and more people utilize high technology to boost the fast development of the national economy (Tang Wan-min, 2006; Yin Li-hui, 2006). Since the project of agricultural science and technology to farmer was put into practice by Ministry of Agriculture in 2005, 212 representative counties have been set, 200,000 typical households have been added, and 4 million peasants have enhanced production and increased income (http://www.gov.cn/jrzg/2006-10/17/content_415658.htm, 2006). But the Information Management System of agricultural science and technology to farmer has not been developed adequately yet in each of domestic provinces, lacking effective management and analysis. Under this background, the topic of scientific management and analysis to the information of science and technology to farmer is put forward. And 3S(RS, GPS and GIS), which has great predominance of spatial management and analysis, has been widely applied in all kinds of domains, such as transportation, military affairs, agriculture, forestry and so on (Alexander Köninger, 1998; Yan Tai-lai, 2005; Wei Fu-quan, 2004; Li Li-wei, 2006; Gong Jian-ya, 2004). In view of these factors, the item closely followed the project of the science and technology to farmer put into practice by Ministry of Agriculture, researched, designed and developed the information management system of agricultural science and technology to farmer using GIS technology and computer technology.

2. SYSTEM DESIGN

Using the virtues of friendly interface, strong simulation, rapid inquiry localization, good expansibility and special spatial analysis of GIS (Wu Lun, 2001; Gong Jian-ya, 2004; Chen Zheng-jiang, 2005) and according to the criterion of software engineering, the article collected the information of Agricultural science and technology to farmer and the geography information of all villages and towns in HUA county, designed the system structure of Agricultural science and technology to farmer, and implemented the query and management of the information of agricultural science and technology to farmer and the special topic analysis of the information distribution by the tools of OOP, the components of GIS and network database, integrating GIS and MIS smoothly.

2.1 System Structure

The system introduced Client/Server/DBMS which separates the logic service from the user connection (Zheng Ke-feng, 2005). Fig. 1 shows the structure of the system. Controlled by unification database interface that is the foundation platform of database, all kinds of spatial data and attribute data are stored by using ArcSDE & MS SQL Server, which are programmed by using VB and MapObjects.

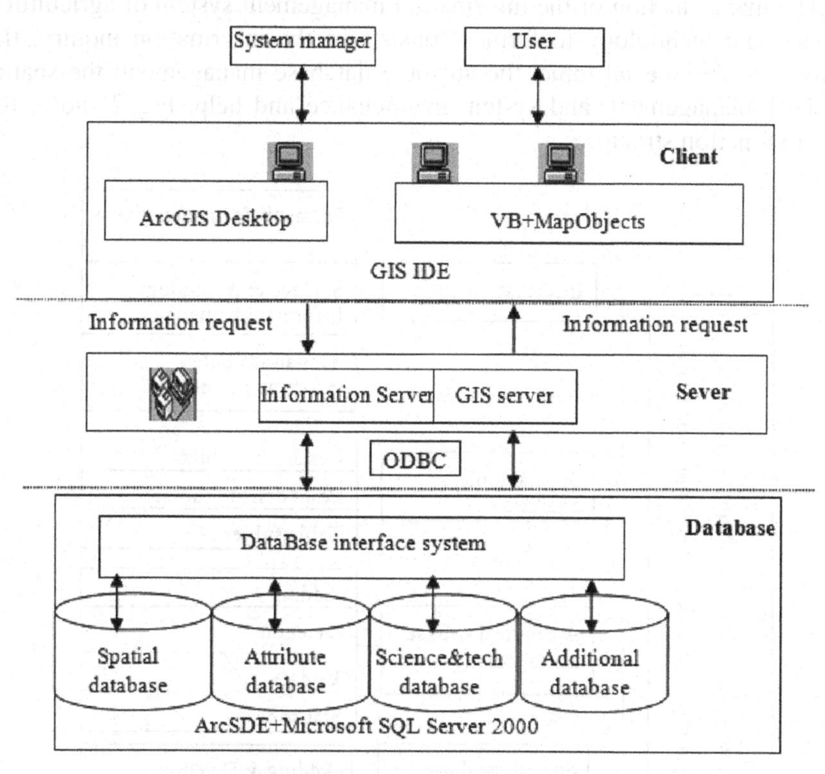

Fig. 1. The structure of the system

Client end is responsible for display of the data and communication with user, which requests information from the server in order to implement all kinds of function, such as information browsing, inquiry, adding, deletion, update and so on. Server end mainly realizes data sharing and data transmission to client. The database platform uses GeoSpatial Database and SQL Server 2000, whose duty is to accept server request operation to the database and transmit data. The C/S pattern needs to install server software in the server end and client software in the client end. The server data such as spatial data, the attribute data, the information of agricultural science and technology to farmer and other information are saved in the server end, the client data are put on the Client end. When client end needs to request data of server end, it will send out the request through the local network to server, server end confirms connection, accepts and processes the request information, and returns the processed result to client end.

2.2 System Function

The main function of the information management system of agricultural science and technology to farmer consists of the information inquiry, the analysis to the special topic, the attribute database management, the spatial database management, and system maintenance and help. Fig. 2 shows the system function structure.

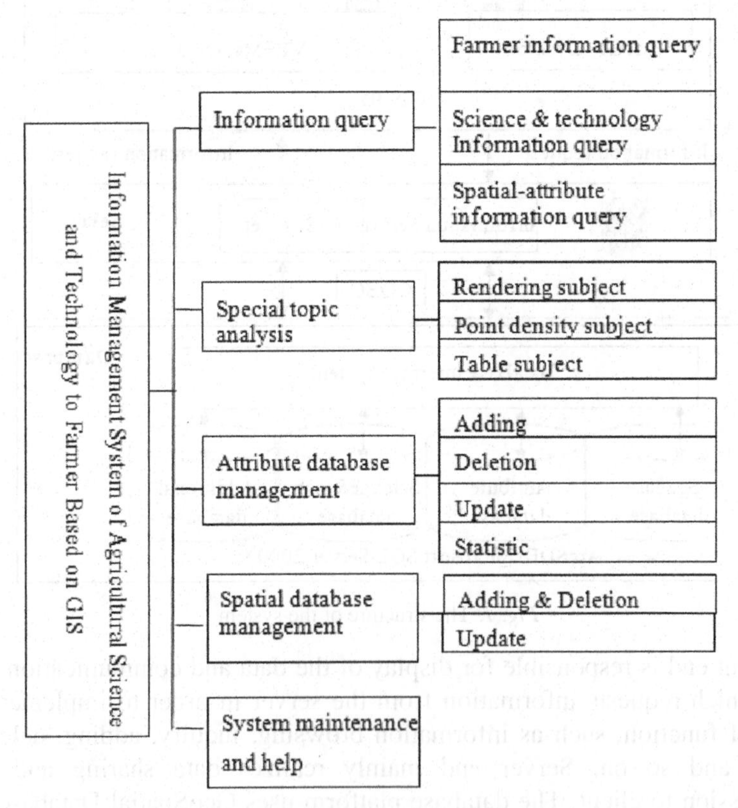

Fig. 2. The function of the system

(1) Information inquiry. It includes farmer information inquiry, the information inquiry of science and technology to farmer, the spatial and attribute information inquiry and so on.

(2) Special topic analysis. It makes dynamic color topic, point density topic, table topic and so on, and carrying on kinds of statistic, analysis and the decision-making according to the special data chosen by user.

(3) Attribute database management. It includes adding, deletion and update to attribute data.

(4) Spatial database management. It includes adding, deletion and update to spatial data.

(5) System maintenance and help. It includes system information maintenance, system operation manual and so on.

3. SYSTEM REALIZATION

System realization includes database design and the application of key technology.

3.1 Database Design

Database design is the key of effective working and function implement of GIS and MIS (Yang Bao-zhu, 2005). This system database designed by adopting E-R model mainly includes the spatial character database, the attribute characteristic database, the database of agricultural science and technology to farmer, the peasant information database, the system maintenance information database and so on. Fig. 3 shows the database structure of agricultural science and technology to farmer. Fig. 4 shows the

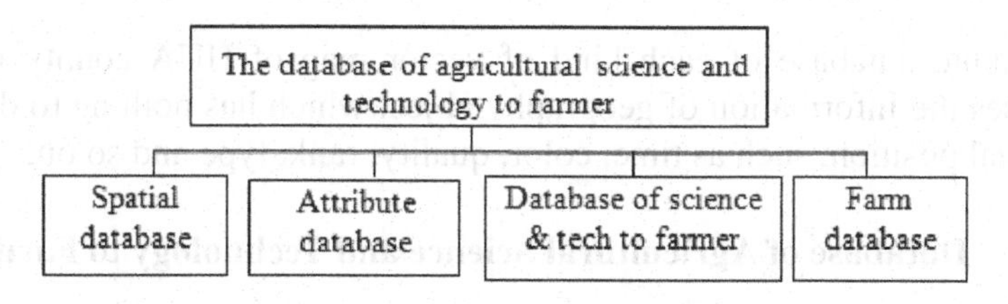

Fig. 3. The database structure of agricultural science and Technology to Farmer

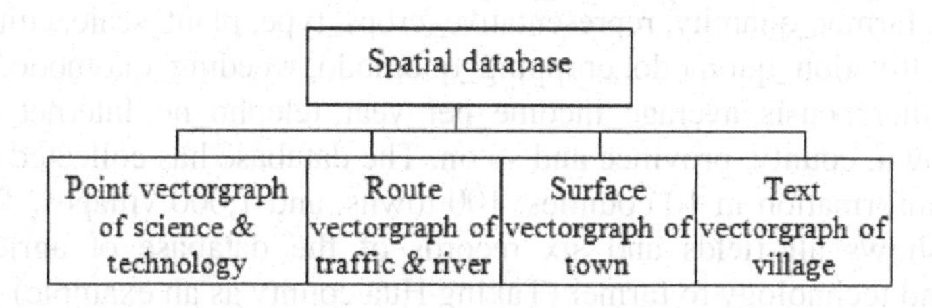

Fig. 4. The database structure of spatial character

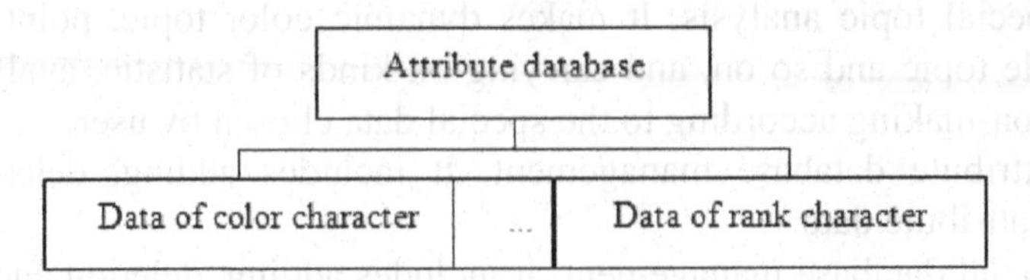

Fig. 5. The database structure of attribute character

database structure of spatial character, and Fig. 5 shows the database structure of attribute character.

The content classification and the structural design of Attribute database and the information database of agricultural science and technology to farmer are the successful and unsuccessful key factor of system development.

3.1.1 Spatial Database

Spatial database is created by using 1:10,000-scale electronic map. The content of spatial database mainly includes: the point vector map of the information distribution of science and technology to farmer, the line vector map of transportation and rivers, the surface vector map of villages and towns, the village text vector map and so on. The information distribution of science and technology to farmer are collected according to the unit of village and town.

3.1.2 Attribute Database

Attribute database of each kind of vector map of HUA county mainly comprises the information of geography object which has nothing to do with the spatial position, such as time, color, quality, rank, type and so on.

3.1.3 Database of Agricultural Science and Technology to Farmer

The content of the database of agricultural science and technology to farmer includes: serial_number, name, sex, birthday, culture_level, population _quantity, farmer_quantity, representative_crops_type, plant_scale, cultivation _scale, Cultivation_quomodo, cropping_quomodo, weeding_quomodo, plant_ di_sease_therapeusis, average_income_per_year, telepho_ne, Internet_or_not, village, town, county, province and so on. The database has collected 10,000 farmers' information in 10 counties, 100 towns, and 1,000 villages. Table 1-Table 5 shows all fields and six records of the database of agricultural science and technology to farmer (Taking Hua county as an example).

Table 1. The information of Agricultural Science and Technology to Farmer

serial_number	name	sex	birthday
11	Hongsheng Lv	man	Jul-54
12	Yuejin Lv	man	May-56
13	Qunli Lv	man	Sep-63
14	Xianbing Sun	man	May-70
15	Guobao Lv	man	Mar-63
16	Junqiang Lv	man	Jul-63

Table 2. The information of Agricultural Science and Technology to Farmer

culture_level	population	farmer_quantity	representative_crop
junior	10	7	wheat
senior	6	4	wheat
junior	5	4	wheat
junior	3	2	wheat
senior	5	2	wheat
junior	5	4	wheat

Table 3. The information of Agricultural Science and Technology to Farmer

plant_scale	cultivation_scale	Cultivation_quomod	cropping_quomodo
22	15	machine	machine
11	10	half-machine	half-machine
10	8	half-machine	half-machine
6	6	handwork	handwork
7	6	half-machine	half-machine
15	9	half-machine	half-machine

Table 4. The information of Agricultural Science and Technology to Farmer

average_income	weeding_quomodo	disease_therapeusis	telephone
1950	herbicide	pesticide	0372-8425009
2450	herbicide	pesticide	0372-8425016
2100	herbicide	pesticide	0372-8425013
2250	handwork	pesticide	0372-8425159
2870	herbicide	pesticide	0372-8425153
2150	herbicide	pesticide	0372-8425262

Table 5. The information of Agricultural Science and Technology to Farmer

Internet_or_not	village	town	county
not	Xi_yuan	Ba_li_ying	Hua Xian
not	Xi_yuan	Ba_li_ying	Hua Xian
not	Xi_yuan	Ba_li_ying	Hua Xian
not	Xi_yuan	Ba_li_ying	Hua Xian
not	Xi_yuan	Ba_li_ying	Hua Xian
not	Xi_yuan	Ba_li_ying	Hua Xian

3.1.4 Additional Database

Additional database includes farmer information database and system maintenance database and so on. The farmer information database saves and manages the basic farmer information. The basic farmer information content includes: serial number, name, sex, birth, ID_card, culture_level, political_faction, married_or_not, spouse_name, native_place, address, zip_code, telephone, village, town, county, province and so on. The system maintenance database mainly includes: parameter_initialization and user_information, diary_information and so on.

3.2 Key Technology

Key technology includes OOP (Object Oriented Programming), GIS components, net database operation and so on.

3.2.1 Redeveloping By VB and MapObjects

By MapObjects component loaded in Visual Basic and a series of operations to MapObjects, the system realized map cruise, zoom in, zoom out, whole map display, localization inquiry and so on, and rendered the created special topic of the information of science and technology to farmer. It is advantageous to make analysis and decision to the distribution information of agricultural science and technology to farmer by GIS. The development steps with VB and the MapObjects are introduced in a lot of related books, such as MapObjects-GIS Programming (Xue Wei, 2004) and Getting Started with MapObjects (Mchael Zeiler, 1999). Fig. 6 shows the interface of the system.

3.2.2 Database Operation

There are all kinds of operations to database in Visual Basic, such as inquiry, adding, deletion, update, statistical classification and so on. The

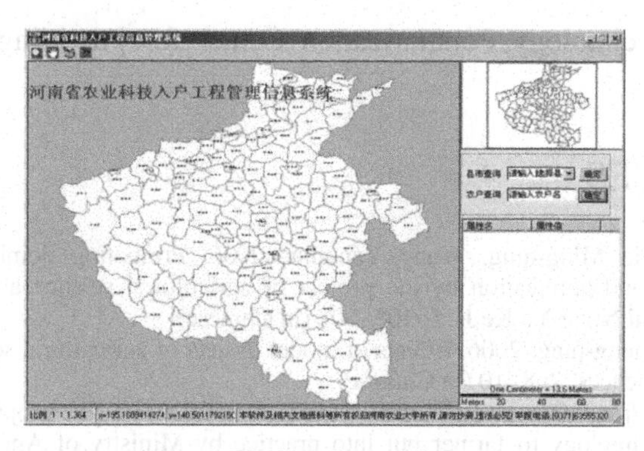

Fig. 6. The interface of the system

system took ADO component to realize operations to SQL Server 2000. Besides three components that are used to operate database: ADO component, DATA component, DAO component, user may also research and develop special components to operate database (Julia Case Bradley, 2003; Microsoft Corporation, 1999). Using SQL command, system realized inquiry and update to the information database of agricultural science and technology to farmer.

4. CONCLUSION AND FUTURE WORKS

System design has followed: scientific and solid system structure, practical and extensible function, artistic interface and so on (Huang Liuqing, 2005). The system has been applied to the information management of agricultural science and technology to farmer in HUA county, and facilitated information management, analysis and decision-making to the agricultural science and technology to farmer. The system may be popularized in all over the county, also the country.

Along with the continuous development of computer technology, Internet and WebGIS (Shang Wu, 2006; Wu Yun-chao, 2007; Liu Yi-jun, 2007), GIS will be applied and spread out deeply in the domain of "agriculture, country and farmer". In the following work, the system will take B/S and use network programming technology to realize the information management of agricultural science and technology to farmer based on WebGIS.

ACKNOWLEDGEMENTS

This study is supported by Henan Education Department tackling key problem of Science and technology Projects (Contract Number: 200510466005). Sincerely thanks are also due to the Hua County

Agricultural Technology Popularization Center for providing the data for this study.

REFERENCES

Tang Wan-min, Jia Ming-qing, Wang Yan-chao. 2006. Promoting dominant technology popularization and application by the project of agricultural science and technology to farmer, Xian Dai Nong Ye Ke Ji, 5: 105–107 (in Chinese).

Yin Li-hui, XIA Sheng-ping. 2006. Effects of model project of agricultural sci & tech, Hunan Agricultural Sciences, 2: 8–10 (in Chinese).

Http://www.gov.cn/jrzg/2006-10/17/content_415658.htm. 2006, 10. The project of agricultural science and technology to farmer put into practice by Ministry of Agriculture drives 4 million peasant households to increase production and income.

Alexander Köninger, Sigrid Bartel. 1998. 3d-Gis for Urban Purposes, GeoInformatica, 2(1): 79–103.

Yan Tai-lai, Zhu De-hai, Zhang Xiao-dong. 2005. To apply "3S" technology and carry on the concept of scientific development for agriculture, Journal of China Agricultural University, 10(6): 16–20 (in Chinese).

Wei Fu-quan, Xie Fang, Huang Tian-zhou, Sun Duan. 2004. Application of 3S technology in forestry, Forestry Prospect and Design, 1: 45–47 (in Chinese).

Li Li-wei, Xiao Ya-li, Liang Bao-song. 2006. Study on the GIS-based forest resources management information system, Henan Agricultural College Press, 5(40): 503–505 (in Chinese).

Gong Jian-ya. 2004. Review of the progress in contemporary GIS, Geomatics & Spatial Information Technology, 27(1): 5–11 (in Chinese).

Wu Lun. 2001. Theory, method and application of GIS, Science publishing House, Beijing.

Gong Jian-ya, Du Dao-sheng, Li Qing-quan. 2004. Modern GIS, Science publishing House, Beijing.

Chen Zheng-jiang, Tang Guo-an, Ren Xiao-dong. 2005. Design and development of GIS, Science Press, Beijing.

Zheng Ke-feng, Zhu Li-li, Hu Wei-qun. 2005. System design and realization of agriculture geography information system, Zhejiang Agricultural Science, 4: 244–246 (in Chinese).

Yang Bao-zhu, Liu Feng, Li Xiang. 2005. Design and implement of crop management system based on WebGIS, Agricultural Network Information, 4: 18–25 (in Chinese).

Xue Wei. 2004. MapObjects-GIS Programming, Defense industry press, Beijing.

Mchael Zeiler. 1999. Getting started with MapObjects, (USA) ESRE Press.

Julia Case Bradley, Anita C. Millspaugh. 2003. Highlevel programming about Visual Basic 6.0, Tsinghua University Press, Beijing.

Microsoft Corporation (USA), 1999. The component reference manual of Microsoft Visual Basic 6.0, Beijing Hope Electron Press.

Huang Liu-qing. 2005. Becoming a component-oriented designer, Programmer Journal, 9: 133–135 (in Chinese).

Shang Wu. 2006. WebGIS:status and prospects, Geologcal Bulletin of China, 25(4): 533–537 (in Chinese).

Wu Yun-chao, Wang Wen, Niu Zheng, Song Guo-jun. 2007. Integrating ajax approach into WebGIS, Geography and Geo-Information Science, 23(2): 43–46 (in Chinese).

Liu Yi-jun, Hu Xiang-yun. 2007. Application of XML in WebGIS, Science Technology and Engineering, 7(6): 1095–1097, 1106 (in Chinese).

MULTICAST IN MOBILE AD HOC NETWORKS

Zhijun Wang [*], Yong Liang, Lu Wang
** Corresponding author, Information Science & Engineering, College of Shangdong Agricultural University, Taian Shandong, China 271018, Email: wzj@sdau.edu.cn*

Abstract: Multicast is a very efficient technology in one-to-many communication scenarios. With the popularity of mobile devices, and demanding group information exchange, multicast in mobile ad hoc networks attracts much research attention. This paper reviews the state-of-art multicast protocols and classifies them into two categories: tree-based and mesh-based. We review one classic protocol closely for each category and briefly describe others. Then some open problems were discussed such as scalability and reliability.

Keywords: ad hoc Networks, Multicast, MAODV, ODMRP

1. INTRODUCTION

Multicast is a one-to-many communication strategy, in which the source sends a copy of data to multiple members of a multicast group. The packet is duplicated only when necessary, that is, at the branch point. Thus the minimum numbers of copies per packet are used to disseminate the data to all receivers. Compared with unicast communication, multicasting saves much bandwidth and achieves high efficiency. In wired networks, multicast is a very well studied research topic. A myriad of papers and RFCs have been published in this area. However the emerging and popularity of wireless ad hoc networks brings new life and challenges to the multicast strategy. In ad hoc networks, mobile nodes are resource constrained, especially when mobile devices like PDAs and hand phones are internetworking. The constraints

Wang, Z., Liang, Y. and Wang, L., 2008, in IFIP International Federation for Information Processing, Volume 258; Computer and Computing Technologies in Agriculture, Vol. 1; Daoliang Li; (Boston: Springer), pp. 151–164.

include limited battery capacity, limited computation capability and storage. Also, due to nodal mobility, the underlying topology changes often, which introduces new challenges to the multicast problem. In this situation, how to establish a multicast underlying structure efficiently becomes an essential issue to the lifetime of a whole ad hoc network.

Other issues such as scalability and reliability are critical to the success of multicast applications in wireless ad hoc networks. Wireless transmission is more error-prone than wired counterparts. Thus, reliability is another important issue in multicasting here.

The rest of the paper is organized as follows. Section 2 discusses the multicast support in wired networks and challenges of deploying multicast in mobile ad hoc network environment. We survey existing multicast protocols in Section 3 followed by a comparison of protocols and other multicast issues in MANET. Finally, we conclude the paper in Section 5.

2. RELATED WORK

In this section, we discuss multicast protocols in wired network, followed by the challenges introduced by the mobility and characteristics of terminal nodes in ad hoc networks during the process of applying traditional multicast protocols directly.

2.1 Multicast support in wired network

In this section, some typical multicast protocols are briefly reviewed. The concept of multicast was proposed by Steve Deering in his dissertation in 1988. It was driven by the observation that much bandwidth could be saved if the data could be delivered to all receivers at one time instead of using multiple individual transmissions. Through the years, many research efforts focused on Internet multicast, and after a test in wide scale of "audiocast" in 1992, a multicast Internet (now called MBone) was setup for experiment use. A new type of IP address is reserved for multicast, and Internet Group Management Protocol (IGMP) was proposed to support dynamic joining and leaving of a group. The up-to-now multicast protocols could be classified into two categories: one category of multicast protocols works at the network layer, and the other works over the transport layer (but below application layer). The first category covers Distance Vector Multicast Routing Protocol (DVMRP) (S.E. Deering et al., 1990), Multicast Extension to OSPF (MOSPF) (J. Moy, 1994) and Protocol-Independent Multicast-Spare Mode (PIM-SM), Protocol-Independent Multicast-Dense Mode (PIM-DM) (S.E.

Deering et al.,1996). DVMRP is based on distance vector routing protocol, and uses reverse path multicasting algorithm to build a spanning tree for each multicast group. If a leaf router finds no nodes in its domain belonging to the group, it sends prune messages to the multicast source, which leads the leaf pruned from the multicast spanning tree. MOSPF is an extension of Open Shortest Path First (OSPF). It requires the information obtained by IGMP to build a multicast forwarding tree on demand for each multicast group. MOSPF, like DVMRP, is source-based multicast protocols. Instead PIM-SM is a core-based multicast protocol that maintains a rendezvous point. The rendezvous point is responsible for forwarding all packets for the multicast group. And each of the multicast domains selects a designated router, which handles multicast group messages in its domain. PIM-DM multicast protocol is very similar to DVMRP.

The second category of multicast protocols works over the transport layer. The classic protocols like Scalable Reliable Multicast (SRM) (S. Floyd et al., 1997) and Reliable Multicast Transport Protocol (RMTP) (S. Paul et al., 1997) fall into this category. SRM provides reliable multicast delivery service. It delegates the responsibility for recovery of packet loss to members in the multicast group. Through clever use of randomized timers, the numbers of feedbacks (replies) are effectively suppressed, and repair locality problem could be alleviated. RMTP makes use of logic tree structure to solve repair locality problem and refrain the feedback implosion problem. Specialized receivers located at the root of the sub-trees of the logic tree receive requests and initiate retransmission only to their own children in the tree. Note that this category multicast protocols does not require multicast support from router. Some of them impose some requirements on receivers instead.

2.2 Challenges of multicasting in mobile ad hoc network

Unlike wired networks, mobile ad hoc networks have no fixed underlying infrastructure. Nodes/terminals are free to move arbitrarily, thus the underlying topology may change randomly in an unpredictable manner (S. Corson et al., 1999). This makes the task of multicast group maintenance more difficult, and packet forwarding more challenging. Also, these mobile terminals/nodes are more resource-restricted compared to the counterparts in wired networks. These resources include, but are not limited to, bandwidth, energy (most cases terminals are run by battery instead of main), and link quality (wireless link are more error-prone than wired link). Thus one possibility is to require using multicast in order to save resources when multiple receivers exist. Another possibility is that careful design is required to consider in the sake of avoiding waste precious resources. For example, network-wired broadcast operations should be used less frequently in finding paths if it could not be avoided. How to balance between the efficiency and robustness is a big challenge for multicasting in mobile ad hoc network.

3. MULTICAST PROTOCOLS IN MOBILE AD HOC NETWORK

There are two approaches to categorize the existing multicast protocols in mobile ad hoc networks. One approach is to group together protocols that evolved from a similar chronological path. Multicast protocols in MANET evolved through three paths: extending existing multicast solution from wired network to MANET; extending existing MANET unicast protocols to support multicast; and proposing new multicast protocols. The (Figure 1) shows the relationship among varying protocols. It would be very interesting to review the protocols in this way. The other approach is to classify the protocols based on the structures the protocols used, tree-based or mesh-based. In this paper, we prefer surveying these literatures in this more natural and technical viewpoint.

Although there exist some protocols (e.g. hierarchical structure employed like (Y.J. Yi et al., 2000)) that do not fall into the following structure, the two very important classes of multicast routing protocols in mobile ad hoc network are reviewed in following subsections. Section 3.1 discusses the tree-based multicast protocols and Section 3.2 reviews the complicated mesh-based multicast protocols. We acknowledge that with hierarchical structure, multicast protocols are more scalable than without it.

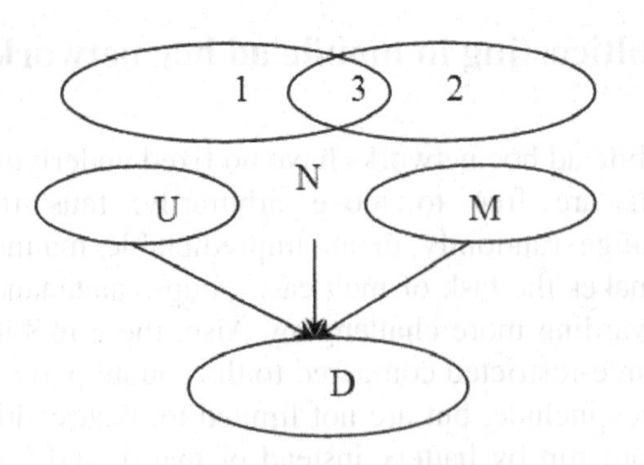

Figure 1. Evolution of multicast protocols in MANET

3.1 Tree-based multicast protocol

Large families of multicast protocols for ad hoc networks are based on a tree structure. One reason is that tree-based multicast protocols are well studied in wired network, thus more researchers tried to extend those feasible solutions to the mobile ad hoc environment. These sets of protocols usually establish a shared multicast delivery tree before multicasting packets in the group. The protocols are Multicast Operation of the ad hoc On-demand Distance Vector (MAODV) (E.M. Royer et al., 1999), ad hoc Multicast Routing Protocol Utilizing Increasing ID Numbers (AMRIS) (C.W. Wu et al., 1998) (C.W. Wu et al., 1999), On-Demand Associatively-Based Multicast Routing for ad hoc networks (ABAM) (C.K. Toh et al., 2000), Adaptive Demand-Driven Multicast Routing (ADMR) (J.G. Jetcheva et al., 2001) etc. AMRIS dynamically assigns each participant an id-number that reflects the "logic height" in the multicast delivery tree. A multicast tree starts to grow after receivers express interest in joining the multicast session. In ABARM, the concept of association stability (such as spatial, temporal, connection, and power stability of a node with its neighbor) is used to establish a multicast tree. Because the link quality and relations to neighbors are considered in an early stage, the tree structure tends to be very stable and does not require frequent reconfiguration in low mobility scenarios. ADMR creates a source-based forwarding tree when a multicast group starts. Receivers adapt to the traffic patterns of the multicast source application for efficiency and maintenance. Passive acknowledgements are used for efficient branch pruning instead of explicit pruning messages. Some other researchers published some tree-based multicast solutions instead of a full set of multicast protocols, such as (Sajama et al., 2003).

Some similarities are shared among these tree-based protocols. They work in two phases: tree establishment and tree maintenance. Tree establishment usually involves starting a multicast group and building a multicast forwarding tree. The phase of tree maintenance consists of adding a branch when a receiver requests to join the multicast group and pruning when no receivers exist in a tree branch. Instead of reviewing each protocol in detail, the classic protocol, MAODV, is reviewed closely on how it creates a multicast group, processes the join/leave request, and maintains the multicast tree.

MAODV is a naturally extension to ad hoc On-demand Distance Vector Routing (AODV) for providing multicast capabilities. Therefore, during the process of tree establishment, unicast is also used to disseminate some information, for example, Multicast Activation (MACT). For this functionality, each node maintains two tables pertaining to routing, and a third table called request table for optimization purposes. The first is route table, which is used to record the next hop for routes to other nodes. The second routing table that a node maintains is multicast route table. The following information is stored in each entry of a multicast route table:

```
Entry Multicast Rt {
IP_t ipGroup; //multicast addr
IP_t ipLeader;//leader addr
Seq_t seqNo; //group seq
int hopCnt; //to group leader
HopList nextHops;
Time_t Lifetime;
}
```

In MAODV, the first member of the multicast group becomes the group leader, and it remains the leader until it leaves the group. This leader takes responsibility of maintaining a multicast group sequence number and disseminating this number to the entire group through a proactive Group Hello Message. Members use the group hello message to update its request table and its distance to the group leader.

Once the group is setup, it is ready to accept join requests from others. When a source node broadcasts RREQ for a multicast group, it is expected to receive multiple replies. Only one of RREPs causes a branch to connect to the existing tree in order to avoid loops. The source node unicasts MACT to determine the next hop. The next hop propagates the MACT further until the node sending out the RREP if it is not a member of the multicast trees. Otherwise, it just updates its multicast route table when necessary. The multicast tree is created in this manner. (Figures 2, 3, and 4) show the process of multicast join operation.

During normal network operation, a multicast group member may decide to terminate its membership in the multicast group. As usual, leave operation leads multicast tree pruning. If the node is not a leaf node of the tree, it may revoke its membership status but may continue to serve as a router for the tree. Otherwise, if the node is a leaf node, it unicasts MACT messages with flag prune being set to next hop, thus it prunes itself from the tree.

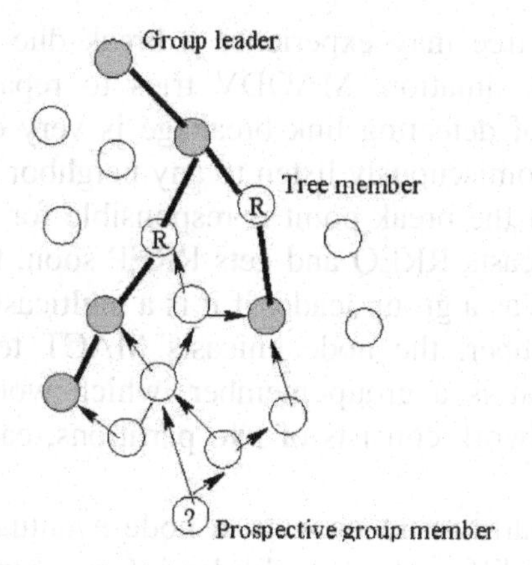

Figure 2. Route Request propagation

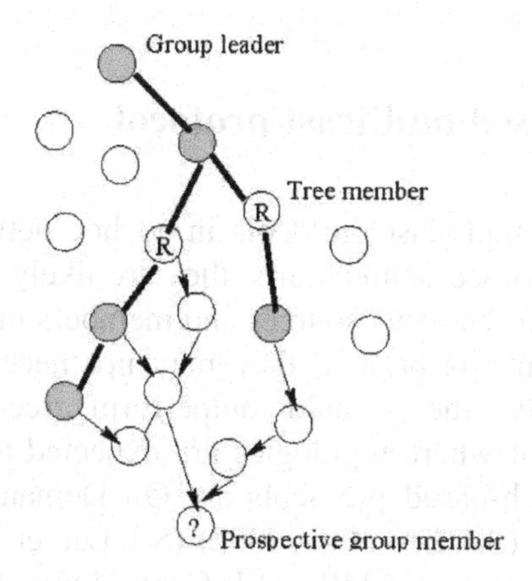

Figure 3. Route Reply sent back to source

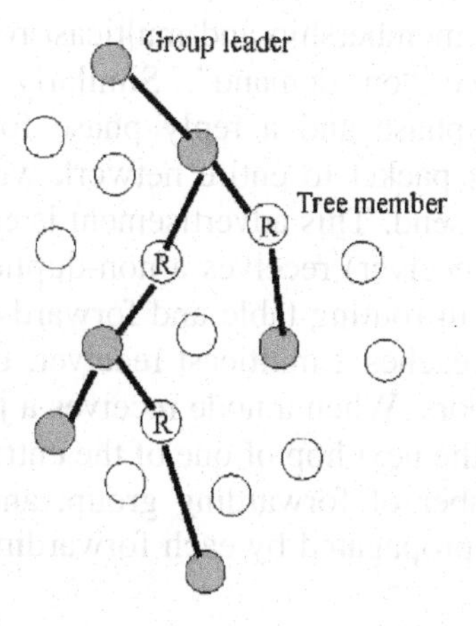

Figure 4. Tree branch growth

Multicast group tree may experience a break due to node mobility or dysfunction. In this situation, MAODV tries to repair the broken links. However, the cost of detecting link breakage is very expensive because it requires nodes to promiscuously listen to any neighbor's transmissions. The node downstream of the break point is responsible for repairing the broken link. Either it broadcasts RREQ and gets RREP soon, thus the link is fixed soon, or it has to act as a group leader if it is a multicast group member. If it is not a group member, the node unicasts MACT to the next hop until reaching a node that is a group member, which would become a group leader. Thus the network consists of two partitions, each one with a group leader.

If the network partition reconnects, a node eventually receives a group hello message with different group leader information. The node unicasts RREQ to each group leader to get permission of rebuilding by grafting a branch on the tree.

3.2 Mesh-based multicast protocol

Another class of multicast protocols in ad hoc network is mesh-based. Compared with tree-based counterparts, they are likely more robust because keeping multiple paths between sources and members in the multicast group. In the case that a link is broken, they may not necessarily initiate route discovery. Intuitively, they would outperform tree-based protocols in MANET environment where topologies are expected to change frequently. Typical existing mesh-based protocols are On Demand Multicast Routing Protocols (ODMRP) (S.J. Lee et al., 2002) (S.J. Lee et al., 1999), and Core-Assisted Mesh Protocol (CAMP) (J.J. Garcia-Luna-Aceves et al., 1999) (J.J. Garcia-Luna-Aceves et al., 1999).

In ODMRP, group membership and multicast routes are established and updated by the source "on demand". Similarly the protocol operation consists of a request phase and a reply phase for join. Sources flood a member advertisement packet to entire network with piggybacked payload when it has packets to send. This advertisement is called join query. A node (not necessarily be a receiver) receives a non-duplicate join query, it stores the upstream node ID in routing table and forwards the packet by flooding. When the join query reaches a multicast receiver, the receiver broadcasts a join reply to its neighbors. When a node receives a join reply, it checks if its own ID matches with the next hop of one of the entries. In the case of match, it marks itself a member of forwarding group, and broadcasts join reply. Thus the join reply is propagated by each forwarding group member until it reaches the multicast source. This process constructs (updates) the routes from source to receivers, and build a mesh of nodes. (Figure 5) shows the example of mesh.

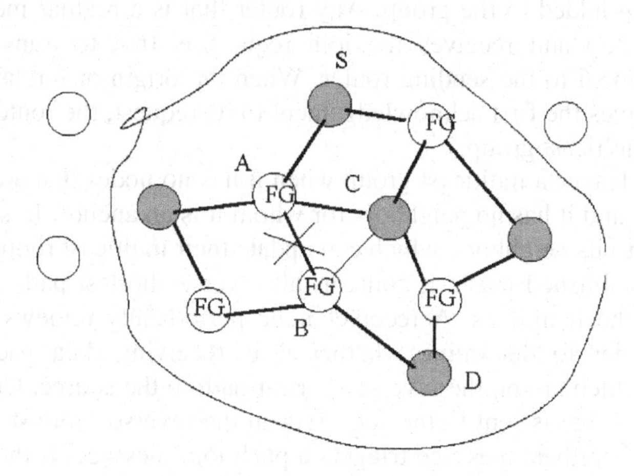

Figure 5. Example of mesh and the concept of forwarding group in ODMRP

ODMRP has several features worth of mentioning. The "on-demand" is source based, which means it does not require receivers to send leave explicitly when they are not interested in the group. It is different from MAODV and other multicast protocols. Secondly, the concept of forwarding group is very similar to the role of "forwarding nodes" in MAODV. Thirdly, the use of mesh configuration enables high connectivity thus its feature of robustness. For example, in (Figure 5) if the link between A and B is broken, the packet transmission from source S to receiver D is not affected because the redundant path S →A→C→B→D could be used instead.

S.J. Lee, et al. proposed some enhancements to ODMRP in (S.J. Lee et al., 2001) (S.J. Lee et al., 2002). The enhancements include adapting the refresh interval via mobility prediction, reliability, and elimination of route acquisition latency by flooding data instead of Join requests when the source does not know any multicast route. Another improved version of ODMRP is proposed in (H. Dhillon et al., 2005). It consolidates join queries in intermediate nodes, thus reducing the total number of control packet transmissions. Compared with ODMRP, simulation results show that it increases multicast efficiency and improves the packet delivery ratio.

CAMP is also a mesh-based multicast protocol. It borrows concepts from core-based tree (CBT), but unlike CBT where all traffic flows through the core node, the core nodes in CAMP are used to limit traffic flow through the core node. CAMP uses a receiver-initiated method for routers to join multicast groups. A node first determines the address of the group it is interested in. Then it uses this address to ask its attached router to join the multicast group. Upon receiving a host request to join a group, the router then determines whether to announce its membership in the group or to request being added to the group. Any router that is a regular member of a multicast group and receives the join request is free to transmit a join acknowledgment to the sending router. When the origin or a relay of a join request receives the first acknowledgement to its request, the router becomes part of the multicast group.

A router leaves a multicast group when it has no nodes that are members of the group and it has no neighbors for whom it is an anchor. It issues a quit notification to its neighbors, which can update their multicast routing tables.

In the established mesh, it contains all reverse shortest paths between a source and the recipients. A receiver node periodically reviews its packet cache in order to determine whether it is receiving data packets from neighbors, which are on the reverse shortest path to the source. Otherwise, a heartbeat message is sent to the successor in the reverse shortest path to the source. The heartbeat message triggers a push join message. If the successor is not a mesh member, the push join forces the specific successor and all the routers in the path to join the mesh. The requests only propagate to mesh members. To date, CAMP is the only multicast routing protocol based on the mesh topology and without using flooding of data or control packets.

The protocol CAMP requires the support of unicast and Domain Name Service (DNS). Additionally, the unicast routing protocol must provide correct distance to known destinations within a finite time. These requirements are difficult to meet in the current MANET environment.

CAMP is improved by unified multicasting through announcement (PUMA) in ad hoc networks from the same researchers in (R. Vaishampayan et al., 2004). Like its preceding work, "core-based mesh", this protocol also establishes and maintains a shared mesh for each multicast group. It is based on a novel idea of using simple multicast announcements to elect a core for the group, inform all routers of their distance and next-hops to the core join and leave the multicast group.

4. COMPARISON OF PROTOCOLS AND OTHER MULTICAST ISSUES IN MANET

Multicast technology was invented/developed at the early internet-working time when the bandwidth was a very precious resource. With the advances of wired technology and reduced costs, it is not an issue in wired networks. The situation of multicast residing in wireless ad hoc networks is very similar to that in wired networks of early stage. Due to the natural characteristics of MANET, multicasting is a very promising technology desired in the scenarios where multiple receivers exist at one time. In this section, we compare multiple multicast protocols, followed by a discussion of various open problems.

Table 1 shows the comparison of multicast protocols in MANET in terms of several evaluation metrics. From the table, we can see all of them could provide the mechanism of avoiding loops. Those protocols that evolved from unicast protocol in MANET usually depend on the support of unicast in the network. In contrast, the newly designed protocols often borrow some ideas from multicast protocols in wired network or others. For example, CAMP borrowed the concept of Core from CBT. ODMRP borrows the concept of the forwarding group from FGMP, and ABAM utilizes the concept of associativity from Associativity-Based Routing (ABR), a unicast protocol in MANET. Generally, the designs of tree-based protocols are not as complicated as mesh-based protocols. However, they are less robust than mesh-based protocols due to the connectivity of each node. This table also shows that all existing proposals have not been tested in large-scale networks. The largest network simulated consists of 100 nodes in AMRIS (C.W. Wu et al., 1999). From the survey and comparison, it is not difficult to summarize that on-demand is a desired property of all multicast protocols with the complement of periodical messages to keep structure or information updated.

Most active research of multicast in MANET is focused on the protocols itself, which mainly propose mechanisms of how to process join/leave request and how to establish the underlying packet forwarding structures. If we compare them with the peers in wired networks, it is interesting to find that no existing work matches to higher layer multicast protocols/

Table 1. Comparison of multicast protocols

Protocols	Underlying Structure	Loop Free	Dependence	Flood	Evolution	Simulation Size
MAODV	Tree	Yes	Unicast	Yes	From AODV	50 nodes
AMRIS	Tree	Yes	Unicast, beacon & broadcast	Yes	New protocol	100 nodes
ABAM	Tree	Yes	Beacon, scoped broadcast	No	From ABR	40 nodes
ADMR	Tree	Yes	No	Yes	From DSR	50 nodes
ODMRP	Mesh	Yes	No	Yes	New protocol	20 nodes
CAMP	Mesh	Yes	Unicast, DNS, etc	No	New protocol	30 nodes

frameworks (e.g. SRM, MFTP, etc) in wired networks. Maybe it is because so far no "killer" applications in MANET multicast scenario has driven the research toward this direction. Also, there are still many important topics of multicast in MANET that require further investigation, such as experiments, scalability, reliability and power consumption, etc.

Simulations could be used to evaluate the performance of proposed multicast protocols. However, further experiments with testbeds are still necessary. The largest ad hoc evaluation testbed (APE) consisted of up to 37 physical nodes (H. Lundgren et al., 2002). Also, no multicast protocol is supported in the testbed yet. To our best knowledge, none of the proposed multicast solutions perform experiments in a testbed. With recent advance in low-power supply, and reduced cost in mobile terminals (handsets, PDAs, laptops, etc), it is feasible to build a large-scale testbed and test multicast protocols in MANET.

The benefit of multicasting turns out to be tremendous only when a large number of receivers exist simultaneously. Therefore, scalability is one of the most important merits that should be provided by proposed multicast protocols. Although those published literatures claim the scalability of their proposals through simulations, further experiments in testbed are required to verify it. Scalability has two-fold meanings: one involves how large a multicast group could be processed, and the other one is how many multicast groups could be processed in the multicast group. A scalability proposed multicast protocol appears in (C. Gui et al., 2004). This paper studied the relationship of the protocol state management techniques and the performance of multicast provisioning. In order to address scalability and enhance performance, domain-based hierarchical and overlay-driven hierarchical routing are proposed. In domain-based hierarchical routing approach, large multicast group is divided into many sub-groups, and in each sub-group a node is selected as a sub-root and these sub-roots maintain the protocol states. The second approach is to use overlay multicast as the upper layer multicast protocol built upon low layer stateless small group multicast.

And reliability is also a very important issue of multicast in MANET. So far, few researchers emphasize this problem. The only paper is (J. Luo et al., 2003). The proposed router driven gossip in this article could achieve probabilistic reliability. Its main idea is based on a partial view for each group member. The spread of information is propelled mainly by a gossiper-push (each group member forwards multicast packets to a random subset of the group), but complemented by gossiper-pull (multicast packets piggyback negative acknowledgement of the forwarding group member). Three sessions are defined, join, leave and gossip. The dissemination of a leave indication relies on the gossip session.

The fourth important aspect that no literature mentioned is the power consumption problem for these proposed multicast protocols in MANET. Some multicast protocols rely on eavesdropping neighbors to detect link breakage while others periodically flood messages to refresh a multicast group. These are undesirable features in a MANET environment. Therefore how to minimize the power consumption and how much benefit could be achieved remain unanswered.

5. CONCLUSION

Multicasting can efficiently support many applications in mobile ad hoc networks. However, the characteristics of MANET, such as frequent topology changes and resource constraints bring many challenges to deploy multicast solutions. In this paper, we discuss the multicast protocols in MANET. The multicast protocols are classified into tree-based and mesh-based mechanisms. In each class of protocols, at least one of classic proposals is reviewed in detail. So far, the research for multicast in MANET is far from exhaustive. Some very important issues, such as scalability, reliability, and power consumption, are not yet investigated thoroughly. Also, existing multicast proposals are not convincing enough without running simulations in large-scale networks and performing experiments in a testbed.

REFERENCES

S. Corson, J. Macker, available online: http://www.ietf.org/rfc/rfc2501.txt, 1999.

E.M. Royer, C.E. Perkins, Multicast Operation of the ad hoc On-demand Distance Vector Routing Protocol, ACM MOBICOM, Aug.1999.

Y.J. Yi, X.Y. Hong and M. Gerla, Scal able Team Multicast in Wireless Ad hoc Networks, available online: http://www.cs.ucla.edu/NRL/wireless/uploads/ngc-yjyi.pdf.

C.W. Wu, Y.C. Tay, and C.K. Toh, ad hoc Multicast Routing Protocol Utilizing Increasing ID Numbers, Internet draft, 1998 (Working in progress).

C.W. Wu and Y.C. Tay, AMRIS, A Multicast Protocol for ad hoc Wireless Networks, Proceedings of IEEE MILCOM'99, Atlantic City, NJ, Nov. 1999.

C.K. Toh, G. Guichal, and S. Bunchua, ABAM: On-Demand Associativity-Based Multicast Routing for ad hoc Mobile Networks, Proceedings of IEEE Vehicular Technology Conference, pp. 987–993, 2000.

J.G. Jetcheva, D.B. Johnson, Adaptive demand-driven multicast routing in multi-hop wireless ad hoc networks, MobiHoc 2001, pp. 33–44.

Sajama, Z.J. Haas, Independent-tree ad hoc multicast routing, Kluwer Academic Publishers, Mobile Networks and Applications 8, pp. 551–566, 2003.

S.J. Lee, W. Su, and M. Gerla, Wireless ad hoc Multicast Routing with Mobility Prediction, Kluwer Academic Publishers, Mobile Networks and Applications 6, pp. 351–360, 2001.

H. Dhillon, H.Q. Ngo, CQMP: A Mesh-based Multicast Routing Protocol with Consolidated Query Packets, Proceedings of IEEE WCNC'2005.

S.J. Lee, W. Su, M. Gerla, On-demand multicast routing protocol in multihop wireless mobile networks, Kluwer Academic Publishers, Mobile Networks and Applications 7, Issue 6, 2002.

J.J. Garcia-Luna-Aceves, E.L. Madruga, The Core-Assisted Mesh Protocol, IEEE Journal on Selected Areas in Communications, Vol. 17, no. 8, AUGUST 1999.

J.J. Garcia-Luna-Aceves, E.L. Madruga, A Multicast Routing Protocol for ad hoc Networks, Proceedings of IEEE INFOCOM 1999.

R. Vaishampayan, J.J. Garcia-Luna-Aceves, Efficient and Robust Multicast Routing in Mobile ad hoc Networks, IEEE MASS 2004.

C. Gui, P. Mohapatra, Scalable Multicasting in Mobile ad hoc Networks, Proceedings of IEEE INFOCOM 2004.

J. Luo, P. Th. Eugster, J.P. Hubaux, Route Driven Gossip: Probabilistic Reliable Multicast in ad hoc Networks, Proceedings of IEEE INFOCOM 2003 .

S.E. Deering, D.R. Cheriton, Multicast routing in Datagram Internet works and Extended LANs, ACM Transactions on Computer Systems, Vol. 8, no. 2, 1990.

J. Moy, Multicast routing extensions for OSPF, communications of the ACM, Vol. 37, no. 8, 1994 .

S.E. Deering, D.L. Estrin,etc, The PIM architecture for wide-area multicast routing, IEEE/ACM Transactions on Networking, Vol. 4, no. 2, 1996.

S.J. Lee, M. Gerla, etc, On-demand multicast routing protocols, Proceedings of IEEE WCNC 1999.

S. Floyd, V. Jacobson, C.G. Liu, S. McCanne, L. Zhang, A reliable multicast framework for light-weight sessions and application level framing, IEEE/ACM Transactions on Networking, 1997.

S. Paul, K.K. Sabnani, J.C.H. Lin, S. Bhattacharyya, Reliable Multicast Transport Protocol (RMTP), IEEE Journal on Selected Areas in Communications, 1997.

H. Lundgren, D. Lundberg, J. Nielsen, etc, A Large-scale Testbed for Reproducible ad hoc Protocol Evaluations, Proceedings of 3rd annual IEEE Wireless Communications and Networking Conference (WCNC 2002).

THE DESIGN OF FRUIT AUTOMATED SORTING SYSTEM

Pingju Ge [*], Qiulan Wu, Yongxiang Sun
College of information Science & Engineering, Shandong agricultural university, Taian 271018, China
* *Corresponding author, Address: College of information Science & Engineering, Shandong agricultural university, Taian 271018, Shandong, P. R. China, Tel: +86-538-8242497, Email: gepj2@163.com*

Abstract: The article introduced the flow and design scheme of fruit automated sorting system, the computer technology and automatic technology is used in agriculture production and management. In this system, the application of some technology, such as the detection and control technology, the digital image processing technology, the computer peripheral device and interface technology, Pattern Recognition technology etc, is used in the fruit automated sorting system.

Keywords: the fruit, automated sorting system, automation, the digital image processing, Pattern Recognition

1. INTRODUCTION

With the development of modern agriculture & forestry and the enhance of industrial standard, the output of fruits from inner country and abroad are being raised swiftly. On the one hand, the fruits from different district are sold in local area, on the other hand, more plenty of fruits are sold to other district even abroad, and these fruits demand being packed according to different quality. In order to package these fruits, we must to design a fruit automated sorting system, because if these fruits were packed by hand, one the one hand, The task is hard and it is inefficient; on the other hand, it is not meet the demands of modern fruit package, which is need to be Multi-layer or high standard or high sanitary standard.

Ge, P., Wu, Q. and Sun, Y., 2008, in IFIP International Federation for Information Processing, Volume 258; Computer and Computing Technologies in Agriculture, Vol. 1; Daoliang Li; (Boston: Springer), pp. 165–170.

To advanced country after second war, the automated sorting system is one of the essential equipment of Logistics centre, Distribution Centre, or circulation centre (Huang, 2002), it was used in American and Europe in the 1960s, automated sorting system was used in Japan in 1970s (Xu, 2002). At the present, automated sorting system is also success used in post office besides above field. With the improvement of modernization standard, it is widely used in more and more field. at the present, almost no success using about fruit automated sorting system because it concern a lots of knowledge and subject, at the same time, it demand more higher level of artificial. At present, we can only see the news about Fruitonics company, which have designed a set of fruit automated sorting system.

2. THE HARDWARE DESIGN OF FRUIT AUTOMATED SORTING SYSTEM

In the course of designing of the System, all kinds of factor are comprehensively thought when fruits are sorted. All kinds of technology, such as detection and control technology, the computer peripheral equipment and interface technology, the digital image processing technology and Pattern Recognition technology, is used, at the same time, advanced algorithm is used in the course of pattern identification, and it improve the accurate rate of fruit sorting.

2.1 The hardware consists of fruit automated sorting system

There are all sorts of automated sorting system, and its standard is different. In general, an automated sorting system is consists of seven partition (Xu, 2002), it is: collection Conveyor, feed Conveyor, the Setting Device Of sorting instruction, confluence device, sorting conveyer, sorting and discharge cargo port, computer controller (Fig. 1).

2.2 The consists of hardware and work principles of fruit automated sorting system

The fruit automated sorting system is based of the traditional automated sorting system, but it is different of the traditional automated sorting system, it is an automated sorting system which is high intelligent, real-time, Multi-layer, multi knowledge fields and multi class. This system has several parts (Xu, 2002), such as: conveyer device, the weight sorting device, the appearance sorting device, and the quality sorting device etc. (Fig. 2).

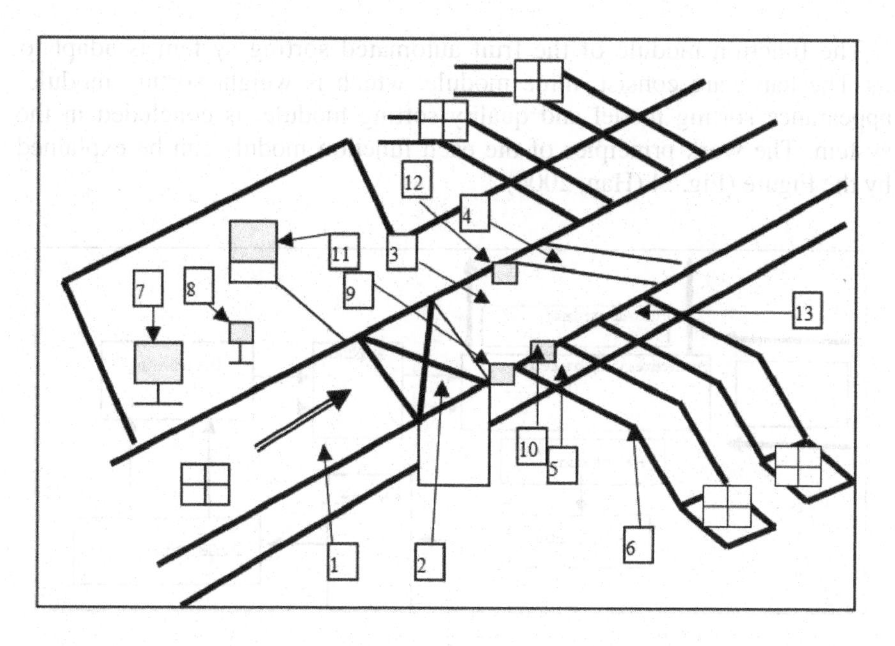

Fig. 1. The hardware diagram of fruit automated sorting system (1) input conveyer belt; (2) fill cargo conveyer belt; (3) steel conveyer belt; (4) Diverter; (5) output Rollers; (6) sorting port; (7) signal giver; (8) laser ISBN reader; (9) pass detector; (10) magnetism signal generator; (11) controller; (12) magnetism signal reader; (13) full scale detector.

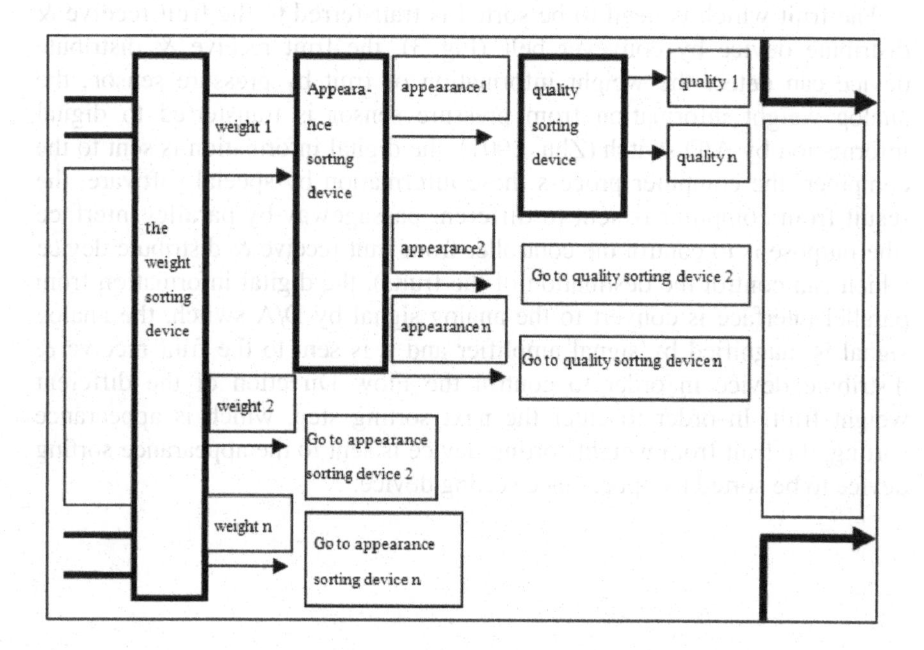

Fig. 2. The hardware consists of fruit automated sorting system

The function module of the fruit automated sorting system is adapt to its The hardware consist, three module, which is weight sorting module, appearance sorting model and quality sorting module, is concluded in the system. The work principles of the each function module can be explained by the Figure (Fig. 3) (Han, 2004).

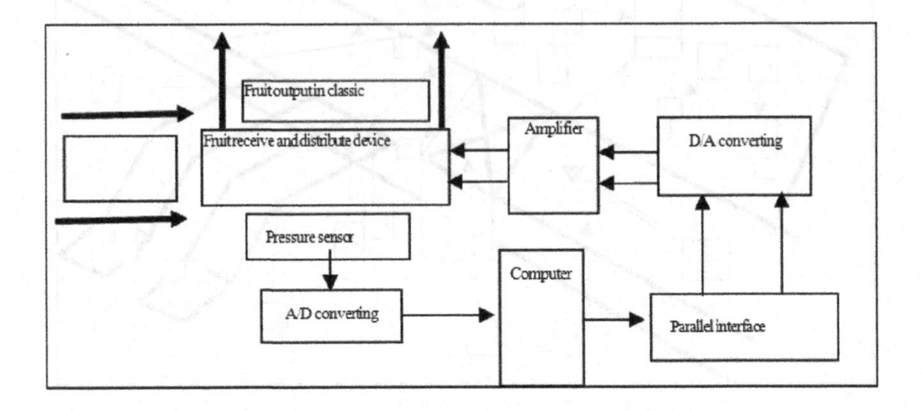

Fig. 3. The hardware consists of the weight sorting device

2.2.1 The weight sorting device and its work principle

The fruit which is need to be sorted is transferred to the fruit receive & distribute device by conveyer belt (Fig. 3), the fruit receive & distribute device can detect the weight information of fruit by pressure sensor, the analog weight information from pressure sensor is transferred to digital information by A/D switch (Zhu, 2001), the digital information is sent to the computer, the computer process these information by special software, the result from computer is sent to different passageway by parallel interface (the purpose is to control the controller from fruit receive & distribute device which can control the destination of the fruits), the digital information from parallel interface is convert to the analog signal by D/A switch, the analog signal is magnified by signal amplifier and it is sent to the fruit receive & distribute device in order to control the Flow Direction of the different weight fruit. In order to enter the next sorting step, which is appearance sorting, the fruit from weight sorting device is sent to the appearance sorting device to be sorted by appearance sorting device.

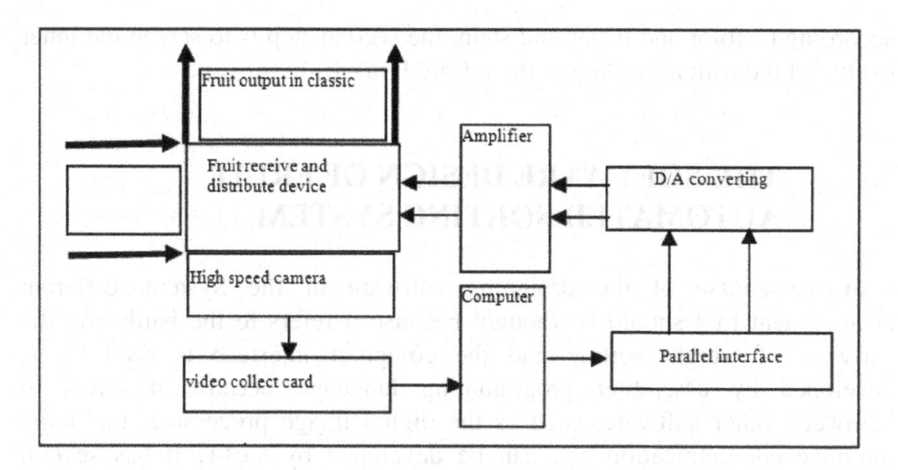

Fig. 4. The hardware consists of the weight sorting device

2.2.2 The appearance sorting device and its work principle

The fruit from weight sorting device is sent to the fruit receive & distribute device of the appearance sorting device (Fig. 4), the high speed camera shoot fruit from various angles, the image from camera be sent to the computer by the video capture card, the computer process the video information by special software, the result from computer is the classified signal of fruit, the signal open the different sorting passageway by parallel interface, and then the signal is magnified by the signal amplifier. In order to control the fruit receive & distribute device, the fruits are sent to the different passageway, and at last the fruit can be sent to the package device.

The image from camera should be processed by edge detection and noise removing (even it must be curve fitting) (Guo, 2004), and then it be matched to different fruit appearance so that the fruit can be sorted by appearance.

2.2.3 The quality sorting device and its work principle

The two devices, which is the quality sorting device and appearance sorting device, have the similar work process (Fig. 4). there are two different: one different is that the high speed camera must be infrared high speed camera; another different is that they have different process software .the software of the quality sorting device can process the fruit appearance quality (such as color and luster, stain etc) and inner quality (it is rot or not) according to the image from camera. It sorts manly according to the image from the infrared high speed camera and the trade standard. In the process of sorting, it makes full use of the digital image processing technology and Pattern Recognition technology. In general, the first step is to screen according to color and luster and stain, the second step is to screen the inner quality of the fruit according to the infrared image.

3. THE SOFTWARE DESIGN OF FRUIT AUTOMATED SORTING SYSTEM

In the course of the designing software of the System, different development tool should be thought because it refers to the hardware. the software of weight sorting and the computer interface is need to be developed by assembler programming language because it refers to hardware, other software, such as the digital image processing, the serial interface communication etc, can be developed by Vc++, It has several special Control in Vc++ about serial interface communication (Dong, 2005).

4. SUMMARY

The fruit automated sorting system refers to different technology form different subject, such as automatic technology, the software and hardware technology of computer. The digital image processing technology, the artificial intelligence technology etc. the paper introduced the design thought of fruit automated sorting system, the consists of hardware, work principle and the design of software. The final application of the system can bring about large social benefit and economic benefit to agriculture production and management.

REFERENCES

Dong S. L. Application of PLC in the Automatic Sorting System, Machine Tool & Hydraulics, 2005, 5:136-147.

Guo X. F., Liu W. T. An Automatic Feather Selecting System by Image Processing, Modern Electronic Technique, 2004, 3:30-33.

Han Y. M., Wang D., Wei H. H., Tan C. Q. The Design of the Parcel Automatic Sorting System, Mechanical & Electrical Engineering Magazine, 2004, 8:11-12.

Huang Q. M. Automatic Sorting System and Analysis of application prospect, Logistics Technology, 2002, 5:7-15.

Xie L. X., Yang H. Y. Design and Implementation of Statistic and Analysis System for Automatic Luggage Sorting System, Computer Engineering (Supplementary Issue), 2005, 31:3-6.

Xu S. Y. Automatic Sorting System and the application, Logistics and Material Handling, 2002, 3:33-39.

Zhu Z. T. The design of computer auto-sorting system, Computer Engineering and Design, 2001, 6:66-88.

CONSTRUCTING AN INFORMATIONAL PLATFORM FOR FAMILY AND SCHOOL USING GPRS

Yongping Gao *, Yueshun He
School of Information Technology East China Institue of Technology, FuZhou, JiangXi, China
** Corresponding author, School of Information Technology East China Institue of Technology, FuZhou, JiangXi, China, Email: ypgao_ypgao@163.com*

Abstract: The article is mainly about how to employ IC card and mobile phone short messages to construct a convenient and real time information communication platform in the educational field, and to quicken the transmission of educational information so as to realize benefits of communication to education. It introduces the whole schematic design, and the designs of the software mainly include Web serving software, the desktop application software, background service software, two types of machines, controlling software, short messages gateway and so on.

Keywords: schematic design, IC card, two types of machines, ActiveX controls, short message gateway

1. INTRODUCTION

With the advent of information age and the development of communication, there are more and more demands for information. Internet and cell phones are becoming important tools for people to obtain information and to keep in touch. Their applications can be seen in different fields, and they play an important part in these fields. The article is mainly about how to construct a multi-functional platform of short messages, sounds

Gao, Y. and He, Y., 2008, in IFIP International Federation for Information Processing, Volume 258; Computer and Computing Technologies in Agriculture, Vol. 1; Daoliang Li; (Boston: Springer), pp. 171–178.

and Web by combining IC card, GSM cell phones with short messages. The platform is to realize a real time information communication and educational communication among schools, parents and pupils, and to realize the modernization of family education with elementary school information (Jinhua, 2004).

2. THE WHOLE SCHEMATIC DESIGN

The system consists of hardware and software. Hardware is composed of IC card terminals and IC card phones. They achieve the collection and transmission of IC card data, and accept the results and commands from upper layer software. Software is responsible for dealing with IC card data and answering the requests of hardware (Ruhong Gong, 2004) [2]. Hardware and software work together to realize the whole systematic goal and functions. The topological structure of network in the system is as the following fig. 1.

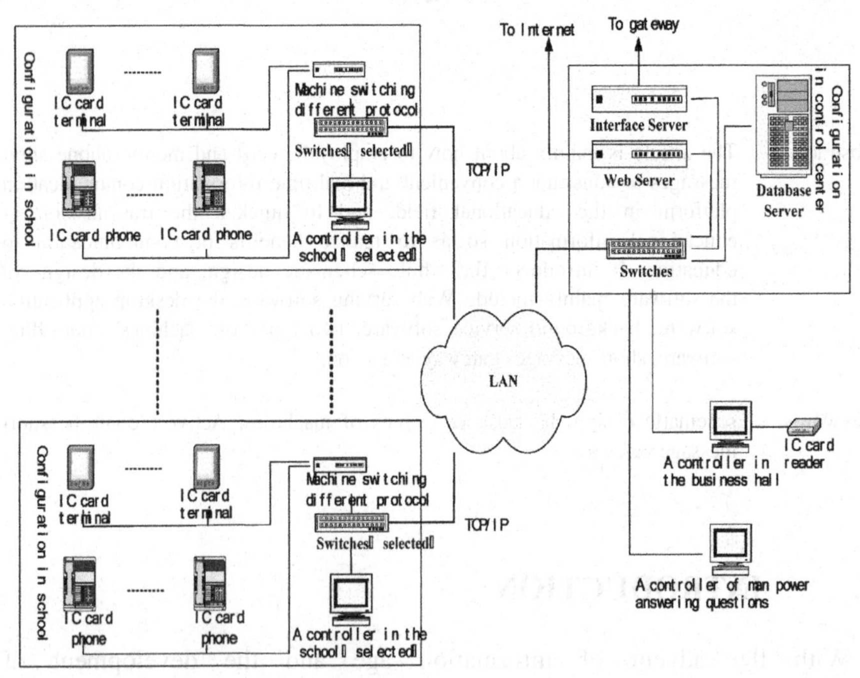

Fig. 1. Based on the GPRS about the informational platform project of family and school

Database server is connected to every IC card terminal and phone in a local network. The collection and transmission of data have been done by placing LAN lines connected with IC card terminals and phones in schools. A machine equipped with an Ethernet controller transmits the data to the database server over the LAN. The database server manages the data, and the results will be transmitted to the IC card terminal or phone.

The functions of management interface and web server can be done by a PC. The software in the interface controller reads the message(s) data which needs to be sent to the server database and send the message(s) to parent cell phones through the short message gateway. Parents can send enquiry messages to short message gateway from their cell phones. The software will search the database and inform parents of the results via the short message gateway. There are two network cards. One is on the same subnet as the PC and the database server. The other is set as the Internet IP address and it provides web server functionality and can send message to the short messages gateway.

A controller in the business hall runs software which deals with matters such as card management and is responsible for IC cards business of all schools.

A controller of man power answering questions makes use of 1860 information service to provide terminal operation software to accept the man power service hot lines.

A controller in the school mainly runs the software for school management.

3. THE DESIGN OF SOFTWARE STRUCTURE OF SYSTEM

Nowadays, three-layer (multilayer) structure has become the mainstream in developing software. It is derived from traditional two-layer (Client/Server) structure, and represents the future of enterprise applications. The software structure chart of system follows in fig. 2.

The software structure of system has some features that it is combined model B/S (Browser/Server) with model C/S (Client/Server), and that it adds business logic layer between application layer and database layer to form three-layer client/server framework (Wei Chen, 2004). The Web server answers the requests from the browser. If the requests include database operations, the Web server will switch the requests to the business logic layer, and then the business logic layer will perform the database operation and return the results to the Web server. Correspondingly, the Web server will produce and return the results to the browser. Similarly, if there are database requests from application software within the application platform, the operation will be performed by the business logic layer. In order to hide the lower communication details, the service software of the service platform needs to communicate with IC cards and the short message gateway, so it uses ActiveX Controls which provide events and methods by which the communication can be accomplished. The main design thoughts and implementation functions are as follows:

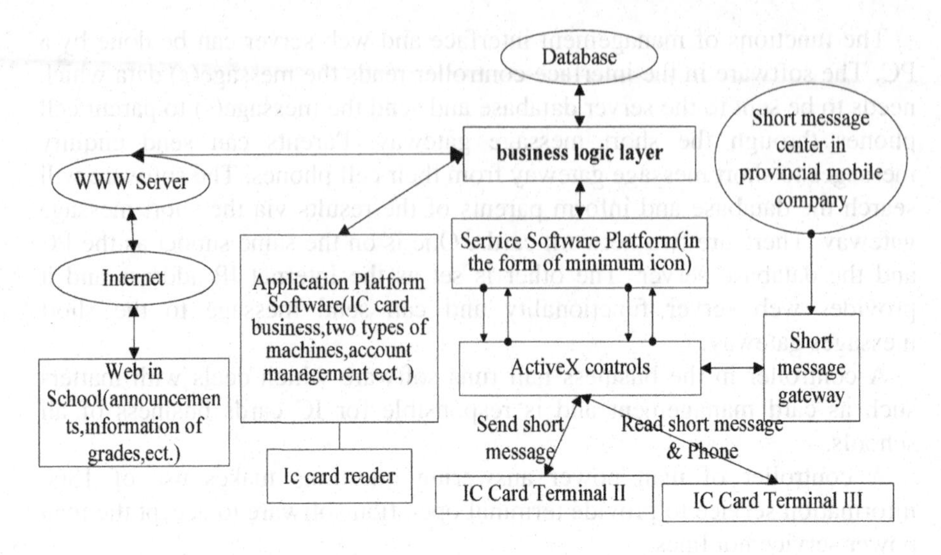

Fig. 2. The software structure chart of system

3.1 The Design of Web Service Software

The Web service is implemented in the Web server and the Web pages can be accessed by the browser on Internet or local network. Teachers can use their own IDs and passwords to login in and send announcements to parents of school activities, how pupils perform in school and so on. Teachers and parents can know whether pupils have arrived at school safely. They can communicate with parents online, and when their pupils have any difficulty, and they can work out a better way with parents. And parents can use pupils IDs and passwords to login in and communicate with teachers.

3.2 The Design of Application Platform Software

The application platform software includes IC card business management software, school management software and 1860 service software.

IC card business management software is installed on the PC in the business hall reception desk, and only the person who has the access rights can operate it. This software mainly manages IC card business transactions, information management for the IC cards, management of equipment, statistics and general system management.

School management software is installed on the PC in the schools, and it is operated by the administrator(s) within the schools. The software provides the following functions: binding teacher cards with pupil cards, looking into whether pupils are in school or not, and sending announcements to teachers.

1860 service software is installed in the information service platform at the mobile corporation and is used by 1860 operators. The software has the main functions of leaving messages, sending short messages between teachers and pupils when they dial man power hot lines.

3.3 The Design of Service Software Platform

The service software platform, in the form of minimum icon, runs on the PC which acts as database server and administrating interface. It consists of three function modules: IC card service, short messages service and accounts management.

The communication software, accepting IC card data and answering requests between every IC card terminal and phone at the schools, is the most important function of the IC card service module. After running this program, the local terminal 5009 will be opened to take over the data and demands of IC cards through the Ethernet. The system employs Command/Response communications mechanism: When a user puts an IC card in IC card terminal, the data on the IC card will be transmitted to the IC card service module. At this time, the terminal is waiting for a response from the IC card service module. If the answer is received in a certain time, the communication succeeds. Otherwise the communication fails. The state of communication can be examined in real time by this method.

The main function of the short message service module is to set up connection with the mobile short message gateway, and transmit and receive short messages to and from the short messages gateway on the Internet. The program received the messages to be sent at any time, and sends the messages one by one to the short message gateway until all have been completed, and it waits for a second, before repeating the whole process.

The account management module is to create accounts information every month, and to update the status of IC cards with the status of parent cell phones. The program can automatically produce all the accounts information of IC cards in the last day per month, and transmit this data to BOSS charging system which will deduct appointed fees from parent's cell phone account. In addition, the program downloads the updated information of parent cell phones from BOSS charging system every day, and renews data of IC cards according to different information of cell phones.

3.4 The Design of Two Types of Machines—the Design of IC Card Terminal and IC Card Phone

IC card reader is to initialize IC cards and there are these kinds of readers in market.

IC card terminal is designed by us, which includes MSC, screen and circuit etc. when a user puts an IC card close to the filed of IC card reader terminal, the machine can read the data of this card, and transmit the data to the service platform software[4].

An IC card phone has been added LCD, headphone, keyboard, GSM card on the basis of IC card terminal. Its appearance resembles a phone. Its working model is request—answer, and users can operate it using menus. It has the functions of dialing stored numbers, reading short messages from parent cell phones and sending the stored short messages.

3.5 The Design of ActiveX Controls

We have designed an OCX Control to accomplish the communication and transmission among IC card terminals, IC card phones and the server. The control comprises many events and methods. It deals with the primitive data and hides the details of communication in the lower layer, and provides a normal interface to the upper layer software. The main events and methods are as follows:

OnICRWRegist Event: This is called when a registration request is received from an active card in an IC Card Terminal.

OnConnect Event: This is called when a command is sent to IC Card Terminal and IC Card Terminal gives response correspondingly.

OnGetUser Event: This is called when data are received from IC card terminals.

ReqLeaveWords Event: This is used to process requests to receive short messages from the IC Card.

OnReqForRTAndCN Event: Used to retrieve phone numbers and balances from IC Card Phones

OnFeeRemainTime Event: Updates the balances on the IC Card Phone

FSInitPort Method: Initialises communication between the server and IC Card Terminal or Phone

AckICRW Method: This is used to deal with a registration by the upper software after a registration request is received from an active card in an IC Card Terminal.

ShotTSN Method: This is used by the upper layer software to switch the pupils account numbers and their names to IC card terminals or phones.

ShowLeaveWord Method: This is used by the upper layer software to transmit total short messages, their sequence numbers and contents to IC card phones.

ShowLeaveWord Method: this is used by the upper layer software to transmit total short messages, their sequence numbers and contents to IC card phones.

3.6 The Design of the Short Messages Gateway

We have developed an OCX control as an interface (Jian Guo, 2005) between the server and the short message gateway based on the CMPP3.0 (China Mobile Communications Corporation, 2003) protocol of China Mobile which includes CMPPAPI.dll and MFC40.dll (Qiang Ji, 2004). The control consists of the following:

SMSArrive Event: Handles the reception of messages from the short message gateway.

AffirmSMS Event: Handles the status of short messages from the short message gateway.

SMSStart Method: Establishes communications with the short message gateway.

EmitMsg Method: Sends short messages to the short message gateway.

Sconn Method: Reports the status of communications with the short message gateway.

4. CONCLUSION

Today the system has successfully been installed in the branch of the China Mobile of Fuzhou City in Jiangxi Province and its application has been extended to every school. The system has been stable for a year. It can send to parents the information whether their children have gotten to school safely in a very short time and can receive short messages from parents without any error. It has been received good feedback from pupils, parents, schools and the branch of the China Mobile of Fuzhou City.

REFERENCES

Jinhua Tang, Jian Cao. Information Service Platform Based on Short Message. Computer Engineering, 2004.12:238-240

Ruhong Gong, Bin Tang, Renbo Wang. Application of Campus IC Card in Wireless Communication Field, Journal of Henan University of Science & Technology (Natural Science), 2004, 4:55-57

Wei Chen, Caopeng Yi, Peixiao Luo. The Implement of SMS Value-Added Applications Platform, Microelectronics & Computer, 2004, 12:121:67-69

Renbo Wang, Gangyong Lin. SMS Controller Using MCU, Science Mosaic, 2004, 10: 7-8

China Mobile Communications Corporation. China Mobile Peer to Peer (CMPP) v3. 3. 3 [EB/OL]. http://104061.playicq.com/1/18435.html, 2003206

Jian Guo, Lijuan Sun. Design and Implementation of Short Message Service System Based on CMPP, Jiangsu communication technology, 2005, 1:26-29

Qiang Ji. SMPP and CMPP interface of SMS application system, Computers and Applied Chemistry, 2002, 6:786-788

REDUCTION IN AGRICULTURAL TAX AND THE INCOME GROWTH OF RURAL RESIDENTS: AN EMPIRICAL STUDY

Ruiping Xie [1,*], Fanling Sun [1]
[1] *Department of Public Economics, Xiamen University, Xiamen 361005, P. R. China*
* *Corresponding author, Address: P. O. Box 79, Department of Public Economics, Xiamen University, Xiamen 361005, P. R. China, Tel: +86-13385926856, Fax: +86-592-2182136, Email: rpxie@163.com*

Abstract: Based on the Chinese statistics from 1976 to 2004, this paper conducts an empirical study on the relationship between reduction in agricultural tax and income growth of rural residents by using econometric methodologies of co-integration theory and error correction mechanism. The result reveals that the policy of reducing agricultural tax has a positive effect on the whole level of peasant's income, but this measure won't resolve the problem of a long-term and continuous growth in peasant's income. In the short run, reduction in agricultural tax obviously increases the direct income of farmers; while in the long run; household operational income and wage income are key elements of peasant's income growth.

Keywords: agricultural tax, reduction, income growth, empirical research

1. INTRODUCTION

During the 1990s, with high economic growth rate in China, the growth rate of the peasant's income was relatively low and the problem of peasant's economic burden was increasingly worsening. In order to raise peasant's income, the central government conducted the rural tax & fee reform from the beginning of 2000, aiming at straightening out the distributive relations among government, collective and farmers and actually slashing peasant's burden. From 2000 to 2003, China started to undertake experiments in Anhui

Xie, R. and Sun, F., 2008, in IFIP International Federation for Information Processing, Volume 258; Computer and Computing Technologies in Agriculture, Vol. 1; Daoliang Li; (Boston: Springer), pp. 179–186.

Province and then gradually generalized to the whole country. The government has conducted trials on reduction and exemption in agricultural tax since 2004. "Central No. 1 document" puts forward that "agricultural tax rate will be cut down by one percentage point in average", and at the National People's Congress it was announced that the rate of agricultural tax would be reduced yearly by more than one percentage point until it was cancelled five years later. One year after, the reducing term was shortened to three years. On December 29, 2005, the 19th Meeting of the Tenth Standing Committee of the National People's Congress passed the decision concerning abolishing the agricultural taxation regulation by an overwhelming majority vote. This policy is branded as the most important reforming measures after the implementation of family-contract responsibility system in China. In order to understand the effects of this reform on peasant's income growth, an empirical analysis, through the econometric methodologies of co-integration theory and error correction mechanism, is carried out, covering the relevant data from 1976 to 2004.

On the relations of reduction in agricultural tax and the income growth of rural residents, there exist three kinds of standpoints. The first standpoint is that the rural tax-fee reform is the efficient path to raise peasant's income (David E. Sahn et al., 1996; Alex Winter-Nelson, 1997; Christine A. Wilson et al., 2002; Yifu Lin, 2003; Li-an Zhou & Ye Chen, 2005). The second standpoint puts forward that the rural tax - fee reform isn't as important as expected, even though the short-term result is significant, the long-term result is uncertain (Robert G. Chambers, 1995; Hui Qin, 2003; Qiyun Fang et al., 2005). The third standpoint reveals that the rural tax - fee reform does little help to the farmers, so maybe agricultural tax shouldn't be reduced or exempted (Mahmood Hasan Khan, 2001; David M. Newbery, 1992; Daiyan Peng, 2004; Junchu Zhu, 2005).

According to these standpoints above, we will carry out an empirical study to reveal whether this policy really reduce the peasant's burden and increases their income. At the same time, we will also try to analyze the scope of change in peasant's income and the time span of this effect.

2. BASIC MODELING

In the yearbook entitled The Agricultural Statistical Annual of China, the peasant's net income consists of wage income, household operational income, property income, transfer income and other incomes. Because, relatively speaking, property income, transfer income and other incomes are just a fraction of the total income, so to reduce the loss of the freedom degree, we combine these three parts into one variable. Therefore we have $Y_0 = X_1 + X_2 + X_3$, where Y_0 represents the peasant's net income, X_1 represents wage income, X_2 represents the peasant's household operational income, X_3 represents the total of property income, transfer income and other incomes. The distribution of the peasant's net income includes the profit deduction and reservation for the collective, national tax and peasant's household income. To reveal the relationship between agricultural tax and the peasant's income growth, we introduce the variable T which stands for agricultural tax. Hence we have $Y_0\text{-}T= X_1 +X_2 +X_3\text{-}T$. Set $Y=Y_0\text{-}T$, thus Y represents the after-tax net income, therefore $Y= X_1 +X_2 + X_3\text{-}T$. So Y then can be rendered as the function of X_1, X_2, X_3 and T: Y=F($X1$, X_2, X_3, T). Differentiate it with respect to X_1, we get:

$$dY = \frac{\partial Y}{\partial X_1} dX_1 + \frac{\partial Y}{\partial X_2} dX_2 + \frac{\partial Y}{\partial X_3} dX_3 + \frac{\partial Y}{\partial T} dT \quad (1)$$

Divide (1) by Y, then we obtain:

$$\frac{dY}{Y} = \frac{X_1}{Y}\frac{\partial Y}{\partial X_1}\frac{dX_1}{X_1} + \frac{X_2}{Y}\frac{\partial Y}{\partial X_2}\frac{dX_2}{X_2} + \frac{X_3}{Y}\frac{\partial Y}{\partial X_3}\frac{dX_3}{X_3} + \frac{T}{Y}\frac{\partial Y}{\partial T}\frac{dT}{T} \quad (2)$$

where: $\frac{Xi}{Y}\frac{\partial Y}{\partial Xi}$ (I=1, 2, 3) represents the after-tax net income Y's elasticity of X_1, abbreviated as α_i, $\frac{T}{Y}\frac{\partial}{\partial T}$ denotes the after-tax net income Y's elasticity of the agricultural tax T, abbreviated as β, $\frac{dY}{Y}$ can be taken as the growth rate of Y, abbreviated as RY, $\frac{dXi}{Xi}$ can be seen as the growth rate of X_i, abbreviated as RXI, $RXII$, $RXIII$ respectively, and $\frac{dT}{T}$ can be regarded as the growth rate of T, abbreviated as RT. So model (2) can be rewritten as:

$$RY = \alpha_1 RXI + \alpha_2 RXII + \alpha_3 RXIII + \beta RT \quad (3)$$

3. STATISTICS CHECKING AND EMPIRICAL RESULT

Statistics checking: We get the sample annual data from 1976 to 2004 of peasant's net income per capita, peasant's household operational net income per capita, peasant's average wage income per capita, peasant's average property income per capita and average transfer income per capita from the yearbook entitled The 2005 Rural Statistical Annual of China and the agricultural tax data from the yearbook entitled The 2005 Financial Statistical Annual of China, the population data of the rural areas comes from the yearbook entitled The 2005 China Population Statistics Annual, the amount of average burden per farmer equals to the ratio of total agricultural tax and the agricultural population. All variables above are real ones. We get the real values adjusted to GDP deflation index with 1978 as the base year.

3.1 The stationary Test of stochastic series

With EVIEWS5.0, through which all the following results are obtained, we get the Unit Root Testing results of each variable as shown in Table 1:

Table 1. Unit Root Testing

Variables	Testing Form	ADF Value	P Value
RY	0,Y,N	-2.906	0.0588
RY	1,Y,N	-6.166	0.0000
RXI	0,Y,N	-2.657	0.0100
RXII	0,Y,N	-2.865	0.0060
RXIII	0,Y,N	-4.117	0.0037
RT	0,Y,N	-4.192	0.0002

Note: the sign patterns in the testing form from left to right respectively mean difference orders, constant item and time trend item. 0 means no difference and 1 means difference order is one and so on. Y means that there has a constant item or time trend item, while N means the opposite.

The results of the unit root test show that, under the significance level of 0.01, variables *RXI*, *RXII*, *RXIII* and *RT* have no unit roots and they are all stationary time series; while variable *RY* is a non-stationary series even with significance level of 0.05, actually it is integrated of order 1. Therefore we could proceed to analyze whether there's a co-integration relationship between dependent variable and independent variables.

3.2 Co-integration test

The model (3) is regressed, and the results are as follows:

$$RY_t = 0.4627RXI_t + 0.3581RXII_t + 0.0653RXIII_t - 0.0431RT_t + u_t \quad (4)$$

t statistic (12.49) (16.24) (2.32) (-1.53)

P value 0.0000* 0.0000 * 0.0299 0.1394

R^2=0.8703 DW=2.3029

Taking a unit root test on model (4)'s residual error gives the results as shown in Table 2.

Table 2. Co-integration test

ADF Value	Testing Form	Critical Value		
		0.01 Level	0.05 Level	0.10 Level
-6.6231	0,N,N	-2.6570	-1.9544	-1.6093

Under the significance level of 0.01, the ADF value of the residual, which is -6.6231, is less than the critical value (-2.6570). So the residual of the model (4) is reposeful and the equation (4) is co-integration regression, which means there is a co-integrated relationship among variables *RXI, RXII, RXIII, RT* and *RY*. And model (4) is the function of long-term relation among these variables. Each regression coefficient stands for the long-term elasticity. Although the parameter values of *RXI, RXII* are statistically significant under the significance level of 0.01 and *RXIII* under the significance level of 0.05, while the parameter value of *RT* is not statistically significant even with the significance level of 0.05, so relevant examination and revision to model (4) are carried out.

The Test for Heteroscedasticity: Because sample data has only 28 observations (a small sample), so we adopt Goldfeld-Guandt Test to test the Heteroscedasticity. After sequencing the data according to the *RT*, depriving the middle 4 observations, we do OLS regression on the top 12 observations and the bottom 12 observations respectively and get their residual sum of squares (rss1 and rss2), therefore we know $F = \dfrac{rss1/df1}{rss2/df2} = 50.40$, with significance level of 5% which is more than the critical value under the freedom degree of 7 for the denominator and 7 for the numerator, that is to say, the residual has the quality of conditional (Heteroskedasticity) relation. Using White Test, we will get the same conclusion.

To eliminate the Heteroskedasticity, we use White heteroscedasticity correction and get the following result:

$$RY_t = 0.4627RXI_t + 0.3581RXII_t + 0.0653RXIII_t - 0.0431RT_t + u_t \quad (5)$$

t statistic	(16.37)	(20.17)	(2.89)	(-2.55)
P value	0.0000	0.0000	0.0083	0.0180

R^2=0.8703 DW=2.3029

After the White heteroscedasticity correction, there is no difference for the value of each coefficient, but they are more statistically significant. Variables *RXI*, *RXII*, *RXIII* are all statistically significant with significance level of 0.01, so is the variable RT with significance level of 0.05. Therefore model (5) makes a more accurate simulation about the long-term relationship between the independent variables and dependent variable.

Model (5) expresses that during the sample period, the growth rate of the peasant's wage income (*RXI*), the growth rate of peasant's household operational income (*RXII*) and the growth rate of peasant's property income and transfer income (*RXIII*) all have positive effects on peasant's after-tax income RY. Among them, the growth rate of the wage income and household operational income are the predominant factors. If *RXI*, *RXII* and *RXIII* change 1% respectively, accordingly, the peasant's after-tax income will change 0.4627%, 0.3581% and 0.0653% in the same direction. However the coefficient of variable *RT* (the agricultural tax) is negative, which means the rise in the agricultural tax will bring negative influence to peasant's income. This also indicates that the policy of reducing agricultural tax is advantageous to the growth of peasant's income per capita. Compared to several other factors in the model, the agricultural tax is the one over which the government can exert the most direct control. So reducing agricultural tax has become one of the most efficient means for the government to work on the peasant's income in recent years.

The coefficient of *RT* is 0.0431, which reveals if the government reduces the agricultural tax by 1% each time, the peasant's real net income will increase by 0.0431%. Therefore, the effect on the growth of peasant's income, brought about by the reduction in the agricultural tax, depends on the degree of the reduction, the larger the reduction is, the more significant the effect is; the shorter the time span is, the more remarkable the effect will be. If the government exempted the agricultural tax, peasant's net income per capita would increase by 4.31%, which makes peasant's overall income rise to a new level. If the agricultural tax is cut down year by year, the effect is also apportioned to the same period. For example, if we carry on a calculation to the sample data, we can get that, during the sample period, peasant's real net income per capita will increase by 7.5% every year. Therefore, if the agricultural tax is reduced within one year, the direct contribution rate to peasant's income growth rate is 4.31/7.5*100% = 57.5%; If this period is increased to 5 years, then the contribution rate will decrease to 12.5%. Actually, many provinces have started reducing the agricultural tax since 2000 and adhered to such measure until the nation declares to abolish the agricultural tax in the whole nation in 2006.

3.3 The error correction mechanism

We have already proven that there exists a co-integrated relationship between *RXI, RXII, RXIII, RT* and *RY*, which means they have an equilibrium relationship in the long term. However, in the short term, variations in the dependant variable will generate dependant variable's deviation from equilibrium. And the error in model (5) can be regarded as "equilibrium error". With this error and ECM, we can relate the short-term behavior with its long-term equilibrium value. Making use of the relevant time series of model (5), an appropriate Error Correction Mechanism and White heteroscedasticity correction, we get the following result:

$$\Delta RY_t = 0.3541\Delta RXI_t + 0.4614\Delta RXII_t + 0.0548\Delta RXIII_t - 0.0399\Delta RT_t - 1.1881u_{t-1} + \varepsilon_t$$

t statistic	9.3735	8.9481	2.2074	1.9040	-5.7472
P value	0.0000	0.0000	0.0447	0.0106	0.0000

R^2=0.854 DW=1.9274 (6)

Where: Δ means a One order Difference, ε_t is the stochastic error, derived from model (5), which equals to $u_{t-1} = RY_t - 0.4627RXI_t - 0.3581RXII_t - 0.0653RXIII_t + 0.0431RT_t$. Equation (6) expresses the Δ *RY* is decided by Δ *RXI*, Δ *RXII*, Δ *RXIII*, Δ *RT* and equilibrium error. If the latter is not zero, the model then deviates from the equilibrium state. The coefficient of u is negative, so if there is a positive deviation in the former period, then the deviation will be corrected by the negative error item, which brings the model back to the equilibrium states, so is the negative deviation. The absolute value of the coefficient of u_{t-1} is to 1.188, so the deviation will diminish in a very short time. Just as the results given by the analysis above, we will come to the conclusion that the short-term changes in the peasant's household operational income per capita, the average wage income per capita, the average property income per capita and the average transfer income per capita data all have positive influence on the peasant's after-tax income per capita, while the rise of the agricultural tax per peasant has a negative impact. At the same time, the absolute value of the short-term effect coefficient (-0.0399) is smaller than the long-term effect coefficient (-0.0431), as given by equation (5). One explanation to this maybe that there exists a time lag before the policy takes effect and the peasants change their production plans. While in the long run, peasants have adequate time to make adjustments to avoid this time lag.

4. CONCLUSIONS

After the empirical analysis on the relationship between reduction in agricultural tax and peasant's income growth, we get conclusions as follows:

1. There exists a long-term equilibrium relationship between the peasant's net income per capita and the reduction of agricultural tax. Reduction or exemption in agricultural tax has long-term effect on the growth of peasant's income. If agricultural tax were abolished, peasant's overall net income per capita would increase by 4.31% compared to the agricultural tax case.
2. The short-term impact of reducing agricultural tax is relative to the degree of reduction and time span. If the agricultural tax is reduced to zero in a very short time, e.g. one year, then the effects will be very remarkable.
3. After the abolishment of agricultural tax, there is no continuing effect from the reduction. Therefore, in the long-term perspective; such policy can't guarantee the continuous growth of peasant's income. While peasant's wage income and household operational income plays a key place in their income growth, especially their wage income.

To sum up, reducing or abolishing agricultural tax has a positive effect on increasing peasant's whole income in the short term; however, this effect is not sustainable in the long-term. Therefore, establishing an effective, long-term mechanism to increase peasant's wage income and household operational income continuously and steadily is the foundation to the growth of peasant's income.

ACKNOWLEDGEMENTS

This paper has been funded by the project of National Social Science Foundation of China (NO. 07CJL027).

REFERENCES

Alex Winter-Nelson 1997. Rural taxation in Ethiopia 1981-1989: a policy analysis matrix assessment for net producers and net consumers, Food Policy, 5: 419–431

Christine A. et al. The effects of a federal flat tax on agriculture, Review of Agricultural Economics, 2002, 24: 160–180

Daiyan Peng. The rural tax-fee reform, the polarization of village leaders and the transformation of social contradiction - according to the analysis about findings in rural areas of Hubei, Journal of Nanjing Administrative Institute and The Nanjing Party School, 2004, 4: 58–62 (in Chinese)

David E. Sahn et al. Exchange rate, fiscal and agricultural policies in Africa: Does adjustment hurt the poor. World Development, 1996, 4: 719–747

David M. Newbery. Agricultural pricing and public investment, Journal of Public Economics, 1992, 2: 253–271

Hui Qin. "Huang Zhongxi laws" and system foundation of tax - fee reform: historical experience and the selection of reality, Tax Administration Research, 2003, 7: 2–8 (in Chinese)

Junchu Zhu. Unscramble reducing agricultural tax and peasants' income growth, Beijing Youth Daily, 2003, 1: 22 (in Chinese)

Li-an, Zhou, Ye Chen. The policy effect of tax - and - fees reforms in rural China: a difference-in-differences, Economy Research, 2005, 8: 44–53 (in Chinese)

Mahmood Hasan Khan. Agricultural taxation in developing countries: a survey of issues and policy. Agricultural Economics, 2001, 3: 315–328

Qiyun Fang et al. An empirical study of the influences of rural tax - fee reform on peasant's income, Chinese Rural Economy, 2005, 5: 36–38 (in Chinese)

Robert G. Chambers. The incidence of agricultural policies, Journal of Public Economics, 1995, 2: 317–335

Yifu Lin. Some opinions relevant to current rural policy, The Problem of Agricultural Economy, 2003, 6: 4–7 (in Chinese)

EXPLORE A NEW WAY TO CONVERT A RECURSION ALGORITHM INTO A NON-RECURSION ALGORITHM

Yongping Gao *, Fenfen Guan
School of Information Technology East China Institue of Technology, FuZhou, JiangXi, China
* *Corresponding author, School of Information Technology East China Institue of Technology, FuZhou, JiangXi, China Email: ypgao_ypgao@163.com*

Abstract: This article discusses how to use queue to make non-recursion algorithm of binary link tree. As for a general binary tree, if we adopt sequence storage, firstly we should extend it first into a complete binary tree, secondly we store it to a temporary queue according to the sequence of up-down and left-right. On the basis of the properties of the complete binary tree and the queue, if we can confirm every element in the queue, then we can find the left and right children of the element in the queue. When we recur these steps, we can create a binary link bit tree. This algorithm enriches the method from recursion to non-recursion.

Keywords: sequence storage, bit link tree, queue, recursion, non-recursion

1. INTRODUCTION

There are two main problems in the designing program of recursion algorithm. On the one hand, not every language backs up recursion. On the other hand, it will take a longer time to finish a recursion program than a non-recursion one, and when there are too many layers of recursion, the stack overflow will appear in the running results (Xiaoyun Guo, 2004).

Gao, Y. and Guan, F., 2008, in IFIP International Federation for Information Processing, Volume 258; Computer and Computing Technologies in Agriculture, Vol. 1; Daoliang Li; (Boston: Springer), pp. 187–193.

If we convert a recursion algorithm into a related non-recursion one, the program can be implemented more efficiently. There are many different methods to transform recursion algorithm to non-recursion one. There are some familiar methods to solve different problems. For example, a recursion algorithm can be converted into a non-recursion one by stacks (Zhenyuan Zhu, 2003), depth-first search and N order Legendre polynomials (Zhong Li, 2001). As for the tower of Hanoi, we usually adopt Inorder-traversing binary tree to convert a recursion algorithm into a non-recursion one (Changbao, 2002).

In order to convert a recursion algorithm into a non-recursion one of a bit link tree, the article attempts to adopt another method—using a temporary queue and two points to finish the conversion.

2. PROBLEM FORMULATION

2.1 Introduction of the problem

If we apply the binary link tree to practical problems, we have to know how to produce a binary link tree. The algorithm adopted by many related books is to extend a binary tree first, and then input some information of node according to preorder traversal to create a binary link tree by recursion (Yuli Yuan, 2006). The detailed function is as following (Weimin Yan, 1997):

```
PBinTreeNode createRest_BTree()
/* To create a bit tree from root node by recursion */
{ PBinTreeNode  pbnode;
char ch;
  scanf("%c",&ch);
  if(ch=='@') pbnode=NULL;
  else
  { pbnode = (PBinTreeNode )malloc(sizeof(struct BinTreeNode));
    if(pbnode==NULL)
    {  printf("Out of space!\n");
       return pbnode;
    }
  pbnode->info=ch;
   pbnode->llink=createRest_BTree();  /*to create left bit tree*/
   pbnode->rlink=createRest_BTree();  /* to create right bit tree */
  }
  return pbnode;
}
```

As it is known to all, a recursion function is called by a stack data structure (Naixiao Zhang, 2002). When we teach such recursion algorithm, we find it difficult for many students to understand the process of creating a binary tree (Chunbao Li, 2002). To help students to understand the algorithm, we have to adopt a non-recursion algorithm to create a binary link tree. Meantime, we can break the rule—recursion of a function should be realized by a stack data structure (Fuying Wu, 2003), but we can change recursion algorithm into non recursion one by other data structures such as queue.

2.2 An analysis of the problem

As it is known by us, we store a complete binary tree according to the sequence of up-down and left-right, by this way; we use a series of memory in order to store the binary tree nodes. The fig. 1 and fig. 2 illustrate this.

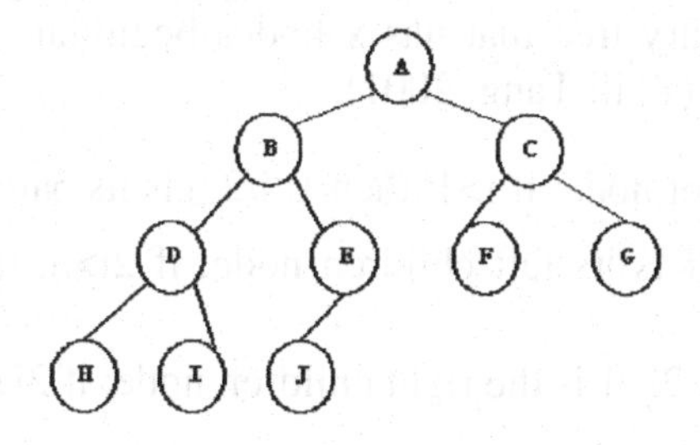

Fig. 1. A complete binary tree

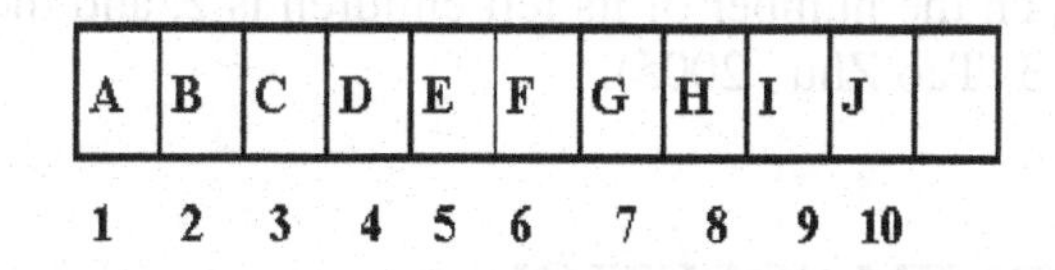

Fig. 2. In order to store a complete binary tree

As for a general binary tree, if we adopt sequence storage, we should extend it first into a complete binary tree, and then store it according to the sequence of up-down and left-right as the following fig. 3.

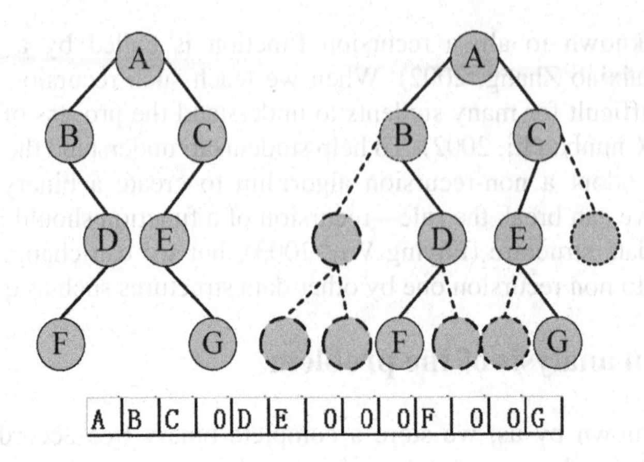

Fig. 3. In order to store a general binary tree

The binary tree stored by this way has the qualities of a complete binary tree and the detailed description is the following (Yongping Gao, 2005).

If we number a binary tree that has x nodes beginning with 1, then any node i (1<= i<=n) will (Zhili Tang, 2001):

1) if i=1, then i is root node; if i>1, then $\lfloor i/2 \rfloor$ is its parent node.

2) if 2i <=n, then 2i is its left children node; if 2i>n, then it has no left children.

3) if 2i+1 <= n, then 2i+1 is the right children node; if 2i+1 > n, then it has no right children.

From these qualities we can know that we can find the positions of one node's left and right children if we are given the node's position in an array. For example, node A is the first number in an array, if its left and right children exist, then the number of its left children is 2, and the number of its right children is 3 (Tao Zhu , 2005).

3. PROBLEM SOLUTION

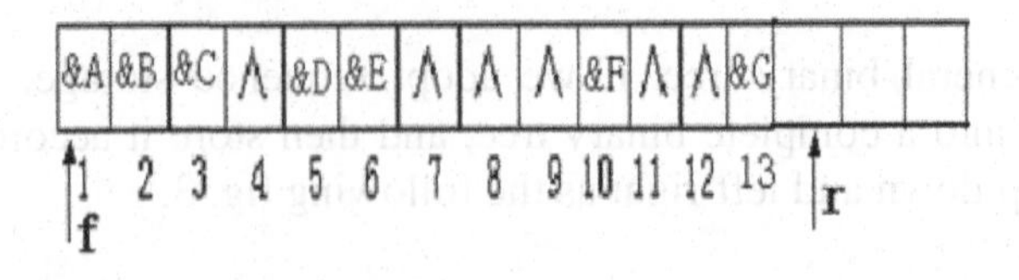

Fig. 4. In order to store a binary tree after which is extend a complete binary tree

From the above analysis, we propose that we can create a binary link tree by the sequence storage of a binary tree.

We can regard the sequence storage of a binary tree as a queue, and add two pointers — one pointer f refers to the head pointer of the queue, the other pointer r refers to the tail one of the queue (Shuqun Gong, 2007), as what the fig. 4 illustrates.

1) Extend a binary tree into a complete binary tree.

2) Store the node address of the complete binary tree, rather than value of the node, in the unit of memory by the sequence of up-down and left-right. If the node is empty, then its address is null. When we store a node, we will add 1 to the value of tail pointer of the queue.

3) When the number element i comes out of the queue, we will add 1 to the value of head pointer and find out the left and right children of this element (if their children exist, the value of left children will be 2*i and the value of the right children will be 2*i+1). For instance, when the first node A comes out of the queue, the value of its left children is 2 (or the element B) and the value of its right children is 3 (or the element C).

4) Repeat the third step until there is no any element in the queue, then the head pointer equals to the tail pointer.

We can realize the function by the following codes:

```
typedef struct BiTNode /*the definition of node type*/
{   elemtype data;
    struct BiTNode *lchild, *rchild;
}BiTNode,*BiTree;
BiTree CreateBiTree()
{ /*to create a binary tree by using a queue according to the sequence of the up-down and
left-right*/
   char ch[20];
BiTree head, p, tm[20];/*head stands for storage of the head pointer of a binary tree, and
tm stands for a queue*/
int i=0,n,f=1,r=1; /*f is the head pointer of the queue and r is the tail pointer*/
    gets(ch);  /*store the extended complete binary tree according to the sequence of up-
down and left-right in the variable ch*/
n=strlen(ch);
  while(1)
  { if(ch[i]=='\0' ) break;
```

```
if(ch[i]=='*') p=NULL; /*if the value of ch[i] is empty node, then we will store null in the queue*/
else
 {p=(BiTree)malloc(sizeof(struct BiTNode));
   if(p==NULL)
 {printf("not enough space!"); return NULL; }
     p->data=ch[i]; p->lchild=NULL; p->rchild=NULL;
 }
 tm[r]=p;  /*we put the first address of the node in the queue*/
  r++; /*we add 1 to the tail pointer*/
 if(i==0) head=p; /*we use head to store the address of head pointer*/
   i++;
 }
while(f!=r) /*we recur the following steps if the queue is not empty*/
{ p=tm[f];
if(p!=NULL)
{ if(2*f<=n) /*we can judge whether the left children exist*/
 p->lchild=tm[2*f] ; /*then p->lchild refers to the left children*/
   if(2*f+1<=n) /*we can judge whether the right children exist*/
  p->rchild=tm[2*f+1]; /* p->rchild refers to the right children*/
}
f++; /*the head pointer of the queue will be added 1*/
}
 return head;
}
```

We can get the result after we run the program:

Input: ABC*DE***F**G

The output of preorder traversal will be (Wanlan Tian, 2003): A B D F C E G

The output of in order traversal will be (Zhong Li, 2003): B F D A E G C

Based on the results of the preorder and in order traversal, we can get only one binary tree (Chuanhong Chen, 2005), and the binary tree is in the fig. 4 and on the other hand, the algorithm we provided above is right (Yuansong Li, 2003).

4. CONCLUSION

In this paper, we have changed successfully a recursion function into non-recursion one by a queue. Thus it is very helpful for students to understand and master a binary link tree. This shows that in some occasions we can convert a recursion function into a non-recursion one by some others instead of by stacks, and this enriches the ways of converting a recursion algorithm into a non-recursion one.

REFERENCES

Changbao Shu, Zhenhai Liu, A Non-recursion Algorithm about Tower of Hanoi, Computer Development & Applications, Vol. 15, No. 11, 2002, pp. 33-34.

Chuanhong Chen, Wuying Shen, The Research and Application of Traversing Binary Tree, Journal of Xiaogan University, Vol. 25, No. 3, 2005, pp. 72-73.

Chunbao Li, Hui Zeng, Zhimin Zhang, Praxis of Program of Data Structure, The Press of Qinghua University, 2002.

Fuying Wu, Luosheng Tan, Mingwen Wang, Nonrecursive Algorithm of Inorder Traversing Sequential Storage Full Binary, Journal of Jiangxi Normal University (Natural Sciences Edition), Vol. 27, No. 4, 2003, pp. 372-375.

Naixiao Zhang, Algorithm and Data Structure, Higher Education Press, 2002.

Shuqun Gong, Yu Ren, Weiwei Chen, The Front and Rear Pointer Design of Circular Queue GONG, Modem Computer, No. 2, 2007, pp. 17-20.

Tao Zhu. Judging Fully Binary Tree on the Basis of Traversing Binary Tree [J]. Journal of Honghe University , Vol. 3, No. 6, 2005, pp. 47-48.

Wanlan Tian, The Discussion of Inorder-traversing and Postorder-traversing Binary-tree with Recursive Algorithm, Journal of Liangshan University, Vol. 5, No. 3, 2003, pp. 3-3.

Weimin Yan, Weimin Wu, Data Structure (C Language), The Press of Qinghua University, 1997.

Xiaoyun Guo, Non-recursive Simulation on Recursive Function, Journal of Xuzhou Normal University (Natural Science Edition), Vol. 22, No. 1, 2004, pp. 40-42.

Yongping Gao, Shumin Zhou, Use Stack to Make the Non-recursion Algorithm of Bit Link Tree, Computer Era, No. 11, 2005, pp. 24-25.

Yuansong Li, New Method of Traversing Binary Tree, Journal of Sichuan University of Science & Engineering (Natural Sicence Edition), Vol. 16, No. 4, 2003, pp. 45-46.

Yuli Yuan, Ling Hu, The Teaching Analysis of Inorder Traversing Binary Tree in the Data Structure, Journal of Neijiang Teachers College, Vol. 21, No. 4, 2006, pp. 109-111.

Zhenyuan Zhu, Cheng Zhu, Non-recursive Implementation of Recursive Algorithm, Mini-Micro Systems, Vol. 24, No. 3, 2003, pp. 567-570.

Zhili Tang, Methods for uniquely determining a tree or a binary tree based on its traversal sequences, Mini-Micro Systems, Vol. 22, No. 8, 2001, pp. 985-988.

Zhong Li, Recursive Algorithm Transform into Non-recursive Algorithm, Computer Science, Vol. 28, No. 8, 2001, pp. 96-98.

Zhong Li, Lin Meng, Dehui Yin, A discussion of postorder-traverse binary tree with no-recursive algorithm, Journal of Southwest University for Nationalities (Natrual Science Edition), Vol. 29, No. 5, 2003, pp. 537-538.

THE CONSTRUCTION AND IMPLEMENTATION OF DIGITIZATION PLATFORM FOR PRECISION FEEDING OF COMMERCIAL PIG

Shihong Liu*, Huoguo Zheng, Haiyan Hu, Yunpeng Cui, Xi Su
Agricultural Information Institute of CAAS, 100081, Beijing, China
** Corresponding author, Address: 12# Zhongguancun South Street, 100081, Beijing, P. R. China, Tel: +86-10-68975098, Fax: +86-10-68975098, Email: lius@mail.caas.net.cn*

Abstract: In this paper, we described the solution method, system implementation, critical technology, etc. of "The Construction and Realization of Digitization Platform for Precision Feeding of Commercial Pig". What's more, we pointed out the architecture main functions of the platform and critical technology taken to achieve the system in detail. The platform includes the following modules: production management, feedstuff management, illness management and production monitor & statistic, implants the digital management of the commercial pig form.

Keywords: Precision Feeding, Platform, J2EE, Component Technology

1. INTRODUCTION

Modern enterprises utilize information technology, and make use of information resources, in order to improve the efficiencies and levels of production, management, and decision making.

Compared with the rural economy, agriculture enterprises, as a main force of Chinese agriculture production, own lots of advantages in human resources, technologies, management etc. The digitization of agriculture enterprises posses the importance of reality and operation and is being

Liu, S., Zheng, H., Hu, H., Cui, Y. and Su, X., 2008, in IFIP International Federation for Information Processing, Volume 258; Computer and Computing Technologies in Agriculture, Vol. 1; Daoliang Li; (Boston: Springer), pp. 195–204.

applied widely. The agriculture enterprise digitization technologies have been diffusely applied in agriculture production management systems, agriculture decision supporting systems, agriculture production monitor systems, agriculture production inspection systems, and agriculture product e-Commerce systems (Shihong Liu, 2003).

There are a couple of problems existing in agriculture enterprises, especially in animal husbandry including expanding production scale, soaring intensive level and exaggerating difficulties in production management. Aimed at the peculiar problems in Chinese feeding enterprises and taken the commercial pigs feeding enterprises as an example, it is applicable to establish a comprehensive technology platform which incorporates information collection, information transmission, information optimization, and information monitor. Based on this platform, the enterprises would digitally and visually express, design, monitor and manage all the objects and production processes. Consequently utilizing of computer technologies would support the decision making, and improve the effects of production management. Information technology, as an important factor in agriculture production, is becoming the indispensable part applied in each phase of the feeding process.

Platform for Precision Feeding of Commercial Pig utilizes modern network technology, web technology, modeling technology and component technology. Moreover, it focuses on three significant factors: information collection, information procession and information application. The construction of unified digital platform would effectively avoid the data redundancy, and realize the different functional modules' completely integration, and by implementing network technology and across-platform technology to eliminate the information-isolates, and provide reliable support to breeding and breed selection, scientific feeding, research analysis.

2. SYSTEM SOLUTION

2.1 System Technology Scenario

The system introduces the enterprise architect frame – J2EE, which is comparatively excellent and low-cost. J2EE is designed to provide a component based solution to design, develop, assembly and deploy enterprise application and offer a multi layers distribution application model which could support component based reuse solution, XML based data exchange, unified security patter and agile transaction monitoring (Xiaohua, Liu, 2003). J2EE has superiority in scalability, deployment, operation system independence, platform independence, high performance and total cost. Thus, it could achieve a system design goal with the availability, maintenance, manageability, high performance, high reliability, scalability, and security. Implementation of J2EE could reduce the cost, speed up system design and application development significantly (Huajun Chen , 2002).

2.2 System Framework

From the point of view of information procession, the technology platform for precision breeding of commercial pig could be divided into 3 layers: information selection, information procession and information application, as shown in figure 1 (Renchu Gan, 2000).

Within layer of information collection, the major goal is to acquire knowledge of the domain knowledge and the enterprises' specific requirement of the platform, and prepare for the system analysis and design.

Within layer of information procession, the major goal is to build up the commercial pig's production record database, production inspection database, feedstuff prescription database, and animal disease diagnostic

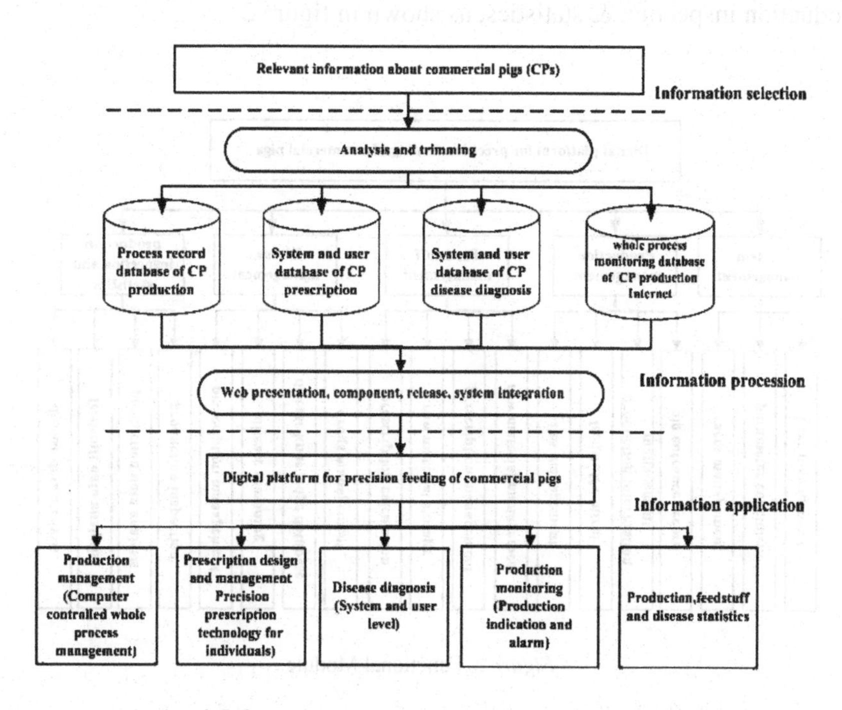

Figure 1. Information procession and information application

knowledge base. With the support of four databases, the core business process logic layer would be realized by using web development, component based and graphic & report and etc technologies.

Within layer of information application, through the web UI interfaces which are designed to orient to the mass system users, users could promote system level domain knowledge base to accomplish the breeding enterprises' specific business processes. In the module of system statistics analysis, intuitionistic graphics and reports would exhibit the situation of enterprise production and management, thus give support to enterprise decision making. Meanwhile, users could also take their own enterprise' practical need into account, and add data and related knowledge into the database to proliferate the platform's resources (Weidong Zhou et al., 2002).

2.3 System Functional Module

The technology platform for precision feeding of commercial pig includes functions following: system management, production management, feedstuff management, prescription management, animal disease management, and production inspection & statistics, as shown in figure 2.

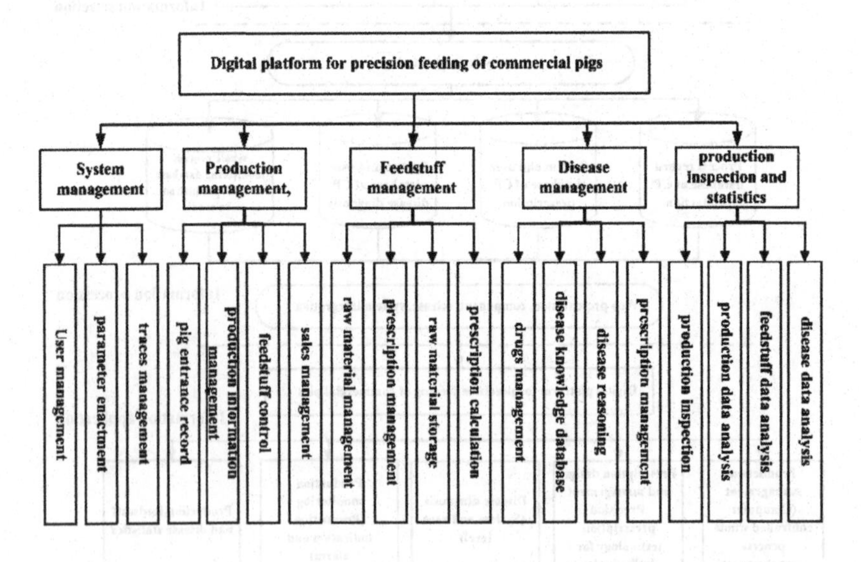

Figure 2. Functional Module

3. SYSTEM IMPLEMENTATION

3.1 Major Function Descriptions

The technology platform for precision feeding of commercial pig introduces J2EE framework, MVC design pattern, client layer/business layer/data layer – three independent layers' component based development method (Chuan Chen, 2001). The 5 major system functions in the platform are as follows:

(1) System Management Sub System: It is responsible for system parameters configuration, user registration, user information maintenance, user authorization, system visit record, database operation record's query, and maintenance. This part is the fundamental premise of the entire platform.

(2) Production Management Sub System: This module monitors the entire processes in commercial pig feeding, and includes several parts: phase management of pig feeding, pig house moving management, feedstuff putting in management, cost management, feedstuff modifying management and sales record management.

In this module, through comparative analysis of feeding patterns in Chinese pig feeding enterprise, couples of common production patterns were found, including 4PF3M (4-Phase-Feeding-3-Moving), 5PF4M (5-Phase-Feeding-4-Moving), and 6PF5M (6-Phase-Feeding-5-Moving). Finally, we chose the 6PF5M pattern which is standard, professional and easy to implement All-In and All-Out streamline technical process to achieve systematic production management.

(3) Feedstuff Management Sub System: It processes the feedstuff, prescription and other related issues, and includes feedstuff resources management, prescription management, feedstuff warehouse management, and prescription computing.

Besides the basic warehouse management functions, this module implements the feedstuff prescription import interfaces, and put forward a java based linear programming minimum cost model to achieve optimism feedstuff computing. Therefore, users could modify and gain their featured prescription according to the enterprise's characteristics.

(4) Illness Management Sub System: It primarily handles pig illness and other related issues, includes illness knowledge base management, illness knowledge query, medicine information management, medicine warehouse management, and epidemic disease management.

This module capsules lots of illness information, meanwhile providing users with accessible interfaces to expand their own illness knowledge base.

It also realizes the FAO AgroVoc based light weight ontology to achieve ontology based pig diseases reasoning algorithm.

(5) Production Monitor & Statistics Sub System: It partly fulfills the alerts and forecasts to production management items in all above sub systems. Besides, the items would view all the forecasts in the module. Moreover, the module partly realizes the numbering statistics of pig, feedstuff employed situation statistics, illness incidence situation statistics to facilitate the managers monitoring the enterprise production operation conditions.

3.2 System Operation

3.2.1 System Runtime Environment

The system's development and operation pattern introduces Tomcat + Java + XML + Crystal Reports + DBMS. The system's specific runtime environment is as follows:

Operation System: Windows 2000 Server (SP4)/Windows XP (SP1)
Java Runtime Environment: J2SDK 1.4.2
Web Container: Tomcat 5.0.19
Database: MS SQL Server 2000 (SP3) Enterprise Edition
Report Server: Crystal Enterprise Report Application Server 9.2 (SP2)

The Client must be compatible of MS IE, and IE 6 (SP1) which are recommended.

3.2.2 System Operation

Put URL into explorer's address, for example http://localhost:8080/pig-login.jsp, and the system would turn into a system login page, as was shown in Figure 3.

Figure 3. System login page

If the username and password are validated, the system turns into a portal page, as was shown in Figure 4.

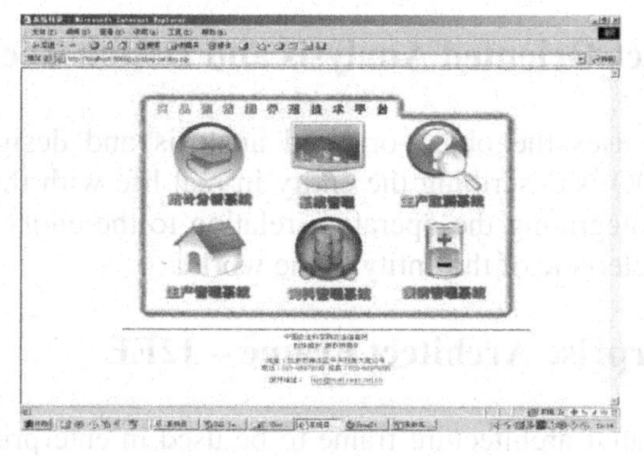

Figure 4. Portal page

On the portal page, choosing different functional menus could enter corresponding functional modules, as shown in Figure 4.

The feedstuff management interface is as shown in Figure 5. The statistic interface is as shown in Figure 6.

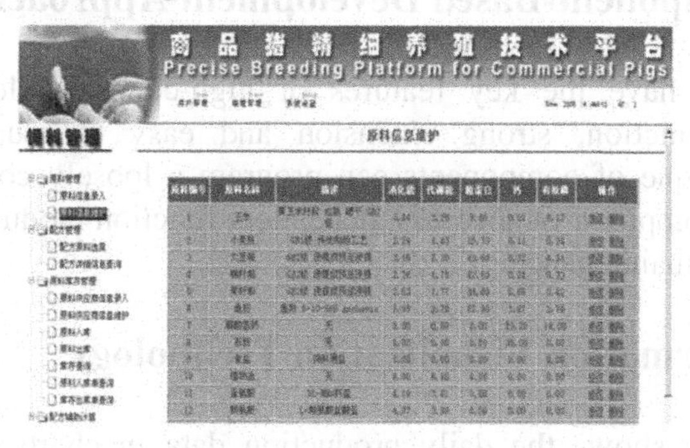

Figure 5. Feedstuff management interface

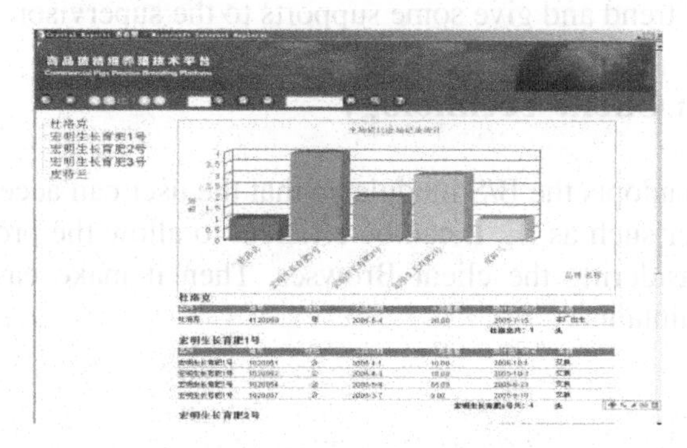

Figure 6. Statistic interface

4. KEY TECHNOLOGY

4.1 Object-Oriented Analysis and Design Technology

The research uses the object-oriented analysis and design technology. Feature of the OO is describing the entity in real-life with the properties of the object and integrating the operation relation to the entity. It is better to reflect the characteristic of the entity in the world.

4.2 Enterprise Architect Frame – J2EE

J2EE is a popular architecture frame to be used in enterprise, key feature of which includes better extension, Cross-platform, strong integrating, OO whole, easy to reuse and deploy, independent of operation system, high performance. This frame can offer the whole support to design enterprise application software, so it can be used widely (Ming Li, 2001).

4.3 Component-Based Development Approach

Components have the key features of high-cohesion, loose-coupling, independence-function, strong extension and easy to reuse. Based on program technique of components can program a loosely coupled system, the system is composed of several independent function module so that easy to upgrade and maintenance.

4.4 Information Visualization Technology

The platform shows the daily production data in charts and graphics. Those data relating to statistic and analyze can reflect the disadvantage and advantage directly during the management of enterprise; what's more, it also can forecast the trend and give some supports to the supervisor.

4.5 Thin Client Technology

The platform adopts the B/S module so that the user can access the system through Browser such as IE. It can be efficient to allow the programmer get away from considering the client Browser. Then it make easy to deploy, upgrade and maintain.

5. SYSTEM IMPLEMENTATION FEATURES

(1) We implementing digitization management in precision feeding of commercial pigs including production management, prescription computing, feedstuff management, illness diagnostics, production monitor & statistics.

(2) We introduced J2EE based system platform into commercial animal breeding domain, and developing robust, scalable and high maintenance system.

(3) We built FAO AgroVoc based light weight ontology of pig illness, introducing the ontology based reasoning mechanism and consequently enhance the accuracy of disease diagnostics, and apply the ontology concept and related methods on animal disease diagnostics.

(4) We adopted UML modeling, object oriented database modeling and etc design methods, based on fully understand of domain knowledge and users requirements building system business models, accomplishing the high availability, rationality, reusability and scalability.

(5) We adopted Java technology to develop component based platform of precision feeding of commercial pig. We developed our own warehouse management, production monitor, illness reasoning, feedstuff management, medicine management, prescription management, database encapsulation, user management with 8 high aggregations, loose coupled, high flexibility, and independent function components reusable.

(6) Through data convergent procession, apply data mining, graphic reporting analysis, and tendency analysis technologies, users can realize visual management in feeding enterprise's decision making. Managers can get qualitative judgments from quantificational analysis, consequently improving the pertinence of decision supporting and rationality of decision making.

(7) We Utilized Java technology to design feedstuff and prescription management function based on linear programming and its related principles based design.

REFERENCES

Chuan Chen, etc. Building JSP/Servlet + EJB Web Application With MVC Based Design Pattern [J]. Computer Engineering, 2001

Huajun Chen. Building Enterprise Application Solution with J2EE [M]. Beijing: Post and Telecom Press, 2002

Ming Li. Java Based Application Server Design and Implementation [J]. Computer Research and Development, 2001

Renchu Gan. Information System Development [M]. Beijing: Economy Science Press, 2000

Shihong Liu. Study on Digital Agriculture and Information Technology Application in Agricultural Enterprise Management. Computer and Agriculture, 2003

Weidong Zhou, Jin Zhou. Component Based Design of New Distributed Client/Server System [J]. Journal of Shandong University, 2002

Xiaohua, Liu. J2EE Enterprise Application Development [M]. Beijing: Publishing House of Electronics Industry, 2003

STUDY ON THE APPLICATION OF DIGITAL IRRIGATION AREA SYSTEM

Yong Liang [1,2,*], Jiping Liu [1], Yanling Li [2], Chengming Zhang [2], Mingwen Ma [3]

[1] *Chinese Academy of Surveying and Mapping, Beijing 100039, China*

[2] *Department of Surveying and Mapping, School of Information Science and Engineering, Shandong Agricultural University, Taian 271018, Shandong, China*

[3] *XueYe Reservoir of Laiwu, Laiwu 271100, Shandong, China,*

[*] *Corresponding author, Address: School of Information Science and Engineering, Shandong Agricultural University, Taian 271018, Shandong, China, Tel: 86-0538-8249322, Fax: 86-0538-8249322; Mobile: 86-13605383139; Email: yongl@sdau.edu.cn*

Abstract: The aim and meaning of study on digital irrigation area are discussed at first. Then the author puts forward the need and feasibility. Taking XueYe reservoir as an example, the software and hardware of digital irrigation area are analyzed and designed, while researching and developing on the key techniques and automatic regulation supervision system. The digital irrigation area software is mainly composed of five modules, which are floodgate control, agriculture-use water management, life-use water management, flood discharge and electricity generation. The hardware system includes the hardware which are installed at the information control and transaction center of chief regulation center, and the systems of floodgate on-off remote sensors, floodgate on-off degree remote sensors, water depth, water quantity and water level remote sensors, electricity generation quantity remote sensors and industry water supply remote sensors. We also designed the software function framework, the hardware constitution, general floodgate flux and distribution of electricity flux, flood discharge flow chart, floodgate on-off degree remote sensors structure chart, floodgate on-off remote sensors structure chart and water depth water quantity remote sensors charts. On this base, the key techniques of digital irrigation area are discussed. Also, we bring forward multi-goal case evaluation method, auto make and inter make method, database technology and modularization management functions. With dynamic supervision on rain and project instances, we can issue alarm information ahead of schedule, and provide supports for scientific decision-making, as while as saving abundant water resource and improving management level.

Liang, Y., Liu, J., Li, Y., Zhang, C. and Ma, M., 2008, in IFIP International Federation for Information Processing, Volume 258; Computer and Computing Technologies in Agriculture, Vol. 1; Daoliang Li; (Boston: Springer), pp. 205–214.

The project was applied in XueYe reservoir of LaiWu in Shandong Province and achieved good effects.

Keywords: digital irrigation area, data auto-collection, regulation supervision

FOREWORD

It has been a long time since reservoir became main source of agricultural water. But with rapid development of country economy and society, the conflict of water supply and consume stands out increasingly. The lack of water resource has touched our foodstuff directly. The foodstuff output reduces 700 hundred million to 800 hundred million every yeah due to lack of water (Science Publishing Company, 2004). As population increase, industrialization and city boost, the contradiction of water lack will be more outstanding. So saving agricultural water is imperative under the situation, while rebuild conventional management measures and irrigation establishments are essential instruments to realize water saving.

Excessive irrigation water waste, great manpower expenditure and civil bothers due to dated management measures have been difficulties all along. The advent of information age affords opportunity for reservoir irrigation area management. The concept of digital irrigation area is put forward under such background. The aim of digital irrigation area is to actualize dynamic, real time optimize collocation and regulation with the technologies of computer, multimedia, modern communication and scientific computation, starting with auto collection, transmission, storage and disposal of weather, rain, water and water supply (Cui Weihong, 1999). To actualize digital irrigation area construction is certain result from water conservancy acclimation of age development, is main content of information-based water conservancy, is also the only way from traditional water conservancy to modern water conservancy and continuable development (Liang yong, 2002).

In digital irrigation area, digital management measures are used from farmland weather information collection, to irrigation water flux information collection, regulation and operations management etc. By solution of key technical problems for auto inspection and control, real time and dynamic watch is processed on real time rain and water circs and water supply. According to real time circs, alarm information is issued ahead, to provide a platform and supports for experts and leaders decision (LuoYunqi, 1999).

We take Xueye reservoir irrigation area as an example in the study. Xueye reservoir irrigation area is one of the large irrigation areas in Shandong Province. It is in the northwest of Laiwu city, span 9 villages and towns and offices. It is 21.6 km long in south-north, 20.8 km wide in east-west. The designed irrigation area is 305 thousand mu, effectual area is 223 thousand mu. With the bound of Yingwen river, it is divided into two parts: one is from northern hill to Yingwen river, controlled by the west channel, ground slope 1/30–1/600; the other is from Yingwen river to Changbu mountain range of Muwen river, controlled by the east channel, ground slope 1/50–1/200. The constructions include chief channel, east channel, west channel, branch channel, dou channel and farming channel, in which the chief channel is 3.1km long, with 14 constructions; east and west channel are 61.2km long, with 401 constructions; the 10 branch channeles are 48.63km long, with 541 constructions. There are 7 Baxieer water weirs, 8 non-throat water weirs, 2 standard water weirs, 1 farming diffluence water measure meter. The entrance of chief channel is flume. While east and west channel, Hebei management station, the area from Qiguanzhuang management station to Heguanzhuang management station, the area from Heguanzhuang management station to Huzhai management station, the area from Wenshi management station to Gushan management station, the area from Gushan management station to Yuchi management station are all Baxieer water weirs. The area from Kouzhen management station to Qiguanzhuang management station and the area from Yangshan management station to Wenshi management station are all standard water weirs. Other management stations and dou channel entrance are all non-throat water weirs.

1. THE SOFTWARE DESIGN OF DIGITAL IRRIGATION AREA SYSTEM

The software of digital irrigation area is composed of five function modules which are floodgate control, agriculture water management, life water management, flood discharge and electricity generation. By the interface with the hardware of the system, the software can realize flux measure, flux control to every floodgate and water weir in the irrigation area. And it may analyze and collect flux data gathered, and realize the aims of agriculture water scientific control, water rate income with flux analysis and statistic modules. Figure 1 is the general framework of the function modules of the software.

1.1 Floodgate control management

The floodgate control management module includes chief floodgate control and floodgate status. In chief floodgate control module, water flux is controlled by unlock degree controller of chief floodgate, at the same time, the corresponding electricity generation sets are turned on according to water flux. In floodgate status module, the open-close command is sent by the software to floodgate according to real time supervision on all floodgates in the area. The distribution of water flow between water gates and corresponding electricity is showed in figure 2.

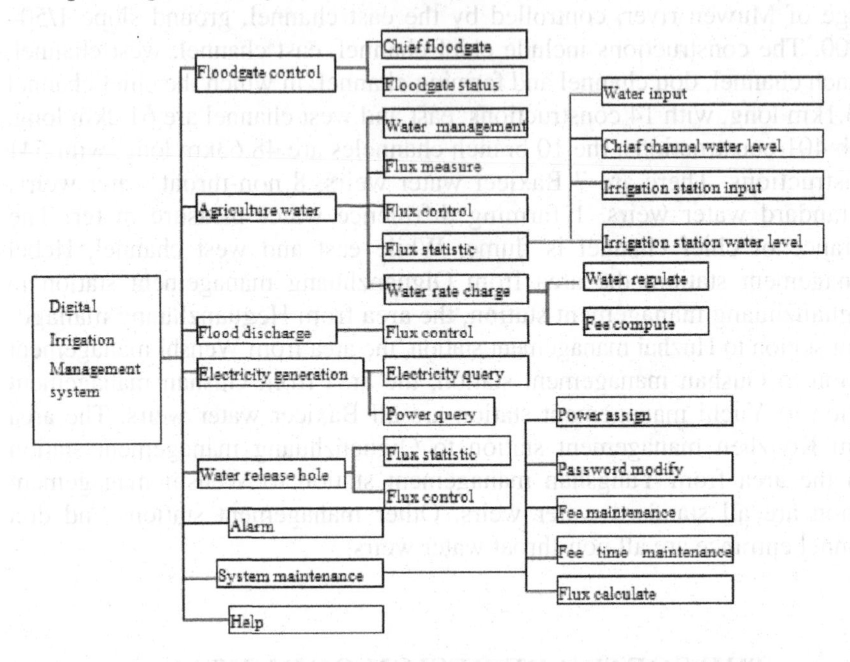

Fig. 1. General framework of the function modules of the software

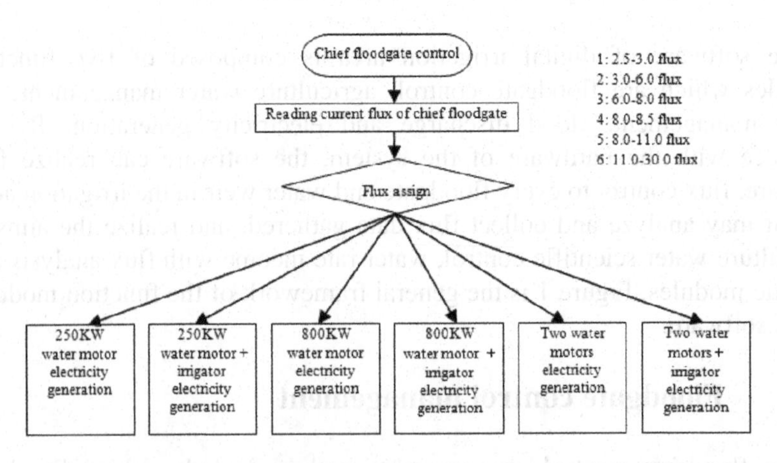

Fig. 2. Distribution of water flow between gater and corresponding electricity

1.2 Agriculture water management

The module of agriculture water management is made up of five child modules which are water administration, flux measurement, flux control, flux statistic and water rate charge. Water administration module supervises the times of agricultural irrigation water every year, start time and end time of each water discharge. Flux measurement module can measure current water depth and flux of water weirs in every irrigation station, and statistic current water wastage in the responsible area of every station. Flux control module accommodates current flux of each station by control the floodgates of east and west channels. Flux statistic module carries out statistic and analysis with graphs on total water input, output, wastage, water level and water flux in each irrigation station of the area. Water rate charge module statistics and charges water rate of each irrigation station in the area, according to its wastage in the year and interrelated national policies.

With agriculture water management module, we can measure, control, statistic and analyze agriculture water so as to realize scientific regulation, reasonable distribution and accurate water rate charge.

1.3 Life water management

Life water management module mainly means statistic on city life water flux. Life water flux statistic module summarizes daily water supply and total water supply, according to water consumed which is recorded automatically.

At present, the city of Laiwu is main life water customer of Xueye reservoir, which is about 20% of total input of the reservoir. Life water is principal in water regulation. To analyze history data and arrange the plan reasonably is important measure to ensure life water.

1.4 Irrigation area flood discharge management

Irrigation area flood discharge management module is realized by flood discharge flux control module. The flood discharge flux control module supervises current height of flood discharge floodgates and abstention floodgates. And regulate the height of flood discharge floodgates rationally according to total flood discharge flux of the chief floodgate, to insure that

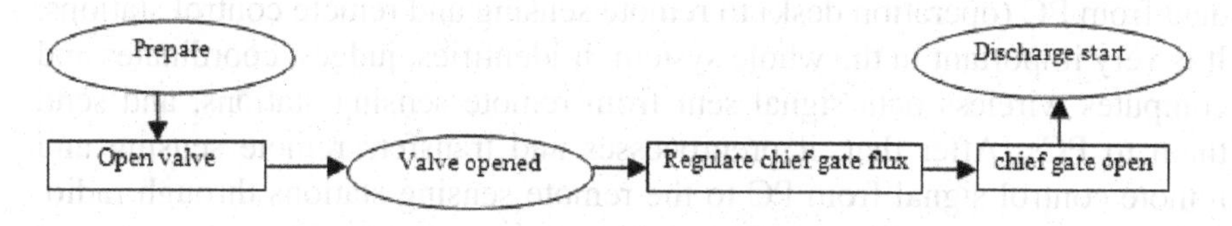

Fig. 3. Sketch map of flood discharge flow

flood discharge works successfully. Figure 3 indicates the flow of flood discharge.

1.5 Electricity generation management

Electricity generation management module includes two child modules, electricity quantity and power enquiry. The module can record electricity quantity of the generators in the area, so that customers can enquire and statistic in time electricity quantity of every generator according to period of time.

The communication part of the software is developed with VC + + language. It can realize effective, reliable rock-bottom data communication and command sending. With wireless data transmission device, the remote sensing points and remote control points are measured and supervised real time. The user interface, which is designed with GUI, expresses the status of floodgate, current water depth and flux of remote sensing points. The interface is kind and convenient for users. Especially, the system is very good at statistic and analysis, so it can query, collect, statistic and analyze water consumed in all periods, and provide supports to scientific and reasonable water regulation.

2. THE HARDWARE DESIGN OF DIGITAL IRRIGATION AREA SYSTEM

The hardware system includes the hardware which are installed at the information control and transaction center of chief regulation center, and the systems of floodgate on-off remote sensors, floodgate on-off degree remote sensors, water depth, water quantity and water level remote sensors, electricity generation quantity remote sensors and industry water supply remote sensors. Data is transmitted with wireless signal between information control center and remote sensors. The composition of hardware is showed in figure 4.

Information control and disposal center is composed of prepositive machines, PC and operation desk. The prepositive machine in the center of information receiving and process is processor which transfers in-between data from PC (operation desk) to remote sensing and remote control stations. It is very important in the whole system. It identifies, judges, coordinates and computes wireless data signal sent from remote sensing stations, and send them to PCs. After that, it preprocesses and transfers remote sensing and remote control signal from PC to the remote sensing stations through radio.

PCs mainly complete data store, query and water quantity auto regulation. Operation desk is specially designed for manual operations, on which Chinese LCD screen and operation buttons are fixed. All survey data of the remote sensing and remote control stations can be displayed on the LCD screen to the life. When it needs human to regulate open-close degree of floodgate, all we need to do is to press up or down button. Then we can watch open-close degree of floodgate and change of water level, water flux. At the same time, the change is transferred to PC by prepositive machines so as to deal with the change data.

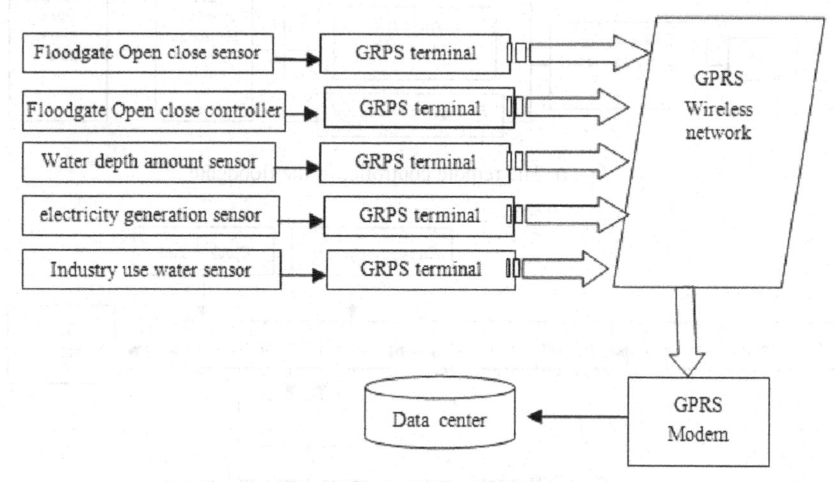

Fig. 4. Composition of hardware

The main function of remote sensor on open-close degree of floodgate is to convert open-close information detected to digital signal by anti-jamming circuit, and send the data to data center through GPRS wireless data terminal after disposal of data processing circuit. The remote sensor is composed of floodgate unlock sensor, floodgate lock sensor, anti-jamming circuit, data processing circuit, data store circuit, clock circuit, data terminal and etc. Figure 5 is the graph of remote sensor on open-close degree of floodgate.

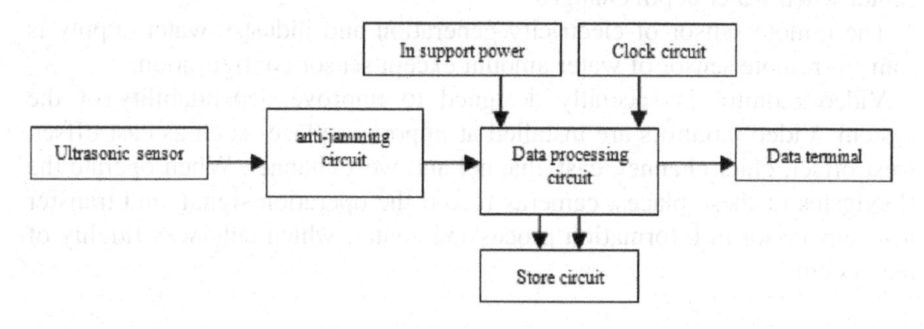

Fig. 5. The remote sensor of the floodgate

The main function of remote controller on open-close degree of floodgate is to receive floodgate control signal from data center and open and close floodgate according to control information. The remote controller on open-close degree of floodgate is composed of GPRS wireless data terminal, data analysis and processing circuit, floodgate up and down circuit, direction shift delay circuit, over load and over flow protection circuit. Figure 6 is the graph of remote controller of floodgate.

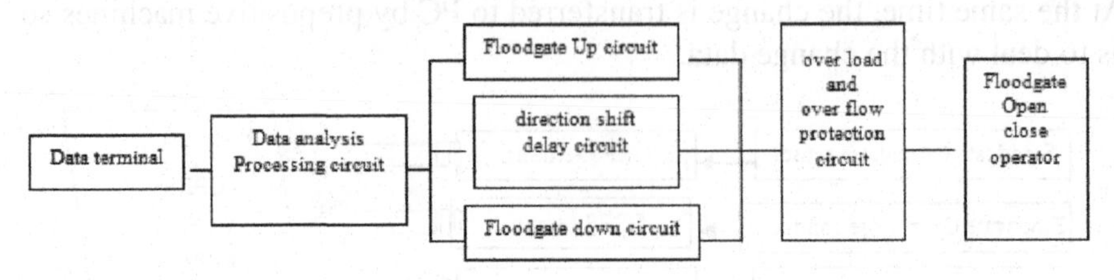

Fig. 6. The remote controller of the floodgate

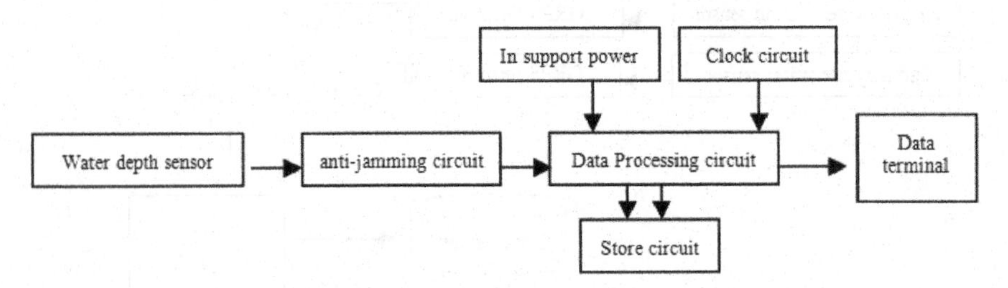

Fig. 7. Remote sensor of water depth and amount

Remote sensor of water depth and amount is made up of water depth sensor, anti-jamming circuit, data processing circuit, store circuit, clock circuit, GPRS wireless data terminal and etc. The structure of remote sensor of water depth and amount is showed in figure 7. The main function of remote sensor of water depth and amount is to measure water depth of the channel by ultrasonic sensor, and calculate flux according to prepared depth-flux curve. It measure every 6 second, and send the change data to data center when water depth changes.

The remote sensor of electricity generation and industry water supply is same as remote sensor of water amount except sensor configuration.

Video monitor is specially designed to improve dependability of the system. Video monitors are installed at important places such as east offset, west offset, chief channel, east channel and west channel. When operate the floodgates of these places, cameras record the operation signal, and transfer it to supervisor in information processing center, which enhances fidelity of the system.

3. KEY TECHNOLOGIES OF DIGITAL IRRIGATION AREA SYSTEM

3.1 Multi-aim plan evaluation method

Multi-aim plan evaluation method adopts multi-aim fuzzy optimized decision-making model. Considering experiences, regulation decision-making people confirm power of qualitative aim with duality, and evaluate multi-aim regulation project quantificationally with the model. The model combines traditional optimize technology and the new theory fuzzy mathematics so as to consider multi-aim in decision-making.

Multi-aim project regulation decision-making model, considering the complexity of different water situation and customers of auto control and regulation for the irrigation area, analyzes water supply aims in all phases of auto control and regulation to confirm power value of the phases. After that, it synthesizes the supply aims in all phases as one aim, and gets optimal regulation plan. The main characteristic of the method is that the decision-making model is more theoretical and effective, because it confirms power according to fuzzy theory.

3.2 Auto created and alternating created regulation project

The system investigates two ways, auto created and alternating created, to provide auto inspection regulation project. Auto created project is created automatically according to real time course of water supply. Alternating created project can regulate according to water supply in period of time and alternating open or close water supply devices. It is very convenient and visual to create a project. When a new regulation project is created, regulator only needs to fill a form on the screen. When a regulation project needs to be modified, regulator only needs to fill a form on the screen too.

3.3 Database technology and modularization management function

The irrigation area auto supervision and regulation system is on the base of database. With multi user system in network, the system should be able to suit all complex situations. Modularization management can be realized only when the problem of water supply regulation is solved effectively. While the only effective way to solve the problem of water supply regulation is to design and manage database aboratively. When designing database, we should consider real banausic need fully and make full use of potential of database function.

4. CONCLUSION

At present, the auto supervision and regulation system of reservoir irrigation area, designed and researched independently according to above advanced technology and theory, is put into practice and achieved good effect. Nowadays, the technologies of information auto collection, database, simulation, computer and other high techs are developing rapidly. Taking the characteristics of water supply into account, we design and research the hardware and software according as standardization. Aiming at influence of real time water supply error, the system can modify real time in self adapt way. The efficiency of system design is improved. The measure precision of the system is assured. The system has important effect on scientific and reasonable regulation decision-making of reservoir irrigation area.

REFERENCES

Continuable development research team of CAS. 2004 China Continuable Development Strategic Report [M]. Beijing: Science Publishing Company. 2004.

Cui Weihong, Li Xiaojuan. Digital Earth [M]. Beijing: Chinese Environment Science Publishing Company, 1999.

Liang Yong, Lu Xiu-shan et al. Study on the Framework System of Digital Agriculture [J]. Chinese Geographical Science, 2003, 13(1):15-19.

Liang Yong, Lu Xiu-shan et al. The Main Content, Technical Support and Enforcement Strategy of Digital Agriculture [J]. Geospatial Information Science, 2002, 5(1): 68-73.

Luo Yunqi. GIS Construction and Mapinfo Application [M]. Beijing: Tsinghua University Press, 2003, 100-201.

GIS COMBINED WITH MCE TO EVALUATE LAND QUALITY

Fengchang Xue[1,*], Zhengfu Bian[1]

[1] *School of Environment Science and Spatial Informatics of China University of Mining and Technology*

** Corresponding author, Address: School of Environment Science and Spatial Informatics of China University of Mining and Technology, Xuzhou, Jiangsu province, China, 221008, Email: xfc9800@126.com, Tel: 0516-83521039*

Abstract: The evaluation of land quality plays an important role in the study of land resources. Some software of GIS was applied to evaluate land quality in recent years. There is some limitation in Appling GIS to evaluate land quality directly. GIS combined with MCE (Multi-Criteria Evaluation) can eliminate limitation existing in using GIS singly. In this thesis key technology of evaluating land quality with GIS and MCE is studded, including methods of confirming evaluation criteria, plotting spatial cells, diffusing spatial attribute value and combining spatial data in numeric way. Technology above was applied in plotting out requisition blocks of land in Tongshan country. The research indicated that the technology of GIS combined with MCE can integrate multiple source information associating with land quality commendably, and achieve measurable land quality evaluation.

Keywords: GIS, MCE, Land, Evaluation

1. INTRODUCTION

The study of land resources in China and abroad since the 1970s has evolved into an important and comprehensive field of applied research in earth sciences (Leng Shuying, 1999). The evaluation of land quality plays an important role in this field. More attention is centralizing land quality indicators and how to measure land quality (Su Biyao, 2001). Some software of GIS was applied to evaluate land quality based on Overlay analysis in recent years.

Xue, F. and Bian, Z., 2008, in IFIP International Federation for Information Processing, Volume 258; Computer and Computing Technologies in Agriculture, Vol. 1; Daoliang Li; (Boston: Springer), pp. 215–222.

Geographic Information System (GIS) emerged in 1960s, which functions of data managing, map showing and spatial analysis are widely used in land evaluation. Overlay analysis of GIS can mix attribute value distributing in different layers, which leads to a new layer containing some new attribute value. There is some limitation in appling overlay analysis to evaluate land quality directly (Ye Jia-An, 2006): (1) General overlay analysis can not comparatively consider weightiness of layers; (2) There is not effective way to confirm threshold value of consecutively distributing variable.

Multi-Criteria Evaluation (MCE) is one of applications of multiple criteria decision making (MCDM). Generally, it is composed of two forms: multi-object evaluation and multi-attribute evaluation. As an application of multi-attribute evaluation, land quality evaluation is the process analyzing land quality differentia based on multiple attributes associated with evaluation region.

GIS combined with MCE can eliminate the limitation existing in using GIS singly and achieve measurable evaluation of land quality.

2. THE PRINCIPLE OF GIS COMBINED WITH MCE

GIS combining with MCE is a process combining multiple source information associating with geographical location to expose relations hiding in resource information. Main content of it are as follows:

(i) Confirming the goal of evaluation;

(ii) Choosing criteria: to confirm effectual factors using for evaluating, which founds a standard for judgment;

(iii) Confirming weights of factors: to confirm comparative weightiness of factors;

(iv) Calculating factor's value diffused in every spatial cells: to diffuse attribute value of factors to make them distributing spatially and numerically based on GIS;

(v) Calculating combined value of spatial cells: to achieve combined value of spatial cells through numeric overlay calculation;

(vi) Classing spatial cells according to their combined value to achieve evaluation results.

3. KEY TECHNOLOGY OF APPLYING GIS AND MCE TO EVALUATE LAND QUALITY

3.1 Founding Criteria and Calculating Weights

The criteria used in MCE must be fit for analyzing the particularity of the question, requirements of which as follows:

(i) Which can mapping the aim of the evaluation synthetically;
(ii) Which can be expressed measurably;
(iii) Factors in criteria are independent.

Weight of factors is used to judge the weightiness of factors in criteria. Generally, methods used to calculate it as follows:

(i) Subjective method: such as Delphi Method, Pairing Comparison, Priority Order Decision and so on.
(ii) Objective method: such as Principal Component Analysis, Entropy Technology, Multiobjective Programming Model and so on.

All of methods above are based on determinate information, but uncertainties widely exist in objective world, human cognition and data processing (Shi Wen-Zhong, 2006). To fit for practical application, it was request that relations of spatial data must be expressed under the condition of lacking determinate and accurate information. Rough set is a novel mathematical tool for dealing with uncertainties and imprecision of geographic information, which can be used to represent uncertainties in multi-levels spatial knowledge (Deng Min, 2006). The follow is a method based on condition and decision attributes of a given knowledge expression system to acquire criteria using useful data and useful characters existing in observed data.

A given knowledge system can be expressed as $S=<U,C,D,V,F>$,where U is universe, $A= C\cup D$ is the set of attributes on U, C and D are the sets of condition and decision attributes, V is the value set of A, F: $U\times A\rightarrow V$ is information mapping from $U\times A$ to V.

In the attributes set $A=C\cup D$, $x\in X$ is an attribute. Resolution enhancement of X after x is joined into X can be taken as a weightiness of x belonging to X. The resolution is more enhanced, x for X is more important.

The concept of the attribute weightiness degree (Miao Duo-qian, 2002) in rough sets can be used to evaluate the importance of the factor.

Let $X \subseteq C$ be a attribute set and $x\in C$ be a attribute. Marking Sigx as the attribute weightiness degree of x for X:

$$\mathrm{Sin}(x) = 1 - \frac{|X \cup \{x\}|}{|X|},\quad |X| = |IND(X)| = \sum_{i=1}^{n} |X_i|^2$$

Where IND(X)is undistinguished relations on X.

The attribute weightiness degree represents the attribute's affecting on decision-making and reflects the different factors' importance in evaluation, so that objective weight of attribute in evaluation can be confirmed.

3.2 Spatial Cell Plotting

The object of plotting spatial cells is to form units to process spatial analysis and evaluation based on GIS. To fit for GIS, spatial cells must possess two characters as follows at least (Paul A. Longley, 2004):

(i) The model of plotting spatial cells should be repeated infinitely so that it can be used to map views in any scale.

(ii) The model of plotting spatial cells can be divided into infinitely subtle ones so that spatial elements in any resolution can be created and represented.

Generally, there are two models to plot spatial cells: shapely model and unshapely model. Shapely model refers to plot spatial cells in equal shapes, such as Archimedes spatial plotting. Unshapely model refers to plot spatial cells according to boundaries of some irregularly distributing spatial attributes, such as administration boundaries, Voronoi map and so on.

In examples of this article, to evaluate land quality, the shapely model and the unshapely model were applied synthetically: for attributes only associated with region, such as local economic development, land output and so on, plotted spatial cells according to administration boundary and land block boundary; for attributes that not only associated with distance but also with orientation, such as effect degree of towns, advantage degree of roads and so on, plotted spatial cells according to regular grid. Furthermore, attributes value of all kinds of spatial elements were integrated to a new sets of appropriate grids. In process of plotting spatial cells, units of interiorly equal and exteriorly dissimilar were obtained by controlling grid size. These units composed of basal cells to carry on spatially analyzing to achieve land quality evaluation.

3.3 Diffusing Spatial Attribute Value

Factors associated with land quality are being forms of scattering dot, line and field in space, but these factors affect more area around them. In order to represent above affection, spatial interpolation was employed, such as Kriging Interpolation, spline interpolation, polynomial interpolation. In example of the article, linearly attenuate function and exponential attenuate function were employed, which as follows:

(i) Linearly attenuate function: $e_{ij} = f_i \times (1 - r)$;

(ii) Exponential attenuate function: $e_{ij} = (f_i)^{1-r}$.

where, e_{ij} is diffused attribute's functional value of factor i in cell j; f_i is attribute's functional value of factor i; r is distance from j cell to i factor's center.

3.4 Numeric Overlay Calculation of Spatial Data

Overlay analysis is basal function of GIS, including arithmetical overlay, logic Overlay, fuzzy Overlay and so on. There is some limitation in general overlay analysis, such as not comparatively considering weightiness of layers, no way to confirm threshold value of consecutively distributing variable and not be able to evaluate all of spatial elements roundly, for example, not be able to order results of evaluation.

Because of limitations above, general overlay analysis can not meet for the requestment of evaluating land quality based on GIS and MCE.

Numeric overlay calculation of spatial data refers to that taking attribute value of spatial object as consecutive variable, and transforming them into some new value which falling into some special numeric zone, after these, factors' new value is weighting and linearly combined. All of these are based on spatial cell plotting and spatial attribute value diffusing. Spatial cell plotting divides consecutive space into equal spatial cells and spatial attribute value diffusing transforms consecutive attribute value to discrete dot value which belongs to different spatial cells. After depositing data above in date table, numeric overlay calculation can be carried on as follows:

$$S = \sum_{i}^{n} D_i \times P_i$$

where, S is value of evaluation, D_i is measurable value of factor, P_i is weight of factor.

4. EXAMPLE

We applied technology above to plot out requisition blocks of land in TongShan country. It succeeded to plot land in TongShan country into some blocks in equal condition and achieve grades of land requisition blocks. Table 1 describes factors in criteria and their weights, Figure 1 is demonstration of diffusing spatial attribute value, Figure 2 describes plotting spatial cell in evaluation region, Figure 3 is grades map of land requisition blocks.

Table 1. Factors in criteria and their weights

Factors in criteria	Land type	Land output	Land condition	effect degree of towns	advantage degree of roads	Local economic development
Weight of factors	16.22	19.68	15.11	17.32	14.23	17.44

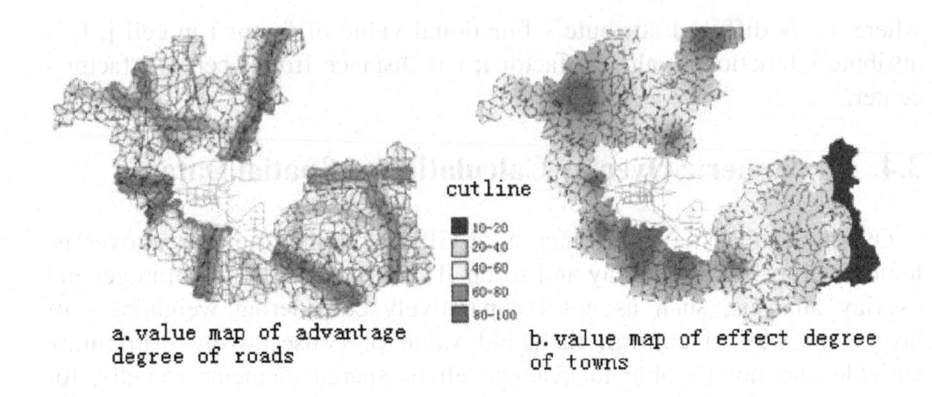

Figure 1. The demonstration of diffusing spatial attribute value

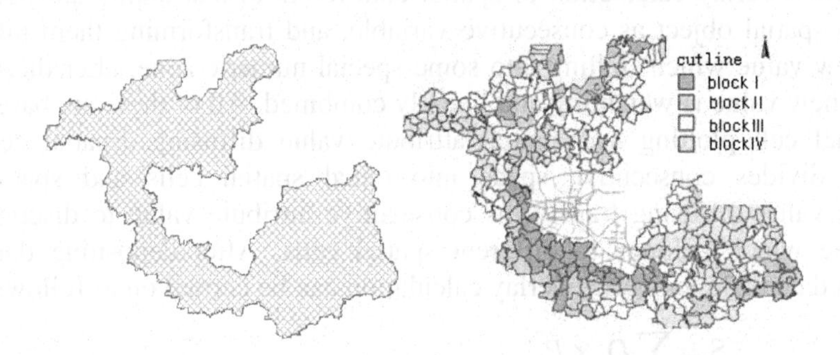

Figure 2. Plotting Spatial cell in evaluation region

Figure 3. Grades map of land requisition blocks

5. CONCLUSIONS

Land quality holds many connotations, such as land environment quality, land economy quality and land management quality, which are involved in extensive spatial data and social statistical data. To combine multiple source information associating with land quality, functions of GIS must be expanded. MCE increased the decision ability of GIS.

Key technology of evaluating land quality with GIS and MCE was studed and is summed up as follows:

i. confirming evaluation criteria;
ii. plotting spatial cells;
iii. diffusing spatial attribute value;
iv. combining spatial data in numeric way.

Technology above was applied in plotting out requisition blocks of land in Tongshan country. The research indicated that the technology of GIS combined with MCE can integrate multiple source information associating with land quality commendably, and achieve measurable land quality evaluation.

ACKNOWLEDGEMENTS

Authors are grateful to the Land Management Bureau of Tongshan Country for its financial support for this research.

REFERENCES

Leng SY, et al. New Progresses of International Study on Land Quality Indicators (Lqis). Acta Geographica Sinica, 1999, 54(2): 177-186.

Su BY, Land Qualities Evaluation of Cangwu. Journal of Nanjing Normal University (Natural Science Edition), 2001, 24(2): 104-110.

Jin XB, et al. Land Quality Evaluation of Land Reclamation Project Based on Farmland Gradation. Journal of Nanjing Forestry Universty (Natural Sciences Edition), 2006, 30: 93-97.

Zhang L, et al. Preliminary Study on Concept and Measurement of Land Quality. Journal of Nanjing University (Natural Sciences), 2004, 40: 378-390.

Tian SM, et al. Land Quality Evaluation Aided By Gis – A Case Study In Quzhou County, Hebei Province. Journal of China Agricultrural Resources and Regional Planning, 2002, 123(13): 16-20.

Shi CY, et al. Evaluation of Land Quality Based on Gis – A Case Study on Paddy Field in Suzhou. Acta Pedologica Sinica, 2001, 38(3): 248-256.

Nie QH, et al. The Quality Evaluation and Site Assessment of Agricultural Land Based on GIS: A Case Study in Liangxiang Town, Fangshan District, Beijing. Scientia Geographica Sinica, 2000, 20(4): 307-314.

Wang XF, et al. Preliminary Study on Landscape-Scale Land Quality Quantitative Assessment in Luan Mining Area. Journal of Soil and Water Conservation, 2007, 21(1): 197-201.

Xu MJ, et al. A Preliminary Study of Land Quality Indicators in Local Scale. System Sciences and Comprehensive Studies in Agriculture. 2006, 122(13): 201-206.

Wang J, et al. Review on Spatial Variability and Scale Effects of Land Quality. Progress in Geography. 2005, 24(4): 28-36.

Li XY, et al. Land Quality Evaluation of Land Reclamation Project Based on Farmland Gradation. Journal of Nanjing Forestry University. 2006, 20(2): 99-104.

Ouyang JL, et al. Agricultural Land Evaluation Based on Different Crop Planting at County Level and Household Behavior. Resources Science. 2003, 125(15): 58-66.

Ye JA, GIS and Programming Support System. Science Press, 2006.

Paul AL, et al. Geographic Information System. Publishing House of Electronics Industry, 2004, 464-473.

Shi WZ, Indeterminacy Theory of Spatial Data and Spatial Analysis, Science Press, 2006, 7-11.

Deng M, et al. Rough-Set Representation of GIS Data Uncertainties with Multiple Granularities. Acta Geodaetica et Cartographica Sinica, 2006, 35(1): 64-69.

Miao DQ, et al. The Calculation of Knowledge Granulation and its Application. Systems Engineering – Theory & Practice, 2002, 48-56.

Paul AL, et al. Geographic Information System. Publishing House of Electronics Industry, 2004, 474-495.

Xu JH, Mathematic Methods in modern geography. Higher education Press, 2002, 35.

Zheng HL, Rough Sets Theory and Applications. Chong Qing university publisher, 1996, 44-55.

SOIL WATER CONTENT FORECASTING BY SUPPORT VECTOR MACHINE IN PURPLE HILLY REGION

Wei Wu[1,2], Xuan Wang[1,2], Deti Xie[1,3], Hongbin Liu[1,3,*]

[1] *Chongqing Key Lab of Digital Agriculture, Southwest University, Chongqing, China, 400715*

[2] *College of Computer and Information, Southwest University, Chongqing, China, 400715*

[3] *College of Resources and Environment, Southwest University, Chongqing, China, 400715*

* *Corresponding author, Address: College of Resources and Environment, Southwest University, 216 Tiansheng Road, Beibei, Chongqing, 400715, P. R. China, Tel: +86-23-68251069, Fax: +86-23-68250444, Email: wuwei_star@163.com*

Abstract: Soil water distribution and variation are helpful in predicting and understanding various hydrologic processes, including weather changes, rainfall/runoff generation and irrigation scheduling. Soil water content prediction is essential to the development of advanced agriculture information systems. In this paper, we apply support vector machines to soil water content predictions and compare the results to other time series prediction methods in purple hilly area. Since support vector machines have greater generalization ability and guarantee global minima for given training data, it is believed that support vector machine will perform well for time series analysis. Predictions exhibit good agreement with actual soil water content measurements. Compared with other predictors, our results show that the SVMs predictors perform better for soil water forecasting than ANN models. We demonstrate the feasibility of applying SVMs to soil water content forecasting and prove that SVMs are applicable and perform well for soil water content data analysis.

Keywords: support vector machines, soil water content, statistical learning, prediction, forecasting, time series

Wu, W., Wang, X., Xie, D. and Liu, H., 2008, in IFIP International Federation for Information Processing, Volume 258; Computer and Computing Technologies in Agriculture, Vol. 1; Daoliang Li; (Boston: Springer), pp. 223–230.

1. INTRODUCTION

Soil water, though very small in volume, provides valuable information for water resources planning and management. Soil water transferring model and prediction are important in agriculture, hydrology, and meteorology. In agriculture, accurate forecasts of future soil water conditions can be helpful in water quality monitoring, irrigation scheduling, and yield forecasting. In hydrology, information about soil water is required for understanding rainfall/runoff generation processes and managing water resources. Similarly, in meteorology, soil water measurements can be helpful for modeling surface/atmospheric interactions. Over the past several years, various attempts have been made to produce soil water content estimates by using different statistical models, such as Artificial Neural Networks (ANNs) (Liou et al., 2001; Baghdadi et al., 2002; Liu et al., 2004; Liu et al., 2003) and Auto Regression (AR) (Liu et al., 2003).

Recently, support vector machines (SVMs), developed by Vapnik and his co-workers (Vapnik et al., 1996), have become a very active research area with machine learning. Motivated by statistical learning theory, SVMs have been successfully applied to solve various problems, among others in data mining, classification, regression, density estimation and times series prediction (Cao et al., 2001; Flake et al., 2002; Mukherjee et al., 1997; Zhang 2003; Vapnik, 1995). SVMs implement the structural risk minimization principle and Vapnik-Chervonenkis (VC) dimension. Based on this principle, SVMs achieve an optimum structure by striking a right balance between the empirical error and the VC-confidence interval. Eventually, this results in better generalization performance than other models (Vapnik et al., 1996). Furthermore, the SVMs deal with non-linear tasks by mapping the input space into high dimensional feature spaces, and then use a kernel function instead of high dimensional inner product. This means that the solution of SVMs is unique, optimal and absent from local minima (Vapnik, 1995; Xian et al., 2005; Yan et al., 2000).

In this paper, we use support vector machines to predict the soil water content and show that SVMs are applicable to soil water prediction and outperforms ANN in purple hilly area. The SVMs are based on statistical learning theory and can be used to predict a quantity forward in time based on the results of "training" that uses past data. This paper is organized in six sections. Section 2 presents the study area and data preparation. Machine learning scheme and kernel function are described in Section 3. Section 4 and Section 5 include the prediction methodology, error measurements and experiment results. Concluding remarks are presented in Section 6.

2. SITE AND DATA DESCRIPTION

The study site is located in Southwest University in Chongqing (long. 106° 26'E, lat. 30° 26'N). The hilly with an area 0.1hm^2 was selected to test the approach presented herein. Climatic conditions are semi-tropical wet with a mean annual temperature of 18.3°C and a precipitation of about 1150.7mm/ year. The soil is classified as purple.

All the data in this study are taken from the Hydrology Experiment. Monitoring has been done about every 5 days, from Dec, 2001 to Apr, 2004. Samples were collected from a depth of 30 to 40 cm, placed in a can, sealed and transported to a lab, where they were weighed before and after oven drying.

3. SUPPORT VECTOR MACHINE

3.1 Theory of SVMs in regression approximation

Compared to other neural network regressors, there are three distinct characteristics when SVMs are used to estimate the regression function. First of all, SVMs etimate the regression using a set of linear functions which are defined in a high dimensional space. Secondly, SVMs carry out the regression estimation by risk minimization where the risk is measured using Vapnik's e-insensitive loss function. Thirdly, SVMs use a risk function consisting of the empirical error and a regularization term which is derived from the structure risk minimization principle.

Given a set of data points $G = \{(x_i, d_i)\}_i^n$ (x_i is the input vector, d_i is the desired value and n is the total number of data patterns), SVMs approximate the function using the following:

$$y = f(x) = w\phi(x) + b \tag{1}$$

where $\phi(x)$ is the high dimensional feature space which is non-linearly mapped from the input space ***X***. The coefficients w and b are estimated by minimizing

$$R_{SVMs}(C) = C\frac{1}{n}\sum_{i=1}^{n} L(d_i, y_i) + \frac{1}{2}\|w\|^2 \tag{2}$$

$$L\varepsilon(d, y) = \begin{cases} |d - y| - \varepsilon & |d - y| \geq \varepsilon \\ 0 & otherwise \end{cases}$$

In the regularized risk function given by Eq. (2), the first term $C(1/n)\sum_{i=1}^{n} L\varepsilon(d_i, y_i)$ is empirical error (risk), and measured by function $L\varepsilon$. The second term $0.5\|w\|^2$, is the regularization term. C is referred to as the regularized constant and is determines the trade-off between the empirical risk and the regularization term. Increasing the value of C will result in the relative importance of the empirical risk with respect to the regularization term to grow. ε is called the tube size and it is equivalent to the approximation accuracy placed on the training data points.

To obtain the estimations of w and b, Eq. (2) is transformed to the primal function given by Eq. (3) by introducing the positive slack variables ζ_i and ζ_i^* as follows:

$$\text{Minimize } R_{SVMs}(w, \zeta^{(*)}) = \frac{1}{2}\|w\|^2 + C\sum_{i=1}^{n}\left(\zeta_i + \zeta_i^*\right) \tag{3}$$

$$\text{Subjected to } \begin{aligned} & d_i - w\phi(x_i) - b_i \le \varepsilon + \zeta_i \\ & w\phi(x_i) + b_i - d_i \le \varepsilon + \zeta_i^*, \zeta_i^* \ge 0 \end{aligned}$$

Finally, by introducing Lagrange multipliers and exploiting the optimality constraints, the decision function given by Eq. (1) has the following explicit form Zhang (2003):

$$f(x, a_i, a_i^*) = \sum_{i=1}^{n}(a_i - a_i^*)K(x, x_i) + b \tag{4}$$

The detail computation procedure can be found in Vapnik (1996) and Flake et al. (2002).

3.2 Kernel function

$K(x_i, x_j)$ is defined as kernel function. The value of the kernel is equal to the inner product of two vectors $\boldsymbol{X}_i$ and $\boldsymbol{X}_j$ in the feature space $\varphi(x_i)$ and $\varphi(x_j)$, that is, $K(x_i, x_j) = \varphi(x_i) * \varphi(x_j)$. The elegance of using the kernel function is that one can deal with feature spaces of arbitrary dimensionality without having to compute the map $\varphi(x)$ explicitly. Any function satisfying Mercer's condition (Flake et al., 2002) can be used as kernel function. The typical examples of kernel function are the polynomial kernel $K(x, y) = (x*y+1)^d$ and the Gaussian kernel $K(x, y) = \exp(-1/\delta^2(x-y)^2)$ where d is the degree of polynomial kernel and δ^2 is the bandwidth of the Gaussian kernel. The kernel parameter should be carefully chosen as it implicitly defines the structure of the high dimensional feature space $\varphi(x)$ and thus controls the complexity of the final solution.

4. METHODOLOGY

4.1 Prediction methodology

Suppose the current time is t, we want to predict $y(t+l)$ for the future time $t+l$ with the knowledge of the value $y(t-n)$, $y(t-n+1)$,..., $y(t)$ for past time $t-n$, $t-n+1$, ..., t, respectively. The prediction function is expressed as:

$$y(t+l) = f(t, l, y(t), y(t-1), \ldots, y(t-n)) \quad (5)$$

4.2 Error measurements

In addition, we examine the soil water content time series of different prediction methods. Relative Mean Errors (RME), Root Mean Square Error (RMSE) and Coefficient of Variation (CV) are applied as performance indices. The computational methods are described as follows:

$$RME = \frac{1}{n}\sum_{i=1}^{n}\left|\frac{y_i - \hat{y}_i}{y_i}\right|$$

$$RMSE = \sqrt{\frac{\sum_{i=1}^{n}(y_i - \hat{y}_i)^2}{n}} \quad (6)$$

$$CV = \frac{\sqrt{\sum_{i=1}^{n}(y_i - \hat{y}_i)^2 / n}}{\bar{y}}$$

where n, y, $\hat{y}$ and $\bar{y}$ represents the number of test data, the observation value, the predicted value and the average of the observation samples, respectively.

All the mentioned procedures above are carried out in MATLAB 6.5.

5. RESULTS

The experiment samples are classified into two groups. The first one includes the soil water content every 5 days, and the monthly average soil water measurements consist in the second group. For the first group, we use data from the first 160 samples as the training set and use the last 21 samples as our testing set. Meanwhile, for the second group, the first 21 samples are selected as the training set and the last 8 samples as the testing set.

SVMs combined with three kinds of kernel functions, that are linear (SVM_Linear), polynomial (SVM_Poly) and radius base function (SVM_Rbf), are applied to soil water content forecast. In addition, the application presented in this paper is also compared to a very well known machine learning tool used in hydrology, Artificial Neural Networks (ANNs). The ANN model was set up in such a way as to have one output node, one hidden layer, and one input layer. Moreover, a tan sigmoid function was employed in the hidden layer neurons and a linear transfer function was used at the output node. ANNs use a least squares loss function, unlike SVMs, which use an ε-insensitive loss function as a fitness measure. Moreover, the ANN predictions are not stable and depend on the averages from various network initializations, which may give a different result each time a model is trained. On the other hand, the SVM results are stable and unique.

The experimental results are summarized in Table 1 and Table 2. It is observed that the SVMs regression outperform the ANN predictors, especially for the monthly average soil water content forecasting. The results in Table 2 show that the radius base function SVM predictor reduces relative mean errors, the root mean squared errors and the coefficient of variation than those achieved by the linear, polynomial SVMs and the ANN predictor.

Furthermore, this experiment examines the errors greater than 5% which are produced by SVMs and ANN prediction methods for the different soil water content. For the first group, the results in Table 1 shows that 30.17% portion of total errors produced by linear and polynomial function SVM predictor are greater than 5% whereas radius base function SVM and ANN predictor produce the number of 78.83% and 70.76% to total errors which are over the 5% RME threshold. Moreover, as far as the monthly average soil water content concerned, Table 2 shows that the bad parts (the portion of errors exceed 5%) of the linear, polynomial function SVMs and ANN prediction errors occupy 10.90%, 11.12% and 9.87% of total errors, respectively. However, for the radius base function SVM predictor, there are only 1.33% of the errors belongs to the bad portion.

Table 1. The forecast performance of different models for the soil water content every 5days

Model	RME	RMSE	CV	Proportion
SVM_Linear	1.0573	0.2200	-2.0938	30.17%
SVM_Poly	1.0573	0.2200	-2.0938	30.16%
SVM_Rbf	58.9678	3.2695	-31.1202	78.83%
ANN	52.4897	2.9878	-28.4386	70.76%

Table 2. The forecast performance of different models for the monthly average soil water content

Model	RME	RMSE	CV	Proportion
SVM_Linear	3.7727	0.3023	-3.4011	10.90%
SVM_Poly	3.7790	0.3024	-3.4017	11.12%
SVM_Rbf	1.2057	0.2440	-2.7448	1.33%
ANN	3.2326	0.2891	-3.2523	9.87%

Moreover, the performance indices of the radius base function SVM for the monthly average soil water content forecasting improve remarkably than that of the soil water content every 5 days. It is can be conferred that to some extent the noise in monthly average samples reduce to less than those of every 5 days.

6. CONCLUSION

Support vector machine and support vector regression have demonstrated their success in time-series analysis and statistical learning. However, little work has been done for soil water content analysis. Prior knowledge of soil water content behavior can not only help in better management and understanding of hydrological systems but also result in improved forecasting, especially for agricultural basins. In this paper we examine the feasibility of applying support vector regression to soil water content time series prediction in purple hilly area. The application presented here uses measured soil water data to predict future soil water. After numerous experiments, we propose a set of SVR parameters that can predict soil water content time series very well. The results show that the SVM predictor significantly outperforms the other baseline predictors. This evidences the applicability of support vector regression in soil water content analysis.

ACKNOWLEDGEMENTS

Authors are thankful to Chongqing key lab of digital agriculture, Southwest University for financial help. We are also thankful to the anonymous referees for their helpful comments in improving the manuscript.

REFERENCES

Baghdadi N, Gaultier S, King C. Retrieving Surface Roughness and Soil Moisture From Synthetic Aperture Radar (SAR) Data Using Neural Networks. Canadian Journal of Remote Sensing, 2002, 5: 701-711

Cao L J, Tay F E H. Financial forecasting using support vector machines. Neural Comput. 2001, Appl(10): 184-182

Flake G, Lawrence S. Efficient SVM regression training with SMO. Machine Learning. 2002, 1/3: 271-290

Liou Y A, Liu S F, Wang W J. Retrieving Soil Moisture From Simulated Brightness Temperature by a Neural Network, IEEE Transactions on Geoscience Remote Sensing, 2001, 8: 1662-1672

Liu H B, Wu W, Wei C F, Xie D T. Comparison of autoregression and neural network models for soil water content forecasting, Transactions of the CSAE, Vol. 19, No. 4, 2003, pp. 33-36 (in Chinese)

Liu H B, Wu W, Wei C F, Xie D T. Soil water dynamics simulation by autoregression models, Journal of Mountain Science, 2004, 1: 121-125 (in Chinese)

Liu H B, Wu W, Wei C F. Study of soil water forecast with neural network, Journal of Soil and Water Conservation, 2003, 5: 59-62 (in Chinese)

Mukherjee S, Osuna E, Girosi F. Nonlinear prediction of chaotic time series using support vector machine. In: NNSP'97: Neural Networks for Signal Processing VII: Proceedings of the IEEE Signal Processing Society Workshop, Amelia Island, FL, USA. 1997, 243-254

Thissen U, van Brackel R. de Weijer A P, Melssen W J, Buydens L M C. Using support vector machines for time series prediction. Chemometrics and Intelligent Laboratory Systems. 2003, 69: 35-49

Vapnik V B, Golowich S E, Smola A J. Support vector method for function approximation, regression estimation, and signal processing. Advances in Neural Information Processing Systems. 1996, 9: 281-287

Vapnik V N. The nature of statistical learning theory. New York: Springer, 1995

Xian G L, Luo X C, Xiao Y F. Statistics learning theory and support vector machine. China science and technology information. 2005, 12: 178-181 (in Chinese)

Yan H, Zhang X G, Li Y D. Support vector machine methods in pattern recognition of sedimentary facies. Computing techniques for geophysical and geochemical exploration. 2000, 2: 158-164

Zhang G P. Time series forecasting using a hybrid ARIMA and neural network model. Neurocomputing. 2003, 50: 159-175

AN EFFICIENT NEW METHOD ON ACCURATELY ESTIMATING GALILEO VPL

Ying Guo [1,2], Xiushan Lu [1]
[1] *Geoinformation Science & Engineering College, Shandong University of Science and Technology, Qingdao, China 266510*
[2] *The Geomatics and Applications Laboratory, Liaoning Technical University, Liaoning, FuXin, China 123000*

Abstracts: The future GALILEO system will use multi-frequency data to process, which may eliminate preferably the influence of ionosphere relay. So the VPL estimation is mainly to consider how to efficiently use the satellite Signal-In-Space (SISE). To efficiently utilize SISE in GNSS and to rationally estimate the user Vertical Protection Level, it is the key to ensure normal application of GNSS and user's safety. For the investigation requirement of GALILEO VPL algorithm. Based on the character which SISE influences on users' positioning, this paper provides an estimating model of equivalent range error of SISE which influences users efficiently. On the bases of the integrity theory of VPL and the principle of synthetizing and assessing the uncertainty of measurement errors, it also gives a new algorithm to estimate VPL rationally to provide references for the development and the application of the integrity theories of GNSS.

Keywords: Galileo; integrity theory; SISE; VPL

1. INTRODUCTION

The satellite broadcast ephemeris error is an important influence on the reliability of user's positioning and the safety of user's life. It includes the orbit error and the clock error, which is named as Signal-In-Space (SISE). GNSS enhances the reliability of the satellite broadcast ephemeris by predicting and monitoring SISE integrity (Veit Oehler, 2004; Carlos Hernandez Medel, 2002, 2005).

Guo, Y. and Lu, X., 2008, in IFIP International Federation for Information Processing, Volume 258; Computer and Computing Technologies in Agriculture, Vol. 1; Daoliang Li; (Boston: Springer), pp. 231–238.

User applies the integrity parameters of SISE, which are broadcasted by GNSS, and local integrity information, which affects on the broadcast of satellite signal, for example, delay correction and correct error of ionosphere and troposphere, receiver noise, multi-path effect and so on, to estimate user vertical protection level (VPL) to predict and monitor the safety state of navigation system (J.T. Wu, 2002; Curtis A. Shively, 2005; Stephane, 2002). If it is estimated more than nominal range, VPL includes some false data, and reduces the integrity function of system; If it is less, it leads to some mislead data to influence normal application of system. So the ideal VPL algorithm is the key to accurate application of the navigation system.

The appearance of the integrity theory of the GALILEO system, which has its own characteristics, improves the integrity theory of GNSS. Because the integrity principle of the GALILEO SISE varies from that of EGNOS/WAAS UDRE (User Different Range Error), which is used to name as satellite ephemeris error, the VPL algorithm of the two systems are also different. At present, the estimating method of GALILEO VPL investigated by ESA is kept secret. Based on the algorithm of EGNOS/WAAS VPL, Stephane and Bruno in reference (Stephane, 2002) provide two selective methods under the assumed condition that the satellite exists in the two cases.

Aim at the need of the development and the application of the integrity theories of GALILEO system and GNSS, this paper studies the estimating formula of equivalent range error of SISE which influences users efficiently. On the bases of the integrity theory of VPL, and the principle of synthetizing and assessing the uncertainty of measurement errors, it further provides a new method of estimating VPL and illustrates its reliability.

2. PRINCIPLE OF ESTIMATING VPL

Based on the linear model of user position estimation

$$y = G\,x + \varepsilon \tag{1}$$

Here

The vector y is the vector of observed differences between the receiver measured pseudo range, and the distance between satellite and initial known approximate user position. G is the geometry metric consisting of one row for each monitored satellite in the view. The 4-vector x represents user position (east, north, up) and clock bias, ε represents the variance metric of error term.

According to the least-square law, weight of remain-covariance and assessed principle of measurement uncertainty to compute VPL. Its primary model is

$$VPL = K_{vpl}\sqrt{\sum_{i=1}^{n} S_{up,i}^{2}\cdot\sigma_i^2} \tag{2}$$

$$\sigma^2{}_i = \sigma^2_{i,SISE} + \sigma^2_{i,ion} + \sigma^2_{i,trop} + \sigma^2_{i,mp+noise}$$

$$S = (G^T W^{-1} G)^{-1} G^T W \tag{3}$$

$$S = (S_{east}, S_{north}, S_{up}, S_{clock})$$

Where

K_{vpl} is the coefficient factor which is estimated by VPL, W is the error-covariance weighting metric, σ_i represents the total mean error which No. i satellite arouses user position, including $\sigma_{i,SISE}$, which is the mean error of SISE, $\sigma_{i,ion}$ which is the remain mean error of ionosphere delay correction, $\sigma_{i,trop}$ which is the remain mean error of troposphere delay correction, and $\sigma_{i,mp+noise}$ which the mean error of receiver noise and multi-path effect.

Because the error between SISE and other local factors are relative independent, their effects are considered as the simple accumulation of each part. And SISE of formula is computed in term of mean-error.

3. EXISTING VPL ALGORITHM

Formula (2) is the primary model of VPL, and is the usual one that is used to estimate VPL of EGONS and WAAS. Based on that model, both Stephane and others provide two selective algorithms:

Algorithm 1: the VPL estimating algorithm based on WAAS/EGNOS (Stephane, 2002)

$$VPL = \sum_{i_{SAT}} \left| M\ (3, i_{sat}) * SISE \right| + K_{VPL}\sqrt{\left(G^T W_{u-local}^{-1} G\ \right)^{-1}_{(3,3)}} \tag{4}$$

Here

$$M\ = \left(G^T W^{-1} G\ \right)^{-1} G^T W^{-1} \tag{5}$$

$$W_{u-local} = \begin{bmatrix} \sigma_{u-local,RX}(1)^2 & 0 & \cdots & 0 \\ 0 & \cdots & \cdots & 0 \\ 0 & 0 & \sigma_{u-local,RX}(i)^2 & 0 \\ 0 & 0 & \cdots & \sigma_{u-local,RX}(n)^2 \end{bmatrix}$$

$$\sigma^2_{u-local,RX}(i_{sat}) = \sigma^2_{i,ion} + \sigma^2_{i,trop} + \sigma^2_{i,mp+noise}$$

Where

$W_{u-local}$ represents the error-covariance metric of user location, $\sigma_{u-local}(i_{sat})$ represents the total error of No. i satellite, except SISE.

The assumed condition of formula (4) is that all satellites in the view are in the state of WUL, the probability seldom occurs. And the errors which both SISE and others effect on are estimated respectively first, and then added together. So, Stephane studies another algorithm again.

Algorithm 1: the estimating model of GALILEO VPL is expressed as

$$VPL = \max_{i_{sat}} \left\{ \left| M_u(3, i_{sat}) * SISE \right| + K_{VPL} \sqrt{\left(G_u^T W_u^{-1} G_u \right)^{-1}_{(3,3)}} \right\} \qquad (6)$$

$$W_u = \begin{bmatrix} \sigma_{u,RX}(1)^2 & 0 & \cdots & 0 \\ 0 & & \cdots & \cdots & 0 \\ 0 & & 0 & \sigma_{u-local,RX}(i)^2 & 0 \\ 0 & & 0 & \cdots & \sigma_{u,RX}(n)^2 \end{bmatrix}$$

$$\sigma^2_{u,RX}(i) = \sigma_i^2$$

This method is assumed that only one satellite is located in WUL. SISE of this satellite is solely computed the influence, others are created weighting metric together with local factors. In the formula of the weighting, SISE is computed in term of mean-error. But broadcast SISE is estimated with uncertainty principle, and user can not obtain its error. So, this method need to be fully studied and illustrated its reliability.

4. A NEW ESTIMATING VPL METHOD

In the user navigation system, there is only certain probability of the both assumed conditions of the above two algorithms and it can not generalize all cases better. Algorithm 1 respectively considers SISE and other effects, and the final effect is considered to be adding up simply to increase the error of positioning; Algorithm 2 needs to provide mean-error of SISE.

Based on the satellite positioning law, the uncertainty law of measurement error, and the integrity theory of VPL, the equivalent range error of SISE accurately effecting user is analyzed, and a method of rational computing VPL is provided, which is as follows.

4.1 Accurate Equivalent Range Error of SISE

In GALILEO system, SISE, which includes predicting parameter SISA and monitoring parameter SISMA, is the value that the satellite locates in

WUL. So, at one epoch, in the location of user receiver, all satellites in the view can not be located in WUL at the same time, and can not ensure that only one or some locate in WUL. But, at this time, both satellite position and SISE may be obtained, and the position of the user receiver may be approximately computed. So, the angle between the SISE vector direction and the one form satellite to receiver may be computed, furthermore, $SISE_{AC}$ which is the equivalent range error of SISE effecting on user may computed, its estimating program is following.

Because satellite clock error has the same influence on all users in view, but the influence that the satellite orbit error affects varies with different location. These two errors will be respectively considered.

(S_x, S_y, S_z) are assumed as predicted vector parameter of satellite orbit error, b_S is its scale, and its unit vector is $e_{SV} = (e_x, e_y, e_z)$, b_{clock} represented satellite clock error, the vector form satellite to user is $e_{SV-u} = (e_{x,u}, e_{y,u}, e_{z,u})$. So the angle ϑ between e_{SV} and e_{SV-u} is computed by

$$\vartheta = arc\,(e_{SV} \bullet e_{SV-u}) \tag{7}$$

b_{AC} is accurate equivalent range error created by satellite orbit error, is expressed by

$$b_{AC} = b_S \cdot \cos\vartheta \tag{8}$$

Assumed that the observable vector G^i is expressed by

$$G^i = (G_1^i, G_2^i, G_3^i, 1) \tag{9}$$

So b_{AC} is

$$b_{AC} = \sqrt{(G_1^i \cdot S_x)^2 + (G_2^i \cdot S_Y)^2 + (G_3^i \cdot S_Z)^2} \tag{10}$$

That is to say, $SISE_{AC}$ is represented as

$$SISE_{AC} = \sqrt{b_{AC}^2 + b_{clock}^2} \tag{11}$$

VPL is computed with $SISE_{AC}$ without considering the position of satellite at some epoch. This estimating method possesses preferable applicability and reliability.

4.2 Estimating VPL

Based on the relative principles, this paper introduces into $SISE_{AC}$, estimating VPL model of GALILEO system is expressed by

$$VPL = \sqrt{\sum_{i-1}^{n} \left(S_{up,i}^{2} \cdot \sigma_{G,i}^{2} \right)} \tag{12}$$

Here

$$\sigma_{G,i}^{2} = SISE_{AC}^{2} + K_{vpl}^{2} \cdot \left(\sigma_{iono,i}^{2} + \sigma_{trop,i}^{2} + \sigma_{mpth,i}^{2} \right) \tag{13}$$

5. CASE STUDY

It is rational in theory to estimate GALILEO VPL based on equivalent range error accurately effect on SISE, furthermore to illustrate feasibility and reliability by simulating data.

5.1 Data Source

(1) Ephemeris datum: They are broadcast ephemeris of Sept. 19–21 in 2005, which are updated with the software of BERNESE GPS, and the epoch interval is five minutes.

(2) SISE: Satellite orbit errors are obtained by updating and comparing broadcast ephemeris and precise one, and then transformed into the value of WUL. Clock error is described as 0.3m[5]; the interval of epoch is r5 minutes, both SISE and σ_{SISE} are estimated with existing methods.

(3) Positioning error: The relative parameters are obtained with formulas which are provided to compute remain of ionosphere and troposphere in the document [5]. The vertical grid ionosphere error is 0.7m.

5.2 Simulated Result and Analysis

According to the above-simulated data, the results which are estimated by algorithm 1 and the new illustrated method are described in the following figures. Where, series 1 represents the computed value with algorithm 1, and series 3 is estimated protect error.

By analyzing and comparing with different values (fig. 1), it can conclude that in the normal condition, VPL estimated by the new method is obviously less than that computed by algorithm 1, and greatly more than the real error

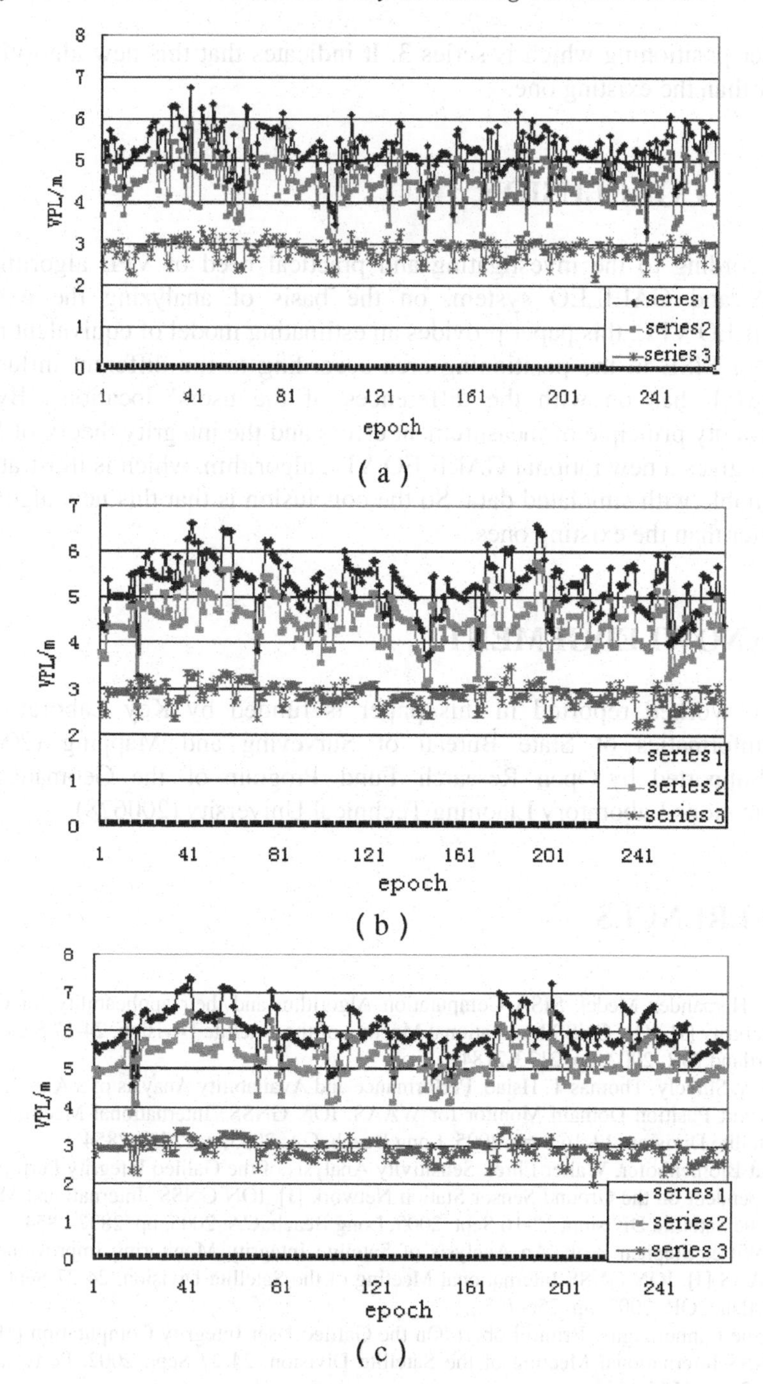

Fig. 1. The comparison of the VPL and VPE estimated by different algorithm

of user positioning which is series 3. It indicates that this new algorithm is better than the existing one.

6. CONCLUSION

According to the investigating and practical need of VPL algorithm in GNSS and GALILEO system, on the basis of analyzing the existing GALILEO VPL, this paper provides an estimating model of equivalent range error of SISE in the positioning area according to the different influences that SISE has on with the differences of the users' locations. By the uncertainty principle of measurement errors and the integrity theory of VPL, it also gives a new rational GALILEO VPL algorithm, which is illustrated to be reliable with simulated data. So the conclusion is that this new algorithm is better than the existing ones.

ACKNOWLEDGEMENT

The worked reported in this paper is funded by Key Laboratory of Geo-informatics of State Bureau of Surveying and Mapping (200607) and Supported by Open Research Fund Program of the Geomatics and Applications Laboratory Liaoning Technical University (200628).

REFERENCES

Carlos Hernandez Medel. SISA Computation Algorithm and their applicability for Galileo Intergrity [J]. ION GNSS International Meeting of the Satellite Division, 24-27 Sept. 2002, Portland, OR. 2002, pp. 2173-2184.

Curtis A. Shively, Thomas T. Hsiao. Performance and Availability Anaysis of a Aimple Local Airport Position Domain Monitor for WAAS. ION GNSS International Meeting of the Satellite Division, 13-16 Sept. 2005, Long Beach, CA. 2005, pp. 2837-2854

Helmut Blomenhofer, Walter Ehret. Sensitivity Analysis of the Galileo Integrity Performance Dependent on the Ground Sensor Station Network [J]. ION GNSS International Meeting of the Satellite Division, 13-16 Sept. 2005, Long Beach, CA. 2005, pp. 2837-2854

J. T. Wu & Stephen Peak. An Analysis of Satellite Integrity Monitoring Improvement for WAAS [J]. ION GNSS International Meeting of the Satellite Division, 24-27 Sept. 2002, Portland, OR. 2002, pp. 756-765.

Stephane Lannelongue, Bruno Lobert. On the Galileo User Integrity Computation [J]. ION GNSS International Meeting of the Satellite Division, 24-27 Sept. 2002, Portland, OR. 2002, pp. 1547-1556.

Veit Oehler. The Galileo Integrity Concept [J]. ION GNSS 17th International Meeting of the Satellite Division, 21-24 Sept. 2004, Long Beach, CA. 2004, pp. 604-615.

CONTROL SYSTEMS IN OUR DAILY LIFE

Irsan Suryadi[1], Rohani Jahja Widodo[2]
[1] *Electrical Engineering Department – Maranatha Christian University, Bandung – Indonesia irsan.ss@gmail.com & 0022088@eng.maranatha.edu*
[2] *President of MASDALI (Masyarakat Sistem Kendali Indonesia) rjwidodo@yahoo.com & masdali@yahoo.com*

Abstract: This paper presents development and applications of Control Systems (CS). Several characteristics of CS can be linked to human behavior. CS can "think" in the sense that they can replace to some extent, human operation. CS can distinguish between open-loop and closed-loop CS and it is a concept or principle that seems to fundamental in nature and not necessarily peculiar to engineering. In human social and political organizations, for example, a leader remains the leader only as long as she is successful in realizing the desires of the group. CS theory can be discussed from four viewpoints as: an intellectual discipline within science and the philosophy of science, a part of engineering, with industrial applications and Social Systems (SS) of the present and the future. In global communication, developed countries and developing countries should build several attractive and sound symbiosis bridges, to prevent loss of universe balances. CS applications have social impacts not only in developed countries but also in developing countries. A new work force strategy without denying the existing of CS is established by retooling the work forces, thus the challenges of social impacts could be answers wisely and would be bright opportunities to improve human standards of living.

Keywords: CS, SS, social impacts, human standards of living

1. INTRODUCTION

Control System (CS) is used to control position, velocity, and acceleration is very common in industrial and military applications. They have been given the special name of servomechanisms. With all their many advantages, **CS** in advertently act as an oscillator. Through proper design,

Suryadi, I. and Widodo, R.J., 2008, in IFIP International Federation for Information Processing, Volume 258; Computer and Computing Technologies in Agriculture, Vol. 1; Daoliang Li; (Boston: Springer), pp. 239–251.

however, all the advantages of **CS** can be utilized without having an unstable system.

Several characteristics of **CS** can be linked to human behavior. **CS** can "think" in the sense that they can replace to some extent, human operation. **CS** can distinguish between open-loop and closed-loop **CS** and it is a concept or principle that seems to fundamental in nature and not necessarily peculiar to engineering. In human social and political organizations, for example, a leader remains the leader only as long as she is successful in realizing the desires of the group. **CS** theory can be discussed from four viewpoints as: an intellectual discipline within science and the philosophy of science, a part of engineering, with industrial applications and social problems of the present and the future. In global communication, developed countries and developing countries should build several attractive and sound symbiosis bridges, to prevent loss of universe balances. **CS** applications have social impacts not only in developed countries but also in developing countries.

2. HUMAN CONTROL SYSTEMS (CS)

The relation between the behavior of living creatures and the functioning of **CS** has recently gained wide attention. Wiener implied that all systems, living and mechanical are both information and **CS**. Wiener suggested that the most promising techniques for studying both systems are Information theory and **CS** theory.

Several characteristics of **CS** can be linked to human behavior. **CS** can "think" in the sense that they can replace to some extent, human operation. These devices do not have the privilege of freedom in their thinking process and are constrained by the designer to some predetermined function. Adaptive **CS**, which is capable of modifying their functioning in order to archive optimum performance in a varying environment, have recently gained wide attention. These systems are a step closer to the adaptive capability of human behavior.

The human body is, indeed, a very complex and highly perfected adaptive **CS**. Consider, for example, the human actions required to steer an automobile. The driver's object is to keep the automobile traveling in the center of a chosen lane on the road. Changes in the direction of the road are compensated for by the driver turning the steering wheel. The driver's object is to keep the input (the car's desired position on the road) and the input (the car's desired position on the road) as close to zero as possible.

Fig. 1. Illustrates the block diagram of the **CS** involved in steering an automobile. The error detector in this case is the brain of the driver. This in turn activates the driver's muscles, which control the steering wheel. Power amplification is provided by the automobile's steering mechanism, which controls the position of the wheels. The feedback element represents the human's sensors (visual and tactile). Of course, this description in very crude, any attempt to construct a mathematical model of the process should somehow account for the adaptability of the human being and the effects of learning, fatigue, motivation, and familiarity with the road.

CS process as that found in physical, biological, and social systems. Many systems control themselves through information feedback, which shows deviations from standards and initiates changes. In other words, systems use some of their energy to feedback information that compares performance with a standards and initiates corrective action.

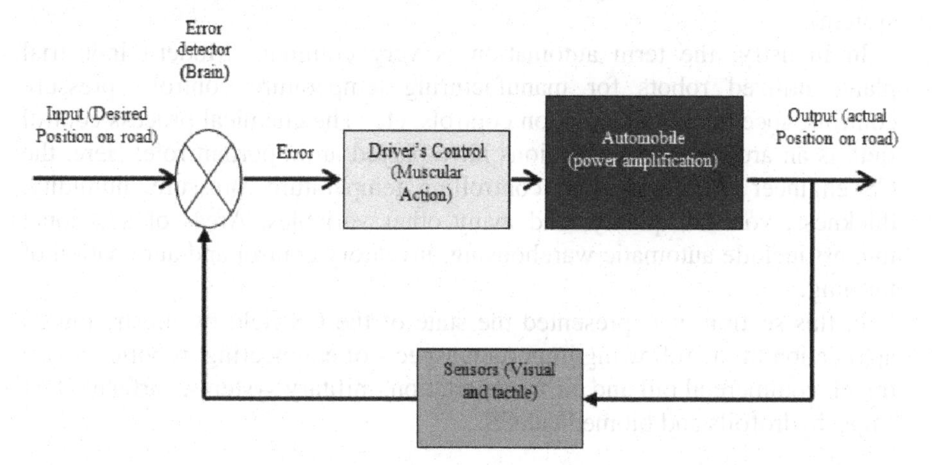

Fig. 1. Steering of an automobile: a feedback control system involving human capability

The house thermostat is a system of feedback and information control. When the house temperature falls below the preset level, an electric message is sent to the heating system, which is then activated. When the temperature increases and reaches the set level, another message shut off the heater. This continual measurement and turning on and off the heater keeps the house at the desired temperature. A similar process activates the air-conditioning system. As soon as the temperature exceeds the preset level, the air-conditioning system cools the house to the desired temperature. Likewise, in the human body, a number of CS control temperature, blood pressure, motor reactions, and other conditions. Another example of feedback is the grade a student receives on a midterm test. This is intended, of course, to give the student information about how he or she is doing and, if performance is less than desirable, to send a signal suggesting improvement.

3. CONTROL SYSTEMS (CS) IN PHYSICAL SYSTEMS

CS is to be found in almost every aspect of our daily environment. In the home, the refrigerator utilizes a temperature-control system. The desired temperature is set and a thermostat measures the actual temperature and the error. A compressor motor is utilized for power amplification. Other applications of control in the home are the hot-water heater, the central heating system, and the oven, which all work on a similar principle. We also encounter **CS** when driving our automobile. **CS** is used for maintaining constant speed (cruise control), constant temperature (climate control), steering, suspension, engine control, and to control skidding (antiskid system).

In industry, the term automation is very common. Modern industrial plants utilized robots for manufacturing temperature controls, pressure controls, speed controls, position controls, etc. The chemical process control field is an area where automations have played an important role. Here, the CS engineer is interested in controlling temperature, pressure, humidity, thickness, volume, quality, and many other variables. Areas of additional interest include automatic warehousing, inventory control and automation of farming.

In this section, it is presented the state of the CS field by illustrating its application in the following important aspects of engineering: robotics, space travel, commercial rail and air transportation, military systems, surface effect ships, hydrofoils and biomedical CS.

4. CS IN INDUSTRIAL ROBOTS

A new work force strategy without denying the existing of **CS** is established by retooling the work forces, thus the challenges of social impacts could be answers wisely and would be bright opportunities to improve human standards of living.

In manufacturing plants in several countries, there has been a large-scale increase in the usage of **CS** for industrial robots, which are programmable machine tools designed in many cases to accomplish arduous or complex tasks. Although there has been some opposition to the fact that robots often replace human labor, but the trend toward robotics will continue, and on balance, be beneficial to the national economy.

CS developments of the last decade are likely to have as profound a potential impact on productivity, labor markets, working conditions, and the quality of life in the developed countries as the introduction of robot into workplace. The conclusions can be reached based on four factors:

First, the estimate of the number of jobs that could be performed by is relatively small.

Second, almost all of these workers would be spared forced unemployment because of retraining and in some cases the job attrition that occurs through normal retirement.

Third, total employment is a function of real economic growth: robots can have a positive effect on real economic growth and, therefore, a positive effect on total employment.

Fourth, in 10 years, retraining programs can adequately shift displaced workers to new careers. In fact, the main challenge posed to policymakers by increased use of robots is not unemployment but need for retraining.

History shows that labor-saving techniques have led to improved living standards, higher real wages, and employment growth. In large measure, the robotics revolution is merely a continuation of a centuries-long trend that has resulted in enormous material progress. Protection from job loss can come through retraining programs. Working condition and job safety will improve as robots take over dangerous and undesirable forms of works.

Technological advances in computers and microprocessors are increasing the sophistication of robots, giving them some "thinking" capacity that increase potential uses. The key to usage in such area as office work depends in large part on the ability to develop "intelligent" robots capable of performing tasks that vary somewhat over time. Some industry observers believe breakthroughs may allow for extensive introduction of robotics in non-manufacturing tasks within a few years.

Three important dimension of the growth of robotics are subjects to economic analysis:

The first is the determinants of the magnitude of the growth of the robotics industry.

The second is the impact robotics unemployment.

The third is the impact that robots will have on wages, profits and prices.

There are two reasons for the growth in the use of robots, one related primarily to supply and the second primarily to demand. In the long run, robots will be increasingly utilized because the cost of traditional labor-intensive techniques is rising over time, while the cost of the capital-intensive robotic techniques is falling relative to prices generally. These costs decline because the technological advances in robotics lower the capital costs of robots per unit output.

On the other hand, some government policies may speed robotic introduction. For example, where environmental regulations lower worker productivity or raise capital costs associated with the traditional **CS**, the traditional technique cost line will shift upward, advancing the date at which robotic adoption becomes profitable.

Important changes in the composition of the work force have occurred over the past four decades and, in some opinions, even more massive changes lie ahead as many thousands of low-skill jobs are eliminated while at the same time large numbers of new jobs are created to meet the demands of technological advance. If serious employment displacement effects are to be avoided, development of broad-scale training programs in which the private sector plays a key role, in concert with various governmental bodies will be required.

5. CONTROL SYSTEM (CS) THEORY

CS theory is needed for obtaining the desired motion or force needed; sensors for vision and computers for programming these devices to accomplish their desired tasks. Just what is **CS** theory? Who or what is to be controlled and by whom or by what, and why is it to be controlled? In a nutshell, **CS** theory, sometimes called **automation**, **cybernetics** or **systems theory** is a branch of **applied mathematics** that deals with the design of machinery and other engineering systems so that these systems work, and work better than before.

As an example, consider the problem of controlling the temperature in a cold lecture hall. This is a standard engineering problem familiar to us all. The thermal system consists of the furnace as the heating source, and the room thermometer as the record of the temperature of the hall. The external environment we assume fixed and not belonging to the thermodynamic system under analysis. The basic heating source is the furnace, but the control of the furnace is through a thermostat, the thermostat device usually contains a thermometer to measure the current room temperature and a dial on which we set the desired room temperature. The control aspect of thermostat is that it compares the actual and the desired temperatures at each moment and then it sends an electric signal or control command to the furnace to turn the fire intensity up or down. In this case, the job of the **CS** engineer is to invent or design an effective thermostat.

Let us next look at a **CS** problem from biology. Parts of the world are being overrun by an increasing population of rats. Here the system consists of the living population of rats and the environmental parameters that affect that population. The natural growth of the rat population is to be controlled to near some desired number, say, and zero. Here the job of the **CS** engineer is to build a better mouse-trap.

From this viewpoint **CS** theory does not appear too sinister. On the other hand, it does not seem too profound. So let us elaborate on the structure of **CS** theory to indicate the reasons why many scientists believe this subject is important. To organize these ideas, it shall be discussed **CS** theory from two viewpoints:

(1) as an intellectual discipline within science and the philosophy of science,

(2) as a part of engineering, with industrial applications and as a force in the world related to social problems of the present and the future.

5.1 CS Theory is a Teleological Science

First consider the philosophical position of the discipline of **CS** theory. Within the framework of metaphysics, **CS** theory is a teleological science. That is, the concepts of **CS** involve ideas such as purpose, goal-seeking and ideal or desirable norms. These are terms of nineteenth century biology and psychology, terms of evolution will and motivation such as were introduced by Aristotle to explain the foundations of physics, but then carefully exorcized by Newton when he constructed a human geometric mechanics. So CS theory represents a synthesis of the philosophies of Aristotle and Newton showing that inanimate deterministic mechanisms can function as purposeful self-regulating organisms. Recall how the inanimate thermostat regulates the room temperature towards the agreed ideal.

5.2 CS is an Information Science

Another philosophical aspect of **CS** theory is that it avoids the concepts of energy but, instead, deals with the phenomenon of information in physical systems. If we compare the furnace with the thermostat we note a great disparity of size and weight. The powerful furnace supplies quantities of energy: a concept of classical physics. Thus **CS** theory rests on a new category of physical reality, namely information, which is distinct from energy or matter. Possibly, this affords a new approach to the conundrum of mind versus matter, concerning which the philosophical remarked,

"What is matter? - Never mind
What is mind? - No matter"

But what are the problems, methods and results of **CS** theory as they are interpreted in modern mathematical physics or engineering? In this sense **CS** theory deals with the inverse problem of dynamical systems. That is, suppose we have a dynamical system, for example many vibrating masses interconnected by elastic springs. Such a dynamical system is described mathematically by an array of ordinary differential equations that predict the evolution of the vibrations according to Newton's laws of motion.

5.3 CS in Nature and in Humanity

CS can distinguish between open-loop and closed-loop **CS** and it is a concept or principle that seems to fundamental in nature and not necessarily peculiar to engineering. In human social and political organizations, for example, a leader remains the leader only as long as she is successful in realizing the desires of the group. If she fails, another is elected or by other means obtains the effective support of the group. The system output in this case is the success of the group in realizing its desires. The actual success is measured against the desired success, and if the two are not closely aligned, that is, if the error is not small, steps are taken to ensure that the error becomes small. In this case, control may be accomplished by deposing the leader. Individuals act in much the same way. If our study habits do not produce the desired understanding and grades, we change our study habits so that actual result becomes the desired result.

Because **CS** is so evident in both nature and humanity, it is impossible to determine when **CS** was first intentionally used. Newton, Gould, and Kaiser' cite the use of feedback in water clocks built by the Arabs as early as the beginning of the Christian era, but their next references is not dated until 1750. In the year Markle invented a device for automatically steering windmill into the wind, and this was followed in 1788 by Watt's invention of the fly-ball governor for regulation of the steam engine.

5.4 Development of CS Theory

However, these isolated inventions cannot be construed as reflecting the application of any **CS** theory. There simply was not theory although at roughly the same time as Watt was perfecting the fly-ball governor both La Place and Fourier were developing the two transform methods that are now so important in electrical engineering and in **CS** theory in particular. The final mathematical background was laid by Cauchy, with his theory of the complex variable. It is unfortunate that the readers of this text cannot be expected to have completed a course in complex variables, although some may be taking this course at present. It is expected, however, that the reader is versed in the use of the La Place transform. Note the word use. Present practice is to begin the use of La Place transform methods early in the engineering curriculum so that, by the senior year, the student is able to use the La Place transform in solving linear, ordinary differential equation with constant coefficients. But no until complex variables are mastered does a student actually appreciate how and why the La Place transform is so effective. In this text we assume that the reader does not have any knowledge of complex variables but does have a working knowledge of La Place transform methods. Although the La Place transform is the mathematical language of the **CS** engineer, in using this book the reader will not find it necessary to use more transform theory.

Although the mathematical background for **CS** engineering was laid by Cauchy (1789-1857), it was not until about 75 years after his death that an actual **CS** theory began to evolve. Important early papers were "Regeneration Theory," by Nyquist, 1932, and "Theory of Servomechanisms," by Haze, 1934. World War II produced an ever-increasing need for working **CS** and thus did much to stimulate the development of a cohesive **CS** theory. Following the war a large number of linear **CS** theory books began to appear, although the theory was not yet complete. As recently as 1958 the author of a widely used control text stated in his preface that "**CS** are designed by trial and error."

In the early 1960s a new **CS** design method referred to as modern **CS** theory appeared.

This theory is highly mathematical in nature and almost completely oriented to the time domain. Elementary conventional linear system and subsystem modeling (again using computer tools) and approaches to loop design: a comparison of traditional and "intelligent" techniques; notions of self-tuning and adaptive controllers.

5.5 Establishment of Standards

Because plans are the yardsticks against which managers devise controls, the first step in the **CS** process logically would be to establish plans. However, since plans very in detail and complexity, and since managers cannot usually watch everything, special standards are established. Standards are, by definition, simply criteria of performance. They are selected points in an entire planning program at which measure of performance are made so that managers can receive signals about how things are going and thus do not have to watch every step in the execution of plans.

There are many kinds of standards. Among the best are verifiable goals or objectives, as suggested in the discussion of managing by objectives.

5.6 Measurement of Performance

Although such measurement is not always practicable, the measurement of performance against standards should ideally be done on a forward-looking basis so that deviations may be detected in advance of their occurrence and avoided by appropriate actions. The alert, forward-looking manager can sometimes predict probable departures form standards. In the absence of such ability, however, deviations should be disclosed as early as possible.

If standards are appropriately drawn and if means are available for determining exactly what subordinates are doing, appraisal of actual or expected performance is fairly easy. But there are many activities for which it is difficult to develop accurate standards, and there are many activities that are hard to measure. It may be quite simple to establish labor-hour standards for the production of a mass-produced item, and it may be equally simple to measure performance against these standards, but if the item is customs-made, the appraisal of performance may be a formidable task because standards are difficult to set.

5.7 Correction of Deviations

Standards should reflect the various positions in an organization structure. If performance is measured accordingly, it is easier to correct deviations. Managers know exactly where, in the assignment of individual or group duties, the corrective measure must be applied.

Correction of deviations is the point at which control can be seen as a part of the whole system of management and can be related to the other managerial functions. Managers may correct deviations by redrawing their plans or by modifying their goals. This is an exercise of the principle of navigational change or they may correct deviations by exercising their organizing function through reassignment or clarification of duties. They may correct, also, by that ultimate re-staffing measure-firing or, again, they may correct through better leading-fuller explanation of the job or more effective leadership techniques.

6. SUMMARY

CS is to be found in almost every aspect of our daily environment. The human body is, indeed, a very complex and highly perfected adaptive **CS**. Consider, for example, the human actions required to steer an automobile. **CS** is highly multidisciplinary, with issues and features that are distinct from those of other branches of engineering. These issues are numerous and subtle, and often the most important aspects depend on the seemingly most insignificant details. Historically, the subject has advanced by employing abstraction to extract principles that are potentially applicable to a broad range of applications. Unfortunately, this abstraction often obscures the practical ramifications of important ideas. A more concrete approach to the subject an rejuvenate and reinvigorate education in this exciting and important area of technology. Wiener suggested that the most promising techniques for studying both systems are information theory and **CS** theory.

CS process as that found in physical, biological, and social systems. Likewise, in the human body, a number of **CS** control temperature, blood pressure, motor reactions, and other conditions. The human body is, indeed, a very complex and highly perfected adaptive **CS**. Consider, for example, the human actions required to steer an automobile.

Let us next look at a **CS** problem from biology. Parts of the world are being overrun by an increasing population of rats. Here the system consists of the living population of rats and the environmental parameters that affect that population. The natural growth of the rat population is to be controlled to near some desired number, say, and zero. Here the job of the **CS** engineer is to build a better mouse-trap.

CS is to be found in industry, the term automation is very common. Modern industrial plants utilized robots for manufacturing temperature controls, pressure controls, speed controls, position controls, etc. The chemical process control field is an area where automations have played an important role. The philosophical position of the discipline of **CS** theory within the framework of metaphysics, **CS** theory is a teleological science. That is, the concepts of **CS** involve ideas such as purpose, goal-seeking and ideal or desirable norms. Another philosophical aspect of **CS** theory is that it avoids the concepts of energy but, instead, deals with the phenomenon of information in physical systems. In this sense **CS** theory deals with the inverse problem of dynamical systems. Because **CS** is so evident in both nature and humanity, it is impossible to determine when **CS** was first intentionally used. Newton, Gould, and Kaiser' cite the use of feed-back in water clocks built by the Arabs as early as the beginning of the Christian era, but their next references is not dated until 1750. In the year of 1788 by Watt's invention of the fly-ball governor for regulation of the steam engine.

In the early 1960s a new **CS** design method referred to as modern **CS** theory appeared. This theory is highly mathematical in nature and almost completely oriented to the time domain. Elementary conventional linear system and subsystem modeling (again using computer tools) and approaches to loop design: a comparison of traditional and "intelligent" techniques; notions of self-tuning and adaptive. Because plans are the yardsticks against which managers devise controls, the first step in the **CS** process logically would be to establish plans. However, since plans very in detail and complexity, and since managers cannot usually watch everything, special standards are established. Standards are, by definition, simply criteria of performance. Although such measurement is not always practicable, the measurement of performance against standards should ideally be done on a forward-looking basis so that deviations may be detected in advance of their occurrence and avoided by appropriate actions.

The alert, forward-looking manager can sometimes predict probable departures form standards. In the absence of such ability, however, deviations should be disclosed as early as possible. Standards should reflect the various positions in an organization structure. If performance is measured accordingly, it is easier to correct deviations. Managers know exactly where, in the assignment of individual or group duties, the corrective measure must be applied. Correction of deviations is the point at which control can be seen as a part of the whole system of management and can be related to the other managerial functions. Managers may correct deviations by redrawing their plans or by modifying their goals.

CS applications have social impacts not only in developed countries but also in developing countries. A new work force strategy without denying the existing of **CS** is established by retooling the work forces, thus the challenges of social impacts could be answers wisely and would be bright opportunities to improve human standards of living.

REFERENCES

Activities Report, Department of Automatic Control and Systems Engineering, The University of Sheffield, UK, January 1999 to December 1999.

B.J. Habibie, B.J. Science, Technology and Nation Building, Vol. I & II, Technology Idonesia & The Agency for The Assessment and Application of Technology, Jakarta, 1991.

Charles L. Philips, Royce D. Harbor. Feedback Control Systems. Prentice Hall International, Inc, 2000.

Dato' Lee Yee Cheong, The President of World Federation of Engineering Organizations, Keynote Address: Current Activities of the World Federation of Engineering Organizations, CAFEO-21, Yogyakarta, Indonesia, 22-23 October 2003.

Douglas Bullis. Write a Winning Business Plan. Time Books Int., Singapore, 1996.

G.F. Franklin, J.D. Powell, A. Emami-Naeni: Feedback Control of Dynamic Systems, 3rd edition, Addison-Wesley Publishing Co., Reading Massachusetts, 1994.

Heinz Weihrich, Harold Koontz. Management A Global Perspective. McGraw-Hill, Singapore, 1994.

P. Choate, Retooling The American Work Force, Northeast-Midwest Institute, 1982. [2] S. Zuboff, Computer-Mediated Work: A New World, The President and Fellows of Harvard College, 1982.

R.C. Dorf, R.H. Bishop: Modern Control Systems, 9th edition, Prentice-Hall, Upper Sadle River, New Jersey, 2001.

R.J. Widodo, Automatic Control for Reducing Energy Consumption and Improving Energy Conservation, (CAFEO-10), Manila Phillipines, 5-6 November 1992.

R.J. Widodo, R.J. Control Education at Bandung Insitute of Technology, (CAFEO-11), Singapore, 18-19 November 1993.
R.J. Widodo, Development of Control Applications in Electrical Power Systems, PSDC'95, Bandung Institute of Technology, Bandung, 14-16 March 1995.
R.L. Phillips, R.D. Harbor & R.J. Widodo, Feedback Control Systems, Third Edition, (R.J. Widodo Alih Bahasa, Sistem Kontrol Lanjutan), Prenhallindo, Jakarta 1998.
Ricard A. D'aveni, Robert Gunther. Harper - Competition Managing the Dynamics of Strategic Maneuvering. Macmillan Inc, New York, 1994.
Wayne C. Booth, Gregory G. Coulomb & Joseph M. Williams, The Craft of Research, The University of Chicago Press, Chicago & London, 1995.
www.—, European Control Conference - ECC 2001.
www.—, NSF/CSS Workshop on New Direction in Control Engineering Education, Coordinated Science Laboratory, University of Illinois at Urbana-Campagne October 2-3, 1998. k 1975.
www.—, The 4th Asian Control Conference - ASCC Singapore 2002.
www.iasted.com, IASTED International Conference on Control and Applications CA, USA, 2002.

APPLICATION OF GIS AND GEOSTATISTICS TO CHARACTERIZE SPATIAL VARIATION OF SOIL FLUORIDE ON HANG-JIA-HU PLAIN, CHINA

Zhengmiao Xie*, Jing Li, Weihong Wu

Institute of Environmental Science and Engineering, Hangzhou Dianzi University, Hangzhou 310018, P. R. China

** Corresponding author, Tel: +86-571-86878583; Fax: +86-571-86919155; Email address: zhmxie@sina.com*

Abstract: Spatial variability of soil fluoride in the plough layer (0–20cm) of paddy soil from Hang-Jia-Hu Plain of Zhejiang Province in China was studied using geostatistical analysis and GIS technique. The results of Semivariograms analysis showed that two forms of soil fluoride were correlated in a given spatial range, and total fluoride (T-F) was controlled by intrinsic factors of parent material, relief and soil type, whereas water-soluble fluoride (Ws-F) was greatly affected by extrinsic factors such as fertilization and soil management. Kriging method was applied to estimate the unobserved points and their distribution maps were obtained, which indicated that the concentrations of soil T-F and Ws-F had a close relationship with parent material, pH value, organic matter, cation exchange capacity content and soil texture. The main contents distribution of T-F and Ws-F were 200–300mg kg^{-1}, 0.5–1.0mg kg^{-1} in the studied area, respectively. And what is more, the range of T-F contents in soil was as low as less than 100 mg kg^{-1} in Yu-hang area accounting for 23.7% area scale. The range of fluoride contents in the soils from central and eastern parts of Hang-Jia-Hu Plain was higher than that from the western part. The accumulation of fluoride contents in soil was lower in the whole studied area, suggesting that local fluoride epidemic such as dental caries due to lack of fluoride should be prevented by using fluoride-containing toothpaste.

Keywords: Fluoride, Geostatistics, GIS, Spatial variation, Kriging method

Xie, Z., Li, J. and Wu, W., 2008, in IFIP International Federation for Information Processing, Volume 258; Computer and Computing Technologies in Agriculture, Vol. 1; Daoliang Li; (Boston: Springer), pp. 253–266.

1. INTRODUCTION

Fluoride (F) is regarded as an essential trace element, primarily because of its benefits to dental health and its suggested role in maintaining the integrity of bone (Underwood and Mertz 1987; Wheeler and Fell 1983). A small amount of fluoride is beneficial in the prevention of dental caries. It has also been used to treat osteoporosis (Fung *et al.* 1999). However, excessive fluoride is built up in the apatite ctystals in teeth and bones and reduces their solubility (Fejerskov *et al.* 1994; Fung *et al.* 1999). Traditionally, excessive fluoride has been connected with high intake of fluoride through drinking water and food (Marian *et al.* 1997; Singh and Dass 1993), but water and food take up fluoride from soil and accumulate it in human body finally though food web (Marian *et al.* 1997), unbalance of fluoride in the human body can cause diseases of teeth and bones (Fung *et al.* 1999; Xie *et al.* 2001). Therefore, increasing attention should be paid to soil fluoride quality.

In recent years, geostatistics has been proved as a successful method to study distributions of soil heavy metals (Atteia *et al.* 1994; Steiger *et al.* 1996; White *et al.* 1997; Yu *et al.* 2001; Romic and Romic 2003) and soil nutrient (Tsegaye and Robert 1998; Fisher *et al.* 1998; Cahn *et al.* 1994). However, most of the previous geostatistical studies were focused on data at small scale (Wang 1999; Goovaerts 1999). With the development of Geographical Information System (GIS), GIS can integrate attribute data with geographical data of system variables, which makes the application of geostatistics technique for large spatial scale more convenient (Steiger 1996; Bai *et al.* 1999; Mendonca Santos *et al.* 2000). Geostatistics and GIS are becoming indispensable in characterizing and summarizing spatial information in large regions to provide quantitative support to decision and policy making for soil, agricultural and natural resources management (Wang 1999; Guo *et al.* 2000; Liu *et al.* 2003).

However, the papers on soil fluoride was less relatively, and that according to local fluoride epidemic has evolved in response to high soil concentrations of fluoride in contaminated sites (Horner and Bell 1995), so previous study had stressed on the fluoride contents in contaminated soil too and there is minimum information on spatial distributions of soil fluoride in paddy fields, and less information in a large scale (Geeson *et al.* 1998; Li *et al.* 2004). In this paper, we applied geostatistics combined with GIS to (1) analyze the spatial dependency and explain the variation mechanism of soil fluoride in the paddy soils; (2) map the spatial distribution of soil fluoride in the soil; (3) provide information for environmental monitoring and evaluation in Hang-Jia-Hu Plain.

2. MATERIALS AND METHODS

2.1 Study area

Hang-Jia-Hu Plain is in the center of Hangzhou-Jiaxing-Huzhou in the North of Zhejiang Province in the southeast of China, including Jiaxing, Pinghu, Tongxiang, Haining, Jiashan, Haiyan, Hangzhou, Yuhang, Deqing, Changxing, Anji, Huzhou and part of Lin'an, 13 regions altogether. It boders the Hangzhou Gulf, a part of the East China Sea (Fig. 1). It is an coastal and lacustrine alluvial plain with an altitude of 3–7m above sea level. The climate of the area is subtropical humid monsoonal climate and has abundant rain capacity, the average rang of temperature and rainfall density are 16–19°C and 1200–1300mm every year, respectively. It is densely dotted with drainage ditches that form a network waterway. Rice (Oryza Satiya) has been dominant crop in the studied area, a large part of the area has acidic paddy soil.

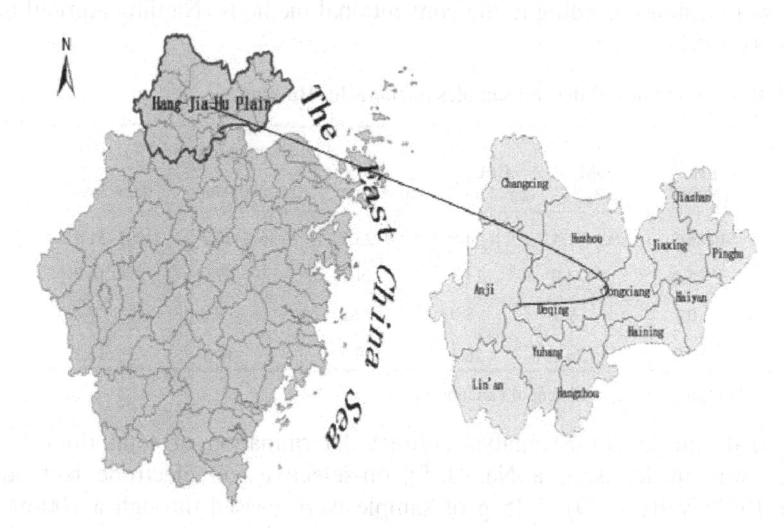

Fig. 1. Location of the study area

2.2 Soil sampling and analysis

Soil samples were taken from over 460 locations within Hang-Jia-Hu Plain in April 2000. Sampling points are presented in Fig. 2. Because there are more low mountains and hills in Anji, Lin'an and Deqing region in the western of Hang-Jia-Hu Plain, the sampling points were sparse comparatively.

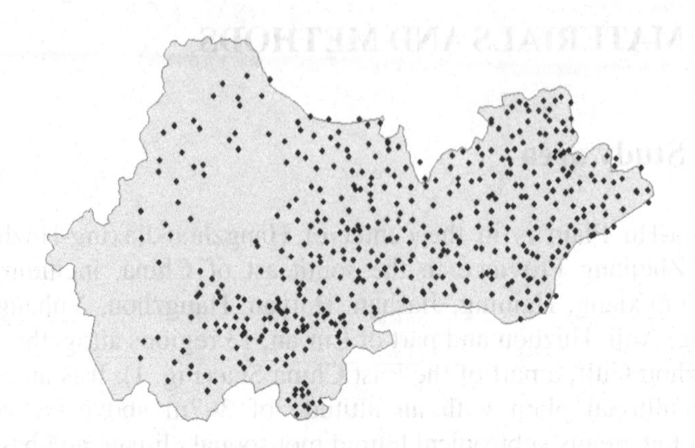

Fig. 2. Distribution of sampling sites in Hang-Jia-Hu Plain

Some characteristics of the soils are presented in Table 1. Pipette method was used to determine the particle composition according to the International System. Soil pH value, organic matter (OM) and cation exchange capacity (CEC) were tested according to the conventional methods (Nanjing Agricultural University 1981).

Table 1. Basic properties of the soil samples in Hang-Jia-Hu Plain

	pH (H_2O)	OM (g kg^{-1})	CEC (cmol kg^{-1})	Particle composition (%)		
				<0.002mm	0.002–0.05mm	0.05–2mm
Range	4.1–8.3	10.9–61.4	5.3–24.8	4.9–23.5	36.4–80.2	5.1–51.4
Mean	5.8	34.0	14.8	14.2	69.0	16.7
S.D.	0.7	9.1	3.9	2.8	7.0	7.6
CV%	12.0	26.7	26.3	19.9	10.1	45.2

S.D., standard deviation; CV, coefficient of variation

For total fluoride (T-F) analysis, direct determination of total fluoride in samples was made using a NaOH fusion-selective ion electrode technique (Baker 1972; Villa 1979). 0.25 g of sample were passed through a 100-mesh sieve and put into a 50-ml nickel crucible 3.0 ml of 16.75 mol/l NaOH solution, then placed in an oven at 150°C for 1 h until dry. The crucible with sample was then placed in a Muffle furnace. The temperature was raised to 600°C. The samples were fused after 30 min at this temperature. After the samples had been removed from the muffle furnace and cooled, 5 ml of de-ionised water was added and then heated slightly to facilitate the dissolution of the fused soil with sodium hydroxide. Then, 4 ml of concentrated HCl were added slowly, with stirring, to adjust pH to 8–9 (checked with pH test paper). The cooled dissolved sample was transferred to a 50-ml volumetric flask, diluted with distilled water to volume, and then filtered through dry filter paper. The filtrate was used for the determination of fluoride. A reagent blank was produced.

Water-soluble fluoride (Ws-F) was extracted by ratio of 1:5 soil to water. Ten grams of soil passed through a 60-mesh sieve and 50-ml distilled water were placed in 60-mesh sieve and 50-ml distilled water were placed in polyethylene bottles, shaken for 0.5 h on an end-over-end shaker, then centrifuged. Then fluoride levels were measured by ion-specific electrode potentiometer (Xie *et al.* 2003).

2.3 Data analysis

Distribution of soil fluoride element were characterized using the Kolmogrov-Smirnov (K-S) test for goodness-of-fit (Sokal and Rohlf 1981) to ensure that the distribution could be validly applied to data sets. Descriptive statistics, including the range, mean, standard deviation (SD) and coefficient of variation (CV), were determined for each set of data using the statistical analysis system (SPSS) and correlation analysis was conducted.

Geostatistics were used to estimate and map soils in unsampled areas (Goovaerts 1999). Among the geostatistical techniques, Kriging is a linear interpolation procedure that provides a best linear unbiased estimation for quantities that vary in space. The procedure provides estimates at unsampled sites. Kriging's estimates are calculated as weighted sums of the adjacent sampled concentrations. That is, if data appear to be highly continuous in space, the points closer to those estimated receive higher weights than those farther away (Cressie 1990).

In this study, spatial patterns of soil fluoride element were determined using the geostatistical analysis. Semivariograms were developed to evaluate the degree of spatial continuity of soil fluoride element among data points and to establish a range of spatial dependence for each soil soil fluoride element using GS+3.1 software. Information generated through variogram was used to calculate sample-weighted factors for spatial interpolation by a Kriging procedure (Isaaks and Srivastava 1989) using Arc/Info8.1 and Arcview3.2 software based on GIS technique from ESRI company.

3. RESULTS AND DISCUSION

3.1 Summary statistics

Fig. 3 displays the histograms (a) on the orginal scales and (b) as common logarithms ($\log_{10}$) of T-F and Ws-F. The distributions of T-F and Ws-F had long upper tails, and there are several data that might be considered as outliers. Some transformation was desirable for further analysis. Taking logarithms achieved approximate symmetry (Fig. 3(b)) and allowed a confident comparison of mean value for different forms of land use. It also brought the apparent outliers of T-F and Ws-F within the distributions and showed that they should not be treated as exceptional. The statistical results using the Kolmogrov-Smirnov (K-S) test indicated that the soil T-F was more nearly normally distributed than logarithm transformed, but the distribution of Ws-F remained more strongly peaked than normal (leptokurtic). Taking logarithms for Ws-F brought the skewness to only 0.23 and the kurtosis was –0.14. Clearly, its distribution is close to logarithm normal distribution.

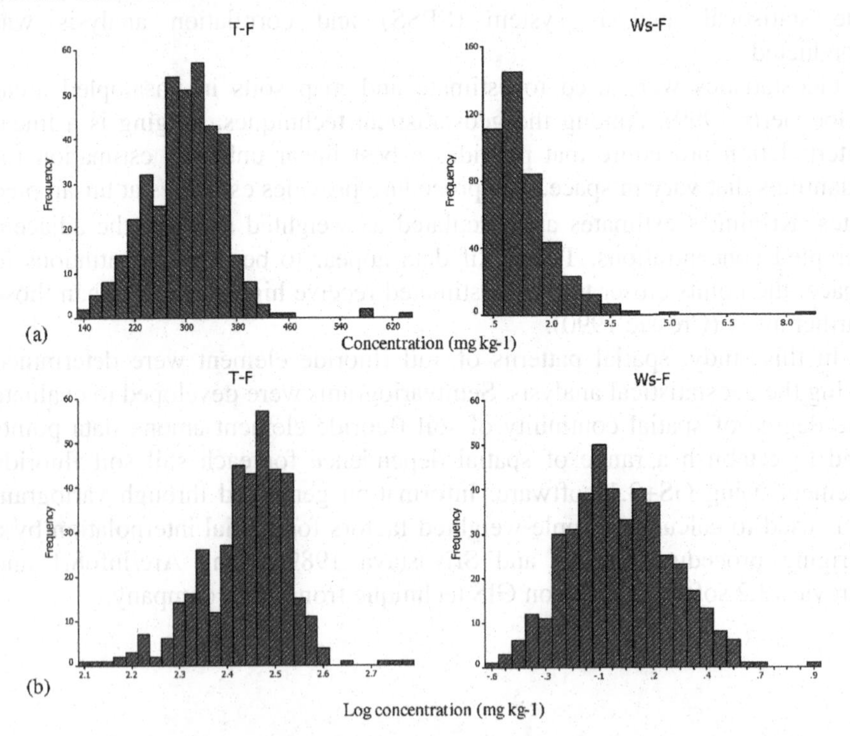

Fig. 3. Histograms (a) on original scales and (b) after transforming to logarithms

3.2 Geostatistical analysis

The concentrations of T-F and Ws-F were transformed to standard normal deviates by Hermite polomomaials, as described above. For T-F and Ws-F all the data were included. Fig. 4 presented the semivariogram and fitted models for fluoride element. The attributes of the semivariograms for soil fluoride were summarized in Table 2.

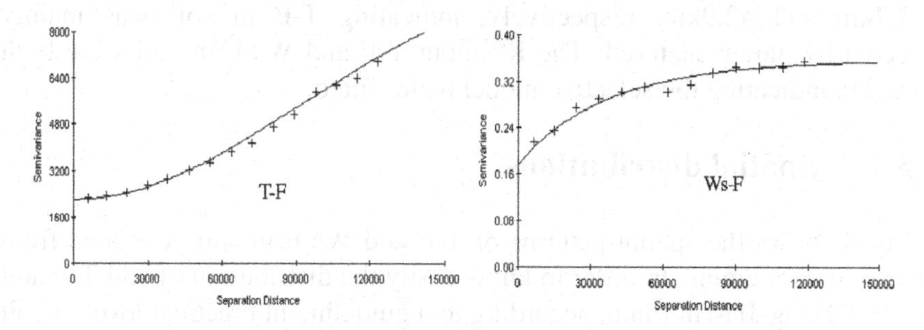

Fig. 4. Experimental semivariograms of soil fluoride element with fitted models

Table 2. Best-fitted semivariogram models of soil F and corresponding parameters

F Element	Model	C_0	C_0+C_1	Range (km)	$C_0/(C_0+C_1)$	R^2
T-F	Gaussian	2200	9510	632.7	23.1%	0.995
Ws-F	Exponential	0.178	0.357	42.9	49.9%	0.960

Nugget variance (C_0) represents the experimental error and field variation within the minimum sampling spacing. The Nug/Sill ratio ($C_0/(C_0+C_1)$) can be regarded as a criterion to classify the spatial dependence of soil properties. If the ratio is less than 25%, the variable has strong spatial dependence; between 25% and 75%, the variable has moderate spatial dependence; and greater than 75%, the variable shows only weak spatial dependence (Chien *et al.* 1997). The spatial variability of soil properties may be affected by intrinsic (soil formation factors, such as soil parent materials) and extrinsic factors (soil management practices, such as fertilization). Usually, strong spatial dependence of soil properties can be attributed to intrinsic factors, and weak spatial dependence can be attributed to extrinsic factors (Cambardella *et al.* 1994). Range is the distance over which spatial dependence. Regression coefficient (R^2) provides an indication of how the model fits the variogram data. The higher the regression coefficient, the better the model fits (Hu *et al.* 2004).

The semivariograms results suggested that the semivariagrams of T-F was well fitted for the gaussian model, while semivariagrams for logarithm conversion value of soil Ws-F was well fitted for exponential model. And their Nug/Sill ratios were 23.1% and 49.9%, respectively. The results suggested that T-F had strong spatial dependence, its spatial variabilities were mainly controlled by intrinsic factors such as parent material, relieves and soil types; Ws-F had moderate spatial dependence, mainly controlled by intrinsic factors and extrinsic factors. The ranges for T-F and Ws-F were 632.7km and 42.9km, respectively, indicating T-F in soil was mainly affected by parent material. The R^2 about T-F and Ws-F in soil were both over 0.9, indicating the selective model better fitted.

3.3 Spatial distributions

Fig. 5 shows the spatial patterns of T-F and Ws-F in soil generated from their semivariagrams. In order to know easily the distribution of soil T-F and Ws-F in Hang-Jia-Hu Plain, according to a guideline in practical level of soil and the area proportion were analysed (Table 3). To understand the effect of Parent material and soil property on T-F and Ws-F content, main parent materials with corresponding average soil property and soil F contents in 13 regions of Hang-Jia-Hu Plain were listed (Table 4), the correlativity between them and soil properties was analyzed (Table 5 and Table 6).

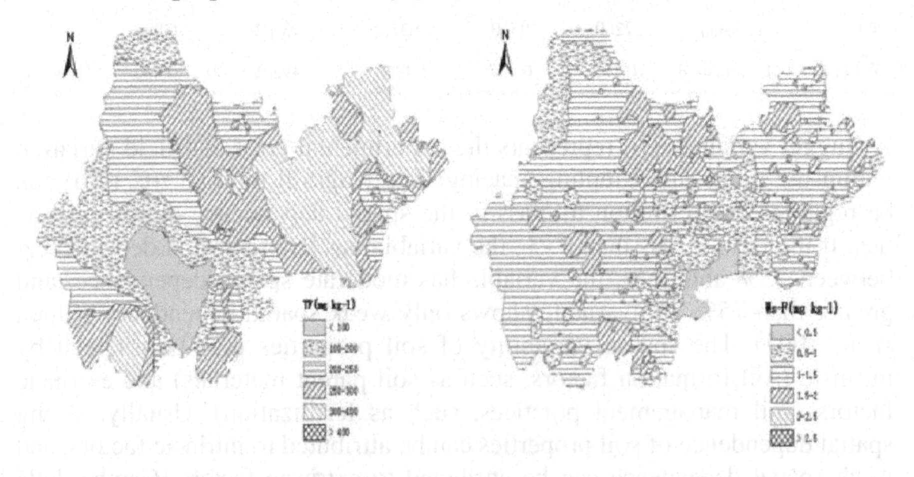

Fig. 5. Filled contour maps produced by ordinary Kriging of T-F and Ws-F in soil

Table 3. The guideline and the area ratio of T-F and Ws-F produced by ordinary Kriging

T-F			Ws-F		
Guideline	Area (km^2)	Ratio (%)	Guideline	Area (km^2)	Ratio (%)
< 100	1019.0	8.1	< 0.5	213.8	1.7
100-200	1959.0	15.6	0.5-1	6475.5	51.4
200-250	3606.3	28.6	1-1.5	4419.4	35.1
250-300	2902.6	23	1.5-2	1319.5	10.5
300-400	3108.4	24.7	2-2.5	167.1	1.3
> 400			> 2.5		

Table 4. Main parent material and corresponding average soil property and soil F contents in 13 regions of Hang-Jia-Hu Plain

						Particle composition(%)				
Code	Region	Parent material	pH (H_2O)	OM (g kg^{-1})	CEC (cmol kg^{-1})	<0.002mm	0.002–0.05mm	0.05–2mm	T-F	Ws-F
1	Jiaxing	River deposit, lake warp	5.80	36.92	18.65	14.58	70.32	15.19	332.94	1.07
2	Pinghu	lake deposit, offing deposit	5.61	35.06	18.01	14.55	73.07	12.38	329.05	1.33
3	Tongxiang	Fluvio-marine deposit, River deposit	6.32	28.44	15.67	12.85	72.77	14.36	338.51	1.78
4	Haining	River deposit, marine deposit	6.05	24.01	14.04	13.63	73.74	12.64	294.16	1.50
5	Jiashan	River deposit, lake warp	5.72	37.08	18.33	14.61	67.37	18.04	338.92	1.41
6	Haiyan	Ancient lake warp, fluvio-marine deposit	5.52	35.27	15.69	14.32	72.61	13.07	298.07	1.49
7	Hangzhou	Old-river alluvium, shallow-sea alluvium	6.07	32.80	11.83	12.19	68.26	19.55	259.81	1.64
8	Yuhang	Red slope deposit, transported redeposit	5.64	33.83	12.29	14.56	63.98	21.46	255.34	1.08
9	Lin'an	Diluvial alluvium	5.84	33.87	10.22	17.18	56.72	26.13	316.32	0.88
10	Deqing	Lagoonal lake wrap	5.67	34.52	14.41	15.23	69.31	15.45	288.00	0.92
11	Changxing	Lagoonal lake wrap	5.41	31.05	13.24	15.84	65.06	19.14	223.06	0.74
12	Anji	Yellow and red soil redidual deposit	5.19	33.42	9.20	15.81	63.26	20.93	211.97	0.57
13	Huzhou	Lagoonal lake wrap, Fluvio-marine deposit	6.03	40.58	14.63	13.77	70.97	15.26	277.44	1.36

Table 5. Correlation coefficients among pH, organic matter, CEC, clay and F elements

		OM	CEC	Particle composition (%)			T-F	Ws-F
	pH(H_2O)	(g kg^{-1})	(cmol kg^{-1})	<0.002mm	0.002–0.05mm	0.05–2mm	(mg kg^{-1})	(mg kg^{-1})
pH	1							
OM	0.019	1						
CEC	0.113*	0.388**	1					
<0.002mm	-0.140**	0.359**	0.359	1				
0.002–0.05mm	0.142**	-0.142**	0.241	0.023	1			
0.05–2mm	-0.079	-0.003	-0.355	-0.396**	-0.927**	1		
T-F	0.193**	0.186**	0.580	0.274**	0.054	-0.160**	1	
Ws-F	0.612**	-0.050	0.070	-0.234**	0.159**	-0.060	0.169**	1

*p <0.05; **p <0.01

Table 6. Stepwise regression of F elements in tested soil in Hang-Jia-Hu Plain

Regression equation of T-F and Ws-F	Parameters of regression equation		
	Multiple correlation coefficient	F value	Significant level
$Y_{T\text{-}F} = 26.584 + 19.253 X_1 - 6.473X_2 + 8.372 X_3 + 3.989 X_4$	0.661	52.37	$p < 0.05$
$Y_{Ws\text{-}F} = -2.551 + 0.661 X_1 - 0.049X_4 + 0.010 X_5$	0.628	86.26	$p < 0.05$

X_1, pH value; X_2, organic matter; X_3, CEC content; X_4, clay fraction (<0.002mm); X_5, silt fraction (0.002–0.05mm); X6, sand fraction (0.05–2mm)

The content distribution of T-F in the plough layer (0–20cm) of paddy soils in Hang-Jia-Hu Plain was the eastern part > central part > western part. As a whole, however, the concentration of soil T-F in Hang-Jia-Hu Plain was comparatively lower, had 2978 km^2 area accounting for 23.7% in the studied area and was lower than the fluoride content in soil which is 200mg kg^{-1} on average in the world (China Environmental Monitoring General Station, 1990). The main range of T-F content was 200–300mg kg^{-1} accounting for 51.6% in the studied area and lower than the fluoride content in soil which is 478mg kg^{-1} on average in China (China Environmental Monitoring General Station, 1990). The average range of soil T-F concentration in Anji, Deqing, Hangzhou zone was 100–200mg kg^{-1}, especially the range of T-F contents in soil was as low as less than 100 mg kg^{-1} in Yu-hang area. The range of T-F contents in soil in Jiaxing, Jiashan, Pinghu zone and part area of Lin'an was exceed 300mg kg^{-1}. The characteristic distribution of soil T-F was mainly associated with each parent material distribution in the 13 regions of Hang-Jia-Hu Plain, the contents of soil T-F developing from river deposit and lake warp exceeded others, yellow and red soil redidual deposit had the least soil T-F contents. And the distribution rule of soil T-F was as same as pH value, organic matter and clay fraction (<0.002mm), which was in positive correlation with them, and stepwise regression of soil T-F indicating the concentration of soil T-F was mainly affected by pH value and CEC contents. The studied results were the same as the previous studied papers. Although fertilization and irrigation can also add up soil fluoride contents, the less effect on it for the less source.

The content distribution of Ws-F in the plough layer (0–20cm) of paddy soils in Hang-Jia-Hu Plain was the central part > eastern part > western part. The main range of Ws-F content was 0.5–1.0mg kg^{-1} in the whole studied zone accounting for 51.4% and centralized in the western area. The average range of soil Ws-F concentration in part area of Anji and Lin'an was less than the water-soluble fluoride content in uncontaminated soil surface layer which is 0.5mg kg^{-1} on average in the world (China Environmental Monitoring General Station, 1990) and only had 213.8km^2 area, where had the lowest pH value and CEC content in the whole area. The content of

Ws-F was comparatively higher in the central and eastern zone, especially in part area of Tongxiang and Hangzhou, the range of Ws-F contents in soil was 2.0–2.5mg kg^{-1}, where had the highest pH value and the lowest organic matter in the whole zone. Which was mainly connected with different parent material distribution in this 13 region of Hang-Jia-Hu Plain too. And the content distribution rule of Ws-F was as same as that of pH value and CEC content and which was in positive correlation with them, in contrast to that of organic matter and clay fraction (<0.002mm) which was in negative correlation with them, and stepwise regression of soil Ws-F indicating the concentration of soil Ws-F was mainly affected by pH value.

The results of this study confirm previous reports indicating that the fluoride concentration has connected with a number of factors such as soil type, parent material, pH value and farm management. Further research should be carried out to study the relationship between the fluoride level of the soil and the fluoride content of agricultural products, et al., and, likewise, the relationship between the content of the water and food and the bioavailability of soil fluoride should be elucidated.

4. CONCLUSIONS

1. The distribution of T-F contents in the plough layer (0–20cm) of paddy soils was normally distributed, and Ws-F was fitted for logarithm normal distribution from Hang-Jia-Hu Plain in China. The semivariagrams of T-F was well fitted for the gaussian model, while semivariagrams for logarithm conversion value of soil Ws-F was well fitted for exponential model, respectively. And their Nug/Sill ratios were 23.1% and 49.9%, respectively. The results suggested that T-F had strong spatial dependence, its spatial variability was mainly controlled by intrinsic factors such as parent material, relives and soil types; Ws-F had moderate spatial dependence, mainly controlled by intrinsic factors and extrinsic factors. The order of range was T-F > Ws-F, indicating T-F in soil had correlation among the whole area and which was mainly affected by parent material.

2. Spatial distribution of the main contents of soil T-F and Ws-F were among 200–300mg kg^{-1} and 0.5–1.0mg kg^{-1}, respectively in the studied area. And what is more, the range of T-F contents in soil was as low as less than 100 mg kg^{-1} in Yu-hang area accounting for 23.7% area scale. The range of fluoride contents in the soils from central and eastern parts of Hang-Jia-Hu Plain was higher than that from the western part. The

soil fluoride concentrations had a close relationship with soil parent material, pH value, organic matter, CEC and soil texture.

3. The accumulation of fluoride contents in soil was lower in the whole studied area, suggesting that local fluoride epidemic such as dental caries due to lack of fluoride will occur easily in Hang-Jia-Hu Plain in China, so where should be prevented by using fluoride-containing toothpaste and other effective measure.

ACKNOWLEDGEMENTS

This research was supported by the Program for Changjiang Scholars and Innovative Research Team in University (PCSIRT) IRT0536 and Foundation of Hangzhou Dianzi University.

REFERENCES

Atteia. O, Dubois JP, Webster R. 1994 Geostatistical analysis of soil contamination in the Swiss Jura. Environment pollution 86, 315-327.

Bai YL, Li BG, Shi Ych. 1999 Research of the soil salt content distribution and manatement in Huang-Huai-Hai plain based on GIS. Resour Sci 21(4), 66-70.

Baker RL. 1972 Determination of fluoride in vegetation using the specific ion electrode. Anal. Chem 44, 1326.

Cahn MD, Hununel JW, Brouer BH. 1994 Spatial analysis of soil fertility for site-specific crop management. Soil Sci. Soc Am J 58, 1240-1248.

Cambardella CA, Moorman TB, Novak JM, Parkin TB, Turco RF, Konopka AE. 1994 Field-scale variability of soil properties in central lowa soils. Soil Sci Soc Am J 58, 1501-1511.

Chien YL, Lee DY, Guo HY, Houng KH. 1997 Geostatistical analysis of soil properties of mid-west Taiwan soils. Soil Sci 162, 291-297.

China Environment Monitoring General Station. 1990 The background values of Elements in Soils in China. China Environment Science Press, Beijing (in Chinese).

Cressie C. 1990 The origins of kriging. Math Geo 22(2), 239-252.

Fejeskov O, Larsen MJ, Rrchards A, and Baelum V. 1994 Dental tissue effects of fluoride. Adv. Dent. Res 8, 15-31.

Fisher E, Thornton B, Hudson G. 1998 The variability in total and extractable soil phosphorus under a grazed pasture. Plant and Soil 203, 249-255.

Fung KF, Zhang ZQ, Wong JWC, Wong MH. 1999 Fluoride contents in tea and soil from tea plantations and the release of fluoride into tea liquor during infusion. Environmental Pollution 104, 197-205.

Geeson NA, Abrahams PW, Murphy MP, Thornton I. 1998 Fluorine and metal enrichment of soils and pasture herbage in the old mining areas of Derbyshire, UK. Agriculture, Ecosystems and Environment 68, 217-231.

Goovaerts P. 1999 Geostatistics in soil science: state-of-the-art and perspectives. Geoderma 89, 1-45.
Guo XD, Fu BJ, Ma KM, Chen LD. 2000 Spatial variability of soil nrttients based on geostatistics combined with GIS – A case study in Zunhua City of Hebei Province. Chinese Journal of Applied Ecology 11(4), 557-563 (in Chinese).
Honrner JM, Bell JNB. 1995 Evolution of fluoride tolerance in Plantago lanceolata. The Science of the Total Environment 159, 163-168.
Hu KL, Zhang FR, Lv YZ, Wang R, Xu Y. 2004 Spatial distribution of concentrations of soil heavy metals in Daxing county, Beijing. Acta scientiae circumstantiae 24(3), 463-468 (in Chinese).
Isaaks EH, Srivastava RM. 1989 An introduction to applied geostatistics. Oxford University Press, New York.
Li YH, Wang WY, Luo KL. 2004 Distribution of selenium and fluoride in soils of Daba mountains. Acta pedologica sinica 41(1), 61-67 (in Chinese).
Liu XM, Xu JM, Zhang MK, Shi Z, Shi JC. 2003 Study on spatial variability of soil nutrients in Taihu lake region – a case of Pinghu City in Zhejiang Province. Journal of Zhejiang University (Agric. and Life Sci.) 29, 76-82 (in Chinese).
Marian Kjellevold Malde, Amund Maage, Elizabeth Macha, Kare Julshamn, and Kjell Bjorvatn. 1997 Fluoride content in selected food items from five areas in East Africa. Journal of food composition and analysis 10, 233-245.
Mendonca Santos ML, Guenat C, Bouzelboudjen M, Golay F. 2000 Three-dimensional GIS cartography applied to the study of the spatial variation of soil horizons in a Swiss floodplain. Geoderma 97, 351-366.
Nanjing Agricultural Unviersity. 1981 Analytical Methods for Soil and Agricultural Chemistry. Agriculture Press, Beijing, pp. 29-39 (in Chinese).
Romic M, Romic D. 2003 Heavy metals distribution in agricultural topsoils in urban area. Environ Geo 43, 795-805.
Singh B, Dass J. 1993 Occurrence of high fluoide in groundwater of Haryana. Bhujal News 8(1), 28-31.
Sokal RR, Rohlf FJ. 1981 Biometry, 2nd Edn. WH Freeman and Company, New York.
Steiger B von, Webster R, Schulin R, Lehmann R. 1996 Mapping heavy metals in polluted soil by disjunctive kriging. Environ Pollu 94(2), 205-215.
Tsegaye T, Rovert L Hill. 1998 Intensive tillage effects on spatial variability of soil test: plant growth and nutrient uptake measurements. Soil Science 13(2), 155-165.
Underwood EJ, Mertz W. 1987 Introduction. In Trace Elements in Human and Animal Nutrition (W. Mertz, Ed.), 5th ed. Academic Press, San Diego, pp. 1-19.
Villa AE. 1979 Rapid method for determining fluoride in vegetation using an ion-selective electrode. Analyst 104, 545-551.
Wang ZQ. 1999 Geostatistics and its application in ecology. Science Press, Beijing, China, pp. 162-192 (in Chinese).
Wheeler SM, Fell LR. 1983 Fluorides in cattle nutrition. Commonwealth Bureau of Nutrition, Nutition Abstracts and Reviews, Series B, 53, No. 12. Commonwealth Agricultural Bureaux, Farnham Royal, Bucks, pp. 741-767.
White JG, Welch RM, Norvell WA. 1997 Soil zinc map of the USA using geostatistics and geographic information system. Soil Sci Soc Am J 61, 185-194.
Xie ZM, Wu WH, Xu JM. 2003 Study on fluoride emission from soils at high temperature related to brick-making process. Chemosphere 50, 763-769.

Xie ZM, Ye ZH, Wong MH. 2001 Distribution characteristics of fluoride and aluminum in soil profiles of an abandoned tea plantation and their uptake by six woody species. Environment International 26, 341-346.

Yu Pin Lin, Tsun Kuo Chang, Tung Po Teng. 2000 Characterization of soil lead by comparing sequential Gaussian simulation, simulated annealing simulation and kriging methods. Environ Geo 11 41, 189-199.

DEVELOPMENT FOR BREEDING PERFORMANCE MANAGEMENT SYSTEM ON PIG FARMS

Benhai Xiong*, Qingyao Luo, Jianqiang Lu, Liang Yang

Institute of Animal Science, Chinese Academy of Agricultural Sciences, State Key Laboratory of Animal Nutrition, Beijing 100094, China

** Corresponding author, Institute of Animal Science, Chinese Academy of Agricultural Sciences. No. 2 Yuanmingyuan West Road, Beijing 100094, China, Tel: +86-10-62811680, Fax: +86-10-62815988, Email: Bhxiong@iascaas.net.cn or Xiongbenh@263.net.*

Abstract: The study was conducted to supply systemic and dynamic analysis data to support a better operation on a breeding pig farm with process management, especially in reproduction parameters. A full simulation model on a breeding pig farm running was proposed in the study, and a series of definitions of process parameters related to service performance, farrowing performance and weaning performance was put forward. Some of them are described on the calculating models. The relationship structural database was designed and a set of digital management information system was developed, based on proposed definitions and models by using Visual Basic 6.0. Access databases and Crystal report combined with genetic characteristic of different pig breeds. The System supplies a series of convenient, intelligent input interfaces of original datum, and all different reproduction data can be counted, analyzed and graphically shown, based on different performances in a specific duration, and it can dynamically derive out all sows history card that shows a complete reproduction performances including some important indexes such as farrowing rate, farrowing interval, average gestation days and average weaned weight et al. in terms of parities, which can be used to decide whether a female needs to be fell into disuse. Therefore, with the help of system analysis and software design techniques, the system made it possible to realize information management and intelligence analysis for a breeding pig farm based on whole digital management of reproduction process from services through weaning and among different categories of breeding pigs and parities.

Keywords: Pig farms, Breeding sows, Information management, Farrowing performance

Xiong, B., Luo, Q., Lu, J. and Yang, L., 2008, in IFIP International Federation for Information Processing, Volume 258; Computer and Computing Technologies in Agriculture, Vol. 1; Daoliang Li; (Boston: Springer), pp. 267–275.

1. INTRODUCTION

With the help of advanced science and management technologies, pig farming has gained fast advances, especially in production rate of pigs and utilization efficiency of nutrients around the world. Currently, the most efficient method to reduce cost of pig farming is to enhance reproduction performance of sows by strengthening the whole breeding process. (Rothschild MF, 1996). Besides reproduction techniques themselves, application of information technologies is becoming more and more important due to timely data gathering and just-in-time decision making. Therefore, many information management system have been developed and applied in pig farms, especially those farms with breeding sow (www.pigwin.com; www.agritecsoft.com/en/porcitec/; http://psru.marc.usda.gov). Among these systems, the most core task focus on female process management from mating, gestation, farrowing to weaning. The objective of managements is not only to calculate reproduction performances on individual pigs, but also to know the whole performance of a specific herd so that the producers can find out existing problems.

Although there are different information systems on pig farms abroad, the most important is that we should have own system with our intellectual property rights, and that the differences among language and culture as well as management patterns bring a requirement for designing similar system by using Chinese language. The study consideration based on breeding pig farming will be introduced and a short description was given: (1) a general model of lifecycle for pigs to express a production system of pigs, (2) identification technology adopted for individual pigs, (3) the required recording for reproduction process documents, (4) reproduction performance analysis for females.

2. FORMULATION OF PROBLEM

Before developing the management system on breeding pig farms, it is necessary to know what key figures for a breeding pig farm is. With the help of analyzing the key figures, deviations of different female individuals can be found out quickly. If reproduction problems in a farm can timely be recognized, losses will be prevented or reduced to a minimum; especially it can supply basic decision making for culling and updating of basic sows herd.

2.1 The categories and production system model on a farm

Six categories of pigs are distinguished in the reproduction management system, that are suckling piglets, weaned piglets, gilts, empty and pregnant sows, lactating sows and fattening pigs. The key figures can be calculated separately for each of these categories. Fig. 1 shows a whole production system or lifecycle model for pigs and the above categories are located in different processes. Based on the model, their production data can be recorded and integrated in terms of the categories of pigs respectively.

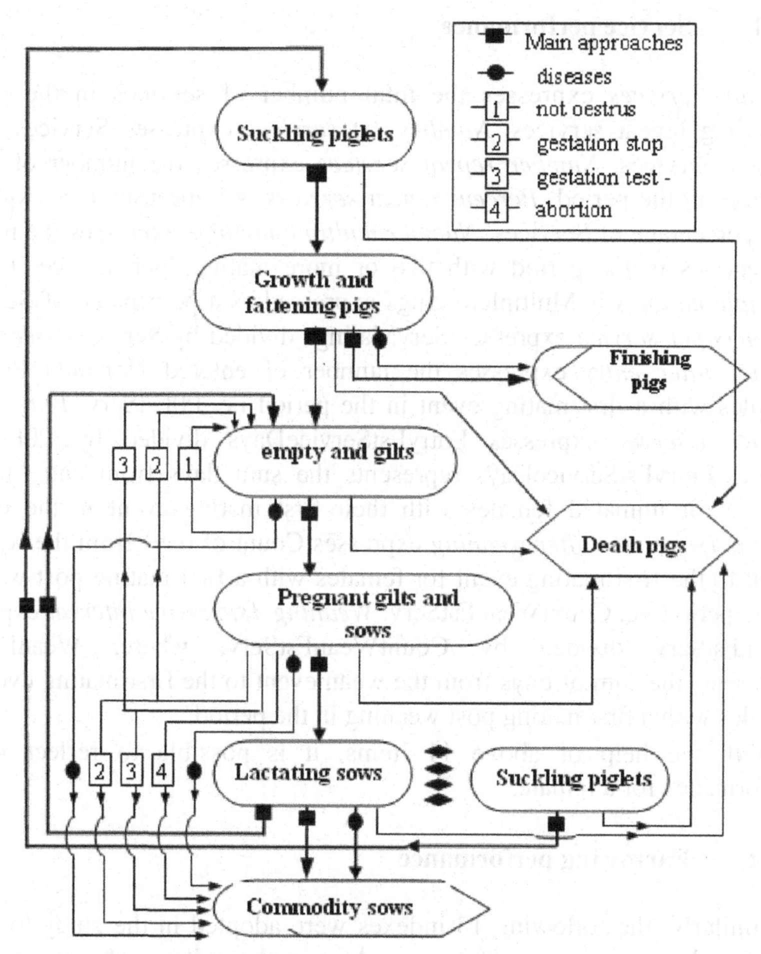

Fig. 1. Production system model sketch map for pigs

2.2 Definition of main data items and required recordings

As known, there are many data items involved in a management information system for a breeding pig farm. Considering length limitation of this paper, the following comprehensive performance analysis indexes for reproduction traits on pigs are shown only. The whole reproduction traits are divided into three parts: service performance, farrowing performance and weaning performance and all indexes are defined in the parts is in the context of one specific period. In the following description, those italic items represent parameters required and to be recorded.

2.2.1 Service performance

Total services expresses the total number of services in the period, including repeat services. *Number 1st services* expresses Services minus Repeat Services. *Number repeat services* expresses the number of repeat services in the period. *Percent repeat services* is RepeatServices expressed as a percentage of Services. *Number multiple matings* expresses the number of services in the period with two or more matings per service. *Percent multiple matings* is MultipleMatings expressed as a percentage of Services. *Matings per service* expresses ServMatings divided by Services. *Served 1st service after entry* expresses the number of entered Unmated Parity 0 females with a first mating event in the period i.e. P0FstServ. *Entry to 1st service interval* expresses EntryFstServiceDays divided by P0FstServ, where, EntryFstServiceDays represents the sum days from entry to first service for unmated females with their first mating event in the period. *Served 1st service after weaning* expresses Count of days from the weaning event to the first mating event for females with a first mating post weaning in the period i.e. CountWeanFstServ. *Weaning-1st service interval* expresses WeanFstServ divided by CountWeanFstServ, where, WeanFstServ represents the sum of days from the wean event to the first mating event for females with a first mating post weaning in the period.

With the help of above 11 items, it is possible to reflect service performance for a female.

2.2.2 Farrowing performance

Similarly, the following 13 indexes were adopted in the study to figure out female farrowing performance. Among these items, *Farrowing rate, Average gestation and farrowing interval* are main parameters that reflect female genetic traits as well as the farm management level.

Farrowings expresses the number of birthing events in the period i.e. births. *Average parity farrowed* expresses total birth parity divided by farrowings in the period. *Total born per farrow* expresses Born divided by Birthings, where, Born means the sum of born alive to all birthing events in the period, Birthings means the number of birthing events in the period. *Liveborn per farrow* expresses Liveborn divided by Birthings. *Stillborn per farrow* expresses Stillborn divided by Birthings. *Stillborn percent* is Stillborn expressed as a percentage of Born. *Mummies per farrow* expresses Mummies divided by Birthings. *Mummies percent* is Mummies expressed as a percentage of Born. *Farrowing rate* is Birthings expressed as a percentage of ServicesPeriodBirth, where ServicesPeriodBirth means the number of females served to farrow in the period.

Average gestation length is SumGestationLen divided by CountGestationLen, where SumGestationLen means the sum days of gestation lengths of females with a birthing event in the period, CountGestationLen means the count of gestation lengths in the period, used for statistical purposes. *Farrowing interval* expresses days between two consecutive farrowing dates for a mated breeding female. *Litters/mated female/year* is calculated as GestationDays multiplied by 365 and divided by both 115 and MatedFemaleDays, where GestationDays means the sum of days of all females gestating in the period. A female contributes one gestation day for each day that is between conception and birthing in the period. MatedFemaleDays means the sum of all mated female days. A breeding female is considered as a part of the mated breeding female inventory, effective of its first service date. Lastly, *total born/mated female/ year* equals to Born multiplied by 365 and divided by MatedFemaleDays.

2.2.3 Weaning performance

Compared with farrowing, weaning performance analysis actually reflects the situation of a farm management level. In general, the better a farm management does, the earlier a herd weaning is. These parameters are as follows:

Total pigs weaned expresses the sum of pigs weaned from sows with weaning events in the period, including weaning events of nurse litters. Pigs weaned with a partial weaning event are counted on the date of the following weaning event. *Litters Weaned* expresses the number of litters weaned in the period, including litters weaned from a nurse on event. *Pigs weaned per litter* are Total Weaned divided by LittersWeaned. *Females weaned or nursed off* expresses the number of females with a last wean or nurse off event in the period. *Pigs weaned per female* expresses LastWeaned divided by FemalesWeaned, where, LastWeaned means total number of animals weaned from sows with their last weaning or nurse off event in the period. *Net fostered* expresses pigs fostered on with a foster event minus pigs fostered off with a foster event plus pigs nursed on with a nurse on event minus pigs nursed off with a nurse off event in the period.

Avg. weaning age is SumWeanLitterAges divided by CountWeanLitterAges, where SumWeanLitterAges means the sum of age at weaning for natural litters weaned in the period, and CountWeanLitterAges means Count of wean litters with weaning ages. *Avg weight per weaned pig* is SumWeanWeight divided by CountWeanWeight, where SumWeanWeight means the sum of litter weights at weaning for litters weaned in the parity, includes nurse litters, CountWeanWeight means Number of animals weighed at weaning.

Recorded preweaned deaths expresses the sum of preweaned deaths recorded with the PwDeath event in the period. *Preweaning mortality rate* means pigs that die between live birth and weaning expressed as a percentage of those at risk.

Weaned/mated female/year is TotalWeaned multiplied by 365 and divided by MatedFemaleDays, where TotalWeaned is the sum of pigs weaned from sows with Weaning events in the period, includes Weaning events of nurse litters. Pigs weaned with a Partial Weaning event are counted on the date of the following Weaning event.

Weaned/female/year is TotalWeaned multiplied by 365 and divided by FemaleDays, where femaleDays is the sum of all active female days during the period. A female contributes one female day for each day she was active during the period, beginning at her entry date into the breeding herd and ending with her removal.

Based on the above definition and analysis, as well as computer program design, it is possible to develop a digital and intelligent analyzing module on female breeding performance.

3. PROBLEM SOLUTION

3.1 System program development

It is divided into two kinds of situations to design this management information system, non-network system and network system. For the first case, access database and VB language were adopted to develop this system and an integrated version is finally supplied for users. For another case, windows server 2003, SQL server 2005 and Visual C#. Net language were depended upon and it is finally issued it by intranet or internet.

Before running a pig farm management system, some basic female production theory parameters must be supplied in advance as the references of analysis. The input interface including required setting for growth and reproduction data, and an operator editing these parameters from either theory or management objectives are shown in Fig. 2.

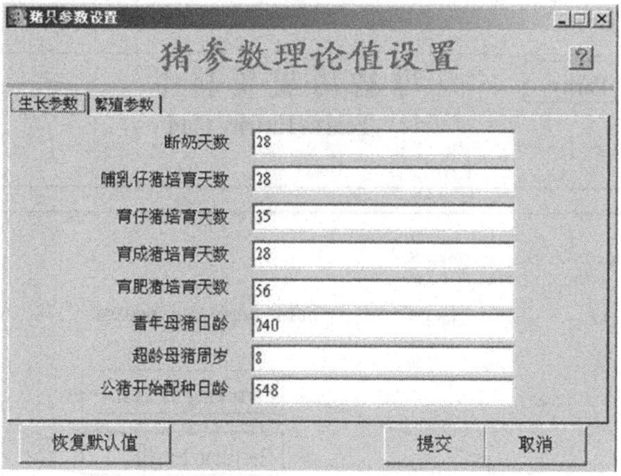

Fig. 2. Setting for theory parameters of a kind of pig

3.2 Identification to pig individuals

The identification of individual pigs depends on a national animal labeling program, which has been implementing by Ministry of Agriculture, China. According to this program, every animal (pig, cattle and sheep/goat) has a unique labeling code (http://www.agri.gov.cn/blgg/t20060628_638621.htm, 2006). The labeling code is composed of 15 numeric digits. The first digit expresses livestock category, i.e. 1 swine; 2 cattle, 3 sheep/goats. The following 6 digits (2nd to 7th) express the administrative region code that obeys to GB T2260-1999, a national standards of People's Republic of China (http://www.stats.gov.cn/tjbz/xzqhdm/t20070411_402397928.htm,2006), which represents a nature county and is updated on December 31, 2006. The last 8 digits (8th to 15th) express a ranking number for a livestock in a specific county in terms of different species respectively. In this system, information or data for individuals is stored in terms of its labeling code.

3.3 Main output results (example)

Based on the female performance analysis including service, farrowing and weaning performance, one of the important statistic outputs is female history card. From a specific female history card (Fig. 3), a whole breeding result with a specific female can be found out from parity 0 to current parity related to service, farrowing and weaning data. In terms of dynamic changes of performance parameters with a female in different parities, an operator can decide when it shall be fallen into disuse.

Using pre-designed statistic module, the system can analyze the whole female herd and print out results calculated at any time. Formulation of different reports is from both crystal-reporting technology and own designed module.

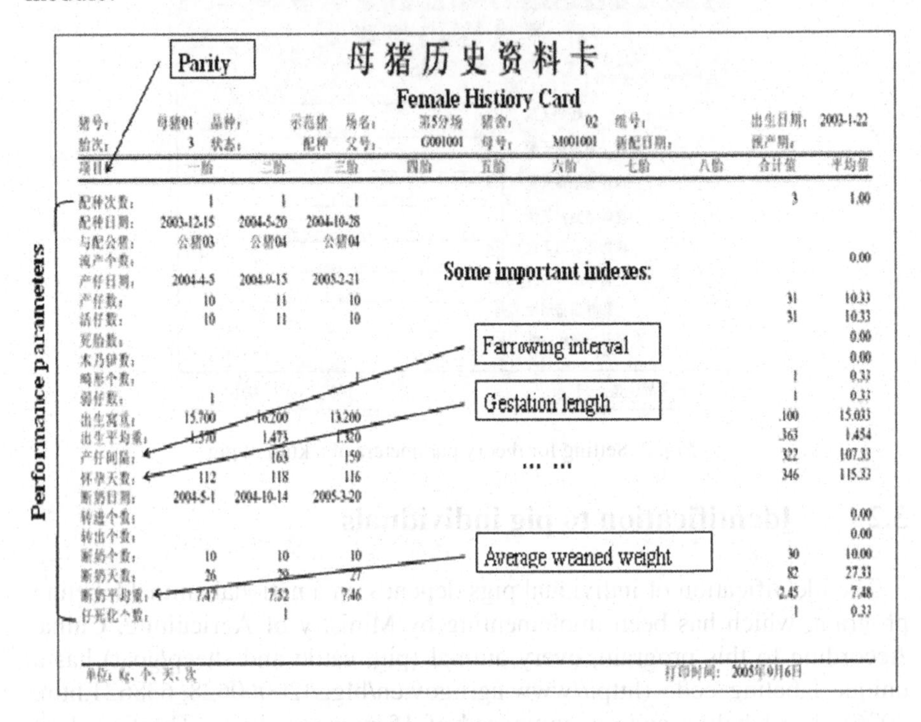

母猪历史资料卡

Female Histiory Card

猪号：母猪01　品种：示范猪　场名：第5分场　猪舍：02　组号：　出生日期：2003-1-22
胎次：3　状态：配种　父号：G001001　母号：M001001　断配日期：　预产期：

项目	一胎	二胎	三胎	四胎	五胎	六胎	七胎	八胎	合计值	平均值
配种次数：	1	1	1						3	1.00
配种日期：	2003-12-15	2004-5-20	2004-10-28							
与配公猪：	公猪03	公猪04	公猪04							
流产个数：										0.00
产仔日期：	2004-4-5	2004-9-15	2005-2-21							
产仔数：	10	11	10						31	10.33
活仔数：	10	11	10						31	10.33
死胎数：										0.00
木乃伊数：										0.00
畸形个数：			1						1	0.33
弱仔数：	1								1	0.33
出生窝重：	15.700	16.200	13.200						.100	15.033
出生平均重：	1.570	1.473	1.320						.363	1.454
产仔间隔：		163	159						322	107.33
怀孕天数：	112	118	116						346	115.33
断奶日期：	2004-5-1	2004-10-14	2005-3-20							
转进个数：										0.00
转出个数：										0.00
断奶个数：	10	10	10						30	10.00
断奶天数：	26	29	27						82	27.33
断奶平均重：	7.47	7.52	7.46						2.45	7.48
仔死亡个数：		1							1	0.33

单位：kg、个、天、次　　打印时间：2005年9月6日

Fig. 3. A female history card (example)

4. CONCLUSIONS

The need for a fully operational and economically visible pig farm production management system that is capable to record all process data is highlighted. In order to explore an integrated data gathering and analyzing as well as output of calculated results, a lot of definitions on those data items related to female production performance is presented in the paper. Among these described performance indexes, although most of them are used for individuals of swine, it is also possible to do some herd traits analysis. A practical system focusing on analysis of individual female reproduction process data is proposed also.

Although such a system has realized prime reproduction information management, there are a few aspects to be perfected in the future. Firstly, regarding to the method to record process data, an intelligent and mobile PDA embedded system is been developing to record and edit data just in time and upload those gathered data into table-server database system by wireless or GPRS/CDMD (Yao W., 2003; Liu yanbo, 2006) based on network management system that is intranet or internet. The following is the precision feeding for females including gestation sows and lactating sows. The nutrition situation of sows affects up growth of fetuses and growth of piglets. A perfect method is to construct dynamic predicting models to calculate nutrient requirements for individual females in terms of its genetic traits, body conditions and environment factors. All above analysis shows that a good information management system on breeding pig farms needs a lots of professional models on reproduction and nutrition (Xiong B H, 2005).

ACKNOWLEDGEMENTS

During this paper's writing, Professor Qian Ping from Institute of Agricultural Information of Chinese Academy of Agricultural Sciences gave careful reading of the manuscript and his fruitful comments and suggestions. The author thanks him for his distinguish work.

REFERENCES

http://psru.marc.usda.gov/

http://www.agri.gov.cn/blgg/t20060628_638621.htm. 2006. No. 67th, order of the ministry of agriculture of the people's republic of china. Animal labeling and their feeding documents establishment.

http://www.stats.gov.cn/tjbz/xzqhdm/t20070411_402397928.htm. November, 31, 2006. National Bureau of Statistics of China. Codes of administrative regions of the People's Republic of China.

Liu yanbo, Hu yan, Ma qi. Platform Application and Development on Windows Mobile. Posts & Telocom Press. 2006.

Rothschild MF. Genetics and reproduction in the pig. Animal Reproduction Science, 1996 42:143-151.

www.agritecsoft.com/en/porcitec/

www.pigwin.com

Xiong B H, Qian P. A comprehensive technology platform for precision feeding of dairy cattle [M]. China Agricultural Science and Technology Press. 2005

Yao W. Visual Basic database Development and project examples [M]. Posts and Telecom Press. 2003.

RESEARCH ON DYNAMIC CHANGE OF GRASSLAND IN WEST JILIN PROVINCE BASED ON 3S TECHNOLOGY

Zhiming Liu, Nanyan Ling
Northeast Normal University, Changchun, China, 130024

Abstract: Grassland had been interpreted with the method that combined supervised and non-supervised classifications using Landsat data in autumns 1986, 1996 and 2000. The amelioration of algorithm promoted the accuracy. With the support of RS and GIS, spatial information of grassland landscape in west Jilin province from 1986 to 2000 was extracted, spatio-temporal dynamics of grassland and change of grassland landscape patterns were analyzed. The results show: 1) The grassland environment in the west Jilin is exacerbating and the area of grassland is decreasing by an average of 4.5×10^4 hectares a year, moreover, the grassland degeneration is severe and the area of degeneration is 81.1% of the total. 2) The degeneration of the grassland is heavier and the proportions of moderate and heavy degenerated grassland are heightening obviously, so their evaluative trends are remarkable. 3) The degenerated grassland is mainly converted into cropland and saline—alkalized land, converted area reaches 992351.4hm^2 and 563031.1hm^2. 4) The total dynamic degree of grassland is a minus value. During 1986–1996, high coverage grassland has a strong effect on change of whole grassland. The fastest degraded area is low coverage grassland from 1996 to 2000.

Keywords: grassland, spatio-temporal change, dynamic degree, remote sensing images, geography information system, west Jilin province

1. INTRODUCTION

Remote Sensing is provided with excellent characteristics such as the exact, macroscopical, dynamic, frequent large area observation. On the other hand, Geographic Information Systems has the advantage on data processing

Liu, Z. and Ling, N., 2008, in IFIP International Federation for Information Processing, Volume 258; Computer and Computing Technologies in Agriculture, Vol. 1; Daoliang Li; (Boston: Springer), pp. 277–285.

and dynamic analysis. The practices indicate that study on the spatio-temporal changes of grassland resources with the support of RS and GIS is very effective. The paper analyzes and reveals the spatio-temporal regulations and momentum of grassland development in west Jilin province since 1986.

2. GENERAL SITUATION OF THE INVESTIGATED REGION

The west Jilin province (121°38′–126°11′E, 43°59′–46°18′N), which is located in the east Kerqin Steppe, one of the biggest prairies in the world, and the south-central Songnen Plain. The dimensionality acreage of region amounts to 47671km^2 that occupies 25.5% in total domanial acreage in Jilin province. It is located in Semiarid Agro-pastoral and belongs to temperate zone continental monsoon climate whose characteristics are long cold winter, short-lived sweltering summer, plenty wind and little rain. The annual average temperature is 3–6°C and the annual precipitation is 400–500mm, otherwise, the annual evaporation is 1842mm in this region.

Firstly, on division, it is a transitional area between the region where agriculture is dominant and the region where mixed pastoral farming area is leading. Secondly, on natural zonation, it lies between the transitional area of temperate zone semi-humid meadow grassland zone and temperate zone semi-dry grassland which also called ecological and climatic fragile zone chestnut soils. Thirdly, on landform, it is the transitional area from Songnen Plain to the downstream plain of Liao river. Finally, on climate, it is located in the transitional area between humid East Asian monsoon region and arid continental area. The investigated region is so sensitive to the environmental changes which happen as a result of the transformation of climate and human activities that it has been concerned by domestic and overseas scholars as an area that responds to global change exceedingly and is the ecological and climatic fragile zone (Chen Da-Ke, 1995; Wang Dongyan et a1., 2002; Qiu Shanwen et a1., 2003).

3. THE METHOD OF STUDY

3.1 Data processing

The data processing which included normal false color composing and accurate correction as well as registration and enhancement of images was accomplished by using Landsat data in 1986, 1996 and 2000 and 1: 100 000 land-use map. The control points were selected to carry through quadratic polynomial fitting correction based on the recently land-use map. The treated TM images matched with the digital border line well, and the error was two pixels at most.

According to the actual situation of the investigated area, the grassland was divided into three categories: high coverage density area, moderate coverage density area and low coverage density area in accordance with the national land use classification system. High coverage density grassland denotes the natural grassland, improved grassland and mowing grassland whose coverage densities are more than 50%. Besides that, the moisture conditions of this kind of lawn are generally good, and the grass spreads sparsely. Moderate coverage density grassland denotes the natural grassland and improved grassland whose coverage densities are between 20% and 50%. It is deficient in water, where the grass is also relative sparse. Low coverage density grassland is the natural grassland whose coverage density is between 5% and 20%. Such lawn is lack of water badly, where the grass is quite sparse. Therefore, the conditions applied to animal husbandry are very poor.

3.2 The model of grassland dynamic change

The land dynamic degree is quantitative changes of some or other type of land-use within a period of time in a certain study area. Using dynamic model to analyze the spatial-temporal changes of regional grassland can truly reflect the exquisite degree of its changes. The expression is:

$$LC=(U_b—U_a)\cdot U_a^{-1}\cdot T^{-1}\cdot 100\% \qquad (1)$$

where: LC represents dynamic degree of a certain type of land-use within study time; U_a and U_b represent the number of the certain land-use type at the beginning and at the end of the research; T represents the time that the study covered.

4. INTERPRETATION OF GRASSLAND RESOURCES BY REMOTE SENSING TECHNIQUES

4.1 Identification of grassland

We identified and extracted the grassland information on the TM images achieved by using the method of computer classification discrimination based on pixel. This method was to apply remote sensing spectral information, space information and time information to identify and classify the target. The non-supervised and supervised classifications were the general methods of classification (Guo Defang, 1987). We combined both methods and improved the algorithm in order to improve the accuracy of remote sensing satellite images automatic recognition (Li Xia; Jiao Weili, 1994).

4.1.1 Application of classification method

It was difficult to select training area when the classifying process because of regional various distribution types and sporadic size of the upper features. Thus, first we get the initial spectral information type map in the area by using dynamic clustering method contained by non-supervised classification, and then compared it with the recently land-use map to determine the relationship between the actual type of objects and spectral information. Artificial visual interpretation process was imported to this step that was the key classification process. The dynamic clustering means that the cluster center can be modified and changed continually, and the cluster number can be also changed and adjusted continuously in the classification process to make the classification more reasonable. As the classification emphasis was more on particular classification for grassland, we merged farmland, woodland and other water bodies correspondingly. Standard deviation means were used to establish the initial category center, and 13 categories spectral information type maps were obtained after iterative convergence. Such classification had better effect on large categories of features, but it was difficult to accurately differentiate some categories which had similar spectra value such as different levels of degradation of grassland.

The results of non-supervised classification were contradistinguished with the actual object categories, and then the grassland was further classified through surveying the samples, selecting the training area, and using supervised classification with expression (1) to deal with the results. Supervised classification adopted maximum likelihood classification method, which was based on Byes principle that pursues the minimum loss of average classification of the entire data sets, so that each data can be classified into the most similar category. The above method was called algorithms I.

Discriminant function:

If $G_i(x) \geq G_j(x)$ $i \neq j$ $(i, j = 1, 2, \ldots, N)$ (2)

then x/w_i

In the expression, i and j represent sequence numbers of categories; w_i represents sign of categories; $P(w_i)$ represents prior probability of w_i; x represents gray value of identified pixel; $G_i(x) = P(x/w_i)\,P(w_i)$ represents set of function for identifying; $P(x/w_i)$ represents conditional probability of x for w_i; N is the total number of categories.

The x was introduced into the various types of discriminant function when a certain pixel needed to be identified. Then the classification results were checked to determine whether the training area needs to be corrected or reclassified or not according to different types of image color and the results of unsupervised classification, we selected 10 training areas whose according categories of the surface features.

4.1.2 Improve classification accuracy

The entire classification process above all was completed with support of Geographic Information Systems (GIS). Regarding some complicated categories which were difficult to distinguish we used the algorithm II:

$$F_i = \sum_{j=1}^{N} \left| R_i - R'_j \right|$$

$$R_j = n_j / n_i \quad (i, j = 12, \ldots, N) \tag{3}$$

where: i and j represent sequence numbers of categories; R_j represents normalization probability of the training sample data for the j type; R'_j represents normalization probability of the local area for the j type; n_i represents the number of pixels which i type of training samples belonged to the j category; N represents the total number of categories.

Algorithm II is different from algorithm I because that it is not the classification of a single pixel as object but the classification of comprehensive analysis and judgment which based on a particular pixel and the surrounding linked pixels. On the basis of supervised classification with algorithm I, we can dispose the data with algorithm II, to further improve the classification accuracy by utilizing spatial information.

5. RESULTS AND ANALYSIS

The desertification, salinization and alkalization of grassland in the region were intensified along with the rapid growth of population, the weak consciousness of protection of grassland. Therefore grassland vegetation decreased or disappeared, the grassland biomass reduced sharply and grassland degenerated at last. Based on the results of remote sensing interpretation that were achieved by interpreting the TM images in 1986, 1996 and 2000, the degrees of regional grassland degradation were classified into mild, moderate and heavy three levels (Northern Grassland Resources Survey Office, 1986). Then in support of GIS, the grading statistics and subarea statistics were accomplished. At last, the situation and development trend of grassland degradation in west Jilin Province were analyzed.

5.1 Obviously decreased area of grassland

5.1.1 The analysis of time-series change on grassland area

According to the spatial information of remote sensing images in the studied region in the different periods, we analyzed the statistics of high, moderate and low coverage density grassland area Table 1 and area change of different types of grassland in 15 years. From Table 1, we can find out that the area of grassland had reduced 42.1% since 15 years ago, and the average annual reduction was 2.81%. The decreasing trend was obvious, which showed that grassland degradation was from bad to worse. High coverage density area continued to decrease with large quantity change of the area, and the rate of net reduction reached 53%, which indicated that the degradation of this grassland category was most serious. Moderate and low coverage density areas didn't change a lot from 1986 to 1996, but from 1996 to 2000 both areas represented a reducing trend overall respectively.

Table 1. Changes of grassland area (hm^2) and dynamic degree in west Jilin province

Type of grassland	Area of grassland in 1986	Area of grassland in 1996	Area of grassland in 2000	Dynamic degree 1986–1996	Dynamic degree 1996–2000
The high coverage density grassland	1052433.0	557218.7	494455.5	-4.7%	-2.8%
The moderate coverage density grassland	631459.8	623523.9	466486.3	-0.1%	-6.3%
The low coverage density grassland	41407.2	94357.4	37958.2	12.8%	-14.9%
Summation	1725300.0	1275100.0	998900.0	-2.6%	-5.4%

5.1.2 The analysis of grassland dynamic degree

The grassland dynamic degree can be used to quantify the instance of grassland degradation, so the grassland dynamic degrees in the studied region from 1986 to 2000 were calculated according to expression (1). Table 2 showed that from 1986 to 1996, grassland dynamic degrees changed drastically. Among them, the dynamic degrees of high coverage density grassland were negative, and the values of moderate coverage grassland density were very small though they were also negative. Otherwise, low coverage density grassland dynamic degrees were positive value and the total dynamic degree accorded with the one of high coverage density grassland which were both -2.6%. It indicated that in this period changes of high coverage density grassland have greater influenced on the unitary grassland change whereas the change of the total grassland area was smaller. According to the analysis of absolute values of dynamic degree, the low coverage grassland density grew the fastest because of its biggest value. The values of moderate and high coverage density grassland were relatively small and they developed comparatively slowly. From 1996 to 2000, various types of grassland and the total dynamic degree were all negative, which suggested that all types of grassland were degenerating, and the absolute values of low coverage density grassland dynamic degree were most. Thus the degradation of low coverage density grassland was the fastest.

5.1.3 The grassland area transfer analysis

In order to more accurately realize the degraded states of different types grassland, the values of high, moderate and low coverage grassland area transfer were acquired by using the land-use map and distribution map in 1986, 1996 and 2000 in support of spatial analysis function of ArcGIS software. The study showed that from 1986 to 1996, the total number of grassland changed to other land-use types reached 1136843.3hm^2. The grassland most converted to cropland, saline-alkali land and woodland, and especially the converted area of the high coverage density one reached 619765.9hm^2 which was the largest. It showed that the grassland in this period was so exploited in excess by human that caused large areas of grassland to be changed to farmland and woodland, at the same time, grassland degradation and salinization were more serious. Undoubtedly, all above all leaded to the further deterioration of ecological environment in the studied area and seriously affected the development of stockbreeding in Western Jilin, so we should pay more our attention to them. From 1995 to 2000, the grassland degradation represented an increasing trend, and 769268.7hm^2 grassland translated into other land-use types. More grassland degenerated into saline-alkali land compared with the previous period, and the trend aggravated. Especially, the high coverage density grassland degenerated most seriously. Despite the areas of grassland which changed to farmland and woodland were reduced, they still can not be ignored for their considerable proportions occupied.

5.2 The exacerbated degree of grassland degradation

The regional grassland degenerated seriously which also represented at the change of degradation degree. According to the grassland resources survey results which were achieved by using of the data of TM in 1986 and 2000, the areas of grassland degradation and the distribution of degradation grade Table 2 that belonged to the western counties (cities) were calculated. Since 1986, the area of grassland degradation had increased 47.8×10^4 hm^2 and the proportion had reached 86%. The light degenerated grassland decreased 66.4%, moderately degraded grassland increased 137.8%, and the heavy degenerated grassland increased most, up to 360.9%. All data above showed that the degree of grassland degradation had increased. In view of proportion, we can see that the grassland area in this region accounted 36.2% for the total area of regional land in 1986 and the degenerated grassland area accounted 32.2% for the area of grassland, however, the regional grassland area accounted 26.7% for the total regional land area while the area of degenerated grassland occupied to 81.1% of the grassland area in 2000.

Table 2. The degeneration of grassland in the West Jilin (10^4hm^2, %)

Type of grassland	In 1986		In 2000	
	area	proportion	area	proportion
Non-degenerated grassland	116.93	67.8	24.16	18.9
Degenerated grassland	55.60	32.2	103.35	81.1
Light degenerated grassland	25.78	46.4	8.65	8.4
Moderate degenerated grassland	19.16	34.5	45.57	44.1
Heavy degenerated grassland	10.66	19.2	49.13	47.5

REFERENCES

Chen Da-Ke. Introduction to Economic Ecology [M]. Harbin: the Northeast Forestry University Press, 1995:63–68.

Guo D F. Computer image processing of remote sensing and pattern recognition. Beijing: Press of electronic industry, 1987:243–390.

Li X, Jiao W L. A study on improving the classification accuracy on TM remote sensing image. Remote Sensing Technology and Application, 1994, 9(3):30–35.

Northern Grassland Resources Survey Office. Technical Rules of Grassland Resources Survey [M]. Beijing: China Agricultural Press. 1986.

Qiu S W, Zhang B, Wang Z C. Status, features and management practices of land desertification in the west of Jilin [J]. Geography Science, 2003, 23(2):188–192.

Wang D Y, Xu W L, Feng H, et al. Study on the element abundances and their characteristics of soil in grassland from western Jilin province. Scientia Geographica Sinica, 2002, 22(6):763–768.

RICE SHAPE PARAMETER DETECTION BASED ON IMAGE PROCESSING

Hua Gao[1,*], Yaqin Wang[1,2], Pingju Ge[1]
[1] *College of Information Science & Engineering, Shandong Agricultural University, Taian, China, 271018*
[2] *College of Geo Info. Science and Technology, Shandong University of Science and Technology, Tsingdao, China, 266510*
* *Corresponding author, Address: College of Information Science & Engineering, Shandong Agricultural University, Taian, 271018, P. R. China, Tel: +86-538-8242497, Fax: + 86-538-8249275, Email: gaoh@sdau.edu.com*

Abstract: Based on image processing technology, the detection, classification and feature extraction for plant grain shape are performed in this paper. Taking rice grain as an example, the shape detection and description method of similar round object are studied firstly. Then a grain shape description method based on 8 feature points of rice grain boundary is proposed. Aiming at rice seed detection, a simple image size calibration method based on black-white grid is put forward too. Finally, an extraction algorithm for 8 feature points is presented.

Keywords: rice grain, shape description, feature extraction, image processing

1. INTRODUCTION

The grain shape of plant seed is one of the important parameters for seed quality detection and classification (Ren et al., 2004). It is a hot issue all over the world to detect and classify plant grain shape based on image processing technology. At the same time, it has long been recognized as a difficult problem. Parameters describing plant grain shape are determined by grain shape feature and research objective. For some researches, only one point or several points shape features of object are needed to describe the shape (Van Eck et al., 1998). While for some other researches, the whole shape features

Gao, H., Wang, Y. and Ge, P., 2008, in IFIP International Federation for Information Processing, Volume 258; Computer and Computing Technologies in Agriculture, Vol. 1; Daoliang Li; (Boston: Springer), pp. 287–294.

of object are described in order to distinguish this kind of object from others or to find shape parameters related with its own special features (Heinemann et al., 1994; Wang et al., 1995). Here some parameters such as area, eccentricity, tightness, percentage of elongation, inertia center, ratio of grain length to width, ratio of grain width to thickness, grain sphericity, grain roundness and Fourier descriptor, are used to describe the object. For example, Ma xiaoyu selects four wheat shape correction coefficients to describe variety difference (Ma et al., 1999). Zhang cong proposes an ellipse matching-based object (rice grain) azimuth location and boundary description method (Zhang et al., 2006). Huang xingyi et al. use area, perimeter, roundness degree and rectangle degree to describe rice shape (Huang et al., 2003). In order to find suitable rice grain shape description method and to improve detection and classification precision, it is imperative that some key parameters should be found to describe grain feature exactly. Taking rice grain as an example, the shape detection and description method of similar round object are presented in this paper. Moreover, based on 8 feature points of rice boundary, the grain shape description method is proposed and its realization algorithm is given.

In computer visual technology-based measurement, image geometric distortion caused by camera lens may result in an inaccurate surveying result, so it is necessary to calibrate the camera before shooting object. Aiming at rice seed detection, a simple image size calibration method based on black-white grid is put forward too.

2. RICE IMAGE ACQUISITION AND IMAGE PIXEL-SIZE CALIBRATION

In order to decide rice image pixel size, it is necessary to calibrate camera before rice image acquisition. At present, most camera calibration algorithms calibrate internal and external parameter at the same time to obtain the relations among three coordinate systems (Huang et al., 2002; Zhong et al., 2005). But in seed detection, the most interested thing to be concerned by people is the relationship between image size and actual size of seed instead of internal or external parameter of camera. Therefore, it is no need to use traditional complicated calibrating method when the seed is detected.

In this paper, the image distortion model is simplified. It is supposed that there is only image size distortion and no image angle distortion. Based on this supposition, a simple image size calibrating method based on black-white grid is proposed.

The specific method is described as follows:

1. Install and fix the camera vertically above the object desk (See figure 1).
2. Shoot black-white grid image before seed image shoot (every black-white grid is a one-inch square, see figure 2).

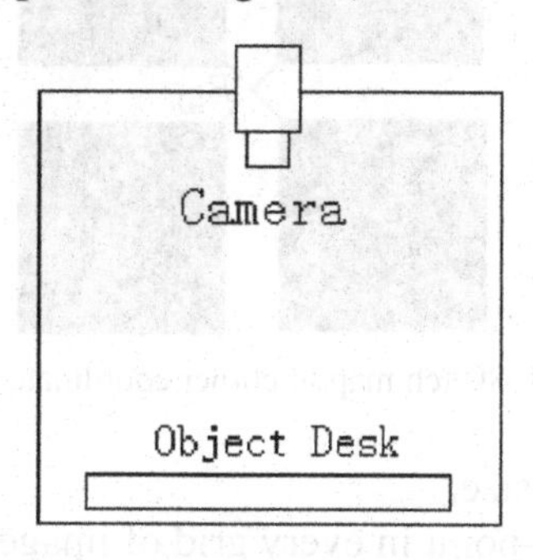

Figure 1. Sketch map of camera installation structure

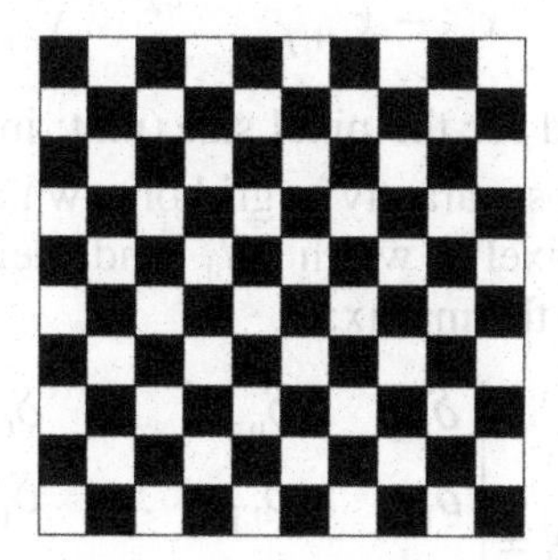

Figure 2. Black-white grid image used for size calibration

3. Search corner point coordinate of grid and get corner point coordinate matrix:

$$G = \begin{vmatrix} g_{0,0} & g_{0,1} & \cdots & g_{0,n-1} \\ g_{1,0}, & g_{1,1} & \cdots & g_{1,n-1} \\ \cdots & \cdots & \cdots & \cdots \\ g_{n-1,0} & g_{n-1,1} & \cdots & g_{n-1,n-1} \end{vmatrix}$$

Here, $g_{i,j} = (x_{i,j}, y_{i,j})$ is corner point coordinate except boundary corners. Because the boundary of black-white grid image is illegible, when calculating the corners, grid's region image is obtained firstly on the basis of region growing algorithm. Then four region inner points are searched. Finally, the cross point of four inner points are calculated. This cross point is the needed corner coordinate (See figure 3).

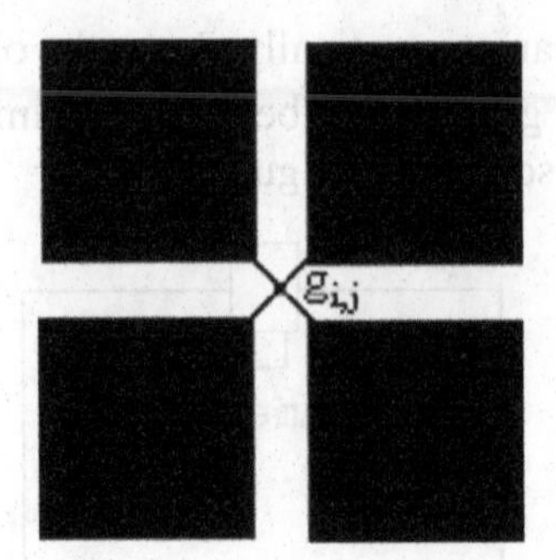

Figure 3. Sketch map of corner coordinate calculation

4. Calculate the pixel size.

The pixel size of core-point in every grid of image is:

$$W_{i,j} = \frac{1}{x_{i,j} - x_{i+1,j}} \qquad H_{i,j} = \frac{1}{y_{i,j} - x_{i,j+1}}$$

Here $W_{i,j}$ and $H_{i,j}$ stand for the pixel size (unit: inch) of core-point along x direction and y direction separately in grid of row i and column j.

5. Calculate every pixel's width W_{ij} and height H_{ij} based on liner difference, and establish the matrix:

$$\Delta = \begin{vmatrix} \delta_{0,0} & \delta_{0,1} & \cdots & \delta_{0,m-1} \\ \delta_{1,0}, & \delta_{1,1} & \cdots & \delta_{1,m-1} \\ \cdots & \cdots & \cdots & \cdots \\ \delta_{n-1,0} & \delta_{n-1,1} & \cdots & \delta_{m-1,m-1} \end{vmatrix}$$

Here $\delta_{i,j}=(W_{ij},H_{ij})$ is the width and height of pixel.

With this matrix we can calculate rice image pixel size accurately. In order to simplify the algorithm, we can substitute rice grain centroid pixel's coefficient for the whole grain image pixel' coefficient when the rice grain size is calculated.

6. Shoot the rice grain: In order to establish the mapping relationship between rice image and pixel size coefficient matrix, it is imperative to put the rice grain on the center of the desk equably.

3. DESCRIPTION OF RICE GRAIN SHAPE

Rice grain shape is very complex and is affected by many factors. There are many methods to describe grain shape at present, but they do have some restrictions. For example, some methods, such as area method, eccentricity methods, percentage of elongation, inertia center, ratio of grain length to width, ratio of grain width to thickness, grain sphericity, and grain roundness, have a lower precision. Fourier descriptor and some other methods are more accurate but they need to do a large amount of calculations and have a bad visualizability. A new method is presented in this paper, which describes grain shape with 8 selected points on grain boundary. As is shown in figure 4(d). This method has several advantages when it is used to describe grain shape, such as simple calculation, small amount of data, accurate description etc. The specific algorithm is described as follows:

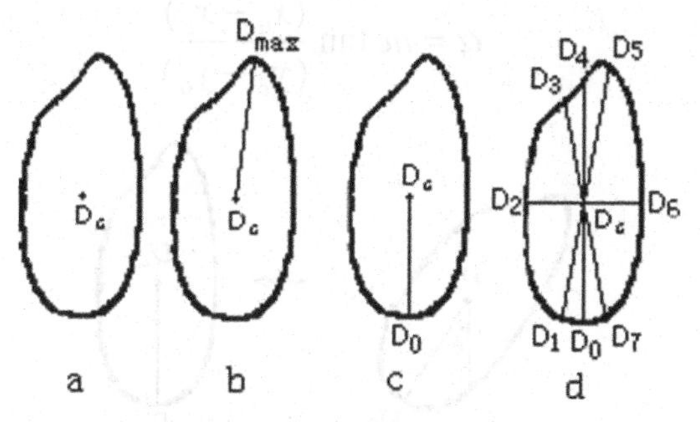

Figure 4. Searching of 8 feature points

1. Calculate centroid coordinate D_c

Centroid coordinate can be expressed by either average of rice boundary coordinate or region barycenter. The former one is used in this paper. See figure 4(a).

$$x_c = \frac{1}{M}\sum_{i=0}^{M-1} x_i \quad y_c = \frac{1}{M}\sum_{i=0}^{M-1} y_i$$

2. Find point D_{max} in the boundary which has the longest distance from centroid.

L_i is the distance from boundary point D_i to centroid D_c

$$L_i = \sqrt{(x_i - x_c)^2 + (y_i - y_c)^2}$$

Find maximum L_{max} along the boundary to get the coordinate of D_{max}. See figure 4(b).

3. Find the longest distance point D_0 within boundary local region, which is on the opposite side of maximum distance point.

Find maximum value along boundary within the scope of $\pm\pi/4$ opposite to D_{max} to get the coordinate of D_0. This point is the first one of 8 points. See figure 4(c).

4. Start from D_0 and search the boundary point clockwise to find D_1 to D_7 one by one. The interval angle between every two feature points are $\pi/16$, $7\pi/16$, $7\pi/16$, $\pi/16$, $\pi/16$, $7\pi/16$, $7\pi/16$. See figure 4(d). The angle can be calculated according to following formula:

$$\theta = ac\tan\frac{(y_c - y_f)}{(x_c - x_f)} - ac\tan\frac{(y_c - y_a)}{(x_c - x_a)}$$

5. Image rotation

When shooting rice image, the direction of grain is random. In order to standardize image, it is needed to rotate the image boundary. The rotation angle is the include angle between line D_0D_c and perpendicular line. Here D_c is centroid point.

$$\alpha = ac\tan\frac{(x_0 - x_c)}{(y_c - y_0)}$$

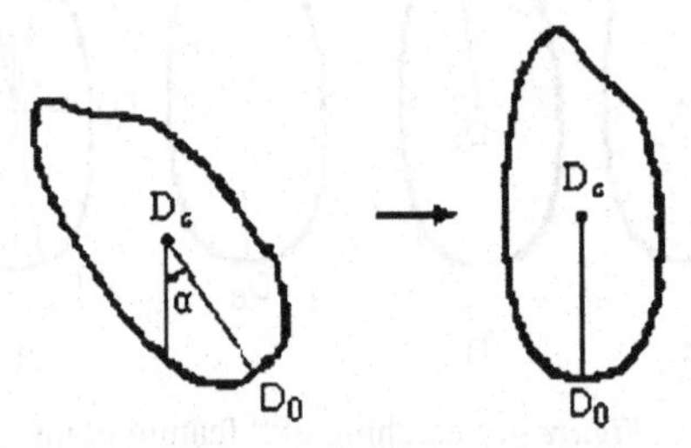

Figure 5. Image rotation

Rotation transformation is performed as follows. See figure 5.

$$\begin{bmatrix} x \\ y \\ 1 \end{bmatrix} = \begin{bmatrix} \cos\alpha & -\sin\alpha & 0 \\ \sin\alpha & \cos\alpha & 0 \\ 0 & 0 & 1 \end{bmatrix} \begin{bmatrix} x \\ y \\ 1 \end{bmatrix}$$

6. Flip horizontal

For the same rice grain, two opposite placement may result in two different images, see figure 6. In order to let 8 describing points have the same sequence, all the embryos are on the top-left in the image. For every nonstandard grain image, flip horizontal should be done along the perpendicular line crossing centroid point.

Supposed that L_3 is the distance from point D_3 to centroid point, and L_5 is the distance from point D_5 to centroid point. Let's compare L_3 with L_5, if $L_3 > L_5$, then to flip grain image horizontal, otherwise, no flip. The flip formula is:

$$\begin{bmatrix} x \\ y \\ 1 \end{bmatrix} = \begin{bmatrix} -1 & 0 & fWidth \\ 0 & 1 & 0 \\ 0 & 0 & 1 \end{bmatrix} \begin{bmatrix} x \\ y \\ 1 \end{bmatrix}$$

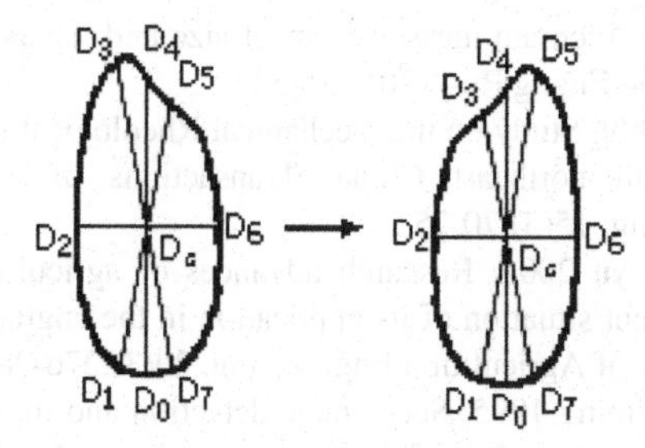

Figure 6. Flip image horizontal

7. Measurement unit conversion

All the measurement units above are pixel. While in practice, the measurement unit should be conversed to length unit. Therefore, the coordinate of every grain's centroid point D_c is recorded, and element $\delta_{i,j}$ is used as grain size coefficient. Here $\delta_{i,j}=(W_{ij},H_{ij})$, is coming from pixel size coefficient matrix. Then, the measurement unit conversion will be performed.

4. CONCLUSIONS

Too fine description of agricultural product grain shape is very complicated. Sometimes it will make trouble for analysis and recognition. In this paper, the method to describe rice grain shape with 8 feature points is not only simple, but also high efficient. Some key feature parameters of rice are contained in the eight data, such as length, width, perimeter, area as well as ellipse approximate degree. Combined with neural network in the experiment, this method has a good effect on the recognition and classification of rice.

The simple method to calibrate rice image size proposed in this paper is also fit to every fixed distance shoot image size calibration.

REFERENCES

Heinemann P H et al. 1994, Grading of mushrooms using a machine vision system. Transactions of ASAE. 37(5):1671-1677.

Huang Fengrong, Liu Jiaomin, Sun Zhuangzhi 2002, Distance measurement system based on new method of determining the three-dimensional position and orientation of object from a single view. Computer Engineering and Applications. 38(4):236-238.

Huang Xingyi et al. 2003, Inspection of chalk degree of rice using genetic neural network. Transactions of The Chinese Society of Agricultural Engineering. 19(3):137-139.

J W Van Eck et al. 1998, Accurate measurement of size and shape of cucumber fruits with image analysis. J Aagric Engng Res. (70):335-343.

Ma Xiaoyu, Lei Detian 1999, Study on the mechanical Rheological properties of soybean and wheat grain grown in northeast China. Transactions of The Chinese Society of Agricultural Engineering. 15(3):70-75

Ren Xianzhong, Ma Xiaoyu 2004, Research advances of agricultural product grain shape identification and current situation of its application in the engineering field. Transactions of The Chinese Society of Agricultural Engineering. 20(3):276-280.

Wang Fengyuan, Zhou Yiming 1995, Seed shape detection and measurement with computer image processing. Transaction of the Chinese society of agricultural machinery. 26(2):52-57.

Zhang Cong, Guan Shuan 2006, Rice figure identification based on an image analysis. Cereal & Feed Industry. 6: 5-7

Zhong Zhiguang, Yi Jianqiang, Zhao Dongbin 2005. A geometric approach for camera calibration based on point pairs. Robot. 1(27):31-35

REPRESENTATION AND CALCULATION METHOD OF PLANT ROOT

Hua Gao[1,*], Yaqin Wang[1,2], Zhijun Wang[1]
[1] *College of Information Science & Engineering, Shandong Agricultural University, Taian, China, 271018*
[2] *College of Geo Info. Science and Technology, Shandong University of Science and Technology, Tsingdao, China, 266510*
* *Corresponding author, Address: College of Information Science & Engineering, Shandong Agricultural University, Taian, 271018, P. R. China, Tel: +86-538-8242497, Fax: + 86-538-8249275, Email: gaoh@sdau.edu.com*

Abstract: In order to quantitatively analyze plant root growth and construct plant root growth model, a root system representation model based on tree-form data structure is proposed in this paper. After processing and analyzing root image, the information of root, such as length, diameter, number of child root, position of child root, angle of child root etc., is stored in a node of this tree-form data structure. The accurate representation of root system is realized, which provides a foundation for the extraction and analysis of various root parameters. The specific algorithm is presented too.

Keywords: root image, representation model, tree-form data structure, image processing

1. INTRODUCTION

The architecture of plant root is the spatial configuration and distribution during its growth. It has an important effect on the nutrition and water absorption for plant from environment and soil. Studying the characteristics of plant root architecture can help people to deeply recognize plant root distribution, configuration, function, and to estimate the ability as well as to describe the process of plant root acquiring nutrition from soil in detail.

Root configuration includes stereo and planar configuration. Stereo configuration is a 3-D spatial distribution of different types of roots in

Gao, H., Wang, Y. and Wang, Z., 2008, in IFIP International Federation for Information Processing, Volume 258; Computer and Computing Technologies in Agriculture, Vol. 1; Daoliang Li; (Boston: Springer), pp. 295–301.

medium. Planar geometric configuration is a 2-D planar distribution of various roots of the same root system along the root axis (Ge, 2002). In order to understand actual situation of growth and distribution of plant root in the soil, many measuring methods for root system have been studied (Luo, 2004). This kind of research has begun since 18 century.

The research of computer recognition for plant root growth characteristic has been paid a significant attention all over the world. Each country has made a different degree of progress (Zoom FC, 1990; Shuman LM, 1993; Hu XJ, 2003). But the algorithm is not mature and the effect is not good.

In order to quantitatively analyze plant root growth and construct plant root growth model, an image processing-based plant root tree-form representation method is proposed in this paper. By using this method, the whole plant root architecture is expressed with a tree-form data structure. In this data structure, a root is represented with a node, and a child root is expressed with a child node. Experimental results show that this method can represent plant root architecture simply, naturally and accurately.

With above tree-form data structure, the root data can be stored into this structure. Based on image processing, the analysis and calculation of plant root can be implemented. Furthermore, an image thinning-based root system search algorithm is proposed in this paper. On the basis of binarization, filtering and thinning of original image, this method searches the whole root image with a breadth first search. After corresponding calculation and judgment, the calculation result information is stored into tree-form data structure. As a result, the representation of plant root is performed, which provides a solid foundation for the retrieval and analysis of various plant root parameters.

2. PLANT ROOT TREE-FORM REPRESENTATION DATA STRUCTURE

A multi-way tree structure is used to express a whole plant root. Here, main node expresses main root, while every child node expresses each child root. Main node and child node are represented with the same data structure, which includes description information and two chained lists. One chained list records detail information of current root, while the other records child root information.

Data structure of "expression tree" node:
<child root ID>;
<start point coordinates>;
<length>;
<included angle with *parent root>;*
<expression chained *list>;*
<number of child root>;
<chained list of child root>.

Data structure of expression chained list:
<*coordinates*>;
<*width*>.

Data structure of child root chained list:
<*child root ID*>;
<*start point* coordinates>.

With above data structure of expression tree, root information can be stored into this structure. In this paper, the measurement and calculation of plant root is realized based on image processing.

3. ACQUISITION OF ROOT IMAGE

It is the first step or even the very critical step for image processing and analysis to acquire appropriate and high quality image. The obtained root image should satisfy following conditions:

(1) It includes all the external figure information of measured root;
(2) It should be separated easily from background image;
(3) Useless background information should be reduced as much as possible.

Based on above three requests, it is needed to take following four measures when root image shooting.

(1) In order to separate root image from background easily, blue A4 paper is used as background, which has a bigger color difference with root system.
(2) In order to ensure measured root system has an even light requirement along shooting direction, so as to catch high quality image, incandescence lamp with diffusion property is used as light source.
(3) The measured root system should be laid as a plane surface when shooting. It is easy to realize. Generally speaking, common plant root is flexible. When the root is laid on the desk, it is nearly planar. Moreover, it is imperative that the root should be laid without crossover.
(4) The end of main root should be laid on the topside so as to be searched easily.

4. IMAGE PREPROCESSING

Firstly, the original image is processed by binarization, filtering and thinning to obtain a search image with only one single pixel width. See figure 1 to figure 3.

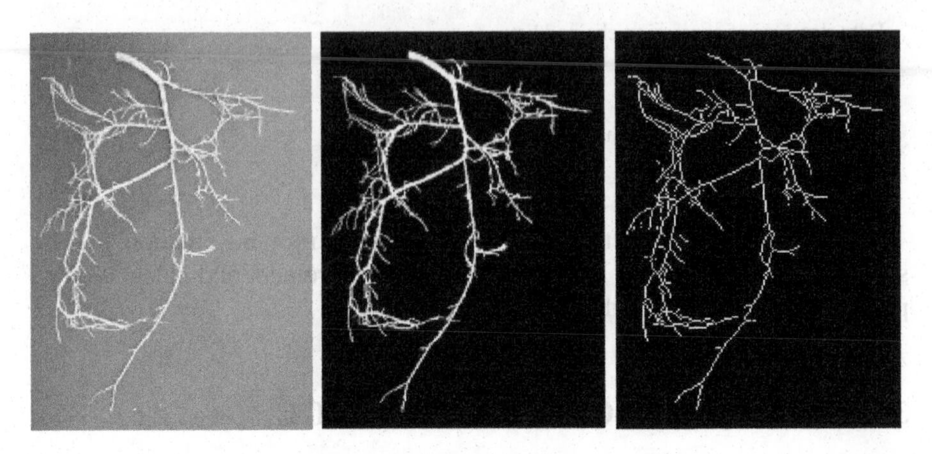

Figure 1. Original root image *Figure 2.* Binary root image *Figure 3.* Thinning root image

5. IMAGE SEARCH ALGORITHM

The first step of this algorithm is finding start point of main root: When thinning image is scanned, the first point encountered is the start point of main root. Then, the root node of expression tree can be generated. The data is initialized as follows:

<*start point coordinate*> = start point coordinate;
<*length*>=0;
<*included* angle *with parent root*> =0;
<*expression* chained *list*>=empty;
<*child root* number>=0;
<*child root chained list*>=empty

The search is begun. The algorithm searches the unlabeled point all the way. When a point is found, it will be labeled and be judged whether it has branches. See figure 4. In figure 4, ▣ represents labeled pixel point, □ represents unlabeled pixel point, and ⊠ represents current pixel point to be processed. In order to determine branch point, the number of 8-neighborhood pixel point except labeled point of current pixel point is used. The method is: if current pixel point has no 8-neighborhood pixel point except labeled point, this point is end point. If there is one 8-neigborhood point, it has no branch. If there are two or more than two 8-neighborhood points, it is a branch point. There are different processing methods according to different number of branches.

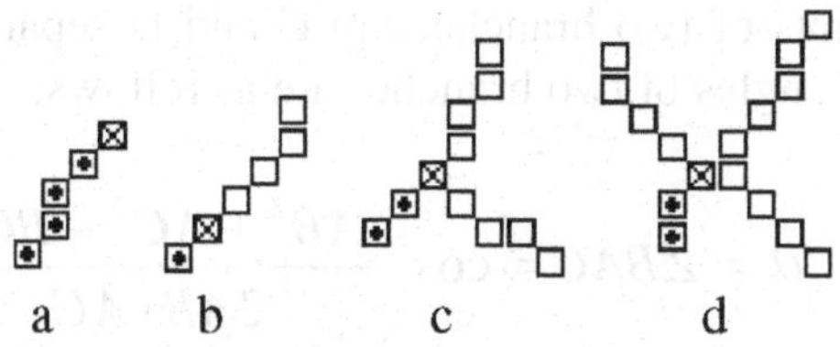

Figure 4. Sketch map of branch situation

(1) If current point is end point:

Here, current pixel point has no 8-neighborhood pixel point except labeled point. See figure 4(a).

Let's label this point;
<*length*>=<*length*>+1;
Add a node in <*expression chained list*>;
<*coordinate*>=*current* point coordinate;
<*width*>=1;

After this node is processed, the brother node or child node will be processed. If all the nodes in this layer have been processed, the next one to be processed is child node. Otherwise, the next one is brother node.

(2) There is only one 8-neighborhood point:

It is obvious that this point has no branch. See figure 4(b).

Let's calculate the root diameter of this point. That is to say to calculate the root image width of this point. According to the coordinate of this point, the two boundary points on the binary image are searched. Then, the distance D between these two boundary points represents diameter of root in this point.

Let's label this point;

<*length*>=<*length*>+1;
Add a node in <*expression chained list*>;
<*coordinate*>=current point coordinate;
<*width*>=D

The neighborhood point is used as current point.

(3) If there are two 8-neighborhood points:

It is obvious that there are two branches in this point. Then this point has a child root. See figure 4 (c).

At this time, it is needed to search the third point behind neighborhood point along two neighborhood points separately (if there are less than three points after neighborhood point, it seemed as noise and should be removed). Then the included angle between each branch and main root are calculated. Supposed that current point is A, the third point before current point is B,

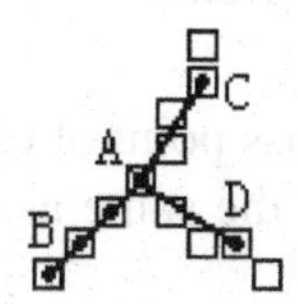

Figure 5. Main root and child root distinguishing in two branches

and two third points of two branches are C and D separately. See figure 5. The two included angles of two branches are as follows:

$$\theta_1 = \angle BAC = \cos^{-1}\frac{AB^2 + AC^2 - BC^2}{2AB \cdot AC}$$

$$\theta_2 = \angle BAD = \cos^{-1}\frac{AB^2 + AD^2 - BD^2}{2AB \cdot AD}$$

Here AB, AC, BC, and BD represent distance between two points.

Let's compare θ_1 with θ_2, the bigger one is main root, and smaller one is child root.

Let's label this point;
<length>=*<length>*+1;
Add a node in *<expression chained list>*;
<coordinate>=current point coordinate;
<width>=D;
<child root number>=*<child root number>*+1
Add a node in *<child root chained list>*;
<child root number>=previous *<child root number> +1;*
<start point coordinate>= current point coordinate;

The neighborhood point with smaller included angle is used as current point.

(4) If there are three 8-neighborhood points, it shows that this point has three branches. See figure 4(d).

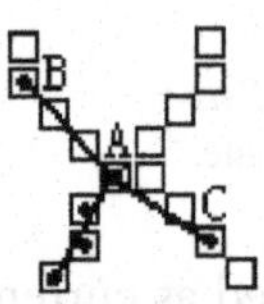

Figure 6. Three branches distinguishing

Now there are two situations. The first situation is that this point is the cross point of two roots. The middle path is current root. The other two paths form anther root. The second situation is that there are two child roots except main root. This can be judged according to the included angle of two paths. See figure 6.

$$\theta = \angle BAC = \cos^{-1}\frac{AB^2 + AC^2 - BC^2}{2AB \cdot AC}$$

If $\theta > 165°$, this point is a cross point of two roots. The other root need not be processed. The procedure is the same as step □. But the tag of this point should be removed. If $\theta \leq 165°$, this point has two child roots except main root. The node data should be modified (add two child roots).

To repeat above procedure until end point is found. After main root is processed, the child root of main root will be searched one by one. Because there may exist loop during the search, the breadth first search method instead of deep first search is used.

6. CONCLUSIONS

The root system is fully represented by tree-form data structure. The information such as root length, root diameter, number of child root, position of child root, angle of child root etc. is stored in a node. The accurate representation of root system is realized, which provides a foundation for the extraction and analysis of various root parameters.

However, the research of root system is on a plane surface in this paper. Further research can be finished on the stereo image of root system to realize analysis of stereo spatial configuration of root system.

Moreover, the algorithm proposed in this paper can only detect root configuration parameters of tap root system. It needs to be perfected in future, so as to realize the measurement of root configuration parameter for every kind of root system.

REFERENCES

Ge Zhenyang, Yan Xiaolong, Luo Xiwen 2002, Simulation Models of Plant Root System Architecture and Application: A Review. Trans. CSAE. 18 (3):154-160.

Hu Xiujuan 2003, The Application of Digit Image Treatment Technique in Plant Root Study. Forestry Machinery & Woodworking Equipment. 31 (3):29-31.

Luo Xiwen, Zhou Xuecheng, Yan Xiaolong 2004, Visualization of Plant Root Morphology in situ Based on X-ray CT Imaging Technology. Transactions of the Chinese Society for Agricultural Machinery. 35 (2):104-106.

Shuman L.M, Ramseur E.L, Wilson, D 1993, Video image method compared to a hand method for determining root lengths. Journal of Plant Nutrition. 16 (4):563-571.

Zoom F.C, van Tienderen P.H 1990, A rapid quantitative measurement of root length and root branching by microcomputer image analysis. Plant and Soil. 126:301-308.

STUDY ON LINEAR APPRAISAL OF DAIRY COW'S CONFORMATION BASED ON IMAGE PROCESSING

Dongping Qian[1,*], Wendi Wang[1], Xiaojing Huo[1], Juan Tang[1]

[1] *College of Mechanical and Electrical Engineering, Agricultural University of Hebei, Baoding, China, 071001*

[*] *Corresponding author, Address: College of Mechanical and Electrical Engineering, Agricultural University of Hebei, Hebei, Baoding, 071001, P. R. China, Tel: +86-0312-7526450, Email: qdp@mail.hebau.edu.cn*

Abstract: Image processing technology is more commonly applied to such fields as virtual reality, biological medicine and permeates rapidly through agricultural scientific research. This paper described a linear appraisal system for analyzing dairy cow types based on image processing, which consists of hardware design, software design, realization of characters click points operating and the interactive measurement of type character parameter for dairy cow. We use NI Lab windows/CVI and NI IMAQ Vision to develop the system software. The key purpose of image preconditioning is to clear up noise, and we use median filtering to eliminate noise during this process. Adopting image measurement method to make dairy cow type linear appraisal is convenient and swift, precision satisfying the request, being able to replace manual appraisal. The system conduces greatly to application research of image processing technology in the type linear appraisal for dairy cows.

Keywords: image processing, linear appraisal, median filtering, Dairy cow

1. INTRODUCTION

Dairy cow type linear appraisal, which was used formally in dairy cow type appraisal in 1983, is a type appraisal method that was developed in America in 1980s (Li et al., 2002; Liu et al., 1994; Chu et al., 1996). The

Qian, D., Wang, W., Huo, X. and Tang, J., 2008, in IFIP International Federation for Information Processing, Volume 258; Computer and Computing Technologies in Agriculture, Vol. 1; Daoliang Li; (Boston: Springer), pp. 303–311.

application of dairy cow type linear appraisal has been studied in China since 1987. The trial standard and demand of appraisal was released in 1995, which stipulates fifteen first-class trait indexes for appraisal (Dao, 2002). The information of type conformation and appearance was used in cattle's inheritance appraisal in China from the result of linear appraisal. Because the items of dairy cow appraisal and data processed are too much, it is necessary to develop a rational and efficient dairy cow linear appraisal system based on computer image processing instead of manual appraisal.

At present, image processing technology is more and more applied to such fields as industry measurement, control and guide, virtual reality and biology medicine so on, and permeates rapidly through agriculture scientific research and produce technology field. Image measuring method makes dairy cow type linear appraisal, which not only reduces workload but also gets rid of the shortcoming of manual appraisal standard changing with men and time changing and the disadvantage of lacking impersonal impartiality. Shunsan Chen, one of graduate students in China Agricultural University, has studied image processing technology of dairy cow type linear appraisal, but they only selected monochrome acquisition board so that the information that was acquired was not enough and measurement was inaccurate. Moreover they measured only 9 traits, other 6 traits have been given score by eyeballing, so they can not appraise all traits automatically. The system adopts a camera to acquire dairy cow image, color image acquisition board to convert its signal into digital signal, image processing technology to measure every trait, so that measurement precision improved and workload is reduced.

There are at least several meanings to use image measurement method in dairy cow type linear appraisal as follows:

(1) It is a not touched measurement method and does not disturb object monitored; (2) Workload of manual measurement is not only more but also has definite danger, which is avoided by computer measuring; (3) Get rid of the shortcoming of manual appraisal standard changing with men and time changing and the disadvantage of lacking impersonal impartiality.

2. SCHEME DESIGN OF IMAGE ACQUISITION AND MEASUREMENT SYSTEM

2.1 System hardware design

System hardware is made up of computer, system display, image acquisition board, CCD Camera, printer and so on, as Fig.1 shows. Image acquisition board is DH-CG300 type color video acquisition board made by

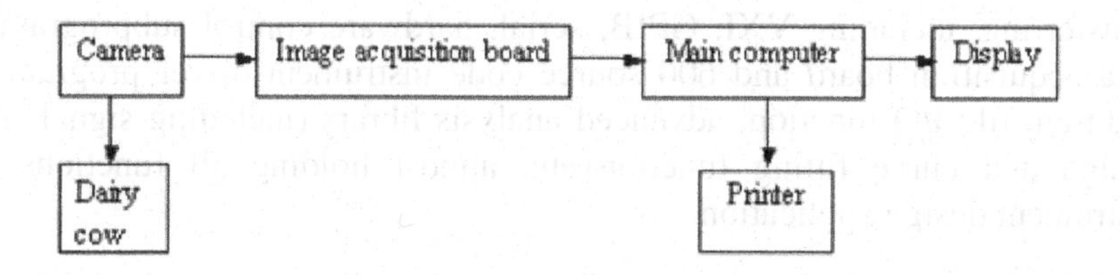

Fig. 1. The block diagram of the hardware structure

Daheng company, image resolution is maximal: PAL: 768×576×24 BIT; NTSC:640×480×24 BIT, supporting single field, single frame, continuous field and continuous frame acquisition way, having red, green and blue three route D/A, color image signal may be inputted to R, G, B three frames deposit body at the same time, and also saved in R, G, B three frames deposit body separately. CCD camera applies Panasonic WV-CP240/G color camera, its pixel is 752 (horizontal) × 582 (vertical), CCD is row transform, its scanning area is 4.89 (horizontal) × 3.67 (vertical) mm (corresponding 1/3inch vidicon scanning area), signal-to-noise is about 50 db.

Dairy cow image may be acquired real-time, and also use dairy cow picture by scanner scanning. Concrete steps as follows: camera acquires dairy cow picture, and converts it into digital image file, image file are inputted computer by color image acquisition board, after that, first converting color image into gray image, then computer processing measuring dairy cow gray image and calculating the score of linear appraisal.

2.2 Software design of the system

The development tool (Deng et al., 2001) of system software is NI IMAQ Vision and NI Lab Window/CVI 6.0. Operating system is Chinese Win2000. IMAQ Vision is a advanced image processing analysis software package. It includes a set of rich MMX optimized library, having gray, color and binary image displaying, processing (statistic filtering and geometry transform) and image morphological processing etc functions, improving rapidly the development course of user image processing and machine vision application. Lab Windows/CVI6.0 (Zhang et al., 2002) is development environment of virtual instruments, holding libraries of powerful function which is used to build the application program of data acquisition and instrument control. Developing program can use developed C language object module, DLL, C static library and instrument driver program. It integrates source code programming, 32 bit ANSIC compiling, connecting, debugging and standard ANSIC library in a interactive development platform. Lab Windows/CVI provides plenty of functional libraries for user transferring, including VXI, GPIB, serial, hardware control subprogram of data acquisition board and 600 source code instrument driver program, in addition, file I/O function, advanced analysis library (including signal filter design and curve fitting function) etc, almost holding all functions for instrument design application.

3. REALIZATION OF CHARACTER CLICK POINTS OPERATING

3.1 Image acquisition and preprocessing

The quality of image acquisition is one of main factors which decide the precision of type appraisal, requiring acquiring image layer to be abundant, detail to be clear, background contrast to be big (Jimenez et al., 1998–2002); so background should be chosen blue one and lighting should adopt reflection light, which can avoid shadow and reduce measurement error. With dairy cow in the standard state of four limbs coexisting, acquire three positive side images, choose the clearest one, similarly positive frontage and positive rear three, and separately choose the clearest one. After passing through window reduced, image of entering inspect view is saved, which will decrease processing time and economize storage space.

Making dairy cow image processing is difficult, because of black and white figure of dairy cow affecting. The image which this research acquires is RGB pattern, it is necessary to change RGB pattern into HSL pattern by extracting Saturation S value, so that color image is converted into gray scale image to realize gray scale processing of dairy cow (Huang et al., 2000; Zhou et al., 2003).

The system applies to median filtering to wipe off noise. In some condition, median filtering method may not only eliminate noise, but also protect image edge to gain satisfactory recover. Basic principle of median filtering is that a point in the digital image or numerical sequence is replaced by the middle value of every point value in a neighborhood of the point.

Assume that there is one dimension array $x_1,x_2,x_3,\ldots,x_n$, according to size order ranged just as $x(1) \le x(2) \le x(3) \cdots \le x(n)$

So that their median filtering output is written as

$y=\mathrm{med}(x_1,x_2,x_3,\cdots,x_n)$

$$= \begin{cases} x(\frac{n+1}{2}) & n \text{ is odd} \\ \frac{1}{2}[x(\frac{n}{2}) + x(\frac{n}{2}+1)] & n \text{ is even} \end{cases} \tag{1}$$

Median filtering is easy to extend two dimensions; here two dimension of some formal can be used. Each point gray scale value of digital image is expressed as $\{x_{i,j}, (i,j) \in I^2\}$, two dimension median filtering is calculated as

$$y_{i,j} = Med\{x_{i,j}\} = Med\{x_{(i+r),(j+s)}, \\ (r,s) \in A, (i,j) \in I^2\} \tag{2}$$

Where A is filtering window

Filtering window chooses 3×3 window.

Fig. 2 shows that dairy cow's edge image is smoothed by mean filtering and median filtering. It can be seen that mean filtering wipes off noise, and blur dairy cow's image edge simultaneously, the effect of median filtering is better, not only wiping off noise availably, but also saving image edge information clearer (Castlman, 2002).

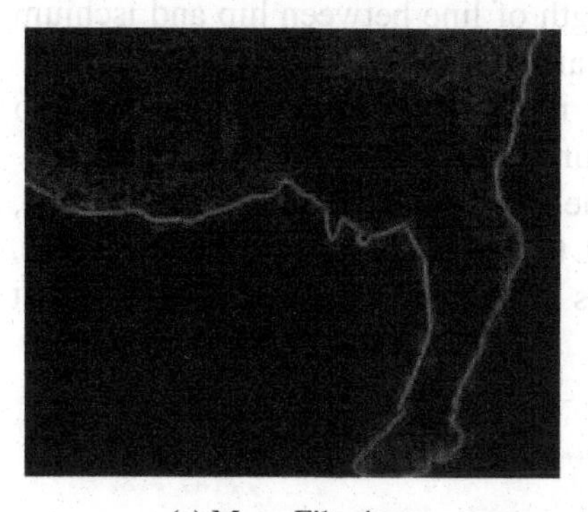

(a) Mean Filtering

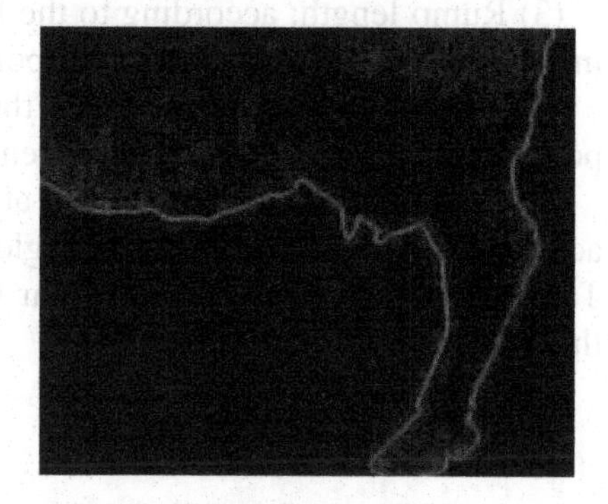

(b) Median Filtering

Fig. 2. Diary cow's edge image

3.2 Calibration

In control of system, click to standard ruler then measure the number of pixel between two points n_1, and input corresponding trim size l_1, consequently work out trim size that each pixel delegates $s_1=l_1/n_1$, measure the number of pixel between character two points n_2, again multiply s_1, educe dairy cow length between two points. According to the measured length of two points, using cosine theorem gains angle among three points (Xu et al., 2002).

4. THE INTERACTIVE MEASUERMENT OF TYPE CHARACTER PARAMETER FOR DAIRY COW

The measurement of most character for dairy cow may convert to measure the distance between two points or angle among three points, therefore make sign for corresponding character points. Fig. 3 shows interface of clicking operation. The measurement way of each type character for dairy cow as follows.

(1) Stature: according to the height of wither point make linear grade. Line out wither point and point of intersection between wither point and ground then gain body height of dairy cow.

(2) Rump angle: according to angle between the line of hip connecting ischium and horizontal make linear grade. Line out hip point and ischium point then gain rump angle.

(3) Rump length: according to the length of line between hip and ischium make linear grade. At the front hip point and ischium point are lined out.

(4) Rump width: according to thurl make linear grade. Line out two points of thurl at rear view figure then gain rump width.

(5) Rear legs side view: look at the posture of rear legs from side, according to the angle of hock angle at the tarsal joint make linear grade. Tag three points on the hock of rear legs at the side view fig then work out the angle of hock angle.

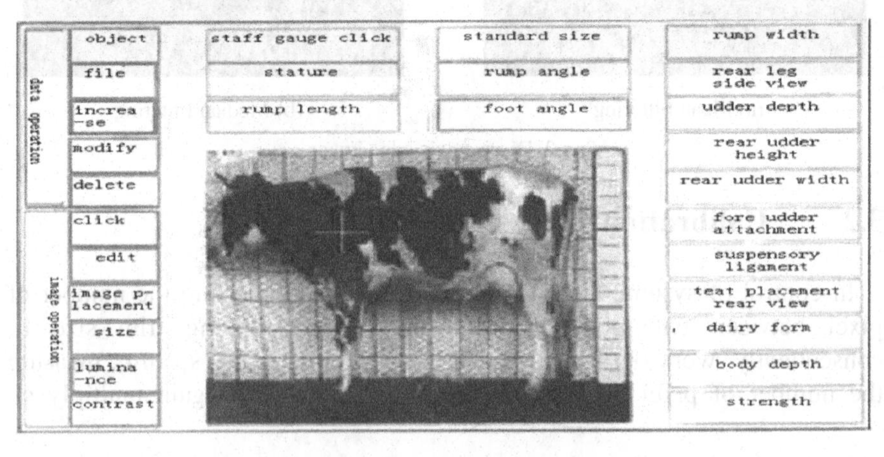

Fig. 3. The operation interface of dairy cow characteristic clicking

(6) Foot angle: according to the angle between hoof parietal and hoof bottom make linear grade. Tag three points on the toes, hoof parietal and hoof bottom then gain foot angle.

(7) Rear udder height: according to rear udder attachment point (transition point of rear legs socket connecting udder) changing between ischium and hock point make linear grade. Tag right and left attachment point, ischium point, hock point then work out rear udder height.

(8) Rear udder width: according to the width of two attachment points of rear view udder make linear grade. Two attachment points are tagged in front.

(9) Udder depth: according to relative position of udder bottom plane and hock make linear, and line out the placement of udder floor at rear view fig then udder depth.

(10) Fore udder attachment: the angle of joint of udder fore hem and abdominal wall.

(11) Suspensory ligament: depth from rear udder basal to median suspensory ligament.

Four fuzzy traits such as strength, body depth, dairy form, teat placement rear view are given to grade by eyes.

5. GRADE TEST

Primary test chose image of six dairy cows to make appraisal, making image click then measure appraisal to eleven traits such as stature, rump width, rump length, rump angle, rear legs side view, udder depth, rear udder height, rear udder width, foot angle, fore udder attachment, suspensory ligament, giving appraisal to four fuzzy traits such as strength, body depth, dairy form, teat placement rear view by eyes. Because linear appraisal is uniform scale size that represents different state between two extremeness of biology, it can not illuminate good or bad of each type trait exactly. To embody better or inferior of type trait, gain linear appraisal of dairy cow, after that, convert them to functional appraisal, finally work out total appraisal. The time which appraising each dairy cow takes is less than 3 min, which is less than 5 min standard regulating, choosing stature that can reflect most measurement precision, appraisal result of computer image appraisal are compared with mutual appraisal, which biggest absolute error is 1.3 cm, relative error 0.9%. The biggest error between computer appraisal and mutual appraisal is 4 point, which is within confessional range. From Table 1, it can be seen appraisal results is gained by linear appraisal method of dairy cow's conformation based on image processing for six dairy cows.

Table 1. Score comparison

Dairy cow	stature/cm		Score		Grade
	Manual	Image	Manual	Image	
1	137.4	136.7	75	76	G
2	140.8	139.9	82	81	G+
3	134.5	134.2	70	72	F
4	137.1	136.4	76	78	G
5	141.3	139.7	84	80	G+
6	132.4	132.2	67	69	F

6. CONCLUSIONS

(1) Making use of Lab Windows/CVI 6.0 and analysis function library of image processing IMAQ Vision put up the development of type linear appraisal system for dairy cow, which is feasible in technology and conduces greatly to application research of image processing technology in the type linear appraisal for dairy cow, and which makes system take on such characters as interface friendly, agility using easily, favorable extensibility.

(2) Adopting image measurement method to make dairy cow type linear appraisal is convenient and swift, precision satisfying request, being able to replace manual appraisal.

(3) Central factor of affecting appraisal precision is the quality of image acquired. The more evident administrative levels of image are, the clearer character points are, the bigger background contrast is, the higher orientation precision is.

ACKNOWLEDGEMENTS

The research work in the paper is supported by the 863 Project (2006AA10Z252) and Hebei Province Science and Technology Department Priorities Program (No. 012134100).

REFERENCES

Castleman K R, Digital Image Processing, Beijing: Machinery Industry Pubulishing House, 2002.

Dao E J, Estimation of Conformation and Appearance Holstein-Friesian Cows Using Linear Classification System, Journal of Agricultural University of Neimenggu, 2002, 21:28-32.

Deng J Z, Zhang T L, Hong T S, Computer Image Processing and Analyzing System Based on Virtual Instruments, Agriculture Organization Research, 2001, 8:30-32.

Huang X Y, Wu S Y, Fang R M, et al., Research on Application of Computer Vision in Identifying Rice Embryo, Journal of Agricultural Machinery, 2000, 31:62-65.

Jimenez, A R, Ceres, R, Pons, J L, A Survey of Computer Vision Methods for Locating Fruit on Trees. GRICOLA, 1998-2002/09.

Li Y F, Wang X F, Li M Q, et al., Relativity Analysis of Dairy Cow Linearity Characters and Milkability for Holsteins, PasturageFarrier of Gansu, 2002, 34:15-16.

Mingxing Chu, Shoushen Shi, Study on Data Variation and Correlation for Type in Dairy Cattle, Journal of China Agriculture University, 1996, 1:113-118.

Xiaojun Liu, Longxiang Wu, Yingxiang Huang, et al., Path Analysis for Linear Traits in Holsteins, Journal of Shanxi Agriculture University, 1994, 14:402-404.

Xu G L, Mao H P, Hu Y G, Measuring Area of Leaves Based on Computer Vision Technology By Reference Object, Journal of Agriculture Engineering, 2002, 18:154-158.

Zhang Y G, Qiao L Y, Virtual Instruments Software Development Setting, Lab Windows/Cvi6.0, Machinery Industry Publishing House, 2002.

Zhou S Q, Ying Y B, Color models and their applications in color inspection and farm produce grading, Journal of Zhejiang University (Agric & Life Sci.), 2003, 29: 684-68.

RESEARCH OF RFID MIDDLEWARE IN PRECISION FEEDING SYSTEM OF BREEDER SWINE

Weiwei Sun, Xuhong Tian*, Minjie Jiang
Department of Computer Science and Engineering, South China Agricultural University, GuangZhou, P. R. China
** Corresponding author, Address: Department of Computer Science and Engineering, South China Agricultural University, GuangZhou, 510642, P. R. China, Tel: +86-20-85282091, Fax: +86-20-85285393, Email: tianxuhong@scau.edu.cn (TIAN Xuhong)*

Abstract: In Precision Feeding System of Breeder Swine, RFID is used to collect some important information about the body status of individual swine, IT is banded together with the adjusting of nutrition model, so as to actualize the precision feeding based on the individual status of livestock. The special swine entity is recognized in hoggery locale and its individual information is prompted, so the corresponding breeding measures can be validated for implementing, and the individual data can be inputted immediately in locale. The RFID middleware that satisfies the requirement of enterprise is designed and actualized; RFID system is integrated seamlessly with the existing ERP system.

Keywords: breeder swine, precision feeding, RFID, middleware

1. INTRODUCTION

RFID (Radio Frequency Identification) is a new technology that can collect and process information quickly and correctly. It identifies the target object with radio signal automatically and obtains the unique identification and other information of a special entity (Harry, 2006). Thereby, the projects of adopting RFID tag to identify and track animal are more and more popular in recent years. The advantages of RFID tag compared with barcode ear-tag are anti-dirt, perdurable, wireless recognizing in a long distance,

Sun, W., Tian, X. and Jiang, M., 2008, in IFIP International Federation for Information Processing, Volume 258; Computer and Computing Technologies in Agriculture, Vol. 1; Daoliang Li; (Boston: Springer), pp. 313–320.

recognizing the moving object, storing and modifying certain data in the tag (Lu CH et al., 2006).

The research and practice of applying RFID in precision feeding of livestock has been carried out for almost twenty years internationally, and mainly focused on dairy cattle feeding. For example, RFID tags are used to recognize individual cow, determine the milking interval and reckon the milk production, determine the formula of concentrate supplement feed according to the body status of cow, determine the position of milker-machine according to the position of teats of different cow, monitor the change of milk quality and locomotion quantity of individual cow, etc (Tan XQ, 2006).

In P. R. China, RFID in agriculture are used in dairy cattle precision feeding (Xiong BH, 2005), disease resistance in dairy cattle (Liu PH, 2005), safety tracing of factory pork production (Lu CH, 2006), pork quality monitoring and food safety (Liu Y, 2006), etc.

RFID was researched to apply precision feeding of breeder swine in our project. Concretely, there are the following requirements or goals:

1) The feeder can acquire his intraday tasks with PDA and wireless network and recognize individual breeder swine with RFID tag, so that the feeding, mating, delivery, epidemic prevention and other tasks can be validated and implemented correctly;

2) The feeder can input individual data immediately in the hoggery with PDA and wireless network so as to update individual status of swine in management system faster and more correct. The RFID system is integrated seamlessly with existing management system and ERP (Enterprise Resource Planning) system of enterprise, the producing flow and schedule can be managed and monitored in real time. This requirement is based on the disadvantage of current mode, i.e. producing data is written on paper at first, and then inputted into computer. Not only data can't be updated on time, but also be mistakable.

3) RFID tags are used to identify the swine, feeders, containers and other important material, and the passive tag readers are installed at appropriate sites in hoggery, so that the system can ascertain and track the position of swine, feeders and so on;

4) According to the swine's individual status of physiology and health, the series of foodstuff formulas for swine are optimized in the goals of nutrition balance and lower cost, and the swine's daily nutritional requirement for routine foodstuff and micronutrient is ensured.

So far, the enterprise's existing Hoggery Management System and ERP System has been running for more than three years, thereby, the RFID middleware is needed to research that integrate RFID system and existing ERP System.

2. USUAL STRUCTURE OF RFID MIDDLEWARE

RFID middleware is a kind of Message-Oriented Middleware, information is transmitted from one module to other in form of message. RFID middleware is situated between RFID reader and enterprise application (such as SCM and ERP), it is the software supporting for development and running of RFID Application (Luckham DC et al., 1998). The main roles are filter, synthesize, compute the events and data related to the tags read by readers, thus it can reduce original data transmitted from readers to enterprise applications and increase meaningful information abstracted from original data (Ding ZH, 2006) (Wang W et al., 2005). RFID middleware shields Operating System and database from the RFID hardware, such as tag, chip, antenna and reader, and it harmonizes the running of enterprise application (Astley, 2001) (Clark S et al., 2003).

Presently, the providers of RFID middleware include IBM, Microsoft, Oracle, Sybase, Sun etc (Wu M, 2007). Fig. 1 shows the structure of IBM WebSphere RFID Middleware. This middleware include two parts, Edge controller and Premises Server (Liu FG, 2006).

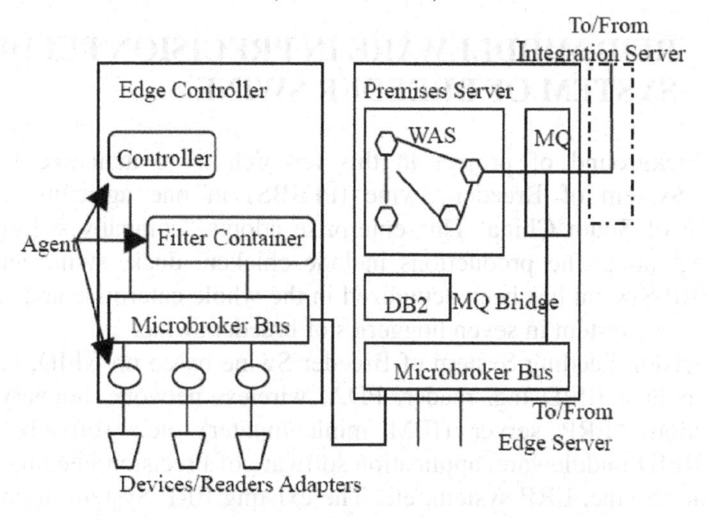

Figure 1. The structure of IBM RFID middleware

Edge controller is responsible for the communication with RFID hardware, which filters and synthesizes the data reading by reader and provides to Premises Server. Premises Server serves as the center of information collected by all RFID devices, it stores data and integrate with the management system of enterprise. Edge controller communicates with Premise Server in the mode of Publishing Topic and Subscribing Topic. The topics are published to and subscribed from Microbroker Bus.

After readers reads tag data, Filter Container will filter tag data, and then publish to Microbroker Bus. Premises Server acquire tag data from Microbroker Bus, the Message-driven Bean in WAS will filter and clean up the data farther, then the data will be provided by MQ to enterprise application in format of XML.

Microsoft's BizTalk RFID provides an open interface based on XML and Web Services, various hardware (such as RFID, bar code, IC card) compatible with DSPI (Device Service Provider Interface) can be Plug and Play in Microsoft Windows, besides that OM/API is provided, the application program can be coded in Managed Code (such as C++.Net, VB.Net, etc.). Oracle has designed Oracle Sensor Edge Server embedded in Server 10g; it actualizes the data collecting, grouping, rule filtering, data packing and routing, organizing and managing of internal data queue before transmission (Wu M, 2007). The RFID middleware provided by Microsoft, IBM, Oracle usually are based on their own kernel production and technology, and may has more dependence and less expansibility (Ding ZH, 2006).

3. RFID MIDDLEWARE IN PRECISION FEEDING SYSTEM OF BREEDER SWINE

The background of project in this research is to actualize Precision Feeding System of Breeder Swine (PFSBS) in one agricultural leader enterprise of South China. This enterprise adopts "company + farmer" as producing mode, the productions include chicken, duck, swine and dairy cattle. ERP System has been actualized in the whole enterprise and Hoggery Management System in seven hoggeries of breeder swine.

In Precision Feeding System of Breeder Swine based on RFID, hardware mainly include RFID tag, reader, PDA, wireless network, hoggery server (workstation), ERP server (IBM minicomputer) etc.; software mainly include RFID middleware, application software of Precision Feeding System of Breeder Swine, ERP system, etc. The existing ERP System in enterprise adopts Oracle as Database Manage System; Sun Jsdk was selected to develop RFID middleware.

The structure of RFID middleware designed in PFSBS includes five modules, i.e. Reader Communication Interface, Event Manager, Process Manager, Mobile User Interface and Application Software Interface (Brock D, 2001) (Harry KH Chow et al., 2006). Fig. 2 shows the structure of RFID middleware and the relationship with other software and hardware.

The roles of Reader Communication Interface include configuring and monitoring and starting Passive Tag Reader, reading RFID tag data and transmitting to Event Manager. Additional functions include cooperation

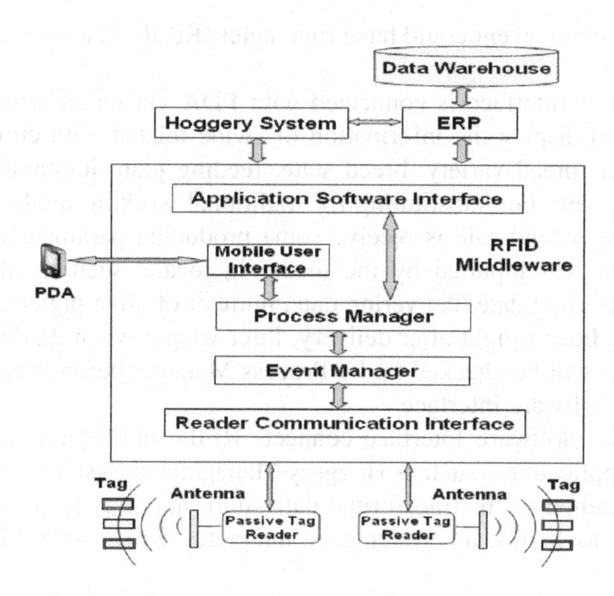

Figure 2. RFID middleware in PFSBS

between readers, anti-collision reading and writing, etc (Daniel W, 2002). The whole RFID system's usability and reliability will be depended on correct tag data provided by this module.

Event Manager has following main function (Britton C, 2001) (Ding ZH, 2006):

1) Data filtering: the wrong or redundant tag data read will be filtered, so that useful and important events will be found out, the number of event transmit to Process Manager will be reduced. Overfull data can also be buffered. Effective operations of filtering include grouping, counting, deleting redundant data, etc.

2) Data routing: different data will be transmitted to different applications according to pre-setting.

3) Event integrating: based on the actual event occurred in system, the subsets of event matching some pattern are extracted, and high-level events that accord with user's definition are produced and outputted. Such event usually has abundant semantic information, and is easy to be comprehended by applications and users.

Process Manager will ensure that processing sequence and logic of data/event is correct in terms of relevant rules. For example, reminding the feeder when he inputs an out-range data. After receiving the data inputted by feeder, Process Manager will update the processing state of the event accordingly, every event could have four states: Ready, Executing, Complete or Problem.

Mobile User Interface is connected with PDA via wireless network, the first role is to display the information of swine related with current tag on PDA, such as breed variety, breed state, feeding plan, immunity measure, intraday task, etc. For substituting for traditional working mode of writing-on-paper, the second role is receive some producing parameters of current swine, which was inputted by the feeder in locale, such as mating date, semen ID, aborting date, delivering date, number of alive piglets, number of dead piglets, litter weight after delivery, litter weight when 21 days old, etc, all these data will be checked up by Process Manager before transmitting to Application Software Interface.

Application Software Interface connects RFID middleware with various high-level applications (such as Hoggery Management System, ERP System) or data warehouse. It transforms data into appropriate format before transmitting to high-level systems, it integrates RFID with ERP System finally.

Level by level up through these five modules, RFID tag data is transformed to match the applications' requirement, and is introduced into Hoggery Management System and ERP System. The goal of integrating RFID with existing ERP System has been achieved.

Fig. 3 and Fig. 4 are the diagrams of data flow. Fig. 3 is Tag data reading, and Fig. 4 is Tag data updating.

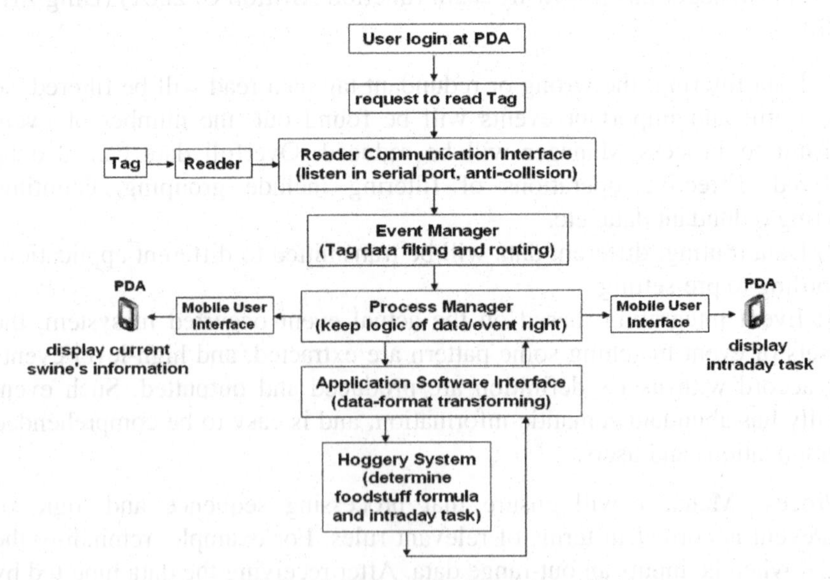

Figure 3. Data flow by Tag data reading

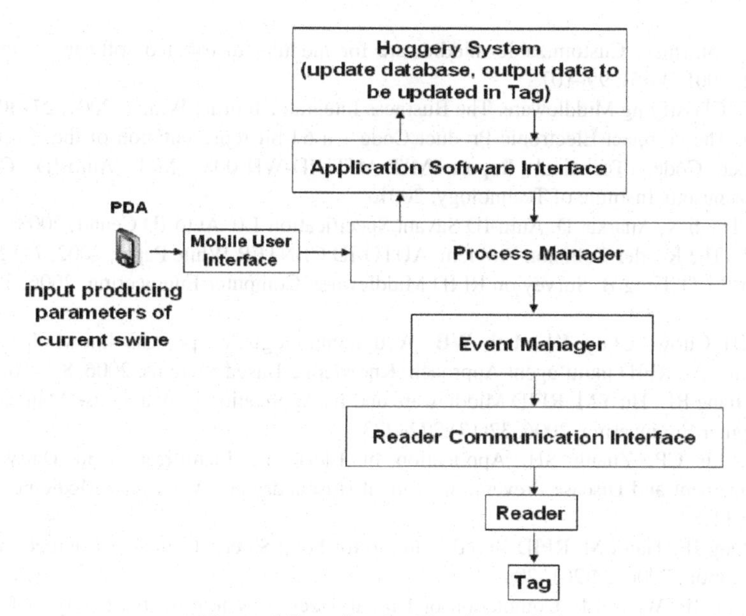

Figure 4. Data flow by Tag data updating

4. CONCLUDING REMARKS

RFID middleware provides the interface with RFID devices as well as the interface integrated with enterprise applications, but the research about RFID middleware is just a rising domain internationally, there is still not enough acknowledged production and comprehensive application.

In Precision Feeding System of Breeder Swine based on RFID, RFID was used to collect individual information of swine, every swine has its special tag. IT is banded together with the adjusting of nutrition model, so as to actualize the precision feeding based on the individual status of livestock. The special individual swine is identified in hoggery locale and its individual information is prompted, so the corresponding breeding measures can be validated for implementing.

For integrating RFID system with ERP System, it is necessary to study RFID middleware, so the RFID middleware is designed and actualized that accord with the requirement of enterprise. The following work will be debugging and perfecting current RFID Middleware System in practical running, and extent to other livestock feeding systems in enterprise gradually.

REFERENCES

Astley M, Sturman. Customizable middleware for modular distributed software. Commun. ACM, 2001, 44(5): 99–107.

Britton C. Classifying Middleware. The Business Integrator Journal Winter, 2001, 27–30.

Brock D. The compact Electronic Product Code – a 64-bit representation of the Electronic Product Code. Technical Report MIT-AUTOID-WH-008, MIT Auto-ID Center, Massachusetts Institute of Technology, 2001.

Clark S, Traub K, Anarkat D. Auto-ID Savant Specification 1.0. Auto-ID Center, 2003.

Daniel W. The Reader Collision Problem. AUTO-ID CENTER White Paper. 2002, 7–12.

Ding ZH, Li JT Feng B. Survey on RFID Middleware. Computer Engineering, 2006, 32(21): 9–11.

Harry KH Chow, Choy KL, Lee WB. A dynamic logistics process knowledge-based system – An RFID multi-agent Approach. Knowledge-Based Systems, 2006, 8:1–16.

Liu FG, Jiang RL, Hu YM. RFID Middleware and Its Application in Warehouse Management. Computer Engineering, 2006, 32(13):272–273.

Liu PH, Lu CP, Zhang SH. Application of Electronic Identification in Dairy Cow Management and Disease Prevention. Animal Husbandry and Veterinary Medicine, 2005, 37(5):32–34.

Liu Y, Zeng JF, Tian LM. RFID Based Solution for Food Safety. Computer Engineering and Application, 2006, 24:201–203.

Lu CH, Xie JF, Wang LF. Completion of Digital Tracing System for the Safety of Factory Pork Production. Jiangsu Journal of Agricultural Science, 2006, 22(1):51–54.

Luckham DC, Frasca B. Complex Event Processing in Distributed Systems. Stanford University, Technical Report: CSL-TR-98-754, 1998.

Tan XQ. Advice from others may help one overcome one's shortcomings – The 8th Japan Automatic Identification Expo and sidelights of visiting Japan. China auto-identify technology, 2006, 2:32–39.

Wang W, McFarlane D, Brusey J. Timing Analysis of Real-time Networked RFID Systems. Cambridge Auto-ID Lab, Cambridge, UK, 2005.

Wu M 2007. Scan RFID Middleware productions. China auto-identify technology, 2003, 1:66–71.

Xiong BH, Qian P, Luo QY. Design and Realization of Solution to Precision Feeding of Dairy Cattle based on Single Body Status. Transactions of the CSAE, 2005, 21(10): 118–123.

SIMULATING LAND USE/COVER CHANGES OF NENJIANG COUNTY BASED ON CA-MARKOV MODEL

Baoying Ye[1], Zhongke Bai[2]
[1] *China University of Geosciences, College of Land Science & Techniques, No. 29 Xueyuan Rd. Beijing, 100083, People's Republic of China, Email: woodfish_ye@yahoo.com.cn*
[2] *China University of Geosciences, College of Land Science & Techniques, No. 29 Xueyuan Rd. Beijing, 100083, People's Republic of China, Email: baizk@cugb.edu.cn*

Abstract: Nenjiang County experienced intense land use/cover changes during 1985–2000.The remote sensing and GIS methods were used to find the changes temporally and spatially. The result indicates: the forests were fallen in a large area, from 49.46% to 39.03% of total land area. Simultaneously, the croplands were increased rapidly from 26.02% to 37.42%. The conversion of forests and croplands were the main activities of landuse. Oppositely, Urbanization resulted in the decrease of the croplands in Southeast China during this period. In order to predict the landuse in 2015 and 2030 in this region, the CA-Markov model was taken. The predicting result indicates: From 2000 to 2015, 2000 to 2030, the croplands would increase 2.53% and 2.85% respectively, which account that the croplands exploitation reached a peak, only a small area of land can be used in croplands.

Keywords: Nenjiang County; Remote Sensing; GIS CA-Markov

1. INTRODUCTION

Land use/cover change is one of important factors that resulted in global warming (Turner et al., 1994). Recently, China is susceptible to land degradation on account of its climate, its geography and considerable population pressure on the land. Land use/cover changes during the past decades have arguably been the most widespread and intense in China's history. In Southeast China, the urban extending and cultivated land

Ye, B. and Bai, Z., 2008, in IFIP International Federation for Information Processing, Volume 258; Computer and Computing Technologies in Agriculture, Vol. 1; Daoliang Li; (Boston: Springer), pp. 321–329.

shrinking are the main activities. Oppositely, cropland increased in the Northeast China. The Nenjiang County is the typical area that the croplands increased and the forest decreased (YE. Baoying, 2003).

The objectives of this study are (1) to analyze the temporal and spatial changes of Nenjiang county during 1985–2000; (2) to simulating and predict the land use of Nenjiang region in 2015 and 2030 based on CA-Markov model.

2. DATA AND METHODS

2.1 Study area

Nenjiang County is located in the West of Heilongjiang province, China (E124°44′ 30″–126°49′ 30″, N48°42′ 35″–51°00′ 05″, Fig. 1). It belongs to North Temperate Zone and continental monsoon climate where the annul temperature is –0.1°C, and annul average rainfall is 550–600mm. The forest is the one of main land use types, mostly covering the low hill in the north. The cropland was planted in the south, which is a part of Nenjiang alluvial plain that is the main food supplying base of China.

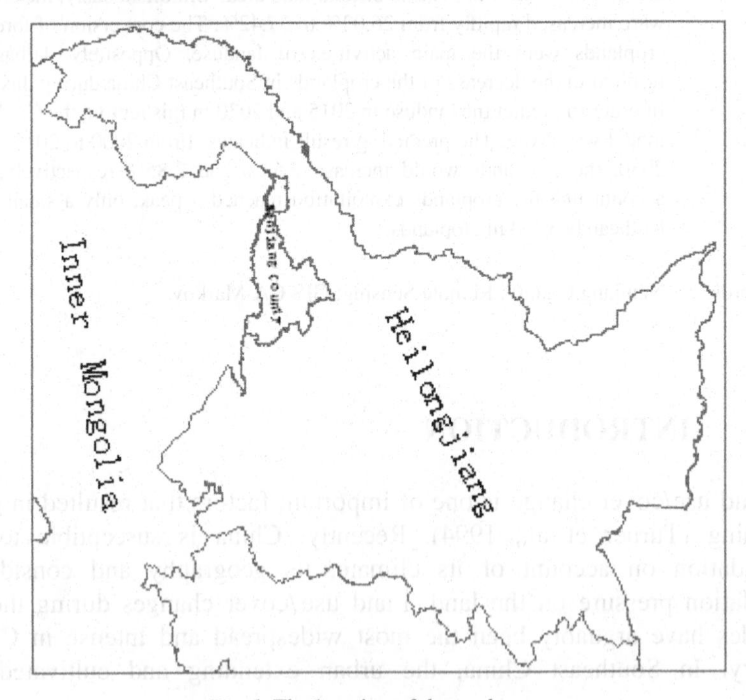

Fig. 1. The location of the study area

2.2 Image and process

Four TM(1985) and 4 ETM+(2000) scenes covering the region were used. The images were processed in the Erdas 9.1 (Leica, 2006), the process included band combination, color enhance and geometric correction, which the 20–30 GCPS were collected from the topologic map at a scale of 1:100000, and a RMS less than 1 pixel. The images were resampling to 30m with the bilinear interpolation.

The artificially and interactively interpreting method was taken to extract the land use/cover classifying information. We used 6 main classes and 24 subclasses for this region. The main classes consist of cropland, forest, grassland, waters, urban/built-up land, and swamp. The ArcGIS 9.2 (ESRI, 2006) were used to interpret and extract the vector data of land use/cover from the images of 1985 and 2000.The classified precision were evaluated is 88.95% (YE Baoying, 2002).

2.3 Markov chain

Markov chain is a series of random values whose probabilities at a time interval depend on the value of the number at the previous time. A given parcel of land theoretically may change from one category of landuse, to any other, at any time. Markovian analysis uses matrices that represent all the multi-directional landuse changes between all the mutually exclusive landuse categories.

The Markov chain equation was constructed using the landuse distributions at the beginning (**Mt**) and at the end (**Mt**+1) of a discrete time period as well as a transition matrix ($\mathbf{M_{Lc}}$) representing the landuse changes that occurred during that period. Under the assumption that the sample is representative of the region, these proportional changes become probabilities of landuse change over the entire sample area and form the transition matrices. The three matrices created above were then assembled to form a 'link' in the Markov chain using the following equation (Michael R et al.,1994):

$$\mathbf{M_{LC}} \quad * \quad \mathbf{M_t} \quad = \quad \mathbf{M_{t+1}}$$

$$\begin{bmatrix} LC_{uu} & LC_{ua} & LC_{uw} \\ LC_{au} & LC_{aa} & LC_{aw} \\ LC_{wu} & LC_{wa} & LC_{ww} \end{bmatrix} \begin{bmatrix} U_t \\ A_t \\ W_t \end{bmatrix} = \begin{bmatrix} U_{t+1} \\ A_{t+1} \\ W_{t+1} \end{bmatrix}$$

Where **U t** represents the probability of any given point being classified as urban at time *t*, and LC_{ua} represents the probability that an agricultural point at t will change into urban land by $t+1$ and so on. Iteration of this matrix equation derives the equilibrium matrix Q which by definition occurs when the multiplication of the column vector (landuse distribution) by the transition matrix yields the original column vector, *i.e.*, $\mathbf{M}_{LC} * \mathbf{M\,t} = \mathbf{Mt+1} = \mathbf{Mt}$.

2.4 CA model

Cellular automata (CA) were firstly used by Von Neumann (Von Neumann J, 1966) for self-reproducible systems. CA is discrete dynamic systems in which the state of each cell at time t+1 is determined by the stated of its neighboring cells at time according the pre-defined transition rules. CA as a method with temporal-spatial dynamics can simulate the evolution of things in two dimensions. This method has been widely applied in many fields of geography. Especially, it was used earlier to modeling the city growing and the land use evolution (Tobler WR, 1970; Couclelis H, 1985, 1989; White R, 1993)

CA is inherently spatial and dynamic, which makes them ideal choices for the representation of spatial and dynamic processes. Because of these properties they have been widely used in land use simulations. Furthermore, they are simple and computationally efficient and therefore make it possible to model land use dynamics at high resolution.

The similarities between CA and raster GIS data structure have led to the implementation of CA models inside GIS by many researchers. CA serves as analytical engines to enable dynamic modeling within GIS. Integrated GIS and CA models for the dynamic simulation of land use changes can generate realistic simulations of land use patterns and spatial structure.

2.5 CA-Markov model

Markov chain and CA both are the discrete dynamic model in time and state. One inherent problem with Markov is that it provides no sense of geography. The transition probabilities may be accurate on per category basis, but there is no knowledge of the spatial distribution of occurrences within each landuse category. We will use CA to add spatial character to the model. The Idrisi 15 (Clark Labs, 2006) integrates CA with Markov very well. In this paper, we take the Idrisi GIS software to simulate the land use in this region. This model needs some parameters as follows:

(1) Landuse data of 2000 is specified as the initial image. The transition probabilities areas from the years 1985 to 2000 were used for Markov conditional probability matrix.

(2) Generate the suitability maps: according to the underlying landuse change dynamics between the years 1985 and 2000, a series of suitability maps consisting of cropland, forest, grassland, urban/built-up land were empirically derived from the land cover images, which values are

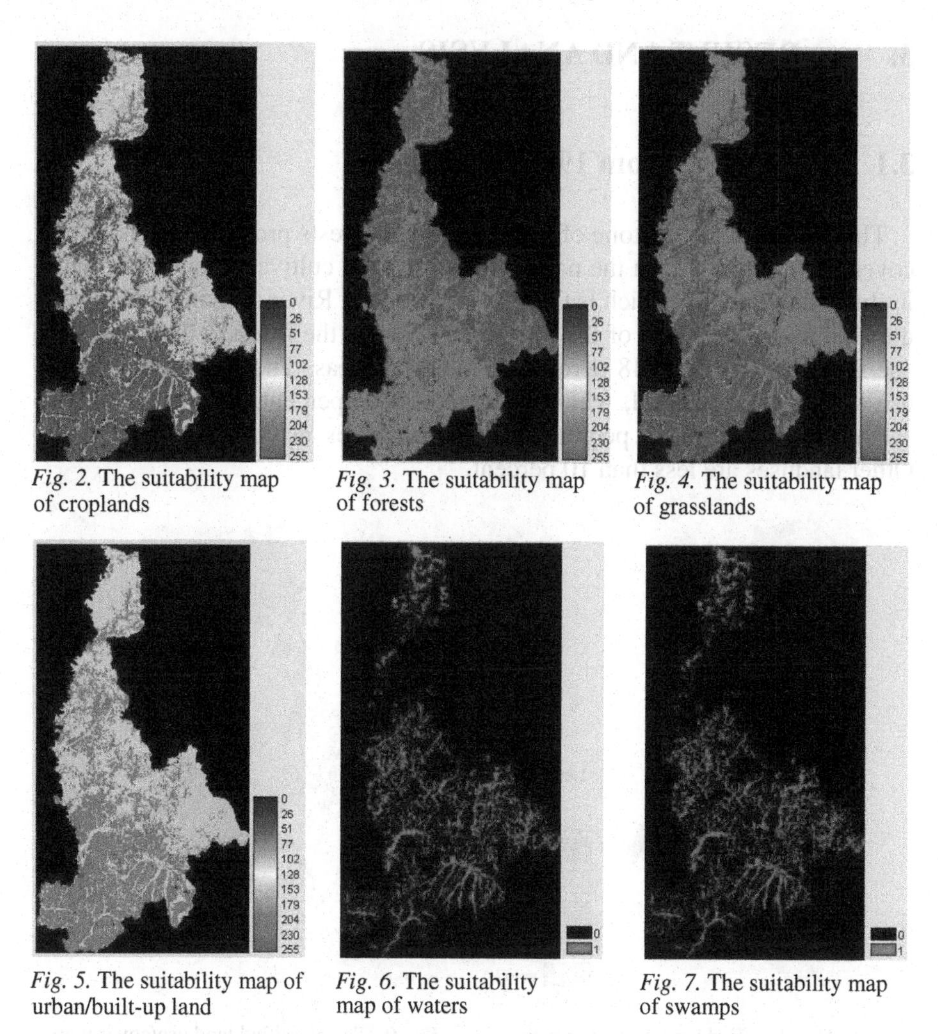

Fig. 2. The suitability map of croplands

Fig. 3. The suitability map of forests

Fig. 4. The suitability map of grasslands

Fig. 5. The suitability map of urban/built-up land

Fig. 6. The suitability map of waters

Fig. 7. The suitability map of swamps

standardized between 0 and 255 (Figs. 2–5). The waters, swamp are conditional constraints, which is a bool value (Figs. 6–7).

(3) A CA filter is used to generate a spatial explicit contiguity-weighting factor to change the state of cells based on its neighbors. The filter is a 5 × 5 contiguity filter as follows:

0	0	1	0	0
0	1	1	1	0
1	1	1	1	1
0	1	1	1	0
0	0	1	0	0

(4) CA loops times: over a 15-year period to 2000 is used to predict the landuse in the 2015, and a 30-year period to 2030.

3. RESULT AND ANALYSIS

3.1 Changes from 1985 to 2000

This region is the ecotone of cultivation and forests production. The forest covers on the low hill in the north of county. The cultivated land distributes in the plain of south, which is the part of Nenjiang River alluvial plain. From the land use/cover maps of the year 1985 (Fig. 8), the forest is the main land use type, covering the 48 percent of the total areas. The second main land use type is the cropland, which occupied 26.12 percent. Subsequently, the swamp occupied 17.53 percent, which distributes near the sides of river. Other landuses are less than 10 percent.

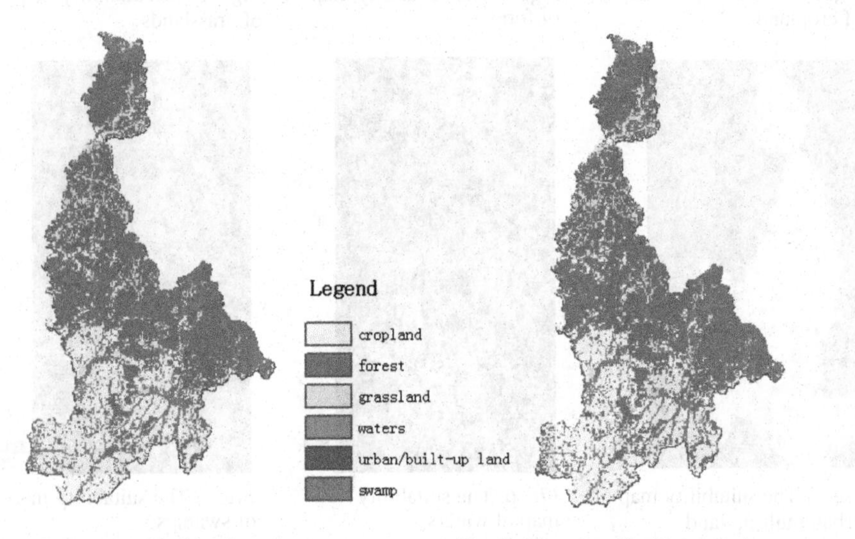

Fig. 8. The classified land use/cover map of Nenjiang County in 1985

Fig. 9. The classified land use/cover map of Nenjiang County in 2000

From the year 1985 to 2000, the cropland increased rapidly from 26.12 percent to 37.42 percent (Table 1, Figs. 8–9). The forested lands have however, decreased substantially from 48.46 percent in 1985 to 39.03 percent in 2000, a record loss of 9.43 percent. Other landuses changed slightly, which grassland decreased 2.22 percent, waters increased 0.6 percent, urban and built-up land increased 0.04 percent and the swamps decreased 0.73 percent. An analysis of the changes in table and figure shows: The forested lands have given way mainly to the expanding agriculture. The forests were felled took place on the foot of hill near the sides of river. An area of 147,231 hm^2 forests were converted into croplands during the year 1985 and 2000. Only 1,661 hm^2 were converted into grasslands, and 609 hm^2 replaced by the swamps. Besides of forests, the grasslands and swamps were also partly converted into croplands, which the converting areas are 19,693 hm^2 and 13,955 hm^2 respectively.

Table 1. Major land use/cover conversions from 1985 to 2000 (unit:hm^2)

1985 \ 2000	Cropland	Forest	Grassland	Waters	Urban/built-up	Swamp
Cropland	0	1263	298	0	170	1666
Forest	147231	0	1661	0	405	609
Grassland	19693	686	0	194	28	679
Waters	0	0	0	0	0	0
Urban/built-up	0	0	0	0	0	0
Swamp	13955	0	37	661	19	0

3.2 Land use/cover simulation

Making use of the CA-Markov model, the land uses in the year 2015 and in 2030 were predicted. From Figs. 10–12, the cropland increased from 37.42 percent in 2000 to 39.95 percent in 2015, an increment of 2.43 percent. In 2015. The forested lands decreased to 36.51 percent in 2015, a loss of 2.52 percent. In the year of 2030, the cropland would account for 40.27 percent and the forested 36.48 percent. From this trend, the cropland increased is limited since 2000. No more land is suitable to cropland.

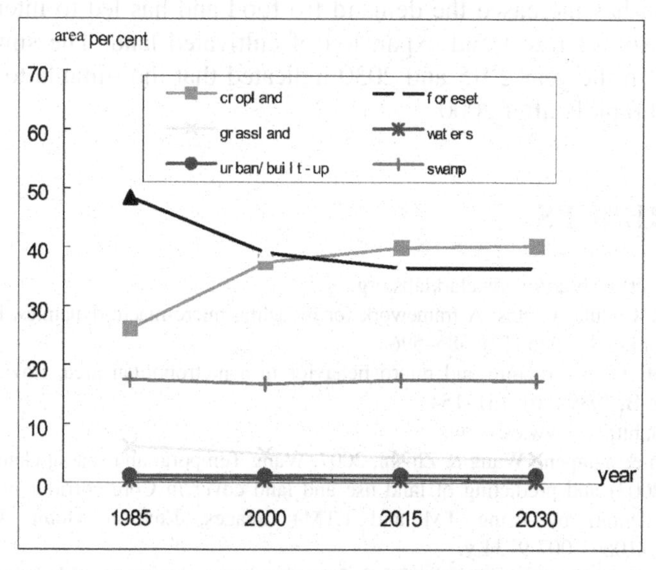

Fig. 10. The land use/cover changes in 1985, 2000, 2015 and 2030

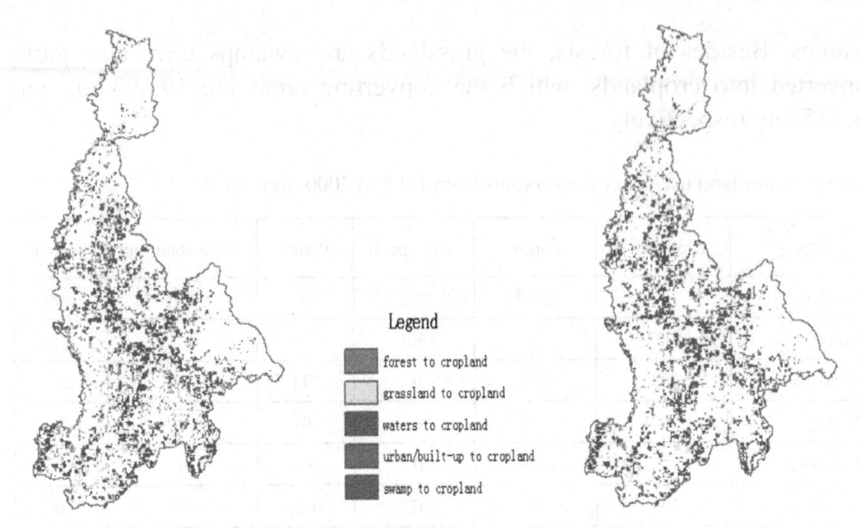

Fig. 11. The predicted increase cropland in 2015

Fig. 12. The predicted increase cropland in 2030

4. CONCLUSION

This paper presents the results for analyzing and predicting land use change by using remote sensing data, CA-Markov model. From this study of land use/cover changes in Nenjiang county between 1985 and 2000, some conclusions were obtain: The increase in agricultural areas from 26.12 percent to 37.42 percent is closed related to the increase in population. The population has increased the demand fro food and has led to intensification of the shrink of forest and expansion of cultivated land. The simulation of land uses in the year 2015 and 2030 indicated that the suitable to croplands decreased rapidly after 2000.

REFERENCES

Clark Labs, 2006, http://www.clarklabs.org.

Couclelis H. Cellular worlds: A framework for modeling micro-macro dynamics. Environment and Planning A, 1985 (17): 585–596.

Couclelis H. Macrostructure and micro behavior in a metropolitan area. Environment and Planning B, 1989 (16): 151–154.

ESRI, 2006, http: //www.esri.com.

Fenglei Fan & Yunpeng Wang & Zhishi, 2007, Wang Temporal and spatial change detecting (1998–2003) and predicting of land use and land cover in Core corridor of Pearl River Delta (China) by using TM and ETM+ images, Environ Monit Assess, DOI 10.1007/s10661-007-9734-y.

Leica Geosystems Geospatial Imaging LLC, 2006, http://www.idasnet.com/idas_site/idasnet_eng/products/gis_software/ leica_erdas. Htm.

Michael R. Muller and John Middleton, 1994, A Markov model of land-use change dynamics in the Niagara Region, Ontario, Canada Landscape Ecology vol. 9 no. 2 pp 151–157 (1994).

Tobler WR. A computer movie simulating urban growth in the Detroit region. Economic Geography, 1970 (46): 234–240.

Turner BL, Meyer WB, Skole DL. 1994. Global land use/land cover change: towards an integrated study. Ambio, 23 (1): 91–95.

Von Neumann J, Burks AW. 1966, Theory of self-reproducing automata. Urbana: University of Illinois Press, 35–39.

White R, Engelen G. Cellular automata and fractal urban form: A cellular modeling approach to the evolution of urban land-use patterns. Environment and Planning A, 1993 (25): 1 175–1 199.

YE Baoying, Zhang Y-Z, Zhang S-W, et al. 2003. Effect of land cover change in Neijiang watershed on runoff volume. Bull Soil Water Conser, 23(2):15–18 (in Chinese).

YE Baoying, 2002, The land use/cover change in Nenjiang watershed and its driving forces analysis. Ph.D. Dissertation, Changchun Institute of Geography, CAS (in Chinese).

DIFFERENTIALLY GENE EXPRESSION IN THE BRAIN OF COMMON CARP (*CYPRINUS CARPIO*) RESPONSE TO COLD ACCLIMATION

Liqun Liang [1,2,*], Shaowu Li [2,3], Yumei Chang [2], Yong Li [2], Xiaowen Sun [2], Qingquan Lei [1]

[1] *Harbin University of Science and Technology, Harbin, China, 150080*

[2] *Heilongjiang River Fisheries Research Institute, Chinese Academy of Fishery Sciences, Harbin, China, 150070*

[3] *College of Life Sciences and technology, Shanghai Fisheries University, Shanghai, China, 200090*

* *Corresponding author, Address: Heilongjiang River Fisheries Research Institute, Chinese Academy of Fishery Sciences, 43 Songfa Road, Daoli District, Harbin, 150070, P. R. China, Tel: +86-451-84861314, Fax: +86-451-84604803, Email: llq-1019@163.com*

Abstract: There are a variety of approaches to identify groups of genes that change in expression in response to a particular stimulus or environment. We here describe the application of suppression subtractive hybridization (SSH) for isolation and identification genes in the brain of common carp (*Cyprinus carpio*) under cold temperatures. The materials were prepared through cooling the hybrid F2 of purse red carp (cold-tolerant strains) and bighead carp (cold-sensitive species) to different regimes of temperatures. A subtracted cDNA library containing 2000 clones was constructed. About 60 positive clones were identified to express differentially by dot blotting in screening 480 clones. Sequencing 26 clones and aligning in GenBank/EMBL database using blastn searching engine, 15 genes showed higher similarities with 85-98%. These annotated genes contained (1) genes for transcription factors and gene products involved in signal transduction pathways such as zinc-finger protein, brevican; (2) genes involved in lipid metabolism such as Acyl-CoA synthetases, and (3) genes involved in the translational machinery such as cytochrome c oxidase, ependymin glycoprotein. In addition, real-time PCR was also conducted to validate these genes. To sum up, we believe this study will make an important contribution to elucidate the possible mechanisms on fish cold tolerance at a molecular level.

Keywords: cold tolerance, suppression subtractive hybridization (SSH), real-time PCR, *Cyprinus carpio*

Liang, L., Li, S., Chang, Y., Li, Y., Sun, X. and Lei, Q., 2008, in IFIP International Federation for Information Processing, Volume 258; Computer and Computing Technologies in Agriculture, Vol. 1; Daoliang Li; (Boston: Springer), pp. 331–339.

1. INTRODUCTION

Temperature has been recognized as a major environmental factor at the molecular, cellular, tissue, organism and ecosystem levels of biological hierarchy. Low temperature may make the cell threaten by a number of physiological and developmental changes. For aquatic ectotherms, visually observed changes associated with decreased temperature include changes in behavior and coloration. As the decline of temperature, fish exhibits a general decline in activity, respiration capacity, immune capacity and an abrupt loss of equilibrium. In addition, many fish will cease feeding and its muscle become more rigid until to death. It is very necessary to improve the ability of cold tolerance in fish, which may be expected to enlarge the culture areas of some thermophilic fish and improve their disease-resistant ability. It is also important for the study of cold tolerance in all other organisms.

Common carp (*Cyprinus carpio*) is a major kind of fish in the aquaculture industry around China, which can survive at the temperature ranging from 0□ to 35□. Many documents reported common carp could alter enzyme activities, membrane fluidities and behaviors to adapt to lower temperature. It is reported that lower temperature had an effect on lipid composition of cell membrane (Yeo et al., 1997). With the decline of temperature, fish produce lots of long-chain unsaturated fatty acid, which will promote membrane fluidity of cells. Cossins et al. (2002) also verified the importance of desaturase induction in the inducible cold tolerance of carp. As for the enzyme activities, Brown (1960) brought forth that there are two pathways of glucose metabolism and activities of LDH isoenzyme increased response to cold acclimation in carp liver. Few studies reported the effect of temperature on behavioural decision-making under predation risk. Fraser et al. (1993) found that Atlantic salmon juvenile use low water temperature as a cue to switch from diurnal to nocturnal foraging. At present, many cold-induced genes have been isolated in teleost fish, including *cytochrome c oxidase*, *cytochrome P450*, *antifreeze protein*, *methallothionein-1*, *carnitine palmitoyltransferase I*, *adenine nucleotide translocase* and *ependimin* (Hardewig et al., 1999; Kloepper-Sams & Stegeman, 1992; Pickett et al., 1983; Beattie et al., 1996; Rodnick & Sidell, 1994; Roussel et al., 2000; Tang et al., 1999). The brain is the organ that senses temperature and makes instructions to cold acclimation. It is possible that cold-induced genes in the brain contribute to the control and regulation of the acclimation responses. Besides of the cold-induced alternations in the composition of the phospholipids, the release of neurotransmitters and hormones in the brain has also been reported (Yeo et al., 1997; Tiku et al., 1996; Poli et al., 1997).

Moreover, studies on the activities of *AchE* in brain showed that *AchE* of some thermo fish becomes unstable at low temperature and possibly exists two isomers at different temperature (Baldwin & Hochachka, 1997). However, little knowledge are known on the molecular mechanism under temperature acclimation.

In the present study, we set out an experimental fish system and describe the application of suppression subtractive hybridization (SSH) on isolation and identification of the cold tolerance-related genes in the brain of common carp in order to further understand the molecular process involved.

2. MATERIALS AND METHODS

2.1 Construction of experimental fish system

Bighead carp (♀, cold sensible) and red purse carp (♂, cold tolerance) were selected to be the parents of experimental fish. Considering that bighead carp cannot survive safely while red purse carp can survive in winter, segregation in F_2 of the crosses may exhibit cold tolerant and cold sensitive respectively. Therefore, F_2 was chosen as the experimental fish system in this study.

2.2 Cold acclimation of experimental fish

The hybrids of F_2 weigned ranging from 58-70mg were maintained in a temperature-controlling aquarium at 16□ for one week. Then water temperature was decreased to 10□ and 4□ at a rate of 2□ per hour, and maintained for 5 days separately (Fig. 1). Brain tissues were collected from 5 fish of each temperature-controlling point under 10□ and 4□ and stored in liquid nitrogen quickly.

2.3 RNA preparation

Brain tissues were ground with a mortal and pestle, and then homogenized in Trizol reagent (Invitrogen Life Technologies, Carlsbad, CA). Total RNA was extracted according to the manufacturer's instructions. The cDNA of cold-treated tissues was synthesized by long distance PCR method using SMART PCR cDNA Synthesis Kit (Clontech).

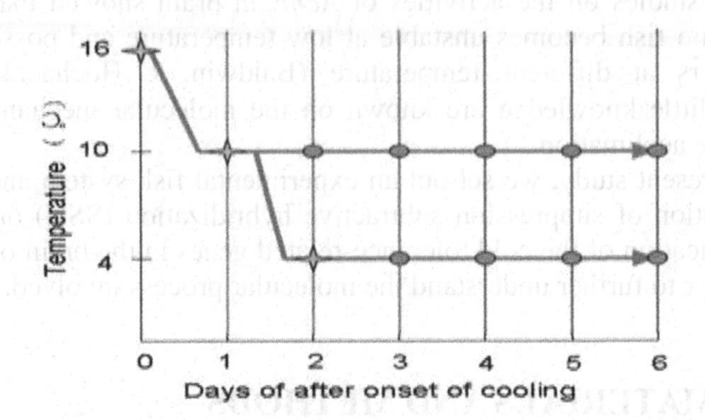

Fig. 1. Schematic diagram showing the cooling time course and sampling regime used
Notes: Samples collected from temperature-controlling point represented by solid asterisk

2.4 Construction of subtracted cDNA library

The mixed brain samples from 10□ and 4□ were selected to be the object of study. SSH was carried out using Clontech PCR-Select cDNA Subtraction Kit and Advantage Klen-Taq Polymerase Mix (Clontech). One microgram each of prepared tester and driver ds cDNA were digested by Rsa □enzyme,and the tester cDNA was separated into half parts and then each was ligated to adaptor 1 and adaptor 2R in separated ligation reaction mixture at 16□ overnight. The reaction was stopped by EDTA/glycogen and ligase was inactivated by heating the sample at 72□ for 5min. For the first hybridization, an excess amount of driver cDNA was added to each tester cDNA (ligated with adaptor 1 and 2R) in separate sample. After denaturation at 98□ for 1.5min, the first hybridization was performed in a hybridization buffer at 68□ for 8h. The two samples from the first hybridization were mixed and fresh denatured driver cDNA were added. In the second hybridization, new bybrid molecules corresponding to differentially expressed cDNA with different adaptors on each end were formed. Two PCR reactions were performed using different primers to selectively amplify the differentially expressed sequences. The PCR products were cloned into pMD 18 T-vector (Takara). The subtracted cDNA library was constructed after transformation into DH5α *Escherichia coli* strain.

2.5 Screening of differentially expressed clones by dot blotting

After purification by phenol-chloroform extraction, PCR products of forward- and reverse-subtracted cDNAs were used to be the probes labeled by α-^{32}P dATP separately. 480 clones were selected at random from the forward subtracted library enriched for cold-related sequences and the cDNA inserts were amplified. The PCR products were then blotted onto membranes together with control cDNAs and probed with the forward and reverse subtracted cDNA pools. 26 of positive clones screened were sequenced and performed alignment in the GenBank/EMBL database using blastn searching engine.

2.6 Real-time PCR

Another group of fish produced through inducing the gynogenesis of bighead carp was conducted the same temperature decreasing experiments described above. The purpose of this experiment is to further validate the genes isolated by SSH. Two tissues of livers and intestines were sampled and prepared single strand cDNA from three temperature points of 16°C, 10°C and 4°C respectively. Parts of sequences obtained by SSH were used to design primer pairs with software Beacon Designer (version 4.0). Real-time PCR was carried out in the Rotor-Gene 3000 PCR (CORBETT) using QuantiTect^{TM} SYBR^{Ro} Green PCR kit (QIAGEN). In this study, double standard curves of comparative quantitation method was used to analyze the differences of gene expression at different temperature points.

3. RESULTS AND ANALYSIS

3.1 Selection of the experimental fish

In this study, the mixed brain samples from 10□ and 4□were selected to be the object of study. It is reported that low temperature firstly have an effect on the central nervous system of fish. In addition, carp can live at 10-16□ normally while abnormally even to death at 4□. Therefore, it is hopefully to find cold-related genes through comparing these two temperature points.

3.2 Experimental key steps: the quality of total RNA, adaptor ligation efficiency and subtracted hybridized efficiency

Total RNA was extracted by Trizol reagents and visualized on 2% agarose gel. The value of OD260/280 was more than 2.0. It is very important to perform the ligation well before subtracted hybridization. The results show a successful ligation and subtraction using α-tubulin gene as the control (Fig. 2).

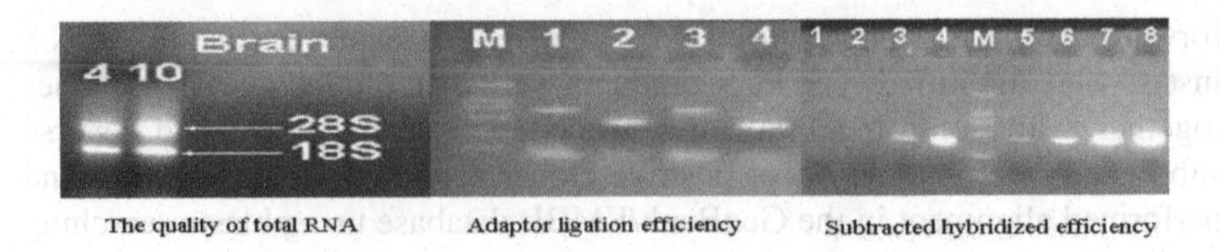

Fig. 2. The results of three key steps of this study

Notes: PCR was performed on subtracted cDNA(Lane1-4) and unsubtracted cDNA(Lane5-8); M: DNA Marker DL2000; Lane 1,5: 18 cycles; Lane 2,6: 23 cycles; Lane 3,7: 28 cycles; Lane 4,8: 33 cycles

3.3 Dot blotting analysis

Sixty of the 480 clones were screened positive by differential screening with forward and reverse subtracted probes and 26 cDNA clones were selected for sequencing. As figure 3 showed, duplicate dot blots hybridized with forward (A, C) and reverse subtracted cDNA probes (B, D), respectively.

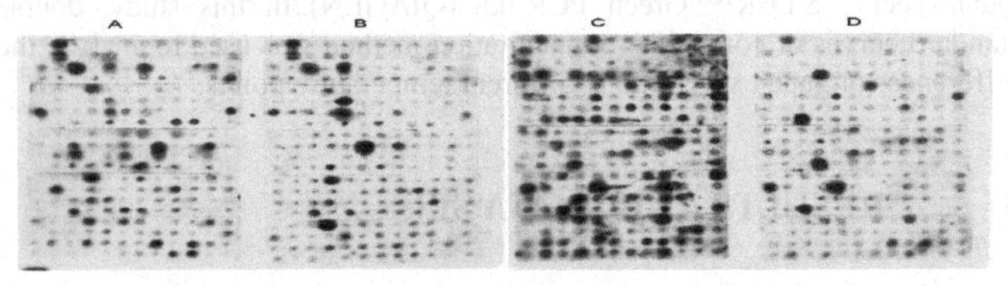

Fig. 3. Differential screening of SSH-selected cDNA clones with forward and reverse subtracted probes

Notes: A,C: probe was prepared by the forward SSH PCR□products; B,D: probe was prepared by the reverse SSH PCR□products

3.4 Sequence analysis

To characterize the 26 differentially expressed genes, we used blastn searching engine to find homological genes in GenBank/EMBL database. Fifteen genes show more than 85% homology to known genes, the remaining genes are unkown function. The major categories of differentially expressed genes in this study included (1) genes for transcription factors and gene products involved in signal transduction pathways such as zinc-finger protein, brevican; (2) genes involved in lipid metabolism such as Acyl-CoA synthetases and (3) genes involved in the translational machinery such as cytochrome c oxidase, ependymin glcoprotein.

3.5 Validation of real-time PCR

Two putative genes (4°C) isolated by SSH named LKE-25 (col-t 22) and LKE-62 (col-t 29) were applied to amplify the target genes, 18srRNA of common carp was used to amplify the housekeeping gene (Table 1). Double standard curves of real-time PCR were constructed individually and showed that the values of relative coefficients (R) are >0.99, most values are even beyond 0.999. Melt curves were also conducted and showed that the primer pairs are very specific in this study. The results of real-time PCR demonstrated that the concentration of LKE-25 decreases generally with the temperature decreased in different tissues, while, the concentration of LKE-62 is on the rise in livers and intestines with the temperature increased. According to our results, we considered LKE-62 is a up-regulated gene and LKE-25 is a down-regulated gene at lower temperature.

Table 1. Characterization of primers for real-time PCR

Primers	Primers sequence (5'→3')	Annealing temperature	Melt temperature	Cycles
LKE-25	F:CATAGCCGATCAACGAACC R:TAGAAACTGACCTGGATTGC	55	65-95	35
LKE-62	F:CTTCGTGGAGTGTGGCTAATC R:CGGTTACATAGGAATGGTCTGAG	55	65-95	35
18srRNA	F:CCTGTCGCCGCTGAATACC R:TCGCTTTCGTCCGTCTTGC	55	65-95	35

4. DISCUSSIONS

There are many studies reported the influences of low temperature in common carp, but mainly focued on the changes of membrane fluidity, lipid composition and enzyme activities. Brain is the most important organ during cold acclimation that can sense temperature and make instructions. Thus, it is necessary to do some research on the response to low temperature in the brain. By cold acclimation, the concentration of unsaturated fatty acid in the membrane increases, thereby promotes membrane fluidity (Roy et al., 1997). In addition, it is reported that ependymin (EPD) expressed increasingly in the brain and play an important role in fish to cold acclimation at an early stage. Up to now, a system analysis of differentially expressed genes in the brain is not available.

In the present study, some cold-induced genes in the brain of common carp were examined using SSH and validated by real-time PCR. It is resulted that 12.5% clones of the subtracted library showed differentially expression at low temperature. Among 26 differentially expressed genes, fifteen genes showed more than 85% homology to known genes while the remaining were still unknown. Because of the large numbers of unknown genes involved in cold acclimation, it is not yet possible to draw a clear picture of the molecular events that lead to adaptation or tolerance to adverse environmental conditions. Ju et al. (2002) hypothesized that a number of molecular events must occur to prepare the organism for environmental stresses including a cascade of signal transduction, activation of transcriptional factors that direct synthesis of new proteins to cope with low temperature. Many of these molecular events are rapid and transitory, but some of them may be persistently induced.

Brevican is a brain-specific proteoglycan that has been suggested to play a role in central nervous system fiber tract development (Gary et al., 1998). Besides preventing the formation of new synapses, it has been speculated that brevican might function as an insulator, sealing the synapses and preventing loss of transmitter substances from the synaptic cleft to the periphery. As the decline of temperature, brevican expressed up regulated that may prevent the formation of new synapses and lead to the low immunity of organisms. Arnab et al. (2004) isolated an intronless gene from rice encoding a zinc-finger protein and found this gene over-expressed after cold stress. However, here we characterized zinc-finger protein down regulated. It is necessary to make a further study.

Recent studies suggest that the long-chain acyl-CoA synthetases (ACS) may play a role in channeling fatty acids either toward complex lipid synthesis and storage or toward oxidation. Acyl-CoA binding protein (ACBP) was involved in fatty acid elongation and membrane assembly and organization (Gaigg et al., 2001). The detection of this gene showed that there must be some changes of cell membrane exposure to low temperature.

As for the genes involved in the translational machinery, several studies have shown that acclimation to low temperature provokes a compensatory increase of cytochrome c oxidase activity in fish tissues (Battersby et al., 1998; Thillart & Modderkolk, 1978; Wodtke, 1981). Ependymin encoding glycoprotein (EPN) is probably associated with collagen fibrils and has the capacity to bind calcium and results in a conformational transition. It is speculated by Tang et al. (1999) that EPN plays an important role in the cold acclimation of fish. The increase of EPN under cold stress not only prevents the nerve system from damage but also promotes its regeneration.

ACKNOWLEDGEMENTS

We would like to thank Zhang QT for his hard work in this research. This work was supported by grants 2004CB117404 from the National Key Basic Research and Development Programs, Republic of China.

REFERENCES

Arnab M, et al. Overexpression of a zinc-finger protein gene from rice confers tolerance to cold, dehydration, and salt stress in transgenic tobacco. PNAS, 2004, 101:6309–6314

Baldwin J, Hochachka PW. Functional significance of isoenzymes in thermal acclimatization-acetylcholinesterase fromtrout brain, Bioch J, 1970, 116:883–887

Battersby BJ, et al. Influence of acclimation temperature on mitochondrial DNA, RNA, and enzymes in skeletal muscle. Am. J. Physiol., 1998, 275:R905–R912

Beattie JH, et al. Cold-induced expression of the metallothionein-1 gene in brown adipose tissue of rats. Am. J. Physoil., 1996, 270:R971–R977

Brown WD. Glucose metabolism in carp. J. Cellular Comp. Physiol., 1960, 55:81–85

Cossins AR, et al. The role of desaturases in cold-induced lipid restructing. Biochem Soc. Trans., 2002, 30(6):1082–1086

Fraser NHC, et al. Temperature-dependent switch between diurnal and nocturnal foraging in salmon. Proc. R. Soc., 1993, 252:135–139

Gaigg B, et al. Depletion of acyl-coenzyme A-binding protein affects sphingolipid systhesis and causes vesicle accumulation and membrane defects in Saccharonmyces cerevisiae. Mol. Biol. Cell., 2001, 12:1147–1160

Gary SC, et al. BEHAB/brevican: a brain-specific lectican implicated in gliomas and glial cell motility. Curr. Opin. Neurobiol., 1998, 8(5):576–581

Hardewig I, et al. Temperature-dependent expression of cytochrome-c oxidase in Antarctic and temperate fish. Am. J. Physiol., 1999, 277:R508–R516

Ju Z, et al. Differential gene expression in the brain of channel catfish (Ictalurus punctatus) in response to cold acclimation. Mol. Genet Genomics., 2002, 268:87–95

Kloepper-Sams PJ, Stegeman JJ. Effects of temperature acclimation on the expression of hepatic cytochrome P4501A mRNA and protein in the fish Fundulus heteroclitus. Arch. Biochem. Biophys., 1992, 99:38–46

Pickett MH, et al. Seasonal variation in the level of antifreeze protein mRNA from the winter flounder. Biochim. Biophys. Acta., 1983, 739:97–104

Poli A, et al. Neurochemical changes in cerebellum of goldfish exposed to various temperatures. Neurochem. Res., 1997, 22:141–149

Rodnick KJ, Sidell BD. Cold acclimation increases carnitine palmitoyltransferase □activity in oxidative muscle of striped bass. Am. J. Physiol., 1994, 266:R405–R412

Roussel D, et al. Increase in the adenine nucleotide translocase content of ducking subsarcolemmal mitochondria during cold acclimation. FEBS Letters, 2000, 477:141–144

Roy R, et al. Regulation of membrane lipid bilayer structure during seasonal variation: a study on the brain membranes of Clarias batrachus. Biochim. Biophys. Acta., 1997, 1323(1):65–74

Tang SJ, et al. Cold-induced ependymin expression in zebrafish and carp brain: implications for cold acclimation. FEBS Letters, 1999, 459:95–99

Thillart V, Modderkolk J. The effect of acclimation temperature on the activation energies of state III respiration and the unsaturation of membrane lipids of gold fish mitochondria. Biochim. Biophys. Acta., 1978, 510:38–51

Tiku PE, et al. Cold-inducible expression of desaturase by transcriptional an post-translational mechanisms. Science, 1996, 271:815–818

Wodtke E. Temperature adaptation of biological membranes. Compensation of the molar activity of cytochrome c oxidase in the mitochondrial energy-transducing membrane during thermal acclimation of the carp (*Cyprinus carpio L.*). Biochim. Biophys. Acta., 1981, 640:710–720

Yeo YK, et al. Ether lipid composition and molecular species alterations in carp brain (*Cyprinus carpio L.*) during normoxic temperature acclimation. Neurochem. Res., 1997, 22(10):57–64

AN IMPROVED FAST BRAIN LEARNING ALGORITHM

Shuo Xu[1], Xin An[2], Lan Tao[3*]
[1] *College of Information and Engineering, China Agricultural University, Qinghua Donglu 17, Haidian, Beijing, 100083, P. R. China, Email: pzczxs@gmail.com. Tel: +86-10-62736755*
[2] *School of International Trade and Economics, University of International Business and Economics, Huixin Dongjie 10, Chaoyang, Beijing, 100029, P. R. China, Email: anxin927@gmail.com*
[3*] *College of Information Engineering, Shenzhen University, Nanhai Avenue 3688, Shenzhen, Guangdong, 518060, P. R. China, Email: taolan@szu.edu.cn. Tel: +86-755-26535078, Corresponding author*

Abstract: In this paper, an underlying problem on the fast BRAIN learning algorithm is pointed out, which is avoided by introducing the quantity *count* (·, ·). In addition, its speed advantage can still be enjoyed only at a cost of a little additional space. The improved fast BRAIN learning algorithm is also given.

Keywords: BRAIN, Numerical Computation

1. INTRODUCTION

Given a labeled dataset (training dataset) $(\mathbf{x}_i, y_i)$, $i = 1, 2, \ldots, l$, $\mathbf{x}_i \in \{0, 1\}^n$, $y_i \in \{-1, +1\}$, where these data points are drawn randomly and independently according to some underlying but unknown probability distribution. We assume this dataset to be self-consistent, i.e., an instance cannot be positive and negative at the same time. The goal is to find a classification rule (hypothesis or function): f: $\{0, 1\}^n \rightarrow \{-1, +1\}$ using this

Xu, S., An, X. and Tao, L., 2008, in IFIP International Federation for Information Processing, Volume 258; Computer and Computing Technologies in Agriculture, Vol. 1; Daoliang Li; (Boston: Springer), pp. 341–347.

dataset such that f will correctly classify a new instance ($\mathbf{x}$, y), that is, $f(\mathbf{x})$ = y for this new instance, which is generated from the same underlying probability distribution as the training data. For convenience, we denote $\mathbf{x}_i^+ = (x_{i,1}^+, x_{i,2}^+, \cdots, x_{i,n}^+)^t$ for positive instances, where $x_{i,k}^+ \in \{0,1\}$, i = 1, 2, ..., l^+, k = 1, 2, ..., n, and similarly, denote $\mathbf{x}_j^- = (x_{j,1}^-, x_{j,2}^-, \cdots, x_{j,n}^-)^t$ for negative instances, where $x_{j,k}^- \in \{0,1\}$, j = 1, 2, ..., l^-, k = 1, 2, ..., n. And t denotes the transpose of a vector. Of course, $l = l^+ + l^-$.

Indeed, there are many approaches that can solve this problem, such as NN (Neural Network) (Haykin, 1999), SVM (Support Vector Machine) (Vapnik, 1998; 1999), and many others. However, from an entirely different perspective, Rampone (1998) put forward the BRAIN (Batch Relevance-based Artificial INtelligence) learning algorithm. The aim of the algorithm is to infer a consistent DNF (Disjunction Normal Form) classification rule of minimum syntactic complexity from a set of instances, i.e., training dataset. Here, the minimum syntactic complexity means the minimum number of *clauses*, each one with the minimum number of *literals*. A Boolean classification rule g is consistent with a training dataset if, and only if, it matches every positive instance and no negative instance in the set. That is, g is consistent with a training dataset if, and only if, $g(\mathbf{x})$ = 1 for all positive instances, $g(\mathbf{x})$ = 0 for all negative instances. Once such Boolean classification rule is found, then our final classification rule is $f(\mathbf{x}) = 2g(\mathbf{x}) - 1$.

The major advantages of this algorithm are the low error rates and high correlation coefficient, the explicit classification rules description as a DNF formula, a polynomial (cubic) computational complexity, and robust and stable "one shot" learning. However, the space and time complexity of this algorithm are very high, which heavily limit the range of its application in real world. On the other hand, by many reasons, errors may be present in the training dataset. That is, the training dataset may be contaminated by noise to some extent. The structural risk minimization (SRM) principle (Vapnik, 1998; 1999) tells us that a function which makes a few errors on the training set might have a better generalization ability than a larger function (with more *literal*s and more *clause*s) which makes zero empirical error.

Soon, Rampone (2004) realized this point, and gave a fast BRAIN learning algorithm with an error tolerance, which will be described in next section. From theoretical viewpoint, there are no problems, but from the viewpoint of numerical computation, there may be a problem, which will be analyzed in section 3. Finally, an improvement will be given in section 4.

2. FAST BRAIN LEARNING ALGORITHM

Rampone (2004) found that building the sets $S_{i,j}$ was a main computational drawback, whose time complexity is O (n × $l^+ \times l^-$), and space complexity is O ($l^+ \times l^-$). By definition of $S_{i,j}$, the sets $S_{i,j}$ can be derived from the given positive and negative instances. When a new *literal* $e_k \leftarrow \arg\max R(e_k)$ (If there is a tie, that is, the *literal*s that reach the maximum value are not just one. At this time, we prefer the *literal* with lower subscript and the one with true form.) is selected, the following two steps are performed:

(1) Delete the $S_{i,j}$ sets for $j = 1, 2, \ldots, l^-$ if $e_k \notin S_i$;
(2) Delete the $S_{i,j}$ sets if $e_k \in S_{i,j}$.

In fact, the $S_{i,j}$ update step (1) can be done by deleting $\mathbf{x}_i^+$ having 0 in position k if e_k is in true form, or $\mathbf{x}_i^+$ having 1 in position k if e_k is in negated form, i.e., the positive instances whose indices belong to

$$II = \left\{ i \left| \begin{array}{c} x_{i,k}^+ = 0, e_k \textit{ is in true form} \\ \mathbf{x}_i^+ \in S, i = 1, 2, \cdots, l^+ \end{array} \right. \right\} \cup \left\{ i \left| \begin{array}{c} x_{i,k}^+ = 1, e_k \textit{ is in negated form} \\ \mathbf{x}_i^+ \in S, i = 1, 2, \cdots, l^+ \end{array} \right. \right\} \quad (1)$$

And the $S_{i,j}$ update step (2) can be done by deleting $\mathbf{x}_j^-$ having 0 in position k if e_k is in true form, or $\mathbf{x}_j^-$ having 1 in position k if e_k is in negated form, i.e., the negative instances whose indices belong to

$$JJ = \left\{ j \left| \begin{array}{c} x_{j,k}^- = 0, e_k \textit{ is in true form} \\ \mathbf{x}_j^- \in S, j = 1, 2, \cdots, l^- \end{array} \right. \right\} \cup \left\{ j \left| \begin{array}{c} x_{j,k}^- = 1, e_k \textit{ is in negated form} \\ \mathbf{x}_j^- \in S, j = 1, 2, \cdots, l^- \end{array} \right. \right\} \quad (2)$$

In this way, we can substitute the $l^+ \times l^-$ sets $S_{i,j}$ for a set S containing at most $l^+ + l^-$ instances. The space complexity can be dramatically reduced.

Now, the extended relevance can be evaluated by

$$R(e_k) = \sum_{i \in I} \sum_{j=1}^{l^-} \frac{\delta_{i,j}(e_k, S)}{d_{i,j}} \quad (3)$$

where $I = \left\{ i \middle| \mathbf{x}_i^+ \in S, i = 1, 2, \cdots, l^+ \right\}$, $d_{i,j}$ is the Hamming distance between $\mathbf{x}_i^+$ and $\mathbf{x}_j^-$, which can be calculated once and used for all, and

$$\delta_{i,j}(x_k,S)=\begin{cases}1, \text{ if } x^+_{i,k}=1 \text{ and } x^-_{j,k}=0\\ 0, \text{ otherwise}\end{cases}$$
$$\delta_{i,j}(\overline{x}_k,S)=\begin{cases}1, \text{ if } x^+_{i,k}=0 \text{ and } x^-_{j,k}=1\\ 0, \text{ otherwise}\end{cases} \tag{4}$$

These quantities can be calculated just once for each *clause*. In fact, by using Eq. 3, it is easy to see that, when we update the sets $S_{i,j}$, the corresponding extended relevance update is for step (1):

$$R^{new}(e_k)=R^{old}(e_k)-\sum_{i\in II}\sum_{j\in J}\frac{\delta_{i,j}(e_k,S)}{d_{i,j}} \tag{5}$$

where $J=\left\{j\middle|\mathbf{x}^-_j\in S, j=1,2,\cdots,l^-\right\}$, and for step (2):

$$R^{new}(e_k)=R^{old}(e_k)-\sum_{i\in I\wedge i\notin II}\sum_{j\in JJ}\frac{\delta_{i,j}(e_k,S)}{d_{i,j}} \tag{6}$$

3. A PROBLEM ON FAST BRAIN LEARNING ALGORITHM

From theoretical viewpoint, iterative formula Eq. 5 and Eq. 6 can work well. But from the viewpoint of numerical computation, there may be some problems, especially when there is a tie for $e_k \leftarrow \arg\max R(e_k)$. In what follows, we will give the analysis.

Though there are several different representations of real numbers (Matula and Kornerup, 1985), by far the floating-point representation is widely used in computer system, from PCs to supercomputers. However, most floating-point calculations have rounding error anyway. So the IEEE standard (IEEE, 1987) requires that the result of addition, subtraction, multiplication and division be exactly rounded. That is, the result must be calculated exactly and then rounded to the nearest floating-point number (using round to even). According to theorem 2 in (Goldberg, 1991), the relative rounding error in the result for addition and subtraction with one guard digit is less than or equal to 2ε, where ε is machine epsilon. That is, each addition or subtraction operation can potentially introduce an relative rounding error as large as 2ε, then a sum involving thousands of terms can have quite a bit of rounding error. The iterative formula Eq. 5 and Eq. 6 introduce much more floating-point addition or subtraction operations than necessary (see below), that is, they introduce quite a bit of rounding error.

More specially, let's consider a simple example $\sum_{i=1}^{n} x_i$, assuming the calculation is being done in double precision. If the naive formula $\sum_{i=1}^{n} x_i$ is utilized, then the computed sum is equal to $\sum_{i=1}^{n} x_i(1+\delta_i)$, where $|\delta_i| \le (n-i)\varepsilon$, that is, each summand is perturbed by as large as $n\varepsilon$ (Goldberg, 1991). Though there is a much more efficient method which dramatically improves the accuracy of sums, namely, the Kahan summation formula, the relative rounding error is still related to the number of operations. Since at this time the calculated sum is equal to $\sum_{i=1}^{n} x_i(1+\delta_i) + O(n\varepsilon^2)\sum_{i=1}^{n}|x_i|$, where $|\delta_i| \le 2\varepsilon$ (Goldberg, 1991). In this way, it is very possible that for the two *literals*, say e_1, e_2, it should be $R(e_1) = R(e_2)$ (or $R(e_1) \neq R(e_2)$), but it becomes $R(e_1) \neq R(e_2)$ (or $R(e_1) = R(e_2)$) after several update calculations using Eq. 5 and Eq. 6. Eventually, it will possibly result in the Boolean classification rule obtained by the fast BRAIN learning algorithm (Rampone, 2004) is not same as the one derived by the origin BRAIN learning algorithm (Rampone, 1998).

4. AN IMPROVEMENT ON FAST BRAIN LEARNING ALGORITHM

According to above analysis, let's consider how to avoid this underlying problem. Assume the training dataset is self-consistent, the Hamming distance between $\mathbf{x}_i^+$ and $\mathbf{x}_j^-$ must be at the interval $[1, n]$, i.e., $1 \le d_{i,j} \le n$. Thus we can count the number of each Hamming distance for each *literal* e_k and denote it as $count(e_k, d_{i,j})$, $i \in I$, $j = 1, 2, \ldots, I^-$. Now, the extended relevance can be evaluated by

$$R(e_k) = \sum_{i=1}^{n} \frac{count(e_k, i)}{i} \tag{7}$$

It is not difficult to see that $count(\cdot, \cdot)$ can also be calculated just once for each *clause*. In fact, Eq. 5 can be replaced by Eq. 8 and Eq. 7, and Eq. 6 by Eq. 9 and Eq. 7.

$$count^{new}(e_k, d_{i,j}) = count^{old}(e_k, d_{i,j}) - 1, i \in II,\ j \in J \tag{8}$$

$$count^{new}(e_k, d_{i,j}) = count^{old}(e_k, d_{i,j}) - 1, i \in I \wedge i \notin II,\ j \in JJ \tag{9}$$

Because the results of integer addition and subtraction calculations are exact so long as operands and result are not out of range represented by computer system, and it is nearly impossible to reduce further the number of floating-point operations in Eq. 7, the underlying problem on the fast BRAIN learning algorithm can be overcome. Compared with the version of

Rampone (Rampone, 2004), the cost that we pay is the additional space (to be precise, $2n^2$) for *count* (·, ·). But in general, for many applications, e.g., splice sites prediction, the order of magnitude of l^+, especially l^- is usually several orders of magnitude of n. That is, the additional space for *count* (·, ·) is negligible. In addition, since Eq. 8 and Eq. 9 are similar to Eq. 5 and Eq. 6, respectively, and Eq. 8 and Eq. 9 are only involving integer addition or subtraction operations, we can still enjoy the speed advantage of the fast BRAIN learning algorithm.

In what follows, the improved fast BRAIN learning algorithm can be sketched:

Improved Fast BRAIN Learning Algorithm

Input: $X^+ = \{\mathbf{x}_1^+, \mathbf{x}_2^+, \cdots, \mathbf{x}_{l^+}^+\}$ and $X^- = \{\mathbf{x}_1^-, \mathbf{x}_2^-, \cdots, \mathbf{x}_{l^-}^-\}$ for positive instances and negative instances, respectively, where $\mathbf{x}_i^+ \in \{0, 1\}^n$, $\mathbf{x}_i^- \in \{0, 1\}^n$, and ε^+, ε^- for the error tolerant parameters;

Output: a Boolean classification rule or a consistent DNF formula $g(\mathbf{x})$;

Initialize: $g(\mathbf{x}) \leftarrow FALSE$, $l_r^+ \leftarrow l^+$;

Calculate the Hamming distance $d_{i,j}$, $i = 1, 2, \ldots, l^+, j = 1, 2, \ldots, l^-$;

While $\left(l_r^+ / l^+\right) > \varepsilon^+$

$S \leftarrow X = X^+ \cup X^-$;

Count the number of each Hamming distance for each *literal* e_k, i.e., *count* $(e_k, d_{i,j})$, $i \in I, j = 1, 2, \ldots, l^-$;

$c \leftarrow TRUE$;

$l_r^- \leftarrow l^-$;

Build the *clause* c: *While* $\left(l_r^- / l^-\right) > \varepsilon^-$

Calculate the extended relevance $R(e_k)$ by Eq. 7;

$e_k \leftarrow \arg\max R(e_k)$;

$c \leftarrow c \wedge e_k$;

Let II the set of indexes [Eq. 1], and update *count* (·, ·) by Eq. 8;

Delete from S the instances $\mathbf{x}_i^+$, $\forall i \in II$;

Let JJ the set of indexes [Eq. 2], and update *count* (·, ·) by Eq. 9;

Delete from S the instances $\mathbf{x}_j^-$, $\forall j \in JJ$;

$l_r^- \leftarrow l_r^- - |JJ|$;

End

$g(\mathbf{x}) \leftarrow g(\mathbf{x}) \vee c$;

Delete from X^+ the positive instances matching c, and update l_r^+;

End

5. CONCLUSION

In this paper, we analyze the reasons that an underlying computational problem on the fast BRAIN learning algorithm may occur, and give an improved algorithm, which is numerically more stable. Furthermore, its speed advantage can still be enjoyed only at a cost of a little additional space. In the end, since the algorithm will give an explicit classification rule description as a DNF formula, we think it can be utilized for feature selection with binary value data, thus the data will be compressed heavily in some cases.

AVAILABILITY

The source code implementing the improved fast BRAIN algorithm is available from the authors upon request for academic use.

ACKNOWLEDGEMENTS

This research is partially supported by the National Natural Science Foundation under Grant No. 60673122.

REFERENCES

Aykin S. Neural Networks: A Comprehensive Foundation, 2nd Edition, Prentice-Hall, Inc., 1999.

Goldberg D. What Every Computer Scientist Should Know about Floating-Point Arithmetic. ACM Computing Survey, 1991, 23(1): 5-48.

IEEE. IEEE Standard 754-1985 for Binary Floating-Point Arithmetic, IEEE, 1985. Reprinted in SIGPLAN, 1987, 22(2): 9-25.

Matula D.W. and Kornerup P. Finite Precision Rational Arithmetic: Slash Number Systems. IEEE Transaction on Computers, 1985, C-34(1): 3-18.

Rampone S. An Error Tolerant Software Equipment for Human DNA Characterization. IEEE Transactions on Nuclear Science, 2004, 52(5): 2018-2026.

Rampone S. Recognition of Splice Junctions on DNA Sequence by BRAIN Learning Algorithm. Bioinformatics, 1998, 14(8): 676-684.

Vapnik V.N. Statistical Learning Theory, Wiley, New York, 1998.

Vapnik V.N. The Nature of Statistical Learning Theory, 2nd Edition, Springer Verlag, New York, 1999.

ONTOLOGY-BASED AGRICULTURAL KNOWLEDGE ACQUISITION AND APPLICATION

Nengfu Xie*, Wensheng Wang, Yong Yang
Agricultural Information Institute, The Chinese Academy of Agricultural Sciences, Beijing, China, 100081
** Corresponding author, Address: Agricultural Information Institute, No. 12 Zhongguancun South St., Haidian District Beijing 100081, P. R. China, Tel: +86-10-68919819, Fax: +86-10-68919820, Email: nf.xieg@caas.net.cn*

Abstract: Agricultural knowledge is a special kind of domain knowledge, and is a significant basis for agricultural knowledge-based intelligent information service system. This paper presents main results of our ongoing project of Agricultural Knowledge Processing and Applications (AKPA): 1) An agriculture-specific ontology, 2) A method for agricultural knowledge acquisition and representation, 3) An experiment of An Intelligent Agricultural knowledge-based knowledge service system. At last, we will conclude the paper.

Keywords: Ontology; Agricultural knowledge; knowledge Service System

1. INTRODUCTION

Agricultural knowledge is a special kind of domain knowledge, and is a significant basis for An Intelligent Agricultural knowledge-based knowledge service system, such as agricultural instructional systems, agricultural language processing systems and agricultural expert systems.

As the starting point of our project AKPA, we have chosen several influential sources of agricultural knowledge and consulted a few agricultural specialists. The knowledge sources includes: 1) the Agriculture Volume of

Xie, N., Wang, W. and Yang, Y., 2008, in IFIP International Federation for Information Processing, Volume 258; Computer and Computing Technologies in Agriculture, Vol. 1; Daoliang Li; (Boston: Springer), pp. 349–357.

the Encyclopedia of China (Cai, 1996), 2) A Dictionary of Agriculture (Com et al., 1998), 3) Illustrated Handbook of Food Crops, Economic Crops and Herbal Plants. Based on the knowledge sources and the advice from our consultants, we started to develop an ontology of agricultural objects, called AgriOnto.

In AgriOnto, each category has a frame-based representation, and each category has a list of slots (attributes or relations) for describing its instances. Categories can be inter-related in several semantic relationships; some are general, such as is-a and have-part(s), and some are more specific to agriculture, such as IsVariantOf(x), immune-disease(x), Harm(x).

In addition to offering a terminology for describing its instances, AgriOnto plays other two significant roles. Firstly, attributes and relationships in a category are knowledge 'place-holders' of the instances of the category, and they are to be filled in during the knowledge acquisition for the instances. Secondly, axioms in an ontology can be used both in knowledge inference and knowledge verification during the knowledge acquisition procedure. When acquired knowledge violates such axioms, the knowledge engineer is alarmed to identify and solve the problems.

We designed a semi-automated method for agricultural knowledge acquisition sources (Cao et al., 2002). In practice, it is a valid and feasible method by building ontology-based base. In essence, the work is primarily divided into the following steps: Firstly, to build AgriOnto hierarchy. Secondly, to formalize the text knowledge on the basis of AgriOnto. Thirdly, to compile and check the knowledge. Fourthly, to built knowledge-based service systems. The primary step is knowledge acquisition, that is to say, knowledge formalization depicted in figure 1.

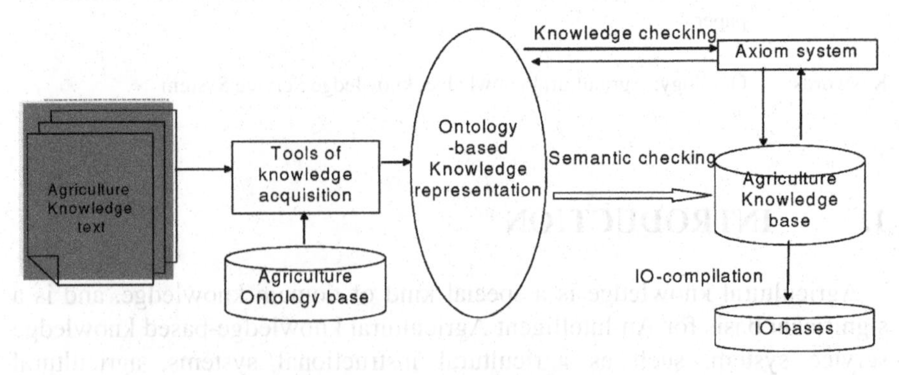

Figure 1. The flow of Ontology-based knowledge acquisitions

The paper is organized as follows. Section 2 presents how to build the hierarchy of agriculture knowledge and the overall description of agriculture-specific ontology. Section 3 gives the method of agriculture knowledge acquisition. Section 4 exemplifies the application of agriculture knowledge. Finally, Section 5 concludes the paper.

2. AGRICULTURAL ONTOLOGY

In general, ontology is an explicit specification of conceptualization (Chaudhri et al., 1997; Cao et al., 2002; Gu et al., 2003). Nevertheless, the term ontology has been controversial in current AI practice, and so far no formal definition exists. In our work, we have selected to use the term of domain-specific ontology (DSO). In practical term, developing AgriOnto includes three steps:

- Building a domain-specific knowledge hierarchy
- Defining slots of the categories and representing axioms
- Knowledge acquisition, this is to say, filling in the value for slots of instances.

2.1 Agriculture knowledge hierarchy

AgriOnto indicates formal definition of agriculture and their relation (Fig. 2). The definition and relations form an integrated hierarchy of agriculture. With labor object as the center of agriculture hierarchy, we divide agriculture knowledge into seven taxa: labor object, production process, production technology, agriculture engineering, agriculture branch, and agriculture environment and agriculture regulation. The labor object as the center of agriculture knowledge hierarchy aims at facilitating the people, who want to know the labor object knowledge, to access to the related knowledge of other taxa.

Agriculture branches are divided into farm branch, forest branch, herd branch, sideline branch and fishery branch. Agriculture engineering is classified as agriculture machine, soil and water conservation, land and land utilization, rural building and structure, agricultural product and by product processing and rural energy resources by agriculture production requirement. Agriculture environment and agriculture regulation are classified by agriculture branch characteristic.

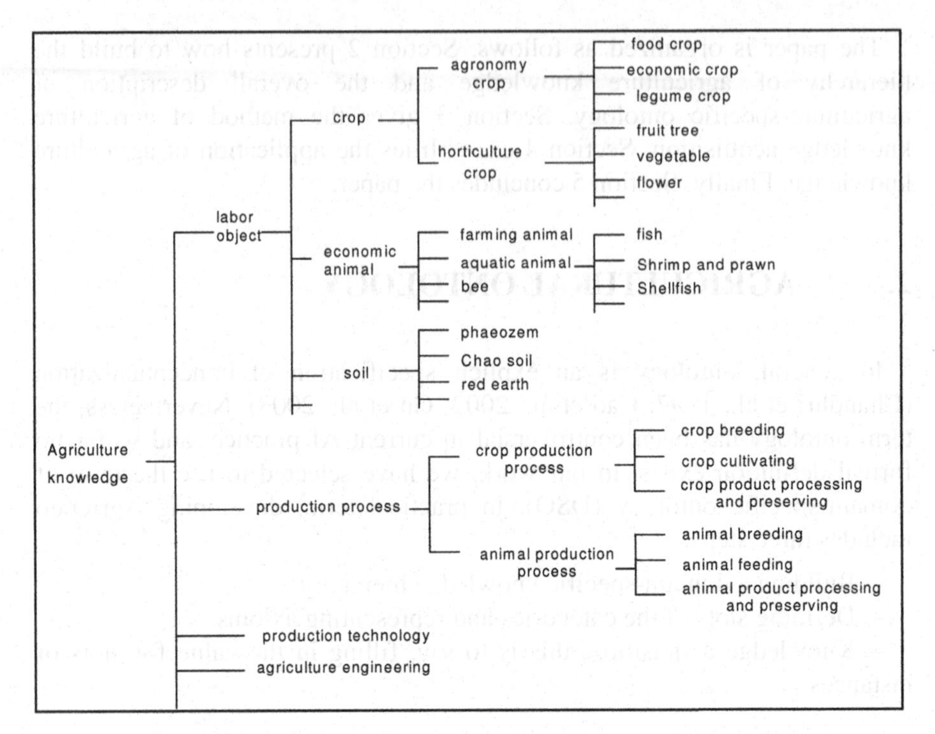

Figure 2. This shows some parts of agriculture knowledge hierarchy

To represent AgriOnto, we have designed a formal language that description the category definition and axiom. The syntax of the language is given in (Gu et al., 2003; Zhang et al., 2002). In practice, to some degree we have classified the attributes of a category when we classify the categories of agriculture knowledge, but this is not yet describe the natural character of a category and need be divided into deeper subclasses so that a concept can be clearly represented.

First, considering the knowledge connection between different domain-specific and the redundancy of NKI base (Cao et al., 1998; Cao et al., 2002), we share many botanical concepts by inherit relations in botanical ontology. In addition to this inheriting knowledge, we must analyze the particular character of crop. The crop knowledge has a close relation with other subject knowledge such as botany, zoology, physics, etc. Firstly, crop knowledge can be look as one part of botany, so it contains the attribute category such as heredity category, form and structure category, distributing category and so on. From agricultural view, there are crop growth category such as plant-disease and insect-pest category, crop environment category. Each attribute category also contains many slot definition like concept category. Figure 3 shows crop attribute hierarchical system. crop-distribute category inherit biological-distribute category and then contains all slots in biological-distribute category (Figure 4).

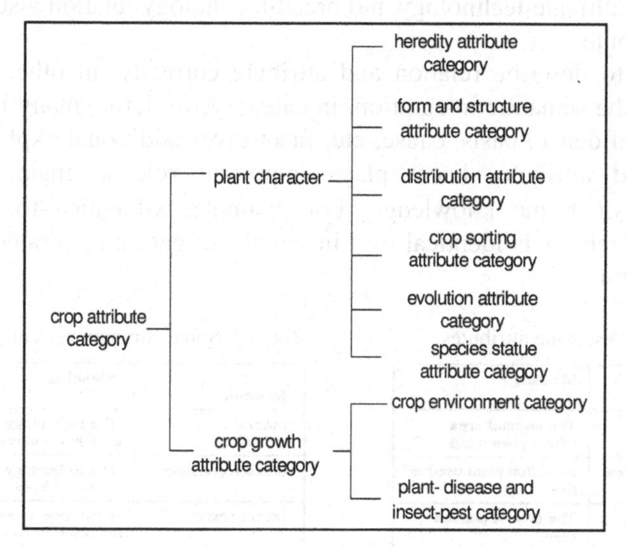

Figure 3. The hierarchical structure of crop- crop attribute categories

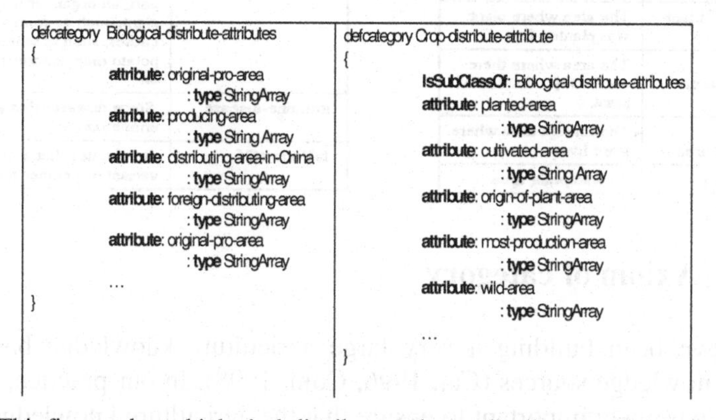

Figure 4. This figures shows biological-distribute and crop-distribute-attribute categories and their relations

2.2 Slots of category

In agriculture domain, Addition to attribute, relation is an most important definition in category. It primarily describes the relationship of entity concept. It indicates a proposition or an assertion so we usually use verb to represent it, but sometimes we also use noun to represent it. Relation can connect knowledge between concepts so that we can get the related knowledge when we look for part of concept knowledge. An important point is that a word as relation is also as attribute, which we call it attribute-relation definition. In table 2, it shows some common relations in the current AgriOnto. Cultivate-technology and breed-technology relation also represent a plant attribute.

In order to describe relation and attribute correctly, in other words, to implement the semantic integration, in category we define many facets such as time, confidence, basis, cause, etc. facet gives additional explanation for relation and attribute. Facet plays important role in maintaining the integration of frame knowledge. For example, x·Identical-to: y {facet: length} indicates x is identical to y in length, so get an integrated semantic representation.

Table 1. Distribute attributes

Attribute	Meaning
original-pro-area	The original area where plant lived
producing-area	Area that plant used to live
planted-area	The area of planted crop
cultivated-area	The area of cultivated crop
origin-of-plant-area	The area where plant was planted first
most-production-area	The area where there are most production of plant
foreign-distributing-area	The foreign area where plant live
wild-area	A wild plant area

Table 2. Some common agriculture relations

Relation	Meaning
cultivate-technology	The technology that a crop usually is cultivated
breed-technology	The technology that a crop usually is bred
Insect-pests	Some insect pests that destroy the plant, Such as budworm, ladybird beetle and so on
plant-diseases	Some diseases that destroy a part, an organ, or a system of plant, such as from various causes, such as wheat ergot, potato mop-top virus and so on.
immune-disease	Some diseases that a plant immunize.
IsVariantOf	It indicates that a species is a variant of another one.

2.3 Axiom of category

We have been building a very large agriculture knowledge base from several knowledge sources (Cai, 1996; Com, 1998). In our practice, we find that it is extremely important to ensure that the agriculture knowledge stored in the knowledge base is accurate and consistent. For each slot defined in a category, we have to specify one or more axioms to constrain their interpretation. These constraints are actually integral components of our categories. We have summarized a list of agriculture-specific axioms both for identifying inconsistency and inaccuracy in the acquired knowledge and for reasoning with the acquired knowledge. They form a first-order axiomatic system, and are an integral part of our whole ontology of crop. When a piece of crop knowledge is stored into the knowledge base, it is first checked by these axioms. If one of the axioms is violated, relevant information is reported to a knowledge engineer.

3. KNOWLEDGE ACQUISITION FORM TEXT: ONTOLOGY-DRIVEN METHODS

In recent years, knowledge acquisition from text has received much attention (Zhang et al., 2002). A key reason is that majority of the knowledge of a domain are presented in domain texts and documents.

In this paper, we utilize two methods to acquire agriculture knowledge from free-structured text. The first KAT system is a frame language for knowledge engineers to formalize text, together with the frame compiler mentioned above in OKEE. After the text is formalized, a frame compiler compiles frames into IO-models based on relevant categories. Although this method is not natural, most of our project knowledge engineers choose the frame language (NKI-FL) in formalizing domain knowledge. The second system is OMKE system which is an ontology-mediated knowledge extractor. The input to this system is semi-structured text (Cao et al., 2002). By semi-structured text, we mean that the syntax of the text is relatively fixed and thus can be easily summarized manually. Experiment in AgriOnto shows it can extract 50,000 Chinese characters per minute, this is, about 40 pages of A4 size per minute.

4. AN INTELLIGENT AGRICULTURAL KNOWLEDGE-BASED KNOWLEDGE SERVICE SYSTEM

In knowledge engineering, ontology is used to share and reuse knowledge and as standard for communication between computer and man. We develop some ontology-based applications on basis of knowledge base. In following, we introduce An Intelligent Agricultural knowledge-based knowledge service system: a Web-based consultant system.

A Web-based consultant system is a platform building on AgriOnto base. It can provide a user with what he wants to know quickly and correctly when he put logical query into its human knowledge interface (Feng et al., 2002). A query may have many expression forms, but they represent the same meaning. The consultant system will answer the query by relating the query's intension with corresponding ontology. For example the two queries "菜豆的栽培技术是什么？" (what's wheat's cultivated technology?) and "菜豆怎样栽培？" (how to cultivate wheat?) can be look as to ask the same question that discuss wheat's cultivated technology, so the answer must be same (Fig. 5). In our system, it provides two means for user to communicate with computer: the first means is keyboard communication. The second is voice communication, that is to say, a user query the computer by

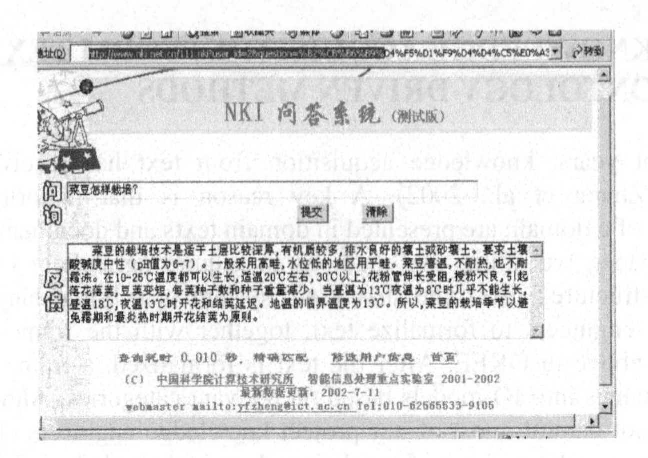

Figure 5. It is Web-based consultant system interface. The "问询" (ask) edit box is used to put the query. The "反馈" (feedback) edit box is used receive the answer the system will give when the "提交" (submit) button is pressed

microphone and the computer will give the answer in voice. But in voice communication, there exists noise and then causes many ambiguous semantic sentences which it is difficult or impossible for the computer to answer them. On basis of AgriOnto base, we restore these sentences to their original before the computer processes them.

5. CONCLUSIONS

This paper presents ontology-based acquisition of agriculture knowledge. In practice, ontology-based knowledge facilitates the sharing and the application of agriculture knowledge. At last we give an example of the application of ontology-base knowledge. With further study of AgriOnto, much future work immediately suggests itself:

– A most acquisition of knowledge is still not automatic for natural language complexity. The rate of acquiring knowledge is very low. the ongoing automatic acquisition tool is still more work.

– We have done some work about knowledge analysis (Xie, 2007). The automatic evaluation algorithm of knowledge is necessary including insistency checking, incompleteness analysis and knowledge redundancy removing.

– Further research on intelligent agricultural knowledge-based service systems.

ACKNOWLEDGEMENTS

This work is supported by Special fund project (2007211) for Basic Science Research Business Fee, AII, the Chinese Academy of Agricultural Sciences: Research on Grid-based Massive agriculture information technology. Thanks to Prof Cao of ICT, CAS for experiment platform.

REFERENCES

Cai SL 1996. The Volume of Agriculture in the Chinese Encyclopedia, China Encyclopedia Publishing House, 1996.

Cao CG 1998. National Knowledge Infrastructure. A Strategic Research Direction in the 21st Century, Computer World, 1-3, 1998.

Cao CG 2002. Progress in the Development of National Knowledge Infrastructure. Journal of Computer Science & Technology, Vol. 17 (No. 5) 2002, 523-534.

Chaudhri VK, Farquhar A. et al. 1997. The generic frame protocol 2.0. SRI International Technical Report, 1997.

Com 1998. The Editorial Committee of Agriculture: An Agricultural Dictionary, China Agriculture Press, 1998.

Feng QZ, Cao CG, Si J, Zheng Y 2002. A Uniform Human Knowledge Interface to the Multi-Domain Knowledge Bases in the National Knowledge Infrastructure. the 22nd SGAI International Conference on Knowledge Based Systems and Applied Artificial Intelligence. Applications and Innovations in Intelligent Systems. pp. 163-176. 2002.

Gu F, Cao CG, Sui YF 2003. A Domain-Specific Ontology of Botany. Journal of Computer Science and Technology, 2003.

Lenat DB 1995. "Cyc: A Large-Scale Investment in Knowledge Infrastructure." Communications of the ACM 38(11): 33-38, 1995.

Liu XH 1994: Automatic Inference based on Resolution Principle, 1994.

Lu RQ, Cao CG 1990. Towards knowledge acquisition from domain books. In: Wielinga, B, Gaines, B., Schreiber, G., Vansomeren, M. (eds.): Current Trends in Knowledge Acquisition. Amsterdam: IOS Press, pp. 289-301, 1990.

Natalya F. Noy, Mark A 1999. Musen. SMART: Automated Support for Ontology Merging and Alignment, Proceedings of the Twelfth Workshop on Knowledge Acquisition, Modeling and Management, Banff, Canada, July 1999.

Si JX, Cao CG, et al. 2002. An Environment for Multi-domain Ontology Development and Knowledge Acquisition, in Proc. First International Conference, EDCIS 2002, LNCS 2480, Springer-Verlag, Berlin, Germany, 104-116, 2002.

Tian W, Gu T, Cao CG 2002. Designing a Top-Level Ontology of Human Beings: A Multi-Perspective Approach, Journal of Computer Science & Technology, Vol. 17 (No. 5) 2002, 636-656

Xie NF 2007. Agricultural Knowledge inconsistency Research. Agriculture Network Information, pp. 11-13, 2007.

Zhang CX, Cao CG et al. 2002: A Domain-Specific Formal Ontology for Archaeological Knowledge Sharing and Reusing, in Proc. 4th International Conference PAKM 2002, LNAI 2569, Springer-Verlag, Berlin, Germany, 213-225, 2002.

RESEARCH ON PRECISION IRRIGATION IN WESTERN SEMIARID AREA OF HEILONGJIANG PROVINCE IN CHINA BASED ON GIS

Q. X. Jiang, Q. Fu *, Z. L. Wang

College of Water Conservancy & Architecture, Northeast Agricultural University, Harbin, 150030, China

** Corresponding author, Address: College of Water Conservancy & Architecture, Northeast Agricultural University, 59 Mucai Road, Harbin, 150030, P. R. China, Tel: +86-451-55191294, Email: fuqiang@neau.edu.cn*

Abstract: Geostatistics and GIS were used to analyze the spatial variability of the soil water characteristics in western semiarid area of Heilongjiang province. Irrigation management zone was established based on spatial distribution of the soil water characteristics and actual condition of the study area. Sampling by layered sampling method and calculating of the rational sampling number by best distribution method for each management zone gave the results, saturated water content with the most rational sampling number 15, whereas natural water content with the least 3. The rational sampling number determined by layered sampling method was 85%–97% less than the measured sampling number. Measuring the natural water content for each management zone and comparing with the corresponding saturated water content, field moisture capacity and wilting point can determine whether management zone need to be irrigated. In addition, different irrigation lower bounds should be adopted under diverse irrigation requires.

Keywords: GIS, precision irrigation, irrigation management zone, rational sampling number, semiarid area

Jiang, Q.X., Fu, Q. and Wang, Z.L., 2008, in IFIP International Federation for Information Processing, Volume 258; Computer and Computing Technologies in Agriculture, Vol. 1; Daoliang Li; (Boston: Springer), pp. 359–370.

1. INTRODUCTION

For the sake of decreasing resource waste, relieving environmental pollution, enhancing land utility efficiency and reducing agricultural production cost, American agriculturists brought forward the concept of precision agriculture (PA) in the 1990s (Peng et al., 2001). PA is based on 3S technique. According to the spatial variability of soil characters, water and fertility status, environmental background and climate condition among crop growing, an integrated series of modern farming operation and management are actualized quantificationally, locationally and timely. PA is also explained as "farming by inch". Different farming modes are taken for diverse soil types in the interest of getting the highest income with the least or the most saving devotion, protecting and improving environment, utilizing various agricultural resources efficiently and obtaining economical and environmental benefit (Cheng, 2004).

PA includes precision fertilization and precision irrigation. PA was started at the research of precision fertilization techniques abroad. Because soil fertility affects crop yield directly, and the technique combing with modern variable rate fertilizer can both decrease funds investment and get high crop yield, scholars in China and abroad put much vigor on it (Shiel et al., 1997; Wang et al., 2004; Hou et al., 2003; Wu et al., 2004). Soil moisture content, water demand status of diverse crops in different periods and variable irrigation technique are all needed for precision irrigation, so it is hard to be actualized. Few papers about precision irrigation are published, and it is just underway in China (Tian et al., 2002). Liu D. J. et al., expatiated on the research significance and prospect of precision irrigation in China (Liu & Feng, 2006). Liu G. S. utilized GPS water saving irrigation to research precision irrigation (Liu, 2000). Sun L. et al., used automatic drip irrigation system under plastic film on cotton farming to irrigate cotton precisely in its total growth periods, and provided hardware support for precision irrigation (Sun et al., 2004). Hu L. studied on real-time irrigation prediction model of the irrigation distinct and provided accurate prediction model for precision irrigation (Hu, 2004). Most scholars in China put attention on the research of precision irrigation system and device, and less on irrigation management zone and the quantification of irrigation water. Geostatistics, GIS and sampling technology were used to study the spatial variability of soil water characteristics in western semiarid area of Heilongjiang province in China, and delineate management zone. The implementation scheme was decided by the measurement of natural water content in each management zone.

2. MATERIALS AND METHODS

2.1 Study Area

The study was conducted on a 1 ha dry farmland of the dry farming science-technology demonstration park of Chahayang Farm locating at Gannan County in western area of Heilongjiang province, China. The area belongs to cold-temperature continent monsoon climate and semiarid agriculture climate region. The study area was divided by a 10 m×10 m grids. One hundred sample points were sampled in the central point of each grid. At each point, three samples were got in diverse depths of the crop growing area extending 30 cm from the surface. Soil water characteristics were measured and their mean values were considered as the value of each sample point. The experiment was conducted on Sep. 25 in 2006 after harvesting crops and before fall plowing. Sample location was decided by GPS device and basic measurement tools.

2.2 Measured Items and Methods

Measured items included natural water content, field moisture capacity, saturation moisture content, wilting point and soil dry bulk density. Soil texture is nearly associated with soil water characteristics, so soil dry bulk density is one of measured items. Drying method was used to measure natural water content, Wilcox method to field moisture capacity, ring knife method to SMC and soil dry bulk density and maximum hygroscopicity method to wilting point.

2.3 Research Methods

2.3.1 Geostatistics

In 1970s, geostatistics was introduced in the field of soil science to overcome the deficiencies of classical Fisher statistics theory in researching the spatial variability of soil characters. The concepts of regionalized variable, random function and stationarity hypothesis are the basis of geostatistics; semi-variance function is its main tool and Kriging interpolation is its means. It is the science used to study nature phenomena which having both randomicity and structure or spatial relativity and dependence (Hou & Yin, 1998). Using semi-variance function and Kriging interpolation can ascertain the variant degree and spatial relative scale and predict the spatial distribution of soil characters to guide precision irrigation and fertilization. Presently using the theory and method of geostatistics to study soil spatial variability quantitatively is the main trend in China and abroad.

2.3.2 3S Technology

Precision irrigation based on 3S technology utilizes the macroscopical control of RS, the collection, memory, analysis and output of GIS to the data in field and the ground precise measurement of GPS. Then cooperating with the transform of ground information and time control system, according to soil moisture and water demand of diverse crops in different growing period, precision irrigation is actualized timely and properly. GIS with its powerful function is dominant in precision irrigation.

ArcGIS produced by ESRI has perfect function including data input, compilation, query and cartography and powerful ability of second development. It is the most representative product in GIS field recently (Tang & Yang, 2006). ArcGIS has mighty spatial analysis function which combined GIS and geostatistics perfectly through the module Geostatistics Analyst providing software system for spatial variability research. In ArcGIS, user can import the geographic coordinates and attribute values and utilize Geostatistics Analyst module to get semi-variance function model of regionalized variable and various spatial interpolation figures. ArcGIS also has the functions of zoning, cutting and splicing for the spatial distribution of regionalized variable; powerful calculation function. User can get the results both in figure and data form (Li, 2006).

2.3.3 Sampling Technology

In a certain acceptance of sampling error, different sampling techniques would be adopted for diverse samples and requires reflecting the general features in the most degree with the least sample number. Simple random sampling and layered sampling are the usual sampling techniques. Random sampling is brief and intuitionistic, but inefficient to estimate population. Layered sampling has high estimating precision and can compute the index of population and layers simultaneously, different layer with diverse sampling method. So layered sampling is used in the study to make sampling survey on soil characteristics (Jin, 2002). Samples quantity is decided by the optimal distribution method one method of layered sampling, and the formula is as follows.

Population sample size:

$$n = \frac{(\sum W_h S_h)^2}{V + \frac{\sum W_h S_h^2}{N}} \tag{1}$$

Distribution between layers:

$$n_h = \frac{N_h S_h}{\sum_{h=1}^{L} N_h S_h} \times n \tag{2}$$

where, $W_h = \frac{N_h}{N}$ and $V = (\frac{d}{t})^2 = (\frac{r\overline{Y}}{t})^2$, W_h is the weight of h layer, N_h is the sample number of h layer, N is the sample number of the population, S_h is the standard deviation of h layer, V is the variance decided by sample estimation, d is absolute error, r is relative error, $\overline{Y}$ is the mean value of the population, t is the double side α fractile of standard normal distribution, while $p = 95\%$ and $N = 100$, $t = 1.984$.

3. RESULTS AND DISCUSSION

3.1 Classical Statistics Analysis of Soil Water Characteristics

The statistic analysis software SPSS 11.5 was used to analyze the soil water characteristics and the results was shown in Table 1.

Table 1. Classical statistic analysis and normality test of soil water characteristics

SWC[a)]	Mean	Max	Min	SD[b)]	CS[c)]	CK[d)]	CV[e)] (%)	TD[f)]
NWC (%)	19.16	22.61	15.79	1.499	-0.06	-0.66	7.82	Normal
FMC (%)	29.77	33.12	25.91	1.632	-0.09	-0.31	5.48	Normal
SMC (%)	43.91	53.79	32.24	4.933	-0.13	-0.69	11.24	Normal
WP (%)	12.59	13.92	11.33	0.625	0.26	-0.52	4.96	Normal
SDBD (g/cm^3)	1.25	1.45	1.10	0.074	0.22	-0.57	5.92	Normal

[a)]SWC: soil water characteristics; NWC: natural water content; FMC: field moisture capacity; SMC: saturation moisture content; WP: wilting point; SDBD: soil dry bulk density; [b)]Standard deviation; [c)]Coefficient of skewness; [d)]Coefficient of kurtosis; [e)]Coefficient of variation; [f)]Type of distribution.

As shown in Table 1, the mean value of soil moisture had an array as saturation moisture content>field moisture capacity>natural water content> wilting point. The natural water content at that time was between field moisture capacity and wilting point. Saturation moisture content had the biggest SD (4.933), whereas soil dry bulk density had the least (0.074), and that of natural water content and field moisture content were close. The CV of saturation moisture content was larger than 10% belonging to medium variability (10%–100%), and the others all belonged to small variability (0–10%). Before the data were analyzed by geostatistics, normality test should be run on them. For the sample size was big, skewness-kurtosis joint test method was used in the study to test data normality (Shi & Li, 2006). According to calculating results, random variable belonged to normal distribution when CS and CK less than 0.48 and 0.94 respectively. Under the test and criterion, all soil water characteristics belonged to normal distribution.

3.2 Semi-variogram Analysis of Soil Water Characteristics

Semi-variogram is implemental function to research spatial variability in geostatistics. It is utilized to represent the spatial variability structure or spatial continuity of random variable. Semi-variogram was calculated and its model was fitted by the geostatistics software GS+. In GS+, the maximal lag and lag distance should be selected appropriately. When lag exceeds a certain distance, the quantity of sampling point pairs will decrease which leads to increase the randomicity of semi-variogram calculating value, play down the precision of model fitting and even possibly distort the regularity of variability, so the maximal lag is commonly half of the maximal sampling distance (Liu & Shi, 2003; Cambardella et al., 1994). The maximal lag was 64 m and lag distance was 10 m in the study, for the maximal sampling distance was 128 m and the minimal distance of sampling by grid was 10 m. According to the results form GS+, semi-variogram values of soil water characteristics expressed the same quality on every orientation under 64 m, therefore they were calculated in term of isotropy. After the semi-variogram was calculated, the model was selected and the parameters were adjusted, the model parameters of semi-variogram of soil water characteristics and their fractal dimensions were shown in Table 2.

Table 2. Semi-variance function model and fractal dimension of soil water characteristics

SWC	Theoretical model	Nugget C_0	Sill C_0+C	C_0/(C_0+C) (%)	Range (m)	R^2	FD[a]
NWC (%)	Exponential	0.68	2.30	29.52	30.3	0.967	1.919
FMC (%)	Spherical	0.57	2.63	21.49	23.4	0.938	1.919
SMC (%)	Spherical	4.80	23.19	20.70	19.1	0.678	1.964
WP (%)	Spherical	0.11	0.53	20.64	94.2	0.934	1.742
SDBD (g/cm^3)	Spherical	0.001	0.005	22.92	19.2	0.829	1.958

[a]Fractal dimension

The theoretical semi-variogram model of natural water content was exponential model, and the others all were spherical models. The fitting precisions (R^2) of all models were high. According to the comparison among the ratios of nugget and sill ($C_0/(C_0+C)$), the variant degree of soil water characteristics can be met. The ratio of nugget and sill of natural water content was 29.52% belonging to medium variability (25%–75%), and the others belonged to low variability (0–25%). There were differences between CV and the ratio of nugget and sill to partition variability, because CV was relative to mean value, it represented the variant degree of random variable to mean value; whereas the ratio of nugget and sill represented the variability among variables. And rather the proportion of random and structural factors affecting system total variability could be described via the ratio of nugget and sill. Compared with CV, it could represent the variant degree of regionalized variable and its affecting factors more directly. The ranges (relative distance) of soil water characteristics were greatly different. Wilting point had the longest range (94.2 m), whereas saturation moisture content had the shortest (19.1 m). The result implied that sampling distance (10 m) in the study could satisfy the estimating need of the soil water spatial variability.

3.3 Ascertainment of Irrigation Management Zone and Reasonable Sampling Number

As the parameters of theoretical semi-variogram model in Table 2 were imported into the Geostatistical Analyst module, the spatial distributions of soil water characteristics were delineated by ordinary Kriging interpolation and shown in Figure 1 (a–e). Field moisture capacity, saturation moisture content and soil dry bulk density had the similar distribution representing banding, and strong negative-correlation existed between soil dry bulk density and field moisture capacity, saturation moisture content respectively. The area that had small soil dry bulk density had large field moisture capacity and saturation moisture content. The coefficients of soil dry bulk density and field moisture capacity, saturation moisture content were -0.98 and -0.85 respectively, which was significant correlation. The distributions of natural water content and wilting point all represented patch.

The tenet of precision irrigation was actualizing different irrigating schemes according to diverse soil moisture. For field moisture capacity, saturation moisture content and wilting point were the important index of deciding irrigation quantity and time, they were main soil water characteristics for delineating irrigation management zone. In addition, the similarity of the spatial distribution and convenient field management were the principle for partition. For the banding distribution of field moisture capacity and saturation moisture content and north-south ridges in study field, the study area was divided into four rectangle subareas (Figure 1(f)).

From the superposition of the spatial distributions and management zone in Figure 1 (a–e), it could be known that the soil water characteristics in each subarea were homogenous and large difference existed among management zones. It accorded with the principles of partition.

The tenet of precision irrigation was actualizing different irrigating schemes according to diverse soil moisture. For field moisture capacity, saturation moisture content and wilting point were the important index of deciding irrigation quantity and time, they were main soil water characteristics for delineating irrigation management zone. In addition, the

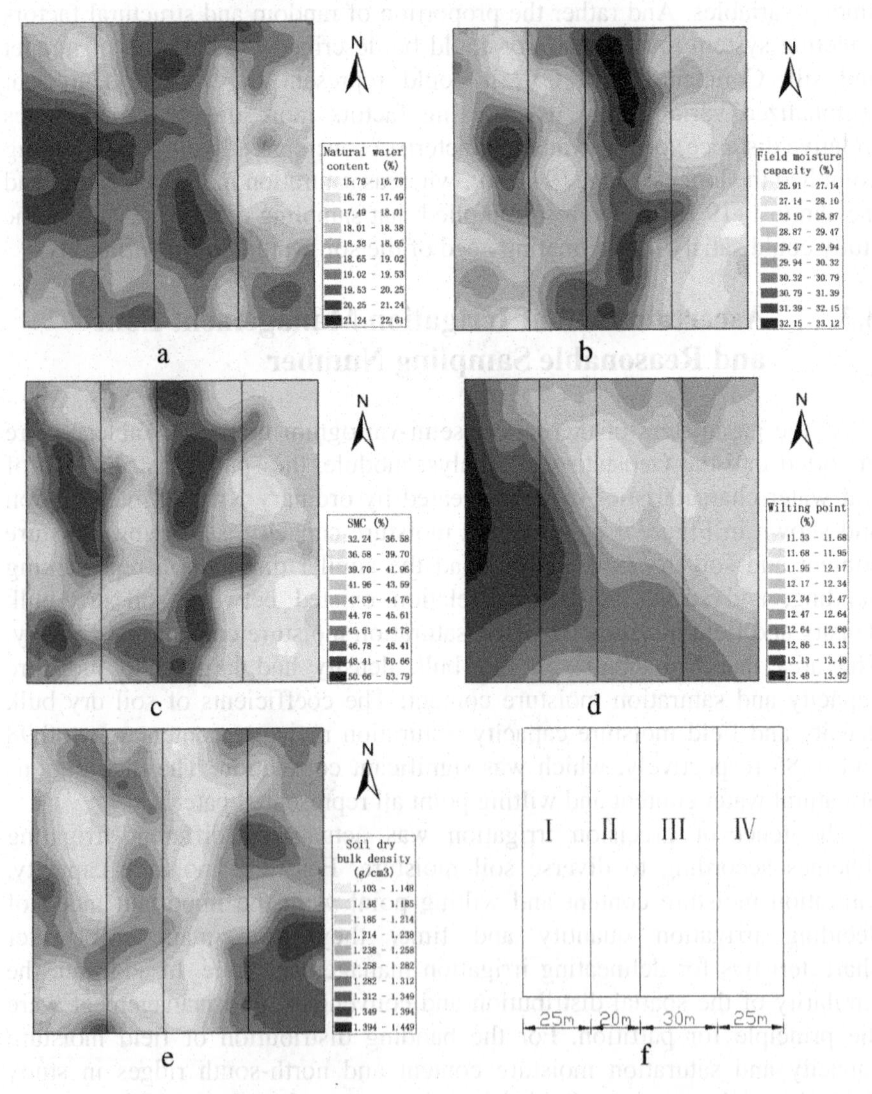

Figure 1. Distribution of soil water characteristics (a–e) and management zone (f)

similarity of the spatial distribution and convenient field management were the principle for partition. For the banding distribution of field moisture capacity and saturation moisture content and north-south ridges in study field, the study area was divided into four rectangle subareas (Figure 1 (f)). From the superposition of the spatial distributions and management zone in Figure 1 (a–e), it could be known that the soil water characteristics in each subarea were homogenous and large difference existed among management zones. It accorded with the principles of partition.

After partition for the study area, the statistical eigenvalues of soil water characteristics in each subarea shown in Table 3 had certain differences. The results indicated that diversity existed among eigenvalues of each subarea. The SD and CV of wilting point in the IV subarea compared with that of total area had been reduced by 31.68% and 29.35% respectively, simultaneously that of other subareas had been reduced by diverse degree. So the two eigenvalues should be considered for the ascertainment of reasonable sampling number.

After the optimal distribution method was used to calculate the reasonable number for each subarea, at the confidence level of 95% and the relative allowable error of 5%, the distributing results of the reasonable sampling number in each subarea were shown in Table 3.

Table 3. Statistical eigenvalue and reasonable sampling number of soil water characteristics for each management zone

Subarea	Sampling number	SE[a)]	NWC (%)	FMC (%)	SMC (%)	WP (%)	SDBD (g/cm^3)
I	25	Mean	19.59	29.59	45.24	13.19	1.24
		SD	1.51	1.71	4.77	0.55	0.08
		CV	7.69	5.78	10.54	4.14	6.06
		RSN	2	1	4	1	1
II	20	Mean	19.17	29.83	43.44	12.57	1.25
		SD	1.35	1.17	3.92	0.56	0.06
		CV	7.07	3.92	9.03	4.43	4.88
		RSN	2	1	3	0	1
III	30	Mean	19.40	30.78	45.87	12.44	1.22
		SD	1.34	1.52	4.74	0.50	0.07
		CV	6.92	4.95	10.33	4.00	5.66
		RSN	2	1	4	1	2
IV	25	Mean	18.43	28.68	40.60	12.17	1.29
		SD	1.59	1.27	4.49	0.43	0.07
		CV	8.64	4.41	11.06	3.51	5.37
		RSN	2	1	4	1	1
Total area	100	Mean	19.16	29.77	43.91	12.59	1.25
		SD	1.50	1.63	4.93	0.62	0.07
		CV	7.82	5.48	11.24	4.96	5.92
		RSN	8	4	15	3	5

a)SE: statistical eigenvalue; RSN: reasonable sampling number

Among the reasonable numbers of each soil water characteristic, saturation moisture content had the most (15) in total area for its biggest SD, whereas wilting point had the least (3). Natural water content had 2 reasonable sampling points in each subarea. The reasonable sampling number decided by layered sampling method was 85%–97% less than that of actual sampling, which greatly retrenched human and material resources for sampling and decreased the investment of precision irrigation.

3.4 Actualization of Precision Irrigation

The spatial distributions of soil water characteristics were transformed into grid figures with the resolution of 0.33 m× 0.33 m by ArcGIS. Then the spatial distributions were cut by management zone to get the grid figures of each subarea. The mean values of soil water characteristics in each subarea were inquired in the attribute table of grids and were shown in Table 4.

Table 4. Mean value of soil water characteristics for each management zone

SWC	I	II	III	IV	MTA[a)]
NWC (%)	19.45	19.27	19.32	18.46	19.12
FMC (%)	29.64	29.80	30.74	28.67	29.74
SMC (%)	45.33	43.50	45.78	40.52	43.88
WP (%)	13.12	12.64	12.38	12.28	12.59
SDBD (g/cm^3)	1.238	1.248	1.216	1.292	1.248

[a)]Mean of total area

For greatly affected by climate, natural water content had high variability in time, and so the mean values calculated in the study just represented the status at the sampling time. The other characteristics were considered as non-variability in time, because they were all relative to soil parent material and particle composition. Therefore, natural water content was the only item requiring to be measured while processing precision irrigation.

In the process of precision irrigation, natural water content needed to be known timely and moisture monitoring devices should be buried in the field to get the data. The quantity of the device was the reasonable sampling number of natural water content in each subarea and the distance between them should be larger than its maximum relative distance (range) 30.3 m. The mean value of the natural water content at the two point in each subarea was compared with the mean value of field moisture capacity, wilting point and saturation moisture content in the same subarea to determine whether it need to be irrigated. When the natural water content was between saturation moisture content and field moisture capacity, irrigation was not required. However, when it was less than field moisture capacity and close to wilting point, irrigation was needed. Field moisture content was the upper limit of irrigation. While water was ample, 70% of field moisture content was the lower limit. If water saving irrigation techniques were utilized, the lower limit of irrigation should be ascertained according to the requirements of diverse crops and the water saving irrigation experiments of main crops in the study area. Usually, the lower limit of irrigation should be the lower limit value of the feasible moisture content of the main crops in different stages (Hu, 2004).

4. CONCLUSIONS

As the maximum lag by 64m and the lag distance by 10 m, the range of wilting point is the longest by 94.2 m and that of saturation moisture content is the shortest by 19.1 m. Field moisture capacity, saturation moisture content and soil dry bulk density have the similar distribution representing banding. Furthermore field moisture capacity and saturation moisture content have strong negative-correlation with soil dry bulk density respectively. The distributions of natural water content and wilting point all represent patch.

For the banding distribution of field moisture capacity and saturation moisture content and north-south ridges in study field, the study area is divided into four rectangle subareas. In all the subareas, the soil water characteristics in each subarea are homogenous and large differences exist among management zones. It accords with the principles of partition.

After the reasonable sampling number of soil water characteristics are calculated by the optimal distribution method of layered sampling method, saturation moisture content has the most sampling number (15), whereas wilting point has the least (3). The reasonable sampling number decided by layered sampling method is 85%–97% less than that of actual sampling, which will greatly retrench human and material resources for sampling and decrease the investment of precision irrigation.

As the natural water content of each subarea is measured, it will be compared with field moisture capacity, wilting point and saturation moisture content in the same subarea to determine whether it needs to be irrigated. When the natural water content is between saturation moisture content and field moisture capacity, irrigation will not be required. However, when it is less than field moisture capacity and close to wilting point, irrigation is needed. Customarily, field moisture capacity is the upper limit of irrigation and 70% of it is the lower limit. When water saving irrigation technologies are adopted, the lower limit of irrigation will be the lower limit value of the feasible moisture of the main crops in different stages.

ACKNOWLEDGEMENTS

The project is supported by Program for Innovative Research Team of Northeast Agricultural University, "IRTNEAU"; Science & Technology Tackle Key Problem Program of Heilongjiang (No. GB06B106-7)

REFERENCES

Cambardella CA, Moorman AT, Novak JM 1994. Field-scale variability of soil properties in central Iowa soils. Soil Science Society of America Journal 58: 1501-1511.

Cheng JC 2004. Technology and application of precision agriculture. Beijing, China, Science Press.

Hou JR, Yin ZN 1998. Practical geostatistics. Beijing, China, Geology Press.

Hou SY, Wang X, Xue XZ 2003. Development of GIS application system for variable rate fertilization in soil precise management. Journal of Hebei University (Natural Science Edition) 23(2): 193-197.

Hu L 2004. Real-time forecast model of irrigation in irrigation districts. Nanjing, China, HeHai University.

Jin YJ, Jiang Y, Li XY 2002. Sampling technology. Beijing, China, Renmin Press. pp. 48-51.

Li RX 2006. Coal thickness spatial variability analysis based on geostatistics module in ArcGIS. Energy Technology and Management 1: 97-99.

Liu DJ, Feng JX 2006. Precision irrigation and its prospect analysis. Water Saving Irrigation 43(2): 43-44.

Liu FC, Shi XZ 2003. Characteristics of spatial variability of soil granules in a typical area of southern Jiangsu province. Chinese Journal of Soil Science 34(4): 246-249.

Liu GS 2000. Research on the GPS water-saving irrigation system. Transactions of the Chinese Society of Agricultural Engineering 16(2): 24-27.

Peng WL, Pierre Robert, Cheng HX 2001. Development of agricultural information technology and precision agriculture. Transactions of the Chinese Society of Agricultural Engineering 17(2): 9-11.

Shi Z, Li Y 2006. Geostatistics and its application in soil science. Beijing, China, China Agriculture Press. pp. 21-22.

Shiel RS, Mohamed SB, Evans E 1997. Planning phosphorus and potassium fertilization of field with varying nutrient content and yield potential. In: Stafford JV. ed. Precision agriculture 97. Processings of the First European Conference on Precision Agriculture. Oxford, UK, Bios Scientific Publication Ltd. pp. 171-178.

Sun L, Wang J, Chen X 2004. Experiment demonstration research on precision irrigation index system of cotton in Xinjiang. China Cotton 31(9): 22-24.

Tang GA, Yang X 2006. Experiment tutorial of geographic information system spatial analysis in ArcGIS. Beijing, China, Science Press.

Tian JC, Han BF, Wang BL 2002. Research on precision irrigation. Journal of Ningxia Agricultural College 23(2): 33-36.

Wang X, Zhao CJ, Meng ZJ 2004. Design and experiment of variable rate fertilizer applicator. Transactions of the Chinese Society of Agricultural Engineering 20(5): 114-117.

Wu CC, Ma CL 2004. Research on reasonable distances of soil sampling and fertilizing based on GIS in precision agriculture. Transactions of the Chinese Society for Agricultural Machinery 35(2): 80-83.

DECISION SUPPORT SYSTEM FOR RISK MANAGEMENT IN AQUATIC PRODUCTS EXPORT TRADE, CHINA

Feng Wang[1,2], Xiaoshuan Zhang[1,2,*], Cheng Tan[1,2], Chuanli Zhuang[1,2], Zetian Fu[1,2,*]

[1] *China Agricultural University, Beijing, P. R. China, 100083*

[2] *Key Laboratory for Modern Precision Agriculture System Integration, Ministry of Education, P. R. China, China Agricultural University, Beijing, P. R. China, 100083*

* *Corresponding author, Address: China Agricultural University (east) 209#, 100083*

Abstract: The fishing industry not only acts as foreign exchange earner but also plays an important role in China's economy. But with the development of technology and liberalization of international trade, the foreign countries adopted in succession trade barriers to limit China's fishery product export, which had make China's fishery product export disproportionated with fishery products production. By the literature analysis, we find practical research aiming at aquatic products export trade is in shortage greatly. So it is necessary to provide risk management for export trade of aquatic products. A decision support system for risk management in aquatic products export trade had been developed by China Agricultural University. Based on questionnaire and interviews, we analyze the decision problems, user needs and the difficulties involved in developing the aquatic products risk management system. The system architecture and its components, such as database, knowledge base and model base are described. At last we discussed on problems we had encountered during development and promotion.

Keywords: aquatic products, export trade, decision support system, risk management, China

Wang, F., Zhang, X., Tan, C., Zhuang, C. and Fu, Z., 2008, in IFIP International Federation for Information Processing, Volume 258; Computer and Computing Technologies in Agriculture, Vol. 1; Daoliang Li; (Boston: Springer), pp. 371–378.

1. INTRODUCTION

In China, aquaculture has been the fastest growing subsector within the agricultural economy, and the fishing industry not only acts as foreign exchange earner but also plays an important role in the China's economy by creating significant employment opportunities in various sectors of the seafood industry (Shekar Bose et al., 2005). China had exceeded over the Thailand in volume and been the world's largest exporter of aquatic products since 2002, and average export volume has increased 17% (Zhang Xiaoshuan, 2005). The export value is now around $9.36 billion at 2006, up 18.7%, and ranks first in bulk agricultural exports all the same.

However, with such tremendous growth in the export volume of aquatic products and the globalization of fish trade, the new problems are emerging. Many foreign countries have tightened food safety controls, imposed additional costs and requirements on imports, and adopted technology barrier to limit China's fishery product export (Zhou Deqing et al., 2003; Lahsen Ababouch, 2006). Secondly, the international market for fish and seafood products is dynamic greatly in nature due to some factors (Shekar Bose et al., 2005). Moreover, China's aquatic products export market is excessively concentrated in Japan, Korea, EU and US. When those countries adopt new trade policy, the export trade of aquatic products will be severely influenced. All these risk contributed to the instability and restrained from exerting advantage of China's aquatic products export trade.

Recent theoretical literature had provided some useful information concerning some of above mentioned problems. The cases and theoretical analysis, which focus on analysis on influence extent, internal and exterior reasons of trade friction, and the market conditions when the cases occur (Lahsen Ababouch, 2000; Zhang Fan, 2006; Chen Li, 2005; Yu Yahong, 2006; Meng Di et al., 2006; Titus O. Awokuse, 2006). Moreover, there are many literatures about fishery forecasting model and information system, which are powerful and useful to support participants of aquaculture industry to predict market risk and related information (Max Nielsen, 2000; Zhang Xiaoshuan, 2004; Helmut Herwartz, 2005).

It can be found that the practical research on risk management of aquatic products export trade is almost inexistent. So it is necessary to provide a useful risk management tool for aquatic products export trade. A decision support system (DSS) for export trade risk management can be typically used to detect any change in international aquatic products market via integration of market segmentation, risk early warning and forecasting, and offer ways to automate decision-making by generating information immediately accessible to all actors in a straightforward way.

Based on the above analysis, this paper introduces and discusses the experience in developing DSS-RMAPET, a decision support system for risk management in aquatic products export trade by China Agricultural University. The following is organized as: section 2 describes the users' need analysis; Section 3 and 4 show the system design and implementation. Then the last section draws some conclusions and further improvement.

2. THE SURVEY AND USERS' NEED ANALYSIS

2.1 The decision-makers and decisions

Table 1 identifies the decision problems that the different decision-makers need to face by analyzing the results of questionnaires and interviews based on aquatic products export trade.

Table 1. Decision-makers and decisions

Decision-makers and decisions
Government at national, provincial, County level, who need evaluate product prices, production, trade and macroeconomic aggregate by species and quality of fish to
• Early-warning risk by trade protectionism (anti-dumping, TBT, and so on.) to avoid potential dumping and protect healthy development of fishery sector. • Set out an enabling policy and regulatory environment at national and international levels with clearly rules and standards. • Establishing appropriate food control systems and programmes at national and local levels. • Collect information about the new aquaculture management and development policies, economical direction, rules and laws of import country, which play an important role in risk prevention.
Export enterprises of aquatic products and some organizations (including fishery institution), who give service to aquaculture production enterprises and farmers to
• Understand the objective reasons and conditions bringing about trade risk, which are gathered for accounting, statistics and forecasting the future potential risk. • Establish the factors which influence the international fish products market, and select appropriate time entrance to target market. • Sum up and classify results and losses caused by export trade risk, such as exchange rate, anti-dump law, TBT etc. • Seek how to shift the pertinent risk to the other memberships.

2.2 The needs for DSS-RMAPET

Table 2 illustrates the key function modules after analyzing the results of questionnaires and interviews based on aquatic products export trade.

Table 2. Modules and their functions

Module	Function
Data management module	Responsible for collecting data involved in import country's economical condition, trade policy, trade risk cases and so on, and analyzing and preparing the data sets for market classification, forecasting and early warning.
Market classification module	Assessing all import countries enables effective comparison, then selecting reasonable export markets.
Forecasting module	Responsible for monitoring and forecasting the export volume of aquatic products with relevant analysis.
Risk early warning module	Evaluating the degrees of trade risk and issue a warning signal.

3. SYSTEM DESIGN

3.1 Database module

The module is with responsibility for preparing the data sets for the model base and knowledge base needed for export trade risk management. It is composed of a database management system (DBMS) and a series of mini-databases. The DBMS provides all required data management capabilities including commands for adding, deleting, updating, browsing, and sorting records as well as importing and exporting (Omar F, 2000). The database is composed of three sections, including: Macro-economy information; Trade barriers information; Trade policy, rules and regulations of aquatic products safety and quality, such as import protection, new safety regulations and standards of import country, etc.

3.2 Knowledge base module

The knowledge base contains both factual and heuristic knowledge. Factual knowledge is that knowledge of the task domain that is widely shared, typically found in textbooks or journals, and commonly agreed upon by those knowledgeable in the particular field. Heuristic knowledge is the less rigorous, more experiential, more judgmental knowledge of performance, which is the knowledge of good practice, good judgment, and plausible reasoning in the field.

In the research, the objective of developing a knowledge base is to provide expert interpretation of available data regarding trade risk, which have an important impact on China's aquatic products export. Therefore, the knowledge base stores abundant knowledge including the forms of international aquatic product market classification, the cases of trade risks, relative risk forecasting and early warning result, an indication of the relative possibility in the accuracy of that result, and the preventive measure, as well as some advice for decision-making.

3.3 Model base module

The model base is the core component of this system, which operates a series of models to provide the managers with important risk decision reference.

1) *The international market classification model*

In order to avoid relying on these markets excessively and improve the ability to resist aquatic products market risks, it is necessary to develop and exploit new markets and reduce make market concentration degree. So the classification and selection of international aquatic products market are the first and most important step in export strategy.

Factors impacting international aquatic products market are fuzzy, uncertain and complicated. A neuro-fuzzy system, which can combine neural networks and fuzzy logic, and provide the advantage of reducing training time. Moreover, such results emphasize the benefits of the fusion of fuzzy and neural network technologies as it facilitates an accurate initialization of the network in terms of the parameters of the fuzzy reasoning system (Muhammad Aqil et al., 2007). Therefore, the Fuzzy Artificial Neural Network Inference System (FANNIS) classification model is established, which can offer the comparatively accurate division of the international aquatic products market and reduce aquatic products export market risk effectively.

FANNIS is not a fixed inference system in advance. It needs to learn according to the regulations which experts have established, then forms a network that can classify international aquatic products market. So there are two processes to ensure reasonable classification for international aquatic products market: learning and classification.

2) *The export volume forecasting model*

In recent years, China's export volume of aquatic products show growing trend, but the strict trade barriers of foreign country are increasing rapidly, which will have a profound impact on export volume of aquatic products. It is a perceivable need to develop forecasting models able to capture export volume changes.

Considering that all the raw sequences about export trade are vibrating and changing in a certain variety trend. To improve the predication capability, the research introduced Markov(1, 1)-GM model by integrating the Markov chain with the GM(1, 1), which can forecast the future by analyzing the inside regulation of development in time to come, and it reflects the influence degree and laws. The GM(1, 1)-Markov model, i.e. a single variable first order grey model integrated with Markov Chain.

3) *The trade risk early-warning model*

The model is responsible for integrating information and operational experience of experts with forecasting results to evaluate the degrees of trade risk and issue a warning signal. When the warning situations emerge, the system need to analyze and evaluate the creditability and impact degree on export trade of aquatic products with domain experts, then provide effective decision advice on dealing with, avoiding and breaking through. According to the procedure of early-warning system proposed by Fulai Huang (2005), the early-warning process include searching the origin/source that warning condition is produced, choosing warning signs, determining warning limits of the warning signs, computing the early-warning index, compartmentalizing warning grade and determining warning degree. In this paper, the early-warning model for export trade risk is developed based on Support Vector Machines (SVM).

4. SYSTEM IMPLEMENTATION

The fishery market data are constantly being updated and the mathematical models are also potentially changeable. So a B/S/S, i.e. three-tier architecture, was adopted as our system's fundamental architecture to provide the transparency among the data layer, business logic layer and the user interface layer.

There are many solutions for developing the web-based systems. The solution, which is Microsoft .NET Framework 1.1+ASP.NET+MS SQL Server 2000 +IIS 6.0 +COM, is adopted to develop this system. Among the development plan:

- Microsoft's Visual Basic .NET is one of the core development languages in the new Visual Studio .NET and it remains a critical component in almost every enterprise wide development effort. It is serving as the main develop language to bridge user interface and web server and database server.
- MS SQL Server 2000, a relational database management system and data warehouse development tool, is serving as the back-end of the workstation to facilitate data storage and retrieval and as a means to preserve analysis methodologies and knowledge.
- IIS 6.0 is serving as a web server to provide information service.

- Microsoft COM technology enables software components to communicate. COM is used to create re-usable software components, link components together to build applications, and take advantage of Windows services.

5. CONCLUSION AND DISCUSSION

From the above analysis, the main outcomes can be outlined as follows:

- Different software modules for data storage, communication, processing, early warning and evaluation as well as forecasting have been successfully developed and combined into DSS-RMAPET software package. We also integrated fuzzy logic, ANN and grey theory to produce a hybrid intelligent algorithm for solving evaluation, forecast and early warning models. The integration has enhanced the system performance. Results of this study have important implications for the refinement of aquatic products risk assessment model and optimization of aquatic products export trade configuration.
- The system provides a comprehensive decision support for users due that it integrates multiple knowledge including macro-economic theory, trade policy, forecasting and early-warning methods and so on.
- By utilizing crossing validation to select anova as the kernel function, indicators of warning situation and warning omen as variables, an early-warning model based on support vector machine is established to study the export trade risk of aquatic products.
- Based on the theory of fuzzy inference system, the author establishes an FANNIS classified model of international aquatic product market by introducing the artificial neural network as fuzzy inference device and non-fuzzy device. This classified model has simplified the procedure structured by fuzzy inference system. It can offer the comparatively accurate division of the international aquatic products market.
- An effective DSS-RMAPET depends on both the model and data. The availability and accuracy of data are major constraints to the usefulness of such models. In China, the fishery market data are scattered among a multitude of producers with dissimilar formats, resolutions. Moreover, it is difficult to acquire relevant trade data about import countries, and new trade risk cases and fish trade policy need to be tracked timely, so the update of track record of system models for risk management is not so good.
- Further research is to integrate more effective forecasting and early warning method into the system to improve the system's intelligent functions, for example, Knowledge Discovery in Database (KDD) can be used to produce the rules automatically for the knowledge base embedded in the DSS.

ACKNOWLEDGEMENTS

This study has been funded by the EU FP6 program (FP6-016333-2).

REFERENCES

Chen Li. Risk assessment and international aquatic products trade. Journal of Fujian Fisheries, 2005, 2: 55-56 (in Chinese)

Fulai Huang and Feng Wang. A system for early-warning and forecasting of real estate development. Automation in Construction, Volume 14, Issue 3, June 2005, 333-342

Helmut Herwartz and Henning Weber. Exchange rate uncertainty and trade growth - a comparison of linear and non-linear (forecasting) models Applied Stochastic Models in Business and Industry, 2005, 21: 1-26

http://www.statsoft.com. Support Vector Machines (SVM)

K. Bryant. ALEES: An agricultural loan evaluation expert system. Expert system with application 2001, 21: 75-85

La Sen, Liu Yadan. Risk analysis and international aquatic products trade. Chinese Fisheries Economy Research, 2004, 4: 33-34 (in Chinese)

Lahsen Ababouch. Assuring fish safety and quality in international fish trade. Marine Pollution Bulletin, 2006, 53: 561-568

Lahsen Ababouch. The role of government agencies in assessing HACCP. Food Control, 2000, 11: 137-142

Max Nielsen. Forecast of Danish ex-vessel seafood prices. Http://esb.sdu.sk/~uho/ime/EAFE2000/papers/maxniel.pdf

Meng Di, Zhou Deqing. Review on Risk Analysis of Seafood. Journal of Chinese Institute of Food Science and Technology, 2006, 1: 390-395 (in Chinese)

Muhammad Aqil, Ichiro Kita, Akira Yano and Soichi Nishiyama. A comparative study of artificial neural networks and neuro-fuzzy in continuous modeling of the daily and hourly behaviour of runoff. Journal of Hydrology, 2007, 337: 22-34

Omar F. El-Gayar. ADDSS: a tool for regional aquaculture development, Aquacultural Engineering, 2000, 23: 181-202

Shekar Bose, Arna Galvan. Export supply of New Zealand's live rock lobster to Japan: an empirical analysis. Japan and the World Economy, 2005, 17: 111-123

Titus O. Awokuse, Yan Yuan. The Impact of Exchange Rate Volatility on U.S. Poultry Exports. Agribusiness, 2006, 22: 233-245

Yu Yahong. Some reflections on aquatic products safety of Chinese export trade. China Fisheries, 2006, 1: 16-18 (in Chinese)

Zhang Xiaoshuan, Hu Tao, Brain Revell and Fu Zetian. A forecasting support system for aquatic products price in China. Expert Systems with Applications, 2005, 28: 119-126

Zhang Xiaoshuan, Zhang Jian, Fu Zetian. A forecasting support system for aquatic products price based agent. Computer engineering, 2004, 8: 65-67

Zhou Deqing, Li Xiaochuan. The quality safety and its promotion of fishery product in China. Fishery modernization, 2003, 1: 3-5 (in Chinese)

DESIGN OF DATA CENTER'S HIGH RELIABILITY IN LARGE AGRICULTURAL ENTERPRISE

Hanxing Liu[1], Yingjie Kuang[1], Caixing Liu[1*], Lei Xiao[1], Guomao Xu[2]

[1] *Department of Computer Science and Engineering, South China Agricultural University, Guangzhou, P. R. China*

[2] *Informatics Center, Guangdong Wen's Food Corporation, Xinxing, Guangdong Province, P. R. China*

* *Corresponding author, Address: Department of Computer Science and Engineering, South China Agricultural University, GuangZhou, 510642, P. R. China, Tel: +86-20-85281371, Fax: +86-20-85285393. Email: liu@scau.edu.cn (LIU Caixing)*

Abstract: It's necessary for the large agricultural enterprise to construct a stabile and reliable network environment and a powerful Data Center. In this paper, the constructing of high reliability in Data Center will be discussed according to the requirements of large agricultural enterprise; an effective solution will be presented that adopts structural redundancy and double VPN in the aspect of network and application. This solution has been carried out effectively in practical application, and proved that it is an exercisable example for the constructing of Data Center in large agricultural enterprise.

Keywords: large agricultural enterprise; data center; high reliability

1. INTRODUCTION

Usually, Data Centers are constructed in large industrial enterprises, banks, telecom, portal sites, etc (Gao GQ, 2004) (Wan XJ et al., 2003), but the application in agricultural enterprises is somehow unusual. The reasons are that the information basis of agricultural enterprise is weak in P. R. China and that the constructer is limited by person, finance and environment etc. One example is

Liu, H., Kuang Y., Liu, C., Xiao, L. and Xu, G., 2008, in IFIP International Federation for Information Processing, Volume 258; Computer and Computing Technologies in Agriculture, Vol. 1; Daoliang Li; (Boston: Springer), pp. 379–388.

Guangdong Wen's Food Corporation, which is a national leader enterprise of agriculture, and has more than one hundred branches or sub-companies, every branch has constructed Local Area Network and application system severally, but the information communication is poor between headquarters and branches. As a result, "Information Island" appeared, the statistic and feedback of enterprise information is delayed, it is not helpful to establish and implement decision.

Thereby, this enterprise decided to construct Data Center based on Internet in order to manage data centrally and to serve branches. In the beginning, Data Center needs to build up a reliable network system, and then construct the application software based on the network. As constructors of this project, the authors think that large agricultural enterprise (especially large livestock-breeding enterprise) has characteristics as following:

1) Wide and interlocal location: large enterprise has many branches and they are located in wide area, even the whole country;

2) Out-of-the-way site for connecting: agricultural enterprises, especially livestock-breeding enterprise, always locate in the rural area where is far away from city or town, and the telecom establishment usually not good enough, this make it more difficult for network connection;

3) Various types of user: although the production bases are always in rural area, but its main market and clients are located in the edge of cities or towns. there are various types of user and connection mode.

The above characteristics make it more difficult to build up Data Center of agricultural enterprise. The main task is to ensure the reliability of network, consequently to ensure the steady running of enterprise business across wide area and out-of-the-way locations.

2. RELIABILITY OF DATA CENTER

The characteristic of Data Center is "centralized data and distributed application", its importance for enterprise is obvious. When designing the Data Center, the requirements of enterprise shall be thought over in aspects of business, management and technology. The design will have certain foresight as well as keeping stabile, so advanced and accredited technology is adopted, the system should be implemented step by step (Gao GQ, 2004).

Arregoces et al. (2003) thought that three aspects should be considered when designing Data Center:

1) Scalability: supporting fast and seamless growth without major disruption;

2) Flexibility: support new services without a major overhaul of its infrastructure;

3) High availability (HA): having no single point of failure and should offer predictable uptime.

Authors of this paper think that the reliability of Data Center includes three aspects as following:

1) Reliability of environment. The environment of computer rooms in Data Center should be satisfy with corresponding national standards or constraints, such as temperature, humidity, dust, fire protection, thunderbolt prevention, grounding, weight bearing, power supply, illumination etc. Main measures include: installing air-conditions to ensure constant temperature and humidity in computer rooms; setting gas fire system in network rooms and server rooms, and water-spray fire system in monitor rooms and test rooms; preparing plenteous backup electric power (Kieffer S et al., 2003).

2) Reliability of network. The reliability of network is the hardware base of Data Center's reliability; it is related not only with performance of devices and topology, but also with environment of communication network (Luo PC et al., 2000). Network with high reliability will be redundant and having no single point of failure, so as to support various applications and ensure system safety (Jiang WJ et al., 2000).

3) Reliability of application. Apart from reliable function of software itself, optimization is needed according to characteristic of hardware during designing, operating system and topology of server cluster are also pivotal for reliability.

Large agricultural enterprise depends on network and relative application system increasingly in routine producing and official business. No doubt, Data Center can decrease cost of producing and increase advantage of competition for enterprise, but in malfunction befallen case, it might confuse the producing and management of enterprise, and cause unnecessary lost, so it is important to pay more attention to reliability of Data Center. The reliability of network and application is analyzed in this paper, and a solution for constructing Data Center with high reliability will be presented.

3. THE DESIGN OF NETWORK RELIABILITY

The principal objective of Network Reliability is fault-tolerance (Jiang WJ, Xu YH 2000). The way of fault-tolerance is to seek regular points of failure, make them robust with redundancy, so as to shorten fault time of network furthest. It has two principles as following: parallel backbone and double network centers. For Data Center of enterprise, the key measures are redundant network topology and connection mode.

3.1 Redundant network topology in data center

In order to ensure business persistent, the network must be persistent, especially in interlocal agricultural enterprise having distributed applications. Network is charged with transferring business data, business will be break down if network broken. In large agricultural enterprise, users connected with Data Center by several types, which were shown in Table 1.

Table 1. User types of Network Connection

User type	Mode of network connection	Bandwidth
Corporation headquarters	Intranet in Corporation	100M
Suburb branch	FDDI	2M, 10M
Out-of-the-way branch	ADSL, modem	<2M
Mobile or sporadic user	ADSL, wireless, modem	<2M

The types of connection are various, and the quality of connection is often not controllable (except Intranet in Corporation). So, the emphasis is made to ensure the reliability of Intranet and response for various connections in order to avoid network breaking, Network topology in Data Center is suggested like in Fig. 1.

To ensure the reliability of connection with Internet, double network redundancy is adopted for fault-tolerance, i.e., two communication links that provided by different telecom providers were used at the same time. Even if one of the redundant links have malfunction, business can connect with Internet by the other link. Synchronously, the pivotal network devices might also be redundant, such as IPS (Intrusion Prevention System), Switches, VPN (Virtual Private Network) Firewall and Main Switch.

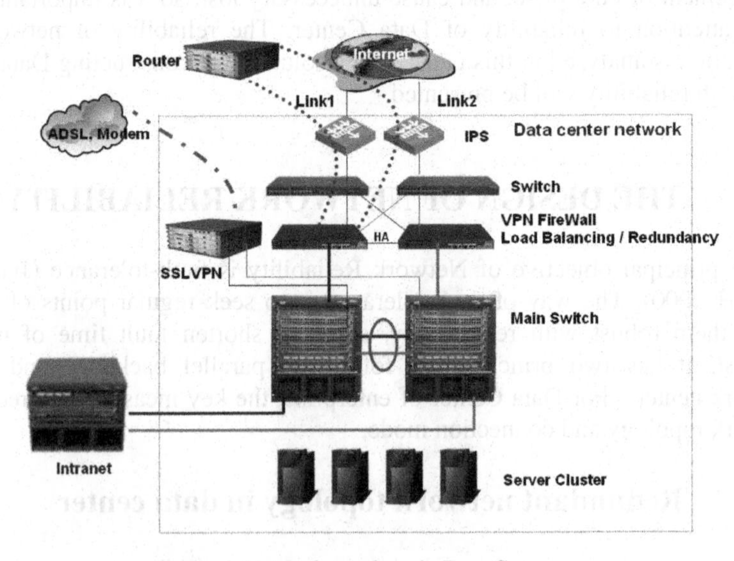

Fig. 1. Network topology in Data Center

IPS up-connects with two communication links, and down-connects with Switches. Switches connect crosswise with Firewall. Firewalls achieve high reliability via VRRP (Virtual Router Redundancy Protocol), Firewalls down-connect with Main Switches. Between Main Switches, the interface of VRRP with TRUNK is set up. In such architecture, Firewall can be load-sharing or redundant backup, and redundant backup is more frequently. When users connecting with VPN, Switch will build a Tunnel with one of the Firewalls that connected crosswise, and switch to another redundant Firewall if the former Firewall faulting. If the former Switch faults, the client VPN device will route to the other communication gateway automatically, and connect with other Firewall through another good Switch. Limited by Firewall's ports, Firewalls don't connect crosswise with Main Switch, and actually, it is not necessary to connect crosswise. Main Switch is used as VRRP, and TRUNK port is used to connect two Main Switches, this TRUNK port will not fail even when one of Main Switches faulting. Thus, Firewalls are ensured to connect with Main Switch reliably (Liang XJ et al., 2004) (Stallings W, 2006). Table 2 shows the strategies of Network Reliability in Data Center.

Table 2. Strategies of Network Reliability in Data Center

Item	Strategy of high reliability
Communication link	FDDI provided by different telecom providers, two gateways, route to the other gateway automatically when former gateway faults;
IPS	Up-connect with communication link and down-connect with Switch, prevent intrusion;
Switch	Down-connect crosswise with Firewall, route to the other gateway automatically and connect with Firewall when the former Switch faults (i.e. former link break down);
VPN Firewall	Up-connect crosswise with Switch, adopt VRRP between two Firewalls, VRRP chooses the other Firewall to connect with Main Switch when the former Firewall faults;
Main Switch	Two Main Switches up-connect with Firewalls respectively, but connect with each other through VRRP and TRUNK, Firewall connect with the other Main Switch through TRUNK when the former Main Switch faults;

3.2 Strategy of double VPN connection

Besides the reliability of device and link in network level, clients' connection might also be redundant. Double VPN is set up, end-user will choose VPN Tunnel and route automatically according to the state of network link, and submit VPN connecting requirement.

In large agricultural enterprise, possible modes of client VPN connection are various and Table 3 shows several connection modes.

Table 3. Mode of client VPN connection

User type	Mode of network connection	Mode of VPN connection	Connection device
Corporation headquarters	Intranet in Corporation	None	None
Suburb branch	FDDI	GRE (Generic Routing Encapsulation), IPSec (Internet Protocol Security)	Router, firewall
Out-of-the-way branch	ADSL, modem	SSL (Secure Socket Layer)	IE, client plug-in
Mobile or sporadic user	ADSL, wireless, modem	SSL	IE, client plug-in

In corporation headquarters, users don't need VPN connection; it is needed by corporation branch and other users. VPN connection of corporation branch is just discussed here. If only IPSec connection is adopted, the high reliability of double VPN will not be realized. The reason is that IPSec itself can't route automatically in common device, i.e., device can't route to another VPN Tunnel when one of VPN Tunnels unavailable. If only GRE connection is adopted, route can be achieved automatically, but security prevention will be lack. Thereby, IPSec is suggested to be integrated with GRE in this paper. GRE Protocol encapsulates message of some network-level protocols (such as IP, IPX), and make these encapsulated message able to be transferred in another network-level protocol. GRE is the third level Tunnel protocol of VPN, and a technology called as Tunnel is adopted between different levels of protocol. Tunnel is a virtual point-to-point link, and can be regarded as a virtual interface supporting point-to-point link, this interface provides a thoroughfare to transfer encapsulated message. The message will be encapsulated and de-encapsulated in two ends Tunnel. An IPX message encapsulated in IP Tunnel has the following format:

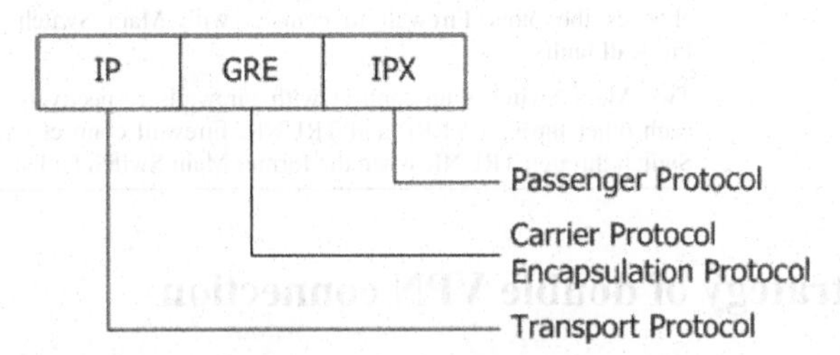

Fig. 2. The format of massage transferred in Tunnel

When actualizing double VPN, two Tunnels are set up, their target addresses are corresponded with certain gateway of different links. One Tunnel is the preferred route, normally it will be chosen automatically. If this Tunnel is detected to be unreachable, the other Tunnel will be switched to, thus double VPN connection is actualized. Actually, target address of message can be recognized when de-encapsulated after GRE encapsulates route data. But GRE itself can't transfer message safely. In order to transfer message safely, the data will be encapsulated by GRE at first, and then the encapsulated message will be encrypted by IPSec. Figure 3 is the description of GRE integrated with IPSec.

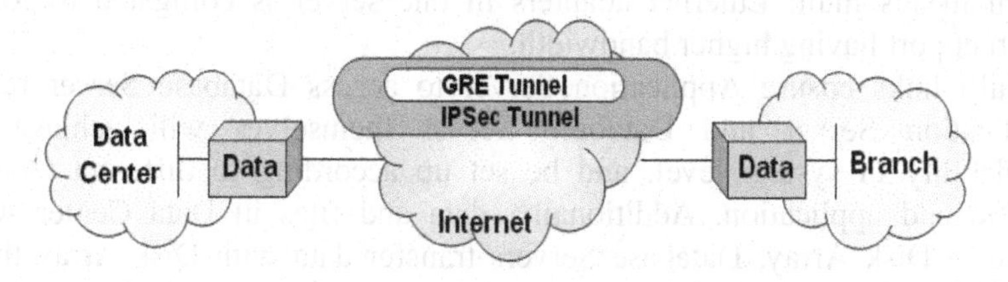

Fig. 3. GRE-IPSec Tunnel

4. THE DESIGN OF APPLICATION RELIABILITY

Data Center based on WEB requires server has high performance of real-time and throughout, it will receive clients' requests reasonably and response in a short time, Cluster is an effective way to implement high performance for WEB server (Lin C, 2000) (Zeng BQ et al., 2004). Server Cluster System consists of multi congener or heterogeneous servers. It offers transparent services and fulfills tasks cooperatively. It has the following advantages: avoiding temporary halt when updating server's software and hardware; joining or exiting of single server will not influence the whole cluster; avoiding single point of failure; high usability, reliability, performance and expansibility; transparent load-balance.

Application Reliability can be guaranteed by Server Cluster in this paper. Fig. 4 shows the topology of Server Cluster, Server Cluster is behind Main Switch usually consist of Database Server, Application Server and Disk Array.

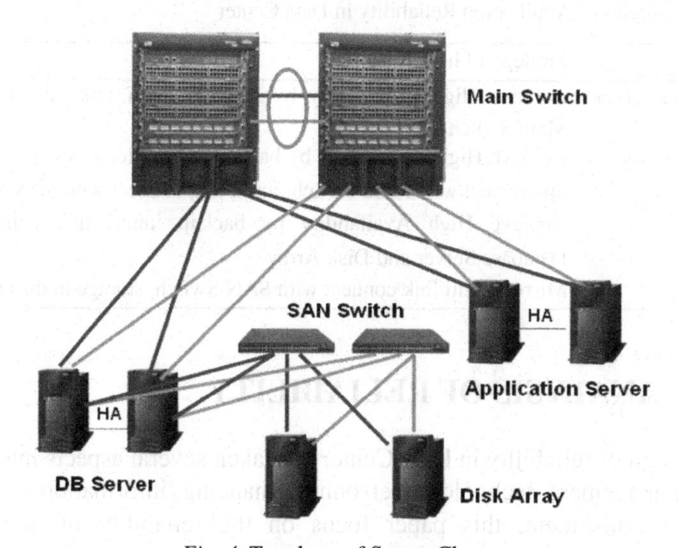

Fig. 4. Topology of Server Cluster

Main Switch down-connects with Application Server and Database Server. For high reliability and high-speed transference, two Switches will connect crosswise with four servers, and network port shall be colligated if possible, which means multi Ethernet adapters in one server is colligated to form an Ethernet port having higher bandwidth.

Multi links ensure Application Server to access Database Server reliably, Application Server and Database Server themselves will achieve High Availability in system level, and be set up according to different Operating System and application. Additionally, data and files in Data Center will be stored in Disk Array, Database Servers transfer data with Disk Array through SAN (Storage Area Network). Two SAN Switches backup each other and down-connect crosswise with the controller of two Disk Arrays, up-connect crosswise with FDDI adapter of Data Server. System-level mirror is implemented in two Disk Arrays. One Disk Array is located in Data Center, and the other in computer room of another building, this strategy of disaster-backup on different locations could be complex but quite effective.

Clients access Application Server, the fault of single Application Server and communication link were considered. Application Servers request data from Database Servers, Database Server itself is high reliable with backup storage. Data in Database store in Disk Arrays, two arrays mirror each other. Data is transferred in independent SAN; there is no single point fault in the path from servers to disk arrays. Under the control of multi-path balance software, data stream will be loaded balancedly when paths no problem, once a path faults, all of I/O will switch to another path for accessing, and after the problem path is resumed, data stream will be loaded balancedly again. SAN is highly reliable under such design. Table 4 figured the strategies of Network Reliability in Data Center.

Table 4. Strategies of Application Reliability in Data Center

Item	Strategy of high reliability
Application Server	Achieve High Availability by backup, multi link up-connect with Main Switch
Database Server	Achieve High Availability by backup or parallel accessing, multi link up-connect with Main Switch, multi link connect with SAN Switch
SAN Switch	Achieve High Availability by backup, multi link connect with Database Server and Disk Array
Disk Array	Mirror, multi link connect with SAN Switch, storage in different site

5. ANALYSIS OF RELIABILITY

The design of reliability in Data Center has taken several aspects into account, such as management, technology personnel, financing, information security and telecom establishment, this paper focus on the reliability of network and application.

Luo PC et al. (2000) thought that there are two evaluations of reliability for communication network: Invulnerability and Survivability. Invulnerability involves the reliability of network under purposive destroying, so it emphasizes particularly on the reliability of topology, always being used to evaluate army's network. For enterprise network, purposive destroying is hardly probable, so Invulnerability is seldom evaluated. Survivability involves the reliability of network under random failure (or being destroyed), it covers the concept of Network Robust, such as dynamic route, fault resume, prevention and redundancy etc, so Survivability was always evaluated. At the present time, hardware devices have been quite reliable, and been easy to inspect even faulting. In real running, system failure was infrequently caused by the malfunction of network devices, so designer of Data Center has concerned about the business performance of network, such as throughout and delay, i.e. how to ensure Application Reliability.

Take example for Guangdong Wen's Food Corporation (a national leader breeding enterprise), the above strategies were actualized when constructing Data Center, and high reliability of system is achieved. The frequent fault of network in routine, including device fault and link fault, were prevented thanks to redundant system framework. Double VPN solves the problem of link intermittence in client level, and keep business 24 hours running. In server level, cluster is adopted reasonably, links are highly reliable, and the speed of cluster switching is fast when single point failure, so the Data Center never intermits business in routine running due to fault. This shows that the above strategy of high reliability is reasonable and feasible.

6. CONCLUDING REMARKS

The primary goal of constructing Data Center is the centralized data management for uniform service, the design of reliability is essential guarantee for stabile running. For large agricultural enterprise, especially breeding enterprise, its characteristics of trade are wide area, distributed branch, out-of-the-way location, various user and complex mode of connection. All of these cause difficulties for constructing Data Center and ensure high reliability of network and client application.

Network Reliability and Application Reliability were analyzed based on practice. Network Reliability was achieved by redundant link, device, VPN and adapter, cooperated with VRRP, GRE, IPSec technologies of Switch, Firewall, router, etc. Application Reliability was achieved by redundant links between servers, server cluster and mirror of Disk Array. Based on these strategies of reliability, the loss caused by single point failure will be reduced effectively.

REFERENCES

Arregoces M, Porotolani M 2003. Data Center Fundamentals. Cisio Press.
Gao GQ 2004. Research and Practice on Data Center's Construction on Bank of China. Tsinghua University Master degree thesis.
Jiang WJ, Xu YH 2000. Research & Design on Architectonic of Network's Reliability. Computer Engineering and Applications 36(12):156-159.
Kieffer S, Spencer W, Schmidt A & Lyszyk S 2003. Planning a Data Center. Denver, USA, Network System Architects, Inc.
Liang XJ, Sun QH 2004. Communication Network Reliability Management. Beijing, China, Beijing University of Posts and Telecommunications Press.
Lin C 2000. Performance Analysis of Request Dispatching and Selecting in Web Server Clusters. Chinese Journal of Computers 23(5):500-508.
Luo PC, Jin G 2000. A Review of Study on Reliability of Communication Network. Mini-Micro Computer Systems 21(10):1073-1076.
Stallings W 2006. Computer Networking with Internet Protocols and Technology. Beijing, China, Publishing House of Electronics Industry.
Wan XJ, Yang JW 2003. Design and Realization of XHNET's Data Center. Computer Applications 23(5):74-76.
Zeng BQ, Chen ZG 2004. Research of Server Cluster System. Application Research of Computers 21(3):186-187.

ROLE'S FUNCTIONS IN CMS

Xianfeng He [1], Mingtian Wang [1,2], Jinlian Zhang [1], Yongkang Luo [1,*]
[1] *Sichuan Rural Economic Integrated Information Center, Chengdu, 610072, China*
[2] *Agronomy College, Sichuan Agricultural University, Yaan, Sichuan 625014, China*
* *Corresponding author, Address: Sichuan Rural Economic Integrated Information Centre, No. 20 Guanghuacun Street, Chengdu, Sichuan, 610072, P. R. China, Tel: +86-28-87313950, Fax: +86-28-87343798, Email: lyk@scnjw.gov.cn*

Abstract: This paper is about role's main functions in the content management system. We respectively discuss the following five points: relationship between role and party; the category of the role within CMS; the influence of roles to the design of Virtual File System; the influence of roles to the design of content category; the role's function in workflow.

Keywords: content management system, role, virtual file system, workflow

1. BACKGROUND

The rural economic information center of Sichuan is a governmental organization which serves the rural economy. It contains a provincial center, 21 subordinate centers in urban district, 188 service centers in county, 3380 service stations in villages and towns, 100 market service stations. Its website is Sichuan Rural Economic Information Net. This website is established on the basis of CMS (Content Management System). All centers' daily routine is managed through this system.

APLAWS+ (Accessible & Personalized Local Authority Websites) (APLAWS, 2004) is the national technical standard which is developed by the government of UK. It is an expanded content management system which

He, X., Wang, M., Zhang, J. and Luo, Y., 2008, in IFIP International Federation for Information Processing, Volume 258; Computer and Computing Technologies in Agriculture, Vol. 1; Daoliang Li; (Boston: Springer), pp. 389–396.

bases on CMS. On the basis of the APLAWS+, Sichuan Rural Economic Information Net has established a content management system to manage organizations of the province at different levels.

Sichuan Rural Economic Information Net is made up of several channels, such as, Chinese version, English version, Calling Center website and 21 sub-sites which represent 21 subordinate centers.

In order to construct such a software system which has organizations with four levels from content management and information service point of view, the following questions must be answered.

How to make the user and its corresponding role have different meanings on different occasions?

How to classify the roles in the view of content management and content service in order to make the content management in order?

What does the functions of the role in Virtual File System?

Does the role and content category (Redhat, 2003) have relation between each other?

What is the relationship between the role and workflow?

2. RELATIONSHIP BETWEEN ROLE AND PARTY

In the content management system, Party is an individual which manages content or accepts content service (Redhat, 2003). It can be a concrete person or the pronoun of a group. The scope of the function of the party is aimed at the entire system–all sections. Because, in APLAWS, all the management of the resource is organized according to the section and each section has a corresponding and exclusive Virtual File System. Therefore, every body can visit all sections (such as, system administrator) or some sections after they are privileged. For example, the privileged "A" can be both the party of Chengdu and Beijing (Chengdu and Beijing are two sections).

In the content management system, role is the grouping to a party in a section (Redhat, 2003). Despite of the system administrator role, the scope of role's function is not aimed at the entire system, but limited to a concrete section. For example, the editor (editor is the role) of Chengdu and Beijing may not only contain different party, but also have different management or visit privileges to their own sections.

A same party can have various roles in the same section. For example, "A" can both be the editor and the publisher. A same party in the different sections can have different roles. e.g. "A" can be editor in Beijing section and it can be publisher in Chengdu section.

3. THE CATEGORY OF THE ROLE WITHIN CMS

From the content management and service angle, the party in the content management system can be divided into two kinds: one is content management user; the other is service object user.

3.1 Content management role

Content management role is usually aimed at the party which is in the corporation's organization. According to their working duty, common roles are (Redhat, 2003):

Author: Creates new content items. The privileges for this role include creating, editing, and previewing items.

Editor: Verifies new content items. The privileges for this role include editing, modification and categorizing items.

Publisher: This role with the privilege of an editor, it also has the privileges of approving and deploying content to the website.

Content administrator: The privileges for this role include role administration, category administration, content type administration, workflow administration and lifecycle administration. Meanwhile, it still has the publisher's privileges.

System administrator: The privileges for this role include content category, account administration, role administration, portal administration, creating the Virtual File System and privileging to it.

3.2 Service object roles

Public user: public@nullhost. In APLAWS, public users are defaulted to read the resource of Virtual File System of all sections. Through assigning privileges to folders, the permission of public users of all folders or a certain sub-folder in a certain section can be cancelled.

Viewer: The privilege of this role is viewing folder in a certain section.

VIP member: A specifically folder in a certain section with privileges of viewing, editing and publishing. It is the member of this non-organizational member.

4. THE INFLUENCE OF ROLES TO THE DESIGN OF VIRTUAL FILE SYSTEM

The Virtual File System is introduced into APLAWS content management system in order to making the following three aspects come true.

The Principle of Humanity is the foundation of content management.

Use the transparent mode to reflect the function and the working state of groups and individuals at all levels of organization in order to restrict each role's function in the way of giving permission.

File system replaces files and catalogs which are stored in the computer. In the system of Linux or Unix, File system stipulates the users, groups and others' access permission to the file and catalog. Through Web, the Virtual File System in APLAWS content management system simulates the computer's file, catalog and their safe management. In the content management system, the content is transparently stored in the Web virtual file management system which is characterized as folder. (Figure 1 virtual File System on the web) All the information is stored in this virtual File management System, including each branch structure in the enterprise; every person of the branch; the content of everybody, everyday, every month and every year. So, the virtual file management system is the basis and the symbol of the corporation's information management.

Based on taking the role and individual as the unit, one of the important management works to the virtual File management System is to assign permissions to folders at all levels.

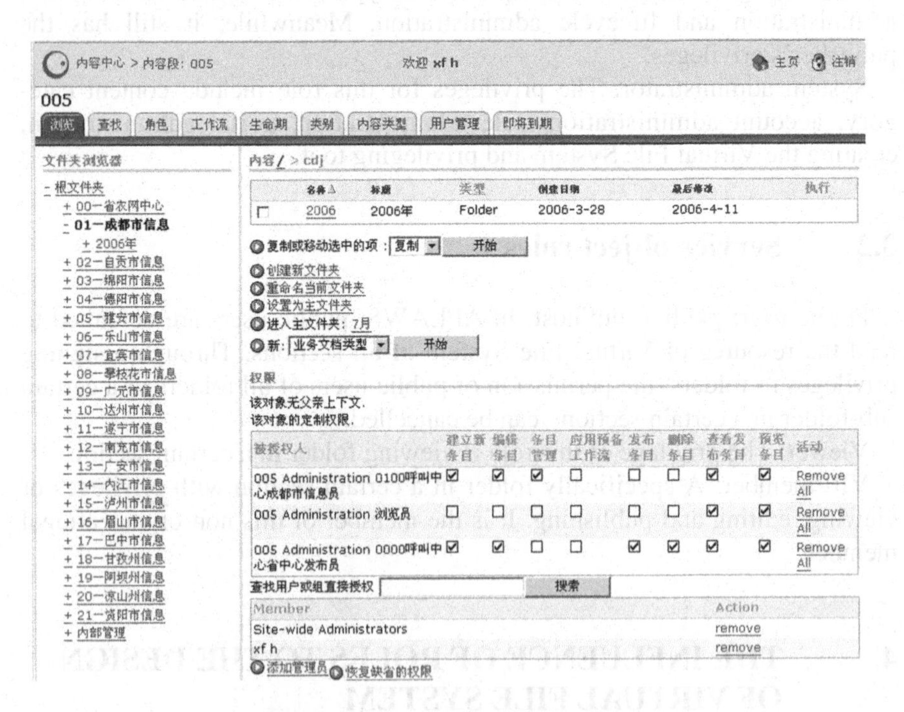

Figure 1. The virtual File System on the web

4.1 Folders of content management roles

Author, editor, publisher, content administrator and system administrator which have the character of content management role can take part in the daily content management working in their own folders. In Figure 4–1, "01-成都市信息" is a folder; "005" is a virtual File System. "01-成都市信息" is one of the sub-folders in the 005 section. We can see that this folder can be managed by "005 Administration 0100 呼叫中心成都信息员" (it has the privileges of creating items–content items, editing items, viewing–search content items, preview content items, excepting publish privilege) and "005 Administration 0100 呼叫中心省中心发布员". The difference between "005 Administration 0100 呼叫中心成都信息员" and "005 Administration 0100 呼叫中心省中心发布员" is that the latter has the publish privilege.

The party with the character of content management role can not only create new documents in their own folder but also create sub-folders under their own folder. e.g. In Figure 4–1, "2006年" is a sub-folder under "01-成都市信息", and all the documents of "2006年" are stored in "01-成都市信息".

4.2 Folders of service object roles

Service object roles and content management roles have the same folders with each other. The advantage of this design is that once "publisher" deploys content to the website, users who enjoy services can immediately get the content service. In (Figure 1), "005 Administration 浏览员" is a service object role, it has the privileges of viewing and previewing the folder of "01-成都市信息" (including contents under this folder). Once "publisher" deploys content to the website, the corresponding users of "viewer" can receive the corresponding voice and text information in their voicemail and website.

5. THE INFLUENCE OF ROLES TO THE DESIGN OF CONTENT CATEGORY

According to the connotation of contents and the pre-designed content category tree, Content Category is a working way of managing contents. In other words, any "document" in the content management system must be attached to one or more content category trees. In form, both of the Virtual File System and content category organize content items according to the treelike diagram and both of them have the meaning of convenient search. But, they have the following essential differences.

The tree node of Virtual File System is the limitation which contains the access permission of a role. But, content category has no this limitation. So, Virtual File System can not be shared between sections, but content category can be shared. Content category which belongs to a section (navigation of the website) can also be shared by other sections. As a result, a content item can not only be deployed in its own website, but also can be deployed in the shared website.

The content item in Virtual File System is located at a confirmed place among the tree nodes. But, content item in the tree of content category or even in the forest of content category can appear time after time, and it also can change freely and flexibly.

The content item actually exists in the physical way in Virtual File System, and it exists in the link way in the tree of content category. So, the content item in the Virtual File System has privileges of copy, move, physical delete. But, there is no such operation in the content category.

In a word, the content can be perfectly shared between or in the sections when the content category is independence from the Virtual File System. Because of aiming at the visit permission and workflow of Virtual File System, roles have no influence on the content category and sharing content.

One needs to be point out is that not all the content management system has the character which content category is independence from the Virtual File System. For example, the widely used MS CMS and Open CMS is a unity which content category is not independence from the Virtual File System.

6. THE FUNCTION OF ROLE IN WORKFLOW

Workflow is a series of mutually linked and automatically worked content management tasks. A workflow contains a group of tasks, their mutually ordered relationship and the description to every task. In (Figure 2), there is a typical content management workflow. The "technical support" phase of content management describes the process which a content item evolves from the type task into publish task.

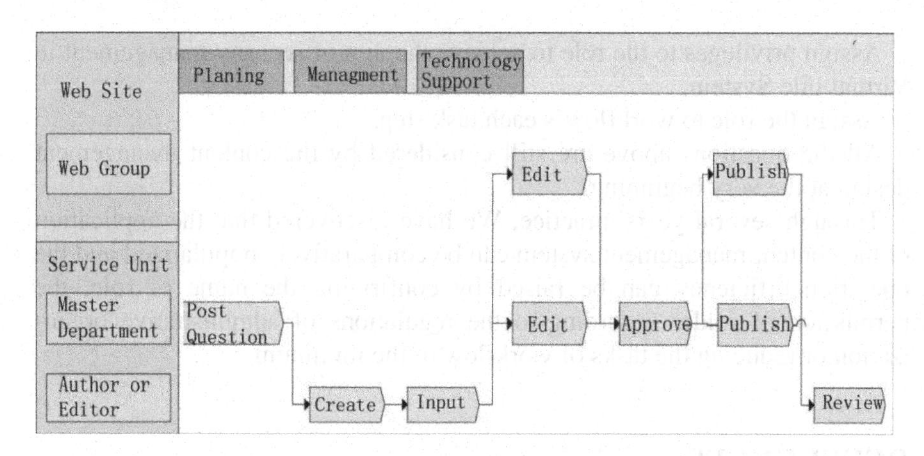

Figure 2. The relationship between workflow and roles

After the workflow is introduced, there are at least three advantages: first, it is not necessary to give the training to the party about the workflow and the change of the workflow can easily be implemented. The second, it is good for the worker to focus their attention to the data which they care about. At last, the worker can go to work either in unit or at home or even at other places.

According to the definition, each task of the workflow can only be implemented by the party who has roles. That is roles and workflow directly linked; the confirmed tasks are accomplished by the confirmed role. In Figure 6–1, "问题" (means Question) and "创建" (Creating) are two documents, they are respectively located at the phase of "规划" (Schedule) and "管理" (Management) and both of them need not the help of the computer and the network. Then, they come into the technical support of content management–workflow phase. In this phase, each task needs the party's participation, and all party must have the confirmed roles. Such as, we need "author" to accomplish the task of "input"; we need "publisher" to accomplish the "approval" task.

7. CONCLUSIONS

This article discusses the following questions at the very beginning:

Confirm the role of each member.
Give proper name and assign corresponding duty to the role.

Assign privileges to the role to achieve the aim of security management in Virtual File System.

Assign the role to workflow's each task step.

All the questions above are still considered by the content management design at the very beginning.

Through several years' practice, We have discovered that the application of the content management system can be comparatively popularized and the operation efficiency can be raised by confirming the name of role, the permission of folder according to the regulations of administrative organization or reducing the tasks of workflow to the minimum.

REFERENCES

E-government plans or standards, please refer to http://www.govtalk.gov.uk
http://www.aplaws.org.uk
In https://www.redhat.com/docs/manuals/waf/pdf/rhea-dg-waf-en-6.1-1.pdf
Red Hat, Inc. WAF Developer Guide.

RESEARCH ON QUALITY STANDARD OF RURAL INFORMATION

Mingtian Wang[1,2], Xianfeng He[2,*], Chunlu Li[2], Yongkang Luo[2], Jinlian Zhang[2], Liping Ou[2], Qin Xue[2], Yingwei Ai[3]

[1] *Agronomy College, Sichuan Agricultural University, Yaan, Sichuan 625014, China*

[2] *Sichuan Rural Economic Integrated Information Center, Chengdu, 610072, China*

[3] *College of Life Sciences, Sichuan University, Chengdu, 610064, China*

[*] *Corresponding author, Address: Sichuan Rural Economic Integrated Information Centre, No. 20 Guanghuacun Street, Chengdu, Sichuan, 610072, P. R. China, Tel: +86-28-87341941, Fax: +86-28-87343798, Email: hxf @scnjw.gov.cn*

Abstract: Researches on information quality standard is few both domestic and abroad, furthermore, there is no research on quality standard of rural information. In order to meet the need of China's rural information construction, promote the development of information science, the research objects and difficulties in the study of quality standard of rural information have been analyzed by the authors. New conceptions such as relativity, essential requirement, external expression, requisite element, optional element have been put forward creatively. Besides, researches on commonality extracted from diversified rural information and standardized treatment on relativity of rural information have been taken. Moreover, a qualified expression method based on the organic integration of essential requirement and external expression has been explored. The dialectical relationship between essential requirement and external expression, internal elements within the essential requirement, requisite element and optional element has been highly concluded in this paper. Consequently, "essential requirement, four necessities; external expression, classified description; pursuing integrity, not inexhaustibility; simple and clear, stressing on utility" as the clew of compiling this standard has been extracted and the grades and standard of rural information quality have been preliminarily brought up.

Keywords: rural information; quality standard; research

Wang, M., He, X., Li, C., Luo, Y., Zhang, J., Ou, L., Xue, Q. and Ai, Y., 2008, in IFIP International Federation for Information Processing, Volume 258; Computer and Computing Technologies in Agriculture, Vol. 1; Daoliang Li; (Boston: Springer), pp. 397–405.

1. INTRODUCTION

"The quality is our life" has become the guideline of so many enterprises and organizations. It goes through the whole process of production, operation and development. Not only in the fields of production and sales of agricultural products or industrial goods, but also covers medical services, culture and education, catering and entertainment, and thus rural information service is without exception.

High quality rural information is the corner stone and premise of improving the effectiveness of rural information service. To enhance the research on rural information quality is one of the most basic and key point in the process of promoting rural information construction. We get troubled when facing an ocean of trash information even though great manpower, material resources and financial strength have been put in. And even the more you put in the more you waste and aftermath follows.

Many people have realized the importance of information quality but most of them are still at the stage of perceptual cognition or verbal expression. Take agricultural websites which produce, operate and spread rural information as an example, if we pay a little attention, a great number of trash information would be found. Website undertakers may definitely know the importance of information quality, but they haven't put it into practice and haven't deeply realized that those trash information comes from their website would be concerned by the government or officials, attacked by their competitors, becomes constraint when negotiate with partners such as communication operators, becomes the fear point when users clicks the web pages again, becomes the legal dispute and speculation of major medias. Finally, trash information develops as deadly poison of ourselves. On the contrary, high quality information would become the shinning point and core competency of our websites.

2. RESEARCH PURPOSE ON QUALITY STANDARD OF RURAL INFORMATION

2.1 Meet the social demand

In a sense, "quality is our life" drives the construction and development of quality standard system worldwide. Standards is a ruler, a basic tool for improving the product quality, enhancing comparability, raising management effectiveness, decreasing management cost, unifying and regulating the market, promoting the production share and exchange capability.

What information to gather, when to publish, how to accurately describe the information need unified and normative quality standard. Furthermore, information transportation mode, information executions such as analyzing, utilizing, exchanging and sharing, and technical trainings on web editors also require the guide of unified and normative quality standard.

If there is no quality standard, we have no way to judge and compare the quality of goods accurately, objectively and effectively. If there is no quality standard of rural information, how can we doubt what kind of information is "trash"?

However, up to now, there is no unified and normative quality standard. Just for this reason, the editing, approving and monitoring of rural information are inefficient; we have to be so cautious on the service, usage and development of rural information, or great loss may happen because of fake information, false information and incomplete information; difficulties come out in integrating, exchanging and sharing resources among websites and relative departments. Rural information service requires us to research on quality standard of rural information.

2.2 Promoting the development of science

Research on information quality should be an important constitution to the construction of information science, and also research and establishment of information quality standard should be an important constitution to the construction of information quality standard system. However, the result after our web search both in Chinese and English is that information quality studies are few and it is almost vacant in the field of quality standard of rural information. Therefore, no matter the development of information science or the construction of rural information needs the research and establishment of quality standard of rural information.

3. RESEARCH ON QUALITY STANDARD OF RURAL INFORMATION

3.1 Making clear the research objective

As the name suggests, rural information implies information which related to agriculture, rural areas and peasants. It is much extended than agricultural information. From the definition we may know that rural information is great in number, with various types, complicated relationships, dispersed sources and diversified styles. For example, rural information could be displayed by narration, exposition and argumentation, or in the way of diagram, sound or image.

3.2 Analysis on difficulties

3.2.1 Research on information classification

Different types of information decide different ways of expression such as news item must be different with price information. Therefore, we have to classify the huge and complicated rural information, and classification itself is a research hotspot and difficulty (authors of this paper have taken research on rural information classification and now is asking for advice. This will be introduced specially in other papers).

3.2.2 Quality commonality extraction from diversified rural information

The types of rural information is numerous and if establish standards respectively to all types may seem ponderous and complicated, hard to remember and carry out. Through years of studying, authors have summarized the ways of expression on rural information quality and characteristics have been put forward: authenticity, accuracy, punctuality and utility. Namely, these characteristics are the fundamental criteria for the judging of rural information quality.

3.2.3 Standardized treatment of information quality relativity

The fit and unfit of information quality is relative. Relativity means the uncertainty of information quality. We often call useless information as "trash information". However, utility itself is a variable. Stock information is useful to those stockholders, but useless to those who are not care about stock market. A piece of information may be useful to me yesterday but useless today. Within the same piece of information, some people just glimpse at the title but others may read the full text without satisfaction. Therefore, information quality varies with people, time and circumstances. This is the relativity of quality information. Relativity may hold back the pace of researching on information quality. Two conceptions: requisite element and optional element are put forward in this paper. By effective collocation between them, the way of standardized treatment would be fulfilled and then the key arduous problem be solved in the course of compiling the standard.

3.3 Expression style of information quality standard

By studying on different types of rural information quality, we take up the position that quality standard of rural information could be described with combination of essential requirement and external expression.

3.3.1 Essential requirement for information quality

Essential Requirement indicates that it can reflect the implicit attribution of description objectives by passing through the external expression. Generally, it contains authenticity, accuracy, punctuality and utility.

Authenticity: information expression is in accordance with objectivity. It requires gathering reliable and unbiased information not figment. Editors should check carefully and not trust too readily on any false news or the information provided would lack fidelity.

Accuracy: totally conforms to the reality or expectation. We should accurately gather or describe objects, truthfully reflect things as they are, fully states the information and use statistics when necessary. For example, a piece of information on fruit, a complete description should be provided such as production history and actuality, variety, amount, shape, flavor, grade, transportation and etc. Readers would gain a full understanding on the goods.

Punctuality: means catch the time, meet the needs; not delay, handle right away; pursue speed in collecting and transporting messages, pay great importance on concept of time. Message out of date is useless and even have side effect. As information which reflects market changes usually passes in a twinkling, we have to catch them promptly and win in the competition.

Utility: as the name suggests the information should be with practical use value. It requires us considering the uses of the information not only from our own perspective but also from the perspective of the public. For example, a supply message of well-known goods which with high authenticity may have small production scale, low amount of goods. This kind of information has low utility. So we have to consider its usefulness when providing information.

Among the four requirements, authenticity is the basis and premise. Both accuracy and punctuality are important guarantee of utility; utility is the starting point and essential destination of information service.

Essential requirement is the commonality of information quality. Namely, any types of information should be a combination of authenticity, accuracy, punctuality and utility. Or it would become “trash”.

3.3.2 The external expression of quality standard of rural information

The external expression: It is the outward appearance, the external relations of things and the surface depiction of information description object.

Different types of information have different quality external expression, and the difference may be huge. For instance, the expression style between price information and news information is apparently different. Therefore, the external expression of quality standard is certainly different. Due to different service objects, time and environment, same types of information may also have different outward appearance. That is the relativity of information quality.

In order to reach a simple, standardized and unified format, improve the universality of the standard and satisfy objective requirements of information quality relativity, we have explored a way to standardize the external expression of information quality standard: requisite element organically collocates with optional element.

Requisite element: Some types of information must have requisite element. The information would have apparent quality defects without any one of the elements. For example, the elements of news information are time, place, person, events and so on. The elements of supply and demand information are supplier and purchaser, the content, quantity, contact information and etc.

Optional element: These elements are not essential, but they are useful for some customers. Without these elements, the information itself has no apparent drawback or it may be also high-quality information. But, some customers may pay their attention to these elements and even these elements are very important. There is a piece of news: "It is published by provincial price bureau that the price index of Sichuan province rose 5% in the first quarter of 2007; because of macro regulation and control, it is predicted that the price index would fall down in the second quarter, but it would keep an increase of 3%." This news has no obvious problems, but some customers may pay more attention to other elements, such as the concrete increase in prices in the first quarter, the measures of macro regulation and control, commodities whose price would go up in the second quarter, regions where price would get higher. Some customers still attach importance to the elements which usually appear in the supply and demand information, such as color, shape of goods and etc.

Different types of information have different requisite and optional elements. For example, the requisite and optional elements of laws and regulations information are apparently different from the requisite and optional elements of supply and demand information. Requisite elements of the same kind of information are the commonality of the quality external expression. As previous stated, news information must have the following elements such as time, place, person, events and etc. Optional elements are the embodiment of individuation and relativity of information quality.

3.3.3 The relationship between the essential requirement and the external expression

There are mainly three aspects: firstly, essential requirement is the fundamental requirement of information quality and it reflects the internal relations of things. It is in the first place; the external expression is the surface depiction of information description object, it reflects the external relations of things. Secondly, essential requirement is the commonality and universal demand of all types of information; the external expression is the special expression of different types of information quality, it is the personality and relativity expression of information quality. Thirdly, the essential requirement and the external expression are unified in the information quality and they are inseparable. The essence is the essence of the external expression and it has the function of regulating the external expression. The external expression is the external expression of essence. It must be embodied and described around the essence. It is the concrete description of the information quality surface, the supplementary illustration and the embodiment of the essential requirement. The essence decides the external appearance and we can see through the appearance to perceive the essence.

4. THE CLEW OF COMPILING THE QUALITY STANDARD OF RURAL INFORMATION

4.1 Essential requirement, four necessities

Authenticity, accuracy, punctuality, utility are the essential requirements of all the information quality. The concrete expression of different types of information quality standard must have four basic attributes mentioned above at the same time.

4.2 The external expression, classified description

Because different types of information quality have different external expressions, the requisite element differs from the optional element. Thus, classified description becomes an inevitable choice.

According to the external expression of information quality, the rural information can be divided into 13 types. They are news, laws and regulations, science and technology, product introduction, organization introduction, character introduction, investment, disaster prevention and reduction, price, supply and demand, training, job seekers, employment.

4.3 Pursuing integrity, not inexhaustibility

Firstly, we should aim for the integrity of the classification object and try to avoid omitting important ones. But we do not require all the information have well-defined quality standard. Secondly, we should also aim for the integrity of the external expression of information quality. The utility of the information will not be influenced without some elements under some conditions. Therefore, it is not necessary to have all elements.

4.4 Simple and clear, stressing on utility

The quality standard of rural information should be studied deeply and systematically. In order to improve the maneuverability and practical value of the standard, we should try to simplify, standardize and unify the standard through extracting the essence and removing the unnecessary details.

5. THE GRADES OF INFORMATION QUALITY

5.1 High-quality information

High-quality information authentically, accurately and punctually reflects events which are concerned by people. This kind of information has good utility, complete requisite elements, outstanding characteristics of optional elements, refined and accurate expression.

5.2 Incomplete information

This kind of information is authentically, punctually reflects events which are concerned by people. It has some practical value, but, it has incomplete requisite elements or roughly described elements. These disadvantages influence the efficiency of information.

5.3 Fake information

This type of information is fictitious and concocted out of thin air; it may be out of date or cribbed from someone else.

6. CONCLUSION

The research of quality standard of rural information is an important component of construction in quality standardization of rural information and information science. As rural information is great in number, with various types, complicated relationships, dispersed sources and diversified styles, the research has many difficulties. Besides, there are few researches and references on quality standard of rural information both domestic and abroad. As a creative study, imperfectness is inevitable. This article intends to start further discussion on this issue.

IMPLEMENTATION SCHEME OF AGRICULTURAL CONTENT MANAGEMENT SYSTEM

Xianfeng He[1], Mingtian Wang[1,2,*], Yuan Yang[1], Yongkang Luo[1], Yingwei Ai[3]
[1] *Sichuan Rural Economic Integrated Information Center, Chengdu, 610072, China*
[2] *Agronomy College, Sichuan Agricultural University, Yaan, Sichuan 625014, China*
[3] *College of Life Sciences, Sichuan University, Chengdu, 610064, China*
** Corresponding author, Address: Sichuan Rural Economic Integrated Information Centre, No. 20 Guanghuacun Street, Chengdu, Sichuan, 610072, P. R. China, Tel: +86-28-87364604, Fax: +86-28-87343798, Email: wangmt0514@163.com*

Abstract: This essay is about agricultural content management system (ACMS) based on project implementation and Internet, which defines; consist of content, system structure and the established foundation of Red hat content management system (CMS) open source council.

Keywords: Agricultural content management system (ACMS), Web Application Framework (WAF)

1. SUMMARY

Information-based is the objective trend of the economic and society development in the world today and it is the mark and key to agricultural modernization. With the development of internet, web-based agricultural information-based has already become the main channel of information exchange. Agricultural content management system based on the World Wide Web has to be produced.

Agricultural Content management system (ACMS) contains digital, network of agriculture and its realization. In this way, the organization departments of

He, X., Wang, M., Yang, Y., Luo, Y. and Ai, Y., 2008, in IFIP International Federation for Information Processing, Volume 258; Computer and Computing Technologies in Agriculture, Vol. 1; Daoliang Li; (Boston: Springer), pp. 407–417.

agricultural academy at all levels can use the modern technology of information and communication, realizing the Optimization and Reorganization of agricultural organization structure and workflow on network. Breaking the limitation of time, space and individual departments, it provides all-around management and service to the society with efficiency, quality and regularity.

The significance of ACMS is of consequence. It can built new Net-connected model, breaking the traditional isolated office model. It may contribute new concept and practices to have management, operation, services education and training of agricultural academy integrated.

The realization of ACMS refers to many aspects of CMS, the foundation of Web Application Framework, ACL (Agricultural Category List) of ACMS, AMS (Agricultural Metadata Standard) of ACMS, e-AIF (Agricultural Interoperability Framework) of ACMS, the framework showed by information of ACMS, CMS (Content Management System) of ACMS, and the Portals management of ACMS and so on. This essay will give a simple project based on concept and practice of CMS in the IT circle, to discuss the implementation scheme of ACMS.

2. HARDWARE BASICS OF ACMS WAF

An ACMS WAF (Web Application Framework) (http://www.redhat.com, 2002) site is a Computer Cluster application system, which is based on one database server, and subjected to several Web servers, and led by Load Balance server. ACMS site demands to offer 24 hours non-stop service with agricultural management, information communication and application. That is to say, Intranet should satisfy the need of agricultural informative communication and various public services. Therefore, it has to introduce Computer Cluster technology for hardware basics of ACMS. Below is the realization of it:

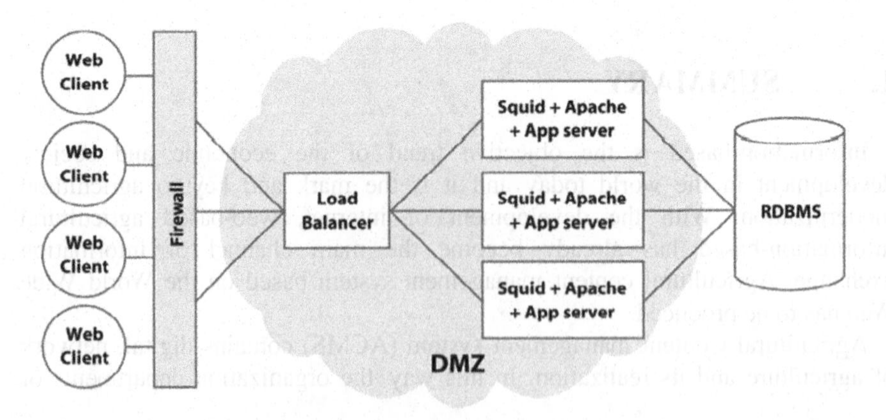

Figure 1. The typical hardware and operational system scheme under Red Hat WAF

A Load Balance receives the request of all Web client-sides, and selects one Web App Server to comply with the request of client-side according to the running state of Front-End Servers.

In order to relieve the stress of Database Server in Web App Server, the Object Cache based on Squid in Apache Web Server is introduced, so as to solve the bottle-neck problem of Database.

At least 2 front-end Web Servers based on Apache Web Server are in convenience to comply several Virtual Webs on one IP.

A stand of Database Server apply to save operation, service and management data.

The essay only lists the hardware basis under the significance of ACMS WAF, but not to take e-AIF of ACMS into consideration. One reason is to avoid the great length, the other is that e-AIF of ACMS is the author's ideal, but no practice. Finally, it prefers to lay emphasis on the key point in the realization of ACMS.

3. WAF (WEB APPLICATION FRAMEWORK) OF ACMS

WAF (Web Application Framework) is not only the basis of ACMS, but also the abstract of all kinds of Web applications. It should follow the principle of open source code, and based on Database, Cross-Platform, and safety, extensible, capable of adapting to the need of various client terminals (e.g. PC, PDA, Mobil-phone, and numerical TV).

The reason why it follows the aforesaid principle is that it saves the cost of Software development and application, and it is also in line with the universal open source and become one of the members in the revolution of open-resource software, and it makes to spread quickly in industry, and it makes CMS standard to follow certain scales of universal standard, of course, it also insures that CMS proceed without interruption through developing of agricultural undertaking.

According to the aforesaid principle, based on the analysis and comparison to nearly 100 CMS systems, Oracle 10g commercial database is selected as foundation, J2EE based and open source Red hat WAF is taken as the application framework to ACMS.

Red hat WAF designed as N layers model, which separates into presentation layer, domain (business logical) layer, data layer, data model layer. Various applications based on WAF (e.g. CMS-content management system, file management, forum…) design according to the layering model.

WAF basic structure as shown in Figure 2, WAF is a Web application management system based on database application, which is built on operating systems, database and J2EE. Above database and J2EE, is CCM system. Many

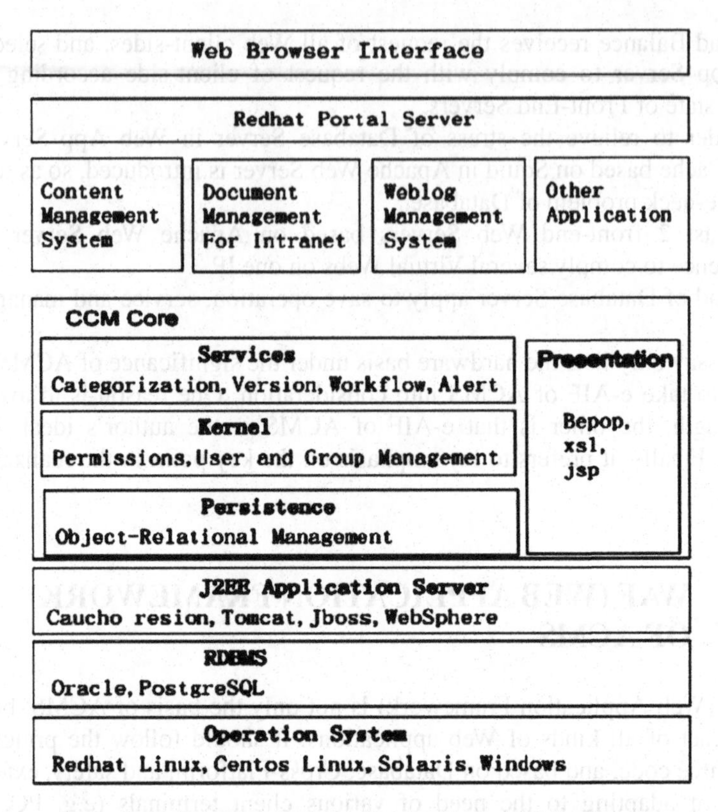

Figure 2. Redhat Linux Web Application Framework basic structure

web-based functions can be achieved on a basis of CCM, including content management, forum, bulletin Board, channel management, theme management and portal management etc.

CCM kernel layer of MAF is completed.

The interaction between database and Java object through successive layers lays a solid foundation for extensible, integrated, and congruity of system. That is to say, users need not to tangle with database directly, but to comply UDCT (User Define Content Type) in database. For example, in order to comply Web application of Doppler radar puzzle, it completes UDCT of puzzle in CMS based on Web application, without ever needing to write Java code and SQL Create Table Script.

Security realization based on Web application. That is to say, member, group, role and authorization are the basis of WAF, and every domain object example works in its own scale of security.

Basic Web services realization. That is, WAF regards version control of data item, workflow control, Category list management, to inform users group by bulletin, to alarm to assigned users as basic Web services, no duplication of effort of research staff.

Presentation and alternation of Domain object. XML, XSL, JSP are the necessary technology to present data item of database. Bepop is also Mark Language to meet XML standard, which is not only to present Web information, but also to transfer the information of event and session between servers and client terminals. The services supplied by ACMS could adjust different client terminals and language environments because it mainly uses XSL technology.

4. AGRICULTURAL CATEGORY LIST OF ACMS

ACL (Agricultural Category List) is the start and end of information organization and category. If it prefer to application of ACMS, it should built security certification system, work standard, and quality system to be up to the demand of Internet. Content management and service of ACMS essentially follow as hierarchy category for clue. It particularly comes down to tabulation and management of agriculture terminology list, agricultural category list, domain list, navigation list and loose-leaf folder list.

Agricultural terminology list is the basis of information organization, communication and ACL. It defines the scope of database and technology standard.

Domain list is the top category of Agricultural terminology list.

ACL is tree-category of Agriculture terminology list.

When some terminologies are both belong to A and B, ACL support link-category concept. For example, civilization contains education, science and technology also contains education, in this way, science and technology is set for the link contained education.

A database admits of several LACL (Local Agricultural Category List), which is aimed at different areas, various departments, different services and management. LACL and ACL can set up mapping relationship. Generally LACL is to further detail of ACL.

Navigation list is URL of ACL.

Folder list is another form of sorting assigned items by users. Different from ACL, every file in folder list corresponds to different access-rights and security precautions mechanism. The particular dividing regulation is determined by soft science achievements.

5. AMS (AGRICULTURAL METADATA STANDARD) OF ACMS

Metadata is first of all data about data, and then it is about the frame information of resources. The metadata standard of ACMS is e-AMS, whose design objective is to adjust to any content and resource in ACMS, and Web site and data recorded management of ACMS. E-AMS complied with Dublin core standards. Besides its own elementary metadata, the support of MCL and ADSC is also required (Agricultural Data Standards Catalogue). E-AMS is the component of E-AIF.

E-AMS is mainly applied to search for the information and resources on agriculture for the public and agricultural staff.

ACL and the element subject category in e-AMS are corresponding.

There will be an assay especially on e-AMS.

6. AIF OF ACMS

Agricultural information is the resources within the agricultural department, but also social assets, which is just like fuel for the national knowledge economy. It should be shared by both the departments and other members in the society. Because of ACMS based on Web, distributed Web service and Database are inevitable features. How do we ensure the consistency of the web content? How do we complete the information exchange among each database? How do we ensure all levels of agricultural organizations follow the regulations admitted by inner department and the society, when agricultural information is shared to public?

E-AIF (Agricultural Interoperability Framework) is the standard of information exchange followed by ACMS members. It makes a seamless linkage to agriculture knowledge, services and management in agricultural departments within or cross provinces. At the same time, It offers a more accordant and broader service mode.

With the need from strategic management, e-AIF selects XML and XSL as the core standards for data integration and presentation data. E-AIF also defines and applies e-AMS, which provides quick queries on agricultural information and resources.

E-AIF structure is as shown in Figure 3. E-AIF on top level is Framework, which is responsible for formulating and action of a higher level policy, system and standard. E-AIF on second level is the exact approach and realization of

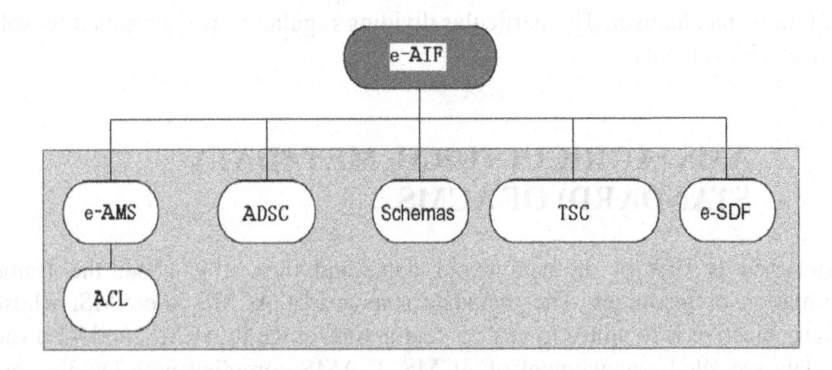

Figure 3. AIF structures

Framework, in other words, the realization content of e-AIF standard, which are e-AMS, ADSC (Agricultural Data Standards Catalogue), XML Plan Schemas, TSC, e-SDF (Service Development Framework).

It is presentation below that e-AIF is framework. There will be an assay especially on all items on the second place.

The main line and principle of e-AIF Framework

Completely follow by Internet and World Wide Web standard

Make XML the chief standard of data integration

Brower is taken as the main interface with other interfaces as additional

Add metadata to agricultural information resources

Develop e-AMS based on Dublin Core Model

Develop and maintain the authority of ACL in all-level administrations

Develop and maintain the authority of e-AIF in all-level administrations

Key evaluated indicators of e-AIF standard

Interoperability: Capability of standard in Net Link, data integration, e-service and content management among systems.

Market support: Capability of standard to win mass support in the market in order to reduce expenses and risk of agricultural department.

Scalability: Capability of standard can be adjusted to system extended. For instance, it can adjust to the expansion of quantity of data, work disposed and quantity of users.

Openness: Capability of standard to be general applicability.

Main application scope of e-AIF standard

Between Chinese agricultural and citizens

Between Chinese agricultural and world business

Among Chinese agricultural organizations

Between Chinese agricultural organizations and the other world agricultural organizations

All new research systems of agricultural organizations and information exchange within the listed above, e-AIF standard is necessary.

E-AIF standard can be adjusted to all present and future channels such as Internet, digital television, WAP phone, PDA and so on.

The technology line is related to at least 4 parts, which are Net link, data integration, e-service and content management.

The realization content of e-AIF standard

XML schemas of ACMS

E-AMS of ACMS

E-Services Development Framework

In this essay, e-AIF framework is rough, and it does not contain the way of practice and management of XML schemas of ACMS, the organization and management of e-AIF etc. A special thesis is planned to write to discuss these issues.

7. INFORMATION PRESENTED FRAMEWORK OF ACMS

In the ideal situation, agricultural department information makes the seamless linkage according to public's habits and the operational rules. That is also the start point of the information presented framework model and technology support of building the unified external image of ACMS.

The CMS Website bases on web technology. It not only has the standard recognized by all the levels, which contain affairs category, sub-category, and sub-sub-category, but also the navigation standard which is in accordant with the Web browse habit.

The CMS focus on the information framework model on the programming of information framework, information displayed template, overall page element, inquiring, landing, dictionary, site map etc.

The basic contents of information framework are:

Title	Explanation
Category Bill	It is the start point of information framework. Generally, the first ten categories are also the basis of navigation guide.
Top Category	It defines a general category over subordinate department categories, and abstracts the same category, as for the public to browse information on different Webs among different departments.
Top guidance	A set of guide label placed on the top of every page.
Z	Information dictionary arranged alphabetically.
List	It is information inquiring based on list like Yahoo, BBC. It resets Category Bill.
A set of Public page element	The page element and style followed by all subdivisions.
A set of URL	A group of standardize resources location, by which the public only need to enter URL to look for the kind of information.

The main function of the information display model is to make the web inf. (information) not only used in the PS, but also in the digital TV, mobile phone and PDA. The CMS mainly uses XML and XSLT technology to assure the information could be exchanged from one platform to another.

Every spec of XSL works with corresponding inf. model.

The element of overall page is made up of common tools on the top of page, category of page top, page position indicator, navigation on the left, page bottom, and related link information on the right.

8. ACMS (AGRICULTURAL CONTENT MANAGEMENT SYSTEM)

CMS (Content Management System) is the start point of agricultural content management and the core part of the whole system. CMS makes creation and processing in all-levels agricultural organizations be in accordant with each other, hence sustainable and extendable. The updating of agricultural information is not limited by the place of work, because CMS allows the individual to create and maintain the content. In addition, management activities focal on content management improved the capabilities from knowledge sharing and communication in organizations, and improved the work efficiencies, removed the communication barriers.

Generally, CMS contains creation and practice of content type, version control, file management, Workflow managing, life cycle of data item, people managing, security management, loose-leaf folder management, file style definition, page model management, category management, file nationalization management, and information exchange with other CMS by RSS.

Content type is the basis of communication and work on the Internet. Because of content type management (create, modify, delete) being contained by Red hat CMS, therefore it is easy to extend to agricultural service system, resources agricultural service system, agricultural management system, financial system, and human resources system and so on.

Version control is completed with managing of work process.

File management is completed with creation, modification and deletion of various content types, and it contains all kinds of editors such as MS Word.

Workflow managing regulates a series of stages similar to creation, modification, deletion, check-up, and the publishing of contents.

Life cycle of data item regulates save time of data item in database.

People managing contain the assigning of roles and security authority of members besides creation, deletion and modification of members.

Security management contains role, user, group management and authorized management.

Folders management contains creation, deletion, modification, and security management of file and sub-file (hierarchically authorized to work group and staff).

Category management determines that the clues of CMS are terms and the classification of them. The determinate term is correspondent with the type of the determinate content. A category contains one or more terms. Every category and terminology has its own URL label.

9. PORTALS MANAGEMENT OF ACMS

Generally, ACMS is made up of several sub-Webs. Every sub-Web has its own theme page which is consisting of several xsl, css, and jsp. The portals management model realizes developing, testing and releasing of various theme pages, and applies it to the allocation of sub-Web.

10. TESTS

Website for 96999 Call Center: http://96999.scnjw.gov.cn
Website for Meteorological Bureau of Sichuan, China: http://www.scqx.gov.cn

They are all bases on Redhat WAF and CMS. They can also be viewed as tests for the systematic construction of ACMS. Steps are as follows:

It forms as the working group contained Sichuan Provincial Agricultural Academy, city-level, and county-level branches. The service object is the public;
Install Java environment, Oracle date base, set Apache server, set Resin, install Red hat CMS;
Establish terms, terms category and navigation address;
Confirm labor division. Establish members, assign roles and authority;
Build content type, content life cycle and workflow;
Build homepage and theme-page template.

Sichuan Provincial Agricultural Website has been completed according to WAF and CMS technology, and optimized on that base. The test shows that ACMS systems engineering in each information center is practicable.

At present, the system has been applied in the construction of Sichuan Rural Economic Information Net with 6000 working staff and English version of it, Sichuan Meteorological Intranet. It can be seen that those websites share one Database and work in different sections of CMS.

11. DISCUSSION

The origin of ACMS: Through the study of e-government, many aspects of e-government can be applied in the agricultural information management in the terms of information management. Therefore, it gets the concept of Agricultural Content Management System.

The relationship between technology and strategy: When construction of ACMS implements, first of all, it has to face the relationship between technology and strategy. The meaning of “Science and Technology is the First Impetus” has to be reviewed. Although ACMS solves a series of technology problems, it is the basis of founding and practicing the strategy. In other words, the technology determines the management regulations, methodologies, service functions and implementation of strategy. On the other hand, strategy both affects and boosts the development of technology. The application of ACMS construction is up to the orientation to strategic development.

The relationship between technological engineering and soft-science: When implementing the construction of ACMS, there is a problem of coordination between technology and management. ACMS is involved with establishment and implementation of a large amount of standards, terms, management regulations, workflow and etc. On one hand, administrators should understand the terms in ACMS and form a complete management system. On the other hand, obstacles in management and trainings should be cleared in the process of technological realization.

Formation of overall objectives: There are three steps in the construction of ACMS. Firstly, the development and application of technology are under the basis of open source code Web CMS. Secondly, it builds standards and guides for Web site of province-level ACMS. Finally, it builds e-AIF (Agricultural Interoperability Framework). The overall objective is to found a complete set of nationwide Web agricultural standard and general framework.

Tasks in the near future: The construction of ACMS cannot be achieved in one move, but a long and tough process of spiral. Application tests and technology development with CMS as the core will be carried out in the near future. Meanwhile, it should be paid close attention to the establishment of standards with category as clue. The most complicated thing is to build e-AIF before which technology guide should be realized.

REFERENCES

Related to the information of building plan and standard of electronic affairs, refer to http: //www.govtalk.gov.uk

Related to the information of development of CMS and OpenCMS, refer to http: //www.opencms.org

Related to the information of Red hat CMS, WAF and Portal Server, refer to http: //www.redhat.com/ccm

Related to the information of technology realization of APLAWS and CMS, refer to http: //aplaws.sourceforge.net

CONTENT SHARE IN CMS

Xianfeng He[1], Chunlu Li[1], Mingtian Wang[1,2,*], Yongkang Luo[1], Yingwei Ai[3]
[1] *Sichuan Rural Economic Integrated Information Center, Chengdu, 610072, China*
[2] *Agronomy College, Sichuan Agricultural University, Yaan, Sichuan 625014, China*
[3] *College of Life Sciences, Sichuan University, Chengdu, 610064, China*
[*] *Corresponding author, Address: Sichuan Rural Economic Integrated Information Centre, No. 20 Guanghuacun Street, Chengdu, Sichuan, 610072, P. R. China, Tel: +86-28-87364604, Fax: +86-28-87343798, Email: wangmt0514@163.com*

Abstract: This paper is on the background of content share in CMS implemented in Sichuan Rural Economic Information Net, discussing the intentions and significances of content share. Instances during application are demonstrated.

Keywords: Content Management System (CMS), Content Share, Navigation Category

1. BACKGROUND

Sichuan Rural Economic Information Center is a government organization dedicated in the information service to agriculture, rural areas and peasants. It consists of one provincial center, 21 municipal branches, 188 county-level service stations, 3380 township-level service stations, and 100 market service stations. Sichuan Rural Economic Information Net is a web portal under Sichuan Rural Economic Information Center and its daily information is managed by the Content Management System (CMS) which is based on the web.

APLAWS+ (Accessible & Personalized Local Authority Websites) (http: //aplaws.sf.net, 2004) is a national technological standard launched by the UK government for the project of e-government (http://www.govtalk.gov.uk, 2004). It is a system which develops and enlarges itself on the basis of

He, X., Li, C., Wang, M., Luo, Y. and Ai, Y., 2008, in IFIP International Federation for Information Processing, Volume 258; Computer and Computing Technologies in Agriculture, Vol. 1; Daoliang Li; (Boston: Springer), pp. 419–424.

Redhat CMS. Sichuan Rural Economic Information Net has built a content management system at all levels in Sichuan Province by applying the open source of APLAWS.

Sichuan Rural Economic Information Net consists of websites such as Chinese version, English version, and Call Center at the province-level and 21 second level websites operated by municipal branches.

Many challenges are faced in the process of building a 4-level application system in the angle of content share:

How can one content item shared among different "navigation categories". In other words, once a content item created, how to make it appear in several categories or subcategories;

How can one "navigation category" shared between province-level and city-level websites. In other words, once a content item created, how to make it appear in different websites;

How to establish its own particular features while share the same content among different channels and websites—"navigation categories" are not contained in content share;

How to make the contents be directly acquired or understood by other websites with the same or different structures;

How to realize the exchange of contents between two same (different) structured content management systems by RSS. For example, to exchange contents between APLAWS and OpenCMS (http://www.opencms.org, 2000).

2. SHARE OF CONTENT ITEM IN NAVIGATION CATEGORIES

Navigation category is a basis of content category in content management system. In public users' view, content category shows the function of navigation. Therefore, navigation category manages information in accordance with the core of content information—standard of content category. It is evident that navigation category is the building line of content management.

In CSM, content item is a byword for document files, images, audio files and videos. Usually, one single content item has connection with one category. However, it is often happened that one single content item belongs to both category A and category B. For example, a news item which belongs to three categories: "domestic news", "provincial news", and "hot spot", it can be easily realized that this item be put into these three different categories in APLAWS CMS. (See Figure 1)

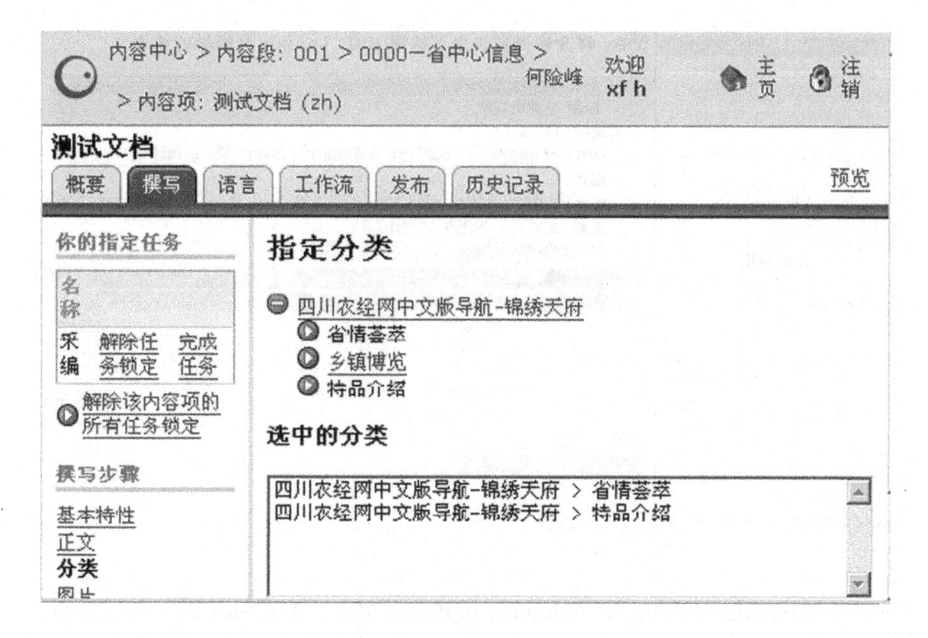

Figure 1. The test item belongs to two different categories

3. SHARE OF CONTENT ITEM IN NAVIGATION CATEGORIES OF DIFFERENT SECTIONS

In order to represent the particular features of different channels or sub sites, keep apart administrators of different websites, ease the pressure of web servers and database servers, the concept of section has been introduced in APLAWS (http://aplaws.sf.net, 2004). In the angle of Web Application Framework (WAF), each section has a corresponding web application. In the CMS, content center is a gathering of sections.

Usually each section has its own navigation category. However, it doesn't mean one navigation category can serve only one section, but much more. In other words, one navigation category can be used in different sections. (See Figure 2)

The example of figure 2 is a tree diagram of category of Institute of Plateau Meteorology in Content Section. (There are first level categories such as "organization structure", "scientific and technical personnel", etc.) Categories in Section 001,002 and 003 are also introduced into Content Section. If the tree diagram of category of Institute of Plateau Meteorology

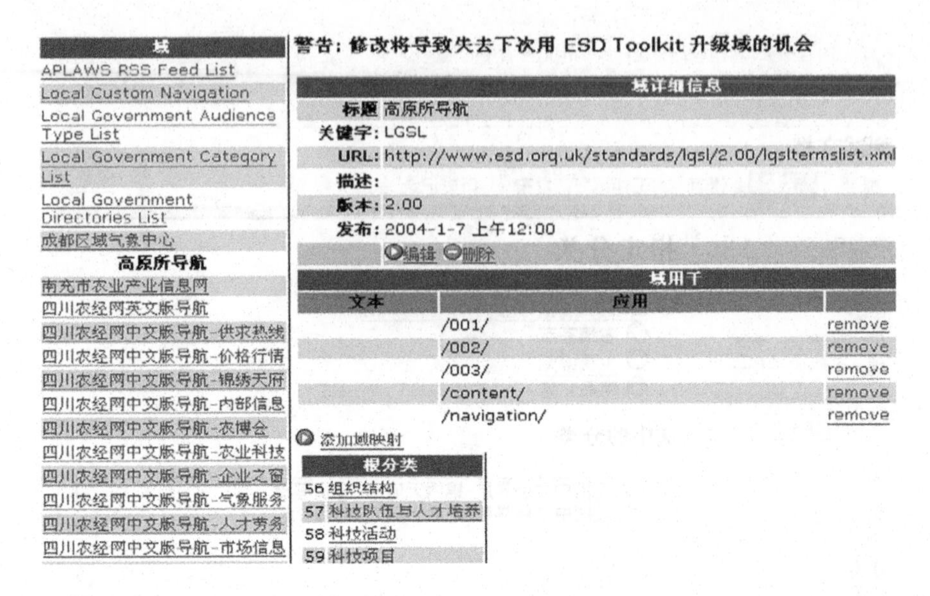

Figure 2. The navigation category of Institute of Plateau Meteorology shared by Section 001, 002 and 003

under Content Section is introduced into other sections, administrator can create a new item in one of the sections and put it into the same category, take "organization structure" for instance, then this newly created item can be seen in all "organization structure" in different website navigations—the realization of sharing content item in navigation categories of different sections is achieved.

4. CATEGORIES WITH PARTICULAR FEATURES

If all sub sites in CMS share the same information with each other and don't have its own information, it will bring us the feeling of similarity and we will lose control in content management. Therefore, the balance between content share and particular feature is very important. For example, sections with the feature of "prices" should not contain information of sections with the feature of "news"; Sections with categories served for "province-level" should not accept categories from "city-level" sections; However, categories in "city-level" sections" should take categories at the "province-level" into consideration in order to make the information be shared in Sichuan Province. Figure 2 shows the featured category of Institute of Plateau

Meteorology in Content Section and at the same time, this section deals with information from Section 001,002,003. The aim of both sharing content in different sections and showing particular features are reached.

5. EXCHANGE OF INFORMATION BY SEMANTIC XML

All web pages contain the information in CMS. Theoretically, all contents can be separated from the web page by the method of program analysis. However, too much unexpected information such as locale, colors is contained in web pages. It seems that other ways are necessary for the exchange of information among websites.

The concept of semantic XML has been introduced in APLAWS. It brings the possibility of acquiring pure content information with architecture description by HTTP with "? output=xml".

```
http://96999.scnjw.gov.cn/ccm/005/znwzx/xxb/zlj/2007/1
  - <cms:item oid="[com.arsdigita.cms.contenttypes.NewsItem:{id=49
      <newsDate>2007-1-2</newsDate>
      <isHomepage>false</isHomepage>
    - <textAsset oid="[com.arsdigita.cms.TextAsset:{id=4931621}]">
      - <content>
          <p>新年第一天，老天来促销蓉城商场赚得欢。<br /> <br />
          的“促销员”。放弃出游的市民蜂拥而入各大卖场，蓉城的消费市
          引顾客。许多市民一头扎进商场，大包小包、手拎肩扛、排起长队
          前，站在商店门口彷佛可以听到市民们大把花钱的声音。</p><p>
          搭着元旦的列车，提前启动了。记者从王府井、好又多、红旗连锁
          品和干货调料等销售区域就不断上货。昨天，各大卖场的生意普遍
          上涨主要靠交易单价的增长拉动。这就意味着蓉城市民的购买力普
        </content>
      </textAsset>
      <title>新年第一天，老天来促销蓉城商场赚得欢</title>
    - <type oid="[com.arsdigita.cms.ContentType:{id=1005}]">
        <label>业务文档类型</label>
        <description>Article</description>
      </type>
      <name>996</name>
      <language>zh</language>
    - <auditing oid="[com.arsdigita.auditing.BasicAuditTrail:{id=4
        <id>4931620</id>
        <creationDate>2007-1-2 上午9:51</creationDate>
        <creationIP>192.168.3.173</creationIP>
        <lastModifiedDate>2007-1-2 上午9:51</lastModifiedDate>
        <lastModifiedIP>192.168.3.173</lastModifiedIP>
      </auditing>
      <objectType>com.arsdigita.cms.contenttypes.NewsItem</objectT
    </cms:item>
  </cms:contentPanel>
- <terms:assignedTerms>
    <terms:term domain="call-center" id="64" name="天府新闻" url=
  </terms:assignedTerms>
</bebop:page>
```


Figure 3. Contents with architecture described by semantic XML

By entering http://96999.scnjw.gov.cn/ccm/005/snwzx/xxb/zlj/2007/1/996.zh?output=xml in the browser and transfer the information into XML, result in figure 3 shows. It contains semantic information such as <language>, <title>, <content>, <auditing>, <objectType>, <terms.assiegnedTerms>. Such semantic XML has established architecture information, so it is not difficult to exchange and share information between two websites.

6. CONCLUSION

Through the content share realized in APLAWS CMS, we can understand that content share happened in different levels. The lowest level is to share content within a section; content share among sections can be divided into two kinds: content share section and section with particular features; the concept of semantic XML has built the foundation of sharing information among websites. Besides, APLAWS has provided service in RSS Feed, application of Dublin Core, input and output of XML, etc. They will be introduced specially in other papers.

REFERENCES

http://www.aplaws.org.uk
E-government plans or standards, please refer to http://www.govtalk.gov.uk

COMPUTER SIMULATION OF THE PESTICIDE DEPOSITION DISTRIBUTION IN HORIZONTAL DIRECTION SPRAY

Wanzhang Wang[1], Tiansheng Hong[2], Wenyi Liu[1], Xiangfu Li[1]

[1] *College of Mechanical and Electrical Engineering, Henan Agricultural University, 95 Wenhua Road, Zhengzhou Henan Province 450002, P. R. China, Tel: +86-371-63558289, Fax: +86-371-63558040, Email: wanzwang@gmail.com*

[2] *College of Engineering, South China Agricultural University Guangzhou, China, 510642*

Abstract: The objective of this study is taken to realize pesticide precision spray of fruit trees and the other crops and reduce the deposition losses outside the canopy when the real time sensing technology was used in the pesticide target spray. In this paper the Pesticide solution deposition distribution experiments were conducted with two different volume median diameter (VMD) hollow cone nozzles fixed in horizontal direction, to investigate the influence of spray pressure and spray ground speed on the spray deposition region. The probability distribution model of the pesticide deposition was constructed based on the experiments, and the pesticide spray distribution range was simulated by using Matlab statistic toolbox. The simulation result showed that the spray pressure and the ground speed had the great influence on the maximum spray distance. With the increase of the spray speed, the spray deposition distribution range decreases gradually, when the nozzle 200 is under the speed above 1.20km/h and nozzle 300 is under the speed above 2.22km/h, the deposition range was reduced greatly. So the computer simulations make a reference for the choice of the spray control parameters.

Keywords: computer simulation, pesticide deposition, lognormal distribution, target spray

1. INTRODUCTION

It's essential that the spray solution distribution range of the level placed nozzle must be given for precision spraying. In order to reduce the pesticide

Wang, W., Hong, T., Liu, W. and Li, X., 2008, in IFIP International Federation for Information Processing, Volume 258; Computer and Computing Technologies in Agriculture, Vol. 1; Daoliang Li; (Boston: Springer), pp. 425–432.

loss in target sprays, the level spray distribution should be controlled by spray pressure and spray ground speed. It is known that level spray range of the droplets has relationship with quality of the droplet, nozzle moving velocity and the air conditions. The flying of the droplets has turbulence flow state, mix flow state and laminar flow state. Flying distance of single droplet can be calculated on the given conditions (Wanzhang Wang, 2005). Phase Doppler particle analysis under different pressure shows that with the increase of the pressure, the speed of big spray droplet increases rapidly and small droplet speed increases slowly. The higher the pressure of pesticide spray droplet, the greater the distinction of speed becomes in the spray droplet spectrum (Giles et al., 1992). The spray ground speed affects the spray solution distribution not only in the direction of the spray, but also in the moving direction. Horizontal movement of the nozzle is equivalent to the airflow moving relative to the nozzle act on the spray droplet, this will change the movement direction of the droplet, thereby the spray scope is reduced and the spray solution distribution is changed (China Agricultural University. 1999). When the speed of the nozzle is low, it only has effect on the movement of small droplet. When the speed of the nozzle is big, it has big effect on the whole spray cloud. But in actual spray, the size and speed of the spray droplet out of the nozzle are different. It is difficult to quantify the scope of the droplet sediment because of the influence of the spray velocity and many other complex factors. This thesis uses the method of pesticide spray experiment to obtain the changing regularities about the deposition distribution by the change of spray ground speed and pressure, and then carry through the computer simulation analysis.

2. PESTICIDE SOLUTION DISTRIBUTION

2.1 Pesticide spray solution distribution test

Two hydraulic hollow cone nozzles are used in experiments. Their models are type 320 and type 200 produced in Ikeuchi, Japan. Their VMD are 210μm and 130μm respectively. The diffuser angle of two nozzles is 80° and operating pressure is from 0–7MPa. Experiments of different spray pressures and spray ground speed have been designed on the spray test stand. The test stand is made of two or more 960 × 2000mm tables, each has 12 "V" shaped groove and the centerline distance of the "V" shaped groove is 80 mm.

Figure 1 shows the pesticide spray solution distribution histogram of the experiments in different pressure and speed with two type nozzles. It can be seen from the pesticide spray distribution histogram that, the pesticide spray distribution shows skew distribution in the spray pressure and spray speed experiments. In order to study the rule of the pesticide spray solution distribution, three parameters lognormal distribution functions are chosen to construct the probability distribution model of it. The three parameters lognormal distribution probability density function is shown below.

$$f[ln(x-d)] = \frac{1}{\sqrt{2\pi}\sigma} \exp[-\frac{[\ln(x-d)-\mu]^2}{2\sigma^2}] \quad (1)$$

Where d is position parameter, which represents the minimum distance of the pesticide deposition; σ is the standard deviation, which shows the centralize degree of the pesticide spray, when σ is small the spray sediment scope becomes narrow, and σ big the scope becomes wide; χ is random variable which represents the spray range.

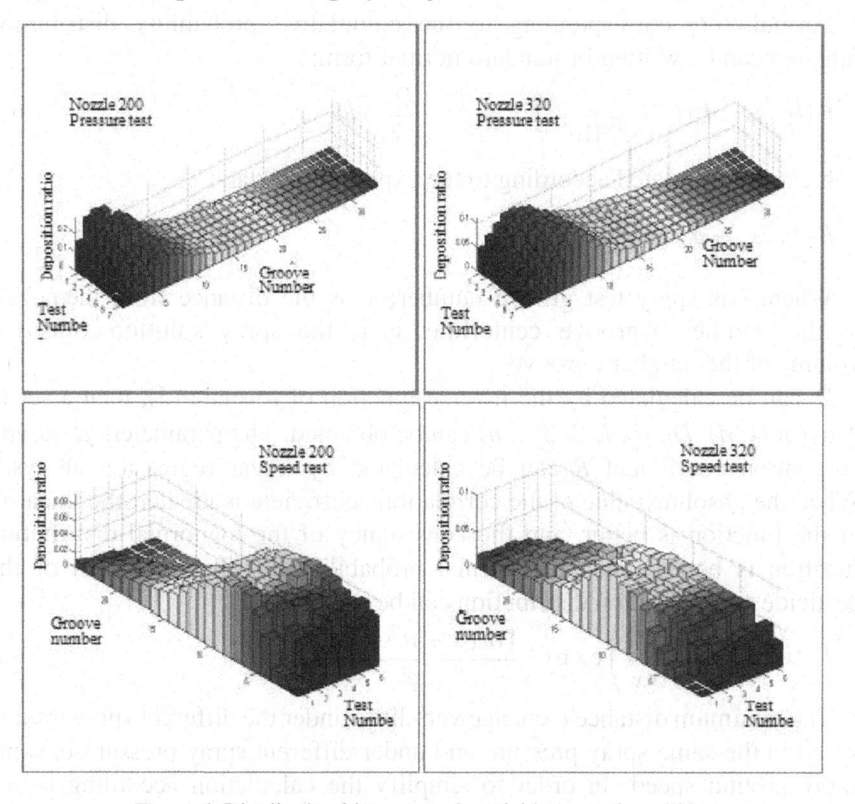

Figure 1. Distribution histogram of pesticide spray deposition

2.2 Lognormal distribution model

The method of quick optimization parameter estimation of three parameters lognormal distribution function is chosen to calculate d according to the test datum (Liu Dianrui et al., 1995).

$$d = \frac{x_{mo}^2 - x_{min} \cdot x_{max}}{2x_{mo} - x_{min} x_{max}} \tag{2}$$

Where x_{mo}, x_{min}, x_{max} are respectively the mode, minimum and maximum of the test sample.

And $2x_{mo}-x_{min}x_{max} \neq 0$

Standardize formula (1) and Let standard normal distribution value as:

$$D = \frac{\ln(x-d)-\mu}{\sigma}$$

Then $\ln(x-d) = \mu + \sigma D$

Formula (1) corresponding to the cumulative probability distribution function can be written in standard normal form:

$$F[ln(x-d)] = \frac{1}{\sqrt{2\Pi}} \int_0^{ln(x-d)} \exp(-\frac{D^2}{2}) dD \tag{3}$$

It can be calculated according to the experimental data:

$$F[ln(x_i - d)] = \sum_{i=1}^{i} v_i / \sum_{i=1}^{n} v_i$$

Where i is spray test groove number; x_i is the distance from the nozzle to the number i groove centerline; v_i is the spray solution collection volume of the number i groove.

D_i can be calculated by the inverse function of formula (3), then a set of data $(ln\,(x_i\text{-}d), D_i)$ $(i=1, 2, 3..., n)$ can be obtained. The parameters μ, σ and correlation coefficient R can be calculated by linear regression analysis. When the absolute value of the correlation coefficient is greater, the linearity of the function is better, and the consistency of the lognormal distribution function is better. Then the normal probability distribution model of the pesticide spray solution distribution can be established:

$$F(x) = \frac{1}{\sqrt{2\Pi}\sigma x} \int_0^x \exp(-\frac{[\ln(x-d)-\mu]^2}{2\sigma^2}) dx \tag{4}$$

The minimum distance d change very little under the different spray ground speed in the same spray pressure and under different spray pressure in same spray ground speed. In order to simplify the calculation according to the experiment datum, d is regarded as constant in pressure experiment and speed experiment. Table 1 and 2 shows respectively the fitting results of pressure and speed experiment for the nozzle 200. Table 3 and 4 shows respectively the fitting results of pressure and speed experiment for the nozzle 320.

Table 1. Nozzle 200 distribution parameters of pressure test d = 584

P/MPa	0.75	0.98	1.18	1.41	1.66	1.94	2.23
R^2	0.959	0.969	0.972	0.955	0.968	0.983	0.964
σ	0.771	0.715	0.727	0.723	0.770	0.694	0.769
μ	5.716	6.077	6.122	6.239	6.318	6.546	6.483

Table 2. Nozzle 200 distribution parameters of ground speed test d = 0

V/km/h	0.51	1.21	1.91	2.60	3.30	4.00
R^2	0.99	0.969	0.959	0.915	0.914	0.897
σ	0.437	0.501	0.497	0.432	0.432	0.399
μ	6.747	6.583	6.564	6.497	6.447	6.393

Table 3. Nozzle 320 distribution parameters of pressure test d = 442

P/MPa	0.67	0.87	1.06	1.29	1.53	1.78	2.06
R^2	0.954	0.960	0.971	0.974	0.973	0.978	0.979
σ	0.652	0.654	0.657	0.639	0.615	0.635	0.625
μ	6.222	6.301	6.420	6.487	6.536	6.578	6.647

Table 4. Nozzle 320 distribution parameters of ground speed test d = 0

V/km/h	0.51	1.21	1.91	2.60	3.30	4.00
R^2	0.996	0.965	0.958	0.951	0.905	0.907
σ	0.379	0.533	0.559	0.555	0.467	0.458
μ	6.906	6.780	6.775	6.767	6.678	6.616

3. COMPUTER SIMULATION OF THE PESTICIDE DISTRIBUTION

Establishing X-axis in the direction of nozzle spraying, and Y-axis in the direction of nozzle moving, the deposition of the spray droplet on the X and Y direction can be viewed as two independent probability event, and the probability density function f(x, y) =f(x)•f(y) can be constructed according to test datum of the horizontal fixed nozzle above the plane area.

The test datum under different spray pressure and speed shows that the pesticide spray solution distribution has normal distribution on the Y-axis. By construct two-dimensional probability density function of the pesticide spray distribution together with the normal distribution function on the Y-axis and lognormal distribution function on the X-axis, the simulation program can be compiled using Matlab6.0 statistic toolbox function Logndf and Normdf (Su Jin et al., 2000). Spray solution deposition distribution on the plane area below the nozzle can be computed by the Monte Carlo simulation with a given reasonable boundaries. A visual representation is obtained by the surface and contour diagram using Matlab drawing function to know where and how much the pesticide deposited. And then get the law of spray distribution change with the spray pressure and speed. Figure 2 shows the pesticide distribution of the static nozzle 200 under pressure 1.2 MPa. Figure 3 shows the pesticide distribution of the nozzle 200 under pressure 1.2 MPa, speed 1.21 km/h, and spray control distance 960mm. Figure 4 shows the pesticide distribution of the static nozzle 320 under pressure 1.2 MPa. Figure 5 shows the pesticide distribution of the nozzle 320 under pressure 1.2 MPa, speed 1.21 km/h, and spray control distance 960mm.

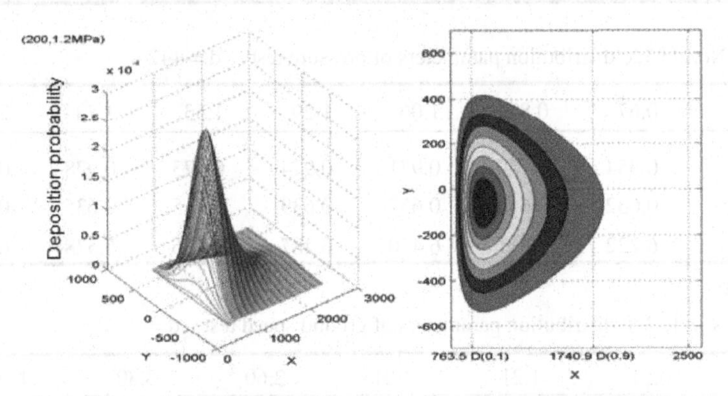

Figure 2. Static nozzle 200 distribution with pressure 1.20MPa

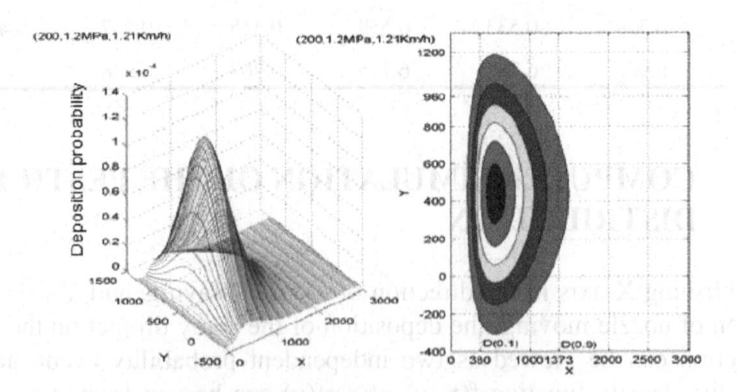

Figure 3. Nozzle 200 distribution with speed 1.21km/h

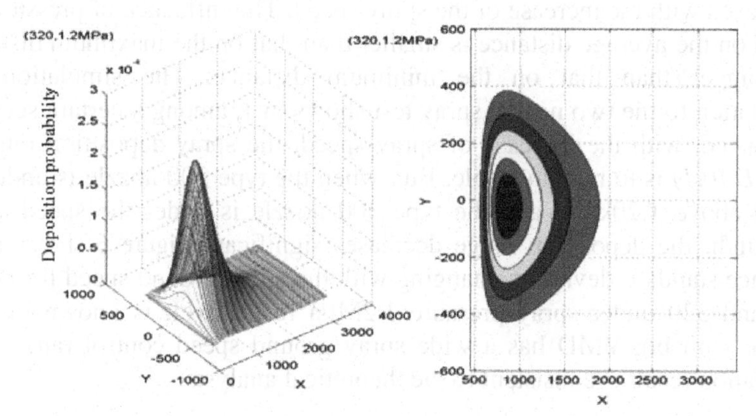

Figure 4. Static nozzle 320 distribution with pressure 1.20MPa

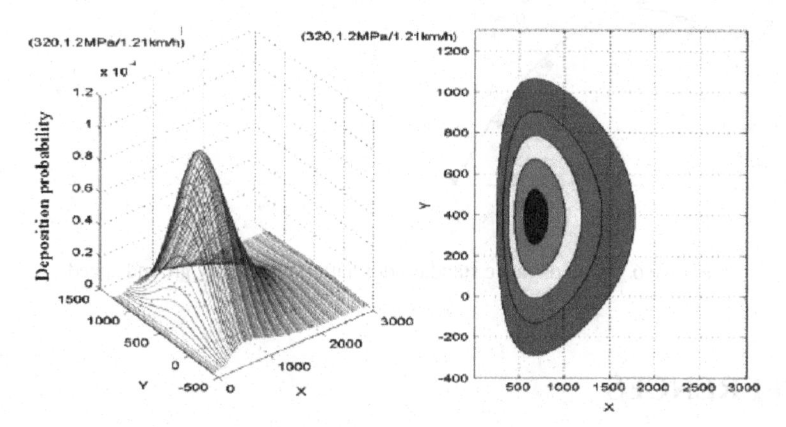

Figure 5. Nozzle 320 distribution with speed 1.21km/h

4. CONCLUSIONS

Based on the spray experiment datum the pesticide deposition distribution can be simulated on computer by using the probability density function and inverse function of cumulative distribution in Matlab Statistic Toolbox, then the minimum spray distance *D (0.1)*, the maximum distance *D (0.9)* and the average distance *D (0.5)* can be computed and the region of the spray solution deposition under different speed and pressure can be calculated.

The simulation result shows that the minimum spray deposition distance *D(0.1)*, the average distance *D(0.5)* and the maximum distance *D(0.9)* have linearity increased with the increase of the spray pressure, and have linearity decreased with the increase of the spray speed. The influence of pressure and speed on the average distance is smaller than that on the maximum distance, but bigger than that on the minimum distance. The simulation and calculation to the two nozzle spray test shows that, among a certain scope of low speed, with the increase of spray speed, the spray deposition range *D (0.9)-D (0.1)* is to remain stable. But, when the type 200 nozzle is under the speed above 1.20km/h and the type 300 nozzle is under the speed above 2.22km/h, the deposition range decreases significant Figure 6 shows spray distance standard deviation changing with the spray ground speed for nozzle 200 and 320 under spray pressure 1.2MPa respectivelt is known that the nozzle with big VMD has a wide spray ground speed control range. This simulation result is consistent to the theoretical analysis.

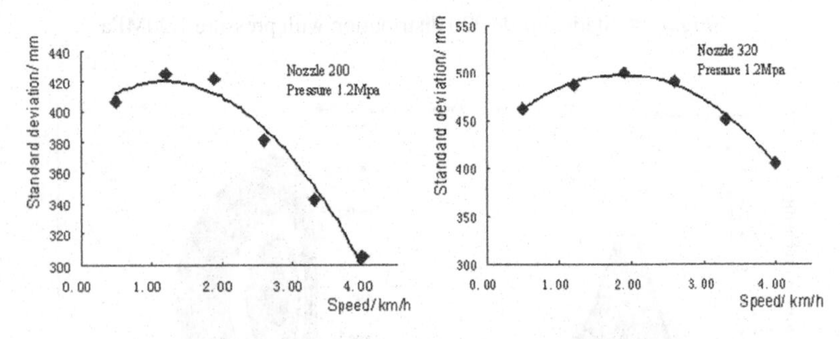

Figure 6. Spray distance standard derviation with spray ground speed

REFERENCES

China Agricultural University. 1999. Agricultural Machinery (1) [C]. Beijing: China Agriculture Publishing House

Gillis K P, Giles D K, Slaughter D. C, et al. 2003. Distance-Based Control System For Machine Vision-Based Selective praying [J]. Transactions of the ASAE. 45(5): 1255-1262

Liu Dianrui, Liu Gang. 1995. Fitting three parameters on the lognormal distribution at the third parameter C optimization method [J]. Geological exploration. 35 (5): 43-48

Molto E, Martin B, Gutierrez A. 2000. Design and testing of an automatic machine for spraying at a constant distance from the tree canopy [J]. Journal of Agricultural Engineering Research. 77(4): 379-384

Su Jin, Zhang Lian, Liu Bo. 2000. Matlab Toolbox Application [C]. Beijing: Electronics Industry Publishing House

Wang Wanzhang. 2005. Pesticide Distribution Test of the Profile Modeling Spray Based on Ultrasonic Sensor for Fruit Tree [D]. Guangzhou, College of Engineering, South China Agricultural University

THE HARDWARE RESEARCH OF DUAL-PORT RAM FOR MAIN-SPARE CPU IN RURAL POWER TERMINAL SYSTEM OF POWER QUANTITY COLLECTION

Ping Yang[1], Shu Dai[2], Xiuhua Wu[1], Yong Yang[1,*]

[1] *College of Information and Electrical Engineering, Shenyang Agricultural University, Shenyang, China, 110161*

[2] *Anshan Electricity Power Bureau, Anshan, China, 114001*

* *Corresponding author, Address: College of Information and Electrical Engineering, Shenyang Agricultural University, 120 Dongling Road, Shenyang, Liaoning, 110161, P. R. China, Tel: +86-24-88487860, Fax: +86-24-88487122, Email: xdxy7128@126.com*

Abstract: With the development of rural power market in our country, how to improve the operation more correctly and timely of power quantity collection, transmission and processing becomes a problem. The requirement of equipment in power quantity collection needs new and higher demands. The advantage of using dual-port RAM and main-spare CPU structure, when main CPU running, spare CPU supply monitoring of main CPU status messages, once main CPU break down, spare CPU could instead of main CPU completely. This paper introduces collectivity design in flow chart, hardware design include theory and application in detail. Using dual-port RAM to communicate between the main and spare CPU not only make sure the transmission efficiency, good anti-jamming performance, improve the speed of disposal, but also reduce the costs, making the operation of rural power network more security, economy and reliability. In rural power terminal system of power quantity collection has broad application prospects.

Keywords: dual-port RAM, main-spare CPU, terminal of power quantity collection, parallel communication, data exchange

Yang, P., Dai, S., Wu, X. and Yang, Y., 2008, in IFIP International Federation for Information Processing, Volume 258; Computer and Computing Technologies in Agriculture, Vol. 1; Daoliang Li; (Boston: Springer), pp. 433–440.

1. INTRODUCTION

To meet he need of power industry change into commercial operation has many requirements, how to improve power quantity collection, transmission, accuracy, reliability and operation in time are all the factors to be considered. With the development of Chinese power industry market, electric power department has already adopted power quantity as standard for expense, test, rewards and punishment. Therefore, higher requests are required on device of power quantity collection. Building a power quantity management system that based on automation is imperative under the situation.

Terminal system of power quantity collection provides multi-route data collection, information storage, inspect, check, balance calculate, prevent analysis of steal electricity, analysis of degradation loss, assess and so on, the most important thing is reliability. Therefore adopt one CPU will have some risks, once CPU has some problems, the data should be lost. If adopt two systems that supply for each other, need two powers, machine interfaces and so on. In this way the cost should be higher. Therefore, this system adopts the structure of main-spare CPU; both of them adopt the chip of ARM that has high performance (Chen et al., 2005). When main CPU running, spare CPU supply monitoring of main CPU's status, once the main CPU appears the problem, the spare CPU can instead the main CPU to carry on the work completely. Moreover, using dual-port RAM to communicate between the main and spare CPU not only make sure the transmission available, improve the speed of disposal, but also good at anti-jamming, making the operation of electric power system more security (Jia et al., 2006).

2. THE MAIN DESIGN PLANNING OF SYSTEM

The main-spare CPU is different from the master-slave CPU. If the system adopt the master-slave of CPU, the relationship of master CPU and slave CPU are leader and subordination, which means master CPU and slave CPU can be seen one CPU to carry out all the functions, just burden the different work in the system (Huang et al., 2004).

This terminal system adopts the structure of the forward plug-in type, each plug-in passes communicates using panel bus, and display and keyboard link the CPU through the cable, this system contains 5 plug-in parts, including copy meter panel, pulse panel, remote control panel, remote supervision panel and CPU panel (Ti et al., 1999). This design of plug-in provides the convenience to the install and debug for spot, any plug-in passes have no differences and can insert any module, but its address are different and each slot contains an unique address, thus any module has its own address after inserting the slot, this method can eradicate stir the wrong address completely (Liu et al., 2006).

The system adopts main-spare CPU, there are isolated in the functions, which means the two CPU can achieve the terminal's functions independently. Therefore, the two CPU have their own connection, including broadcasting station, alternating equipment is interface between man and computer, other function panels (Jia et al., 2006). The configuration of the whole system is shown in figure 1:

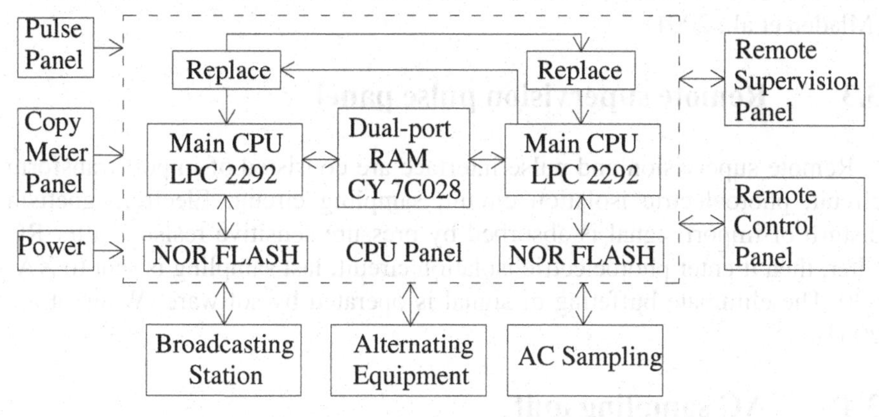

Figure 1. System configuration chart

3. THE DESIGN OF HARDWARE

3.1 CPU panel

The dual-port RAM adopts CY7C028 of CYPRESS corporation. It has 64K x 16 static RAM, the access speed is less than 25ns, it has a true dual-ported memory cells which allow simultaneous access of the same memory location, the both ends has the independent control signal bus, the address bus and data bus. The CY7C028V consist of I/O and address lines, and control signals (CE, OE, R/W) (Cypress Semiconductor Corporation, 2002). These control pins permit independent access for read or write to any location in memory.

Program memory adopt SST39VF160 as NOR FLASH, it is CMOS multi-function FLASH (MPF) machine piece of the SST corporation, the memory capacity is 2MB, 16 bit data width, work voltage is 2.7V–3.6V. NOR FLASH have the same interface as SRAM which provide enough address to lead the pin to seek address, which can access each byte in chip easily and address lines and data lines of the NOR flash are divided, thus the efficiency of transmission is very high and the performance can be implemented in the chip.

3.2 Copy meter panel

Copy meter module adopts ST16C552 and extend RS-485 interface with two photoelectric isolations as the communications to meter. Each RS-485 interface can connect 32 blocks electricity meter with RS-485 interface (Mladen et al., 2001).

3.3 Remote supervision pulse panel

Remote supervision and pulse interface are consisted of import transform circuit, photoelectric isolation circuit, sampling circuit. Electromagnetism disturb of import signal is absorbed by pressure-sensitive resistor, after RC filter, then it enter photoelectric isolation circuit, last sampling is sent to XA-S30. The eliminate buffeting of signal is operated by software (Wang et al., 2004).

3.4 AC sampling unit

AC sampling circuit adopts ADS7864 that controlled by main CPU. AC sampling unit consists of voltage PT, current CT, corresponding analog signals disposal circuit and sampling A/D circuit. Main control panel can achieve all controls of circuit and data collection.

3.5 Broadcasting station

Broadcasting station adopts data transmission radio station. The customer can choose to use MODEM, fiber, GSM, GPRS or other communication methods. This broadcasting station is the vehicle radio stations with 200MHZ which made by New Zealand, it can place 16 pairs frequency of receive, dispatch and different blast-off powers, it also has call function, having the characteristics of channel establishment in a short time and high reliability (Deng et al., 2005).

4. IMPLEMENTATION OF DUAL-CPU FOR PARALLEL COMMUNICATION

Because this system adopts the structure of main-spare CPU, therefore the interfaces of data exchange become important parts that affect the whole system data processing ability. The whole hardware circuit connection is showed in figure 2.

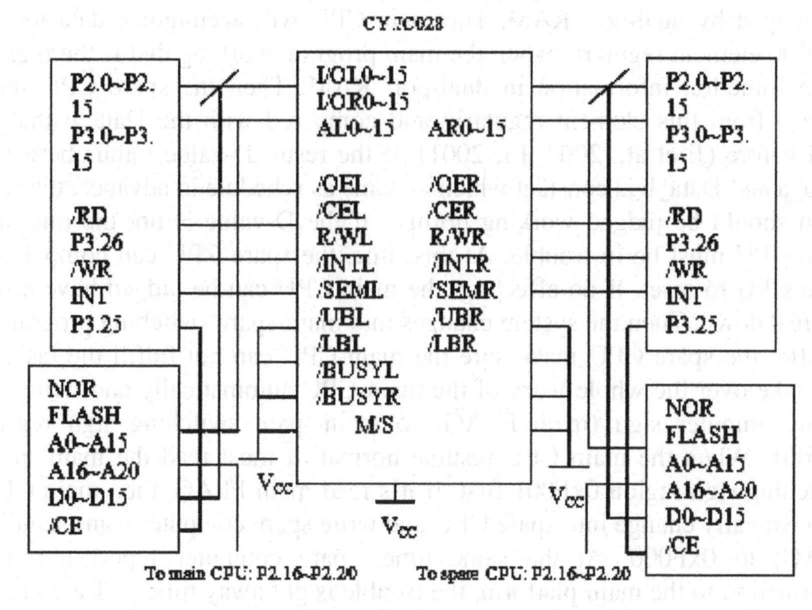

Figure 2. Hardware circuit connection chart

4.1 Data exchange

Using dual-port RAM achieve parallel communication between main-spare CPU means to share memory, which can lead data conflict, how to solve one memory for two sides to use together is the key point. CY7C028 has many solution modes, including hardware logic, interrupt logic, and semaphore logic (Cypress Semiconductor Corporation, 2004). According to the request of this system, when the main-spare CPU switching, need a great deal of data to communicate, and the data exchange is concentrated, therefore this system adopts the interrupt logic mode.

The interruption function of the CY 7C028 chip is use "mailbox" to achieve. For the hardware only need to link between the $/INT_R$ and $/INT_L$ of CY 7C028 and the main CPU and spare CPU respectively (Li, 1999). For the software only needs to write operation program and the interruption program.

4.2 Switching between main CPU and spare CPU

Here is discussing how to implement switching from the main CPU to the spare. Firstly, need to think about how to judge whether the work of main CPU is normal or not. Malfunction information of the main CPU can be exchanged by dual-port RAM. The main CPU will accumulate data to the 0xE000 element regularly when the main program working that is the region of malfunction information in dual-port RAM. Then the spare CPU read Data_a from this element regularly and compared with the Data_b that is read before (Ji et al., 2004; Li, 2001). If the result D-value minus between Data_a and Data_b is constant which as same as schedule in advance, the program should be judged working normal; if the D-value is not the one, the main CPU must be in trouble. At this time, the spare CPU can compel the main CPU to reset. If no effecting, the main CPU can be judged have error or breakdown. Then the system changes into main-spare switching program.

After the spare CPU make sure the main CPU can not fulfill the task, it will take over the whole work of the main CPU automatically and write the main computer sign (main FLAG) to main-spare switching sign region 0xF001. When the main CPU resume normal, it must read the main-spare switching sign region 0xF001 first, if it's read main FLAG, then main CPU automatically change into spare CPU and write spare computer sign (standby FLAG) to 0xF000. At the same time, spare computer report terminal information to the main platform, the trouble is got away timely (Tang et al., 2001; Wang et al., 2005).

Nevertheless, with the development of electronic technology, the mass appear of 32-bit CPU provides conditions for main-spare CPU system. At beginning, because of processing speed of chip accelerating increasingly, only one chip can achieve the functions of system perfectly; moreover, with the chip price reducing, the cost of main-spare CPU system is lower (State Economic and Trade Commission, 2001). Main-spare CPU in a system, but independence with each other, can fulfill the task of system independently. That is say the main CPU in function is equal to the spare one. When the main CPU is in trouble and cannot resume back independently, after judgment system can switch main CPU to the spare one automatically. Then spare CPU can take over all work of main CPU that can prevent system's breakdown from main CPU's problems. In terminal system of rural power quantity collection, this structure can prevent loss of mass instant data from emerging issues of problem CPU and can reduce the cost of the whole system.

5. CONCLUSIONS AND FUTURE WORKS

Using high speed dual-port RAM and main-spare CPU structure to deal with the information, not only in the share data in the parallel network, make sure the information can easy get across the channel, improve the speed of transmission, be good at anti-jamming, but also with the price of lower-end processor reducing, the cost of using main-spare CPU also can be decreased, in order to get more benefit in economy of rural power network.

The dual-port RAM and one-way store are the same in data store ability, simple interface, convenient operation, easy for communication, the communication processing and protocol in the two sides CPU are easily, only need to make sure the store space of data, the two sides CPU operate this space independently, in this way, the whole system can achieve high speed and reliability in the parallel communication. Otherwise, the dual-port RAM has good expansibility and easy for bit and byte expand, make sure the update operation for the future. Therefore, compared with the traditional double CPU communication, dual-port RAM has more superiority in main-spare parallel communication. Based on the dual-port RAM embedded multi-CPU system has good broad prospects in application of rural power enterprises.

To be worthy, when the designing of hardware and software, the distributed of store space, security issues of access in memory, failure handling, recovery and data redundancy should still be considered in detail of this main-spare CPU system, and also need to do some deeper research in the future.

ACKNOWLEDGEMENTS

As with any effort for this paper, there are a number of people who contributed to this in a roundabout way. Without their help, this paper would not exist.

Special thanks to professor Yang Yong, who is my PhD supervisor not only suggested the whole structure of design, but who always gave me good idea, no matter how odd.

Particular thanks are due to Dai Shu, the engineer of the Anshan electricity power bureau for sharing her knowledge and skills environment. The author is also grateful to associate professor Wu Xiuhua for her help, advice, comments, and excellent proofreading skills and gently showed me the errors in my ways.

Finally, the authors would like to thank the countless people who contributed to this paper with informal reviews and suggestions.

REFERENCES

Chen Deming, Xiong Liebin, Application of Dual RAM in Automation System, International Electronic Elements, 2005, No. 4, pp. 20-23.

Cypress Semiconductor Corporation, Interfacing the CY7C028V with Motorola's MPC860 Power QUICCTM Processor, 2002.

Cypress Semiconductor Corporation, 3.3V 64K×16 Dual Port Static RAM CY7C028V Data Sheet, 2004.

Deng Xiaogang, Xu Shiming, Tang Houjun, New Type of Power Amount Acquisition Terminal Based on Insert-style System, North China Electric Power, 2005, No. 11, pp. 42-45.

Huang Bin, Hu Rongqiang, Meng Qinghong, Design and Implementation of Real-Time Supervisory System Based on Dual CPU, Journal of WUT (Information & Management Engineering), 2004, Vol. 26, No. 2, pp.45-48.

Ji Qiang, Liu Liqiang, Application of Dual-port RAM to Data-acquisition System, Applied Science and Technology, 2004, Vol. 31, No. 5, pp. 22-24.

Jia Guixi, Qi Le, Design of Data Supervision System Based on Dual CPU Structure, Engineering Science, 2006, No. 4, pp. 68-71.

Li Xiaoqing, The Application of Dual-Port RAM in the Computer Measurement and Control System of Multi-cpu, Control & Automation, Vol. 15, No. 1, 1999, pp. 54-56.

Li Yufeng, Design of Two Monolithic Computers Parallel Communication System Based on Dual-port RAM, J. Changchun Inst. Tech. (Nat. Sci. Edi.), 2001, Vol. 2, No. 3, pp. 56-58.

Liu Xin, Yin Zhiqiang, Mu Guobao, Ma Shidian, Research on Parallel Communication Based on Dual-port RAM in the Noncompetitive Pattern, Journal of Hefei University of Technology, 2006, Vol. 29, No. 2, pp. 234-237.

Mladen Kezunovic,Yuan Liao, A New Method for Classification and Characterization of Voltages Sags. Electric Power Systems Research, 2001, Vol. 58, No. 1, pp. 27-35.

National Instrument, A New Open Specification for Modular Instrumentation, 1997.

National Instrument, The PXI System Architecture, 1997.

State Economic and Trade Commission, Terminal System of Electrical Energy Collecting for Electric Power Industry, China Electric Power Press, 2001.

Ti Zhaoxu, Li Yuan, Design and Optimization Study of MIS of A Large Power Supply Bureau Client/Server System Being Used Successfully in Guangzhou Power Supply Bureau, Proceedings of the CSEE, 1999, Vol. 19, No. 5, pp. 85-88.

Tang lei, Zhang Boming, Sun hongbin, A Processing Method of Real-time Data in Energy Management System, Power System Technology, 2001, No. 5, pp. 15-19.

Wang Jing, Ji Xinsheng, Zhu Yunzhi, Strategies of Improving the High Usability of Application Software on Embedded OS, Control & Automation, Vol. 21, No. 82, 2005, pp. 42-44.

Wang Junliang, Yuan Yijiang, Zhang Jiang, The Design and Application of Electric Energy Acquisition Terminal Based on GSM Communication, Journal of Henan Mechanical and Electrical Engineering College, Vol. 12, No. 2, 2004, pp. 42-44.

AN AGENT-BASED POPULATION MODEL FOR CHINA

Zhongxin Chen[1,2,*], Einar Holm[3], Huajun Tang[1,2], Kalle Mäkilä[3], Wenjuan Li[2,3], Shenghe Liu[4]

[1] *Key Laboratory of Resource Remote Sensing and Digital Agriculture, Ministry of Agriculture, Beijing 100081, China*

[2] *Institute of Agriculture Resources & regional Planning, Chinese Academy of Agricultural Sciences, Beijing 100081, China*

[3] *Kiruna Spatial Modelling Centre, Umeå University, Kiruna SE98128, Sweden*

[4] *Institute of Geographical Science & Natural Resources Research, Beijing 100101, China*

* *Corresponding author, Address: Room 322, IARRP, CAAS, 12 Zhongguancun Nan Dajie, Beijing, 100081, P. R. China, Tel/Fax: +86-10-68918684, Email: zxchen@mail.caas.net.cn*

Abstract: Human is a crucial factor responsible for local and large scale environmental change and is a key factor in rural management. Accurate simulating and predicting population change and its spatial distribution is of great interests to environmental scientists as well as social scientists. In this paper, we introduced the development of a spatial-explicit agent-based population model for China based on a prototype from SVERIGE model. The model was used to predict the population in China. The validation result shows that agent-based model performances well in population prediction in China.

Keywords: Agent-Based Model, SVERIGE, Population, Prediction, China

1. INTRODUCTION

Our planet has suffered great changes at very strong intensity world wide since industrial revolution. There are a lot of evidences from various observations, investigations and scientific researches showing that the amplitude of global change has been enhanced since the mid-term of last century. The causes for global change are various and very complicated, among which anthropogenic factors contribute a lot (Turner et al., 1995).

Chen, Z., Holm, E., Tang, H., Mäkilä, K., Li, W. and Liu, S., 2008, in IFIP International Federation for Information Processing, Volume 258; Computer and Computing Technologies in Agriculture, Vol. 1; Daoliang Li; (Boston: Springer), pp. 441–448.

Accurate simulating and predicting population change and its spatial distribution is of great significance in various fields of research and application including global changing, land use/cover change, rural management and so on (Ziervogel, 2005).

The world population is about 6.6 billion in 2007 (US Census Bureau, 2007). The world population has been doubled in the past 40 years, and is still increasing at a very rapid pace. The rapid population growth is not only a demographic problem. It also concerns the social and economical development, food security, environment and sustainability and so on in specific regions as well as worldwide. China contributes more than 1/5 to the world's population. And the ratio has been quite stable during the past half century. The population in China increases rapidly. In order to alleviate the side-effect brought by the population growth, family planning program has been applied in China since late-1970s. So there are other demographic problems in China, such as age structure shift, the increasing of aged people, etc. The rapid population growth have great impacts on the following issues, such as economic development, food security, the demand of educational resources, population migration and urbanization, land use and land cover changes and other environmental and social issues, etc. (Brown et al., 1999).

Although there have been 5 national demographic censuses in China and there are a lot of population monitoring and survey data, yet the demand of a robust, accurate demographic prediction tools is very strong. The tools can be used to predict the future trend of the population in China, and it can also be used to aid the decision-making of the population-related issues, such as population migration management policy, land use planning, educational resources configuration, etc. To meet the urgent demands, we decided to set up the China population simulation and prediction model by employing individual based micro-simulation model, i.e. agent-based model. (Bankes, 2002; Bonabeau, 2002; Holm et al., 2002; Nuppenau, 2002; Thomas et al., 2006; Rykiel, 1996; Michael et al., 2006.)

2. DATA PREPARATION

2.1 The national demographical census data in 1990

The 1990 population Census of China was conducted by the State Statistics Bureau of China. This dataset (1% sampling) was prepared taking villages as sampling unit, and contains a record for each household and supplies variables describing the location, type, and composition of the households. Each household record is followed by a record for each individual residing in the household. Information on individual includes demographic characteristics, occupation, literacy, ethnicity, and fertility.

2.2 Aggregation data of 2000 national demographical census

The 5th National Demographic Census in China was committed on November 1, 2000. The Aggregation dataset was compiled by Population Census Office of The State Council and National Bureau of Statistics of China, and it was published in 2002.

2.3 County-level demographical database

There are the frequently used demographic indices in 1990 for more than 2000 counties from 30 provinces, autonomous regions and municipalities. These indices include: total population, sex ratio, size of family, population in cities, population in towns, the ratio of illiterate and semi-illiterate among the population over 15, the number of college-educated people per 10,000 people, the number of middle-school-educated people per 10,000 people, population aged between 0–14, population aged 15–49, population aged 50–65, population older than 65.

2.4 Provincial migration database

This database summarized the population migration data from 1949. It included the outputs from the 4th national demographic census in 1990, and some other large-scale sampling survey since 1980s, such as the migration survey in 74 cities in 1986, 1‰ national sampling of population in 1987, the national sampling survey on anti-pregnancy in 1988, the sampling on pregnancy and delivery in 1992, and other related data from police's registering system.

2.5 County-level agro-eco-environmental database

The database is from Information Center, Ministry of Agriculture of China. It contains the statistics for agriculture production in 2523 counties in China. The data is for 1986 to 2001. The data items include population, cropland acreage, acreage and yield for each crop, GDP, etc.

2.6 County-level administration region's map

The map is from Columbia University's Center for International Earth Science Information Network (CIESIN). The original scale for this administrative map is 1: 1 000 000. The ID for each county for each dataset is based on "Codes for the Administrative Divisions of the People's Republic of China" 1990 version (GB 2260–90).

3. MODEL DESIGN AND IMPLEMENTATION

The prototype for the China population model is the Swedish SVERIGE (System for Visualising Economic and Regional Influences in Governing) model, which was developed by Spatial Modelling Centre (SMC), Department of Social and Economic Geography, Umeå University (Holm et al., 2002). The developing tools for the China population model include Microsoft Visual C++, other data preparation tools include SPSS, SAS, ESRI ArcGIS, etc.

3.1 Overall structure of the model

The China population simulation model is an individual based model. The basic agent in this model is an individual person, but with a household (or a family) as the basic operational unit. There can be one or more individual person(s) in one family. The reason for us choosing household as the basic organizational unit is that the members in a same household usually share a lot of resources such as housing, income, etc. But the household is not static but dynamic. It means that either the members can leave or join a household by migration, death, birth, etc. Even the household itself can migrate, emerge, or disappear. Then each household has an attribute of geographical location. Herein, a county code is assigned to it to represent its geographical location. The household as well as the household members can share the resources and limitations in this geographical region.

3.2 The attributes of individual agent

The attributes for a household and an individual person can be very complex. In this paper, we only consider the critical and essential features for them. The following is the data structure for an individual person in the model.

```
typedef struct chinastruct { // -------------------
  unsigned FID              :24;
 unsigned EducationLevel     :3;  // 0-7
 unsigned EducationSector  :3;
 unsigned Relat              :2;
// Relation in the family //1=Head,2=Spouse,0=Child

 unsigned Sex              :1;
// 0 = male, 1 = female
 unsigned BirthYear            :9;
// Base year is 1870
 unsigned OutOfLabor          :1;
 unsigned InEducation         :1;
   // In school or university
 unsigned Working              :1;      // Working
 unsigned Unemployed          :1;
//Unemployed
 unsigned MaritalStatus    :2;
// (0-3) 0=Single, //1=Married,2=Widowed,3=Divorced
 unsigned LARegion          :12;
//region identifier 1-2500
 unsigned RuralUrban          :1;
 unsigned School            :3;

 unsigned EthnicGroup         :6;
 unsigned ffiller             :26;
} ChinaBuffer;
```

3.3 Parameterization

The demographical attributes of the individuals as well as households are correlated with other demographical parameters and environmental factors. Herein, the China population simulation model is running in a simple way. The parameters for the model are estimated mainly be probability tables and rules. The probability tables are from the census data or other investigation data. The rules are either based on the demographical theories or related studies or investigations. The parameterization for each processes are being explained as follows.

(1) Aging

The simulation step for the China population simulation model is 1 year. In the beginning of each simulation step, each active agent in the model will automatically gain one year in age. Age is also a basis to determine the parameters for other processes, such as education, marriage, migration, giving birth to a child, death, etc.

(2) Fertility

The fertility of the women older than 15-years-old and younger than 50-years-old is estimated based on the 1990 censuses data. The conditional probabilities of the women in each age falling in this age group were given. We considered the following conditions: (a) whether the woman lives in the rural area or the urban area (city or town), (b) different martial status, and (c) if she has already given birth to a certain number of child(ren).

Also the ethnic groups of the wife and the husband can affect the fertility. We used rules in this regard. If either of the couples is from a minority ethnic group, they can have more than one child. If the couples live in rural area and they have a 4-year-older child, they can have the second child.

(3) Mortality

The mortality module in the China population simulation model is employed to terminate the lives of individual agents. A lot of factors affect the mortality. But in order to keep the model simplicity at the beginning, age, gender, and residential type (either in rural area or in urban area) are considered as the main impact factors to determine the mortality for a specific person in the model. The mortality probability is parameterized based on the 1990 and 2000 census data, and the mortality database of China.

(4) Education

The education is estimated based on simple rules and probability. The rules are as follows: (a) A child begins to go to primary school at 6 or 7 at a probability; (b) He or she will spend 6 years in primary school, after that he or she goes to junior high school at a probability; (c) He or she will spend 3 years in junior high school, after that he or she goes to senior high school at a probability; (c) He or she will spend 3 years in senior high school, after that he or she goes to Junior college or university at a probability. The probabilities at the critical ages are from statistical data.

(5) Marriage

Marriage module is used to set up families in the model. It affects the fertility and the number of families. An agent tries to find a partner to get married after certain age. In this model, age, gender, education level, the residential type, and geographical location are considered as the main factors to determine the probability for the potential partners to get married.

(6) Migration

The migration rates between provinces are used to estimate the probability of migration for a specific agent in the model. The data is from 1990 and 2000 national census data.

(7) Retirement

The retirement model is used to let the agents to quit from the labor markets. It is a rule-based module. The women agents retire at 55, while the men agents retire at 60.

4. RESULTS AND DISCUSSION

After running of the model, we can get the year-by-year population distribution map of China like Fig. 1.

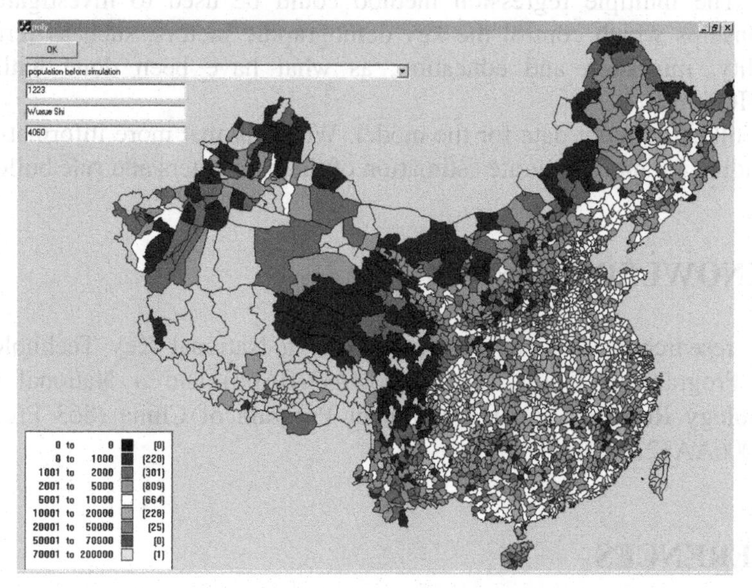

Fig. 1. The population distribution map from the model

The model outputs are evaluated by the existing population datasets during 1990–2001, which includes 2000 census data, provincial population dataset from 1990. The preliminary validation of the total population is listed in the following table. The projected population for China is very close to the statistical figure. The relative errors are less than 0.64%. We can conclude that the model is quite good overall.

We succeeded in setting-up of the spatial-specific population mode for China with employing agent-based technique. The validation result shows

Table 1. The validation of the model

Year	Statistic Population (Million)	Projected Population (Million)	Relative Error (%)
1991	1158.23	1165.70	0.64
1992	1171.71	1177.78	0.52
1993	1185.17	1190.04	0.41
1994	1198.50	1202.48	0.33
1995	1211.21	1215.14	0.32
1996	1223.89	1227.67	0.31
1997	1236.26	1240.14	0.31

that it can accurately simulate the population in China. But there are still a lot works to be done. These works include:

1. More accurate estimation of the parameters for the processes in the agent. The multiple regression method could be used to investigate the relationships which control the key demographic factors, such as fertility, mortality, migration and education, as what have been done with the SVERIGE model.
2. Collecting more data for the model. We think that more information is demanding for more accurate estimation of the parameters and rule building.

ACKNOWLEDGEMENTS

The research in this paper is funded by a National Key Technologies R&D Program of China (No. 2006BAD10A06) and a National High Technology Research and Development Program of China (863 Program No. 2006AA12Z103).

REFERENCES

Bankes, SC, Agent-based Model: Revolution? PNAS, 2002, 99(spp3):7199-7200.

Bonabeau, E, Agent-based modeling: Methods and techniques for simulating human systems. PNAS, 2002, 99(spp3):7280-7287.

Brown, LR, GT Gardner, B Halweil, Beyond Malthus. Norton. 167pp. 1999.

Holm, E, K Holme, K Mäkilä, M Mattsson-Kauppi, G Mötvik, The SVERIGE Micro-simulation Model. Dept. of Soc. & Econo. Geography, Umeå University. 55pp. 2002.

Michael M, A Miguel, C Baird, Coupled human and natural systems: A multi-agent-based approach. Environmental Modelling & Software, 2006, 1-8.

Nuppenau, E, Towards a genuine exchange value of nature: interactions between humans and nature in a principal-agent-framework. Ecological Economics, 2002, 43:33-47.

Rykiel J, Testing ecological models: the meaning of validation. Ecological Modelling, 1996, 90:229-244.

Thomas B, S Pepijn, W Johannes, Multi-agent simulation for the targeting of development policies in less-favored areas. Agricultural Systems, 2006, 88:28-43.

Turner II BL, D Skole, S Sanderson, G Fischer, L Fresco, R Leemans, Land-Use and Land-Cover Change (LUCC), Science/Research Plan. IHDP Report 7/IGBP Report 35, Stockholm and Geneva. 1995.

US Census Bureau, http://www.census.gov. 2007.

Ziervogel, GM Bithell, R Washington, T Downing, Agent-based social simulation: a method for assessing the impact of seasonal climate forecast applications among smallholder farmers. Agricultural Systems, 2005, 83:1-26.

HOW TO DEVELOP RURAL INFORMATIZATION IN CHINA

Junjing Yuan[1], Daoliang Li[2], Hongwen Li[1,*]
[1] *College of Engineering, China Agricultural University, Beijing, China, 100083*
[2] *College of Information and Electrical Engineering, China Agricultural University, Beijing, China, 100083*
* *Corresponding author, Address: P.O. Box 46, College of Engineering, China Agricultural University, 17 Tsinghua East Road, Beijing, 100083, P. R. China, Tel: +86-10-62737631, Fax: +86-10-62737300, Email: lhwen@cau.edu.cn*

Abstract: This paper analyses the status of rural informatization of China from different perspectives, including China's development timeline, info-platform, resources, geography, and income; and puts forward some recommendations of development as follows: (1) Strengthening government promotion and infrastructure construction. (2) Increasing optimism and confidence of farmers. (3) Providing training and facilitating discussion. (4) Solving localization issues. (5) Providing satisfactory goods and services.

Keywords: rural informatization, digital divide, information and communication technologies, ICT, recommendation

1. INTRODUCTION

Global history has shown that science and technology are the main drivers of social and economic development. In the past two decades, information and communication technologies (hereafter abbreviated as ICT) have made rapid progress, which has resulted in science and technology quickly becoming widespread (Wang, 2006). Currently, traditional agriculture is undergoing a process of modernization, the development and dissemination of rural information and the application of ICT in the agricultural field can not only improve rural conditions, but also raise the level of agriculture currently being practiced.

Yuan, J., Li, D. and Li, H., 2008, in IFIP International Federation for Information Processing, Volume 258; Computer and Computing Technologies in Agriculture, Vol. 1; Daoliang Li; (Boston: Springer), pp. 449–456.

From the beginning of the 21st century, ICT have become more common and expanded around the globe. On one hand, computers are more readily available to the ordinary person and the costs (table 1) of computers are declining rapidly (Singh, 2003). On the other hand, the imbalance of the information network development has enlarged the digital divide, which not only exits between developed countries and developing countries, but also between urban and rural areas in developing countries (Li, 2006). Currently, many poor families or/and minority families are living on the less fortunate side of the divide.

Table 1. Falling Costs of Computing (dollars)

Costs of computing	1970	1999
1 Mhz of processing power	7 601	0.17
1 megabit of storage	5 257	0.17
1 trillion bits sent	150 000	0.12

2. THE STATUS OF RURAL INFORMATIZATION IN CHINA

China is now one of the most important developing countries in the world. The nation's rural informatization has been developing for over 20 years, but there are characteristics of China's rural informatization that must be taken into account, such as late entry, rapid development, unbalanced disemination and so on.

From a time perspective, rural informatization of China is currently 10 to 20 years behind that of developed countries, but its progress has been made at a particularly rapid rate. In the 1980s, computer technology was applied in the field of agriculture, and subsequent to the 1990s, agricultural informatization systems were established and rural informatization developed quickly. For instance, the "Golden Agriculture Project," agricultural websites and national agricultural informatization systems were promoted and established by the Ministry of Agriculture and other agricultural institutes at this time. After the turn of the 21st century, developing rural informatization was listed in the Outline of the Tenth Five-Year Plan for National Economic & Social Development of the People's Republic of China and additionally in the Eleventh Five-Year Plan.

From an info-platform perspective, the government has invested a large amount of funds to construct the info-platform, and has established the tele-communication network with relatively advanced technology. Today, the number of both fixed line telephones and mobile phones in China is the largest in the world, and the number of people with access to the internet has reached 123 million on the Chinese mainland (Tong, 2006). According to the Chinese Ministry of Information & Industry, by implementing the "Village Coverage Project," the volume of telephones in rural areas increased rapidly. By the end of 2005, there were 11 provinces and cities where every basic village could be reached by telephone, and 97.1% of all villages in the entire country could be reached by telephone. By the end of 2006, the proportion increased to 98.9%, with 24 provinces and cities where every basic village could be reached by telephone (Yao, 2007). According to the China Statistical Report (NBSC, 2001–2007), the telephone volume per 100 Chinese people and the number of rural fixed line telephone subscribers has increased quickly from the year 2000 (see fig. 1 and fig. 2). Additionally, statistical information from the State Administration of Radio Film and Television shows that, after the extension of the "Village Coverage Project," the proportion of towns and villages covered by radio and television broadcasts is respectively 99.16% and 96.43%.

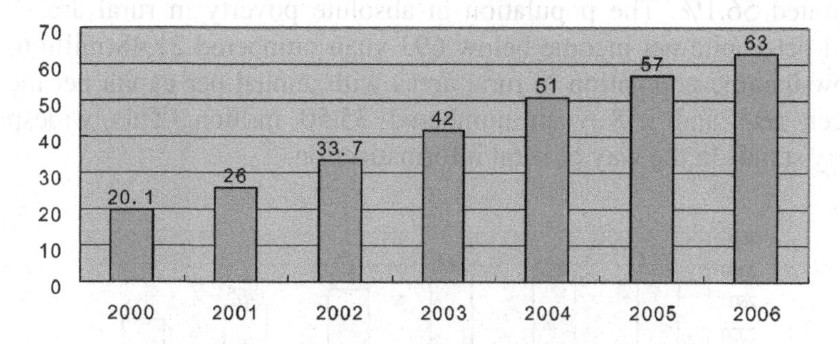

Fig. 1. Telephone volume per 100 Chinese people

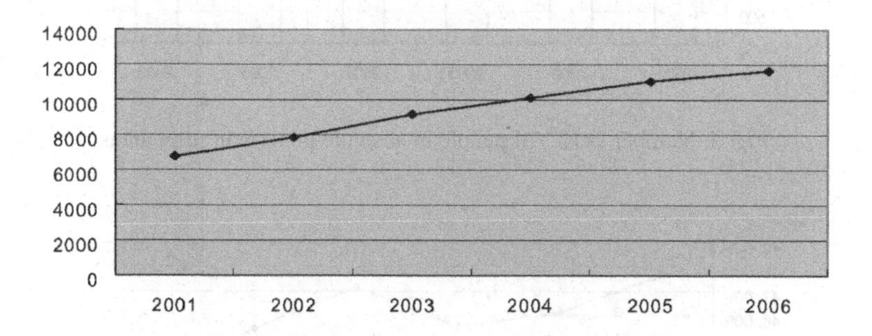

Fig. 2. Number ($\times 10^4$) of rural fixed line telephone subscribers

From a resource perspective, rural or agricultural informatization resources are consistently deficient and many resources are out of date. The 17th statistical report of CNNIC shows that there were 694,200 websites in China by Dec. 31, 2005 with only 0.66% of the total related to agriculture and rural communities.

From a geographic perspective (NBSC, 2000–2007), the majority of agricultural websites and information service stations are mostly based in Beijing and some coastal regions, with only a few located in China's western provinces. The total proportion of agricultural websites in five areas (including Beijing, Shandong, Zhejiang, Jiangsu, and Guangdong) exceeds more than 50% of the total number in the whole country, while in Yunnan, Gansu and other western provinces, the quantity of agricultural websites and libraries is considerably less (Wang, 2006; Wan et al., 2006).

From an income perspective, the proportion of rural population to total population is very high, and the degree of modernization is low. In fig. 4, the decreasing trend of the Engel coefficient for rural households is obvious, while fig. 3 shows a stable reduction in numbers of rural people in absolute poverty. However, absolute poverty and low income are still important problems (NBSC, 2001–2007). For instance, at the end of 2006, the total Chinese population reached 1,314.48 million, of which the rural population constituted 56.1%. The population in absolute poverty in rural areas with annual per capita net income below 693 yuan numbered 21.48 million, and the low-income population in rural areas with annual per capita net income between 694 and 958 yuan numbered 35.50 million. This widespread poverty stands in the way of rural informatization.

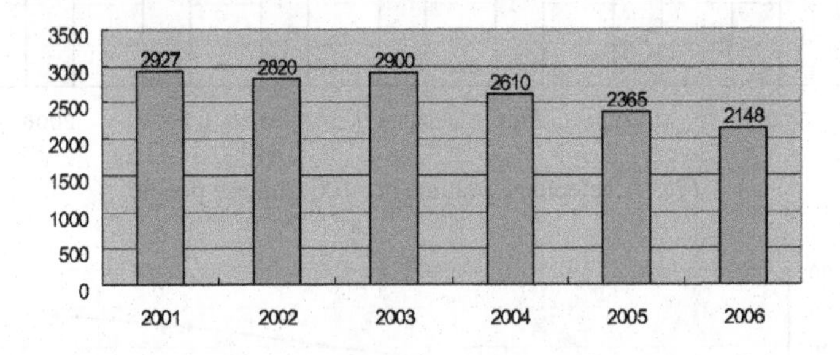

Fig. 3. Number (×10^4) of people in absolute poverty in rural areas

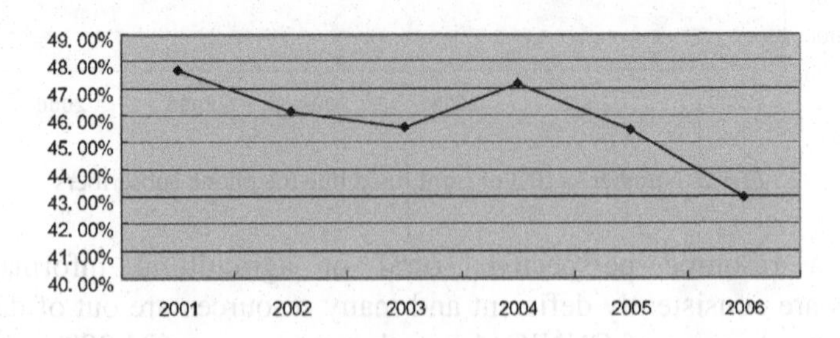

Fig. 4. Engel coefficient for rural households

3. RECOMMENDATIONS

After analyzing the status of developing rural informatization in China, there are some simple recommendations put forward in order to accelerate further rural informatization.

3.1 Government promotion and infrastructure construction

In the construction of villages and dissemination of rural informatization, the government's role as leader should be the provision of funds, development of legislation, regulation of inter-departmental relations, and construction of infrastructure and so on. Bridging the digital divide and solving "the last mile" requires the leadership of government (Guo et al., 2006).

3.2 Optimism and confidence

Although developing quickly in the past twenty years, the internet is still very young. The internet is becoming increasingly more established and individuals around the world are learning to use the web's many functions. For people on the wrong side of the technology tracks, the government and non-profit organizations should provide residents of rural and remote areas with more training opportunities. It is important for them to relinquish doubts and fears and take part in training programs with confidence and optimism.

3.3 Training and discussion

The internet is accessible to everyone, irrespective of age, gender, religion, income, nationality and level of literacy. Training is a key process to assist disadvantaged farmers. After training, farmers should be able to master sufficient expression skills to be able to describe problems clearly and obtain detailed explanations from experts. Without relevant training, farmers are unsure how to put forward a problem, or problems are outlined in terminology that is too general and ambiguous for experts to understand and resolve.

A typical example (Singh, 2003) shows the necessity of training. Before training, experts received this question from a 32-year old resident of a village: "I observed flowers dropping in my castor field, please advise me." The advice of experts to this situation was "We need adequate information to understand the problem." After training, the same question was repeated as follows: "In the 3-month-old castor crop on my 4 acres of land, I have observed two kinds of flowers, red and green; only the red ones turned into fruit while the green flowers fell from the plants. Please advise me." To this more detailed question, the advice of experts was: "Green flowers are male flowers. After fertilization, male flowers fall off and the red female flowers turn into fruit. This is natural and there is no need for taking any measure." (Dileep et al., 2006).

3.4 Solution of localization issues

China is a large country with a vast territory, comprising 56 ethnic groups with their own language or dialect. Localization is a serious issue in the spreading of agricultural informatization, because local names vary from one location to another even within a province. Experts often use scientific names in their discussions (Guntuku et al., 2006). How can this problem be solved? There are two effective measures. One is to popularize agricultural knowledge in all rural areas, which is a long-term task, and the other is to train local personnel at an information station and broadcast information downloaded from the internet to the rural population in the local language, which will quickly solve some problems to a certain extent.

3.5 Satisfactory goods and service

According to American researchers, the consumer will disseminate his/her dissatisfaction at an astonishing speed when he/she is not content with the goods or services provided by enterprises, government or organizations (Chen, 2004). Fig. 5 describes the process of dissemination. For every consumer who expresses his/her discontentment to the providers of the goods or services, there are 26 dissatisfied consumers who remain silent. However, each of the 26 dissatisfied consumers will transmit his/her discontentment to 10 other people, and one-third of each of those 10 will transmit the information to a further 20 people.

In reality, only one consumer is dissatisfied with the good or service but, in fact, many more people become potential consumers affected by negative information. The detailed calculation is as follows:

Level 1 $1+26 = 27$

Level 2 $26\times10 = 260$

Level 3 $1/3\times10\times26\times20 = 1733$

In level 1, there are 27 dissatisfied consumers; in level 2, there are 260 potential consumers affected; and in level 3, there are 1733 potential consumers affected.

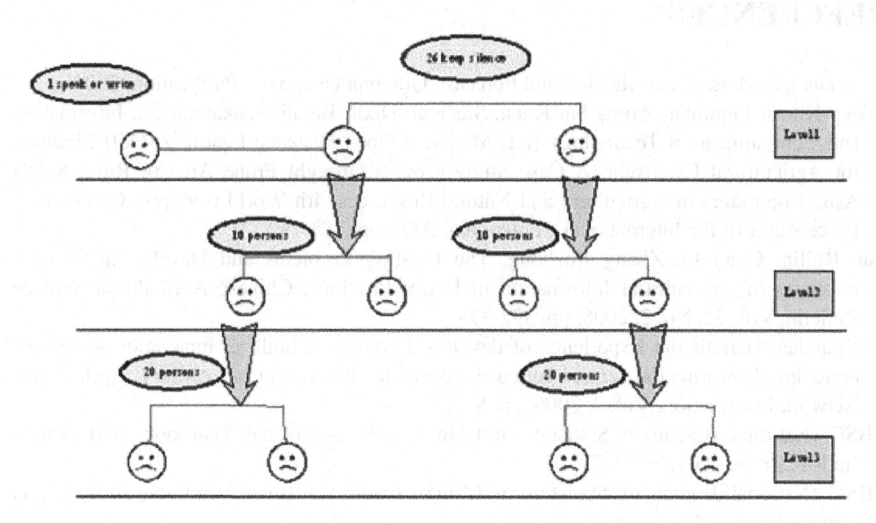

Fig. 5. Dissemination of dissatisfaction

During the process of developing rural informatization, residents in rural and remote areas are sensitive consumers. High quality facilities and high-grade services will attract consumers step by step, and inferior production will hinder the development of rural informatization.

4. CONCLUSION

The aim of developing rural informatization is to improve agricultural yields and improve rural conditions. There are many differences between different areas; it is impossible and unnecessary to purchase a computer for every resident of rural and remote areas. Which model and which technology should be adapted? Although a "one-size-fits-all" solution does not exist: different circumstances may dictate different choices, the following measures are important to develop agricultural and rural informatization: strengthening government promotion and constructing infrastructure, increasing the optimism and confidence of farmers, providing training and facilitating discussion and providing satisfactory goods and services to farmers.

ACKNOWLEDGMENTS

The author wishes to thank Ms. Robyn Chesney and Mr. Wang Qingjie to revise and polish this paper.

REFERENCES

Chen Diange, Market Investigation and Forecast, Qinghua University Publication, 2004.

Dileep Kumar Guntuku, Aruna Sai Kuna, Sreenath Dixit, Balaji Venkataraman. Information and Communication Technology (Ict) Mediated Open Distance Learning (Odl) Methods for Agricultural Extension: A Case Study From a Drought Prone Area of Rural South Asia, Computers in Agriculture and Natural Resources, 4th World Congress Conference, Proceedings of the International Conference, 2006, pp. 188-192.

Guo Ruilin, Guan Li, Zhang Jinzhong, The Existing Problems and Developing Counter measures of Agricultural Information in Henan Province, Chinese Agricultural Science Bulletin, Vol. 22, No. 2, 2006: pp. 392-394.

Li Xianglan, Benefit the Experience of developed countries, built an innovation system of agricultural information service suited the concrete situation in our country, Agricultural Network Information, Vol. 2, 2006: pp. 8-12.

NBSC (National Bureau of Statistics of China), China statistical Yearbook 2001, China statistic Press, 2001.

NBSC (National Bureau of Statistics of China), China statistical Yearbook 2002, China statistic Press, 2002.

NBSC (National Bureau of Statistics of China), China statistical Yearbook 2003, China statistic Press, 2003.

NBSC (National Bureau of Statistics of China), China statistical Yearbook 2004, China statistic Press, 2004.

NBSC (National Bureau of Statistics of China), China statistical Yearbook 2005, China statistic Press, 2005.

NBSC (National Bureau of Statistics of China), China statistical Yearbook 2006, China statistic Press, 2006.

NBSC (National Bureau of Statistics of China), China statistical Yearbook 2007, China statistic Press, 2007.

Nirvikar Singh, India's information technology sector: what contribution to broader economic development, OECD Development center, (Working Paper No. 207), 2003.

Tong Peijie, The Gap of Internet is Shrinking, China Computer Users, Vol. 28, 2006: pp. 13.

Wan Honghui, Wang Junhui, Feng Lu, Study on Difficulties Faced in Agricultural Information Service in Yunnan Province and its Development Counter Measures, Journal of Library and Information Sciences in Agriculture, Vol. 18, No. 01, 2006: pp. 39-41.

Wang Yuzhen, Research on Some Problems of Countryside's informatization in Gansu Province, Agricultural Network Information, Vol. 1, 2006: pp. 37-39.

Yao chunhua, Analysis on the Application and development of VSAT in Village Coverage Project, 2007 http://www.cww.net.cn/article/article.asp?id=76607&bid=2816

STUDY ON SUITABILITY EVALUATION MODEL OF NEW MAIZE VARIETIES

Weili Wang, Xiaodong Zhang*, Xiao He
College of Information and Electrical Engineering, China Agricultural University, Beijing, China, 100083
** Corresponding author, Address: College of Information and Electrical Engineering, China Agricultural University, 17 Tsinghua East Road, Beijing, 100083, P. R. China, Tel: +86-10-62737628, Email: zhangxd@cau.edu.cn*

Abstract: Seed industry is an important basic industry in agricultural, an emerging industry as well. However, the new varieties are driving force and providing source for seed industry development. How extending the new varieties to the most appropriate place are important issues that seed industry needs to address. East of North China as an example, the county for the evaluation unit, on the basis of average daily meteorological data of all meteorological stations in the 50 years and the test data of new varieties of maize in the district, this paper discussed the suitability evaluation model of new maize varieties based on the characteristic values of new maize varieties (temperature, plant diseases and insect pests, lodging and yield) and the corresponding characteristic values of the environment; besides, quantifies the relationship between all characteristic values and their effect on yield; finally, recommends the region that new maize varieties suit and the varieties that fit for regions, providing the premise basis for maize varieties matching.

Keywords: new maize varieties suitability evaluation

1. INTRODUCTION

Seed is the most basic, indispensable and irreplaceable agricultural production and the internal of yield, the carrier of various production technologies as well. Therefore, requirements of improved varieties greatly

Wang, W., Zhang, X. and He, X., 2008, in IFIP International Federation for Information Processing, Volume 258; Computer and Computing Technologies in Agriculture, Vol. 1; Daoliang Li; (Boston: Springer), pp. 457–470.

exceed other productions. Especially the high-yield, high-quality and high-efficiency agriculture establishing, farmers urgently call for stronger competitiveness of the varieties. The dominant of seed production is new varieties. New varieties can be used in production, transformed into real productive forces after regional test, validation and promotion. Vigorously promoting the excellent new varieties of crops is to promote the development of agricultural technology in the most effective measures and the important guarantee.

At present, as a member of WTO, There are higher demands on the corn seed industry development because of international economic integration in China. Promotion of new varieties of maize, as the focus of seed industry development, we must make improvements. However, the current agricultural production, cultivation of maize, some did not fully take into account the suitability and stability, blindly promoted based on past experience and resulted in undue losses. So, we need to create a best suitability assessment model of maize varieties, quantitative calculate the extent appropriate, reduce the choice to promotion region blindness and subjectivity and the promoting risk, improve the average corn yield, and adapt seed market for the new competitive situation.

2. STUDIES ON SUITABILITY EVALUATION MODEL OF NEW MAIZE VARIETIES

New maize varieties suitability evaluation can be divided into two types: (1) Given a new maize variety, study its suitability for promotion of the region and quantify the appropriate extent. (2) Given a study area, determine which new varieties of maize in the region fit, while which not, quantify the appropriate extent as well. The appropriate level will be measured by the average yield per acre. The second method was adopted in the paper, with east of north China as the study area.

2.1 Determining evaluation factors

There are many factors in judging whether a new maize variety suits the study area. Including climate, pests, soil and so on (Liao, 2002; Li, 2002; Liu, 2003; Huang, 2004; Zhang, 2006). This paper mainly discussed the accumulated temperature (Yu, 1997; Wang, 1997), pests and lodging that will impact on the suitability. Thus, evaluation factors include: temperature (°C), big spot disease(level), small spot disease(level), bending fungus(level), gray leaf spot disease(rank), black silk(%), smut(%), stem rot(%), sheath blight(level), MDM(level), rough dwarf virus(%), corn borer(%), lodging rate(%) Weights of each evaluation factor.

The first condition that new maize varieties promotion should be satisfied with is that no less than 10°C environmental temperature will meet the required temperature during the growth period. Once the environmental temperature does not meet, 100% losses will be resulted in. Therefore, the counties can not participate in suitability evaluation if the varieties are not met all the required temperature in growth period in the counties. Since the suitability evaluation is carrying out according to the yield per acre in the end, the weights of each factor are determined by the extent affecting on the yield, using V to stand for.

Kinds of diseases and pests and lodging in the event of a serious will cause the production and the degree of reduction not the same. Put the loss of production rate caused by the most serious reduction as weight of a factor. Such as, big spot in the event of a serious disease in the year, susceptible output maybe about 50%. Thus, put 50% as a big stain on the weight. By analogy, all factors affecting the weight, as shown in table 1:

Table 1. Factors and corresponding weights

Factors	Temperature	Big spot	Small spot	Bending fungus	Gray spot
Weight (V)	1	0.5	0.6	0.6	0.5
Factors	Black silk	Corn borer	smut	Stem rot	Sheath blight
Weight (V)	0.7	0.2	0.3	0.5	0.2
Factors	MDM	Rough dwarf virus	lodging		
Weight (V)	0.1	0.5	0.4		

2.2 Establishing the evaluation factor membership function

2.2.1 Accumulated temperature membership function

According to the required accumulated temperature under new maize varieties growing period and counties no less than 10°C accumulated data in east of north China. Based on the following formula we can calculate temperature suitability of a certain new variety

$$S = \frac{T^*_{sum}}{T_{sum}} \quad (1)$$

T_{sum} is the no less than 10°C accumulated temperature of region in east of north China. T^*_{sum} is the required accumulated temperature under growing period. Under the appropriate size, the results will be divided into appropriate and inappropriate. Classification as shown in table 2:

Table 2. Suitability classification

Level	Suitability
One grade (inappropriate)	>1
Two grade (appropriate)	≤1

S>1, illustrates that the provided temperature does not satisfy with the required data, so it should not promote in this environment. While S ≤ 1, is on the contrary. However, the value of suitability degree is not the same. The value of production was affected by the difference in size. We set the true value of S when S∈ [0.9, 1.0], while 0.9 when S<0.9. This is because when S∈ [0.9, 1.0], environmental temperature can be sufficient for maize varieties growth period. When the large accumulated to a certain extent, more temperature does not engage in other activities in the agricultural. Thus, a formula can be expressed in quantitative the effect upon yield as follow:

$$T = 1 - S \begin{cases} \text{S true value}, & S \in [0.9, 1.0] \\ S = 0.9, & S < 0.9 \end{cases} \tag{2}$$

From this we can see that the smaller the value, the more suitable it is within the scope of 0.9 to 1.0, for which the waste temperature will be less, and the maize varieties mature in the shortest time under the conditions of required temperature.

2.2.2 Lodging and diseases and pests membership function

The evaluation of diseases and pests can be described of two aspects from the qualitative and quantitative. From the qualitative evaluation, four conditions exist as follows:

(1) A resistance of the corn varieties in a low stress environment promotion.

(2) A resistance of the corn varieties in a high stress environment to promote.

(3) Not an anti-corn variety in a low stress environment promotion.

(4) A resistance of the corn varieties not in a high stress environment to promote.

While, from the quantitative evaluation, the relationship between varieties susceptible reliability and environmental stress will be researched and established. This paper attempted to quantitatively evaluate the impact of pest on yield. The formulas can be expressed as follows:

$$\begin{cases} K_i = \dfrac{\overline{D}_i - \hat{D}_i}{D^*_{\max}} \quad (i=1, 2\ldots, 12) \quad R^2 \neq 0 \\ K = -0.01 \text{ or } -0.11 \qquad R^2 = 0 \end{cases} \tag{3}$$

$$\hat{D}_i = W_1 * D_i^* + W_2 * D^*_{i\max} \tag{4}$$

$$D_i^* = a + bx \quad W_1 = R^2 \quad W_2 = 1 - R^2 \tag{5}$$

In formula 5, x is the stress degree of diseases and pests and lodging rate of each county, D^* is stress degree of diseases and pests of some variety in different environment, D_i^* is a linear function about environmental stress, x, D_i^* is a number belonging to [0, 9] or [0, 100] at random, R^2 is a correlation coefficient of the linear function.

In formula 4, D_i^* which amended by the management pessimistic act, represents the forecast of degree of the maize varieties flu disease, a function about environmental stress as well. $D^*_{i\max}$ is the maximum value of a variety of lesions to the disease, Taking into account R^2 of D_i^* smaller, $\hat{D}_i$ is a method when the diseases and pests sensitivity and the degree of environmental stress linear fit is not so good. When R^2 small, only illustrates the extent of fitting not good, and lack of regularity, however, could not explain the resistance of the variety. Then pessimistic assumption $\hat{D}_i$ is mainly decided by the maximum value. On the opposite side, is mainly decided by D_i^*. W_1, W_2 are the respectable weights of D_i^* and $D^*_{i\max}$. When R^2 of D_i^* larger, the weight of D_i^* is larger, otherwise, smaller.

Formula 3 is on behalf of that diseases of a certain type of a variety increase or reduce production levels. i is the type of the diseases and pests and lodging (such as big spot disease, lodging), $\overline{D}_i$ is an average value of a variety of disease extent, the average level of a variety in the study area susceptible also. If $(\overline{D}_i - \hat{D}_i)$ >0, then it expressed the susceptible of the variety in the region less than the average value of the variety, which means the variety is more appropriate in the region to promote, and increases the production, vice versa.

2.3 Model establish

Suitability evaluation of new maize varieties is based on the average yield per acre under climatic conditions, lodging and diseases, pests in the end (Tong, 1997; Wang 2006). Therefore, we need to forecast the new varieties of maize yield per acre after establishing accumulated temperature, lodging and diseases and pests membership function, finally, ascertain the suitability evaluation model of new maize varieties.

2.3.1 Method of forecasting new maize varieties average yield per acre

The method is similar to the method of calculating diseases and pests susceptible. The formulas can be expressed as follows:

$$\hat{C}_i = W_1 * C_i^* + W_2 * C_{i\min}^* \quad (6)$$

$$C_i^* = a + bx \quad W_1 = R^2 \quad W_2 = 1 - R^2 \quad (7)$$

Formula 7 is a linear function based on the environmental average yield per acre within east of north China trial regional and average yield per acre of a variety, R^2 is larger, the fitting of the linear equation is better. x is environmental average yield per acre in each county.

In formula 6, C_i^* which amended by the management pessimistic act, represents a forecast of the maize varieties average yield per acre in every counties, a function on environmental average yield per acre, $C_{i\min}^*$ is the minimum average yield. Pessimistic assumption that when R^2 smaller, that is the extent of fitting not well, the weight of the lowest average yield is larger. W_1, W_2 are the respectable weights of C_i^* and $C_{i\min}^*$.

2.3.2 Suitability evaluation model of new maize varieties

According to the impact factors, the weights of each factors and the membership function, we can establish the evaluation model, which can be used to compare the appropriate level of a maize variety in every county, can be compared the suitable degree of new maize varieties in a particular county. The suitable degree is measured by average yield per acre. The higher the average yield, the better appropriate in the county, the better promotion. Comprehensive above formula, we can get the suitability evaluation model of new maize varieties, as shown below:

$$\hat{Y} = \hat{C} * \left(1 + T + \sum_{i=1}^{12} K_i * V_i\right) \quad \text{(i=1, 2... 12)} \qquad (8)$$

In formula 8, $\hat{C}$ is the average yield per acre of a variety in one county by environmental average yield per acre of every counties in east of north China. Calculation is Similar to K_i, $\hat{Y}$ is average yield per acre forecasting by considering temperature, lodging, diseases and pests. Thus, we can judge a variety suitable for which counties and not by the level of forecasting value, meanwhile, the appropriate level can be judged as well. Different varieties comparison, the model can calculate the average yield per acre level, with a judgment in which the county more suitable.

3. SUITABILITY EVALUATION

Take the new variety 628412 and HAOYU12 for reference, we can judge which will be more appropriate in the east of north China through the suitability evaluation model of new maize varieties. Take the 628412 suitability evaluation as example:

3.1 Introduction to 628412

In 2005, the production of the region test was 760.0kg per acre; increasing 17.70% compared to NONGDA108, The effect was remarkable, occupied 1st, 22 fields increased, and didn't have decreased fields.

The average growth period of the variety is 127 days, earlier 4 days than NONGDA108. The height of the adult plants is about 330 centimeters. The height of the tassel is 120 centimeters, and the length of the tassel is 19.8 centimeters, 16–18 rows. The weight of the single tassel is 245g. The weight of one hundred corn kernels is 39.0g. It resists various diseases and pests through appraising. If diseases and pests appear in the fields, it resists all the plant diseases.

3.2 Process of 628412 evaluation

3.2.1 Calculation of the suitability of the accumulated temperature

The accumulated temperature of growth period of 628412 requires 2434.2 °C. So we can get the suitability of the accumulated temperature of 628412 in the new maize varieties promotion fields in the 223 counties of four provinces in the east of north China according to the method of the suitability of the accumulated temperature. As shown in figure 1:

	A	B	C	D	E	F
1	Counties	ID	Environmental	Growth	Result	Suitabil
2	ChangTu	211224	3025.5	2434.2	0.8045612	0.9
3	HuaDian	220282	3025.5	2434.2	0.8045612	0.9
4	XiFeng	211223	2922.2	2434.2	0.8330025	0.9
5	KangPing	210123	3094.4	2434.2	0.7866468	0.9
6	KaiYuan	211282	3010.8	2434.2	0.8084894	0.9
7	ZhangWu	210922	3149.9	2434.2	0.7727864	0.9
8	FaKu	210124	3129.7	2434.2	0.7777742	0.9
9	WeiChang	130828	2557.9	2434.2	0.95164	0.95
10	FuXin	210921	3217.9	2434.2	0.7564561	0.9
11	TieFa	211281	3108.8	2434.2	0.7830031	0.9
12	TieLing	211221	3088.5	2434.2	0.7881496	0.9
13	QingYuan	210423	2791.5	2434.2	0.8720043	0.9
14	TianPing	211322	3184.1	2434.2	0.764486	0.9

Figure 1. The accumulated temperature of 628412 in the 223 counties

According to the rank of the suitability of the accumulated temperature and the suitability of the 223 counties, 223 counties will be divided into two categories of appropriate and inappropriate, as shown in figure 2:

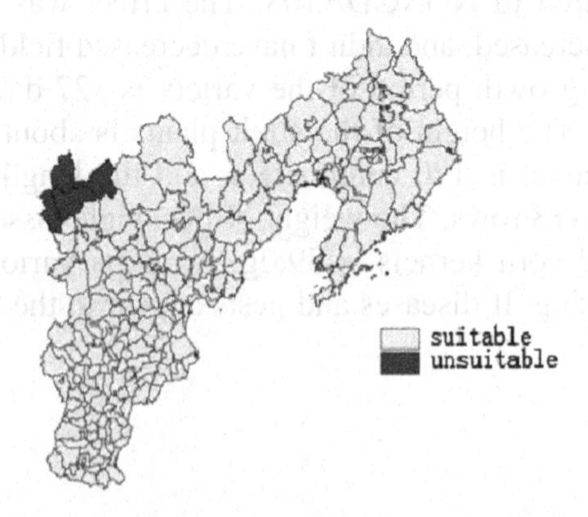

Figure 2. The map of accumulated temperature suitable counties of 628412 in the promotion fields (four provinces) in the east of north China

We can know from the figure, in the new maize varieties promotion regions of the east of north China, 5 counties are inappropriate. The other 218 counties are appropriate for promoting 628412. The inappropriate counties are Kang Bao County, Zhang Bei County, Shang Yi County, Gu Yuan County, and Ping Shan County. The data of inappropriate degree of these counties as shown in Table 3:

Table 3. 628412 inappropriate counties in the promotion fields in the east of north China

County	Kang Bao	Zhang Bei	Shang Yi	Gu Yuan	Ping Shan
Appropriate of accumulated temperature	1.12	1.04	1.09	1.04	1.04

3.2.2 Yield forecasting

The scatter gram on average yield per acre in trial areas of varieties and environmental average yield per acre is shown in figure 3.

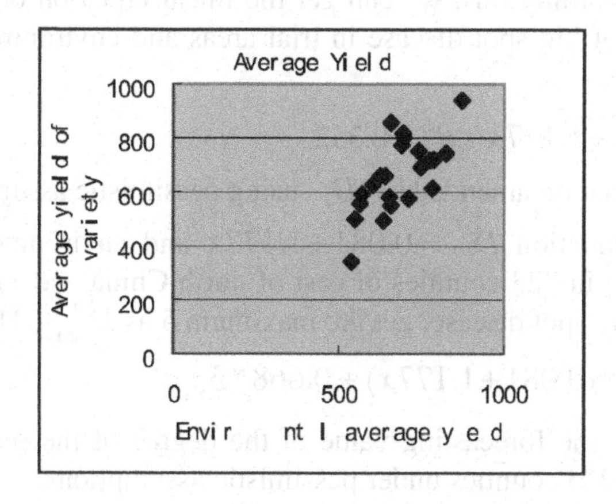

Figure 3. The relationship between 628412 average yield and environmental average yield

From figure 3, we can see the dots distribution showing a certain linear. According this, we can get the linear equation on varieties average yield per acre and environmental average yield per acre.

$$C^* = -37.462 + 1.023x \quad R^2 = 0.462$$

$$\hat{C} = 0.462 * (-37.462 + 1.023x) + 0.538 * 341.1$$

3.2.3 Big spot disease

(1) D_i^* linear equation fitness and R^2 solution

The scatter gram about the degree of the sensitivity of big spot disease in trial areas and environmental stress on big spot disease is shown in figure 4.

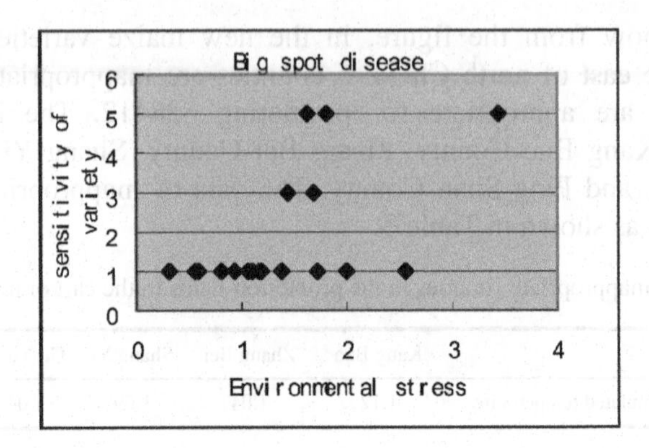

Figure 4. The relationship between 628412 sensitivity of big spot disease and environmental stress

From figure 4, we can see the dots distribution showing scatter, R^2 will be smaller. According this, we can get the linear equation on the degree of the sensitivity of big spot disease in trial areas and environmental stress on big spot disease.

$D_1^* = 0.081 + 1.177x \quad R^2 = 0.332$

(2) $\hat{D}_i$ calculation amended by D_i^* using pessimistic assumptions

By fitness equation $D_1^* = 0.081 + 1.177x$ and environmental stress on big spot disease in 223 counties of east of north China, we can calculate the sensitivity of big spot disease, get the maximum 5 as $D_{i\max}^*$, Thus:

$\hat{D}_1 = 0.332 * (0.081 + 1.177x) + 0.668 * 5$

For $\hat{D}_1$, it is the forecasting value of the degree of the sensitivity of big spot disease in 223 counties under pessimistic assumptions.

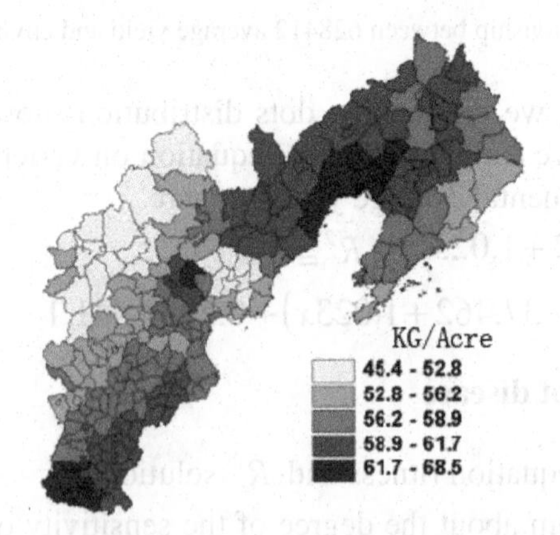

Figure 5. The impact of big spot on yield of 628412

(3) K calculation

We can get the impact of pest on yield by taking the average 1.861 as $\overline{D}_1$

$$K_1 = \frac{1.186 - \hat{D}_1}{9}$$

Similarly, we can calculate the impact of evaluation factors on yield for small spot disease, bending fungus, gray leaf spot disease, black silk, smut, stem rot, sheath blight, MDM, rough dwarf virus, corn borer and lodging rate.

3.3 Calculation by model

According to the formula 8, putting the output of 628412 variety evaluation process into evaluation model, it may be:

$$\hat{Y} = \hat{C} * \left(\begin{array}{l} (1-s)*1 + \frac{1.186 - \hat{D}_1}{9} * 0.5 + \frac{2.097 - \hat{D}_2}{9} * 0.6 + \frac{1.179 - \hat{D}_3}{9} * 0.6 + \frac{1.132 - \hat{D}_4}{9} \\ *0.5 + \frac{1.441 - \hat{D}_5}{100} * 0.7 - 0.01 * 0.3 + \frac{2.615 - \hat{D}_7}{100} * 0.5 + \frac{1.381 - \hat{D}_8}{9} * 0.2 - 0.111 * 0.1 \\ + \frac{1.333 - \hat{D}_{10}}{100} * 0.5 + \frac{1.514 - \hat{D}_{11}}{100} * 0.2 + \frac{3.788 - \hat{D}_{12}}{100} * 0.4 + 1 \end{array} \right).$$

There are 218 counties meeting temperature, so the calculated only carries on 218 counties. We can forecast the average yield per acre under the influence of the weather, pests, lodging integrated in every county, as shown in figure 6:

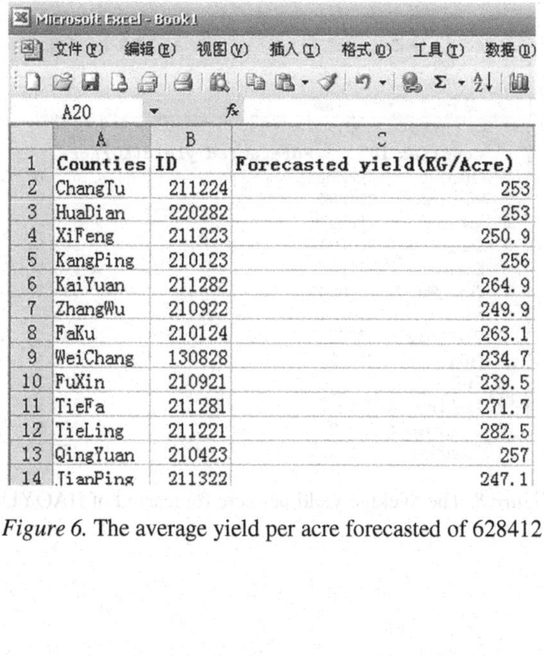

Counties	ID	Forecasted yield(KG/Acre)
ChangTu	211224	253
HuaDian	220282	253
XiFeng	211223	250.9
KangPing	210123	256
KaiYuan	211282	264.9
ZhangWu	210922	249.9
FaKu	210124	263.1
WeiChang	130828	234.7
FuXin	210921	239.5
TieFa	211281	271.7
TieLing	211221	282.5
QingYuan	210423	257
JianPing	211322	247.1

Figure 6. The average yield per acre forecasted of 628412

The higher average yield per acre forecasted the more suitable for 628412 promotions in this county. In the results map of suitability evaluation of 628412, we use different legend shapes and colors to distinguish the fitness level among them.

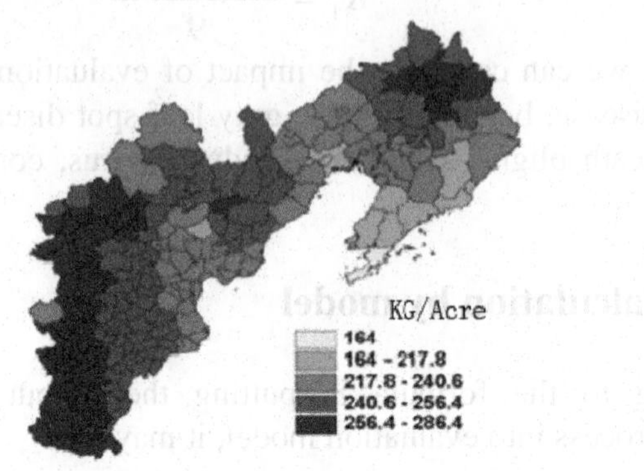

Figure 7. Distribution of average yield per acre forecasted of 628412 in suitable counties

In the same way, HAOYU12 is assessed. We can obtain the formula:

$$\hat{Y}=\hat{C}*\left(\begin{array}{l}(1-s)*1+\dfrac{1.518-\hat{D}_2}{9}*0.5+\dfrac{1.193-\hat{D}_2}{9}*0.6+\dfrac{1.721-\hat{D}_2}{9}*0.6+\dfrac{1.185-\hat{D}_2}{9}\\ *0.5+\dfrac{1.282-\hat{D}_2}{100}*0.7-0.01*0.3+\dfrac{1.363-\hat{D}_7}{100}*0.5+\dfrac{1.570-\hat{D}_8}{9}*0.2-0.111*0.1\\ +\dfrac{1.499-\hat{D}_{10}}{100}*0.5+\dfrac{1.517-\hat{D}_{11}}{100}*0.2+\dfrac{3.780-\hat{D}_{12}}{100}*0.4+1\end{array}\right).$$

Counties	ID	Forecasted yield(KG/Acre)
ChangTu	211224	355.6
HuaDian	220282	355.6
XiFeng	211223	347.3
KangPing	210123	360.5
KaiYuan	211282	375.6
ZhangWu	210922	356.8
FaKu	210124	376.9
WeiChang	130828	340.2
FuXin	210921	388.9
TieFa	211281	414.3
TieLing	211221	339.5
QingYuan	210423	351.5
TianPing	211322	425.3

Figure 8. The average yield per acre forecasted of HAOYU12

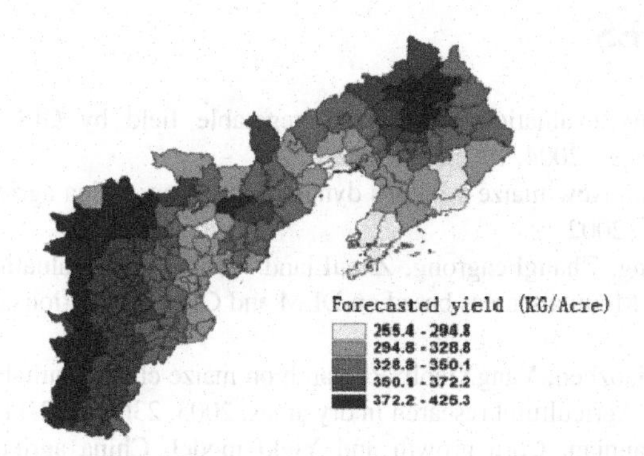

Figure 9. Distribution of average yield per acre forecasted of HAOYU12 in suitable counties

4. CONCLUSIONS

We can draw the following conclusions after the suitability evaluation of 628412 and HAOYU12:

(1) The average yield per acre forecasted of HAOYU12 is higher than 628412 in appropriate counties. It shows HAOYU12 is more suitable for promotion than 628412 in most parts of the east of north China.

(2) Lodging, big spot disease, small spot disease, gray leaf spot disease, sheath blight, bending fungus have greater impact on corn yield in east of north China. In other words, varieties of maize resisting the above diseases should be more suitable for promotion in the east of north China.

(3) The introduction of management of pessimistic assumptions to suitability evaluation model of new maize varieties avoids undervaluing a variety of inappropriate, provides a reliable guarantee to promote the new maize variety.

(4) The suitability evaluation model of new maize varieties can solve the traditional mode of promoted region choosing and make up for the deficiencies in agricultural expert system, assist in raising the accuracy and scientific choice. It is of great practical significance to new maize varieties promotion evaluation from the qualitative evaluation and experience to the quantitative model.

ACKNOWLEDGEMENTS

This research is funded by the Support of Science and Technology Project of the State, Programmed award No. 2006BAD10A01 from March 2007 to November 2009.

REFERENCES

HuangHe, Suitability evaluation of regional vegetable field by GIS, Fujian journal of agricultural sciences, 2004, 19(2): 108-112

LiaoQin, SunShixian, New maize varieties dynamic of China, China agricultural science and technology press. 2002

LiHong, SunDanfeng, ZhangFengrong, ZhouLiand. Suitability evaluation of fruit trees in Beijing Western Mountain areas based on DEM and GIS. Transactions of the CSAE, 2002, 18(5): 250-255

LiuMingChun, LiQiaozhen, YangXiaoli. Research on maize climate suitable for ecological of Gansu province, Agricultural research in dry areas, 2003, 23(3): 112-117

TongPinya, ChenYannian, Corn growth and Yield model. China agricultural science and technology press. 1997

WangBaotang, Skills on extending maize new variety, China seed industry. 2006(3):17-18

WangYongguang, AiWanxiu. Analysis and forecast of ≥10□ effective accumulated temperature in Northeast China, China agricultural meteorology, 1997, 3(18): 39-44

YuRonghuan, ShunMengmei, ZhangLijuan. The re-division about the quantity of heat resources and the accumulated temperature zone of Heilongjiang province, Natural sciences journal of Harbin normal university, 1997, 3(13): 98-102

ZhangJing, FengJinxia, BianXinmin, ZangMin. Variable weight approach in evaluation of corps ecological adaptability, Journal of Nanjing Agricultural University, 2006, 29(1): 13-17

HUMAN-COMPUTER INTERFACE DEVELOPMENT OF WIRELESS MONITORING SYSTEM BASED ON MINIGUI

Zhihua Diao [1], Chunjiang Zhao [1,2], Xiaojun Qiao [2,*], Cheng Wang [2], Gang Wu [1], Xin Zhang [2]

[1] *University of Science and Technology of China, China, 230024*

[2] *National Engineering Research Center for Information Technology in Agriculture, China, 100097*

* *Corresponding author, Address: Shuguang Huayuan Middle Road 11#, National Engineering Research Center for Information Technology in Agricultur, Beijing, China, 100097, Tel: +86-10-51503348, Fax: +86-10-51503449, Email: qiaoxj@nercita.org.cn*

Abstract: As the development of embedded GUI, the human-computer interface is more and more important in embedded system. In order to achieve simple operation and visual display, the better to achieve human-computer interaction, we use MiniGUI to develop human-computer interface of the wireless monitoring system. The hardware of the system includes the S3C2410X, TFT LCD and touch screen with four line resister, the relative software is ARM-Linux. This human-computer interface is used in the wireless monitoring system, provides a visible and friendly platform for customers, and runs steady in the real system.

Keywords: MiniGUI, Human-computer interface, Monitoring system, Wireless communications, Installation agriculture.

1. INTRODUCTION

As the constant development of the facilities agriculture, the requirements to the supervises-control system of agriculture are more and more high. In order to realize the human-computer interface better, set the system procedure and equipment more convenient, look into the system real time data immediately, manipulate correspond of control equipments, such

Diao, Z., Zhao, C., Qiao, X., Wang, C., Wu, G. and Zhang, X., 2008, in IFIP International Federation for Information Processing, Volume 258; Computer and Computing Technologies in Agriculture, Vol. 1; Daoliang Li; (Boston: Springer), pp. 471–477.

human-computer interface with some merits such as easy operation, reliable, and taking up a little resource, is required to realize the control of modernization, intelligence and precise. The user interface is the foundation of human-computer interaction; the user control and computer respond all carry out through the user interface. Unceasing development and wide application of embedded system have injected new vitality to agricultural fields. Embedded system develops from the initial systems in the form of single-chip as the core of the programmable controller to the present embedded system based on Internet. The present hardwares of embedded system include ARM, DSP, Power PC and so on, the operating systems running on them include ARM-Linux, ucLinux, uc/OS, Wince and so on (An et al., 2005). From the 90s of the 20th century, the embedded systems introduced graphical user interface support system, which runs on embedded Linux operating systems are Nano-X, OpenGUI, QT-Embedded and MiniGUI. Because of networking function components without proper adjustment, Nano-X has not too many ready-made application programs to be used. Because the kernel of OpenGUI is achieved by compiled language, its portability will be affected. QT-Embedded has high hardware requirements and the structure is too complex, is hard to be bottom expansion, customization and transplant.

MiniGUI is such a light open source graphic interface support system with the features of less resources occupancy, high performance, high reliability and configurable (Sun et al., 2005). It can not only run on the VxWork, pSOS, uc/OS, but also on the Linux operating environment which the kernel is 2.4 edition. We use ARM9 to realize this wireless monitoring system, and the operating system is ARM-Linux. We can conveniently set system program and parameters of the control devices, display Real-Time system working conditions and environmental information according to human-computer interface developed by MiniGUI. In order to satisfy the real-time requirements of system, MiniGUI is compiled to Threads mode.

2. SYSTEM HARDWARE STRUCTURE

In order to overcome the less coordination, wiring and maintenance disadvantages of the traditional monitoring system as a result of using PC-controller working mode and wire communications, we design the universal wireless monitoring system. It is a new type of monitoring system developed by the combination of wireless communication technology and information-processing technology, adopts monitoring center-front-end wireless module, its whole structure is as shown in Fig. 1. Monitoring center is the core of the entire system, realized by ARM9 system which core is S3C2410X. LCD is a 6.4-inch TFT-LCD display produced by yuan tai company, its type is PD064VT5, its pixel is 640X480. Monitoring center is responsible for receiving the required monitoring information from wireless modules, sending operational orders to wireless modules, to control the action of control devices linked to wireless modules. We use the existing module JN5121 as wireless modules, and the wireless technology is Zigbee. Wireless modules are deployed away from the monitoring center in the monitoring point, are responsible for completion of communication link to monitoring center, front-end monitoring information collection and responding to control orders that the monitoring center sends.

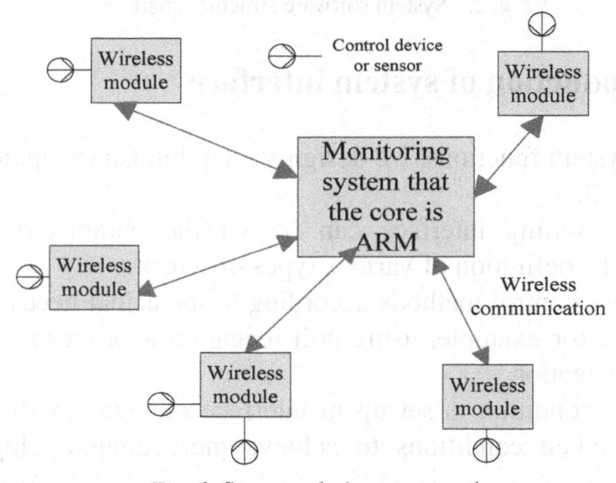

Fig. 1. System whole structure chart

System can not only be set the parameters of control devices and alarm parameters, but also be set logical condition and system operation programs. Then system will control the corresponding devices to work according to the setting program. System can not only store the setting parameters in databases, but also store the data that received from front-end modules. Meanwhile system also supports the control of the most primitive ways: manual control mode.

3. DESIGN OF SYSTEM INTERFACE

According to the system requirements, we design system software structure as shown in Fig. 2. Then we focus on the introduction, realization and development of system human-computer interface module below.

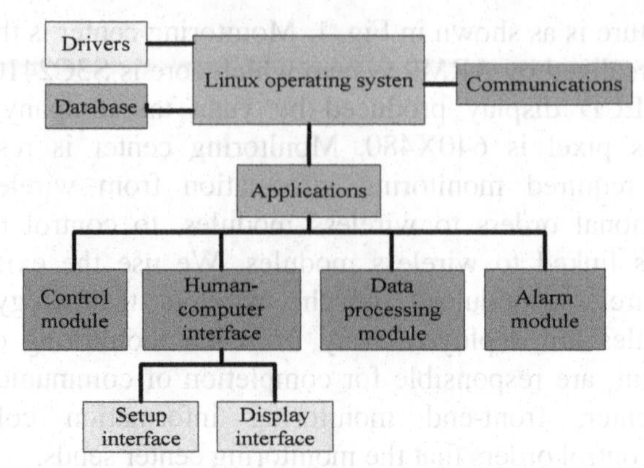

Fig. 2. System software structure chart

3.1 Introduction of system interface

To achieve system functions, we design system human-computer interface as shown in Fig. 3.

System input setting interface can be set the parameters of control equipment and the definition of various types of sensors.

We can choose control methods according to the actual needs in program setting interface, for example, using poll irrigation as a mean of achieving the purpose of irrigation.

Setting Logic condition is set up in interface of logic condition setting, besides using linked conditions to achieve more complex logic control conditions.

Alarm interface can be set up to achieve the output numbers and the setting of alarm time.

Display of real-time data interface can show real-time irrigation program working, sensor information and some information of control equipment.

Manual operation interface can be achieved by controlling the most primitive ways: manual control, including manual suspension and manual operation.

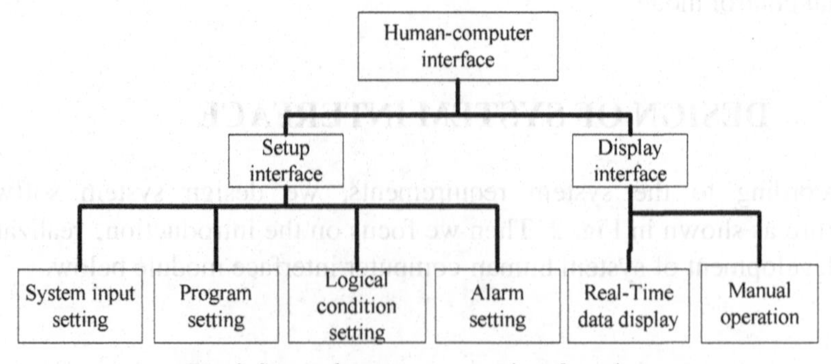

Fig. 3. System human-computer interface chart

3.2 Realization of system interface

MiniGUI provides three kinds of window, Main window, Dialogue box and Control window (Feynman, 2003).

Generally, every MiniGUI application should sets up a main window as the main-interface or start-interface. Main window usually includes some sub-windows, which are usually control window or user-defined window class, and the application can also create other types of windows, such as dialogue box and news box.

In MiniGUI, Dialogue box is a special kind of the main window, which just concerned communications with users – output information to users, but more are for input. It can be interpreted as a kind of main window after sub-classified.

Controls can be understood as the sub-window of the main window, and it acts as the main window. Not only receive the keyboard, mouse and other external input, but also could output in its region – were just all of its activities restricted in the main window.

In this wireless monitoring system, using Modal Dialog create MiniGUI main window as the main interface, and the six sub-pages which representatives of the functional modules all express with a single button in this main interface. All these buttons are taking the form of arrays in the main interface controls layout definition. When opening the main interface, and clicking a button, then a sub-interface will be pop up correspondingly, according to the sub-interface functions, all those setup of system input, irrigation program, real-time-data display and other operations can be carry out.

3.3 Development of system interface

MiniGUI is a graphical user interface support system, and the concepts of GUI programming are also applicable to MiniGUI programming, such as windows and other time-driven programming (Kang et al., 2006). The system is used in Threads model, which is based on messages and window management mechanism of POSIX thread, MiniGUI intercommunicates with outside through receiving messages. Messages are produced by system or application procedures, system produces message for input event and response of input event meanwhile. Applications could complete a particular task through message, or communicate with other application window. System send message to the process of application window which contains four parameters: window handle, message identifier and two 32-bit message parameters. Applications must promptly deal with the messages that delivered to its message queue, and the application generally handles the message queue news through a message cycle in MiniGUIMain function

(Feynman, 2003). Message cycle is the cycle of body. Programs obtain message from the message queue continuously using GetMessage function in the body, then send message to the specified window using DispatchMessage function, which is same to call the designated window process, and impart information and parameters.

The various functional modules program in this system is realized by modular design, and the interfaces of modules are realized in the form of model dialog. Meanwhile system uses information-driven mechanism to complete the realization of the system function. Callback function of the application procedures verdicts incidents occurred on the current interface based on receiving message identifier, then transfers the corresponding processing function to operate. There are six buttons totally in the main interface, and each button has an identifier. When the corresponding button is pressed, the system will receive message and send this message to main system process function, then call the corresponding functional modules and show the corresponding sub window based on different identifiers. The specific system main callback function is as follows:

```
static int MainwindowProc (HWND hDlg, int message, WPARAM wParam, LPARAM
lParam)
{ switch (message) {
    case MSG_INITDIALOG:    return 1;
    case MSG_COMMAND:  {
      int id = LOWORD(wParam);
      int nc = HIWORD(wParam);
if (id == IDC_SYSTEMINPUTSETBUTTON && nc == BN_CLICKED)
 { //enter into system input set sub interface when pushed
  DlgSysteminputset.controls = CtrlSysteminputset;
  DialogBoxIndirectParam (&DlgSysteminputset, hDlg, SysteminputsetProc, 0L);}
if (id == IDC_IRRIGATIONSETBUTTON && nc == BN_CLICKED)
{ //enter into procedure set sub interface when pushed
  DlgIrrigationset.controls = CtrlIrrigationset;
  DialogBoxIndirectParam (&DlgIrrigationset, hDlg, IrrigationsetProc, 0L);}
if(id==IDC_LOGICALCONDITIONSETBUTTON && nc == BN_CLICKED)
{ // enter into system logical condition set sub interface when pushed
DlgLogicalconditionset.controls=CtrlLogicalconditionset;
 DialogBoxIndirectParam(&DlgLogicalconditionset, hDlg, LogicalconditionsetProc, 0L);}
if (id == IDC_ALARMSETBUTTON && nc == BN_CLICKED)
{ // enter into alarm set sub interface when pushed
  DlgAlarmset.controls = CtrlAlarmset;
  DialogBoxIndirectParam (&DlgAlarmset, hDlg, AlarmsetProc, 0L);}
if (id == IDC_DATADISPLAYBUTTON && nc == BN_CLICKED)
{// enter into data display sub interface when pushed
   DlgDatadisplay.controls = CtrlDatadisplay;
   DialogBoxIndirectParam (&DlgDatadisplay, hDlg, DatadisplayProc, 0L);}
if (id == IDC_MANUALLYOPERATEBUTTON && nc == BN_CLICKED)
{ // enter into manual operation sub interface when pushed
DlgManuallyoperate.controls= CtrlManuallyoperate;
DialogBoxIndirectParam (&DlgManuallyoperate, hDlg, ManuallyoperateProc, 0L);}
```

4. CONCLUDING REMARKS

In this system, we have transplanted MiniGUI - 1.3.3 version to the ARM development transplant board successfully, received executable file using ARM-Linux-GCC compiler to cross compile system procedure on the PC. We have copied executable file to the development board and set it as startup operating procedure. Then system human-computer interface have run on the development board and worked well on the ARM-Linux operating system. In order to satisfy the characteristics of operating system easily and adapt it as the habit of operation of touch screen input, therefore, in the choice of man-machine interface controls, the use of button controls, frame composition controls, the drop-down controls will be adopted regularly to make the users operate conveniently. Based on the embedded system MiniGUI developed can achieve beautiful interface, complete functions, good real-time, better scalability and maintainability, and other advantages can be widely used in the development of man-machine interface of embedded system (Yang, 2004).

ACKNOWLEDGEMENTS

This research is supported by the National High Technology Research and Development Program of China (863 Program, Grant 2006AA10A311 and Grant 2006AA10Z253) and Beijing Science & Technology Program (Z0006321001391).

REFERENCES

Beijing Feynman Software Technology Co., Ltd., (2003). MiniGUI Program Guide For MiniGUI Version 1.3.x.

Beijing Feynman Software Technology Co., Ltd., (2003). MiniGUI User Manual For MiniGUI Version 1.3.x.

Chengjin An, Maoyang Sun, Li Po. 2005. Development of Graphical Interface Based on Embedded Linux System, Modern Electronics Technique, 20:108-113.

Jianwu Zhang, Zuopeng Xu, Dong Ping. 2006. Multi-serial Port Communication Based on MiniGUI-Threads and MiniGUI-Lite, Chinese Journal of Electron Devices, 3:985-988.

Shaohua Sun, Lizhong Xu. 2005. Graphics User Interface Based on Embedded Linux System, Microcomputer Development, 10:123-125.

Tieying Kang, Li Wei, Wengang Zheng etc. 2006. Human-computer Interface Design of Irrigation Controller Based on MiniGUI, Journal of Liaoning Normal University (Natural Science Edition), 1:46-48.

Xiaojun Yang. 2004. Research of Embed System Based on Linux in GUI, Modern Electronics Technique, 15:89-91.

Xiaoyan Zhao, Ma Qi. 2005. GUI Support System for Embedded System Based on Linux-MiniGUI, Computer and modernization, 1:10-12.

RESEARCH OF SLUICE MONITORING SYSTEM BASED ON GPRS AND PLC

Qiulan Wu [*], Yong Liang, Xia Geng, Wenjie Li, Yanling Li
School of Information Science and Engineering, Shandong Agricultural University, Taian, China, 271018
** Corresponding author, Address: School of Information Science and Engineering, Shandong Agricultural University, Taian, 271018, Tel: +86-0538-8249755, Fax: +86-0538-8249755, Email: wqlsdau@163.com*

Abstract: The method of sluice control applied to reservoir is executed manually in field at present. To fix this problem, sluice remote monitoring system based on GPRS and PLC is provided. Some advanced technologies used in system are discussed. The system structure and principle are introduced. The system realizes wireless network connections between local control unit and center control unit based on GPRS technology, and it provides remote monitoring of reservoir sluice with a new technique.

Keywords: General Packet Radio Service, Programmable Logic Control, Sluice, Remote Monitoring

1. INTRODUCTION

At present, the field manual operation or field automatic operation is applied to sluice control in XueYe reservoir. Whenever it is of the violent storm or the hot sun, the operator must operate system on the dam, and record the system movement condition. The operator records the open time, position, stop time, stop position, etc of the sluice, and regularly process the data and summarize the information. The manual mode can't satisfy the requirements of development.

With the development of control theory, communication, computer technology and network, putting remote measurement and control, wireless communication and network technologies into sluice's control is the main

Wu, Q., Liang, Y., Geng, X., Li, W. and Li, Y., 2008, in IFIP International Federation for Information Processing, Volume 258; Computer and Computing Technologies in Agriculture, Vol. 1; Daoliang Li; (Boston: Springer), pp. 479–485.

stream of the reservoir sluice control system. The monitoring way of sluice is changed from individual centralized structure control to the hierarchical structure control (Wu et al., 2002).

This paper proposes a sluice remote monitoring system based on GPRS (General Packet Radio Service) and PLC (Programmable Logic Control). Its hardware and work principle are illustrated. The system implements automatic measurement and control of ascending height of the sluice, manual and automatic control of headstock gear, collection, data process and transmission of real-time information. The remote automatic control of sluice is an important part of water conservancy modernization. It provides a better tool to save water and energy, to manage water resources and optionally dispatch accomplished automatic measurement and control of ascending height of the brake (Dai, 2002; Zhang et al., 2002; Han et al., 2003).

2. INTRODUCTION TO GPRS

GPRS transmits data using grouping exchange with high efficiency. GPRS mobile telecommunication network is formed by adding nodes to the public communication telecommunication network, SGSN (Server for GPRS Support Node) and GGSN (Gate GPRS Support Node). It adopts grouping data exchange and provides users connection in the form of mobile grouping IP or x.25. It has same frequency slot, bandwidth, retransmission structure, wireless modulation, frequency-hopping and data frame of TDMA as GSM. GPRS can be divided into two parts: wireless access and the main network. The wireless access part is responsible for communication between mobile node and BBS (Base Station Subsystem) while the main network is responsible for communication between BBS and router of the standard digital communication network.

The advantages of GPRS are:

(1) No extra network needed, only application to be user of the public mobile communication network.

(2) The coverage of the public communication network is huge and the public communication network is extendible.

(3) Fee can be base on the usage or on the monthly plan.

User can be online all the time and fee is charged according to the data amount. When there is no data transmission, even the user is online, no fee is charged. Or, user can pay a monthly fee without usage limitation.

(4) Transmission speed is high.

Grouping data exchange is the necessary condition for data transmission and enhancement of the user capacity.

Thus, wireless communication network making use of the public mobile communication network is low-cost, extendible, non-constraint, low-error and stable system with advancement and standards. GPRS is especially suitable for low data rate with high usage frequency communication like transformation of the data relevant to water quality (Wavecom Company, 2001).

3. INTRODUCTION TO PLC

PLC is a programmable controller. PLC storages instruction in the internals such as logical operator instructions, sequential control instructions, counting instructions and arithmetic operations instructions. PLC can control various types of mechanical equipment or production process through Digital or analog input and output. PLC has the function of network communication. It can exchange information between PLC and PLC, PLC and host computer, PLC and other intelligent devices. It can form an integrated, separately centralized control. Major PLC has the RS-232 interface and the interface which can support individual communication protocol.

PLC can adapt to the bad environment easier than single-chip processor and computer acting as filed controller. It has the advantages of convenient configuration, reliable remote communications and easy maintenance.

4. SYSTEM DESIGN

The layered distributed open architecture is adopted in the sluice Remote Monitoring System. The System consists of three parts: monitoring center, GPRS network and field controller (Fig. 1).

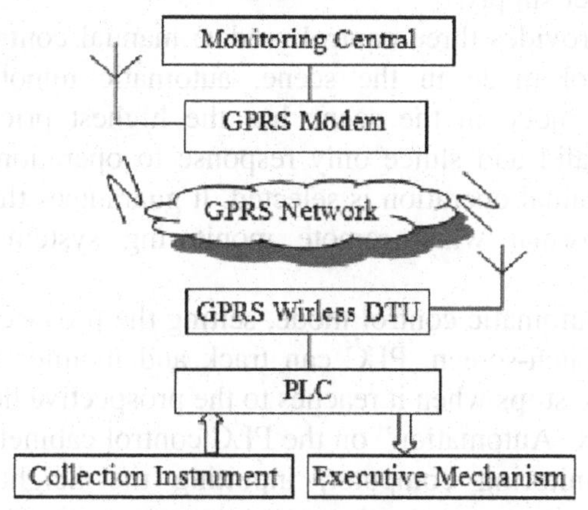

Fig. 1. Function architecture of system

4.1 Field Controller

The main part of the field controller is PLC which collects the information and controls sluice. The data collection instrument of Field controller collects the real-time information such as ascending height of the sluice, water level, temperature and Humidity of sluice chamber. The information of sluice is coded by the coding program in PLC. PLC forwards it to GPRS wireless DTU (Data Terminals Units) through the RS232 communication interface. The data is processed and packaged by Embedded Processor of GPRS wireless DTU, which is transmitted to the data processing and monitoring center through wireless GPRS network.

The Control part is the core of field controller. The control system consists of control procedure of the sluice, man-machine interface of the scene operation, warning devices etc. The system mainly includes PLC, intermediate relays, man-machine interface board (switches, buttons, signs, show devices, alarm devices).

The command coming from sluice monitoring software of monitoring center is transmitted to GPRS wireless DTU through GPRS wireless modem and then retransmitted to PLC through RS232 communication interface. The code of command is translated by translation software of PLC. Then PLC completes corresponding control operation through calling automatically corresponding sluice control program.

Electric control mainly consists of electric equipments such as air switch, AC contactor, and intermediate relay. Automatic (manual) dial switch on the control cabinet can shut off or connect two different circuits. Electric control unit has the function of multiple protections. Motor overload protector can automatically shut off power in order to protect equipment when motor is overload. Limit switch can automatically shut off power of headstock gear in order to protect motor when height of sluice is out of limit location. It can avoid running abnormally of motor because of disorder phase sequence of three-phase power supply.

The system provides three control modes: manual control mode in field, automatic control mode in the scene, automatic remote control mode. Manual control mode in the scene has the highest priority. The remote operation is invalid and sluice only response to operation buttons of PLC cabinet when manual operation is selected. It guarantees that the dispatch of the water is normal when remote monitoring system is failed or in maintenance.

On the filed automatic control mode, setting the prospective height of the sluice on the Touch-screen, PLC can track and monitor the height of the sluice. The sluice stops when it reaches to the prospective height.

Selecting "PLC Automation" on the PLC control cabinet and "remote" on the remote monitoring computer, inputting the height of sluice with keyboard, click increase or decrease buttons with mouse, the sluice automatically runs, and it can stop when it reaches the prospective height. At the same time the operation button is invalid on the PLC control cabinet. The ascending or descending process of the sluice is showed on the image surveillance computer in the monitoring center (Liu et al., 2004; Zhang et al., 2007; Shi et al., 2007; Zhang et al., 2003; Huang et al., 2003).

4.2 Monitoring Center

Monitoring Center consists of server, monitoring computer, manage computer and image surveillance computer.

The server of monitoring center applies for assigning fixed IP address, using of DDN special line provided by mobile communication company to connect with GPRS network. Because DDN special line could provide relative high bandwidth, when the number of data collecting sites increased, monitoring center could meet the need without enlarging the capacity.

After receiving the data transferred from GPRS network, the server will firstly verify, and then transfer the data to the main control computer. The system will restore and process the data.

The function of monitoring center as follows:

(1) Remote Control

Monitoring center can transfer ascending or descending command of sluice.

(2) Surveillance Running

The image surveillance computer browses the website embedded in the video server. The picture of sluice is showed on the website. Camera and scanner can be adjusted remotely.

(3) Statistics and Print

Do statistics on the running of sluice and various performance indicators; print various tables of events and operation.

(4) Accident Analysis and Fault Diagnosis

After accident, accident analysis and anti-accident measures can be done according to a list of accidents, operating records and accident record. At the same time online real-time diagnosis can be done such as the front monitoring diagnosis, communication interface diagnosis, network interface diagnosis and computer equipment diagnosis.

4.3 GPRS Network

The data transmission uses GPRS wireless network. GPRS network communicates with DTU through GPRS modem, a product from Hongdian Company in a single-point to multi-point manner. There are several water data sampling instruments whose data are packed into IP packages and transmitted to the data processing and monitoring center through GPRS wireless network. GPRS network assigns fixed IP addresses within information center, according to the IP address assigned DTU set up data pathway with the information center.

GPRS DTU based on GPRS network of Chinese mobile has advantages of higher reliability and better anti-interference ability. It can connect computer, RTU, PLC, GPS receiver, digital cameras and data terminals through RS232 interface. It can support high-speed data transmission up to 171Kbps. The function of remote diagnosis, testing and monitoring can meet the needs of data acquisition and control between control center and a large number of remote sites (Shenzhen Hongdian Technologies Co., Ltd. 2002).

5. SOFTWARE DESIGN

The software adopts modular design. It is easy to debug, modify, promotion and expand. The program module mainly includes automatic control module of sluice, the acquisition and process module of the sluice's height, the acquisition and process module of water level, the fault diagnosis and Analysis module etc.

6. CONCLUSIONS

After onsite simulation in irrigation district and hydropower plant of the reservoir, the system runs stably and dependably, and meets the requirement of remote monitoring. But it is still not satisfying for control of the flood-releasing sluice. The system needs further improvement.

The sluice remote monitoring system based on GPRS and PLC achieves automatic control of sluice. It can make the better use of water resource and improve the utilization rate of water energy. It is an important measurement to implement automatic and information China's water conservancy projects. It not only greatly enhances the automatic control capacity of sluice and reduces labor intensity and the number of duty, but also receives a lot of economical benefits.

ACKNOWLEDGEMENTS

This study has been funded by Shandong Water Conservancy Bureau. It is supported by technology special fund project of Shandong water resources (Number: 200357), and also supported by Shandong Agricultural University. Sincerely thanks are also due to XueYe reservoir for providing the data for this study.

REFERENCES

Dai Shangan 2001, The Control of Floodgates in Water Conservancy Projects, Computer & Digital Engineering, 29:1-6 (in Chinese).

Han Bing, Li Fenhua 2003, Application of GPRS technology in data acquisition and monitoring system, Electronic Technology, 19:26-29 (in Chinese).

Huang Xu, Zhao Ying 2003, Applications of PLC Programmable Controller and Supersonic Positioner for the Pressure-adjusting Well Gates Control of Hydroelectric Power Plant, Northeastern Electric Power Technology, 42-43 (in Chinese).

Liu Shubo, Xu Zhanguo 2004, Design and Realization of Sluice Automatic Control at the Head of Canal in Doushan Irrigation District, China Rural Water and Hydropower, 21-23 (in Chinese).

Shenzhen Hongdian Technologies Co., Ltd. User manual of H7000 GPRS Wirless DDN data terminal, 2002.

Shi Yitan, Gu Zhong bi 2007, Application of Simons S7-200 in the remote gate control system, Jilin Water Resources, 8-9 (in Chinese).

Wavecom company, GPRS User Guide, 2001

Wu Jianchun, Jiang Zhizhao 2002, Improvement of Water Intake Gate Control System of Generating Sets at Wuxijiang Hydropower Plant, Dam Observation and Geotechnical Tests, 26:55-60 (in Chinese).

Zhang Peng, Liu Qing, Zhang Zhixiu 2007, The reservoir gate monitoring and control system based on Fieldbus, Industrial Instrumentation & Automation, 50-55.

Zhang Rentian, Xiao Jian, Ju Maosen 2002, Key Techniques of Real Time Automatic Control and Optimal Operation System of Irrigation Districts, China Rural Nater and Hydropower, 9-11 (in Chinese).

Zhang Zhixiu, Zhang Peng, Wang Zengyu 2003, A Kinds of the Reservoir Gate Monitoring and Control System, China Instrumentation, 33-35 (in Chinese).

SELF-ADAPTIVE FUZZY DECISION MAP MATCHING ALGORITHM BASED ON GIS BUFFER IN LCS

Yongjian Yang [1] , Xu Yang [2] , Chijun Zhang [1,*]
[1]College of Computer Science and Technology, Jilin University, Changchun, Jilin, China, 130012
[2]School of Electronics & Information Engineering of Shenzhen Polytechnic, Shenzhen, China, 518055
** Corresponding author, Address: Chijun Zhang, College of Computer Science and Technology, Jilin University, No. 2699 Qianjin Road Changchun, Jilin, China, 130012, Tel: +86-0431-85168017, Email: yyj@jlu.edu.cn*

Abstract: The wireless localization service is called Location-Based Services (LCS), is an value-added service provided by mobile communications network. This paper introduces a self-adaptive fuzzy decision algorithm based on GIS buffer. Candidate matching roads information is stored in GIS buffer. Through self-adaptive adjusting and fuzzy decision, the algorithm advances the differentiation degree between the right road and wrong. We can judge the matched road immediately, so it is a real-time algorithm and high accurate still. Changing the Coefficient-Value of measure factor by itself, the algorithm adjusts the weight of position information and direction information in judging the best matching road very well. In this way, the whole algorithm is optimized in many aspects.

Keywords: Location Based Service, Map Matching, Geography Information System, Fuzzy Decision

1. INTRODUCTION

The wireless localization service is called Location-Based Services (LCS), is an value-added service provided by mobile communications network, obtains the location information of mobile subscriber (for example latitude

Yang, Y., Yang, X. and Zhang, C., 2008, in IFIP International Federation for Information Processing, Volume 258; Computer and Computing Technologies in Agriculture, Vol. 1; Daoliang Li; (Boston: Springer), pp. 487–494.

and longitude coordinates data) through some location technologies, provides for mobile subscriber or other people as well as the communications system, realizes each kind of the location relatives services. "Chinese Location-based Service Market Analysis" reported by Bo Tong Zhi Xin which is dedicated in telecommunication consultation indicated that the Chinese LCS service will enter the high speed growth phase in 2006 the second half year. The report thinks that the main reasons of the location service development so slow are that the LCS service technology, the content all quite complex, operation business and the SP preparation are insufficient, and faced with various development barrier, the cognition, the content, the technology, the terminal, the privacy, the backstage, cooperates and roams and so on.

This paper divides the LCS service correlation technology into three parts: Position technologies based on the mobile network, the core network side implementation technology about LCS services and the location service provider side implementation technology about LCS services. Through contrast current positioning accuracy and LCS service demand, no matter which one localization method uses, the mobile network all cannot guarantee stably provides real-time and precise location information. Along with the development of LCS services gradually, some service required high precise location information will be needed more precise requirement inevitably. Like the position accuracy which needed by navigation or track services is 30 meters and the system corresponding time in 5 seconds. There are two methods solve this problem, one method is that from access network side and the core network side adopts more precise position technology, but there is no feasible solution currently. Another solution is that joins a second position module in GIS engine in SP side, which function is implements map matching (Bernstein et al., 2002). The core of this paper is just the second positioning algorithm.

2. MAP MATCHING ALGORITHM

Map matching is a method of modifying location at the basis of software. The main idea is locating the position by connecting the position information with the road network in electronic map.

For different map matching algorithms, they are different only in the model and criterion, but the flow is same.

2.1 Set the area of matching error

The first step of map matching is determining the error area (that is judgment area), so that we can get the information of candidates of matching road from the map database.

2.2 Choose the suitable criterion for map matching

After step (1), we have already got the candidates of matching road from map database. In this step, we begin matching with the defined matching criterion. It's the core of the whole matching algorithm, and different criterion will lead to different matching results.

2.3 Judge the matching results synthetically

GIS buffer technique is the basic function of GIS. We can get the geography information in buffer being in a distance to the positioned object. The core is GIS topological relation and advanced database links and query technique. The prior condition of buffer analysis is that the electronic map has complete topological relation and powerful database engine. Buffer analysis can extract the information of different layers within the buffer, which is useful for map matching. For example, we can analyze the road traffic and other geographic information in the circle of some point, some line and some area with buffer technique. For map matching, the task is searching the roads and nodes information near the location results in the system, and finding out the best matching point and matching it.

Above-mentioned algorithm is a typical map matching algorithm based on GIS buffer. We can see that it is suitable for the straight sparse highway. The solution for turning or intensive highway is to wait for presenting only one same value of highway, then position the current location using position data ahead. The flaws of this algorithm are as below:

The performance of real-time is not very well for solving turning or intensive highway section, even it would not respond for a long time.

The adaptability and the performance of dynamic adjusting are also not very well (George et al., 1999).

The reason for above is because of low positioning precision now, for example, GSM cell-phone positioning depends on the methods of TOA/TDOA or CELL-ID mainly, but the positioning error may reach tens of meters even more than one hundred meters. In this condition, it is very difficult to recognize the object turning real timely, and it is more difficult to match when it happens in intensive highway.

3. SELF-ADAPTIVE FUZZY DECISION MAP MATCHING ALGORITHM BASED ON GIS BUFFER

The paper gives some supposing for introducing more explicitly.

First of all, the algorithm demands for more precise electronic map and position information. From the introduction at the head of the paper we can see that the precise of positioning has been improved in large extent with the integration of GPS and mobile network (Huang, et al., 2004). In this condition, the precise of positioning can reach within thirty meters. It has been possible to recognize the turning of highway. The special traffic electronic map is also needed to implement this algorithm. The precise of the electronic map must be limited less than 15m2. Additionally, it is important to enrich the information of the electronic map, such as the integrality of the electronic map, the type of turning (right-angle, round square and crossroad), one-way street, left-forbidden and right-forbidden of the road and so on (Sinn et al., 2001).

Second, the positioning information refreshes every one second, and the data is saved in server. Dividing the positioned object into Fast-Moving Object (FMO) and Slow-Moving Object (SMO), Matching for FMO mainly adopts Straight Line Map Matching algorithm (SLMM) and Matching for SMO adopts Turning Map Matching algorithm (TMM). Generally speaking, FMO is more possible moving on the straight and wide highway, it would move longer distance in one second, so the position information showing on the electronic map can be refreshed in one second. SMO frequently runs in the area of turning, it could not move long distance only in one second, we can refresh the position information every three or four seconds, the position information in this interval is used as criterion for judging in what type of motion the object is moving. Then we can locate the object position in the electronic map by combining with the type of highway in the error area of the object.

On the basis of the supposing above, this paper introduces a judging model of candidate matching roads.

The DF (Direction Function) is defined as (Takagi et al., 1995):

$$H(h_c, h_r) = \cos(h_c - h_r) \qquad (1)$$

hc is introduced as the direction of the moving object.

hr is introduced as the direction of each candidate matching rode.

$$\left(|h_c - h_r| \le 90°\right)$$

The range of DF is: (0, 1]

The PF (Position Function) is defined as (Takagi et al., 1995):

$$D(p_c, p_r) = \frac{1}{1 + \|p_c - p_r\|^2 / \sigma^2} \tag{2}$$

$p_c = (x_c, y_c)$ is the position coordinate of the current moving object.

$p_r = (x_r, y_r)$ is the position coordinate of the point that is on the candidate matching road to which if current matching point matches.

σ is defined as standard deviation of the position data. The value of is defined as 30 in this paper according to current positioning precision.

The range of PF is: (0, 1].

The function involved measurement factor of the positioning object is defined as:

$$F = \alpha H(h_c, h_r) + \beta D(P_c, P_r) \tag{3}$$

α, β is coefficient of measurement factor. The value of α, β is relative to the number of the cross roads which are fetched form GIS buffer. If there are several cross roads in GIS buffer, α will be bigger. Because the moving object is more possible to turn in this situation, the proportion of DF should be enlarged in function (3).

Measurement factor α, β is defined as below:

$$\alpha = 1 + \cos^2\left(\frac{h_c - h_r}{M+1}\right) - \cos^2(h_c - h_r) \tag{4}$$

$$\beta = 1 \tag{5}$$

M is the number of candidate matching roads in the GIS buffer. The weight of PF in function (3) of PF is invariable in the algorithm, so $\beta = 1$. In order to improve the robustness of the algorithm, the connectivity amount the roads must be taken into account. So it is necessary to introduce the information of the previous matching points.

DF is modified as:

$$H(K) = \cos\left(\frac{1}{N+1} \sum_{j=K-N}^{K} \left| h_{c\,j} - h_{r\,j} \right|\right) \tag{6}$$

PF is modified as:

$$D(K) = \frac{1}{1 + \frac{1}{N+1} \sum_{i=K-N}^{K} \frac{\|p_{ci} - p_{ri}\|^2}{\sigma^2}} \tag{7}$$

The measurement factor is modified as:

$$F(K) = \left(\frac{1}{N+1} \sum_{i=K-N}^{K} \beta_i\right) \cdot H(K) + \left(\frac{1}{N+1} \sum_{i=K-N}^{K} \alpha_i\right) \cdot D(K) \tag{8}$$

K is the current candidate matching point. N is the number of previous matching points which are used to enrich the information of the current matching information. In order to decrease the complexity of the algorithm, N is always 2 or 3. As a result of the joining of the previous N matching points, the judgment information of current candidate matching point is enriched. The robustness of the algorithm is improved and matching vibration is avoided at the same time.

For example, the matched roads of the matching points in center will switch between road A and road B. Certainly, we can also use the method of filter to solve this problem (Zhang et al., 2003). But the algorithm proposed in this paper solves the problem by the method of adding previous matching information to current judgment, the method is convenient and high-efficient, it also makes the algorithm more universal. Additionally, considering some road is one-way, we define θ as the valid direction of the one-way street. Then the measurement factor of DF is modified as below:

$$\alpha' = \gamma\,\alpha \tag{9}$$

γ is a Step Function, it is defined as below:

$$\begin{cases} \gamma = 0 & |h_c - \theta| \geq 90° \\ \gamma = 1 & |h_c - \theta| \leq 90° \end{cases} \tag{10}$$

Finally, we define the Judging-Function as below:

$$J(K) = \max(F1(K), F2(K), \ldots\ldots FC(K)) \tag{11}$$

Here, C is defined as the count of roads in the GIS buffer. The matched road judged by the algorithm is one of the candidate matching roads which makes J(K) reach maximum. The Self-Adaptive Fuzzy Decision Algorithm based on GIS buffer can be concluded as below:

1) Get the information of the candidate matching roads in GIS buffer (angle, coordinate, one-way street or not);

2) Get the matching information about the nearest N previous matching points;

3) Compute the measure values of every candidate matching roads through formulas of H (K) and J (K);

4) Confirm the matched road according F (K) and J (K).

4. EMULATION TESTING AND DISCUSSION

The following table is the object running logical traveling route in some region of road network:

Table 1. The coordinate value of the object run

Anchor point serial number	coordinate value	Anchor point serial number	coordinate value
1	(2.731, 3.909)	9	(2.631, 3.772)
2	(2.710, 3.907)	10	(2.653, 3.744)
3	(2.680, 3.911)	11	(2.668, 3.717)
4	(2.650, 3.906)	12	(2.681, 3.695)
5	(2.595, 3.882)	13	(2.709, 3.684)
6	(2.588, 3.849)	14	(2.735, 3.678)
7	(2.616, 3.824)	15	(2.758, 3.680)
8	(2.620, 3.800)	16	(2.788, 3.677)

According to the data of position, we can normalize the value of F (K) in every second as membership of each candidate matching road and make in one figure.

Changing the Coefficient-Value of measure factor by itself, the algorithm adjusts the weight of position information and direction information in judging the best matching road very well. In this way, the whole algorithm is optimized in many aspects.

In the test, we find that there are several matching points whose positioning errors became larger when the moving object entered some road, but we can still get the matched road rightly. Obviously, the robustness of this algorithm has been improved after importing the matching information of the nearest N previous matching points.

We can also find that the real-time character of this algorithm is very well too. The membership-values of the right matching road and the wrong matching road change very fast when moving object enters one road from another. That is also attributed to importing the matching information of the forward matching points. Both of the right and wrong information are accumulated because of the previous information. So it advances the differentiation degree about the current matching information, then we can just spend little second in finding out the right road.

ACKNOWLEDGEMENTS

This work was supported by ZhuHai Technologies Plans Projects of China (Contract Number: PC200320007).

REFERENCES

Bernstein, D. and Kornhauser, A. Map matching for personal navigation assistants. 77th Annual meeting. The Transport Research Board, 2002, Jan 11-15, Washington D.C.

George Taylor and Geoffrey Blewitt. Virtual Differential GPS & Road Reduction Filtering by Map Matching. ION GPS'99, 14-17.

Huang Yongfang, Yang Xinyong. A Map-Matching Algorithm Based on Adaptive Fuzzy Decision in Vehicle Navigation System [J]. Journal of Huaiyin Institute of Technology, 2004, 13(1).

Sinn Kim, Jong-Hwan Kim. Adaptive Fuzzy-network-based C-measure Map-matching Algorithm for Car Navigation System [J]. IEEE Transactions on Industrial Electronics, 2001, (4).

T. Takagi and M. Sugeno, Fuzzy Identification of System and its Application to Modeling and Control. IEEE Trans. Syst, Man and Cybern, 1995, 15(1).

Zhang M S. A new self-adaptive Kalman filtering method for GPS kinematic positioning. Journal of Central South University of Technology 2003, 34(5).

CONSTRUCTION OF AGRICULTURAL PRODUCTS LOGISTICS INFORMATION SYSTEM BASED ON .NET AND WAP

Yan Zhang [1,*], Yong Liang [1,2], Chengming Zhang [1], Qiulan Wu [1], Pingjiu Ge [1]

[1] *College of Information and Engineering, Shandong Agriculture University, Taian, China, 271018*

[2] *Chinese Academy of Surveying & Mapping, Beijing, China, 100039*

* *Corresponding author, Address: College of Information and Engineering, Shandong Agriculture University, 61 Daizong Street of Taian, Shandong, 271018, P. R. China, Tel: +86-538-8242497, Fax: +86-538-8249275, Email: zhangyandxy@sdau.edu.cn*

Abstract: Functions and construction of agricultural products logistics system based on .NET and WAP technology are introduced in detail. The problems encountered during the process of system development and corresponding solutions are also illustrated. The Windows 2003 Server and SQL Server 2005 serve as the platform and background database server respectively, and windows are designed using the ASP.NET mobile controls. This system will be beneficial to the circulation of agricultural products in undeveloped area. The information can be released and browsed through WAP mobile phone anytime and anywhere, so a convenient means of exchanging information is provided by this system.

Keywords: WAP, .NET, Agricultural Product Logistics

1. FOREWORDS

Traditional circulation of agricultural products is keeping through fairs or market. This approach suits to the small areas while the circulation information within a large area often lags (Ratnasingam et al., 2006). With the

Zhang, Y., Liang, Y., Zhang, C., Wu, Q. and Ge, P., 2008, in IFIP International Federation for Information Processing, Volume 258; Computer and Computing Technologies in Agriculture, Vol. 1; Daoliang Li; (Boston: Springer), pp. 495–502.

development of Internet technology, network has been applied to all levels of society, and the exchange of information has become more timely and easily (Shu Geng et al., 2006). Computers has become very popular in urban areas, but it is not so realistic to buy a computer in the countryside where the economy is relatively undeveloped (Sun Wei et al., 2004). However, with the increasing improvement of living standard, mobile phone was not a luxury any more, and it has become a reality for farmers to have own cell phones. Meanwhile, the communication costs of mobile phone gradually reduced because the amounts of users increase violently. Both provide a possibility of access to Internet through mobile phones. At present, it has become popular to access to Internet using the WAP-enabled mobile phone with the standardization of wireless access protocol (WAP). In this way, users can visit the websites that provide WAP functions, view information and download resources by cell phone at any time. The WAP-based agricultural logistics information system is submitted according to these considerations.

2. SYSTEM BASIS: .NET AND WAP

2.1 .NET Technology

.NET is a collection of applications supported by Web Services programming and is a technology used for the seamless interoperability between applications and computing equipments and the realization of Web interface, which is proved to personal and commercial users by Microsoft (Hou Yingchun et al., 2003). As a part of .NET framework, ASP.NET is a technology used to establish dynamic Web applications. ASP.NET applications can be written by any .NET-compatible languages such as Visual Basic .NET, C# and J#. Compared to the original Web technology, ASP.NET provides a programming model and structure, and can build flexible, secure and stable applications rapidly and easily. Web Forms can create powerful windows based on the web sites. Once the Web Forms established, the common user interface elements can be built using ASP.NET server controls and the tasks can be accomplished by programming. These controls allow the use of built-in reusable components and user-defined components to quickly build Web Form, and simplify your code. In this system, the ASP.NET mobile controls are used for pages design and websites development.

2.2 WAP Technology

WAP is an open global standard for the communication among digital mobile phones, computer applications, Internet and personal digital assistants (PDA) (Ma Xiaojin et al., 2006). A wealth of information from Internet and a variety of business can be introduced into mobile phones and PDA wireless terminals through this protocol without any restriction on network types, network architectures, business of operators and terminal equipments. Thus, users can get the online information resources expressed in unified format through WAP-supported cell phone anytime and anywhere.

Different from the client/server architecture of Internet, the structure of WAP network consists of three parts, that is, WAP gateway, WAP phones and WAP content server. The network structure is shown in Figure 1.

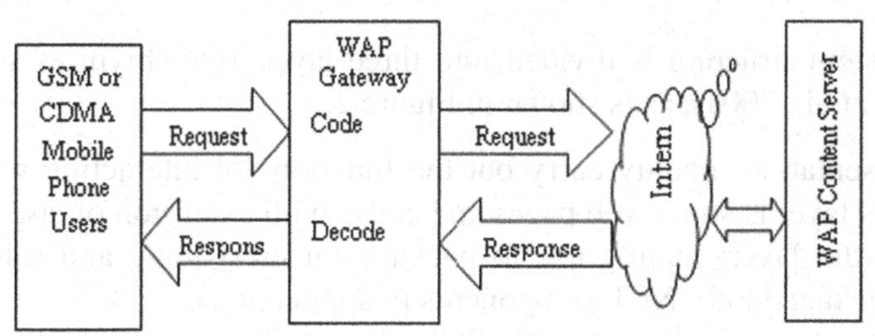

Figure 1. WAP network structure

3. SYSTEM CONSTRUCTIONS

3.1 Platform Construction

The IIS of Windows 2003 Server functions as the WAP server and SQL Server 2005 acts as the background database. Both of them construct the platform of agricultural products logistics system. To enable the server to support WAP functions and to ensure normal operation of the WAP modules, it is necessary to set up the server as follows: (1) Configured WWW services on the IIS server. (2) Add WAP-supported document types on WWW server. To achieve the transmission of WAP documents, supports of MIME that is the unique type of WAP are needed. The specific extensions and content types are shown in Table 1 (Gao Lei et al., 2003; Xu Haoyue, 2005).

Table 1. Document types needed for WAP

Extensions	Content Types (MIME)	Remarks
.wbmp	image/vnd.wap.wbmp	WAP-supported bit maps
.wml	text/vnd.wap.wml	WAP-supported WML web page text files
.wmlc	application/vnd.wap.wmlc	WAP-supported WML application files
.wmlsc	application/vnd.wap.wmlscriptc	WAP-supported WML script application files
.wmlscript	text/vnd.wap.wmlscript	WAP-supported WML script web page text files
.wsc	application/vnd.wap/wmlscriptc	WAP-supported WML script application files

3.2 System Architecture

The system structure is divided into three layers (Qi Zhiyin et al., 2004; Yang Rui et al., 2004), as is shown in Figure 2.

(1) Presentation: mainly carry out the functions of interacting with final users. This layer is some web pages and codes with extension of .aspx;

(2) Middle layer: mainly use to package business logic and rules. This layer is packaged into .NET components in applications;

(3) Data Access: interact with SQL Server Provider through the data access components in intermediate layer.

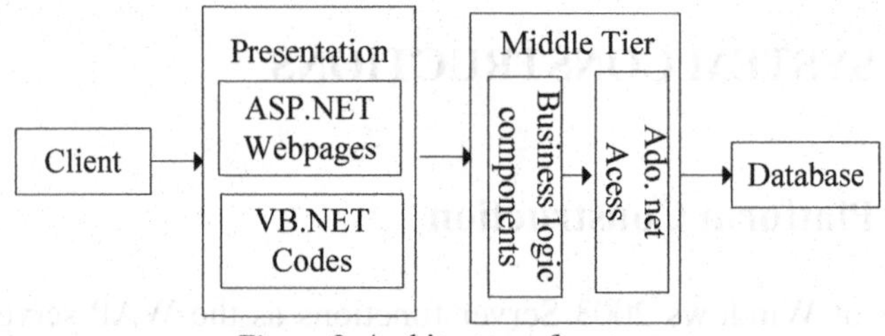

Figure 2. Architecture of system

3.3 Functions Design

The agricultural products logistics system contains the following modules:

(1) The latest developments: access to the agricultural products market, including the latest market news. The information is updated by the system administrator through the background database.

(2) Information Release: mainly release the information of supply and demand.

(3) Information retrieval: retrieve the information of supply and demand according to the product types entered from mobile phones.

(4) The market price: provide the recent prices of agricultural products in different regions. Users can also retrieve the relevant prices of agricultural products according to their requirements.

The main data tables used during the system design are shown in Table 2.

Table 2. Main data tables

Fieldname	Meaning	Field name	Meaning
InfId	ID number	Content	Content of information
Flag	Supply or demand	Sort	Product types
Price	Price	Place	Producing area
Time	Released time	Deadline	Valid before
Contact	Contact person	Phone	Phone number
Fax	Fax number	E-Mail	E-mail address

4. PROBLEMS AND SOLUTIONS

(1) The system needs the support of background data, and some data that used in background database management system SQL Server 2005 are imported from Microsoft Excel. When the data are imported with the DTS Wizard that is built-in SQL Server 2005, any data cannot be found in the required table. On the contrary, a table with the name of imported table plus a "$" is generated in SQL Server, and this table has all the data imported. Although the structure of this table is similar to the original one, the data field types in this table is different from the original data and the constraints to the field length are also not the same. So, the current table is not equivalent to the original one. To import data from Excel into SQL Server 2005 correctly and to generate a table with the same name, some SQL statements are needed. The SQL statements used to import data is shown as follows:

```
EXEC sp_configure 'show advanced options', 1
GO
RECONFIGURE
GO
EXEC sp_configure 'Ad Hoc Distributed Queries', 1
GO
RECONFIGURE
GO
Insert into NCPLeiBie select * from
OPENROWSET('MICROSOFT.JET.OLEDB.4.0','Excel
5.0;HDR=YES;DATABASE=d:\data.xls', NCPLeiBie $)
```

(2) In the process of dealing with information, a lot of IF statements are needed to write in the denotative layer in order to detect the information received by the clients whether accords with norms or not. Thus, the response time of mobile phones will be postponed to some extent because of

their relatively weak capacity. The processing time of wireless terminals can be reduced by setting proper restraints for the related fields in the background database tables.

(3) In the process of writing a code to receive the agricultural products information input from the client and submit it to the background database, the information can be submitted to the Datatable. But, the background database can not be updated using the Update method of SqlDataAdapter object. For this purpose, SqlCommandBuilder object is used to create command and update data sets in data adapters automatically. The concrete realization is shown as follows:

```
'sqladapt agricultural products information data adapters
Dim cb As SqlCommandBuilder = New SqlCommandBuilder(sqladapt)
'tdatatable updated agricultural products data tables
Dim table As Data.DataTable = tdatatable.GetChanges
sqladapt.Update(table)
sqladapt.Fill(tdatatable)
```

5. SYSTEM DEBUGGING

In this system, both the content and the interface of WAP site are written in Chinese. Therefore, encoding command must be used to designate Chinese character sets at the beginning of programs in order that WAP browsers can display correct Chinese characters. Code is written as follows(WAP-Forum, 2002):

```
<? xml version="1.0" encoding="gb2312">
```

The WAP browser M3Gate is adopted for system debugging. The main interface of this system is shown in Figure 3 where the menu can be selected through arrow keys of cell phones in order to access the corresponding link interface. The interface of information retrieval is shown in Figure 4 where the users can retrieve the available supply by inputting the agricultural product's name. The results of retrieval are displayed in a pattern of Figure 5. Only one item can be displayed each time and others can be seen by pressing the "next page" button. The main interface shown in Figure 3 can be returned to from arbitrary interface through the "home" button.

Completing the debugging through simulators on the PC does not mean system can be used normally. After all, there are differences between simulators with real WAP terminals. The view effect of webpage must be debugged to make some adjustment by using the real WAP mobile phones in final stage of WAP development. In addition, the standard implemented by

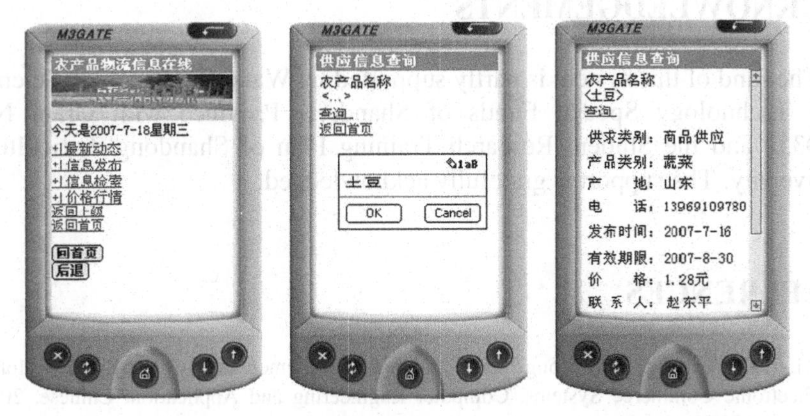

Figure 3. Main interface *Figure 4.* Supply information retrieval interface *Figure 5.* Retrieval results of supply information

different WAP client equipment may have differences on the details. Therefore, constant adjustments are required in the process of WAP development to make the WAP site be compatible with most equipment.

6. CONCLUDING REMARKS

An agricultural products logistics system is constructed by using the ASP.NET of Visual Studio 2005 as the development tools and the SQL Server 2005 as the background database. The architecture of system is given firstly. Then, the overall function modules and the main background data table structures of system are introduced by taking the actual needs of system development into account. Finally, the specific solutions in accordance with the problems encountered in the debugging process are illustrated in detail. The system is debugged and run by using the M3Gate simulator, and part operating results of the system is shown.

The agricultural products logistics information system will be beneficial to the circulation of agricultural products in economically underdeveloped regions, especially in non-urban areas. In this system that consists of WAP gateway, WAP content server and WAP terminals, users can browse agricultural products information released by others or submit his own information through WAP mobile phones anywhere and anytime. The agricultural products logistics system provides a new channel for the exchange of agricultural products information.

ACKNOWLEDGEMENTS

The fund of this system is partly supported by Water Conservancy Science and Technology Special Funds of Shandong Province with Grant No. 200357 and the Student Research Training Plan of Shandong Agriculture University. The support is gratefully acknowledged.

REFERENCES

Gao Lei, Ren Lihong, Ding Yongsheng. Design and Implementation of WAP-based Mobile Electronic Commerce Systems. Computer Engineering and Application, Chinese, 2003, (1):215-217

Hou Yingchun, Geng Baiqiang. Application on .NET Technology in WAP. Journal of Henan Institute of Education (Natural Science), Eng., 2003, 12(3):47-49

Ma Xiaojin, Zhou Yong, Jia Shaorui, et al. Design and realization of score inquiry subsystem based on .NET and WAP mobile. Journal of Hebei Institute of Architectural Science and Technology, Chinese, 2006, 23(3):80-82

Qi Zhiyin, Li Yuchen. Building ePDMS Based on Web and WAP. Computer Engineering, Chinese, 2004, 30(7):66-68

Ratnasingam, P. The role of ecommerce adoption among small businesses: an exploratory study, Eng., International Journal of Cases on Electronic Commerce, 2006, 2(2):39-54

Shu Geng,Tian-zhi Ren, Mao-hua Wang, Technology and Infrastructure Considerations for E-Commerce in Chinese Agriculture, Eng., Agricultural Sciences in China, 2006, 6(1):1-10

Sun Wei, Wan Xiao-ning, Sun Lin-yan. Electronic commerce applied to the structure optimization in agricultural products' supply chains system. Eng., Industrial Engineering and Management, 2004, 9(5):33-41

WAP-Forum (2000a), WAP-199.WTLS (Wireless Transport Layer Security), Version 18-Feb-2000 http://www.wapforum.org/what/technical.htm

Xu Haoyue, Wap-the foundation of the lead-in electronic commerce. Agriculture Network Information, Eng., 2005, (4):34-35

Yang Rui, Guan Xiaohong, Gao Feng, Jiang Lei, Zhao Li. Design and Implement of a Wap-based Mobile Phone Fee Query System. Eng., Microelectronics & Computer, 2004, 21(11):92-95

RESEARCH OF AUTOMATIC MONITORING SYSTEM OF RESERVOIR BASED ON EMBEDDED SYSTEM

Chengming Zhang[1,2,*], Jixian Zhang[3], Yong Liang[2], Yan Zhang[2], Guitang Yin[2]
[1] *School of Geoinformation Science and Engineering, Shandong University of Science and Technology, Qingdao, China, 266510*
[2] *School of Information Science and Engineering, Shandong Agricultural University, Taian, China, 271018*
[3] *Chinese Academy of Surveying and Mapping, Beijing, China, 100039*
* *Corresponding author, Address: School of Information Science and Engineering, Shandong Agricultural University, 61 Daizong Road, Taian, Shandong, 271018, P. R. China, Tel: +86-0538-8242497, Fax: +86-0538-8249275, Email: chming@sdau.edu.cn*

Abstract: The automatic monitoring system of reservoir is an important means to realize modernization of reservoir management. This paper expounds the structure of automatic monitoring system of reservoir firstly. The system consists of three subsystems, which are acquisition subsystem, transmission subsystem and data management subsystem. Secondly, the design of the data collection terminal is studied, which realized the design of hydrological data collection terminal including hardware design based on embedded system and software design. The reason that affects data collection is analyzed and anti-jamming measures are given. And then the system structure of data transmission is offered, and the transmission mechanism of mixed-mode network, in which the elementary channel is wireless mobile communication and the backup channel is wire communication, is achieved. Finally, the data management subsystem is briefly introduced. The system is proved to be useful and efficient by the application on Xueye Reservoir.

Keywords: monitoring system of reservoir, embedded system, data collection

Zhang, C., Zhang, J., Liang, Y., Zhang, Y. and Yin, G., 2008, in IFIP International Federation for Information Processing, Volume 258; Computer and Computing Technologies in Agriculture, Vol. 1; Daoliang Li; (Boston: Springer), pp. 503–513.

1. STRUCTURE OF AUTOMATIC MONITORING SYSTEM OF RESERVOIR

Reservoirs are important water conservancy facilities and water resources protection bases. They are key facilities which ensure industrial and agricultural production and urban people's life. Also, they are foreland of the rapid response to flood prevention, drought control and flood warning. The management level of reservoir is directly related to the normal design efficiency and people's life and property's safety. As the main contents of the reservoir modernization, reservoir automatic monitoring system can realize the automatic collection and delivery of hydrological factors such as rainfall, water level and water scheduling, directly serve the flood forecasting and scheduling and the water resources management, achieve the optimal allocation of water resources, provide the scientific basis for decision-making on the efficient use of water resources, comprehensively upgrade the management level of the reservoir, and is an important means to realize reservoir management modernization (Liang et al., 2005).

Reservoir has become main water source of agricultural. But with rapid development of country economy and society, the conflict of water supply and consume stands out increasingly (Liu, 2004). The lack of water resource has touched our foodstuff directly. The foodstuff output reduces 700 hundred million to 800 hundred million every year due to lack of water. Along with population growth and industry development, the contradiction of water lack will be more outstanding. So saving agricultural water is imperative under the situation, while rebuild conventional management measures and irrigation establishments are essential instruments to realize water saving.

Reservoir automatic monitoring system consists of data collection subsystem, data transmission subsystem and data management subsystem.

Data collection subsystem includes the gate open degrees collection terminals, rainfall collection terminal, water level collection terminal, power output collection terminal, industrial water collection terminals and data center. Data collection subsystem mainly completes the automatic collection of hydrological information including water level, rain and flow, and the data center exchanges data with collection terminals through the GPRS network. Data transmission subsystem is a mixed-mode network, which uses wireless mobile communications (GPRS) as the main channel and cable communications (PSTN) as a backup channel. It is responsible for transmitting the collected data to the database of data center. Data management subsystem uses SQL Server as the data management platform for data inputting, deletion, modification, storage, retrieval, sorting and statistics management. Figure 1 shows the structure of automatic monitoring system of reservoir.

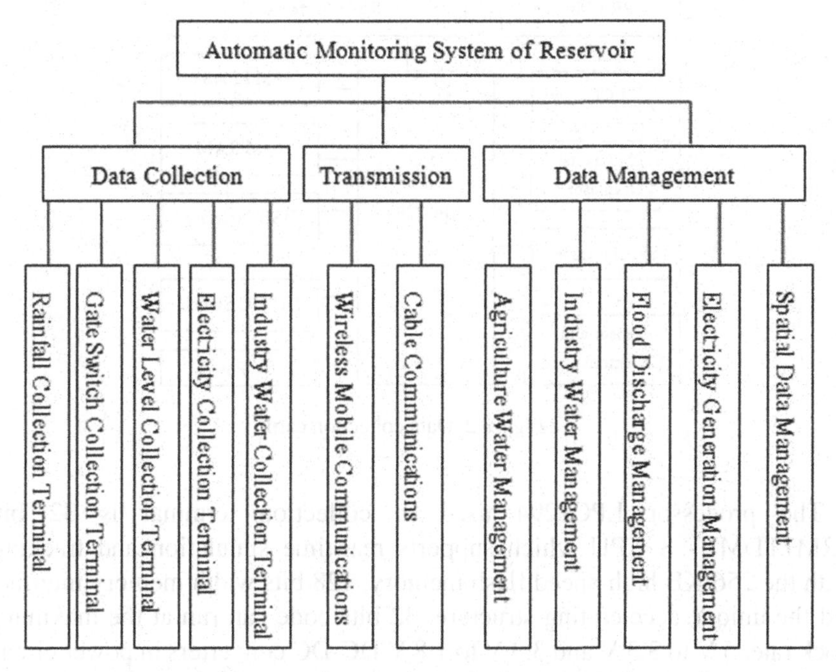

Figure 1. Structure of automatic monitoring system of reservoir

2. THE HARDWARE DESIGN OF SYSTEM

Hardware includes the hardware which is installed at the information control and transaction center, and rainfall collection terminals, irrigation flow collection terminals, reservoir water level collection terminals.

2.1 Data Collection Terminal

Data collection terminals are designed with embedded system. They have the following main features: stable and reliable operation, low power consumption, large storage capacity; high operational speed to deal with complex algorithms and protocols quickly. They also can be connected to Internet and use public networks for data transmission, cost low communication, and realize data on-line monitoring completely (Ye et al., 2007; Shi et al., 2007; Shen, 2007).

LPC2294 processor is used for data collection terminals. An embedded system platform is constructed by making use of the excellent core performance and the abundant external interfaces of LPC2294, which create data collection terminals. The hardware structure of terminals is shown in figure 2.

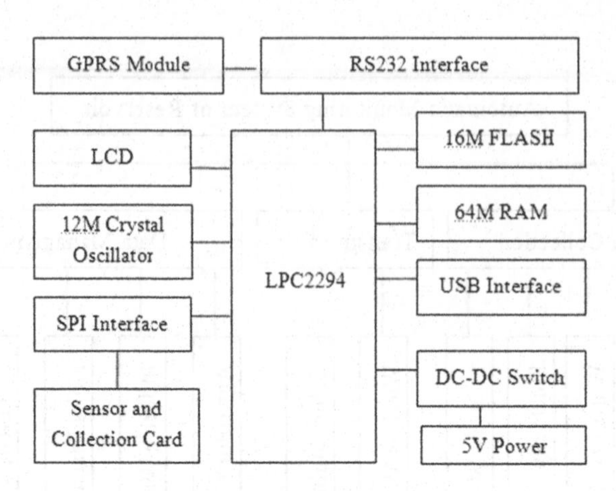

Figure 2. Data collection card

The processor LPC2294 used in collection terminal is 32 bits ARM7TDMI - S CPU which supports real-time simulation and tracking. With the 256 KB high-speed flash memory, 128 bits width memory interface and the unique accelerating structure, 32 bits code can run at the maximum clock rate. 5 V to 3.3 V and 3.3 V to 1.8 V DC-DC converters in power circuit can supply power to LPC2294 and the other external circuits that need 3.3 V.

A 12 MHz oscillator is taken as the system clock for the LPC2294 chip, and the internal clock which controls logic can produce different frequency clock signals that the system needed. FLASH memory can store boot loader, embedded operating system, application procedures and the user data need to preserve after the system restarts. RAM memory is the system's main regions, where the operating system, user data and stacks are located. The system connects to GPRS modules through an RS232 serial link in order to finish wireless data transmission functions. The system expands IO modules through SPI interface, which links hydrological collection equipment sensor and collection card and completes inputting and outputting of digital and analog signals.

2.2 Anti-jamming Technology

Data collection terminals installed in the field are often influenced by electromagnetic fields, lightning, electrostatic, switching power supply, motor starting current and other noise impact. These factors will affect the reliability and safety of whole system, and result in increasing of data collection errors, controlling state failure and procedures disorders. Therefore, these factors must be taken into account during the design of the system.

1. Because the control output circuit load of system is relays, contactors, and other inductive load, high induced electromotive force (EMF) will be produced in inductance coils when they disconnected. This high EMF can not only cause interference electromagnetic induction in circuitry, but also cause spark or arc interference between contacts that affecting the normal work of microcontroller, or even "dies" phenomenon. This problem can be effectively solved by adding discharge diode parallel absorption device at both ends of the relay J coil.

2. Data terminals installed perennially at the field are vulnerable to the wind, the sun, the rain, the lightning and overload factors. The most important issue in terminals design is reliability in the harsh environment. In order to solve the problems, the following measures have been adopted.

(1) All the hardware chips are the wide-temperate chips used in industry which adopted CMOS low-power structure with good anti-interference capability and very low power consumption.

(2) In the layout design of PCB, the signal collection line, the memory data bus and control bus adopt the parallel multi-lane surrounded with a large area of land lines. The critical data lines will be siege by all land lines and TVS tubes are added between the data bus and the large area of land lines, which can absorb the over voltage. A very clear anti-jamming effect has received through above-mentioned measures.

(3) In addition to high power or heating devices, all components used patch components. At this stage, patch components production need a high technological level, so there is little fake patch components. At the same time, patch components are small and easy to integration. The stability and reliability of the system is further enhanced by using hot air returned exclusive automatic welding equipment that can prevent the welding from damaging chips.

(4) A specially designed waterproof aluminum alloy frames is adopted. The outlet lines are four-core shielded twisted paired, the plugs use waterproof jacks and the twisted paired lines use the sensor end grounding method. All the methods can avoid the interference from outside stray electromagnetic field.

3. Sometimes the system program may come into a "cycle of death" due to interference, that is, the system procedures lost control. Software anti-jamming technology such as directive redundancy and software traps can not make these procedures from "death cycle." Surveillance procedures, also known as the "watchdog" technology, can prevent the process from the "death cycle". "Watchdog" technology is constantly monitoring procedures orderly running time. If the time is longer than the known cycle time, the procedure will be considered to be trapped in a "cycle of death" and then

forced it to return to the 0000H procedures entrance where a wrong procedure is arranged that allow the system to run onto the right track.

2.3 Distribution of Data Collection Terminal

Data Collection Terminals include rainfall collection terminals, irrigation flow collection terminals, reservoir water level collection terminals, gate opening collection terminals and industrial water collection terminals.

Rainfall collection terminals distribute in Chaye rainfall stations, Shangyou rainfall stations, Longzi rainfall stations, Luye rainfall stations, Yumen rainfall stations, Xueye rainfall stations, Kouzhen rainfall stations, Qiguanzhuang rainfall stations, Huzhai rainfall stations and Gushan rainfall stations. The rainfall tubes used in these stations are telemetry dedicated rainfall gauges with a resolution of 0.5mm.

Irrigation flow collection terminals distribute in the main channel, the eastern entry, the western entry, Hebei-Sanshan junction, the Sanshan-Kouzhen junction, the Kouzhen-Qiguanzhuang junction, the Qiguanzhuang-Heguanzhuang junction and the Heguanzhuang-Huzhai junction. The ultrasonic sensors are used to measure the channel depth that will be converted into the flow values to calculate the flow.

Reservoir water level collection terminals installed a platform with the altitude of 231.30m in the west side of the dam. A large-range ultrasonic sensor with the measuring range of 20 meters is installed at the height of 231.70m. The minimum measurable height is 211.70m, which is lower than the dead storage water level, so it fully meets the range requirements of water level.

There are six telemetry gate opening collection terminals: the main flood discharge gate, the 1st and 2nd sluice of irrigation area, the 1st gate of irrigation area, the eastern entry, the western entry. Photoelectric rotary encoders are adopted as the gate opening sensors.

Telemetry industrial water collection terminals are Laiwu power plant, Laicheng district power plant and Xueyin Company. Ultrasonic flow meters are installed in these units, and the RS485 interfaces are taken as the data collection terminals which transmit the collected data.

2.4 Remote Controller of Floodgate

The main function of remote controller on open-close degree of floodgate is to receive floodgate control signal from data center to open and close floodgate according to control information. The remote controller on open-close degree of floodgate is composed of GPRS wireless data terminal, data analysis and processing circuit, floodgate up and down circuit, direction shift delay circuit, over load and over flow protection circuit. Figure 3 is the graph of remote controller of floodgate.

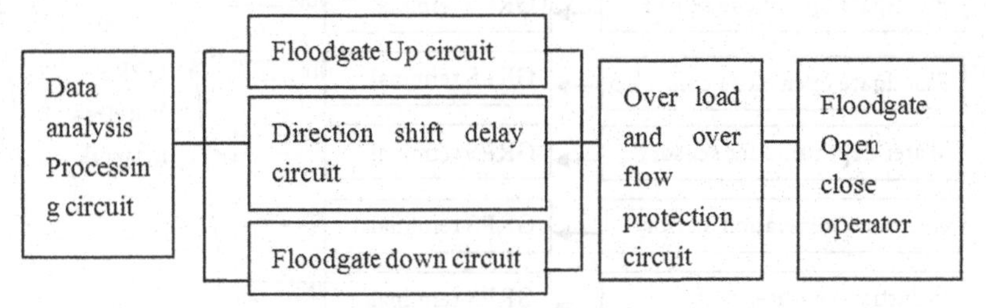

Figure 3. Remote controller of floodgate

3. DATA TRANSMISSION MODE

3.1 Transmission Network

According to the availability of current domestic communication resources and the actual situation and also taking the operational management facilities of the system into account, a mixed-mode network is designed with the use of wireless mobile communications as the main channel and the cable communications (PSTN) as a backup channel. Equipped with the "double channel", the communication can be automatically switched to the backup channel in case of the main channel failure, and return to the main channel to transmit data after the main channel is normal again (Guo et al., 2007; Liu et al., 2006). Because all the stations are situated in the coverage areas of China Mobile's signal, the wireless mobile communications are taken as the main channel. Data are packaged into TCP/IP data packets in the GPRS modules of data collection terminals and are sent to the data-processing and control center through the GPRS wireless network (Li et al., 2004; Cui, 2004). Transmission network is shown in figure 4.

GPRS modules adopted H7000 produced by Hongdian Company in Shenzhen. The unified mobile SIM cards are required to install in each module of collection points, and has the only ID in the mobile network just as the mobile phone. Specialized APN distributed by China Mobile is adopted by GPRS wireless data terminals and control center to access the wireless network. There are four modes for H7000 GPRS wireless DDN System, and different stations can choose arbitrary one mode according to the actual situation (Wu et al., 2007).

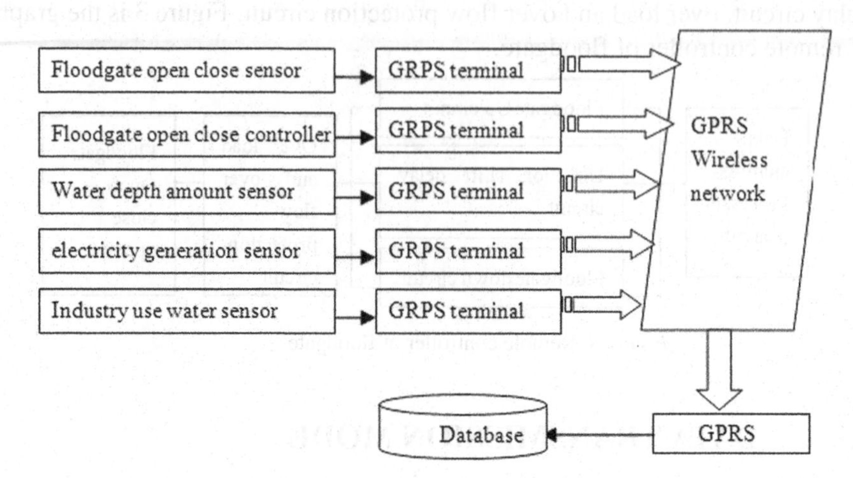

Figure 4. Transmission network

(1) Always-on-line mode

Always-on-line mode will maintain DTU model and data center connections. Its work is: DTU connects to GPRS network automatically when it boots. Connect automatically according to the IP address of data operations center (DSC), and maintain and preserve the connecting link. DTU monitors the operation of link of the network, and automatically re-establish link once the exceptions occur.

Rainfall measuring stations are used to measure rainfall, the water level and water from the upper reaches of the river measuring stations are used to measure the water level in the reservoir and the upper reaches of the river water level. So, they are continuously working methods and GPRS modules in these stations use this mode.

(2) Regular transfer mode

DTU uses regularly transfer mode to send data to the center regularly. Its work is to send and receive data to data center operations (DSC) according to the pre-set interval. After it's over, automatically disconnect.

Pressure tube water level measuring stations are used to measure the water level of pressure tube of the dam every week. GPRS module in these stations can choose this mode for this is a cycle of work.

(3) Data triggered mode

In this mode, the DTU initiates link only when the user need to transmit data. DTU connects to GPRS network and logs data operations center (DSC), and the data are transmitted.

Gate opening measuring stations measure the opened extent of gate and will be used only in agricultural irrigation and flood. Water level measuring stations measure the water level of agricultural irrigation channels in irrigation areas and use only in agricultural irrigation work. These stations can choose this mode.

(4) Center call mode

Center call mode is used for transmitting data to the DTU from data center. The data center sends a data transmission request and the DTU responses and transmits data to operations center (DSC) according to the instructions of DSC.

(5) Dormancy mode

If there is no data transmission, wireless transmission system will be closed and come into a state of dormancy. Apart from rainfall stations, other stations in the system will be in this state if there is nothing.

3.2 Communication Protocol Design

There is a set of stringent response mechanism when transmit data using TCP/IP. The headers of TCP are longer than the telemetry data when the TCP is used for transmission. When the responsive signal sending from the receiving party is not received, the transmission party will be repeated sending until the responses received. If the TCP protocol used for telemetry data transmission, and transmission efficiency will be substantially reduced. However, UDP network broadcasting agreements don't have the response agreement, and not suitable for high reliability telemetry data transmission. Therefore, According to water telemetry system characteristics, the data packets of the system use the first two bytes for station number, then after a number of byte values for telemetry, the last byte for CRC. Before the station sends the data, the first is to connect data center with IP connectivity, and transmit data directly without sending request signals, and then wait for sending the responsive signal from the data center. When the survey station didn't receive response signals from the data center after delaying time, sent again, and then waited for the response from the data center. Such repeated three times. It proves that the agreement is fully in response to satisfy the requirements of the telemetry system.

4. DATA MANAGEMENT

The main function of data management subsystem is to store the attribute data and spatial data, that is, to input, delete, modify, store, retrieve, sort and do statistics.

The attribute database management system mainly managed many data closely relating to the objective function in the form of database, such as text, tables and so on. Attribute data included: water characteristics, such as rivers data, hydrological station distribution and rainfall distribution; reservoir information, such as reservoir characteristics, water-storage capacity curve, inflow, the water level at the dam, the flow out; hydrological information, such as daily flow, the water level information; meteorological information, such as precipitation, temperature, wind speed, air humidity, evaporation, sunshine and vapor pressure; water consumption information, such as water for agricultural, urban water supply, the local industrial water consumption and hydropower; reservoir scheduling information, such as water scheduling information, flood scheduling information, optimum scheduling information.

Geographical information system is based on spatial database. When establishing spatial database, layered technology is used to separate a variety of geographic elements into a number of independent layers, establish the relationship between physical objects and geometric characteristics in order to edit, hide and display, select and analyze. spatial database, such as 1:10 terrain database, the database of geographical names, land used classification map, river map, project distribution maps, digital elevation and so on, is built based on 0.6-meter satellite image shot in 2005 and consulting 10,000 maps, and the input of the spatial data (such as irrigation, drainage and irrigation stations, a field distribution) relating to irrigation management business is completed, and then the function such as the changes of spatial reference, spatial analysis, the inquiry and display of the feature elements, the exports of electronic maps and so on, is achieved.

5. CONCLUSION

The automatic monitoring system of reservoir has been put into use in Xueye reservoir, Laiwu City, Shandong Province. The system completes real-time data collection, transmission and automatic management. It is timely, efficient, accurate, and has low operating cost. Managers can grasp the reservoir information immediately and accurately so as to take proper measures to regulate the water resources reasonably. According to the results, we can see, the system can provide detailed and accurate hydrological information in time, and can provide decision-making basis and reliable information for reasonable water schedule and scientific flood control, which is proved to have significant social benefits and great economic benefits.

ACKNOWLEDGEMENTS

This study has been funded by Special Funds of Water Science and Technology of Shandong (Contract Number: 200357). Sincerely thanks are also due to the Xueye Reservoir for providing the data for this study.

REFERENCES

Cui Ziqian 2004, Data Acquisition System for Industrial Based on GPRS. Mechanization Research. 11:108-109 (in Chinese).

Guo Chunyan, Yu Junqi, Song Xianwen 2007, Design of Wireless Remote System for Reservoir-bedload. Control & Automation. 2:85-88 (in Chinese).

Li Tao, Fu Yongsheng 2004, Data Transmission Based on GPRS. Mobile Communications. 7:76-79 (in Chinese).

Liang Yong, Zhao Ao, Zhang Maoguo 2005, Studies on Digital Reservoir. Journal of Shandong Agricultural University (Natural Science). 2:313-316 (in Chinese).

Liu Chunling, Wang Yanfen, Zhang Shihu 2006, Realization of Embedded Telemetering System of Hydrological Signal Based on ARM for Irrigation District. Water Resources and Hydropower Engineering. 8:74-77 (in Chinese).

Liu Jianbiao 2004, Research of Digital Reservoir. Water Development Studies. 3:44-46 (in Chinese).

Shen Limei 2007, A Novel Platform Design of Data Acquisition Control System. Control & Automation. 2:130-133 (in Chinese).

Shi Yi, Ran Shuyang, Zhang Shengjun 2007, Realization of Network-based and Intelligent Data Collection System. Control & Automation. 5:101-102 (in Chinese).

Wu Qiulan, Liang Yong, Zhang Chengming, Ge Pingju 2007, Design of Wireless Hydrological Automatic Measurement System Based on GPRS. Computer Engineering. 2:280-282 (in Chinese).

Ye Dunfan, Xue Guojiang 2007, Data Acquisition System Based on Embedded Operating System. Control & Automation. 4:6-8 (in Chinese).

LONG-RANGE MONITORING SYSTEM OF IRRIGATED AREA BASED ON MULTI-AGENT AND GSM

Tinghong Zhao [1], Zibin Man [2], Xueyi Qi [1,*]
[1] *College of Fluid Power and Control, Lanzhou University of Technology, Gansu, Chinan, 730050*
[2] *Laboratory, Lanzhou University of Technology, Gansu, China, 730050*
* *Corresponding author, Address: P.O. Box 53, College of Fluid Power and Control Engineering, Lanzhou University of Technology, Gansu, 730050, P. R. China, Tel: +86-931-2976770, Fax: +86-931-2976750, Email: zhaoth2626@163.com*

Abstract: In order to improve irrigated automatic degree of irrigated area, and make the control center and telemeter station to intelligent complete the task, structures the control center and telemeter station Agent union by instruct Multi-Agent theory, and make use of GSM network to finish the communication between the control center and telemeter station, at last make up the long-range monitoring system of irrigated area based on Multi-Agent and GSM. In this system, telemeter station finishes its functions, such as collection, memory, dealing with of the data, and while to carry control out according to the treated data, send the data that the control center needed to the control center Agent union through GSM network; control center Agent union store and deal with the accepted data information from telemeter station at first, then send the treated data to telemeter station Agent union to carry out new regulation and running. The setting-up of this system, not only make the work of the control center and telemeter station have higher intelligent, but also make the communication amount reduce greatly, communication is swifter and high-efficient, which make the real-time character of long-range controls improve greatly.

Keywords: Multi-Agent; GSM network; Long-range monitoring system of irrigated area; Agent union;

Zhao, T., Man, Z. and Qi, X., 2008, in IFIP International Federation for Information Processing, Volume 258; Computer and Computing Technologies in Agriculture, Vol. 1; Daoliang Li; (Boston: Springer), pp. 515–523.

1. INTRODUCTION

With the constant development of the technology of computer and network, it is easier and easier to realize irrigated area long-range monitoring. The wired network is fast and swift in communication, used widely in a lot of fields, but for the scattering of distribution of irrigated area telemeter station, very difficult to wiring, so is less applied in the long-range monitoring system of irrigated area. Modern wireless communication network GSM is relatively suitable in the long-range monitoring system of irrigated area, and it has got certain application in long-range monitoring of irrigated area and has got better effect. (Zhang, 2005)

However, as to long-range monitoring system of irrigated area, GSM network is only a communication media, only play a role in transmitting information of monitoring and controlling, so in the monitoring system, most work need staff member to finish. To reduce the staff member participated and easily control, these works only can finish in monitor center. Thus, telemeter station has accurately and swiftly telemeter, but the course that telemeter station send telemeter message to monitor center through GSM network, which is analyzed by monitor and sent back telemeter by through GSM network, need longer time, make the real-time character of the long-range monitoring system of irrigated area relatively bad.

In order to change this kind of phenomenon, this paper combines Multi-Agent theory and GSM network together, set up a kind of new-type long-range monitoring system of irrigated area. In this system, structure a Agent union respectively in monitor center and telemeter station, through defining structure, function etc. of each one Agent inside the union, to make them have certain intelligence, they can finish one's own task through one's own intellectual behavior. The monitor center Agent union finish such functions as the dispatching calculate of irrigation water of the whole irrigated area, storing dispatching information and sending control information etc. through the negotiation among inside Agent. Telemeter station Agent union is used to finish such functions of monitoring in real time, dealing with and calculation of monitoring information, analysis and storing of calculation result etc. confirm one's own control task in time and real-time control, will not send monitor information to the monitor center again and only send message when telemeter station is unable to finish the task. Thus, GSM network only transmits the task message that telemeter Agent union unable to finish and need monitor center to finish, thereby reduce the communication amount of GSM network greatly, improve communication efficiency; At the same time the telemeter station will not wait the control information of monitor center again, improve the real-time character of the long-range monitoring system of irrigated area greatly.

2. AGENT AND MULTI-AGENT

Generally speaking, Agent is one kind independent calculation entity or procedure that can perceive the environment under specific social environment, and can realize a series of design objects by flexible and independent operate. As independent individual, under urged by certain goal, Agent has certain self- control ability of it's own behavior and inside state, and understand users' true intention as accurate as possible, adopt the positive behavior, effectively utilize various kinds of data, knowledge, information and calculation resource that can utilize in the environment, offer fast, accurate and satisfied services to user. (Zhang, 2004)

Multi-Agent systems is one computing system that one group Agent finishes some tasks or achieve some goals through cooperating, these Agents should cooperate and solve the problem that exceed each individual ability, they independent and distributed to run, and coordination and service each other, the goal and behavior between Agents contradiction and conflict, which is coordinated and solved through the means such as competition or consulting, to finish a task together. The important of Multi-Agent systems is the coordination of intelligent behavior between independent Agent, coordinate one's own knowledge, goal, skill, and scheme between them to produce the corresponding behavior or solve the problem. While the question is solved, for a common goal, these Agents share relevant knowledge of questions and solving method. Multi-Agent system requires system the exchange of every Agent has intelligence or self-organize ability (For instance reasoning, planning, studying, etc.) (Herry, 2006; Mohamed, 2007; Thibaud, 2007)

3. LONG-RANGE MONITORING SYSTEM OF IRRIGATED AREA BASED ON MULTI-AGENT AND GSM

3.1 Systematic structure

The long-range monitoring system of irrigated area is generally make up of the monitor and manage center, telemeter station and communication network between them. The monitor and manage center is a key part of monitoring system, its task can divide into many son tasks, so in the long-range monitoring system of irrigated area based on Multi-Agent and GSM, structured a monitor Agent union formed by a lot of Agent; In same way, the task of telemeter station also can divide into many son tasks, so structured a telemeter station Agent union; To counter the dispersiveness that telemeter station distribute, this system adopt GSM wireless network to realize the communication between monitor center and telemeter station, the concrete system structure is as Fig. 1 shows (Zhao, 2007).

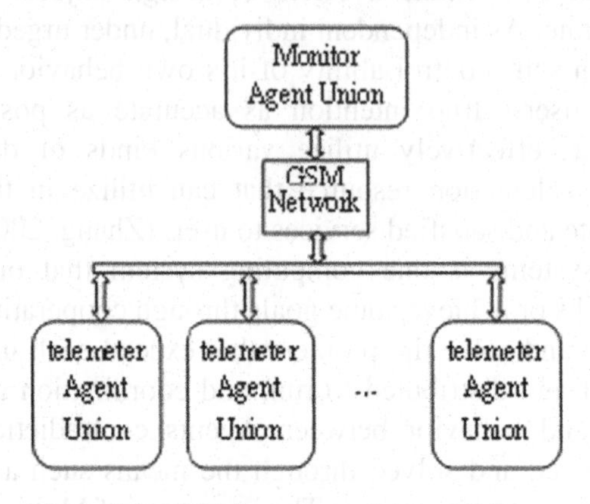

Fig. 1. Monitor system structure chart

3.2 Classification and function of Agent

Can be found from the systematic structure chart, the long-range monitoring system of irrigated area based on Multi-Agent and GSM make up of monitor Agent union and telemeter Agent union these two kinds of Agent unions and GSM communication networks, the important part used to finishing system task is two kinds of Agent unions. Each of These two kinds of Agent unions is composed of a lot of Agents, each one Agent has one's own tasks, and the completion of their task is serves for the completion of the overall task of the union.

3.2.1 Subdivision of the indicators

The monitor Agent union is a monitor center of the whole system, it is make up of one manage Agent and a lot of function Agents, the concrete structure is as Fig. 2 shows. Manage Agent1 (M-Agent1) manage the distribution of monitor center inner task and carries on communication with the external world; Information storage Agent1 (IS-Agent1) is used for storing the concretely information of water proportion of all telemeter station; Calculate Agent1 (C-Agent1) is used to calculate the water quantity distribution among a lot of telemeter stations; Data store Agent1 (DS-Agent1) is used to store the result that C-Agent1 calculated for after using. If there are data in the monitor center need to output (type), it is may add another data output Agent (DO-Agent).

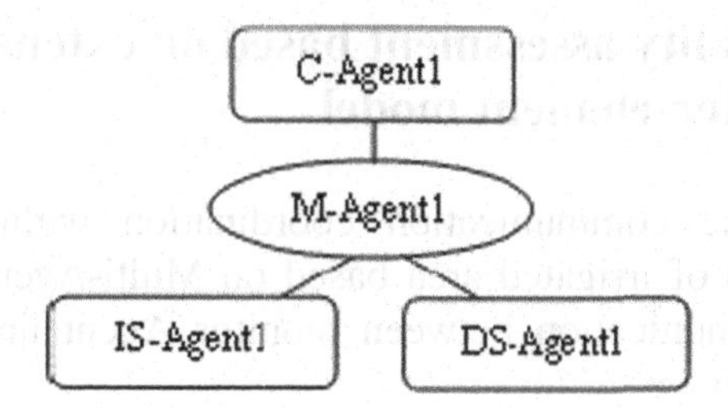

Fig. 2. Monitor agent union structure chart

3.2.2 Telemeter station Agent union

The telemeter station Agent union is make up of one manage Agent and a lot of function Agents too. Manage Agent2 (M-Agent2) responsible to distribute and resolve the task inside telemeter station and carry on communication with the external world. Information store Agent2 (IS-Agent2) is used for storing the relationship curve information of the water level with the gate level; The data store Agent2 (DS-Agent2) is used for storing the concretely relationship data of the water level with the gate level; Calculate Agent2 (C-Agent2) is used for calculating the relation of the water level and the gate level; The water level monitor Agent1 (WLM-Agent1) is used for monitoring the water level and obtaining water level information; Gate monitor Agent1 (BLM-Agent1) is used for monitoring the gate level and obtaining the gate level information; Operate control Agent1 (OC-Agent1) is used for controlling the opening and close of the gate. For the validity of communication, the communication inside the telemeter station Agent union is all finished through M-Agent2. The concrete structure is as Fig. 3 shows

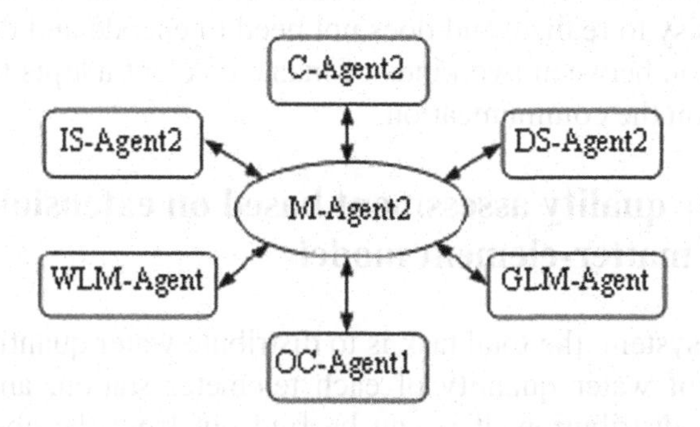

Fig. 3. Telemeter agent union structure chart

3.3 Air quality assessment based on extension of matter-element model

The key of the communication coordination within the long-range monitoring system of irrigated area based on Multi-Agent and GSM is the coordination communication between monitor Agent union and telemeter station Agent union.

Because the telemeter station of irrigated area is distributed more scattered, use the wired network though communication is high efficiency, but it is very difficult to concretely implement, so adopt GSM wireless network to carry on communication. Monitor Agent union and telemeter Agent union all with one manage Agent, each manage Agent is a administrator inside one's own union, responsible for coordinating communication with external besides responsible for distributing the union inside task and communication coordination among Agent, so the communication coordination of the whole big system is the communication coordination between manage Agent1 in monitor Agent union and manage Agent2 in telemeter Agent union. The communication of the wireless network is different from wired network, in order to realize communication, need to set up a communication and control module within each two kinds of management Agent, as the data communication interface of GSM network.

At present, communication agreement interface adopted by mobile communication of GSM is the AT order collection, which norm is had described in detail by GSM07.05 standard and GSM07.07 standard of ETSI standard, in the world all mobile communication equipment terminal such as short news terminal module, GSM cell-phone support above standards at present. Sending and receiving SMS message has two ways: Text Mode (the mode of text) based on AT order collection and PDU (protocol description unit) mode based on AT order collection. Because the mode of the text is simple and easy to realize, and does not need to encode and decoding, so the communication between two kinds of manage Agent adopts the mode of the text to carry on the communication.

3.4 Air quality assessment based on extension of matter-element model

As to this system, the total task is to distribute water quantity according to the demand of water quantity of each telemeter station, and carry on the result of this distribution. It is can be find out from the above discussion, every function Agent finishes the task of subsystem under coordination of manage Agent, and the subsystem and subsystem finish the transmission of the task information between them through the communication of GSM network, the concrete realization step is as follows:

Step1: M-Agent1 receives this task of water amount allotment from the network of rivers (or the message of water amount allotment imported by administrative staff), and send water amount allotment information to DS-Agent1;

Step2: DS-Agent1 inquires in its database according to the information of water amount allotment, and sends the result inquired to M-Agent1;

Step3: M-Agent1 receives the information, and analysis, if there is corresponding information, send information to M-Agent2, carry out Step9; Without, send a request of accessing information to IS-Agent1;

Step4: After received the request, IS-Agent1 accesses the proportion information of the water amount allotment of all gates and send to M-Agent1;

Step5: M-Agent1 send the information of water amount allotment of all gates and the information of overall water amount allotment to C-Agent1;

Step6: C-Agent1 calculates the allocation program and sends to M-Agent1;

Step7: M-Agent1 sends the information of new allocation program of water amount to M-Agent2, carries out Step9; Send this information to DS-Agent1 at the same time;

Step8: DS-Agent1 stores the new allocation program of water amount in its database for accessing next time;

Step9: M-Agent2 send the information of water amount to DS-Agent2;

Step10: DS-Agent2 inquires in its database according to the message received, sees whether there is a corresponding water level - gate level relation data, if it is exist, access this information and send to M-Agent2; If it is not exist, send the message of “without” to M-Agent2;

Step11: M-Agent2 receives the reply message of DS-Agent2, and analysis, if there is corresponding information, send the corresponding relation data of water level - gate level to OC-Agent1, namely carry out Step15; Without, send water amount information to C-Agent1, namely carry out Step12;

Step12: C-Agent2 receives the water amount information, and calculates the relation of water level - gate level, then send the result of calculation to M-Agent2;

Step13: M-Agent2 receives the result of calculation by C-Agent2 and sends it to OC-Agent1 carries out Step15; and send the result of calculation to DC-Agent2, carry out Step14;

Step14: DS-Agent2 stores the result of calculation received in its database for accessing next time;

Step15: OC-Agent1 controls the opening and close of the gate according to the received relation data of water level - gate level;

Step16: After the gate is opened, WLM-Agent1 and GLM-Agent1 begin to run, and the data monitored are sent to M-Agent2;

Step17: M-Agent2 sends the data to C-Agent2;

Step18: C-Agent2 calculates out water level - gate level relation curve and relation data, and returns the result to M-Agent2;

Step19: M-Agent2 sends the relation curve information to IS-Agent2;

Step20: IS-Agent2 compares the new relation curve with already existing, if the result is same, do not send reply information, if does not agree, store new relation curve information in its database, and send the result to M-Agent2;

Step21: After M-Agent2 received the reply of IS-Agent2, send the new relation data of water level-gate level to OC-Agent1, carry out Step15; Send to DS-Agent too at the same time, carry out Step22;

Step22: DS-Agent stores the new relation data of water level - gate level in its databases.

4. CONCLUSIONS

Multi-Agent system is composed of a lot of Agents, finishes the systematic task common and intelligence through the communication coordinating among them; GSM network is the wireless communication networks that is modern generally adopted. The long-range monitor system of irrigate area built by combine Multi-Agent system theory and GSM network together has a lot of advantage: at first, the monitor center forms a Agent union, finish its inside task by a lot of Agent through negotiation, instead of be finished by staff members through coordinated; Secondly, telemeter station (spot) construct a Agent union too, has more functions use it to replace the simple monitor equipment, some problems which can be solved inside the telemeter station no longer send information to the monitor center, thus reduce the communication to bear; Finally, the ones that communicate are not again the monitor center and telemeter station (spot) through GSM, but only manage Agent within them, communication amount reduces greatly, communication is more effective and swift.

It can find out through the test data, the application of the long-range monitoring system of irrigated area based on Multi-Agent and GSM can greatly reduce losses of water quantity, thus can greatly reduce peasants' water expenses burden. This system is low costs, convenient to safeguard, with complete function, and has the functions such as real-time control gate and data gathering, storing, dealing with and automatic yielding water quantity, etc. in real time, overcome the phenomenon such as difficult to macroscopic distributing water quantity, bad to regulating water quantity in real time, reduce work intensity, economize the valuable water resource.

ACKNOWLEDGEMENTS

In the course of research of this system, I have been supported vigorously by the fluid institue of lanzhou University of technology, especially got the help of the teachers of the engineering department of water conservancy and power, such as Xu cungdong, Wangzhijun and Biguiquan etc. In the stage of test and verify, I have got support and help of all staffs of one management office of one administration bureau, express seriously thank to their support and help in here!

REFERENCES

Herry Purnomo, Philippe Guizol, Simulating forest plantation co-management with a multi-agent system, *Mathematical and Computer Modelling*, Vol. 44. No. (5-6). 2006. pp. 535-552

Mohamed Salah Hamdi, A multi-agent approach to information customization for the purpose of academic advising of students. *Applied Soft Computing,* Vol. 7. No. 3. 2007. pp. 746-771

Thibaud Monteiro, Daniel Roy, Didier Anciaux, Multi-site coordination using a multi-agent system, *Computers in Industry*, Vol. 58. No. 4. 2007. pp. 367-377

Zhang jie, Gao liang, Li pei-gen, *The application of Multi-Agent technology in the advanced manufacture,* Science Press of China, 2004.

Zhang Kun-ao, Application of wireless local area network in interconnection of the monitoring system for large irrigating area, *Journal of Xi'an University of Science & Technology,* Vol. 25. No. (4). 2005. pp. 86-87

Zhao ting-hong, Qi xue-yi, Control System of Hydropower Economic Running Based on Multi-Agent Theory, *Journal of Micro-computer information,* Vol. 23. No. (4-3). 2007. pp. 54-56

THE RESEARCH ON GRAIN RESERVE INTELLIGENT AUDIT METHOD AND IMPLEMENTATION IN THREE-DIMENSIONAL STORES

Ying Lin [1,2*], Xiaohui Jiang [1]

[1] *School of Management, Chongqing Jiao Tong University, Chongqing, China, 400074*

[2] *School of Electronic Information Engineering, Tianjin University, Tianjin, China, 300072*

**Corresponding author, Address: No. 66 Xuefudadao, Chongqing Jiao Tong University, Nanan district, Chongqing Municipality, P. R. China, 400074. Tel: 023-66876662. Email: Linyingdyh@yahoo.com.cn*

Abstract: In order to eliminate the drawbacks in grain reserve management, a grain reserve intelligent audit method was designed. It was achieved by using the edge detection technology towards image samples to determine the edge of each object in the photograph, including the grain, the wall and benchmarks, and then separately holding pattern recognition to ascertain the identity of each object. Then according to the basic theories of 3-D reconstruction technology, combining with the location information of these objects in the original photographs, the underlying quantitative information could be dug out. At last, using grain weight measuring algorithm, which had been combined with the perspective error correction method, achieve the real-time and precise audit to the grain in stock and each installment.

Keywords: Grain reserve; Intelligent audit; Image recognition; Perspective error

1. INTRODUCTION

Analyzing the drawbacks in grain reserve management of the China for the last years, such as virtual storage, false discount and so on, it reveals that the supervision and audit system are not effective, mainly because of the contradiction between the high supervisory requirements caused by the

Lin, Y. and Jiang, X., 2008, in IFIP International Federation for Information Processing, Volume 258; Computer and Computing Technologies in Agriculture, Vol. 1; Daoliang Li; (Boston: Springer), pp. 525–532.

geographical dispersion of the granaries and the serious lack of powerful management technologies in reality. Currently, at home or abroad, the research on grain situation's surveillance (mainly including temperature, humidity and pest) have been conducted in depth, achieving several good results (Wei et al., 2004; Li et al., 2003). But on the area of grain audit, because an effective way to determine the weight of a large-scale object hasn't been designed out, this becomes the major bottleneck in grain reserve audit. Therefore, the key of eliminating the drawbacks is to use modern technologies to solve the problem of grain's weight determination in order to make the entire remote supervision comes true, instead of the manned field operations.

This paper mainly introduces an intelligent audit method in grain reserve management, based on the image recognition. As for the grain reserve audit, it includes the audit in two aspects, the weight and the time. So when designing the audit method, we focus on resolving the following three problems. The first is the recognition of a simple-frame image. Because, under the current technologies, it is relatively difficult to achieve dynamic-flow recognition, a feasible idea is resorting to the recognition of numbers of simple-frame static images to refer the dynamic-flow recognition, in order to realize the real-time supervision and recognition. This is the foundation of this intelligent audit method. Further more, the installment judgment, as concerned to grain audit that is the judgment of the starting and ending between which each batch of grain were transport into or out of the stores. The last is perspective error correction. Because of the influence of video camera's view angle, the far object is compress to smaller than the near one. So there must be a certain perspective error in the data dogged from the plane image, which will have seriously influence on the accuracy of recognition (Haven & Betty, 1977). So when designing the grain weight determine algorithm, the interference of the perspective error must be effectively excluded.

2. GRAIN RESERVE INTELLIGENT AUDIT METHOD

Generally, there are two approaches for grain weight measurement, the approach based on the direct measurement and the approach based on recognition. The former is usually to set up a dedicated channel and use correlative facilities to measure the weight, by which we can know the weight of each batch of the grain and the amount in stock. But this law requires a high cost in investment and maintenance. From the perspective of supervision, it asks for a high requirement of standardized operating, so if the grain custodians have illegal motivation, he also could carry out his illegal action easily. Therefore, using this approach to change the weak power of supervision is largely meaningless.

Currently, the approach based on the recognition, has rarely to be seen in reports (in the field of grain management). Image recognition, have two approaches which are worthy of paying closely attention to. The first is based on the distance measuring, which can be come true by using of the infrared or laser scanning to measure the distance between the probe and any point on the surface of an object. Through multi-detection, we can deduce the fitting three-dimensional shape of the object, from which we can calculate the bulk date of the digital object body. This approach asks for a long time and a high cost of equipments, so it is obviously not suitable for such a large amount of dynamic sites in grain reserve audit. Simultaneously, we can know that the key of this 3-D reconstruction, based on pattern recognition, is the third dimension's infers, by our experiential knowledge. (Qing et al., 2005).

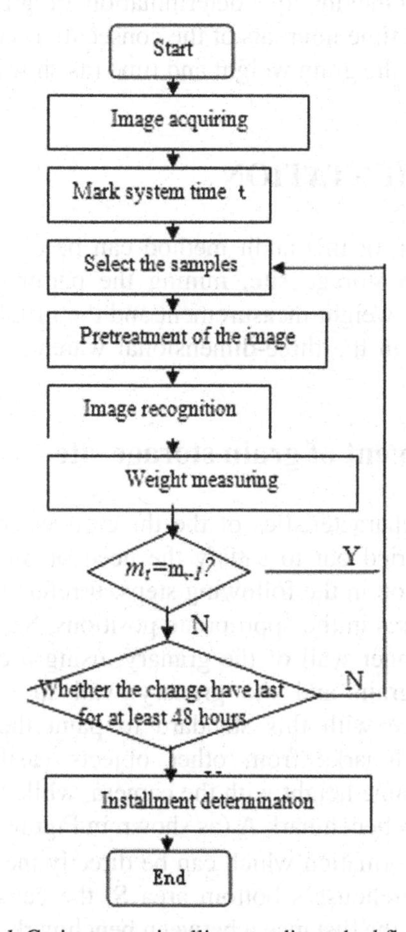

Fig. 1. Grain reserve intelligent audit method flowchart

The other is based on plane image recognition, reaching the identification purpose by pattern recognition. Obviously it can also achieve the dynamic-flow recognition by the contrast between consecutive simple-frames. And this method can achieve remote automatic control just under the supervisory camera in granary, thereby significantly reducing supervisory strength.

Based on the above analysis, we design a grain reserve audit method, whose audit processes are as follows: At first, aiming at the requirement of quantitative information in grain weight measuring algorithm, we should correspondingly do some pretreatment in granary site. Secondly, after the dynamic image flow, which was taken by video camera, was transmitted to the center computer through the Internet, the system will automatically select a simple-frame image as a sample by a fixed time interval. Then the sample will be taken to edge detection and image recognition, in order to dig out the underlying quantitative information. Further more we use the grain weight measuring algorithm, which have been combined with the perspective error correction method, achieving the determination of grain weight. Finally, according to the fixed time intervals of the consecutive samples, we can infer each installment from the grain weight and time (as shown in figure 1).

3. IMPLEMENTATION

The implementation of this audit method can be divided into four steps, pretreatment of grain storage site, mining the parameters of quantitative information, the grain weight measurement and the installment judgment. A experiment was held in the three-dimensional warehouse, a typical way of grain reserve.

3.1 Pretreatment of grain storage site

According to the characteristics of the three-dimensional warehouse, the pretreatment was carried out to satisfy the need of mining the underlying quantitative information in the following steps. It refers to, firstly, setting up the supervisory cameras in the appropriate positions. Secondly, painting two benchmarks on the inner wall of the granary, using a color which is quite different from the grain and the granary wall in color and grayscale features(in accordance with this standard to paint the benchmarks is to distinguish the benchmarks from other objects easily). The height of benchmark A is the same height with the camera, while benchmark B is in a certain distance below benchmark A (as shown in Figure 2).

In this step, the information which can be directly measured out includes: three-dimensional warehouse's bottom area S, the camera and benchmark A's vertical height H, the distance a between benchmark A and benchmark B.

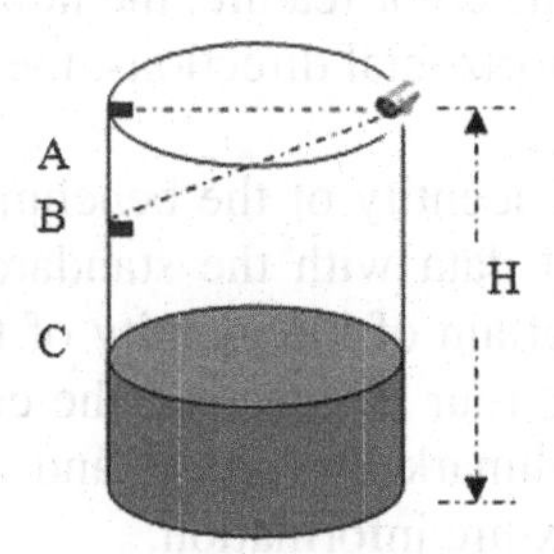

Fig. 2. Picture show of Three-dimensional stores

3.2 Mining the parameters of quantitative information

This step is the key of this intelligent audit method. Concretely, it can be achieve by the following two stages.

Firstly, select a simple-frame image sample from the dynamic flows by a a fixed time interval and evaluate it (the purpose is to exclude the unsatisfactory samples which have too much noise). Then with the computer's system time to mark the time for these simple-frame image samples, intending to be more facilitate to judge each batch's starting and ending of time. Evaluation of the selected samples is accomplished by scanning the image sample point-by-point, so as to form the phase-gray matrix. Then combining the basic principles of edge detection technology (Canny, 1986; Demigny, 2002), contrast the grayscale values of each point with the direction of the gradient near it, by which the image is divided into a set of meaningful regions (Liu, 2001) (each region can be considered to be a kind of object). Therefore, through the contrast of the regions with the number of the objects we have known in advance, can prove whether there is too much noise.

Secondly, to the eligible samples, we hold the image recognition with each meaningful region in the sample, in order to ascertain the identity of each object. Here, we use pattern recognition, which includes two major steps, the establishment of the standard database (that is, all of the information of possible objects and their parameters are established and saved in a database in advance), and the establishment of the test data (Cheng, 2006).

In response to the determination of grain weigh, the main task of recognition includes the identity of two benchmarks and their location in the original image, the identity of the grain surface and its location, from which we can also know the number of the pixel (N_1) between benchmark A and B in the original image, and the number of the pixel (N_2) between benchmark A and grain surface C. Therefore, the information need to be saved in the standard database includes, the data of benchmark and grain surface's features, mainly including the color feature, the number-ration of pixel of the benchmark in vertical and horizontal directions, the grayscale feature and the texture information.

The determination of the identity of the benchmarks (or grain surface) is through comparing the test data with the standard data. For the accuracy close to 100%, we make certain of the identity of the benchmarks (or grain surface) from the following four parameters: the color feature, the number-ration of pixel of the benchmark in vertical and horizontal directions, the grayscale feature and the texture information.

If a test data matches with the standard data in over 95% confidence interval of the above four types of features, we can conclude that it is the benchmark (or grain surface). Based on this, the parameters N_1 and N_2 can be measured out easily.

3.3 The grain weight measurement

Till now, the known information includes the height of benchmark *A* and *B*, the distance *a* between benchmark *A* and *B*, the height *H* of the granary, the bottom area *S* of the granary and density ρ. Then we can calculate out the weight of the grain in stock. The algorithm is as follows.

Step 1: according to the ratio of N_1/N_2 and the length of *a*, we can acquire the length of *AC'*, namely $b=(N_2\times a)/N_1$, which is the length of *AC* in the photograph reflect in the reality, having considered perspective err, and we can know that its length isn't unequal to *AC*.

Step 2: combining with the theory of image forming and plane geometry, we establish Cartesian coordinate system XY, as illustrated in figure 3. Then point *C'* is a point in a circle whose center is point *E* and radius *r* is the length between point *A* and *E*, and we can know that *C, C ', E* has been in a same line; It comes to the equation:

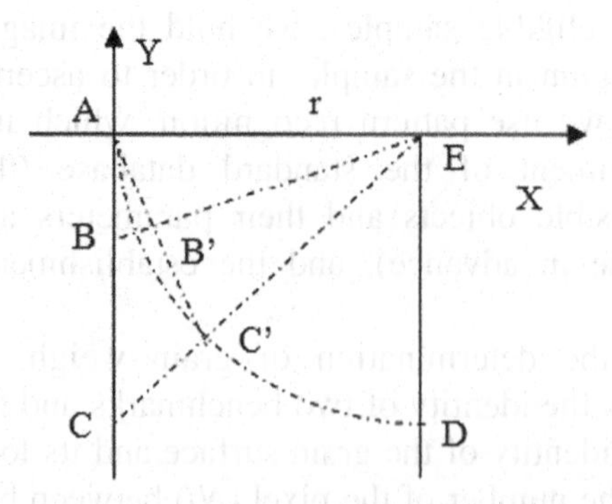

Fig. 3. Three-dimensional stores volume calculation model

$$(x-r)^2+y^2=r^2 \tag{1}$$

Step 3: according to the length of *b*, we can calculate out the coordinate of point *C'*(x, y), and

$$x^2+y^2=b^2 \tag{2}$$

From equation □ and □, we got:

$$\begin{cases} x=b^2/2r \\ y=\dfrac{-b\sqrt{4r^2-b^2}}{2r} \end{cases}$$

Step 4: as we have know the rdinate of point *E*, we can got the equation of line *CE* from two point, point *E* and *C'*:

$$y=\frac{b\sqrt{4r^2-b^2}}{2r^2-b^2}(x-r) \tag{3}$$

Then the coordinate of point *C* can be calculated

$$C=(0,\ \frac{br\sqrt{4r^2-b^2}}{2r^2-b^2})$$

At last, we can obtain the height of the grain in the granary:

$$h=H-\frac{br\sqrt{4r^2-b^2}}{2r^2-b^2} \tag{4}$$

So, the weight of the grain is easy to know by the equation of:

$$m=\rho\times S\times h. \tag{5}$$

Aggregate analysis, the calculating method of the height of the grain can resolve the perspective error problem of the objects in the image due to different shooting angles.

3.4 The installment judgment

The former steps are mainly to solve the problem of the grain weight determination. The grain reserve audit, as we have said before, contains two aspects, the weight and timing. Therefore, the installment judgment is also very necessary. The algorithm is as follows.

As shown in figure 2, according to the samples collected by every fixed time interval, after we have got the data of the grain's weight m_t in the granary, then contrast the data with the former one m_{t-1}. If they are not equal, we think that the amount of grain is changing, this also indicates that this batch of grain haven't not yet complete, so we continue to consider the next sample. If $m_t = m_{t-1}$, then we make a judgment that whether the previous grain bulk hasn't been changed for over at lease 48 hours. If so, we can get the conclusion that this batch has completed, thus combining the initial time when grain weight began to change, the start time and end time of this batch can be inferred. And according to the difference between the two time, we can got the weight of this installment, namely $m = m_t - m_o$. If it is less than 48 hours, we will continue to select next sample and contrast their weight, until the grain weight hasn't changed for over 48 hours.

The accuracy of this intelligent audit method is mainly determined by the precision of the supervision cameras and image recognition, because of the relatively high stability of the images acquired from the close granary, the surrounding environment (such as brightness, humidity) generally will have little influence on it. And the quantity of the objects in the granary is relatively small, so the considering region and its edge of the images will be very clear, which is easier to achieve the edge detection and recognition. Currently, through experimental, the intelligent audit method is proved to be high accuracy, and with the continuous development of digital process technology, the accuracy of this grain reserve intelligent audit will also increasingly high.

ACKNOWLEDGEMENTS

This study has been funded by Finance Bureau of Chongqing Municipality. And sincerely thanks are also due to the Grain Administration of Chongqing Municipality for providing the experiment base for this study.

REFERENCES

Haven, Betty H. (1977). A Method for Minimizing Perspective Error. Journal of Physical Education and Recreation. 48(4), 74-77.

Wei Yong-lu, Zhang Ji-yue, Ji Wen-gang. (2005). Design of distributed control system for grain depot based on MODBUS/TCP Protocol. Beijing Institute of Petrochemical Technology Journal. Chn, 13(2): 2-4.

Li Jian-hua, Sun Hai-bo, Liu Zhan-liang, et al. (2003). The design and realization of supervising on provisions situation system [J]. Journal of the Hebei Academy of Sciences, 3(4): 224-227.

Qiang Wei-zhe, Song Guang-hua, Zheng Yao. (2005). 3D reconstruction based on image segmentation [J]. Computer Engineering and Application, 36: 77-82.

Liu li-bo. (2001). Summarization of segmental way of imagine. Ningxia Agricultural College Journal. Chn, 4(22), 51-53.

Canny, J. (1986). A Computational Approach to Edge Detection [J], IEEE Transations Pattern Analytical Machine Intelligent 8: 679-698.

D. Demigny. (2002). On Optimal Linear Filtering for Edge Detection, IEEE Trans. Image Processing, 2002, 9(11): 728-737.

Cheng Xiao-chun. (2006). A method of shape recognition [J]. Pattern Recognition and Artificial Intelligence, 6: 126-132.

STUDY ON REAL-TIME VIDEO TRANSPORTATION FOR NATIONAL GRAIN DEPOT

Ying Lin[1,2,*], Liang Ge[1]

[1] *College of management, Chong Qing Jiao Tong University, Chongqing, China, 400074*

[2] *School of Electronic Information Engineering, Tian Jin University, Tianjin, China, 300072*

* *Corresponding author, Address: No. 66 Xuefudadao, Nanan district, Chongqing Municipality, 400074, P. R. China, Tel: +86-23-66876662, Email: Linyingdyh@yahoo.com.cn*

Abstract: Because current remote monitor systems can't deal with problems of real-time transmissions in the bad condition of network very well, this article presents a study which is combined with the evaluation of video streams, adjustment and sealing user-defined data pockets and discarding useless data pockets. The goal of this solution is to transmit video information of national grain depot. And its practical use shows that the system has good effect.

Keywords: code rate, adaptive transmission, RTP protocol

1. INTRODUCTION

It's known that, professional companies in China take responsibility of depositing the nation grain. But without strict supervision, someone will, aiming of private benefit, conduct corrupt behaviors, such as making the false bill of document and imitating good grain with bad grain and so on. On the view for this situation, government should construct a series of supervising system which can provide supervision at any time. The most important thing is how to copy with the quality and transmission of real-time videos. They are proof to accuse those criminal behaviors. This monitor system is operated depending on accuracy of real-time video data. However, the quality of video data is associated with conditions of network environment. If the bandwidth is stable and the speed of network transmission is ideal, the packet-losing rate will decrease to make sure video

Lin, Y. and Ge, L., 2008, in IFIP International Federation for Information Processing, Volume 258; Computer and Computing Technologies in Agriculture, Vol. 1; Daoliang Li; (Boston: Springer), pp. 533–541.

transmitted successfully, besides phenomena of frame-losing and time-delay will disappear. But the bandwidth is varying from time to time. If there are so many data pockets in the channel, the increasing rate of packet-losing will impact the transmission of video stream very seriously (B.J. Kim et al., 2000). The media streams compressed by MPEG4 technology become hard to transmit under such bad conditions. It's because that MPEG4 technology split video stream into several layers, losing of key frames will damage the quality of video and the network spending will become larger result from error inspection and retransmission protocol (Talley et al., 1996). In order to solving this kind of problem, this article presents a method. It is based on the speculate speed of network with AMID algorithm and adjusted the speed rate of video transmission automatically and sealed data packets in user-defined format and discarded some useless data packets.

2. GENERAL DESIGN

The system of real-time video transportation is composed by three components: sender, net medium and receiver. This paper concentrates on introducing the design of sender. The overall structure of transportation system is shown in Fig. 1.

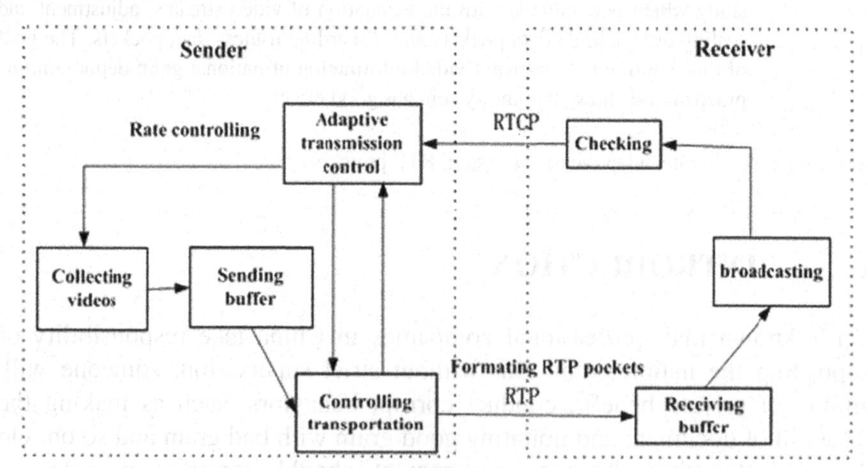

Fig. 1. Overall structure of transportation system

2.1 Obtaining and compressing real-time video streams

Video streams are received at sending-port through cameras devised in grain depots. These videos will be transported to the media server and compressed into MPEG4 video streams. The reason to choose MPEG4 for transporting format is that, it's apparently different with other video compressed technology. MPEG4 is based on objects and it spilt a video file into different objects that are formatted by a special object layer. (The structure of MPEG4 frame is shown in Fig. 2.) Each layer contains much information about figure and texture and other aspects. Moreover, MPEG4 affords extension of time-field and space-field. It will operate some changes on basic layer according with current conditions of network. Furthermore, in order to perfect the quality of video, MPEG4 technology could insert some frame into basic layer and increase or decrease the resolution. In a word, it is most suitable to be used in the field of long distance net TV inspection.

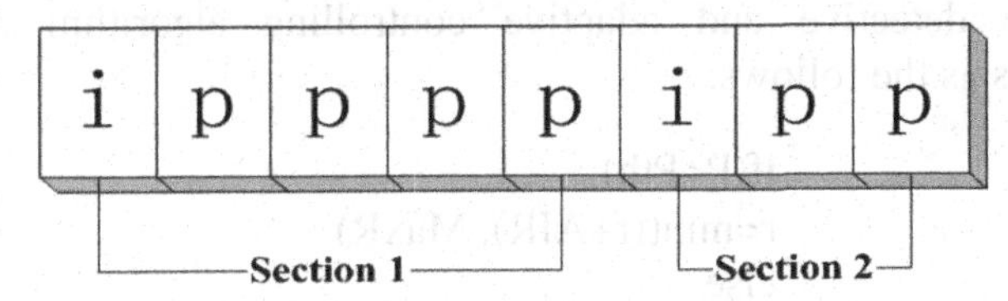

Fig. 2. Map showing the location of the research area

2.2 Design of video transportation

Step1: Choosing RTP/RTCP protocol for transportation.

RTP/RTCP protocol is a real time transmitting method. It is specially used for the transportation of media data. This protocol could achieve stream media data for singleness & group play in internet. It can rearrange the video stream frame by correct order, checks the integrity of frames, and supplies some services such as security guarantee and so on.

RTCP is one kind of controlling telegram that is sent by sender. Its main function is to afford QOS quality feedback. As a part of RTP protocol, it is relevant with stream control supplied by other protocols. And feedbacks operate adaptive code-control directly. The function of feedback is executed by both sender and receiver.

At the conversation of RTP, all members send RTCP control packets periodically. Server can take advantage information to change transmitting speed optionally. The corporation between RTP and RTCP, can perfect the efficiency of transmission by necessary feedback and least cost. Therefore, this protocol is pretty suitable for real-time transmission. Considering information collected from feedbacks, it's easy to make a fitful strategy.

Step2: Evaluating the speed rate of transmission on the foundation of feedback data brought by RTCP pockets, the rate of video stream will adjust by itself.

The solution is based on adaptively adjusting the speed rate of video stream transmission; the sender analyzes the current conditions of network through feedback information brought by RTCP pockets (J.Y. Tham et al., 1998). Bandwidth can be calculated with the loss-pocket rate in the QOS report. After getting approximate value of rate, one can use programs at the sender to adjust transferring rate into a suitable value that is useful to real-time transmission.

There exits two algorithms to adjust the sending rate: the model algorithm and the detect algorithm. The foundation of model algorithm is loss-pocket rate and time for sending and receiving data pockets and maximum number of communicating data pockets, which should be used to calculate the sending rate (D. Mills, 1992). Detect algorithm is that sender evaluates the speed of network by frequent adjustment of sending rate. This solution chooses the detective and adaptive controlling algorithm AMID. Its description is as the follows:

```
if(P<Pth)
r=min((r+AIR), MaxR)
else
r=max(□r, MinR)
```

Where: P stands for the actual loss-pocket rate; Pth stands for a trigger value in a range of time; r means sending rate. MaxR and MinR separately stand for maximum value and minimum value of sending rate that has been set before. AIR means accelerating rate, □ is subtrahend factor and its value is between 0 and 1 (Li, 2006).

The meaning of this algorithm is that, in a defined range, one can increase the sending rate untill the loss-pocket rate is too great to assure affording an accurate play in the receiver. And the next step is to reduce the sending rate to make sure the loss-pocket rate in an acceptable range, following increasing sending rate gradually.

Step3: Analyzing MPEG4 pockets and making a RTP pocket in a particular format.

The chief difference between MPEG4 and other traditional video compressing standard is that MPEG4 is based on objects. This technology splits video data into many different objects and forms a layer for each object which contains figure and texture and other information (S. Palacharla et al., 1997).

Grammatical layers of MPEG4 stream have four aspects: video communication, video objects, video object layer, and video object plane (VOP).

VOP is a frame of video object. MPEG4 separately codes to every VOP and gender three different frame styles: I VOP, P VOP, B VOP. I VOP is very critical to the quality of video and it has no relations with other adjacent VOPs; P VOP needs I VOP in front of it as a consult to compensate movement (J. Shapiro, 1993); B VOP works with adjacent VOPs. Because

VOP is the basic unit for saving video information, so its construction will be introduced by blow picture.

Synchronous code is very important to video play, if once it misses, video will not play well. Report head of video pocket is also crucial, because the receiver need it to sort a correct order for video pockets. Besides as the quantity of information contained by different video pockets, some huge pockets will be lost during the transmission in the network especially under the terrible net conditions, this is one of main reasons that I use the particular defined data pockets to do transmissions.

The particular defined data pocket means that, after analyzing each VOP pocket and getting relative information, one can seal a new pocket by RTP pocket structure which will be introduced in the next section. The focus of this article is to design a new RTP pocket structure which is suitable for special requirement.

The structure of New RTP pocket is shown in Fig. 3.

<table>
<tr><td>V=2</td><td>P</td><td>X</td><td>CC</td><td>M</td><td>PT</td><td>Sequence number</td></tr>
<tr><td colspan="7">Timestamp</td></tr>
<tr><td colspan="7">SSRC</td></tr>
<tr><td colspan="7">CSRC</td></tr>
<tr><td colspan="7">Mpeg4 Video data</td></tr>
<tr><td colspan="5">RSN</td><td colspan="2">CK</td></tr>
</table>

Fig. 3. A new RTP pocket structure is suitable of special requirement

The meaning of each element:

V: version

Extension - X: defined by RTP structure

PT: a introduction of sort of load interpreting the style of code

Sequence Number: the number of each RTP data pockets. It's used to set up a correct order of data pockets and inspect whether there are errors and damage in pockets.

Marker - M: defined by operating structure

SSRC: help receiver identify the adscription of all streams with only one number that sender supplies. SSRC is a strict random number.

CSRC: identify the streams.

Besides (RSN) stands for synchronous code taken by this frame, and critical key (CK) stands that whether this frame is a key frame that takes great responsibility for video quality. This is because, under the conditions that bandwidth is terrible, and some huge pockets will be lost during transmission. And there will be many chances to form congestions. The receiver has to attempt many times to send repeat requests. In this situation, the spending of network will become so large, and the quality of real-time transmission will be impacted seriously (H. Schulzrinne, 1995).

So we can average huge video streams to many RTP pockets, this solution will alleviate the pressure of each pocket and fitful for transmission (B. Paul et al., 1999). Each pocket will take the same RSN to certificate that all pockets come from the same frame and the sequence number will increase by one to record their order.

Step4: Discarding frames which aren't key frames and adjusting the rate, if the condition of network is poor.

Even though the speed rate can be evaluated from Qos report, it's also difficult to operate a perfect transmission because of the abnormal variety of network. If the condition is much better than before, the transmission will be very successful. But if it's worse, the bandwidth will be narrower and loss-pocket rate will increase.

So it's hard to transmit video streams with the speed calculated before. In order to decrease the pressure of network, the reasonable approach is to adjust the rate again by discarding some useless RTP pockets that are not important for video play.

For MPEG4 video stream, I frame is compression of static imagine, P frame is formed depended on previous I frame. I frame is consulting frame whose losing will bring great damage to the quality of video as other frames can be sorted by a correct order. This solution is that discarding some frames that do little impact with video quality, by precondition that video could be play well at the receiver. (A. Kantarci et al., 2000).

Therefore, B frames which are relatively useless and P frames that are far away critical frames are should be discarded. This kind of operation should be done in the buffer. Buffers in this paper are sorted to two aspects, one part is buffer used to real-time transmission at the sender, and another part is used to save data pockets and wait for the command of retransmission.

3. IMPLEMENT

This experiment needs three personal computers to take responsibility of MPEG4 Collecting Server and Net Transportation Controlling Server and MPEG4 Receiver Client. Their functions are different: main functions of MPEG4 Collecting Server are collecting the spotted real-time video, compressing the video stream, making video files and saving some videos to the hard disk; Main functions of Net Transportation Controlling Server are providing sending buffer for stocking real-time video stream, analysis of current net situation, speculating the speed of transportation next time span, producing a RTP pocket in a particular format and checking information coming from receiver; Main functions of MPEG4 Receiver Client are receiving video streams, checking and rearranging the order of video stream pockets, sending requiring messages to sender and playing videos and so on. The structure of the circumstance of experiment is shown in Fig. 4.

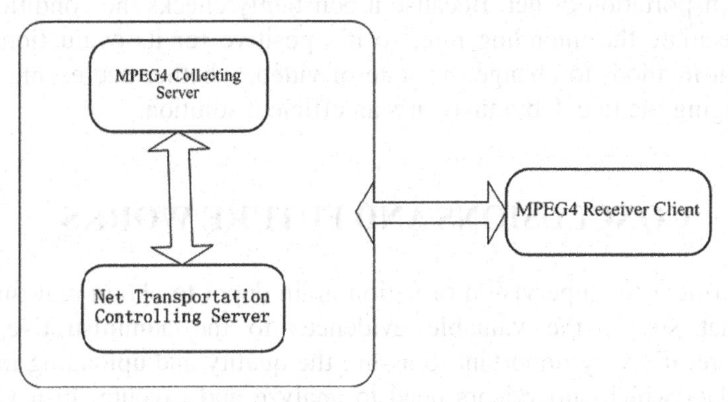

Fig. 4. The structure of the circumstance of experiment

The process of transportation is consisted by several steps as following:

(1) Collection of video streams:

Video streams are collected by cameras settled at grain depots, and sent to the MPEG4 Collecting Server in special line.

(2) Disposal of video streams:

Net Transportation Controlling Server speculate the speed of net transportation with accordance to RTCP data pockets, and send messages of adjusting rate to MPEG4 Collecting Server. According to evaluated result, system has different choices to do the transportation. MPEG4 streams can be transported steadily in the range of rate between 4.8kbit/s and 64 kbit/s (Zhang, 2006). But the situation of network is varying from time to time. So the condition may be very terrible for transportation in some time spans. It's hardly reach to the basic condition for video transportation. At this situation, system can call off these transportations, save video into the data base with unique mark. When the condition is suitable enough for video transportation, system will automatically search whether there are some videos without transportation, and send this kind of videos at prior. If current situation of net reaches the basic condition of transportation, system will adjust the compressing rate and send video streams to the buffer of Net Transportation Controlling Server.

(3) Transportation of video streams:

If there doesn't exist some barriers in the network, system will immediately execute transportation; If the condition is getting worse and the transporting speed is fewer that evaluated speed, the system will remove some useless data pockets in order to reduce the bulk of the whole RTP pockets, and identify each pockets with particular mark to avoid making mistakes in checking process in the receiver.

(4) Receiving video streams:

The receiver will check the order of data pockets. And if there are some phenomenons of disorder, the receiver will send out messages.

Comparing with other solutions dealing with the same problem, this solution has its advantages. This approach use a more positive way to copy with transportation of net. Because it constantly checks the condition of net and speculate the intending rate, so it's positive for its evaluation. It uses different methods to change the state of video, whether decreasing the bulk or changing the rate. Obviously, it's an efficient solution.

4. CONCLUSIONS AND FUTURE WORKS

It's critical for supervision of nation grain depot to obtain real-time video data that supply the valuable evidences to the administrative officer. Therefore, it's very important to assure the quality and uploading in time of video data which supervisors need to analyze and execute. In a view, this paper designs a series of solution which can deal with some difficult problems during the process of the real-time transmission of huge video information. But if the current condition is extreme terrible, it is hard to operate real-time transmission. In this situation, one had better cancel this transmission and save all data pockets into the hardware waiting for next transmission when the condition becomes better. Following with the further development of web technology and hard devices, you will have variety of solutions.

ACKNOWLEDGEMENTS

This work is supported by the financial bureau of Chongqing City.

REFERENCES

Zhang Qing. 2006. The Analysis of the Technology Points and the Prospects for the MPEG4 Applications, Computer Knowledge and Technology, 2:145–147

Li Yun. 2006. Internet Transmission Control Protocol in Streaming Media Proxy Server, Computer and Digital Engineering, 34(7):151–152

A. Kantarci and T. Tunali. 2000. The design and implementation of a streaming application for MPEG videos, in Proc. 2000 IEEE International Conference on Multimedia and Expo, Vol. 2, pp. 1021–1024.

B.J. Kim, Z. Xiong, and W.A. Pearlman. 2000. Low bit-rate scalable coding with 3-D set partitioning in hierarchical tree (3-D SPIHT), IEEE Transactions on Circuits and Systems for Video Technology, Vol. 10, pp. 1374–1387.

D. Mills. 1992. Network time protocol (v3). RFC 1305, Internet Engineering Task Force, April.

S. Palacharla, A. Karmouch, and S. Mahmoud. 1997. Design and implementation of a real-time multimedia pre-sentation system using RTP, in Proc. The 21th Annual International Computer Software and Applications Conference, pp. 376–381.

B. Paul, D. Gibbon, G. Cash, and M. Civanlar, Vtondemand. 1999. A framework for indexing, searching and on-demand playback of RTP-based multimedia conferences, in Proc, IEEE 3rd Workshop on Multimedia Signal Processing, pp. 59–64.

H. Schulzrinne, S. Casner, R. Frederick, and V. Jacobson, RTP: A Transport Protocol for Real-Time Applications. RFC 1889, Internet Engineering Task Force, 1995.

J. Shapiro. 1993. Embedded image coding using zero-trees of wavelet coeffcients, IEEE Transactions on Image processing, Vol. 41, pp. 3445–3462.

T. Talley and K. Jeffay. 1996. A general framework for continuous media transmission control, inProc. 21st IEEE Conference on Local Computer Networks Minneapolis, MN, Oct. pp. 374–383.

J.Y. Tham, S. Ranganath, and A.A. Kassim. 1998. Highly scalable wavelet-based video codec for very low bit-rate environment, IEEE Journal on Selected Areas in Communications, Vol. 16, pp. 12–27. Jianyu Dong received the B.E. and M.S. d

THE KEY OF BULK WAREHOUSE GRAIN QUANTITY RECOGNITION

Rectangular Benchmark Image Recognition

Ying Lin [1,2,*], Yang Fu [1]

[1] *College of management, Chong Qing Jiao Tong University, Chongqing, China, 400074*

[2] *School of Electronic Information Engineering, Tian Jin University, Tianjin, China, 300072*

* *Corresponding author, Address: No. 66 Xuefudadao, Nanan district, Chongqing Municipality, 400074, P. R. China, Tel: +86-23-66876662, Email: Linyingdyh@yahoo.com.cn*

Abstract: According to requests of bulk warehouse grain quantity recognition, we take the scene video as identified object to obtain the object's boundary from the result of edge detection difference iterative analysis. By using region iterative threshold value of gradient operator fitted closely with identified target carries to execute the picture characteristic second-extract and then to carrying on rectangular benchmark judgment using the membership functions of fuzzy recognition, we adopt the Visual C++ realized this recognition algorithm. And the experimental results show that this recognition algorithm effectively enhances the anti-jamming, robustness and the recognition precision and effect.

Keywords: edge detection difference, fuzzy recognition, membership functions, iterative analysis

1. INTRODUCTION

From the last 20 years' practice of storage grain regulatory in China, the quantity of grain reserves supervision and auditing is still manual regulation. Because of the geographical dispersion of reserve granary and the features of whole process and real-time supervising, it leads to the failure of grain reserves effective supervision. So it is necessary to adopt a computerized intelligent recognition technology to achieve the precious quantity of grain reserves automatic monitoring and auditing in a smart method. The key for video-based grain reserves automatic monitoring and auditing technology is

Lin, Y. and Fu, Y., 2008, in IFIP International Federation for Information Processing, Volume 258; Computer and Computing Technologies in Agriculture, Vol. 1; Daoliang Li; (Boston: Springer), pp. 543–551.

based on the result of scenes video image recognition to calculate the quantity of grain granary.

From (Fig. 1), it indicates that the bulk of the reserve granary bottom of the area set S, grain density of ρ reserve granary high set T are known factors, grain surface above the benchmark length is L, M is the quality of grain, then:

Grain quality $M = V \times \rho$

Where $V=S \times H$, $H=T-L$

According to the above statements, the key of grain quantity recognition is the identification and estimation of reserve granary benchmark. The key problem of real-scene recognition of granary reserves benchmark is how to identify the object boundary and discriminate the identified targets. The existing Robert operator partial detection methods can effectively extract the object boundary (M.D. Kelly, 1973). But it has poor adaptive ability for the interference, and also the algorithm efficiency is not high efficiency enough. Therefore, seeking stronger anti-jamming detection methods and target's characteristic judgment method has become an issue that must be addressed. In this paper we introduce a combining intelligent detection method to identify the benchmark by Robert Operator local detection and the overall threshold detection algorithm (Canny J., 1986).

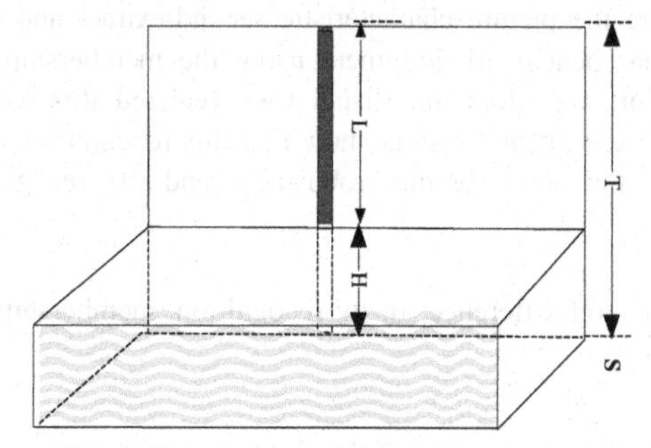

Fig.1. Sketch map of bulk grain warehouse

2. INTELLIGENT DETECTION METHOD

The use of computer recognition is mainly by computer automatic identification and understanding of images. To this purpose, the decomposition of image that contains a large variety of characteristic information as a key step. The accuracy level of intelligent recognition is directly related with the shape, color and specific size of identified targets. A key technology of image detection is how to determine ROI (region of interests), the image divided by Robert operator partial detection operator gradient method and operator of regional iteration threshold image segmentation algorithm combination of images to the second feature extraction, a new intelligent detection method is involved (Pursuing technology, 2006).

This method includes following three major steps.

1. Color extraction: The setting of real-scene should fully consider the issue of the benchmark's color. Firstly image is converted to color space, extracting specific benchmark for the red, tentatively identified ROI region.
2. Edge extraction: Based on the edge of extraction from the red border regions, we use Robert segmentation local edge detection operator and combine with the iterative threshold to extract the pixel of the same difference, to some extent reduce the computational time and the complexity of the recognition (Cheng Xiao-chun, 2006).
3. Rectangular benchmark judgment: Judging from the extracted targets, through certain principles found matching rectangular regional area.

2.1 Color extraction

Based on the specific color of rectangular benchmark, firstly we extract the color from the hue, saturation and intensity model (HSI) color space, initially identifying ROI region. In the actual image acquisition process, from the acquisition of Monitor RGB format for images, since the first need to achieve a conversion to RGB space HSI space (Kim, V. & L. Yaroslavskii, 1986).

First step will be a normalized mode:

$$r = \frac{R}{255}, b = \frac{B}{255}, g = \frac{G}{255} \tag{1}$$

Then proceed into:

$$H = 90° - \arctan\left(\frac{F}{\sqrt{3}}\right) \times \frac{180°}{\pi} + \{0, g > b; 180, g < b\} \tag{2}$$

Where:

$$F = \frac{2r - g - b}{g - b} \tag{3}$$

It is known from HIS color model picture that one can extract the regions of arbitrary color accord to our needs (Tou, J.T. & R.C., 1981). In the benchmark identification process, the design needs extract the red region, the corresponding distribution of red region is 315°–360° and 0°–23°, we can determine the initial ROI region.

2.2 Edge extraction

The main purpose of edge extraction is more accurately to reduce the number of pixels in ROI region, on the basis of color extraction result acquired, in order to effectively improve the accuracy of feature extraction. Edge extraction carried on the source gray image, using the rate of change of intensity and direction of changes in the method Robert boundary segmentation local edge detection operator method checks each pixel point for the neighborhood, and the completion of the pixels in a neighborhood of gray rate of changes, which in the direction of quantifying the identification, then to the border Robert segmentation local edge detection operator method calculate the edge pixels gradient operator basis second operational threshold will have the same rate of difference of gray pixels constitute closure and connectivity region. (Marr, D. & Hildreth, E., 1980)

Then, For two-dimensional image *x, y* respectively, on behalf of the pixels in a two-dimensional pixel-point benchmark of the abscissa, longitudinal coordinates, the position *f(x, y)* of the gradient can be expressed as a vector, using G_x and G_y Specific formula as follows: Gradient vector can be expressed as the following:

$$\nabla f(x,y) = \begin{bmatrix} G_x \\ G_y \end{bmatrix} = \begin{pmatrix} \dfrac{\partial f(x,y)}{\partial x} \\ \dfrac{\partial f(x,y)}{\partial y} \end{pmatrix} \tag{4}$$

Set θ_r Represent gradient direction:

$$\theta_r = \tan^{-1}(f_x / f_y) \tag{5}$$

In the direction of θ_r the rate of change velocity:

$$g(x,y) = |\nabla f(x,y)| = \sqrt{(\frac{\partial f(x,y)}{\partial x})^2 + (\frac{\partial f(x,y)}{\partial y})^2} \tag{6}$$

In the margin calculation gradient operator is equivalent to the following two ways calculated norm:

$$g(x,y) = \left|\frac{\partial f(x,y)}{\partial x}\right| + \left|\frac{\partial f(x,y)}{\partial y}\right| \tag{7}$$

Or to use infinity norm:

$$g(x,y) = \max\left(\left|\frac{\partial f(x,y)}{\partial x}\right|, \left|\frac{\partial f(x,y)}{\partial y}\right|\right) \tag{8}$$

In fact, the use of computers to handle most of the images is targeted at digital images, in the field of digital image processing, using the above difference for alternative differential. The definition of the form is as follows:

$$f_x(x,y) = f(x,y) - f(x-1,y)$$
$$f_y(x,y) = f(x,y) - f(x,y-1) \quad (9)$$
$$g(x,y) = f_x(x,y) + f_y(x,y)$$

The visualization of image threshold processing makes it easy to realize the image segmentation application; this is also the main factor to choose the method of threshold processing (Pavlidis, T., 1982). Adopted the idea of threshold to (i.e. difference value) implement threshold segmentation on the gradient operator, the key issue of the threshold treatment is how to choose appropriate threshold segmentation. Difference is the threshold for the acquisition to take recursive iterative approach is relative to the test selection of simple threshold segmentation (i.e. the average of the overall difference), the accuracy of operational results is greatly improved (J. Kittler, M. Hatef, R.P.W. Duin, and J. Matas, 1998). Specifically operational steps as (Fig. 2) shows:

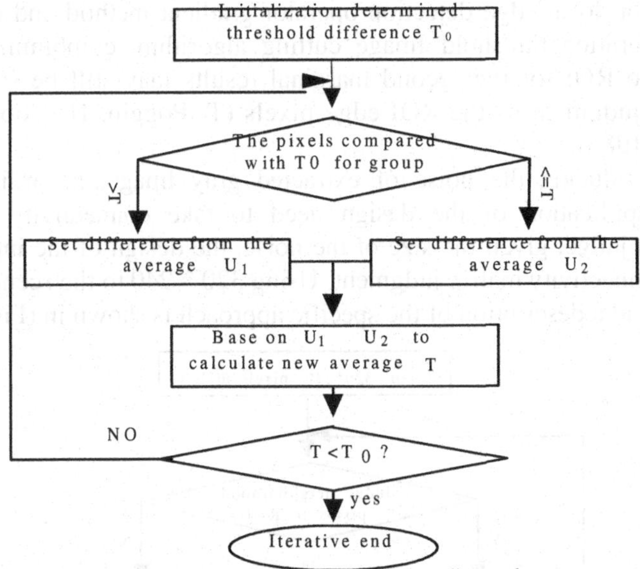

Fig. 2. Process of image division disposal

1. The process of edge calculates the gradient operator's expectations as the initial sub-threshold;

$$E(z) = T_0 = \frac{\sum_{i=1}^{n} g(x,y)}{n} \quad (10)$$

2. Using T_0 for image segmentation, all pixels divided into two groups: V_1 group: $g(x,y) > T_0$ pixels composition; V_2 group: $g(x,y) <= T_0$ pixels composition;

3. The calculation of two different sets of internal difference, on average, and recorded as μ_1 and μ_2 ;

4. And the value of difference for the average operation, to calculate the difference the new threshold: $T_0 = 1/2(\mu_1 + \mu_2)$;

5. Repeating the Step 2 to Step 4 of the calculation process, until successive iteration is less than the value of T_0 prior estimate the parameters T_0 stop iteration. (Gao, J., Zhou, M. & Wang, H., 2001)

2.3 Rectangular benchmark judgment

During the above process, there is likely to be the noise induced by the hardware or in the transmission process generated by the channel, because of the background and there are some isolated the noisy points (J.F. Canny, 1986), in order to improve recognition accuracy in the design, base on the above handled regional ROI pixels. But Robert conducting boundary segmentation local edge detection operator gradient method and operate of regional iteration threshold Image cutting algorithm, combining regional approach to ROI for the second marginal results may still be some noisy points misjudgment as the ROI edge pixels (T. Poggio, H. Voorhees and A. Yuille, 1985).

Firstly, reducing the noise of extracted gray image, according to the practical application of the design need to take connectivity judgment methods to judge, given the size of the noise, the design of the introduction of 2 * 2 connectivity matrix judgment. Using 320 * 240 to the scene graph as template, and a description of the specific approach is shown in (Fig. 3).

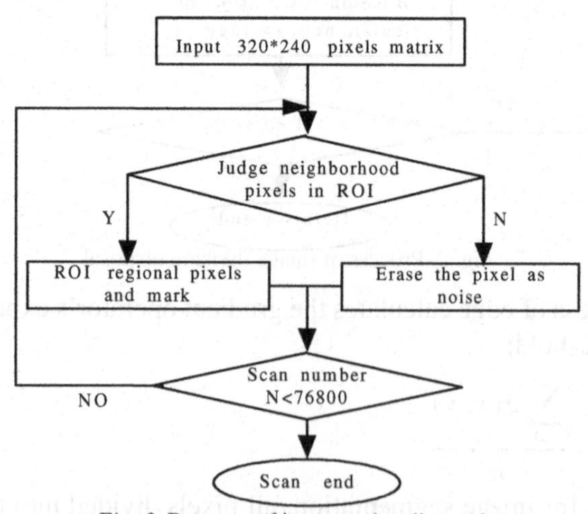

Fig. 3. Remove of image noise disposal

Secondly, conducting the pattern recognition based on the ROI edge region has been labeled, preparing for the conduct of the length calculations. Geometry of the pattern recognition is the key to determine the geometry and its membership function (i.e. fuzzy degrees), the right has been labeled rectangular gray two marginal value of the map edge smoothing and the corresponding pixel matrix conversion (Duda, R.O. & Hart, P.E., 1973).

A, B, C, D quadrilateral as four interior angle, a, b represent two adjacent edges Quadrilateral length, this is the number of pixels, and fuzzy correlation coefficient, ρ_1, ρ_2 are respective the relative factor.

Angle membership functions:

$$(RE)(u_1) = 1 - \rho_1 \frac{1}{90}\left[(A-90°)+(B-90°)+(C-90°)+(D-90°)\right] \qquad (11)$$

Adjacent edges membership function:

$$(RE)(u_2) = 1 - \rho_2 \frac{a}{b} \qquad (12)$$

When had withdrawn the rectangular ROI regional membership meets the following conditions:

$$(RE)(u_1) >= 0.8 \, and \, (RE)(u_2) >= 0.8 \qquad (13)$$

Then the computer will determine the ROI for the region to be identified rectangular benchmark.

3. THE EXPERIMENTAL ANALYSIS AND CONCLUSIONS

From the above implement, we can educe the conclusion of this intelligent recognition, as follows:

(1) Divided by the border Robert partial edge detection operator iterative methods to extract rectangular benchmark is feasible (Fu, K.S. & Mui, J.K., 1981). Advanced intelligent recognition technology has certain value in practical application of the regulatory process of the grain reserves of real-time monitoring.

(2) The methodology used HIS color space conversion, the use of H-hue color space, so it has good robustness under different light circumstance.

(3) Using the design methods in the actual recognition, all of the 10 samples of the rectangular benchmark can be correctly recognized, and there is not misjudgment, that is the correct detection rate is 100%.

Based on the grain reserves automatic video monitoring and auditing, the key technology is through the recognition of real-scenes video image to acquire the quantity of grain storage. In this paper, a boundary segmentation Robert partial edge detection operator methods and iterative threshold regional gradient operator segmentation algorithm are combined to complete the edge of the ROI region extraction, and use a fuzzy membership function of recognition for rectangular benchmark judgment, the scheme is simple and the calculation is not too much, effectively remedy the processing failures that aroused by the edge detection operators to noisy image, and have good effect of edge detection. But the result of some processed image still has a certain edge of the width. For the department of grain management require a high precision for grain quantity calculation, the width of the benchmark length will result in errors for the subsequent calculation of grain quantity, therefore, the further refinement of the image edge extraction still needs further study and discussion.

ACKNOWLEDGEMENTS

This work is supported under the Bulk Warehouse Grain Quantity Recognition, a project (2007) funded by the Finance Bureau of Chongqing Municipality, P. R. China.

REFERENCES

Canny, J. (1986). A Computational Approach to Edge Detection [J], IEEE Transations Pattern Analytical Machine Intelligent 8: 679– 698.

Cheng Xiao-chun (2006). A method of shape recognition [J], Pattern Recognition and Artificial Intelligence 6: 126–132.

Duda, R.O. & Hart, P.E. (1973). Pattern Classification and Scene Analysis [J], New York: Wiley.

Fu, K.S. & Mui, J.K. (1981). A Survey of Image Segmentation. Pattern Recognition [J], IEEE Transactions on Pattern Analysis and Machine Intelligence 13(1): 3–16.

Gao, J., Zhou, M. & Wang, H. (2001). A Threshold and Region Growing Combined Method for Filament Disappearance Area Detection in Solar Images [J], In Proceedings of The Conference on Information Sciences and Systems. The John Hopkins University.

Huertas, A. & Medioni G. (1986). Detection of Intensity Changes with Sub-Pixel Accuracy using Laplacian-Gaussian Masks [J], IEEE Transactions Pattern Recognition Machine Intelligent 8(5): 651–664.

J.F. Canny (Nov 1986), A Computational Approach to Edge Detection [J], IEEE PAMI, Vol. 8, No. 6.

J. Kittler, M. Hatef, R.P.W. Duin, and J. Matas. (1998). On combining classifiers [J]. IEEE Transactions on Pattern Analysis and Machine Intelligence 20(3): 226–239.

Kim, V. & L. Yaroslavskii, (1986). Rank algorithms for picture processing [J], Comput Vision, Graphics, and Image Process 35: 234–258.

M.D. Kelly, Edge Detection in Pictures by Computer Planning [J], Machine Intelligence,Vol. 6 (American Elsevier, New York, 1973), pp. 397–409.

Marr, D. & Hildreth, E. (1980). Theory of Edge Detection [J], Proceedings of the Royal Society London B207: 187– 217.

Pursuing technology. (2006). Visual C++ digital image disposal typical arithmetic and implement [M]. post & telecom press.

Pavlidis, T. (1982). Algorithms for Graphics and Image Processing [J], Computer Science Press, Maryland, USA.

T. Poggio, H. Voorhees and A. Yuille. (1985). A Regularized Solution to Edge Detection [J], May 1985 A. I. Memo 883, M.I.T.

Tou, J.T. & R.C. Gonzalez.Pattern Recognition Principles [M], Addison-Wesley Publishing, Reading, MA, USA, 1981.

POLYMORPHISM OF MICROSATELLITE SEQUENCE WITHIN ABC TRANSPORTER GENES IN PHYTOPATHOGENIC FUNGUS, *MAGNAPORTHE GRISEA*

Lin Liu[1], Chengyun Li[1,*], Jing Yang[1], Jinbin Li[2], Yuan Su[1], Yunyue Wang[1], Yong Xie[1], Youyong Zhu[1]

[1] *Key Laboratory for Agricultural Biodiversity and Pest Management of China Education Ministry, Plant Protection College, Yunnan Agricultural University, Kunming, 650201, China*

[2] *Plant Protection Research Institute of Yunnan Academy of Agricultural Sciences, Kunming, 650205, China*

[*] *Corresponding author, Address: Chengyun Li, Key Laboratory for Agricultural Biodiversity and Pest Management of China Education Ministry, Yunnan Agricultural University, Hei Longtan, Kunming, 650201, P. R. China, Tel: +86-871-5227552, Fax: 86-871-5227552, Email: li.chengyun@gmail.com*

Abstract: Thirteen polyporphic microsatellite markers suitable for population genetic structure analysis and ABC transporter and signal transduction coding genes variation measurement were developed for rice blast fungus, *Magnaporthe grisea*. Polymorphism was evaluated by using forty-six isolates collected from diverse geographical locations and rice varieties. Preliminary results indicated that each locus resolved multiple alleles ranging from two to ten. There results showed that these SSR-containing genes are also polymorphic in the natural population.

Keywords: *Magnaporthe grisea*, ABC transporter, microsatellite

1. INTRODUCTION

Rice blast disease, caused by *Magnaporthe grisea*, is the devastated disease of cultivated rice in most rice-growing regions worldwide. The

Liu, L., Li, C., Yang, J., Li, J., Su, Y., Wang, Y., Xie, Y. and Zhu, Y., 2008, in IFIP International Federation for Information Processing, Volume 258; Computer and Computing Technologies in Agriculture, Vol. 1; Daoliang Li; (Boston: Springer), pp. 553–558.

fungus exhibits a high level of pathotype variation. Potential mechanisms contributing to this variation include mutation, migration, parasexual recombination or an as yet unobserved sexual stage in the field (Ou, 1985). Disease management strategies would greatly benefit from an increased understanding of the amount and distribution of genetic diversity in this pathogen. The completion of the fungus genome sequence project has made it possible to determine not only the total number of genes, but also the exact number of genes of a particular type and analysis their structure and function in details (Ou, 1985; Zeigler et al., 1997). As a consequence, we now know exactly how many regulatory gene are encoded by the blast fungus genome, and how many genes contain simple sequence repeats (SSRs) within protein coding regions. Trinucleotide repeats are clustered in regulatory genes in *Saccaromyces cerevisiae* (Young et al., 2000) and rice blast fungus (Li et al., 2005), but all these SSRs are structurally and functionally polymorphisms, are still unknown.

Microsatellites are founded in both eukaryotes and prokaryotes. It nonrandomly distributes either in expressed sequence tags (ESTs) and genes, including protein-coding, 3'-UTRs and 5'-UTRs, or in introns. The consequences of SSRs repeat-number changes are different in those regions of both prokaryotes and eukaryotes. For example, 14% of protein-coding regions of all known proteins in eukaryotes was proved to containing repeated sequences, and it is three times higher abundance of repeats than in prokaryotes (Marcotte et al., 1998). Characterized with relatively rapid and inexpensive, microsatellites are favored for genetic research, it was not only applied to polymorphic resolve within species but also commonly used to identify specific chromosomal regions consistently across populations.

Genes involved in ABC transporters play a key role in development and pathogenicity of fungal pathogens. The ATP-binding cassette (ABC) superfamily of active transporters is composed of about 50 functionally diverse prokaryotic and eukaryotic transmembrane proteins (Higgins, 1992; Michaelis and Berkower, 1995). The ABC transporters not only carry a variety of substrates into or out of the cell, but also are involved in intracellular compartmental transport. These proteins utilize energy derived from the hydrolysis of ATP to transport the substrate across the membrane against a concentration gradient.

The previous work showed that microsatellite sequences, especially trinucleotide repeats are richness in protein kinase and ABC transporter coding genes of fungus (Keleher et al., 1992). The objective of this study was to determine the polymorphism of these microsatellite loci by PCR assay of loci among natural population in *M. grisea*.

2. MATIERIALS AND METHOD

The DNA sequence, a database of known and predicted open reading frames (ORF) of eukaryotic ABC transporters were obtained from the *Maganaporthe grisea* genome database World Wide Web site: http://www.genome.wi.mit.edu/annotation/fungi/magnaporthe/ on July 14, 2005, and was made sure by *Maganaporthe grisea* genome database World Wide Web site: http://www.broad.mit.edu/annotation/genome/magnaporthe_grisea/ on May 12, 2006. We used the program software tandem repeats finder (TRF) written by Benson (Benson, 1999) with the following options: minimum size =15 bp, 80% matches (namely number of matched bases between two repetitive elements is 80%) and abundance was removed.

Polymorphic loci were detected by screening a subset of 46 isolates of *M. grisea* collected from different regions (including *japonica*, *indica* rice grown regions) and various rice varieties of Yunnan Province, China. The genomic DNA were extracted from mycelia using a simple extraction protocol (Sweigard et al., 1990). Primers were designed for DNA sequence with microsatellite motifs using PRIMER3 (Rozen and skaletsky, 2000) software and synthesized by Invitrongen Biotechnology Co. Ltd. Shanghai, China.

PCR amplifications were performed in 20 μL volumes containing 1 × PCR buffer (10 mM Tris-HCl pH 8.5, 50 mM KCl, 1.5 mM $MgCl_2$, and 0.001% gelatin), 125 μM each dNTP, 5 pmol of each primer, and 0.5 U of *Taq* DNA polymerase (Sino-American Biotechnology Co., Beijing). Approximately 50 ng of genomic DNA was used for each reaction. Amplification were performed in a Eppendoff PCR thermal Mastercycler with the cycling parameters; 2 min and 30 sec at 94°C, 35 cycles of 30 sec at 94°C, 1 min at 55°C and 1 min at 72°C followed by a final extension for 10 min at 72°C. In initial experiments, amplified fragments were visualized by electrophoreses in 1.5% agarose gels stained with ethidium bromide. Those loci appeared polymorphic were further examined by 8% polyacrylamide gel to determine the product size of the PCR product and number of alleles per locus. Fragment size of PCR products were estimated on Bio-Imaging System E5000.

3. RESULTS AND DISCUSSION

Thirteen of the fifteen polymorphic loci produced amplicons from a majority of 46 isolates, and displayed two to ten alleles (Table 1). Observed

Table 1. Polymorphisms of SSRs in ABC transporter genes in *M. grisea.* I, Shannon's information index, Ho, expected homozygosty*, Hg, expected heterozygosity*

	Pirmer(F,5'-3')	Pirmer (R,5'-3')	Gene name	Estimate product size	Super contig	I	Motif	Repeat No.	Product size range	H_O	H_E
SMS1	ACAAGCCAGTCGCAGTCAC	CTAACCCGTCACGCTTCTTC	MGG_00957.5	250	5.194	0.6902	CAGCAA	3	206-216	0.6092	0.3908
SMS2	CATTGCCCTCGATCGTTTC	TGTTGAGCCACTCGATATGC	MGG_02572.5	250	5.193	1.8668	AAG	6	235-263	0.1887	0.8113
SMS3	GGCCGTACGAGGACTATGAC	TCGGTTTCGGGTTTGTATTC	MGG_13490.5	282	5.134	2.0419	GTTGGG	3	245-304	0.1409	0.8591
SMS4	TGCATCCAGGGTAACAGTGA	GTTGGAGCAAGAAGCCTGTC	MGG_05009.5	290	5.175	1.0438	GGTAGC	4	234-324	0.4639	0.5361
SMS5	CCCTGATAGTCGCCCTCATA	GATCCGGACCAGCTTGAGTA	MGG_06707.5	254	5.186	1.7759	CGA	6	243-275	0.1906	0.8094
SMS6	CCGACATTGTTCTCGACCTC	ATCCGAACTGGGCTGAACAC	MGG_06939.5	275	5.186	1.5362	CGCCAT	3	293-331	0.2145	0.7855
SMS7	GAGCTGCTGACGTTTGAGG	TCATGCCCTAACCTTTTTGC	MGG_07375.5	282	5.191	1.6113	GCAGCT	3	281-308	0.2317	0.7683
SMS8	AGCCTGCACACTACACCAAA	CGGGTAAGCTTTTCCATCAA	MGG_07848.5	256	5.183	1.4919	CTC	5	265-339	0.2480	0.7520
SMS9	ATCATACCGCAAGACCCAAC	ATGATCTGTGAGCCCCTGAC	MGG_08309.5	296	5.195	2.0016	GGC	5	328-355	0.1343	0.8657
SMS10	CGTTCACTACGAGCGTTTCA	TACGGGAACCAAGAGAGCAC	MGG_12035.5	260	5.187	1.6648	CAAGGC	3	268-301	0.2059	0.7941
SMS11	ATCGTGGGTTTGATCGAGAG	GGACCTCCACCATTTGATGT	MGG_09931.5	276	5.186	1.9762	AAG	5	248-296	0.1677	0.8323
SMS12	AAGGTCGGGCACCTCTTC	CTCCTCGGGGTTGTAAATGA	MGG_10277.5	273	5.179	1.6606	GCT	5	264-302	0.2699	0.7301
SMS13	GAATTCACCAGCGGATTGTT	GACTCTGAAGCGTTGGAGGT	MGG_11025.5	266	5.187	1.9533	CTCGT	3	237-293	0.1524	0.8476

* Expected homozygosty and heterozygosity were computed using Levene (1949)

heterozygosity and expected heterozygosity values by software GENEPOP (V1.34), were shown in table 1. The results suggested that genes harbored these SSR sequence are also diversity in isolates used.

4. CONCLUSIONS AND FUTURE WORKS

The high degree of polymorphism in this set of microsatellite markers can be used to analysis of population structure and strain distribution in association with particular commodities and locations, as well as complemented for understanding function of regulatory genes in the fungus. With integration of such information into strategies of the functional genomics, it would facilitate SSR functions Study.

ACKNOWLEDGEMENTS

This work is supported by National Basic Research Program of China (2006CB100202), Education Ministry Foundation (307025) and Doctorial Foundation of Education Ministry of China (20050676001).

REFERENCES

Benson G. 1999, Tandem repeats finder: a program to analyze DNA sequences. Nucleic Acids Res 27, 573-580.

Cecconi F., Alvarez-Bolado G., Meyer B.I., Roth K.A., Gruss P. 1998, Apafl (CED-4 homolog) regulates programmed cell death in mammalian development, Cell 94, 727-737.

Dean Ralph A., Talbot Nlcholas J., Ebbole Danlel J. Ebbole, Mark L. Farman, Thomas K. Mltchell, Marc J. Orbach, Mlchael Thon, Resham Kulkarnl, Jin-Rong Xu, Huaqln Pan, Nlck D. Read, Yong-Hwan Lee, lgnazlo Carbone, Doug Brown, Yeon Yee Oh, Nlcole Donofrlo, Jun Seop Jeong, Darren M. Soanes, Slavlca Djonovlc, Elena Kolomlets, Cathryn Rehmeyer, Welxi Li, Michael Hardlng, Soonok Klm, Marc-Henrl Lebrun, Heidl Bohnert, Sean Coughlan, Jonathan Butler, Sarah Calvo, Li-Jun Ma, Robert Nicol, Seth Purcell, Chad Nusbaum, James E. Galagan & Bruce W. Blrren 2005, The genome sequence of the rice blast fungus *Magnaporthe grisea*. Nature 434, 980-986.

Edward M. Marcotte, Matteo pellegrini, Todd O. Yeates and David Eisenberg 1998, A census of protein repeats, J. Mol. Biol. 293, 151-160.

Higgins C.F. 1992, ABC transporters: from microorganisms to man. Annu. Rev. Cell. Biol. 8, 67-113.

Keleher C.A., Redd M.J., Schultz J., Carlson M., Johnson A.D. 1992, Ssn6-Tup1 is a general repressor of transcription in yeast, Cell 68, 709-719.

Li Cheng-yun, Li Jinbin, Zhou Xiao-gang, Zhang Shao-song, Dong Ai-rong, Xu Ming-hui 2005, Frequency and Distribution of Microsatellite sequence in Open Reading Frames of Rice Blast Fungus, *Magnaporthe grisea*. Chinese J. Rice Science 19(2), 167-173, 2005. (in Chinese with English abstract).

Michaelis S., Berkower C. 1995, Sequence comparison of yeast ATP-binding cassette proteins. Cold Spring Harbor Symp Quant Biol 60, 291-307.

Ou S.H. Rice diseases 1985, 2nd Edition. Commonwealth Mycological Institute, Kew, UK, 1985. 380.

Rozen S., Skaletsky H. 2000, Primer3 on the WWW for general users and for biologist programmers. Methods Mol. Biol. 132, 365-386.

Saxena K., Gaitatzes, C., Walsh, M.T., Eck, M., Neer, E.J., and Smith, T.F. 1996, Analysis of the physical properties and molecular modeling of Sec13: a WD repeat protein involved in vesicular traffic, Biochemistry 35, 15215-15221.

Stifani S., Blaumueller C.M., Redhead N.J., Hill R.E., Artavanis Tsakonas S. 1992, Human homologs of a Drosophila Enhancer of split gene product define a novel family of nuclear proteins [published erratum appears in Nat. Genet. Dec2(4), 119-127.

Sweigard, J.A., Orbach, M.J., Valent, B., Chumley, F.G. 1990, A miniprep procedure for isolating genomic DAN from *Magnaporthe grisea*, Fungal. Genet. Newslett 37, 4

Vaisman N., Tsouladze A., Robzyk K., Ben-Yehuda S., Kupiec M., Kassir Y. 1995 The role of *Saccharomyces cerevisiae* Cdc40p in DNA replication and mitotic spindle formation and/or maintenance, Mol. Gen. Genet 247, 123-136.

Young E.T., Sloan, J.S. and Riper K.V. 2000, Trinucleotide repeats are clustered in regulatory genes in *Saccharomyces cerevisiae*. Genetics 154, 1053-1068.

Zeigler, R.S., Scott, R.P., Leung, H., Bordeos, A.A., Kumar, J., and Nelson, R.J. 1997, Evidence of parasexual exchange of DNA in the rice blast fungus challenges its exclusive clonality. Phytopathology 87, 284-294.

POLYMORPHISM OF MICROSATELLITE SEQUENCE WITHIN PROTEIN KINASE ORFS IN PHYTOPATHOGENIC FUNGUS, *MAGNAPORTHE GRISEA*

Chengyun Li[1,*], Lin Liu[1], Jing Yang[1], Jinbin Li[2], Zhang Yue[1], Yunyue Wang[1], Yong Xie[1], Youyong Zhu[1]

[1] *Key Laboratory for Agricultural Biodiversity and Pest Management of China Education Ministry, Plant Protection College, Yunnan Agricultural University, Kunming, 650201, China*

[2] *Plant Protection Research Institute of Yunnan Academy of Agricultural Sciences, Kunming, 650205, China*

* *Correspondence: Chengyun Li, Key Laboratory for Agricultural Biodiversity and Pest Management of China Education Ministry, Yunnan Agricultural University, Hei Longtan, Kunming, 650201, China, Fax: 86-871-5227945, Email: li.chengyun@gmail.com*

Abstract: Eighteen polymorphic microsatellite markers suitable for population genetic studies and protein kinase encoding genic variation measurement were developed for rice blast fungus *Magnaporthe grisea*. Polymorphism was evaluated by using 46 isolates collected from diverse geographical locations and rice varieties. Preliminary results indicate that each locus harbors two to fourteen alleles.

Keywords: *Magnaporthe grisea*; protein kinase; microsatellite

1. INTRODUCTION

Magnaporthe grisea is the most destructive pathogen of rice worldwide and the primary model organism for elucidating the molecular basis of fungal diseases of plants (Valent, 1990) The completion of the genome

Li, C., Liu, L., Yang, J., Li, J., Yue, Z., Wang, Y., Xie, Y. and Zhu, Y., 2008, in IFIP International Federation for Information Processing, Volume 258; Computer and Computing Technologies in Agriculture, Vol. 1; Daoliang Li; (Boston: Springer), pp. 559–563.

sequence for *Magnaporthe grisea* has made it possible to determine the total number of genes as well as to analyze and classify them according to their structure and function (Dean et al., 2005).

The eukaryotic protein kinases comprise one of the largest superfamilies of homologous proteins and genes. There are now hundreds of different members whose sequences are known within this family (Hanks, 2003). Although there are common structural features among protein kinases, differences in structural features, regulation modes, and substrate specificities divide them into separate groups. In the phytopathogenic fungi, components of heterotrimeric G proteins, MAP kinases, and cAMP signal transduction pathway are required for pathogenesis (Muller et al., 2003; Yamada-Okabe et al., 1999; Xu et al., 1996). We previous revealed that many protein kinase genes harbored SSR sequences within their protein coding region (Li C.Y. et al., 2005), but whether these sequences are polymorphic is unclear.

Microsatellites are favored for genetic applications because they are abundant in plant genomes, highly polymorphic within species, relatively rapid and inexpensive to assay, and can be used to identify specific chromosomal regions consistently across populations. Distribution and frequency of SSRs in genomic scale or ESTs have been analyzed extensively, however, reports published to date clearly discussed SSR polymorphism in genes has been limited (Li C.Y. et al., 2005).

The objective of this study was to develop PCR primer pairs targeting previously sequenced genes from *M. grisea* in order to compare the allelic amplification product polymorphism of protein kinase encoding genes among natural populations.

2. MATERIAL AND METHOD

The DNA sequence and a database of known and predicted open reading frames of eukaryotic protein kinases were obtained from the *Magnaporthe grisea* genome database (http://www.genome.wi.mit.edu/annotation/fungi/magnaporthe/) on July 14, 2005. The program software tandem repeats finder (TRF) written by Benson (Benson, 1999) with the following options: minimum size = 15 bp, 80% matches (namely number of matched bases between two repetitive elements is 80%) and abundance were removed.

Polymorphic loci were detected by screening a subset of 46 *M. grisea* isolates collected from different locations and from various rice varieties of Yunnan Province, China. Genomic DNA was extracted from mycelia using a simple extraction protocol (Zhang et al., 1996). Primers were designed for DNA sequence with microsatellite motifs using PRIMER3 (Rozen et al., 2000) software and synthesized by Invitrongen Biotechnology Co. Ltd. Shanghai, China.

PCR amplifications were carried out in 20 μL volumes containing 1 × PCR buffer (10 mM Tris-HCl pH 8.5, 50 mM KCl, 1.5 mM $MgCl_2$, and 0.001% gelatin), 125 μM each dNTP, 5 pmol of each primer, and 0.5 U of *Taq* DNA polymerase (Sino-American Biotechnology Co., Beijing). Approximately 50 ng of genomic DNA was used for each reaction. Amplification were performed in a Eppendoff PCR thermal Matercycler with the cycling parameters; 5 min at 94°C, 35 cycles of 1 min at 94°C, 1min at 54°C and 1 min at 72°C followed by a final extension for 10 min at 72°C. In initial experiments, amplified fragments were visualized by electrophoreses in 1.5% agarose gels stained with ethidium bromide. Loci that appeared polymorphic were further examined by 8% polyacrylamide gel to determine the product size of the PCR product and number of alleles per locus. Fragment size of PCR products were estimated on Bio-Imaging System E5000.

3. RESULTS

Eighteen of the 26 polymorphic loci produced amplicons from a majority of 46 isolates, and displayed anywhere from two to fourteen alleles (Table 1). Gene diversity was estimated with the software program, GENEPOP (V1.32), and are shown in table 1. KMS02, KMS07, KMS18 showed high gene diversity in population used for the study. This suggests that genes harboring these SSR sequences are also highly diverse within the populations.

M. grisea has 11 109 proteins coding ORFs in whole genome, and among these, 85 protein kinase genes, corresponding to – 0.76% of the total number of genes (Dean et al., 2005). More than 30% of these protein kinase encoding genes have SSRs within their protein coding regions. The high degree of polymorphism in this set of microsatellite markers can be used to analyze population structure and strain distribution in association with rice variety and location, adding to the fundamental understanding the function of protein kinase genes of the fungus. These results provide useful information to study possible SSR functions and variation of protein kinases that harbor them.

Table 1. Polymorphsims of SSRs in protein kinase encoding ORFs in *M. grisea.* N_a, number of alleles; G_d: Gene diversity by Shannon's Information inde

Locus Contig*	Primer (F,5'-3')	Primer (R,5'-3')	Gene name	Motif	No of Genotypes	Repeat No.	Na	Percent matches	Product size range (bp)	Gd
KMS012.1190	AGCGAAACAA GAACGCGAGG	TAGCAGCCTG TGCTCGTTCG	MGG 06413.5	ACG	46	5	2	100	219-222	0.5623
KMS022.1299	CGCAAAGAAT TCAAACCCGC	GGAGACGACA CTGGGGTGCT	MGG 07003.5	ACG	46	6	14	100	279-326	2.4256
KMS032.1302	TCCCTTTCGGT CGTTCCAAG	CCCGCTGAGG TAGCCAAAGA	MGG 07012.5	ACG	46	5	4	86	208-220	1.0673
KMS042.1631	CCGAAGAGGT CCTCCAAGCA	GCAGCAGCAT TACCATCCCC	MGG 08643.5	AG	46	6	5	88	216-311	0.654
KMS052.377	CCAAGCCCAG AGCCAGAAAA	CTCGAGGCTG CCCATCTTGT	MGG 01998.5	AGA	46	5	3	85	243-249	0.2826
KMS062.1692	GCATGAAATG CTCGTCGTGG	ATGCAGCGGC CTTCATTAGC	MGG 09000.5	AGAAAA	46	3	4	100	191-209	1.181
KMS072.347	GTTCTCCATCG CCCAAATCG	TAAGTGGGCT CCTTGGCTGC	MGG 01816.5	CAA	46	5	10	86	245-281	2.1018
KMS092.993	GGTCCTATTCC CCTTCCCCC	TTCAACCGAT ACGAGGCCGT	MGG 05397.5	CAAGG	46	3	2	90	200-206	0.5921
KMS112.211	GGGGTAACGA CAGCCAGGTG	CTGCTGCTGCT GTTGCTGGT	MGG 01196.5	CAG	46	7	5	80	218-230	1.1916
KMS122.1645	CGATACCCCT GAGCCACCAC	ATCTGCCGCCT TTTGAGTGC	MGG087 46.5	CAG	46	5	6	100	198-213	1.3926
KMS132.1353	GTCATGACAG GGGTCCTCGG	ATCCCGCTCTC CCTCCATTC	MGG 07291.5	CAGCCC	46	3	8	84	232-292	1.8262
KMS142.1182	GAACCCCGCA GTCCAACAAC	CGCCTGCTGA GAATGGGACT	MGG 06368.5	CAGTCA	46	3	5	92	205-229	1.2688
KMS152.838	GGCCACAGAG GAGAACGGAA	TCCATCAGGA TCGGGGACTG	MGG 04463.5	CGA	46	5	6	83	215-233	1.7452
KMS162.343	TCATGAGCGA GACAATGGGG	TGCTGAACCG ATTCCGCTTT	MGG 01795.5	CGGT	46	4	2	83	206-210	0.5623
KMS182.687	GCAAGTCGCC TCGCCATTAT	TGAACCGACT CGTCGACTGC	MGG 03488.5	GCA	46	6	5	82	244-256	1.4555
KMS192.1360	CAGCACCCAA AAGGAGCCTG	AACATTCCCA GGTGCATCGC	MGG 12406.5	GCA	46	5	2	85	218-221	0.3594
KMS202.1779	CGCCCTCAA AAACCAAGGG	AATTGCGACA AGTCGCTCCC	MGG 14499.5	CAG	46	7	5	100	257-281	1.3841
KMS222.338	AGACGACGAG GCTTCCGATG	GGCATAAGGT TGTCGCGGAG	MGG 00925.5	GCT	46	6	5	100	187-211	0.8856

* Contig is based on the *M. griseea* genome database,
website: http://www.genome.wi.mit.edu/annotation/fungi/magnaporthg/

ACKNOWLEDGEMENTS

We thank Miss J. Krenz, in Department of Botany and Plant Pathology, Oregon State University of USA for helpful advice and revision the paper in detail. This work is supported by National Basic Research Program of China (2006CB100202), Education Ministry Foundation (307025) and Doctorial Foundation of Education Ministry of China (20050676001).

REFERENCES

Benson G. 1999, Tandem repeats finder: a program to analyze DNA sequences. Nucleic Acids Res., 27: 573-580.

Dean R.A., Talbot N.J., Ebbole D.J. et al. 2005, The genome sequence of the rice blast fungus *Magnaporthe grisea*. Nature, 434: 980-986.

Hanks S.K. 2003, Genomic analysis of the eukaryotic protein kinase superfamily: a perspective. Genome Biol., 4, 111.

Li C.Y, Li J.B., Zhou X.G., Zhang S.S., Dong A.R., Xu M.H. 2005, Frequency and distribution of microsatellites in open reading frame of rice blast fungus, *Magnaporthe grisea*. Chinese J. Rice Sci., 19: 167-173. 2005. (in Chinese with English abstract)

Muller P., Weinzierl G., Brachmann A., Feldbrugge M., Kahmann R. 2003, Mating and pathogenic development of the Smut fungus Ustilago maydis are regulated by one mitogen-activated protein kinase cascade. Eukaryot. Cell, 2: 1187-1199.

Rozen S., Skaletsky H. 2000, Primer3 on the WWW for general users and for biologist programmers. Methods Mol. Biol., 132: 365-386.

Valent B. 1990, Rice blast as a model system for plant pathology. Phytopathology, 80: 33-36.

Xu J.R., Hamer J.E. 1996, MAP kinase and cAMP signaling regulate infection structure formation and pathogenic growth in the rice blast fungus *Magnaporthe grisea*. Genes Dev., 10: 2696-2706.

Yamada-Okabe T., Mio T., Ono N., Kashima Y., Matsui M., Arisawa M., Yamada-Okabe H. 1999, Roles of three histidine kinase genes in hyphal development and virulence of the pathogenic fungus Candida albicans. J. Bacteriol., 181: 7243-7247.

Zhang D., Yang Y., Castlebury L.A., Cerniglia C.E. 1996, A method for the large scale isolation of high transformation efficiency fungal genomic DNA. FEMS Microbiol. Lett., 145: 261-265.

APPROACH OF DEVELOPING SPATIAL DISTRIBUTION MAPS OF SOIL NUTRIENTS

Yong Yang[1,*], Shuai Zhang[1]

[1] College of Information and Electrical Engineering, Shenyang Agricultural University, Shenyang, China, 110161

* *Corresponding author, Address: P.O. Box 43, College of Information and Electrical Engineering, Shenyang Agricultural University, 120 Dongling Road, Shenyang, 110161, P. R. China, Tel: +86-24-88036221, Fax: +86-24-88487863, Email: yangsyau@163.com*

Abstract: One of the major components of precision agriculture is the precision fertilization. The basic principle of precision fertilization is to adjust the fertilizer input according to the specific circumstances or properties of soils in each location for the least waste and the highest profit. The paper presents a feasible approach for developing the spatial distribution map of soil nutrients based on a kind of GIS software, the ArcView. According to the field sampling data and localities measured by GPS a database of soil nutrients was set up. Using the semi variance function and the Kriging interpolations algorithm upon geostatistics theory the field data were analyzed, and then the graphic editor of the ArcView was applied to produce soil nutrient spatial distribution map, which describes the precision of the algorithm and distribution range of soil nutrients. This research is a methodological contribution to precision agriculture and lays the ground for precise application of fertilizers.

Keywords: Soil nutrients; Spatial distribution map; ArcView; Geostatistics

1. INTRODUCTION

The ideas for sustainable development of agriculture and precision agriculture theory have been pushed forward. The precision agriculture requires a new theory and technology in terms of reasonable fertilization. The speedy development of the world agriculture was gained under the

Yang, Y. and Zhang, S., 2008, in IFIP International Federation for Information Processing, Volume 258; Computer and Computing Technologies in Agriculture, Vol. 1; Daoliang Li; (Boston: Springer), pp. 565–571.

condition of a large amount of input of chemicals or mineral sources of energy, such as chemical fertilizer and pesticide (Liu Jintong, et al., 2002). But, ecological and environmental problems, for example, increased soil erosion, pollution of agricultural products and groundwater and enrichment of nutrition in water bodies, etc., have caused the extensive concern of international community.

The traditional even fertilizing method is not scientifically suitable and efficient to apply fertilizer in places with different soil nutrients, because soil fertility at different regions differs from place to place significantly. And overuse fertilizers can certainly lead to a waste of fertilizer resources and a serious environmental pollution. The Geostatistics has been proved to be one of the most effective ways to analyze the characteristics of soil nutrient spatial distribution and the pattern of variation. The Kriging interpolation algorithm is the most useful and optimal one in geostatistics that uses of initial data in a region as well as the structural characteristics of variable function to estimate the unknown values by the linear unbiased estimation.

The basic technological principle of precision fertilization is to adjust the fertilizer input according to the specific circumstances or properties of soils in each location for the least waste and the highest profit by fully understanding the variation of soil nutrients. Therefore, understanding the spatial variability of soil nutrient is the first step and the pre-condition for precision fertilization. The spatial distribution map of soil nutrient, developed by using the geostatistics as a principle and the software GIS as a tool, can reflect the spatial variability of nutrient and also make the balanced fertilization possible.

2. COMPUTER ENVIRONMENT

The computer tool of developing spatial distribution maps of soil nutrients is the ArcView GIS. The ArcView GIS is a geography information system software that was developed by the American Environment System Research Institute (ESRI). As a GIS software, the ArcView GIS's key function is the desk mapping and spatial analysis, etc. Owing to its building on the object-oriented data structure, data management and the analysis merit of ArcView can be fairly nimble, and ArcView can read the data taken from the Coverage and Grid of ARC/INFO, also the data from AutoCAD and data outside the base, etc. As far as his flexibility, user can control every element in ArcView's environment because of Avenue's programming language, through using Avenue's script, man can visit the object and class of ArcView's inside, and through its approaches of the internal geostatistical algorithms, one can carry on the geostatistics operations.

3. ALGORITHM

It is a science about studying natural phenomena of the randomness and structure in spatial distribution with the variable theory in region serving as foundation and the variation function as the basic tool. The Geostatistics has been proved to be one of the most effective ways to analyze the characteristic of the soil nutrient spatial distribution and the law of variation (Sun Hongquan, 1990).

3.1 Regional variable theory

The regional variable theory is one kind of real function that possesses numerical values in the space, its every point in the space means a definite numerical value, and when the point moves to the next point, the function value changes (White J. G. et al., 1997). The main characteristic of the region variable is the spatial correlation that it assumes the fixed level in the fixed scope, and after transcending this scope, the correlation becomes so weak that it disappears in the end, this quality is very difficult for the general statistics methods to recognize, but very useful for geostatistics.

3.2 The semi variance function

Semi variance function is a group of functions to describe the spatial variation in soil, which can show the change between the observation values of different distances (Campbell J. B. et al., 1978). The so-called semi variance is the semi variance of the observation value between any two points, the same as:

$$r(h) = Var\,[\,Z\,(X+h) - Z\,(X)\,]\,/2$$

in the equation, Z (X+h) and Z (X) are for the measured value of soil nutrient, r(h) is for semi variance of an interval of h, it is enlarged along with the increase of h in the fixed scope, and this value is stable when the interval is more than the biggest correlation distance.

3.3 The Kriging interpolations

The data between samples need to be estimated. Because the soil is sampled intermittently, and this estimation course is called the interpolation, that is a method used to estimate the unknown soil data in the neighborhood with the data of sample (Gaston L. A., 2001). The Kriging interpolation is the most useful and optimal one in use of the initial data in region and the structure characteristic of variable function to estimate the unknown value by the linear unbiased estimation, the nature is actually a weighted average of partial estimation as below:

$$Z(X_0) = \sum_{i=1}^{n} \lambda_i Z(X_i) \tag{1}$$

in the equation, X_0 is a point waiting for estimation, $Z(X_0)$ is the interpolated estimation value on the point of X_0. $Z(X_i)$ are real measured values of a certain number of observation points near X_0, λ_i are weights which describe spatial variation in consideration of spatial variation weights in semi variance map. Therefore, the estimation of Z value is unbiased.

Because of

$$\sum_{i=1}^{n} \lambda_i = 1 \tag{2}$$

the estimated bias is minimum, which can be worked out by the following equation.

$$\delta^2 = B^T \begin{bmatrix} \lambda \\ \mu \end{bmatrix} \tag{3}$$

where B is semi variance matrix between estimated point and other points, μ is Lagrange's parameter.

4. MAPPING

4.1 Data acquisition

The samples were taken in 34hm^2 of dry farmland. For the difference of size and fertility, a sample (0–20cm) was collected from every 0.49hm^2 in the net (70m to the south and 70m to the north) and 84 samples were obtained in total. The latitude and longitude of sampling sites were recorded by the global position system (GPS), and at the same time, we made a careful investigation about the outcome and manure application circumstances in the place for the recent years. After gaining the sample of soil, we started to analyze them in laboratory to gain each nutrient value of every soil sample.

4.2 The data analysis and mapping

An information database of soil nutrient and fertilizer information was set up by using small-size database tool Microsoft Access. The measured soil nutrient content and coordinate in samples, and all the concerned information of fertilization and yield over the past years were input into the database. Through the extension module Database Access of ArcView and by using ODBC the ArcView is linked to the information database of soil nutrient and fertilizer information. Selecting a table which contains the coordinate and soil nutrient content of samples and using the event subject commands the table will be generated in map form, namely the distribution map of soil sample. Then, it is the time to realize the interpolation algorithm in Avenue's language, and with the ArcView's graphic editor and by way of classifying the soil nutrient values again to define the precision of soil nutrient values and the colors of each region section to produce the spatial distribution map of soil nutrients. Different color represents different range of nutrient value. The corresponding area of color is to be used to examine each nutrient value scope and possibly to know the condition -rich or poor- of the soil nutrient in the map.

4.3 Mapping example

We can reclassify soil nutrients and determine precision of the nutrients and colors of various fields by using graphic editor provided by ArcView, and as the result we can generate the following map.

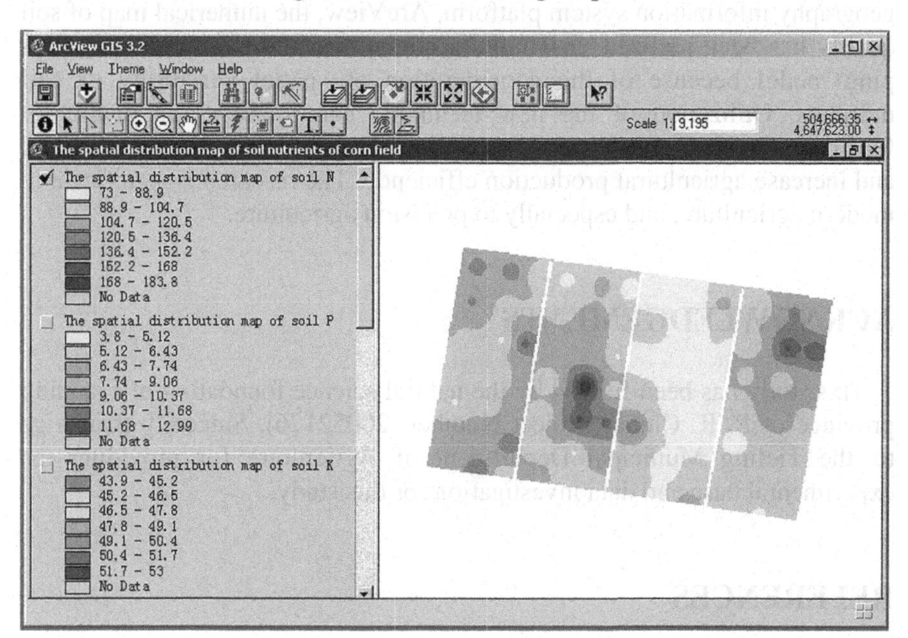

Figure 1. The spatial distribution map of soil nutrients of corn field

The above figure can be divided into two parts, the right part is map area and the left is the legend area. When a legend is selected, a corresponding map will appear in the right part. The different colors in the left part refer to the different ranges of nutrients values, from which one can know the status of a nutrient, while the nutrient level at a certain point can be known by using the "identity tools" of the ArcView.

5. CONCLUSIONS

The core of precision agriculture is to obtain different information in space and time about the crop outcome of farmland and environmental factors in a small area (of soil structure, soil fertility, topography, climate and disease, etc.), to analyze different causes, thus to take some feasible technological measures. It has some disadvantages for wasting fertilizers due to over input and even use of fertilizers. The geostatistics method can be used for quantitative analysis in the research on soil nutrient spatial variation. The semi variance map is the foundation of explaining the spatial variation structure of soil nutrients, while the Kriging's interpolation algorithm among them can predict the soil nutrient values in the unsampled areas with the model of the semi variance utilization map.

By means of generating the soil nutrient spatial distribution map on the geography information system platform, ArcView, the numerical map of soil quality has been realized. It is more accurate than the traditional soil mapping model because of the consideration of spatial variability of soil nutrients. Utilization of the new technique can improve the quality of fertilizing, reduce environmental pollution resulted from over fertilization, and increase agricultural production efficiency. The research is significant to modern agriculture, and especially to precision agriculture.

ACKNOWLEDGEMENTS

This study has been funded by the natural science foundation of Liaoning province of P. R. China (Project Number: 20052126). Sincerely thanks go to the Tieling Municipal Department of Agriculture for providing the experimental base and data investigation for this study.

REFERENCES

Alem M. H., Azari A. B. and Nielsen D. R. 1988. Kriging and univariate modeling of a spatial correlated date. Soil technology. 1(2):133-147.

Baker W. H. and Carroll S. D. 1996. Assessment of rice yield and fertility using site-specific technologies. Better Crops with Plant Food, 80(3):24-29.

Beetz H. F. 1994. Site-specific nutrient management systems for the 1990s. Better Crop, Vol. 78, No. 4:14-19.

Brannon G. R., Hajek B. F. 2001. Update and recorrelation of soil survey using GIS and statistical analysis. Soil Science Society of America Journal. 64(2):679-680.

Burrough P. A. 1991. Sampling designs for quantifying map unit composition. Spatial variability of soils and landforms, International Soil Science Society Working Group of Soil & Moisture Variability in Time & Space/American Society of Agronomy, the Crop Science Society of America & the Soil Science Society of America, 89-126.

Cahn M. D., Hummel J. W. and Brouer B. H., 1994. Spatial analysis of soil fertility for sitespecific crop management. Soil Science Society of America Journal. 58:1240-1458.

Campbell J. B. 1978. Spatial variation of sand content and pH within single contiguous delineation of two soil mapping units. Soil Sci. Am. J. (42):460-464.

Gaston L. A. 2001. Spatial Variability of soil properties and weed population in the Mississippi delta. Soil Science Society of American Journal. (65):470-479.

James D. W. and Hurst R. L. 1995. Soil sampling technique for band-fertilized, no-till fields with Monte Carlo Simulations. Soil Science Society of America Journal, 59:1768-1772.

Liu Jintong, etc. 2002. Refined Precision Agriculture Outline. Beijing. China Meteorological Publishing House (in Chinese).

McBraney A. B. and Webster R. 1986. Choosing functions for semi-variograms of soil properties and fitting them to sampling estimates. Journal of Soil science, 37:617-639.

Mulla D. J. 1991. Using geostatistics and GIS to manage spatial patterns in soil fertility, In Automated agriculture for the 21st century. Proc. Symposium. Chicago, IL. 16-17. Dec. 1991. ASAE, St Joseph, MI. 336-345.

Proc-Hermandez R. 1993. Spatial interpolation in geographical information systems. Proceeding of GIS'93, the Canadian conference on GIS. Energiy mines and resources of Canada.

Rogowski A. S. 1995. Quantifying soil variability for GIS application. I. estimates of position. International Journal of gergraphical Information Systems, 9:81-94.

Schnug E., Murphy D. P. L., Haneklaus S. and Evans E. J. 1993. local resource management in computer aided farming: A new approach for sustainable agriculture. Optimization of Plant Nutrition, M. A. C. Fragoso and M. L. van beusichem (Eds), Netherlands, Kluwer Academic Publisher, 657-663.

Sun Hongquan. 1990. Geology Statistics and Application. Beijing. China Mining Industry University Publishing House (in Chinese).

Tom McGraw, 1994. Soil level Variability in Southern Minnesota. Better Crop, (7894):24-25.

Van Groienigen J. W. 2000. Soil sampling strategies for precision Agriculture research under sahelian condition. Soil Science society of America Journal. 64:1674-1680.

Webster R. and McBraney A. B. 1987. Mapping soil fertility at Broom's Barn by simple kriging. J. Sci. Food Agric. 38:97-115.

White J. G., Welch R. M. and Norvell W. A. 1997. Soil Zn map of the USA using geostatistical and geographic information system. Soil Sci. Soc. Am. J. (61):185-194.

Yost R. S., Uehara G. and Fox R. L. 1982. Geostatistical analysis of soil chemical properties of large land areas. I. semivariograms. Soil Science Society of American Journal. 46:1028-1037.

RESEARCH AND DESIGN ON DOMAIN-AGRICULTURE-CROPS SOFTWARE ARCHITECTURE ORIENTED ADAPTIVE MODEL

Huanliang Xu[1], Haiyan Jiang[1,3], Shougang Ren[1,*], Xiaojun Liu[2,3], Weixin Cao [2,3]

[1] *College of Information Science and Technology, Nanjing Agricultural University, Nanjing 210095, P. R. China, Email: huanliangxu@njau.edu.cn*

[2] *College of Agriculture, Nanjing Agricultural University, Nanjing 210095, P. R. China*

[3] *Hi-Tech Key Laboratory of Information Agriculture, Jiangsu Province*

* *Corresponding author, Address: College of Information Science and Technology, Nanjing Agricultural University, 1 Weigang, Nanjing Jiangsu Province, 210095, P. R. China, Tel: +86-25-84396350, Email: rensg@njau.edu.cn*

Abstract: Composing application with plug & play (P&P) agriculture-crop business component on the domain-agriculture-crops software architecture (DAcSA) is an ideal implementation mechanism to develop the domain-agriculture-crop applied system. A black and white box framework for the adaptive DAcSA is built based on the agriculture-crop business component and hotspot subsystem. According to the domain-agriculture-crops rules, an administer center was designed to realize the plug and play of business component in domain framework by gluing component and hotspot subsystems up, which deposit in component lib and hotspot repository respectively. In line with the domain-agriculture-crops information characteristics, a resource-model-analysis (RMA) data mode and corresponding behavioral model based on the adaptive DAcSA was proposed and a supporting system based on virtual machine architecture was built, which has been applied to the prototype development of wheat growth simulation and decision-making supporting system. The proposed adaptive model proves preferable adaptability and can effectively decrease the cost of development and maintenance.

Keywords: Domain-specific software architecture, Domain-agriculture-crops adaptive software architecture, Business component, virtual machine, Wheat growth simulation, decision-making supporting system

Xu, H., Jiang, H., Ren, S., Liu, X. and Cao, W., 2008, in IFIP International Federation for Information Processing, Volume 258; Computer and Computing Technologies in Agriculture, Vol. 1; Daoliang Li; (Boston: Springer), pp. 573–584.

1. INTRODUCTION

Domain-specific software architecture (DSSA) is the core asset for domain-specific software development (Mei & Shen, 2006). It has the following features (Huang et al., 2006): □ strictly-defined problem area and solution area; □ domain-proper abstraction; □ domain universality applied to specific application development in the domain; and □ fixed, typical and reusable software element in the development process of the domain.

Research on domain-agriculture-crops software architecture (DAcSA) oriented adaptive model is to explore the software system construction and adaptability in the domain of agriculture crop. As the fundamental and principle research in agriculture information, it involves the digitalized expression, design, control and management of domain-agriculture-crops objects (such as biological factor, environment factor, technique factor and social economy factor) and the process (such as growth process, service process as well as management process etc.) (Yu & Cao, 2004). DAcSA mainly designs a universal and agile software architecture oriented to agricultural-crop domain, and eventually creates an application software system effectively and automatically (Cao et al., 2006). Therefore, this article will emphatically study a software process model to adapt for the domain-agriculture-crops and the corresponding architecture to support this process.

The research on numerous domain-specific software architectures (for example, self-adaptive intelligent system, aeronautic electronic equipment system etc.) have been carried out with according achievements abroad. The domestic DSSA research mainly concentrates in EIS (for example, tobacco, insurance, city geology etc.) as well as the intelligent systems (Barbara et al., 1995, Li & Wu, 2005, Shang et al., 2006). At present, it has two kinds of realization methods about the DSSA. That is: □Compile-time method, which creates the executable code according to the software system model automatically; and □Run-time method, which runs the software system model in the Virtual Machine (VM) directly. The former has realized the automatic creation of code, but it is costly and disadvantageous to the supporting system development. Moreover, it is difficult to manage due to the numerous codes created automatically. The latter is implemented based on VM, thus it only involves the description of system script model with less code quantity, which is in favor of the code management and maintenance. In the article a DAcSA-oriented system model called "resource-model-analysis (RMA)" is proposed, taking the digital agriculture-crops as an example. Furthermore, an RMA-based DAcSA is designed to meet the need of changeable, flexible and self-adaptive intelligent agriculture information system by a combination with virtual machine structure.

2. DACSA-BASED ADAPTIVE SYSTEM DEVELOPMENT

2.1 Formal description of the DAcSA-based adaptive model

DAcSA-oriented adaptive model must keep system logic function properly under different environmental conditions, and could covert the passivity of entity unit to initiativeness, the homogeneous coordination mode to diversity and static system evolution to dynamic one. Herein adaptability means software system can adjust its construction and algorithm automatically along with the change of its operating environment, and achieve a balance between them in an evolutionary way. To facilitate the description of DAcSA model, this article gives some descriptions as follows.

2.1.1 Domain-agriculture-crops business component

Domain-agriculture-crops business component is relatively fixed, typical and reusable software element in the domain-agriculture-crops. According to the definition by W. Koyacyuski (Kwozacznski, 1998), business component is the software business objects. Domain-agriculture-crops business component is the domain-agriculture-crops business object component. It is the autonomous domain-agriculture-crops business concept and reusable software unit, formed during domain-agriculture-crops software implementation. And it may be the domain-agriculture-crops object or the set of domain-agriculture-crops objects or the structure involved some domain-agriculture-crops objects, which can implement some specific functions. Hereinto, domain-agriculture-crops business entity component and data component are two basic components in domain-agriculture-crops.

AcEntity component

AcEntity component is the set of business logic functions, used to depict any significant objects such as the static entity, dynamic event and processing logic. AcEntity would be described as follow.

$$\text{AcEntity} ::= \{\text{Name}, \text{Func}, \text{Code}, \text{Interf} \mid (\text{In}, \text{Out})\} \quad (1)$$

Of which, Func means component functional description; Code is the component binary target code entity; Interf is the component interface, including input interface and output interface.

Data Component

Data Component would transform the AcEntity component attribute into the data and store them in the database. That is, Data Component will execute the data processing function including data definition and data manipulating such as data storage, inserting, deleting, updating and so on.

$$\text{Data} ::= \text{Com}(\text{AcEntity}, \text{Val}[, \text{Val}]..) \tag{2}$$

$$\text{Com} ::= \{\textit{Define}, \textit{Store}, \textit{Manipulation}\} \tag{3}$$

2.1.2 Domain-agriculture-crops software architecture, DAcSA

DAcSA is the mechanism or the framework for configuration agriculture-crops business components, and the channel network for connection of the information flow and the control flow among components. Usually, the framework could be divided into white-box framework and black-box framework according to the framework extension and customization technique. To take both of their advantages, the black and white box mixed framework technology is used to construct DAcSA, that is, the fixed part would be designed in the form of component (black-box) and changeable part would be design in the form of hotspot subsystem (white box) based on the design pattern (Xu & Li, 2003, Li & Xu, 2005). It is illustrated in figure 1.

In Figure 1, The DAcSA includes the management & glue center, black box business component library, white box hotspot knowledge library and business rules library. In the "black and white box" mixed framework, the fixed spots in DAcSA framework are designed to generic business component and saved in the business component library and the hotspots designed to hotspot subsystem based on the design pattern and saved in the hotspot knowledge library. The domain-agriculture-crops business rules are extracted and stored in the domain regular library. The management glue center would be responsible for selecting constructing unit from the business

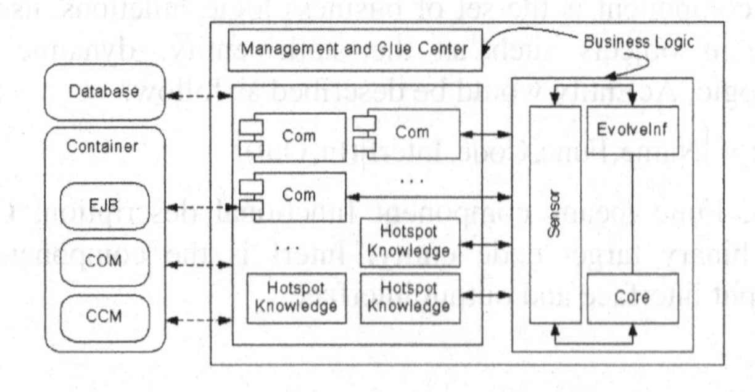

Figure 1. Adaptive model: black and white box mixed DAcSA

component library and the hotspot knowledge library respectively according to the business rules in the domain rules library, and plug & play upon the DAcSA to complete the realization and running of business logic.

2.1.3 Agriculture-crops applied system, AcSytem

In accordance with the above specification, the self-adaptive AcSystem can be manifested by a unit triplet as follows:

$$AcSystem ::= (DAcSA, Businessrule, Controller) \quad (4)$$

In formula (4), (1) DAcSA framework is a set of agriculture-crop fundamental component and hotspot subsystem and management glue center; (2) Businessrule, i.e. agriculture-crop business logic set, is the agriculture business rules set extracted from the agriculture domain analysis; and (3) Controller means to apperceive and adapt business logic change and complete the evolution or instantiation of business component and the hotspot knowledge in DAcSA. Controller may be shown in the following unit triplet,

$$Controller ::= \{Core, \mathrm{SensorInf}, EvolvedInf\} \quad (5)$$

Of which, Core means the logic structure of component for evolution; SensorInf means outside change information or input information. SensorInf deduces conditions triggering component evolution; EvolvedInf means the necessary evolution condition of the component. EvoledInf will justify whether the SensorInf triggers component evolution or designed for hotspot subsystem.

Conceptually adaptive DAcSA has three abstract layers, namely requirement layer, service layer and operation layer, to map the environment change and dynamic property to the software realization. Specifically, user build a requirement model firstly and configure management mechanism as shown in Figure 1 to realize the dynamic mapping of requirement and service.

2.2 Adaptive DAcSA-based RMA model, DAcSA-RMA

2.2.1 RMA model

Adaptive DAcSA design would support the AcSystem production and system evolvement with requirement change. In AcSystem, the system main body is agriculture-crops simulation model. Crops simulation model is the model simulated by means of the crops key composing element during its growth, which the crops get natural resource and grow according to the ecophysiological and zoological rule, as shown in Figure 2. It illustrates the general concept framework of the crops system simulation, key sub-model and simulation element and the interrelation- ship between them. Hereinto, A-F presents the 6 sub-models respectively. A is the development phase sub-model; B the biomass production sub-model; C the partition sub-model; D the organ formation sub-model; E the soil-water balance sub-model; F the soil-nutrition balance sub-model. And they are all restricted and influenced by climate state, conditions of soil, breed, cultivation and management.

Crop simulation model is composed of some independent sub-model, status or velocity variables and process function, which will get a simulation of a given subject (Wang et al., 2002). The ordered set of these concepts forms the whole process of crop simulation. The different crops model construction is the collection of the different algorithmic under the same or similar concept model. The running foundation of these algorithms is the set of various data.

Of course, although the simulation means, methods and the decomposition degree may vary, their goal or destination of simulation is consistent. Taking crop development phase for instance, the aims are all to simulate the growth process and predict phenology. There are different methods and models for different crops and different model-construction staff, but the main concept

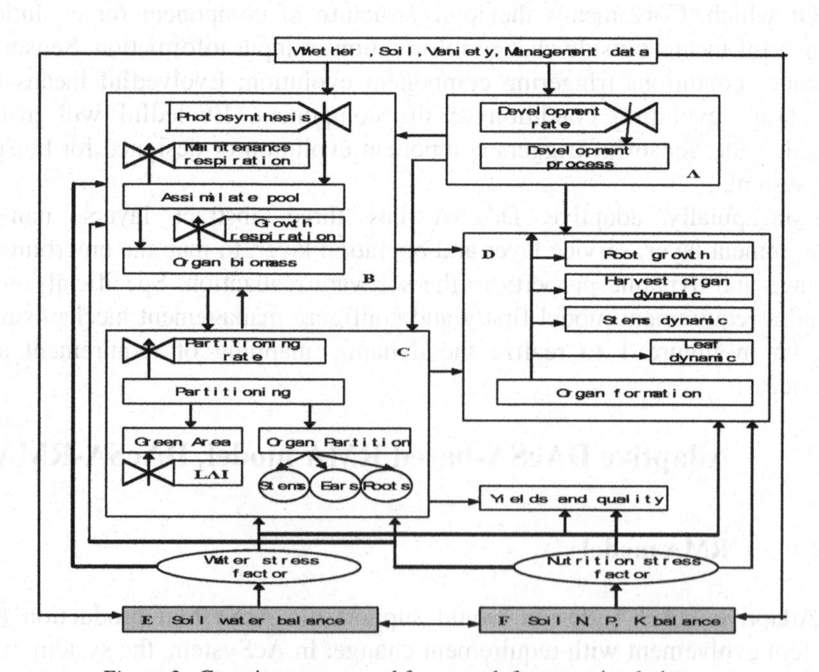

Figure 2. Generic concepts and framework for crop simulation

process and data model are similar (Yan et al., 2000). For example, in the wheat growth duration simulation, the temperature effect, light effect, jarovization effect and the interaction among every influence component are all considered to simulate wheat growth process and to predict phenology. For rice, because of no jarovization effect, only the temperature effect, light effect, growth process and predict phenology are considered to simulate the rice growth process and data. Meanwhile, both of them will also take into consideration of the restriction and influence of climate, breed parameters and cultivation condition.

From the analysis given above, the key to build DAcSA is to extract the relatively fixed business component and the relatively volatile hotspot subsystem from different models for different crops and different model-construction staff. It is operation to data, whatever the abstract mode. From such viewpoint, DAcSA-based AcSystem is a status machine in nature. Therefore, it can also be marked as:

$$AcSystem = \{I, O, S, \delta, \lambda\}$$

Of which, I is input set; O output set; S the inner status set in the AcSystem; δ status mapping function; λ output mapping function, and marked as:

$$\delta : I \times S \rightarrow S$$

$$\lambda : I \times S \rightarrow O$$

To analyze the data characteristics of crops simulation model further, there exists two kinds of data type sets with common character in the crops simulation model, one is describing resource data class (R), and the other is describing resource activity mark data class. The former is an objective resource or experiment-proved public data class, the latter is the data class result from analysis, reasoning and illation activity. In consideration of the characteristics of domain-agriculture-crops, this paper would separate the describing resource activity mark data into two types, simulation model data (M) and analysis data (A). It is called RMA model. In details:

R: resource data, it is the mapping result of outside natural environment entity on AcSystem, including environment data, cultivation and management data and factual measured data of the crops growth, e.g., the data about climate, soil, breed, cultivation condition and water and fertilizer management etc in different area;

M: simulation model data, it is the mapping result of the AcEntity and interrelationship between them on AcSystem, e.g. independent sub-model, status or velocity variables and process function and breed parameters related to simulation model, etc;

A: analysis and evaluation data, it is the mapping result of outside environment, human thinking activity or the change of user demand on AcSystem, e.g. the model prediction result data, the evaluation methods on crops etc.

Obviously, mapping relationship exists in R, M and A data in AcSystem, e.g. there must be R in M and there must be M and R in A. The data relationship among R, M and A can be marked as:

$$\Gamma(R,M,A) = \prod_{M\to R}, \prod_{A\to R}, \prod_{A\to M} \tag{6}$$

Γ means the mapping relationship among data.

$\prod_{M\to R}, \prod_{A\to R}, \prod_{A\to M}$ represent relationship of data M to R, A to R, A to M respectively.

On the basis of the RMA data model, its behavioral model can be further analyzed. Only from the viewpoint of structure, AcSystem behavioral model can be viewed as the set of three parts, i.e.

$$AcSystem = \Lambda\{M_R, M_M, M_A\} \tag{7}$$

Hereinto, $M_R : R \cup M \to R$ means R class data processing part, which would be processed for R class data itself, or M class data would be processed into the R class data;

$M_M : M \to M$ means M class data processing part, which would be processed for M class data itself;

$M_A : (M \times R) \cup A \to A$ means M class and R class data would be processed into the A class data, or A class data would be processed into A class data itself. As shown in Figure 3.

Now take a wheat growth simulation and management decision-making support system as an example to explain RMA model. It can be described as:

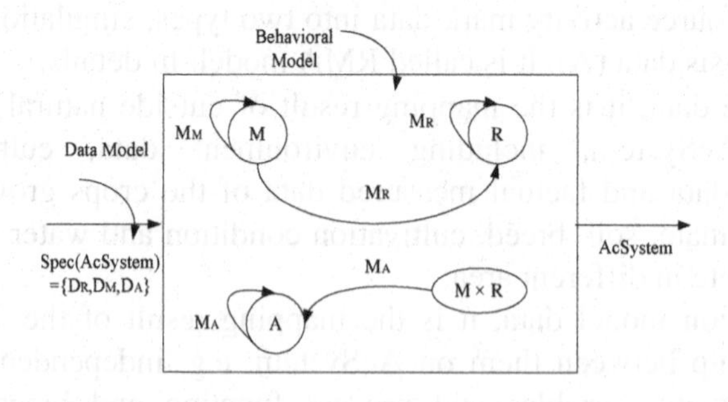

Figure 3. Behavioral model

R = {{soil parameter}, {climate data}, {variety description}, {cultivation condition}, {water-fertilizer management data}, {observed values in field}} ...

M = {{variety parameter}, {wheat development phase sub-model}, {wheat biomass production sub-model}, {wheat dry matter accumulation and yield formation}, {wheat organ formation sub-model}, {wheat water balance simulation}, {wheat nutrient (N, P, K) balance}, (weather environment simulation)... };

A = {{model prediction data}, {index of strategy evaluation}, {project evaluation result}, {observed sensitive values}, {variety parameter debug}, {real-time prediction}, {temporal spatial analysis}, ...}

Well then, the data mapping set among them,

Γ = {{ wheat development phase simulation} → {weather, growth condition, variety parameter, water balance simulation, nutrient balance simulation}, ...{project evaluation result} → {M}, {sensitive observed values} → {model prediction data, observed values in field}, {variety parameter debug} → {R, variety parameter}, {real-time prediction} → {R, M, model prediction data, observed data in field}, { temporal spatial analysis} → {weather, soil, variety description, growth condition}, etc.

Among the above, the behavior model Λ for the project evaluation result can be described as follow.

$$\Lambda = \{M_A | R, M \times R, M_M | R, M\} \tag{8}$$

In formula (8), the behavior model for the project evaluation result includes the processing of relevant resource data and model data. These behavior models embed by domain-agriculture-crops business rules is mapping to software implementation. Management glue center (in Figure 1) would select AcEntity business component or hotspot subsystem to be instantiated from hotspot knowledge and produce AcSystem.

2.2.2 Virtual machine framework-based system implementation

This article uses virtual machine framework to realize RMA. The virtual machine framework is constructed by the hierarchy pattern, and the data mapping layer, business logic layer, interactive control layer, interface presentation layer, the regulation control layer and the inner structure of regulation mapping layer would be design by the interpretation machine pattern (Guo et al., 2004). The inner structure of the virtual machine is as illustrate in Figure 4.

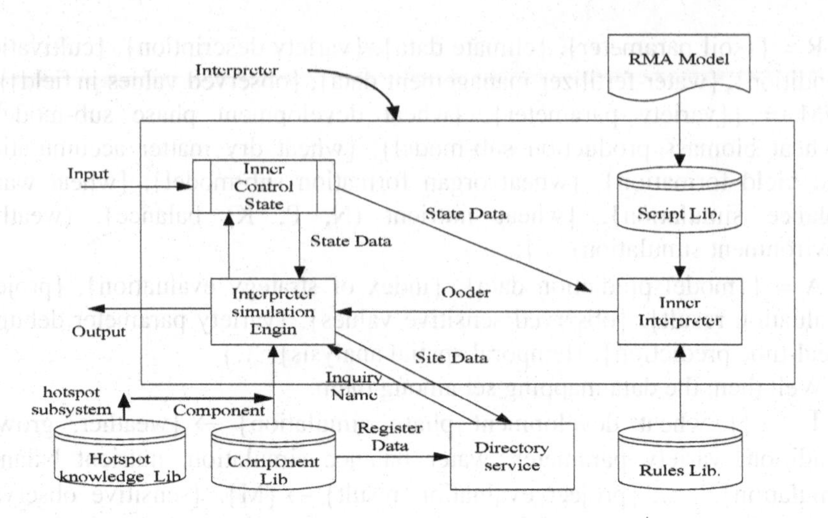

Figure 4. Inner structure of the virtual machine

Implementation process is as follow:

(1) Agricultural experts and software engineers build RMA model together.

(2) RMA model would be transferred into script form. Inner interpreter gets and checks the validity and correctness of the script according to business rules from rules library. Then instruction will be sent to inform the interpreter simulation engine in line with the script semantic content.

(3) According to instruction, interpreter simulation engine inquiry the location of according business component from component directory to obtain component or instantiated hotspot subsystem. The management glue center would be responsible for glue and plug it upon DAcSA. Then these business component and subsystem would run dynamically to implement required functions.

(4) When agriculture-crops business requirement or simulation model change, RMA model will change. Under the revised or new rules, the SensorInf in the Controller will get the input information and transform the condition (EvolvedInf) to evolve the existing component or produce the hotspot library. Based on RMA model and in light of the new rules, the adaptability of DAcSA will be implemented through selecting the new structure block from component library and hotspot directory.

3. CONCLUSIONS & DISCUSSIONS

Research on DAcSA-based architecture adaptability can effectively meet the domain-agriculture-crops business requirement and adapt the system evolution with business requirement change for different simulation model construction and different crops, and improve the productivity of domain software. The proposal of RMA model and virtual machine framework concept is to simplify the domain-agriculture-crop software development, particularly to high efficiently implement the different simulation model or revise model from different domain specialist by software. DAcSA-RMA model has been applied in digital agricultural crops system, which proves preferable adaptability, as shown in Figure 5.

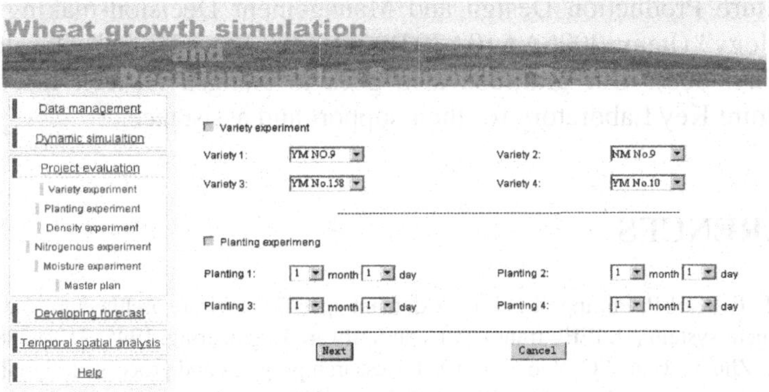

Figure 5. Wheat growth simulation & decision-making supporting system

The simulation model is different with different crops and different agricultural specialist. DAcSA-RMA is put forward to extract the common factors between them, implement the smooth transformation and flexible adaptability from business demand to software implementation. However, more research would be still necessary for different agricultural crops application system. Mainly they are:

DAcSA uses unchangeable black-box component and changeable white-box hotspot subsystem, i.e. DAcSA of the black and white box mixed framework to solve the constructing block. Evidently, change and fixedness are relative. When business requirement changes, the black and white box mixed constructing block will transform and evolve, which requires timely increase and update of component library and hotspot knowledge library. Therefore, the establishment and management of component and hotspot subsystem library must keep abreast of the time and be dynamic.

Extraction of business logic rules and management of rule library is another key. Extraction method of regulations, semantic description of regulations, grammatical structure and management of rule library are all crucial to adaptive model of DAcSA.

Management glue center is the code collection to plug and play upon DAcSA for business component and the hotspot subsystem to be instantiated. Directory service in the virtual machine is use to label the physical location of business components and other components. The common component interface is key technique to achieving seamless connection between constructing blocks, which is subject to further research.

ACKNOWLEDGEMENT

The research and thesis writing was sponsored by "Research on Precise Agriculture Production Design and Management Decision-making Model Technology" (Item: 2006AA10A303) under the State Hi-tech Development Plan (Plan 863). Our gratitude also goes to Jiangsu Hi-tech Information Agriculture Key Laboratory for their support and assistance.

REFERENCES

Barbara H R, Karl P, Philippe L, et al. A domain-specific software architecture for adaptive intelligent system [J]. IEEE transactions on Software Engineering. 1995. 21(4):288-301.

Cao W X, Zhu Y, Tian Y C, Yao X, Liu X J. Research progress and process of digital farming techniques, Scientia Agriculture Sinica, 2006. 39(2):281-288.

Guo Zh, Zhao X B, He F, Gu M, Sun J G, Software oriented to special domain model and architecture for enterprise information system, Computer Integrated Manufacturing System, 2004. 9:1046-1051.

Huang S X, Fan Y Sh, Zhao Y. Research on generic adaptive software architecture style, Journal of software, 2006. 6:1338-1348.

Kwozacznski W. Achitecture framework for business components. 5th Internantional Conference on Software Reuse [C]. Computer Society Press, 1998. 300-307

Li X R, Xu H L. Design and Implementation of Business Component Oriented Black & White Box Domain Framework, Mini-micro system, 2005. 1:64-68.

Li Y, Wu Z H H, Research on architecture of dynamic self-healing system, Journal of Zhejiang University (Engineering Science), 2005. 2:216-222.

Mei H, Shen J R. Progress of research on software architecture, Journal of software, 2006. 6:1257-1275.

Shang J G, Liu X G, Hua W H. Urban geological domain specific software architecture, Earth Science-Journal of China University of Geoscience, 2006. 9:673-677.

Xu H L, Li X R, Business component reengineering based on rule library, Computer Integrated Manufacturing System, 2003. 10:911-915.

Yan M C, Cao W X, Luo W H, Jiang H D. A mechanistic model of phasic and phenological development in wheat [I]: Assumption and Description of the Model. Chinese Journal of Applied Ecology, 2000. 11(3):355-359.

Yu Z H H, Cao Y L, A study on the domain specific software architecture, Micro-electronic and computer, 2004. 21(7):66-69.

GEOSPATIAL COMPUTATIONAL GRID FOR DIGITAL FORESTRY WITH THE INTEROPERABILITY

Guang Deng [1,*], Xu Zhang [1], Quoqing Li [2], Zhenchun Huang [3]

[1] *Research Institute of Resource and Information Technique, Chinese Academy of Forestry, Beijing, China, 100091*

[2] *China Remote Sensing Satellite Ground Station, China Academy of Science, Beijing, China, 100086*

[3] *Department of Computer Science and Engineering, Tsinghua University, Beijing, China, 100084*

** Corresponding author, Address: P.O. Box 33, Chinese Academy of Forestry, No. 1 Dongxiaofu, Xiangshan Road, Beijing, 100091, P. R. China, Tel: +86-10-62889196, Fax: +86-10-62872778, Email: dengg@caf.ac.cn*

Abstract: In the Digital Forestry area, there is more and more requirements for spatial information connection and processing. From a prototype of geospatial computational grid (GCG), the relationship between the geospatial computational grid and OGC Interoperability protocols is proposed in this paper. A prototype of geospatial information grid is given in this paper. Digital Forestry Grid is proposed as a case of implementing of geospatial computational grid with OGC interoperability at the end of this paper.

Keywords: Geospatial Computational Grid; Digital Forestry; Interoperability

1. INTRODUCTION

Earth science is a data-intensive and computation-intensive scientific domain in which the applications always produce and analyze a large volume of distributed heterogeneous geospatial information. The grid concept is revolutionary, because it foresees the future integration of technologies to realize an observing and operating system with scalability and applicability over a broad range of earth surface phenomena. As forest

Deng, G., Zhang, X., Li, Q. and Huang, Z., 2008, in IFIP International Federation for Information Processing, Volume 258; Computer and Computing Technologies in Agriculture, Vol. 1; Daoliang Li; (Boston: Springer), pp. 585–591.

management is a kind of earth resource science, we do the research on the Geospatial Computational Grid for forest management. In this paper, we discuss the relationship between the Geospatial Computational Grid and interoperability based the view of system construction.

As the amount of remote sensing imagery and related GIS database in the forest management area grows, there is an increasing problem of providing sufficient computational resources to assemble and analyze the datasets for environment science research or forestry affairs decision-making supporting.

Grid technology brings together geographically and organizationally dispersed computational resources, such as CPUs, storage systems, communication systems, data, software, instruments and human collaborators, which securely provide advanced distributed high-performance computing to users in one or more Virtual Organizations (VOs) (Foster, et al., 2001; Booth, et al., 2004). The Globus Project proposes the Open Grid Service Architecture (OGSA), Open Grid Service Infrastructure (OGSI) for Globus 3.0 and the Web Service Resource Framework (WSRF) for Globus 4.0 as the guidelines and specifications for system design and implementation to build a geospatial computational Grid (Brunett, 2001; Moore, et al., 1999; Karl, et al., 2004).

Although the grid technology can take an important role in geospatial computational grid design, but it is not enough because of the particularity of geospatial information. As follows, the description of the paper structure is given. First, we discuss the trends and problems which parallel and distributed computing for processing large-scale geospatial data would faced. Then, a prototype of geospatial information grid is given, which is a case study and a simple implementing of our understanding of geospatial computational grid. From this, we discuss the methods to integrate the geospatial interoperability to grid application. The Digital forestry Gird is the case of the methods discussed above. Finally, an opening conclusion is given for further research.

2. A PROTOTYPE OF GEOSPATIAL COMPUTATIONAL GRID

2.1 Aims and structures

A major goal of the geospatial computational grid is to make it as easy and transparent as possible for researchers to move jobs, GIS data and kinds of remote sensing data freely among the machines within the grid scope. The geospatial computational grid would consist of some resource provider (RP) sites. Each site connects through a dedicated geospatial data clearing house network, and provides high-end computing resources totaling more than specifically Floating Point Operations Per Second (FLOPS) of compute power and specifically available storage. Users have the option of storing their data, managing their jobs, and performing their computations on the machine most appropriate for their tasks, using grid technology for access to each resource.

A prototype system of geospatial computational grid is a spatial information grid which designed by our research team and has resolved many problems. By applying WSRF, the spatial information grid is constructed. Fig. 1 shows the structure of geospatial computational grid.

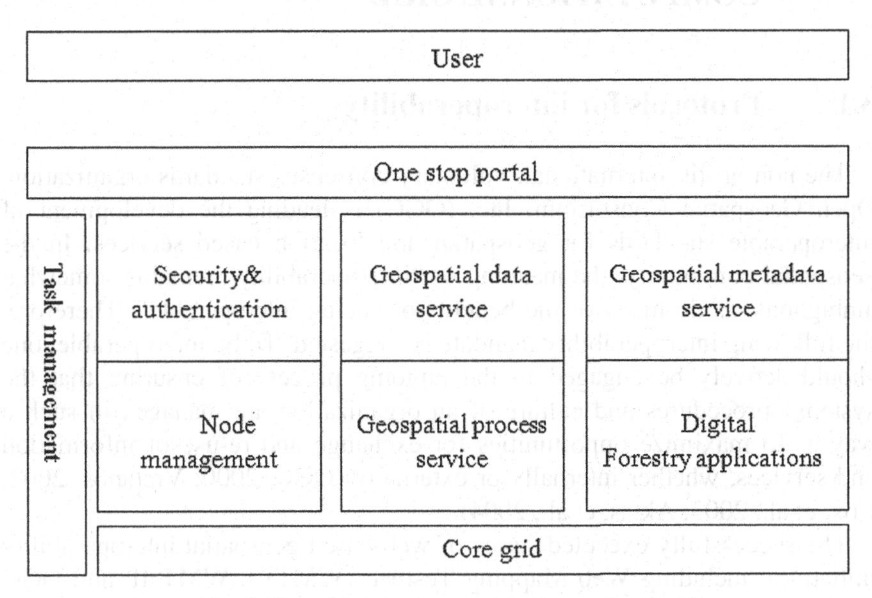

Fig. 1. The structure of geospatial computational grid

2.2 Functions

The geospatial computational grid provides many functional modules such as geospatial data processing and core grid management.

Node management is for monitoring and discovery of node for providing its information and its state, which also provide services through state registration, data modeling and local registry. By globe name service, the service in Geospatial Computational Grid could be understood. Task management is a Resource Allocation Manager for providing a common user interface for submitting a job to the dispersed multiple machines. The security and authentication function provides generic security services such as authentication, authorization and credential delegation for resources that will be run on the Grid. The geospatial metadata service and geospatial process service belongs to Metadata Catalog Service (MCS) for providing a mechanism for storing and accessing metadata of data and service.

Anyway the operational geospatial computational grid must deal with interoperability problems which involved in the related layers.

3. INTEROPERABILITY AND GEOSPATIAL COMPUTATIONAL GRID

3.1 Protocols for interoperability

The non profit, international, voluntary consensus standards organization, Open Geospatial Consortium, Inc. (OGC) is leading the development of interoperable standards for geospatial and location based services. In the geospatial community, the meaning of "interoperability" remains somewhat ambiguous, as do many of the benefits of "being interoperable". Therefore, the following interoperability mandate is suggested. To be interoperable, one should actively be engaged in the ongoing process of ensuring that the systems, procedures and culture of an organization are managed in such a way as to maximize opportunities for exchange and re-use of information and services, whether internally or externally (OGC, 2000; Vretanos, 2002; Cox, et al., 2003; Aktas, et al., 2004).

The successfully executed series of web-based geospatial interoperability initiatives, including Web Mapping Testbed (WMT) I, WMT II, and OGC Web Service Initiative (OWS) 1.1, and OWS 1.2 have produced a set of web-based data interoperability specifications, such as the OGC Web Mapping Service (WMS) specification which allows interactively assembling maps from multiple servers, the OGC Web Coverage Service (WCS) specification which provides an interoperable way of accessing geospatial data from multiple coverage servers, especially those data from remote sensing, and the OGC Catalogue Service-Web (CSW) specification which is based on e-business Registry Information Model (ebRIM) and aims to provide an object-oriented registry system for registering, managing and retrieval of geospatial resources, e.g. services, data and other objects (Chen, et al., 2005).

3.2 Agents for interoperability

How to put OGC interoperable Web Service into the common computing grid is the key technologies to construct geospatial computational grid. We use the agents to integrate the interoperability and Geospatial Computational Grid. Table 1 shows the selected OpenGIS interoperable specification agents used in GCG. The basic methods to construct the agents can get from Li's paper (Li et al., 2005).

4. DIGITAL FORESTRY GRID AS A CASE

Digital Forestry Grid (DFG) is a kind of specialization application grid which supports for forest resource information management and forestry eco-construction projects. A characteristic of the DFG is in specialization grid resource management, business logic development and running engine of grid application. By grid transform of forestry resource information, DFG integrates the data resources of forest and forestry eco-construction projects in four levels (namely state, province, city and county). The data resources have kinds of types, such as remote sensing images, vector data, thematic attribute data and etc. Thematic grid services, such as data resource service, spatial information analysis service, online statistic computing service, is be developed independently, which realizes the sharing and usage of forestry information resources in deeply extents.

Using catalog interface service (Table 1), DFG find the wanted thematic forest resources data service. Using filter encoding and Web Feature Service (Table 1), the forest resources map can be set out on the web. (Fig. 2)

Table 1. OpenGIS interoperable specification agents for GCG

Acronym	Name	Agents at GCG Location
CAT	Catalog Interface	Geospatial metadata service
CT	Coordinate Transformation Services	Geospatial data service
Filter	Filter Encoding	Geospatial data service
GML	Geography Markup Language	Geospatial data service
Common	OGC Web Services Common Specification	Geospatial data service
SLD	Styled Layer Descriptor	Geospatial data service
WCS	Web Coverage Service	Geospatial data service
WFS	Web Feature Service	Geospatial data service
WMC	Web Map Context Documents	Geospatial data service
WMS	Web Map Service	Geospatial data service

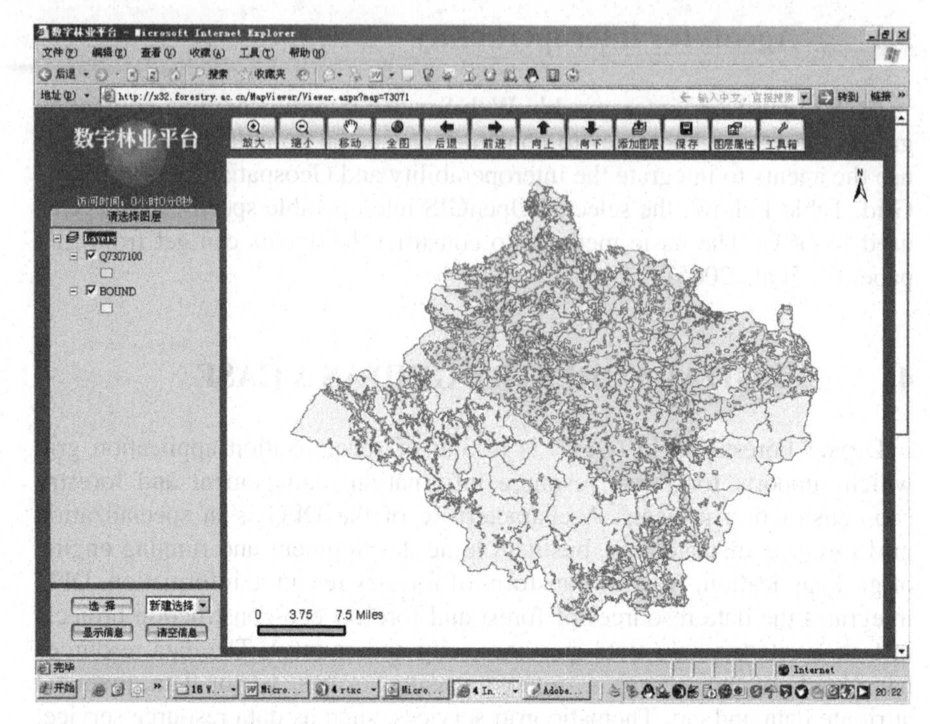

Fig. 2. A Web Feature service result presentation of digital forestry grid

5. DISCUSSION AND CONCLUSION

From above, we give a short presentation of the design of geospatial computational grid in the view of interoperability. The importance of the interoperable features in GCG cannot be ignored in the digital forestry system. Agents can be used to integrate the web service. But more key technologies are not put up in this paper, which is for further research.

ACKNOWLEDGEMENTS

This study has been funded by China National Key Research Project (863) "Research on the Web based Spatial Computation Service" (Contract Number: 2006AA12Z203) and "Research on the distributed management technique of the forest farm faced forest resources heterogeneous data" (Contract Number: 2006AA10Z240).

REFERENCES

Aktas, M., Aydin, G., Donnellan, A., Fox, G., Granat, R., Grant, L., Lyzenga, G., McLeod, D., Pallickara, S., Parker, J., Pierce, M., Rundle, J., Sayar, A., and Tullis, T., Iservo 2004. Implementing the International Solid Earth Research Virtual Observatory by Integrating Computational Grid and Geographical Information Web Services Technical Report, Special Issue of Pure and Applied Geophysics (PAGEOPH) for Beijing ACES Meeting July, December.

Booth, D., Haas, H., McCabe, F., Newcomer, E., Champion, M., Ferris, C., and Orchard, D. 2004. "Web Services Architecture." W3C Working Group Note, 11 February 2004, Available from http://www.w3.org/TR/2004/NOTE-ws-arch-20040211/.

Brunett, S. 2001. Application experiences with the Globus Toolkit, Proceedings of the 7th IEEE Symposium on High Performance Distributed Computing. IEEE Press: 81–89.

Chen Aijun, Di Liping, Wei Yaxing, Liu Yang, Bai Yuqi 2005. Grid Computing Enabled Geospatial Catalague Web Service, SPRS 2005 Annual Conference Baltimore, Maryland March: 7–11.

Cox, S., Daisey, P., Lake, R., Portele, C., and Whiteside, A. (eds) 2003 OpenGIS Geography Markup Language (GML) Implementation Specification. OpenGIS project.

Foster, I., Kesselman, C., and Tuecke, S. 2001. The Anatomy of the Grid: Enabling Scalable Virtual Organizations, International J. Supercomputer Applications 15(3).

Karl Czajkowski, Donald F. Ferguson, Ian Foster, Jeffrey Frey, Steve Graham, Tom Maguire, David Snelling, Steve Tuecke. 2004. From Open Grid Service Infrastructure to WS-Resource Framework: Refactoring & Evolution Version 1.0.Globus Alliance.

Li Guoqing, Liu Dingsheng, Sun Yi. 2005. A Grid Research on desktop type software for spatial information processing.

Moore, R., Baru, C., Marciano, R., Rajasekar, A., and Wan, M. 1999. Data-intensive computing, in Foster, I. and Kesselman, C. (eds) The Grid: Blueprint for a New Computing Infrastructure, Morgan Kaufmann: 105–129.

The Open Geospatial Consortium (OGC): http://www.opengeospatial.org

Vretanos, P. (ed.) 2002. Web Feature Service Implementation Specification, OpenGIS project document: OGC 02-058, version 1.0.0.

SIMULATING SOIL WATER AND SOLUTE TRANSPORT IN A SOIL-WHEAT SYSTEM USING AN IMPROVED GENETIC ALGORITHM

Changshou Luo [1,*], Sufen Sun [1], Junfeng Zhang [1], Qiang Zuo [2], Baoguo Li [2]
[1] *Institute of Information on Science and Technology of Agriculture, Beijing Academy of Agriculture and forestry Sciences, Beijing, 100097, China*
[2] *College of Resources and Environment, China Agricultural University, Beijing, 100094, China*
[*] *Corresponding author, Address: Institute of Information on Science and Technology of Agriculture, Beijing Academy of Agriculture and forestry Sciences, Beijing, 100097, China*
Email: luocsyoupan@163.com

Abstract: An improved genetic algorithm was applied and examined to optimize the weights of a neural network model for estimating root length density (RLD) distributions of winter wheat under salinity stress. Thereafter, soil water and solute transport with root-water-uptake in a soil-wheat system were simulated numerically, in which the estimated RLD distributions were incorporated. The results showed that the estimated RLD distributions of winter wheat using the neural network model combined with the improved genetic algorithm, as well as the simulated soil water content and salinity distributions, were comparably well with the experimental data. The method can serve in modeling flow and transport under salinity or saline water irrigated areas.

Keywords: genetic algorithm; root length density distribution; soil water content; salinity; simulation

1. INTRODUCTION

Because of water resources shortage in China, study on soil water and salinity distributions on the conditions of light-saline irrigation and root-water-uptake was flourish (Khosla, 1997). A lot of root-water-uptake models included root distribution function. Constructing empirical model or fitting measured data were the common methods used for getting distribution

Luo, C., Sun, S., Zhang, J., Zuo, Q. and Li, B., 2008, in IFIP International Federation for Information Processing, Volume 258; Computer and Computing Technologies in Agriculture, Vol. 1; Daoliang Li; (Boston: Springer), pp. 593–604.

parameters (Wu, 1999). Because the model of root distribution in soil profile was not standard and it would be different according to different crop and soil environment, there was more difference between root parameters and real data. Plant roots went hand in hand with canopy (Thornley, 1998). An improved genetic algorithm was applied in this thesis. This method can numerically estimate root distribution and get relative exact parameters of root distribution surrounding salinity in indirect and easy way. Then the parameters combined with water and solute transport model to study water and solute transport. These would be significant for study on crop growth, secondary salinization prevention and light-saline utilization surrounding salinity.

2. EXPERIMENTAL DESIGN OF WATER AND SOLUTE TRANSPORT UNDER CONDITION OF ROOT-WATER-UPTAKE

2.1 Experimental design

This thesis designed winter wheat indoor soil column experiment. Cultivar was NongDa 189 (ND189). Soil was sandy soil. Bulk density was 1.64g/cm3. Field capacity was 0.07cm3/cm3. Soil column was polyethylene tube of 10cm inner diameters. Sandy soil was loaded in hierarchy which was 5cm per layer. The total height was 40cm. There were respectively tensiometers and salinity sensors in 5cm, 10cm, 15cm, 25cm, 35cm. Experimentation which was designed one controlled disposal (no salinity stress), two salinity disposal (salinity disposal 1 and salinity disposal 2) and three repeats was practiced in wheat seedling. There was no NaCl in nutrient solution of controlled disposal. 3g/L NaCl was in nutrient solution of salinity disposal. 1.6g/L NaCl was in nutrient solution of salinity disposal 2. There were four wheat cultivars per soil column (equal to 4500000/hm2 in field). 3cm quartz sand was used to cover sandy soil in order to prevent evaporation after seedling. Water content must be well proportioned and nutrient must be abundant. Gravimetric method was used to control water content. In the process of experiment, soil columns were taken apart once in 6d in order to get concerned data of root and canopy.

2.2 Water and solute transport model under condition of root-water-uptake

On the condition of vertical one-dimensional unsaturated flow, water transport fixed solution problem which included root-water-uptake in this experiment was:

$$C(h)\frac{\partial h}{\partial t} = \frac{\partial}{\partial z}\left(K(h)\frac{\partial h}{\partial z}\right) - \frac{\partial K(h)}{\partial z} - S(z,t)$$

$h(z,0)=h0(z)$, $0 \leq z \leq lz$

$$\left[-K(h)\frac{\partial h}{\partial z} + K(h)\right]_{z=0} = E(t), t > 0$$

$h(lz,t)=h1(t)$, $t > 0$ (1)

h- soil water matrix potential, cm. z- spatial coordinate, cm, adown. C (h)- water capability, cm-1. K(h)- unsaturated hydraulic conductivity, cm·d-1. t- time ordinate, d. h0(z),h1(t)- given function (or discrete points). lz- vertical total depth in simulation zone, cm. S(z, t)- root-water-uptake rate, d-1. E(t)- evaporation or irrigation intensity (evaporation:-, irrigation: +), cm·d-1.

On condition that taking no account of soil static water influence and sorption and ignored influence of soil temperature gradient in this experiment, vertical one-dimensional unsaturated soil water and solute transport equation was:

$$\frac{\partial(\theta C)}{\partial t} = \frac{\partial}{\partial z}\left(\theta D_{sh}\frac{\partial C}{\partial z}\right) - \frac{\partial(qC)}{\partial z}$$

$C(z,0) = C_0(z)$ $0 \leq z \leq lz$, $t=0$

$$[-\theta D_{sh}\frac{\partial C}{\partial z} + qC]_{z=0} = q_0 C_0'(t), z = 0, t > 0$$

$C(L_z,t) = C(t)$ $z=lz$, $t > 0$ (2)

C- soil salinity concentration, g·L-1. Dsh- hydrodynamic dispersion coefficient, cm2·d-1. q- unsaturated DaXi flow velocity, cm·d-1. C0(z)- given profile concentration, g·L-1. C(t)- given concentration, g·L-1. $C_0'(t)$ - irrigating water concentration, g·L-1, the value was 0 on the period of evaporation.

3. ROOT DISTRIBUTION ESTIMATION MODEL BASED ON IMPROVED GENETIC ALGORITHM

On the condition of crop in existence, in the simulating process of water and salute transport model, root-water-uptake source terms was the key for the accuracy of simulating soil water and salute distribution. But the model of root in soil profile was not standard and it would be different according to crop and soil environment. Based on close relationship of root and canopy, this research availed oneself of artificial neural network advantages for modeling information complex question to construct artificial neural network model in order to estimate root distribution parameters. It also took advantage of briefness, robustness and global optimization of genetic algorithm to optimize weights of artificial neural network model with improved genetic algorithm. The artificial neural network model based on improved genetic algorithm could provide root distribution parameters. Moreover, the parameters could combine with root-water-uptake model and water and solute transport model to simulate water and solute transport distribution on the condition of root-water-uptake.

3.1 Construction of artificial neural network model for estimating root distribution parameters

Aboveground dry matter weight and leaf area have close relation with root growth. The two parameters should be used as input variables. Because crop growth time can show crop genetic characters on the whole and soil water and salinity were the main reasons of root growth and distribution difference, crop growth time, soil matrix potential in different depth and salinity should be input variables too. Root length density in different depth was the output variable. This search used feed forward artificial neural network which included three layers (Yuan, 1999). The three layers were input-layer, out-layer and middle-layer (hidden-layer). There were connections between upper layers nerve cells and under layer nerve cells. There were no connections between same layers.

3.2 The process of genetic algorithm optimizing artificial neural network model weights

Genetic algorithm was a computing model which simulating natural Bio-evolution. People were taking more and more attention to genetic algorithm because it was simple, robust, global search and not be limited by search space. Genetic algorithm has simulated propagation, mating and mutation in the process of natural selection and natural inheritance. It expressed questions as chromosomes which would form original chromosomes colonies. Then original chromosomes colonies were parked in question environment. The chromosomes which can adapt environment were selected according to the survival of the fittest and would be operated by copy, across and aberrance. Accordingly the new chromosomes colonies were brought. Circulation would be incessant until the fittest individual was selected (Chen, 1996). Finally the optimal solution was obtained. The main process of genetic algorithm to optimize artificial neural network weights in this thesis was:

1 Real coding strategies was applied in chromosomes gene coding of artificial neural network model weights.

2 Produce the original chromosomes colonies.

3 Calculate the individual fitness value of colonies.

4 Genetic operation which included turntable bet selection as well as across and mutating operation according to certain rules was conducted.

5 Recalculate the fitness value of chromosomes colonies.

6 If the search criteria were met, iteration would stop and gave the optimal solution. Otherwise, turn to step 4.

In order to prevent optimal individual to be destroyed by genetic operation, the optimal individual reservation strategy was implemented in genetic operation.

3.3 Determinate fitness function in model

Fitness function was a criterion to judge individual of chromosomes colonies. Genetic algorithm used fitness value to instruct search direction. The fitness function was used by us in this search as follows:

Input and output variable formed training samples to train neural network which presented by individuals of chromosomes colonies in order to calculate learning error E of each individual. Formula was:

$$E = \sum_{i=1}^{n} E_i , \qquad E_i = \sum_{l=1}^{m} (y_l^i - C_l^i)^2 / 2 \tag{3}$$

n- training samples number. m- output units number. yil-Cil - diffirence between real output and expectation output of No. 1 when No. i sample was training. Fitness function was decided by following formula.

fs = 1/E,

Fitness function assured error less the fitness value more large.

3.4 Improve genetic algorithm

When genetic algorithm was training artificial neural network weights, crossover operator and mutation operator played the great role in optimization process. Chromosomes may be likely to converge or diverge in search space while simple genetic algorithm was optimizing artificial network weights. This caused speediness and global convergence of search under expectation. After improved crossover operator and mutation (Luo, 2000), the time of optimizing neural network weights would decrease and convergence and stability would improve. Realize concretely as follows:

Crossover operator:

If x1, x2 were parent individuals which were uniformly distributing random numbers in interval V=[xmin, xmax]. Progeny individuals z1, z2 after crossover would be produced via formula as follows (Huang, 1999):

$$y_1 = \alpha x_1 + (1-\alpha)x_2, \quad y_2 = \alpha x_2 + (1-\alpha)x_1$$
$$z_1 = MOD(y_1, V), \quad z_2 = MOD(y_2, V) \tag{4}$$

a- integer. MOD- modular arithmetic operator.

Mutation operator conformed to rules as follows (Zheng, 2000):

$$P_m = 0.001 + NG \cdot cof \tag{5}$$

Pm- variance rate of present algebra. NG- continuously un-evolutional algebra since the last evolution. Cof- parametes which decided threshold of chromosomes coercive mutation (100% mutation).

$$\text{var} = rand \cdot w \cdot dyna \tag{6}$$

var- mutation variables. Rand- random number in [0, 1]. w- a fixed value in value range of weight. Dyna- dynamic parameters which decided mutation variables var, and initializing dyna = 1.0. if counter > nochange, dyna = $dyna \times 0.1$ and counter $= 0$, counter was counter which used to cumulate continuously un-evolutional algebra since the last evolution. nochange- constant used to judge alteration of dyna value.

After combination application of crossover operator and mutation operator, on the one hand distribution of progeny individuals was even in search space, on the other hand a self-adaptive mutation mechanism could be constructed. Therefore these could keep upper efficiency in the process of genetic algorithm optimizing neural network model weights all along. Evaluation could refer to reference (Luo, 2005). In conclusion, improved genetic algorithm can improve convergence in the process of genetic algorithm optimizing neural network model and reduce the time of solution. These could gain one's ends to improve efficiency of optimizing root distribution estimation model.

3.5 Obtain parameters of model

In the above-mentioned experimental design, the model parameters can be obtained via following methods:

Canopy parameters measure:

Measuring frequency was once in 6d. Measuring items included dry matter and leaf area. Leaf area measure used Snapscan1236 scanner made by company AGFA of Germany. Then leaf area calculation used WinRhizo pro root analysis software made in Regent Instrument company.

Root length density measure:

Soil columns were taken apart once in 6d. At first cut open soil columns in lengthways. Then sandy soil with root was intercept per 4cm in transverse and put in sifter. The last, water used to wash root cleanly. Snapscan1236 scanner was used to scan root and WinRhizo pro root analysis software calculate root length.

Soil salinity measure:

TYC-□ soil salinity sensor was used to obtain soil salinity very day on time.

Soil matrix potential measure:

Tensiometer was used to measure soil matrix potential very day on time.

Evaporation measure:

Gravimetric method was used to measure evaporation once in 3d.

4. APPLICATION AND ANALYSIS OF MODEL

4.1 Analysis of root distribution prediction result

According to description of constructing artificial neural network model for root distribution parameters estimation and experiment design in this research, winter wheat root distribution estimation model under salt stress was a 13-9-10 network. Input layer was composed of growth time, aboveground dry matter, leaf area, matrix potential in 5, 10, 15, 25, 35 and salinity. Nerve cells number was 13. Output layers were composed of root length density of each 4cm soil column in transverse. Nerve cells number was 10. There were no specific prescribe to design middle layers nerve cells number in theory. The number generally decided by experience. If middle layers nerve cells were less, the model global optimization would weaken. If middle layers nerve cells were more, the model generalization would weaken. This experiment design middle layers were 9. Improved genetic algorithm was applied to optimize neural network weights in this experiment. Moreover, reservation strategy was applied in the experiment process. Colony scale was 31. Training precision was designed under learning error E<0.05. After learning, the samples which had neither part nor lot in learning were estimated as follows:

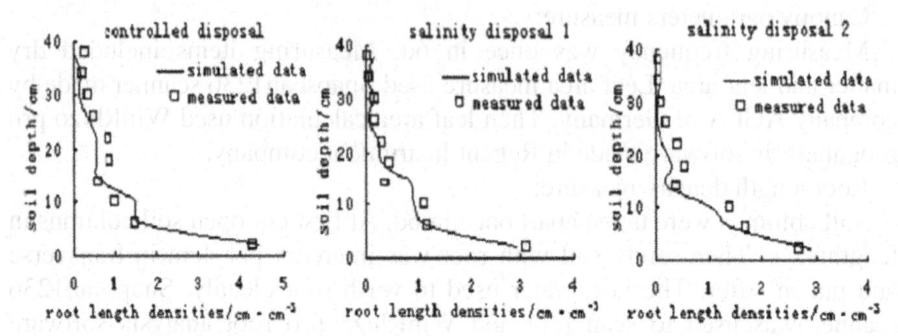

Fig. 1. Comparision between measured and estimated root length densities

Table 1. Error analysis of results between measured and estimated root length densities

Disposal	Average relative error	Correlation factor (n=18)
Controlled disposal	0.24	
Salinity disposal 1	0.26	0.97
Salinity disposal 2	0.27	

Fig. 1 was the comparision between measured and estimated root length densities. Fig. 1 showed holistic estimation was preferable except exceptional individuals. Moreover, while the learning samples were composed of diverse learning samples in the estimation process for diverse samples, the model estimation could be preferable generally and show good convergence and stability. Pictures indicated that root mainly distributed in 25cm under ground. So we analyzed average relative error between root length densities estimated and measured data and calculated correlation factor of theirs. The result refers to table 1. In the process of calculating average relative error, less effect of exceptional points brought on the augment of estimation data average relative error. However, the general correlation factor which was 0.97 showed model estimation generally worked well. As well as Fig. 1 showed that estimation of controlled disposal was better than salinity disposal 1 and salinity disposal 1 estimation was better than salinity disposal 2. Taking complexity and diversification of interaction between crops and environment into account, estimation result of this model was receivable. In conclusion, based on soil water parameters, salinity parameters and canopy parameters which went hand in hand with root distribution and easily be obtained, weights of improved genetic algorithm optimizing neural network model used to quantitatively forecast root distribution was feasible. This method can obtain parameters and root distribution data which was difficult easily.

4.2 Construct water uptake model under salinity stress

On the condition that salinity stress (no water stress), root-water-uptake model which was brought forward by Feddes etc was used broadly to show root-water-uptake rules. The model was:

$$S = \alpha(h_0) S_{max} \tag{7}$$

S- root water uptake rate which denoted root water uptake amount in unit time and unit soil, cm3·cm-3·d-1. h0- soil water osmotic potential, cm. Smax- maximal root water uptake rate, d-1. Smax can be obtained by following formula (Wu, 1999):

$$S_{max} = \frac{T_P(t) L_d(z,t)}{\int_0^{L_r(t)} L_d(z,t) dz} \tag{8}$$

z- soil depth, cm. Tp(t)- potential transpiration rate, cm·d-1. Ld(z,t)- root length densities, cm·cm-3. Ld(t)- maximal root depth, cm. a(h0)–revising coefficient of osmotic potential on the condition that not restrict water, no dimension. On the condition of no water and no nutrient stress, the better practical format of a(h0) as follows (Homaee, 1999):

$$\alpha(h_0) = 1 - \frac{\alpha}{360}(h_0^* - h_0) \tag{9}$$

h_0^*- critical value of osmotic potential, cm. a- the reducing water uptake amount while add unit conductance rate (mS·cm-1), cm-1, a=0.073. The root-water-uptake model under salinity stress (no water stress) can be constructed by Formula (7), (9):

$$S = [1 - \frac{\alpha}{360}(h_0^* - h_0)] \frac{T_P L_d(z)}{\int_0^{L_r} L_d(z) dz} \tag{10}$$

In above root-water-uptake model, root length densities derived from root distribution estimation model based on improved genetic algorithm via aboveground dry matter, leaf area, growth time, matrix potential and salinity parameters. h_0^* was the measured data in reverse and the method as follows:

Given measured water distribution of soil profile, salinity concentration distribution, root length densities distribution Ld(z) and potential transpiration rate Tp in time t1, t2, if h_0^* was given an initial value, root-water-uptake rate since t1 can be calculated by formula (10). Root-water-uptake rate used as source terms to solve water and solute transport equation in order to figure out simulation value of soil water in time t2 and get square sum of the error between simulation value and measured data of soil profile water content in time t2. Then give h_0^* another value, the same method was used to calculate square sum of the error until the square sum of the error was least. The value of h_0^* was the parameter which we want to solve. In this experiment, the h_0^* was -560cm.

4.3 Analysis of water and salinity transport simulation result

Weights of improved genetic algorithm optimizing neural network model used to quantitatively estimate root distribution. The estimation root distribution parameters combine with root-water-uptake model to simulate water and salinity transport distribution. On the condition of salinity stress, water transport solution equation of root-water-uptake used Crank-Nicolson difference schemes to solve problems and salinity solution equation used Bresler (1973) second-order numerical differential method to solve above definite solution problems. The solving process used to simulate soil water and salinity. In the process of simulation, space step-length was Δz=1cm. Time step-length was Δtj+1=1.25Δtj. Subscript (j. j+1) was the ordinal number of time step-length. The control criterion of iterative water potential was ε=0.01. Upper boundary soil surface evaporation was E(t)=-0.038 cm d-1. Because limited by measured data, surface water content was given by linear extrapolation. Lower boundary was 40cm. Water flux and salinity flux were 0. The maximal root depth was L_r=40cm. Fig. 2 was a simulation result of soil water and solute transport model in soil profile. The simulation time was 3d. Average potential evaporation T_p was 0.38cm · d-1. Fig. 2 showed that this method can estimate soil water distribution better under salinity stress. Moreover, except the exceptional points, the salinity simulation was generally good. In general, used as a tentative method, we thought that improved genetic algorithm quantitatively estimated root distribution and the estimation result combined with root-water-uptake model to simulate salinity stress, water and salinity transport distribution worked well.

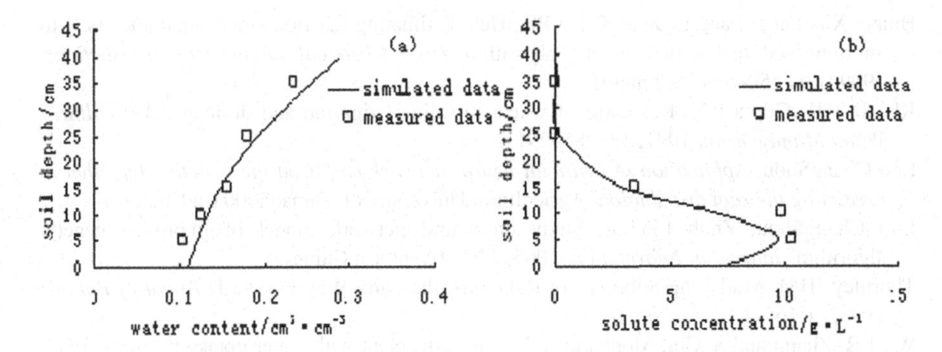

Fig. 2. Comparision between measured and estimated soil water content, salinity

5. CONCLUSION

Crop root-water-uptake was one of important studying problem about water movement in groundwater-soil-plant-atmosphere continuum. As well as it was the absolutely necessary data for soil water dynamic simulation in root area. A simple and practical method to get root distribution parameters was provided by improved genetic algorithm. The parameters which were obtained by this method combined with root-water-uptake model as well as water and salinity transport model can simulate soil water and solute distribution well. This method was of great significance to study relation between root distribution, salinity and crop growth as well as soil water and solute transport distribution regularity. This method can also provide decision-making gist for taking advantage of light-saline and preventing soil secondary salinization.

ACKNOWLEDGEMENTS

The accomplishment of this paper benefits from the enlightenment of professor RenLi. His numerical simulation algorithm furnishes me with inspiring source. I would also like to extend my sincere thanks to professor RenLi of Agricultural University of China.

REFERENCES

Chen GuoLiang, Wang XuFa, *Genetic algorithm and its application,* Post & Telecom Press, 1996 (in Chinese)

Homaee M. *Root water uptake under non-uniform transient salinity and water stress.* Wageningen Agricultural University, the Netherlands, 1999, 41–54, 93–97

Huang XiaoFeng, Pang LiDeng, Chen BiaoHua, Estimating reaction kinetics parameters with an improved real coded genetic algorithm. *High Chemical Engineering transactions,* 1999, 13(1):50–55 (in Chinese)

Khosla BK, Gupta RK. Response of wheat to saline irrigation and drainage, *Agricultural Water Management,* 1997, 32 :285–291

Luo ChangShou, *Application of artificial neural network based on the genetic algorithm in predicting the root distribution.* Agricultural University of China, 2000 (in Chinese)

Luo ChangShou, Zhou LiYing. Study on neural network model of improved genetic algorithm. *Journal of Information,* 2005, 24(5):65–66 (in Chinese)

Thornley JHM, Modelling Shoot: Root Relations: the Only Way Forword? *Annal of Botany,* 1998, 81:165–171

Wu J R. Zhang and S. Gui. Modeling soil water movement with water uptake by roots, *Plant and Soil,* 1999, 215:7–17

Yuan Zeng, *Artificial neural network and its application:* Tsinghua University Press, 1999, 66–70 (in Chinese)

Zheng ZhiJun, Zheng ShouQi. Adaptive genetic algorithm based on real-coded evolve neural network. *Computer Engineering and Applications,* 2000, (9):36–37 (in Chinese)

INFORMATION TECHNOLOGY SPEEDING UP CIRCULATION OF RURAL ECONOMY

Ranbing Yang [1,2], Suhua Liu [3], Jie Liang [2], Shuqi Shang [2,*]
[1] *Engineering School, Shenyang Agricultural University, Shenyang, China, 110161*
[2] *School of Mechanical and Electric Engineering, Qingdao Agricultural University, Qingdao, China, 266109*
[3] *Automation Engineering School, Qingdao Technological University, Qingdao, China, 266109*
** Corresponding author, Address: School of Mechanical and Electric Engineering, Qingdao Agricultural University, Qingdao, 266109, P. R. China, Tel: +86-532-86080842, Fax: +86-532-86080452, Email: yangranbing@163.com (No. of Fund Item: 2006BAD28B06)*

Abstract: Nowadays, the information in the field of Chinese rural economy circulation is not only plentiful but complicated. The current network information technology is insufficient when dealing with the supply-demand relationship, and it cannot fundamentally meet the real needs of rural economy. Based on the current situation of rural economy, this paper puts forward ways to establish, using computers and information technology, a new rural economic information exchange platform. The platform can realize Informatization of rural economy through the database processing technology and Geospatial Information Grid method, thus fundamentally solves Chinese information shortage.

Keywords: rural economy, information technology, database, geospatial information grid

1. INTRODUCTION

With the development of productive forces, rural economy has developed rapidly in recent years; rural areas have undergone tremendous changes, and gradually changed the backwardness of "natural economy". However, at present, most of Chinese rural areas are still no large-scale production, and

Yang, R., Liu, S., Liang, J. and Shang, S., 2008, in IFIP International Federation for Information Processing, Volume 258; Computer and Computing Technologies in Agriculture, Vol. 1; Daoliang Li; (Boston: Springer), pp. 605–611.

belong to the basic self-production and sales of an inward-looking economy (Zhu Naifen, 2007). The main features of this kind of economic activities are:

(1) The economic structure of exchange is miscellaneous. Economic activities of most commodities are only between suppliers and customers. Forms of commodities are diverse, and the commodities are with small quantities of miscellaneous;

(2) Demand information seriously lags behind. As the product quantities of both suppliers and customers are too small, the existing ways of information circulation cannot meet the demand. The primitive market is in a "blind" state, where both suppliers and customers can only try to do unknown commodity activities in prescriptive time, and in this case, the economic information cannot be exchanged timely and effectively;

(3) TV, newspapers and other media are mainly for cost-efficient commodities of large quantities, and do not yet meet the exchange of information in rural areas, which is complex, chaotic and small.

In the global Informatization and Digital background, information technology has become indispensable means to promote agricultural economic development (Su Qizhi, 2007). Only full use of information technology, can fundamentally promote rural economic development and circulation.

2. THE INSUFFICIENCY OF EXISTING NETWORK INFORMATION SERVING FOR RURAL ECONOMIC ACTIVITIES

Taking a panoramic view of various network services systems in China, most of them take diffusive ways to disseminate information. Suppliers and customers directly disseminate the information to whole country and even the whole world through internet.

In this way, there are several problems:

(1) Information waste. If commodities for supply or for demand are too small or too little, it is unnecessary for the overall situation of direct distribution. For instance, farmers in Tianjin want to disseminate information to sell 100 kilograms of apples, which is effective information in some areas of Tianjin. However, this information is redundant for other regions of Tianjin and even the whole country.

(2) Failure to establish effective communication. For example, Guangdong farmers issue a district desires 10 kilograms of strawberries. Strawberry has certain durability, coupled with less demand, so this information can only be effective within 50 km. Chinese Agricultural Information Network, relative to local network, has done a great

improvement in this aspect, as illustrated in Figure 1, but there are still above shortages, particularly in rural areas, where do not facilitate transport. The rage of effective information should be smaller, which is clearly unable to meet everyday needs of the farmers, if only confined to the provinces, municipalities (Pang Jiangang, 2007).

(3) Information region is too broad. Commodity information includes time efficiency and the range, but existing information network platform only pays attention to the time efficiency, ignoring range of information.

Figure 1. Traditional information exchange platform

3. RURAL ECONOMIC INFORMATION EXCHANGE PLATFORM

For rural economic information is cumbersome and complex, rural traffic conditions is relatively poor, and information is with time efficiency and the regional characteristics, a new rural economic information exchange platform is established here, and the whole structure is depicted in Figure 2.

Relative to other agricultural information platform, the new one adds geographic information processing module on the basis of the time information processing module. Figure 3 is the corresponding search engine. The main features of the system are given in the following sections.

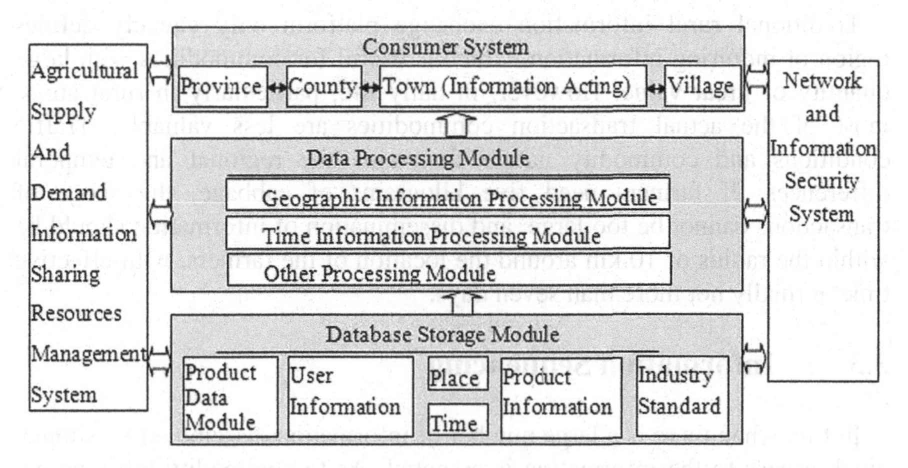

Figure 2. The whole structure of the new information exchange platform

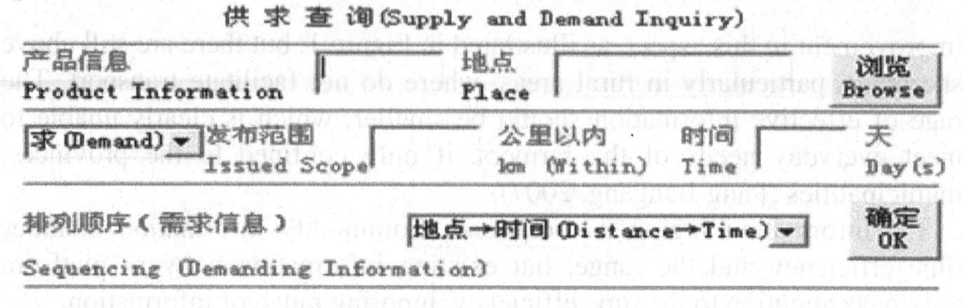

Figure 3. The new search engine

3.1 Information Point

In rural areas, distance factor is the primary factor among product information because of vast territory and inconvenient transportation. Therefore, demand-side should firstly provide geographic information of locations to geographic information processing module, that is to say, after determining the coordinate base point, one can get the product information in a certain area. Geographic information processing module is able to provide the region according to customers' IP addresses, and users can also revise region depending on specific circumstances. In order to get more detailed product information, geographic information should not be too general, and the smallest level is usually village. Time information point is, as the same as traditional rural economic information exchange platform, using the demand time as base point.

3.2 Information Demanding Domain

Traditional rural information exchange platform only vaguely defines region of inquiring information, which is useful for commodities with large quantity or great value. However, in daily life, particularly in rural areas, most of the actual transaction commodities are less valuable. Traffic conditions and commodity nature determine the regional and temporal differences. If farmers need five kilograms of cabbage, the scope of transactions cannot be too large, and dissemination of information should be within the radius of 10 km around the location of the farmers, with effective time normally not more than seven days.

3.3 Information Sequencing

In life, when there is a large number of information dissemination, simple, quick access to the information is essential. As to commodity information, distance and time of supply and demand in rural areas are very crucial factors; therefore, it should be based on different needs to establish different information sequencing. Here are two categories: "distance → time" and "time → distance" (Wang Na, 2007), and the users can determine different priority level to the two factors according to different needs. Traditional rural information exchange platform only pays attention to the time factor, overlooked the distance factor (Chen Wei, 2007). In rural areas, farmers are often based on economic interests, and choose distance factors at the very beginning. Figure 4 is information query results using distance factors as the priority level.

产品(求)Product(D)	距离Distance	时间Time(days)	地点Place	发布时间Issued time
☑ 苹果	0公里	1天	微山县两城乡居桥村	2007-05-07
☑ 苹果	0公里	2天	微山县两城乡居桥村	2007-05-06
☑ 苹果	2公里	1天	微山县两城乡马店村	2007-05-07
☑ 苹果	2公里	3天	微山县两城乡焦庄村	2007-05-05
☑ 苹果	5公里	2天	微山县两城乡西李庄村	2007-05-06
☑ 苹果	5公里	3天	微山县两城乡王庄村	2007-05-05
☑ 苹果	5公里	4天	微山县两城乡王庄村	2007-05-04

Figure 4. Information query results

4. GEOGRAPHIC INFORMATION DIGITAL

As the rapid development of the information technology, geographic information digital (Han Zhenbiao, 2007) has become inevitable. In rural economic activities, different distance is an important factor to be considered. In order to obtain detailed supply and demand information, it is necessary to carry on the digital information communication to the two places of supply and demand.

A vector maps is established in this paper by the ways shown in Table 5, in order to – according to latitude and longitude and taken village as data point basis – erect vector numerical information to the national map. Units here are kilometers, and along with the information development, the units can be even smaller.

Figure 5 is the map of some areas in Ji'ning area of Shandong Province. For example, a farmer in the place of the Bridge Village wants to inquire some commodity information, and his inquiry scope establishment is 5 kilometers, then the demanding scope should take the Bridge Village as the center, within surrounding area of a 5 kilometers circle.

All the issued information, whose location distance D from Bridge Village to be smaller than 5 kilometers, is the farmer's required one. Xilizhuang Village, Zongcundong Village, Madian Village, Jiaozhuang Village, Wangzhuang Village in Table 4 are in this scope. Then the information issued from these villages is classified according to distance D. The distance

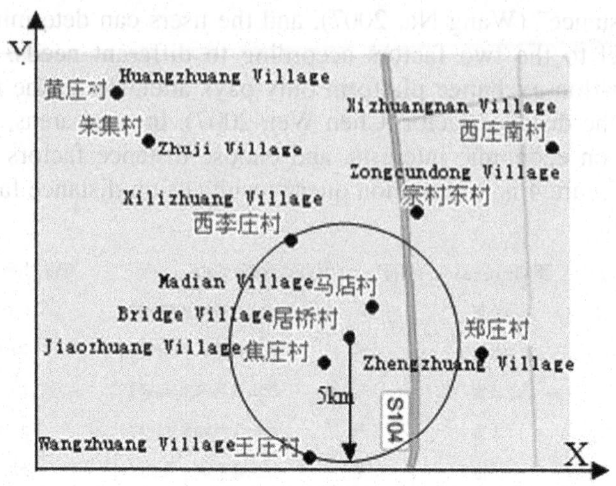

Figure 5. A vector map

is calculated in this platform based on the Dijkstra's shortest searching algorithm. (YueYang, 1999) It has fully used the existing pile of data, reduced the data comparison frequency.

Concrete steps: Each spot has a pair of marking (d, p), D is the most short-path distance from source s to selects j (From the apex to its itself most short-path is the zero road [No arched road] Its length is equal to zero); Pj is in the short-path way from source s to j and is j preceding. To selects j from source s the most short-path algorithm unit process. The shortest path algorithm from source s to j for the basic process:

(1) Initialization. Point of Origin: ds=0, ps empty; All other points: di=∞,pi=∞. Marker source s, mind k =s, all other points as the unmarked.

(2) Testing the distance from all marked point k to the unmarked point j which direct linked, and setting: dj=min[dj,dk+lkj], and lkj is the direct link distance from the point k to j.

(3) Next point selection. From all unmarked node, select the smallest j among the points dj: di=min [dj, All unmarked point j] point i was selected as the shortest path to the point, and has set the marker.

(4) Preceding point in front of point i found. Point j* direct linked to point i is found from marked points, and taken as preceding point,. Set: i=j*.

(5) Point i marked. If all points have marked, the algorithm fully launched, otherwise set k = i, to step (2), to continue.

5. DYNAMIC DATABASE ESTABLISHMENT

It is extremely widespread to establish the network dynamic database in the daily application, and most of existing websites use the dynamic information input. (Zhang Xiao, 2006) Moreover, most database input form

that the majority of farmer information network platform used all may adapt in this network platform, and as to the needed code, this paper will no longer give unnecessary details. The difference is an added module of detailed locations of the user input, to the determination of the inquiry point in vector maps.

6. SUMMARY AND OUTLOOK

Rural information technology is an important way to realize agricultural and rural economic restructuring, and to increase the income of peasants. Based on the actual needs of farmers, this paper puts forward ways to establish a viable rural economic information exchange platform to solve the difficult problems that cannot handle by other media in the flow of rural economy. Along with rural network popularization, people can gradually gain the actual benefit from information technology, achieving the goal of increasing economic returns of agriculture and farmers' incomes.

This search engine can be used not only in agricultural product application, but also on the application promotion of daily life. For instance, if vector map is established for local supermarkets, stores, hotels, and so on, people at home can communicate with each other and exchange information in a timely manner and efficient way.

REFERENCES

Chen Wei, RSS. A technology to syndicate information. Computer Knowledge and Technology, 2007(1): 46–47

Han Zhenbiao. Mapxtreme2004 based query system of webgis vectorgraph. Geomatics & Spatial Information Technology, 2007(2): 36–39

Pang Jiangang. The applied research of information construction in countryside. Market Modernization, 2007, 503: 369–370

Su Qizhi. Strengthening of rural information network to promote rural economic development. China Well-off Rural Technology, 2007(6): 55–56

Wang Na. Discussion about the network information mining. Library Work in Colleges and Universities, 2007, 27(3): 38–40

YueYang. An efficient implementation of shortest path algorithm based on dijkstra algorithm. Computer Applications and Software, 1999, 24(3): 209–212

Zhu Naifen. The countermeasure and suggestion on the strength the agriculture information system. Information Science, 2007, 25(3): 364–367

Zhang Xiao. The Example of PHP+ MySQL+ Dreamweaver, Tsinghua University Press, 2006

NEW FAST DETECTION METHOD OF FOREST FIRE MONITORING AND APPLICATION BASED ON FY-1D/MVISR DATA

Jianzhong Feng[1,3,*], Huajun Tang[1], Linyan Bai[2], Qingbo Zhou[1], Zhongxin Chen[1]

[1] *Institute of Agricultural Resources and Regional Planning, Chinese Academy of Agricultural Sciences, Beijing, China, 100081*

[2] *Institute of Remote Sensing Applications, Chinese Academy of Science, Beijing, China, 100101*

[3] *Department of Surveying and Land Science, China University of Mining and Technology, Beijing, China, 100083*

* *Corresponding author, Address: Key Laboratory of Resources Remote Sensing & Digital Agriculture, Institute of Agricultural Resources and Regional Planning, Chinese Academy of Agricultural Sciences, No. 12, South Zhongguancun Street, Beijing, 100083, China, Tel: +86-13522241414, Email: fengjzh4680@sina.com.cn*

Abstract: In this paper, the authors proposed a new work and process flow algorithm about remote sensing image data to forest fire identification and monitoring, which was greatly different with the traditional approaches. Therefore, a more useful context method was used to detect forest fire spots, banes on statistic rationale, meanwhile the cloud-contaminated pixels were rejected (if any) and the misjudged fire spots excluded by examination of *NDVI*s before and after forest fires occurred; moreover, a simpler and more resultful approach was used to earth-locate forest fire spots, and a forest-fire sub-pixel area situation was taken into account so as to evaluate forest-fire area and enhance accuracy of evaluated forest fire areas. Subsequently, a FFDM prototype system was designed and developed and later tested using the typical FY-1D/MVISR data, performance and running efficiency of which were practically greater than the normal mainstream software systems (e.g. ENVI) in this domain.

Keywords: FY-1D/MVISR, Radiance, vegetation index, geo-referance, workflow

Feng, J., Tang, H., Bai, L., Zhou, Q. and Chen, Z., 2008, in IFIP International Federation for Information Processing, Volume 258; Computer and Computing Technologies in Agriculture, Vol. 1; Daoliang Li; (Boston: Springer), pp. 613–628.

1. INTRODUCTION

Forest Fires are a prominent multi-scale phenomenon, which not only destroy natural vegetation, but also pose enormous danger to wildlife as well as to human life and property, so researchers, multi-level governments and others pay great attention to them. In view of preventing them from occurring in danger areas, there is one of crucial problems that is both smartly and in time to monitoring situation of and achieve environmental change of forest fires, but it is evident that the objectives couldn't be reached, if only using approaches and information of ground-based observation and monitoring, especially on large-scale forest fires. Relying on satellite remote sensing data, a wide-range and real-time monitoring is the most feasible and practical since it is essential for the grasp of fire occurrence situation and useful information that is half-automatically or automatically extracted from those data.

A surprising number of satellite systems are currently available, providing data and other capabilities that can be used for different aspects of forest fire monitoring since the late 1990s (Christopher et al., 2003). For instance, there are very common utilities of the NOAA/AVHRR (Advanced Very High Resolution Radiometer) and EOS (NASA Earth Observing System)/MODIS (Moderate Resolution Imaging Spectroradiometer) data in a large or global scale of satellite forest fire monitoring. The NOAA/AVHRR data are characterized with low cost, high temporal resolution (i.e., two daily passes per day over a given area for each operating satellite, thus two scenes obtained by each operational platform) and pantoscopic scan, as compared with the EOS/MODIS data that are greatly improved in spatial resolution, have more spectral channels (the total of 36 spectral channels), offer a larger dynamic range of radiance values, and enhance derived products and so on (though being somewhat lower in temporal resolution). The MVISR (Multichannel Visible and Infrared Scan Radiometer), onboard the present operating FY-1D satellite (which is a polar-orbiting meteorological satellite, developed and launched by China), is largely similar to the NOAA/AVHRR in aspects of functionalities and performances. It has 10 spectral channels, the corresponding ranges of which are same to the NOAA/AVHRRs', thereby the FY-1D/MVISR data show greatly potential for a large scale or global forest fire monitoring.

In the forest fire monitoring, active fire or hot spot detection is one of underlying tasks using satellite data. So far, methodologies of these detecting researches are abundantly achieved, based on satellite remote sensing systems and their obtained data, especially on the NOAA/AVHRR. Take for examples, Kalpoma Kazi A. et al. proposed a new algorithm of forest fire detection method with statistical analysis using NOAA/AVHRR images (Kalpoma et al., 2006); Galindo I. et al. carried through real-time

NOAA/AVHRR forest fire detection in Mexico (Galindo et al., 2003); it was acquired to study a RS (using NOAA/AVHRR data)-and-GIS-integrated forest fire monitoring (Zhang, 2004). With respect to different satellite data, more effective and efficient algorithms of active fire or hot spot detection and other technologies, however, are still explored by researchers/users, in order to monitor forest fires and prevent their from emerging or reduce their losses in quite broad areas.

In this paper, the authors present a new work and process flow algorithm about remote sensing image data to forest fire identification and monitoring, which is different from traditional approaches. Useful and interesting information of forest fire spots is estimated and extracted from potential forest-fire pixels above all, which lies on corresponding thresholds determined with a statistic manner that validly eliminates subjective influences in a certain context and satisfies real requirements, meanwhile the cloud-contaminated pixels are rejected (if any) and the misjudged fire spots excluded by examination of Vegetation Index (VI) before and after forest fires occur; and forest fire spots are located and assigned exactly geo-coordinates, using an interpolation algorithm of fast location of forest fire spots, rather than all unavailable and forest-fire-free spots; in addition, forest-fire sub-pixel area status is taken into account so as to evaluate forest-fire area and enhance accuracy of evaluated forest fire areas. Subsequently, a FFDM prototype system is designed and developed and later tested using the typical FY-1D/MVISR remote sensing data involved in forest fires, while compared to normal mainstream software systems (e.g. ENVI & ERDAS). In general, this prototype is to certain extent valuable and meaningful for FFDM in theoretic and applicable domains.

2. METHODOLOGICAL MODES

According to Wien's Displacement Law, a wavelength of thermal radiation with a peak is inversed to some absolute temperature of blackbody (Chen, 1985). Thus, there is a peak (about 9.7μm) of thermal value while at the normal temperature (about 300k); meanwhile, related to the FY-1D/MVISR data, there are more thermal radiant fluxes in CH4 (10.3–11.3μm) and CH5 (11.5–12.5μm). As appears a brightness temperature of fire spot (a active fire or hot spot signature), thermal peak values transfers towards CH3 (e.g., a peak value about 5.8μm to the temperature 500k), and there arise energies of observed thermal radiation, as well as are more in CH3&4. Variation ratios of brightness temperatures, therefore, are normally evaluated and their trends depicted in CH3&4, using the Planck function of blackbody radiation. It is obvious that there are distinguishing differences between in CH3&4 (e.g., when temperatures of blackbody arising from 300K to 500K and even to the higher, a several hundredfold increase of variation ratios for the fore, but only over 10 times a increase for the latter) (Che et al., 1900). In FY-1D/MVISR image data, the pixel graylevel value of hot spot (including fire spot) is enormously different from those of its surrounding pixels in CH3, in contrast with the fact that there are only a few graylevel differences for the related pixels in CH4.

Forest fire detection procedures may be based on the fixed threshold value or contextual method, but there are lots of limits in the adoptability of the fore because of hugely relying on empirical analyses. The contextual method is devised to take into account the relationships between observational pixels and its neighboring pixels mainly based on spatial statistics, and then identify fire pixels from around candidate fire pixels, hence the contextual approach is often preferred to be selected (Kudon, 2005) and also adopted in this paper.

2.1 Detection of forest fire spots

In procedure of identifying forest fire spots, the Planck function of blackbody radiation is shown as a fundamental rationale, from which related others are derived. As the following below:

$$T = h \cdot c / (k \cdot \lambda \cdot log(2 \cdot h \cdot c^2 / (I \cdot \lambda^5) + 1)) \quad (1)$$

where T is brightness temperature (K); h, k and c are the Planck, Boltzmann and light-velocity constants respectively, which are $6.6256 \cdot 10^{-34}$ $(J \cdot s)$, $1.3906 \cdot 10^{-23}$ $(J \cdot K^{-1})$ and $2.998 \cdot 10^8$ $(m \cdot s)$; λ radiance wavelength (m) and T I emissivity, whereas I may be obtained from the following function (i.e., the radiometric calibration formula).

$$I = n \cdot g + b \quad (2)$$

where n is pixel graylevel value; g is the gained value and b the bias value, which are the calibration coefficients ($W / m^2 \cdot ster \cdot \mu m$). Formula (1) and (2) hence are jointly evaluated so that the thermal radiances, obtained in CH3&4, are transformed into the corresponding brightness temperatures.

Now it is a key task of detecting fire spots how to identify fire pixels, associated with the real fire spots, from potential fire pixels in satellite image, as below is followed (Kudoh, 2005; Giglio et al., 2003; Zhou et al., 2006):

- ***Definition of fire context window***

A fire pixel, selected by certain conditions (See the next sections), appears to be different enough from its background and meanwhile the brightness temperatures are employed, which are derived from the FY-1D/MVISR in CH3&4, respectively denoted by T_3 and T_4. Statistical information of each potential-fire pixel is calculated for a variable size context window around it, within which at least 25% of neighboring pixels are satisfied with forest fire-free pixels and considered as forest fire background, but not cloud-contaminated (If any, cloud pixels are rejected from the forest fire background because they are obscured. See identification of Cloud below).

- ***Identification of Cloud***

Cloud identification is performed using a technique based on conditions: $Reflectance_1 > 22\%$ and $Reflectance_2 > 23\%$ that are respectively derived from the FY-1D/MVISR data in CH1&2, and $T_4 < 273K$. If the previous conditions are jointly be satisfied, then the pixels in the FY-1D/MVISR may be flagged as cloud (Zhang, 1999).

- ***Identification of potential-fire pixels***

If $T_3 > 320\,K$ and $\square T_{34} > 20\,K$ in daytime, or $T_3 > 315\,K$ and $\square\, T_{34} > 10\,K$ in nighttime, where $\square T_{34} = T_3 - T_4$, then a pixel is identified as a potential fire pixel.

- ***Tentative identification of fire pixels***

If the potential-fire pixels are identified as active fire (including hot spots), the following conditions below are founded on:

$$T_3 > \overline{T_3} + 4\delta T_3\text{, where, if } \overline{T_3} < 2\text{K, then } \overline{T_3} = 2K \quad (3)$$

$$\square T_{34} > \overline{\square T_{34}} + 4\delta T_{34} \quad (4)$$

$$T_4 > 320K \text{ in daytime (or } T_4 > 315K \text{ in nighttime)} \quad (5)$$

$$\square T_{34} > 20K \text{ in daytime (or} \square T_{34} > 10K \text{ in nighttime)} \quad (6)$$

$$T_3 > 360K \text{ in daytime (or } T_3 > 330K \text{ in nighttime)} \quad (7)$$

where $\overline{T_3}$ and δT_3 are the respective mean and mean square deviation of T_3 for the valid potential-fire pixels in the background; $\overline{\square T_{34}}$ and δT_{34}, the respective mean and mean square deviation of $\square T_{34}$ for the valid potential-fire pixels in the background.

In the subsequent sections, the valid potential-fire pixels are tentatively identified as fire pixels using the below conditions, as are described later on:

If {((3) or (5)) and ((4) or (6)) or (7)} is true,
where, if δT_3 or □T_{34} < $2K$, they are assigned with $2K$; or else, they are maintained and invariable,
Then *the above valid potential-fire pixels are figured as fire pixels (fire spots).*

2.2 Location of forest fire spots

To date, fire pixels have been detected in the satellite images. The next task is that identified fire spots are, accurately and quickly, located and assigned with corresponding geo-coordinates, hereby an interpolation algorithm of fast location of forest fire spots, based on the FY-1D/MVISR data, is discussed so as to better serve the direct or derived productions to related departments (e.g. helping for efficient mapping of forest fires).

The FY-1D/MVISR is characterized that there are 2048 counts (i.e. samples) along each scan, each of which is 10-bit binary values (in each spectral channel) and archived into the format of 2 bytes (one per 8-bit byte) in storing system. Based on the 1a.5 file format, earth-location of data is possible using benchmark data provided with each archived sector. The benchmark data consist of line and sample number coordinates of data samples falling on certain span increment of latitude and longitude; furthermore there are latitudes and longitudes of 51 benchmark data points along each scan in the file (bytes 151 to 354), each pair of which are archived into 4 bytes.

The authors thus employ the nearest-neighbor infill pixel interpolation algorithm and WGS-84 Coordinate System to geo-rectify FY-1D/MVISR image datum, and obtain an orthoimage and associated geo-coordinate information file (storing the first langitude and latitude and adjacent pixels' difference values respectively between longitudes and latitudes, which are mapped into the associated orthoimage), in order to obtain the longitude and latitude values of center of an arbitrary pixel in the orthoimage and then earth-locate the corresponding forest fire spots. There is a basic algorithmic despcription of earth location in an associated geo-coordinate information file, as is shown below:

```
FUNCTION fire_spot_location (fnm, i, j)
{/*fnm is an orthoimage file; and i, j are the longitude and latitude values of center of an aimed pixel*/
  map_info = get_map_info (fid = fnm);
  lon_ini = map_info.mc[2]; /*Obtaining the longitude value of center of the first pixel in the orthoimage*/
  lat_ini = map_info.mc[3];  /*Obtaining the latitude value of center of the first pixel in the orthoimage*/
  lon_interval = map_info.ps[0]; /*Obtaining difference values of the most-adjacent pixels between longitudes*/
```

```
lat_interval = map_info.ps[1]; /*Obtaining difference values of the most-adjacent pixels between latitudes*/
lon_j = lon_ini +(j-1)* lon_interval; /*Calculating the longitude value of center of the pixel (i, j)*/
lat_i = lng_ini +(i-1)* lng_interval; /*Calculating the latitude value of center of the pixel (i, j)*/
... ...
}
```

2.3 Examination of Vegetation Index (VI)

When the actual forest fire detection and monitoring are carried out, the observed values just through an infrared channel are inversed to obtain the brightness temperature and determine whether the forest fires occur, but owing to the complexity of land surface, there are often lots of misjudgments (e.g., a high temperature desert area would be misjudged into a forest fire area; and smoking chimneys or other special high heat sources into forest fire spots). It is a good way that the vegetation index (VI) is used to solve the problem. Obviously, the VIs are some significant digital values and indicate vegetation growth status, biomass and so on (Jiang et al., 2001). According to the Satellite image data in the recent period of time before the forest fires occurred, the VIs whose earth-locations are corresponded with the identified potential forest-fire pixels, therefore, are calculated to define identification credibility of forest fire spots and determine whether the potential forest fire spots are real or not: the VIs are higher, and the real probabilities of potential forest fire spots are bigger; furthermore, to evaluate areas of or assess other disaster losses of forest fires and so forth in terms of the VIs changes before and after the forest fires take place.

Up to now, reseachers have studied and developed dozens of different models of the VIs that are widely applied (Xu, 1991). In this paper, a model of Normalized Difference vegetation index (*NDVI*) is expressed below:

$$NDVI = \frac{\rho_{NIR} - \rho_R}{\rho_{NIR} + \rho_R} \tag{8}$$

Where ρ_{NIR} and ρ_R are the reflectances of land surface in the near-infrared and red bands, respectively.

As it is charactered with the design of and data acquisition mode of channels of FY-1D/MVISR, the correponding FY-1D/MVISR data in CH1 (0.58–0.68 μm) and CH10 (0.90–0.96 μm) are preprocessed by a series of ways such as radiometric calibration and solar elevation correction, and then the ρ_{NIR} and ρ_R are obtained to calculate corresponding *NDVI* values and detect forest fire spots or assess forest fire situation and etc.

2.4 Evaluation of forest fire areas

Evaluation of forest fire areas is a very important task of forest fire losses assessment and one intention of detecting and monitoring forest fire. In some cases, a real fire spot is so small that its area is less than a resolution of one pixel, and therefore, a sub-pixel scale should be taken into account necessarily to evaluation of forest fire areas.

Assuming a rate P is a sub-pixel forest fire area over the area of a pixel (i.e. a field of view), there exists $1-P$ that is a sub-pixel forest-fire-free area within a field of view, so the radiant flux density equation below is expressed, saturated in CH3 of the FY-1D/MVISR:

$$E(\lambda,T_3)=p\cdot E(\lambda,T_f)+(1-p)\cdot E(\lambda,T_b) \tag{9}$$

where E is a Planck deformation function; λ a detection wavelength in CH3; T_3 a pixel land surface temperature in CH3; T_f a forest fire temperature for a sub-pixel in CH3; T_b a fire-free area temperature for a pixel (i.e. a background temperature). In the view of mathematic theory, a binary equation group about T_f and T_b may be founded and solved so as to obtain the T_f and T_b values, which is based on the Planck function $E(\square)$; nevertheless, given that $E(\square)$ is non-linear, it is very difficult and even impossible to practically solve the equation group and gain the above rate P. In practice, a logical value T_f is often experientially set on the basis of the actual instances, or, relying on the pertinent relationship, a certain fitting function upon T_f is constructed to obtain the needed T_f a value T_b is derived from the mean temperature for the 8 immediate neighboring pixels around a forest fire pixel (Xu, 1991); and then the rate P is obtained, using Equation (9). Necessarily noting, given that T_3 is obviously more than T_b and often T_f over 500K, and a positive relationship for Variable E with Temperature T in the previous $E(\square)$, a obtained rate P should fall between 0 and 1; otherwise it is invalid, so causes have to be discovered and the values be revised to achieve a sound rate P (Dong et al., 1999). Now a sub-pixel forest fire area can be evaluated by P multiplied by a corresponding pixel forest fire area.

3. PROCESS AND WORKFLOW

According to the previous rationale and methods about forest fire detection, the process and workflow of forest fire detection and monitoring are designed, as illustrated in Figure 1.

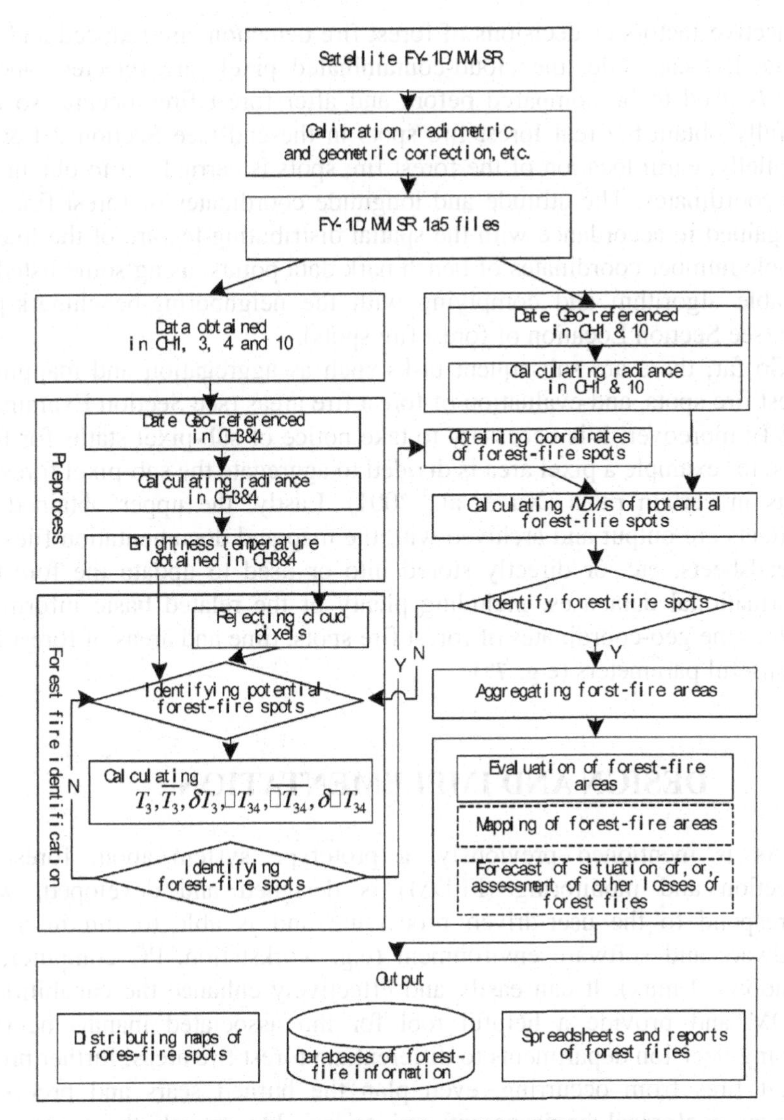

Figure 1. The process and workflow of forest fire detection

After the given raw FY-1D/MVISR data are preprocessed by a series of essential techniques such as calibration, radiometric and geometric corrections, the corresponding data products (i.e. called the *FY-1C/MVISR product data*) may be obtained and ultimately act on the next test of forest fire detection and monitoring. It obviously shows that the forest fire detection approach is a pivotal and complex task that is determined by a series of factors and conditions. The context method, based on statistics, is used to obtain the decision thresholds and identify the potential pixels in the *background*, in order that the great deviations or errors, resulted from the subjective factors or decisions of forest fire detection, are reduced and even avoided; meanwhile, the cloud-contaminated pixels are rejected, and the *NDVIs* used to be compared before and after forest fires occure, so as to usefully obtain the real forest fire spots in the end (see Section 2.1 & 2.3). Parallelly, earth location of the forest fire spots is carried out to obtain their geo-coordinates. The attitude and longitude coordinates of forest fire spots are gained in accordance with the spatial distributing feature of the line and sample number coordinates of benchmark data points, using some listed and suitable algorithm and complying with the neighboring-benchmark-point rule (see Section Location of forest fire spots).

So far, there are subsequent tasks such as aggregation and mapping of forest fire spots, and evaluation of forest fire areas (see Section Examination of VI); moreover, it is necessary to take notice of sub-pixel status for forest fires, for example a pixel area is divided to aggregate the sub-pixel forest fire spots in certain rules (Xu et al., 2001). Lastly, the upper obtained data products are output and archived with the map and documentation files, and spreadsheets, etc, or directly stored into or used to update the forest-fire informational databases, including plenty of the related basic information such as the geo-coordinates of forest fire spots, time and areas of forest fires, and useful parameters (e.g. T_3).

4. DESIGN AND IMPLEMENTATION

As is mentioned previously, a prototype system about forest fire detection and monitoring (FFDM) is designed and developed, which correspond to the user-driven processing and is able to run in a wide hardware and software environment (e.g. workstation, PC computer, and Windows, Linux). It can easily and effectively enhance the capabilities of FFDM and provide a helpful tool for the associated management and making-decision departments to differentiate forest fire areas, further prevent forest fires from occurring, even plan the burned scars and proceed to recover ecological environment and so on. In general, there are a few primary functionalities in the overall system, containing information extraction of forest fires, identification and earth-location of forest fire spots, and visual display of related information and processing results, etc.

The architecture of the FFDM prototype mainly consists of a series of functional subsystems, as presents in Figure 2: (1) the *information extraction subsystem* presents to extract information/data as data sources of next forest fire detection from the FY-1D/MVISR (e.g., the spectral information in CH3&4, the line and sample number coordinates of benchmark data points, and image sampling time); (2) the *threshold determination subsystem* is

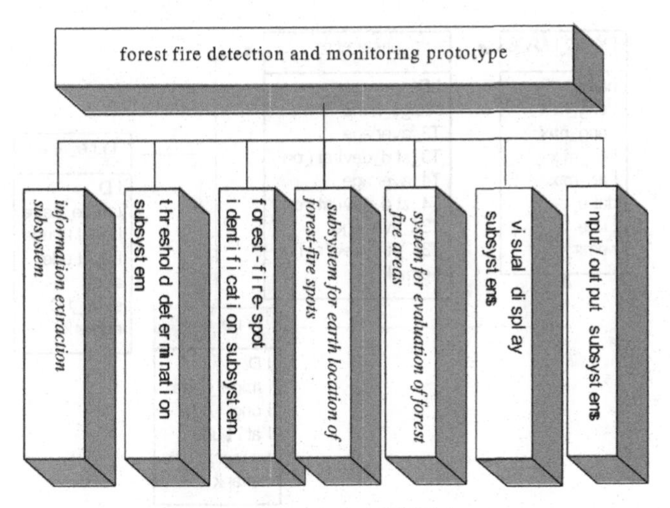

Figure 2. A framework of FFDM system

responsible to the valid thresholds that are obtained in statistic manner (see Section 2.1), using the brightness temperatures transformed from the previous acquired spectral data; (3) the *forest-fire-spot identification subsystem* is responsible to identify the forest-fire spots from the potential forest fires in the background and reject the cloud-contaminated areas (if any) and exclude the misjudged fire spots by the *NDVIs*; (4) the *subsystem for earth location of forest-fire spots* assigns the identified fire pixels with the corresponding geo-coordinates (of course, as well as is available for the forest-fire-free common pixels); (5) the *system for evaluation of forest fire areas*, including the referred sub-pixel areas; (6) the *Input/output and visual display subsystems* present input/output functions with user-friendly man-and-computer interactive mode, especially upon visual display. In addition, the databases subsystem of FFDM is a basic and pivotal component in the overall FFDM system. It can make the related forest-fire information/data very availably stored, accessed managed, and analyzed, etc, thus abilities and levels of analyzing and processing problems in the FFDM system are improved in effect and further all performances of the FFDM system are heightened as well. For instance, Figure 3 shows the structures and relationships of partial spatial database tables about forest fire detection in this system.

Furthermore, parts of the implemented results as to the FFDM system are demonstrated (see Figure 4 and 5). In this system, the data structures are constructed in accordance to the stored scheme of the FY-1D/MVISR 1A.5 files so that the associated available information is extracted. Fig. 5 shows the file-head information extracted from some 1A.5 file, including the ID and orbital parameters and of the satellite, time of data acquirement, and "good"

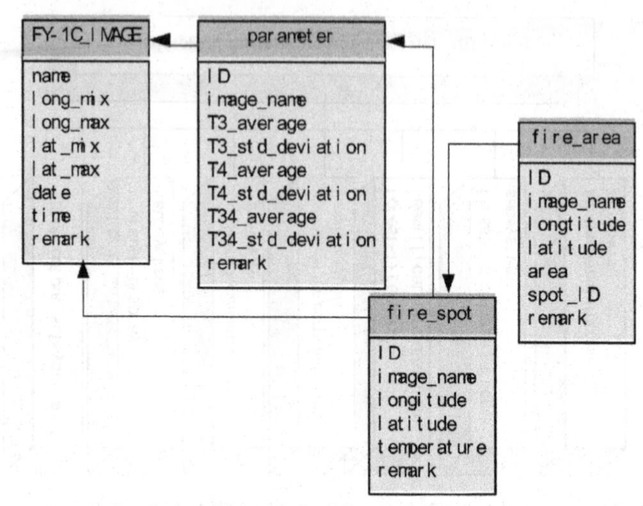

Figure 3. Structures and relationships of partial spatial database tables about forest fire detection in the FFDM system*

*(a) FY-1D_IMAGE table stores the information FY-1D/MVISR Image data (including the related basic information); (b) parameter table stores the corresponding calculated parameters of forest fire detection; (c) fire_spot table records the detected spot information; (d) fire_area table stores the information of forest-fire-spot-aggregated areas (e.g. the remark regarding fire-area features, and evaluated area, etc).

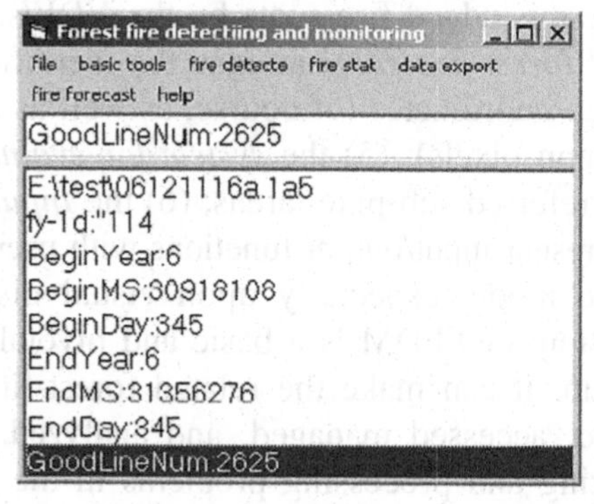

Figure 4. Interface of file-head information extraction for FY-1D/MVISR data

lines (i.e. valid lines) number in the file, the calibration coefficients and standard deviations of data in Channels 1 to 10, and geo-coordinates of 4 edge corners of cover areas, etc. Additionally, the FFDM system presents two patterns of the forest fire automatic and manual detection: if the former selected, an user can obtain the related data in some channels, or, he can manually obtained the needed spectral data (including other basic information) in chose channels (see Figure 5), as well as the obtained related spectral data are stored in the format of *.txt files that contain the spectral bare data file, band parameter file, benchmark-data-point geo-coordinates file and log file, etc, whereas these are only illuminated to data/data-files, although this system are characterized in some other ways.

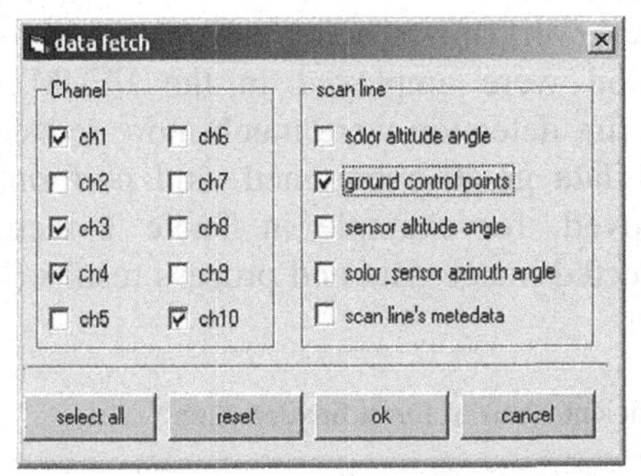

Figure 5. Interface of spectral-band information extraction

5. ANALYSIS AND DISCUSSION

Given the FY-1D/MVISR data of 5 scenes (implicating forest fire information) in China, downloaded from the site (http://www.ers.ac.cn), this FFDM system hence was tested and investigated in the application. By testing, the system demonstrated its high performance and great processing competences and satisfied the valid accuracy and precision requirements of FFDM application with the user-friendly operation interfaces and tools. It was particularly notable that the FFDM prototype offered two operational approaches mentioned above, i.e. the automatic and manual processes: choosing the fore, the user only clicked "File" to "Open", and immediately a window pop up, and the user selected the needed FY-1D/MVISR image file, where after the functions of system were completed such as identification and earth location of forest fire spots, and fire area division, and producing the reports of fire situation, etc, which needed not to be operated by the user, but this system could automatically proceed to operate; if the latter selected, it is necessary to manually complete the related operation. Additionally, the results of processing data were stored into the forest-fire informational databases so as to be retrieved, queried and further processed, and concurrently the associated forest-fire map (see Figure 6, 7 and 8), handing-in spreadsheet and report products and so on.

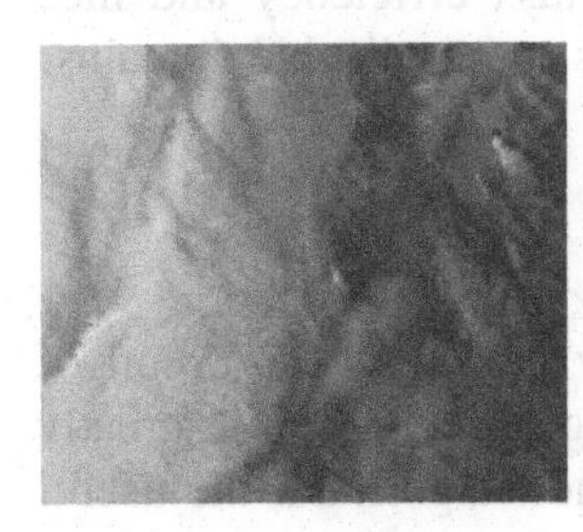

Figure 6. A distribution map of potential forest-fire spots

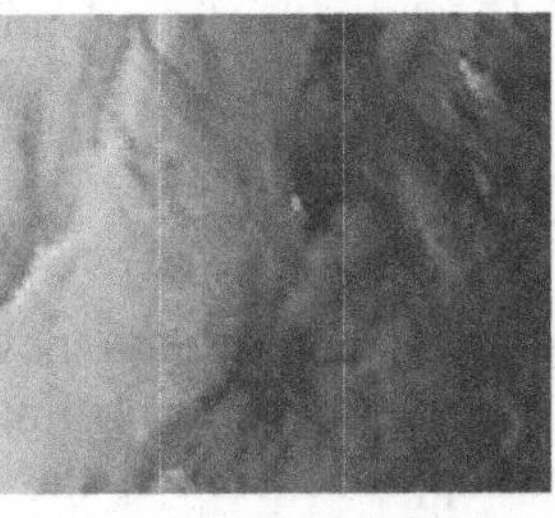

Figure 7. A distribution map of actual forest-fire spots

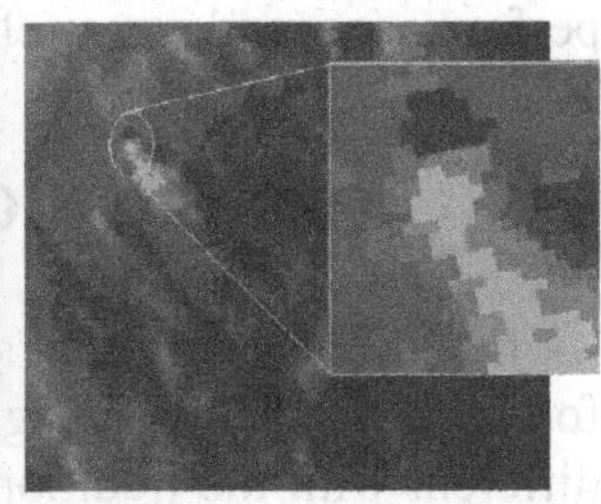

Figure 8. Status of a forest fire area

Because the new workflow scheme and fast earth location algorithm of forest fire detection were employed in the FFDM system, calculation contents of forest fire detection were much more reduced, and cost time of processing image data greatly shortened, and performance of the system enormously improved, for example in Table 1 there lists the relevant information of selected image data and process results (i.e., consuming time of data process).

Table 1. Executing efficient status of forest fire detection*

Name of data files	Samples*lines	Data size (Mbety)	Processing time (Second)
07022607.1a5	2048*3665	155	350
07031005.1a5	2048*3288	138	309
07030919.1a5	2048*2304	97.5	225
07050406.1a5	2048*3584	151	320
07051606.1a5	2048*3606	152	324

*Main parameters of configuration about the computer executing the programs: Intel ® P4 2.4GHz, 512 RAM; the FY-1D/MVISR 1A.5 data source from the site of Remote Sensing of Evironment in China, http://www.ers.ac.cn

At present, there are no special subsystems to process the FY-1D/MVISR data in many mainstream software systems of remote sensing image process (e.g. ENVI & Erdase); therefore, if using them, it is difficult and complex to directly process the remote sensing data in some application ways. Take an example, in some forest fire detection and monitoring there exist complicated steps and users rely on manual operations step-by-step to fulfill related tasks (e.g. many of parameters need to be configured in detail, merely related to opening a FY-1D/MVISR data file, using the ENVI software system; and geometric correction, forest fire identification and so forth are further more perplexing), so efficiency of forest fire detection and monitoring is very lower. In the sake to overcome the mentioned difficulties and faults, in this research a new workflow algorithm of forest fire detection and fast forest-fire earth location scheme were designed and developed. As a result, the FFMD system demonstrated the high task efficiency and nice performance, and was useful and worth of academic and applicable domains.

6. CONCLUSIONS

In this research, the authors employed a new workflow algorithm of forest fire detection using the remote sensing image data, which is greatly different with the traditional approaches. For the new algorithm, the useful context method was used to detect forest fire spots that were identified from potential forest-fire pixels, which relied on corresponding thresholds determined with a statistic manner that validly eliminated subjective influences in a certain context and satisfied real requirements, meanwhile the cloud-contaminated pixels are rejected (if any) and the misjudged fire spots excluded by examination of *NDVI*s before and after forest fires occur; moreover, earth location of forest fire spots was simpler and more resultful because it referred to the available forest fire spots/areas, in which users were interested rather than in unavailable spots/areas, so calculation contents of forest fire detection were more reduced and process running time of system also greatly lowered; additionally, a forest-fire sub-pixel area situation was taken into account so as to evaluate forest-fire area and enhance accuracy of evaluated forest fire areas.

Consequently, a FFDM prototype system was designed and developed and later tested using the typical FY-1D/MVISR remote sensing data involved in forest fires. Practically, performance and running efficiency of the prototype are greater than the normal mainstream software systems (e.g. ENVI & ERDAS) with respect to process and operation of the FY-1D/MVISR data; therefore, it was very beneficial and worth of related departments dealing with fire detection, control, prevention and further burned scars plan and ecological recovery, etc. In short, the system are greatly propitious to forest fire detection, monitoring, management and so the like. Lots of researches, however, need still further to be carried out both theoretically and methodologically in order to be applied to wider application domains.

ACKNOWLEDGEMENTS

This study is supported by the China National High Technology Research and Development Program (863 Program) Project (No. 2006AA12Z103) and China National Key Technologies R&D Program Project (No. 2006BAD10A06).

REFERENCES

Chen H. (1985). Thermal infrared physics. National Defence Industry Press, Beijing

Che N. Z., Yan D. Y. (1900). Radiometry and photometry. Beijing institute of technology Press, Beijing

Christopher O. Justice, Smith R., Gill A. M., Csiszard I. (2003). A review of current space-based fire monitoring in Australia and the GOFC/GOLD program for international coordination. International Journal of Wildland Fire, 12, 247-258

Dong C. H., Zhang G. C. (1999). Xing F. Y. et al., Tool manual of meteorological satellites, China Meteopoiogical Press

Galindo I., Lopez-Perez P., Evangelista-Salazar M. (2003) Real-time AVHRR forest fire detection in Mexico (1998-2000). International Journal of Remote Sensing, 24(1), 9-22

Jiang D., Wang A. B., Yang X. H., Lu H. H. (2001). Ecological Connotation and Application of the Vegetation Index-Surface Temperature Feature Space. Progress in geography, 20(2), 146–152

Kudoh J. I. (2005). Forest fire analysis for several years in Russia by using NOAA satellite http://www2.cr.chiba-u.jp/symp2005/documents/Oralsession/Session3_Dec14/S3-2JunichiKudoh_paper.pdf

Kalpoma K. A., Kudoh J. I. (2006). A new algorithm for forest fire detection method with statistical analysis using NOAA AVHRR images, International Journal of Remote Sensing, 27(18), 3867-3880

Giglio L., Descloitres J., Christopher O. Justice, Yoram J. Kaufman (2003). An Enhanced Contextual Fire Detection Algorithm for MODIS, Remote Sensing of Environment, 87, 273-282

Xu X. R. (1991). Processing of the Remote Sensing of Evironment Monitoring and Crop Yield Estimation, Peking University Press

XU X. R., Chen L. F., Zhuang J. L. (2001). Genetic inverse algorithm for retrieval of component temperature of mixed pixel by multi-angle thermal infrared remote sensing data. Science in China (Series D), 44(4), 363-372

Zhang C. G. (2004). Study on RS-GIS-based Forest Fire Monitoring in Fujian Provinc, Journal of Fujian College of Forestry, 24(1), 32-35

Zhang C. G. (1999). A Preliminary Study of Cloud Detection in remote sensing applications of Polar-orbiting Meteorological Satellites, Journal of Fujian Meteorology, 2, 3-5

Zhou X. C., Wang X. Q. (2006). Validate and Improvement on Arithmetic of Identifying Forest Fire Based on EOS-MODIS Data, Remote sensing technology and application, 21(3), 206-211

A SCHEME FOR SHARE AND EXPLOITATION OF NETWORK AGRICULTURAL INFORMATION BASED ON B/S STRUCTURE

Huitao Liu [1,2], Limei Tan [1,2], Yuan Yao [1,2], Qing Wang [1,2], Hongsheng Zhang [1,2], Guanglu Zhang [1], Jintong Liu [1,*]

[1] *Center for Agricultural Resources Research, Institute of Genetic and Developmental Biology, CAS, Shijiazhuang, China*

[2] *Graduate School, CAS, Beijing, China*

[*] *Corresponding Author: Center for Agricultural Resouces Research, CAS, 286 Huaizhong Road, Shijiazhuang, 050021, China, Tel: 86-0311-85871749, Email: jtliu@sjziam.ac.cn*

Abstract: The features of network agricultural information (NAI) are summarized by analyzing Chinese typical, famous agricultural information websites. The features include the storage-scattered of agricultural informations, high frequency of updating, non-uniform data format and a lots of raw information existing. It is pointed out that the features have disadvantages for sharing and exploiting the NAI efficiently. Then a scheme is proposed to try to overcome the disadvantages of the features, which is mainly based on brower/server structure and information extraction technology from webpages. Also the key technologies to implement the scheme are described in the paper. In the end, an example application of the scheme is carried out to demonstrate the concrete steps to develop practical applications based on the scheme.

Keywords: network agricultural information, information extraction technology, brower/server structure

1. INTRODUCTION

Practices and studies have proved that agricultural information share and development can be an effective way to reduce the pressure of natural resources and promote rational allocation of natural resources (Chen Pinde, 1993). In the face of the massive agricultural information (Gudivada V.N.,

Liu, H., Tan, L., Yao, Y., Wang, Q., Zhang, H., Zhang, G. and Liu, J., 2008, in IFIP International Federation for Information Processing, Volume 258; Computer and Computing Technologies in Agriculture, Vol. 1; Daoliang Li; (Boston: Springer), pp. 629–636.

1997), how to rational and efficient use and sharing of them are becoming increasingly important. Bayesian methods and boosting algorithm has been proposed in respect of data mining (Tang Chunsheng, 2003). And artificial intelligence, mathematical statistics and the sample-specific training methods (O. Etzioni, 1995; Y. Yang, 1999; Pui Y. Lee, 2002) are used in information retrieval. Wai Lam put forward an automatic text categorization according to the information content and it is used in retrieval of text (Wai Lam, 1999). Zheng LiuYue (2003) proposed the W3C DOM (Document object model), metadata and XML-based network information extraction model (Liu Zhengyi, 2003). However, the implementation of those techniques is a complex project and not suitable for general users (T. Radecki, 1997) and these techniques and methods are obviously poor or limit to TXT files or XML. The common defect of the techniuqes and methods is that they cannot accurate positioning information, have no function to exploit the information retrieved to find out new knowledges.

In this paper, an endeavor be done to try to find a scheme, whick not only can efficiently retrieve and share network agricutrual information, but can deeply exploit the retrieved information.

2. FEATURES ANALYSIS OF NAI

2.1 Features of network agricultural information

To gain the features of NAI, this paper select 50 representative, well-knowed agricultural information websites as study sample. By analyzing its producing, storing and dissemination, we get the features as following:

Agricultural Information has a scattered storage. The same type of agricultural information is often distributed in a number of websites.

Agricultural information data have a high frequency of update and the amount of data increases rapidly.

The data formats of agricultural information are not uniform, which vary from website to webite. For prices, a website may use "U.S. dollars/ton", while another may use "Yuan/kg".

Most of NAI is raw information, which is simply stacked into a website, and through which we can't find the internal rules. The deep development is urgent to be implemented. For example, by browsing the items of supply and demand information on the website, it is difficult to get the situation of supply and demand, or to forecast the prices of agricultural products or the seasonal fluctuation of price.

2.2 Disadvantages of the features of NAI

Since the agricultural information storage is dispersed, for a complete grasp of certain type of agricultural information, we have to visit as much websites as possible, so the workload is large and the efficiency is low. Meanwhile, it is a mechanical, monotonous work, for the agricultural information update frequently; and in order to grasp the latest information, we have to do the same job every day (or frequently): visiting web site as much as possible to obtain the latest information. However, it's prone to errors when comparing the non-uniform format data, for example, to convert the "dollar/ton" and "Yuan/kg" to same unit. As NAI is mostly raw information, the use of information is at low-level and cannot meet the high-level demand. For instance, policy maker want to know the price changes with seasons.

In summary, the features of NAI have severely restricted the high efficient share and deep exploitation of NAI.

3. A SCHEME FOR HIGH EFFICIENT SHARE AND DEEP EXPLOITATION OF NAI

To overcome the disadvantages of the featurs, a scheme is proposed here for realization of high efficient share and deep exploitation (HESDE) of NAI, which is mainly based on browser/server structure. The scheme is described as Figure 1.

The contents in dotted-line box are the main body of the scheme. According to its functions, the main body can be divided into three main modules: information extraction module, deep exploitation module and management-query interface module. In fact, it can also be understood as three functional module procedures running on a server.

3.1 Information extraction module

This module is mainly to solve the former three disadvantages of the features of NAI. Further, the model can be divided into three propcedures: data extraction procedure, format standardizing procedure and data store procedure, which respectively completes the automatic information extraction and collection, data format standardization and centralized data repository.

In the light of user's requirement, data extraction procedure extracts specific category information from website1, website2, and so on. At the same time, format standardizing procedure uniform the data formats to a standard one. Then data store procedure saves the uniform datum to database and completes centralized data repository. This module provides basic datum for the next phase of deep exploitation module.

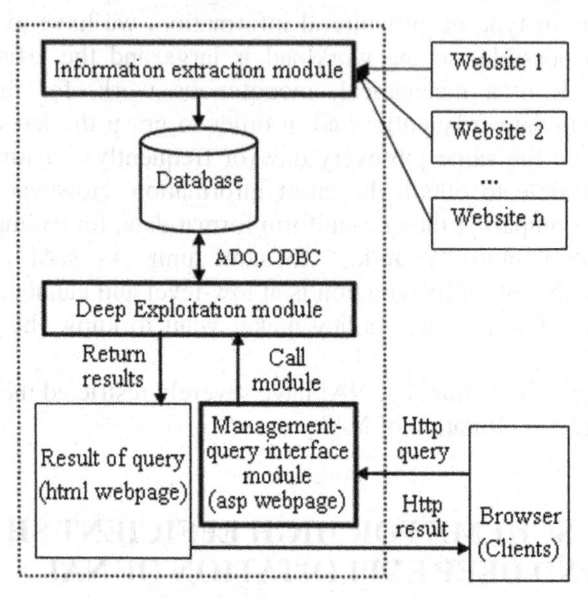

Figure 1. Sketch map for the sheme of HESDE of NAI

As for data extraction procedure, we can use Webbrowser control provided in VB6 (and above) for the browsing the websites, use its Navigate2 method for information location. Then, according to the specific properties of data set, such as the index (related with DOM) of the table and the DOM (Document Object Model) model, we can locate information resources in webpages and get access to the interest data by acquiring object attributes such as InnerText. After uniforming the datum extracted to same format (for example, all to Yuan/kg), data store procedure save them to database by the use of ADO/ODBC database manipulation technology.

3.2 Deep exploitation module

The module mainly implements the deep exploitation and development of raw information of NAI to find out new knowledge, for example, inherent laws or implicit rules. The deep exploitatin module is an open subsystem to the scheme. In other words, users can add, modify and delete deep exploitation programs in the system to meet the needs; also, the deep exploitation programs can be added or deleted remotely. Once a deep exploitation program is uploaded by one user, it can be used by others in the users group. This realizes the sharing of the deep exploitation method and avoids repeat development of procedures with the same function. This module can be viewed as a package of user transaction handling process and can be managed and called from management-query interface module and return results in html webpages.

In fact, deep exploitation program can be developed by Java, PHP, Perl, Javascript, and VC and so on. Also it can be released in many forms, such as scripting language program, .COM, .EXE and .DLL form.

3.3 Management-query interface module

This module mainly provides the interface of background management and foreground call. Through this module users can manage information source websites, set the freqency of data extraction procedure, upload and manage the deep exploitation program, and call certain uploaded deep exploitation program to execute data analysis. Still through this module, the analyzing results will be returned in text, graphics and photographs forms.

Actually, the above three modules are on the server end of brower/server structure. There are many operation systems and information techniques can be choosed to build the interface of the module, the B/S system and backgroud database. For example, we can choose Unix or Windows as the operating system platform, choose Apache or IIS to build the B/S structure, use Oracle or SQL to the develop a database and use different programming languages (VC, Java, VBscript, Jscript, etc.) to develop and produce the deep development programs, system interface and background management procedures.

According the analysis of functions of the above three main modules, we can make a conclusion that the scheme can overcome the disadvantages and is a solution of high efficient share and deep exploitation of NAI. Up to this point, we know the scheme is a framework of utility of NAI, which has the feasibility of implementation by using existing technologies.

4. AN EXAMPLE APPLICATION OF THE SCHEME

In order to further explain the scheme and demonstrate how to build a specific application, an example application of the scheme is given here.

The application is built mainly by Windows 2000 Server, IIS5.0, MS SQL and VB. And it is designed to find out the fluctuations of the price of vegetables and the relationship between supply and demand by analysis of the wholesale prices of vegetables in six markets in Beijing.

In the case, three websites are selected as information source website (see Figure 2), all of which have vegetable price information, but in different format. A simple database with the fields of "data source", "vegetable varieties", "market", "price" and "Date" is build by MS SQL. Data extraction procedure, format standardizing procedure and data store procedure are develped and released to DLL files respectively by using VB and its built-in WebBrowser control, by using the DOM to collect information data and by using ADO/ODBC to visit database. Format standardizing procedure uses "yuan/kg" as uniform unit; "fanqie" and "tomato" are uniformed as "tomato"; publishing date are uniformed as "yyyy-mm-dd". Data store procedure save the uniform format datum into the simple database.

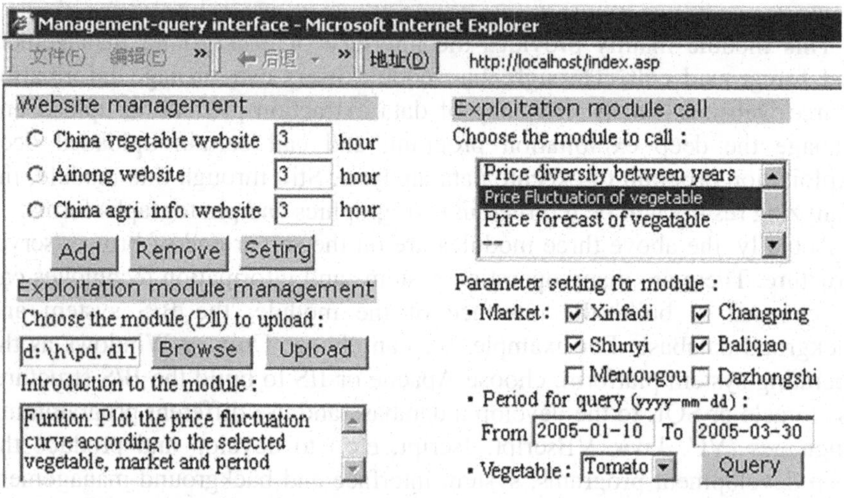

Figure 2. Interface ofr user management and aquery of example application

In this example, we developed three programs (see Figure 2). The program named "Price flactuation of vegatable" can output the given type of vegetable price volatility curve and give the mean price of different market in given period. The program takes a ten-day as a unit to calculate the mean price, and uses DbtoChart components to generate data chart, which be returned to the user in the form of web pages using ASP and VBscript script.

Management-query interface module use Html, VBscript and ActiveX to develop. It is necessary to explain that each deep exploitation program is added to deep exploitation module, there will be a reaction that its name will be listed in textbox under "Exploitation module call" (see Figure 2); Once the user to choose a program in the module list, the bottom of the list will show corresponding parameter options. In this example, this background management procedure is also produced by using VB 6.0, published by DLL files, which are convenient for it's the uploading, remote registration and deletion. The function of uploading file is realized partly by using SA-FileUP components of Artisans. A SA-FileUP component is an ActiveX DLL server component and easily integrated into the ASP website; In ASP pages, we use VBscript to call CreateObject to generate WScript.Shell, and then call regsvr32.exe by WScript.Shell to finish the procedure of uploading and registration module.

Through the interface, the interval time of extracting data from three website is set to 3 hours. Users can upload deep exploitation program in the section of Exploitation module management. Here, three programs have been uploaded, which are listed in the section of Exploitation module call. In this paper, "Price flactuation of vegetable" is called (corresponding parameters had been set and was shown in Figure 2), and the result return to the uers in the form of webpages (see Figure 3).

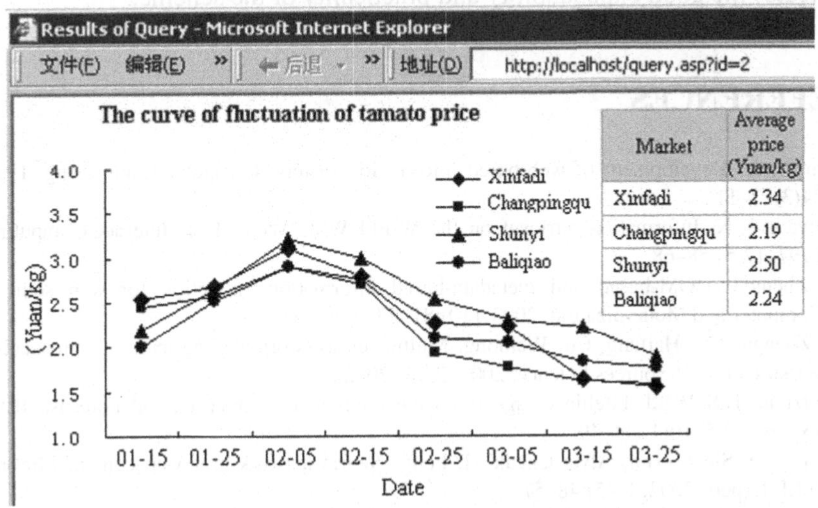

Figure 3. The result of the program of "Price flactuation of vegatable"

Figure 3 indicates that Beijing's vegetable prices in late January and in early February (that is during the Spring Festival) rose rapidly and reached its peak at Feb. 5th, then began to go down steadily. We must notice the information conveyed from Figure 3 cannot be easily to get by simply browsing original imformation in the three websites.

5. CONCLUSION & DISCUSSION

The features of NAI are analyzed and summed up as following: NAI has a scattered storage, have a high frequency of update and the amount of data increases rapidly; the data formats of NAI are not uniform, and most of NAI is raw information.

The disadvantages of the features of NAI are pointed out; and they hinder the realization of high efficient share and deep exploitation of NAI.

In order to overcome the disadvantages of the features of NAI, a scheme is put forward. By analyzing the functions of the scheme, it was theoretically proved that the scheme can greatly promote the high-efficient share and deep exploitation of NAI. In addition, this program is applicable to XML format and a large number of web HTM format.

An example application of the scheme is built up to demonstrate the concrete processes of a practical application of the scheme. Users can build more complex and practical applications by using it as a template.

However, for the limited space, we do not discuss the system building and technical details. In addition, although the scheme is proposed on the basis of NAI, it is equally applicable to other types of network information, which indicates the good expandability and practicality of the scheme.

REFERENCES

Chen Pinde. Development of web-based information system. Computer Engineering, 1993, 24(3): 7–11

Gudivada V.N. Information retrieval on the World Wide Web. IEEE Internet Computing, 1997, 12(5):58–68

Liu Zhengyi. DOM-based and metadata-based information extraction for web sources. Computer and Modernization, 2003, (10):81–83

Niu Zhenguo, Fu Haifang, Cui Weihong. Multilevel-users-oriented agricultural information classification. Resources Science, 2003, 25(2):20–25

O. Etzioni, D.S. Weld. Intelligent agents on the Internet: Fact, Fiction, and Forecast. IEEE Expert, 1995, 10(4):44–49

Pui Y. Lee, Siu C. Hui, Alvis Cheuk M. Fong. Neural networks for web content filtering, IEEE Expert, 2002, 17(5):48–57

Tang Chunsheng, Jin Yihui. A Multiple Classifiers Integration Method Based on Full Information Matrix. Journal of Software, 2003, 14(6):1103–1109

Wai Lam, Miguel Ruiz, Padmini Srinivasan. Automatic text categorization and its application to text retrieval. IEEE Transactions on Knowledge and Data Engineering, 1999, 11(6):865–879

Zhao Chunjiang, Wu Huarui, Yang Baozhu. Development platform for agricultural intelligent system based on techno-componentware model. Transactions of the Chinese Society of Agricultural Engineering, 2004, 20(2):140–143

A THEORY MODEL FOR DESCRIPTION OF THE ELECTRICAL SIGNALS IN PLANT PART I

Zhongyi Wang[1], Lan Huang[1,*], Xiaofei Yan[1], Cheng Wang[2], Zhilong Xu[2], Ruifeng Hou[2], Xiaojun Qiao[2]

[1] *College of Information and Electrical Engineering, China Agricultural University, Beijing, 100083, P. R. China*

[2] *National Engineering Research Center for Information Technology in Agriculture, Beijing, 100097, P. R. China*

[*] *Corresponding author: Lan Huang, College of Information and Electrical Engineering, China Agricultural University, Beijing, 100083, P. R. China, Email address: biomed_hl@263.net, Tel: +86-10-62737778, Fax: +86-10-62737778*

Abstract: The ion mechanism for membrane potential of higher plant was discussed in this paper. A modified Hodgkin and Huxley model was developed for description of the electrical signals in plant. Three individual components of ionic current were formulated in terms of Hodgkin and Huxley model. It include potassium current I_K, calcium current I_{Ca}, and anion current I_{Cl^-}. It model will provide a useful tool to simulate the electrical activity in cell of higher plants, which respond to environmental changes.

Keywords: electrical signals; model; Hodgkin-Huxley equation; higher plant

1. INTRODUCTION

Since 1873 when Burdon-Sanderson discovered bioelectrical activity following stimulation in plant (Burdon-Sanderson, 1873), most investigations were carried out to prove the existence of electrical signals in plant as in animal (Davies, 1987). In fact, environmental stimuli such as spontaneous changes in temperature, light or wounding can induce electrical signals at plant cells. With regard to action potential (AP), the plasma membrane is depolarized by temperature or electrical stimulation as observed in higher plants (Fromm and Spanswick, 1993; Fromm and Bauer,

Wang, Z., Huang, L., Yan, X., Wang, C., Xu, Z., Hou, R. and Qiao, X., 2008, in IFIP International Federation for Information Processing, Volume 258; Computer and Computing Technologies in Agriculture, Vol. 1; Daoliang Li; (Boston: Springer), pp. 637–643.

1994); if the stimulus is up to a certain threshold to depolarize the membrane, an AP is generated. APs are rapidly propagated electrical messages that are well known. APs travel at constant velocity and maintain constant amplitude in animals. They usually present all-or-nothing feature after a stimulus reaches a certain threshold lead resulting in membrane depolarization, but increases in stimulus strength do not change its amplitude and shape. It is well known that the ionic mechanism of APs in animal axons depends on Na^+ channel, K^+ channel and Ca^{2+} channel; the quantitative model of APs was described by Hodgkin-Huxley equation (Hodgkin and Huxley, 1952). Variation Potential of plant is evoked by damaging stimulations (wounding by cutting or burning) and is characterized by decrease in magnitude and as it spreads away from the site of stimulus. The mechanism and pathways of AP or VP transmission in plants have been investigated by several researchers (Davies, 1987; Julien et al., 1991; Stankovic and Davies, 1996; Stankovic et al., 1998; Dziubinska et al., 2003; Dziubinska et al., 2001; Volkov et al., 2005). When electrical signals (fluctuations) are only locally generated and not transferred to the other parts of a plant, they are defined as Local Electrical Potential (LEP) (Lou, 1996). LEP is a subthreshold response induced by natural variations in environmental factors, such as soil water, fertilizer, illumination, air temperature, and humidity (Leng, 1998; Hu, 2003). Although the LEP cannot be transferred, it significantly influences the physiological state of a plant (Ren et al., 1993).

Unlike animals, there are significant differences of bioelectrical activity between the species or individuals in higher plant after the same stimuli, the sensitivity and threshold of the same species varied dynamically with different growth condition. The ion mechanism which channels involve excitation of plant cells was investigated in several higher plants. In summary, ion base of depolarization or repolarization was studied by using traditional electrophysiological methods (intracellular, extracellular recording and ion channel inhibitors) and modern electrophysiological methods (patch clamp technique). Some results support the view that Ca^{2+}, Cl^- and K^+ ions channel and proton pump via cytoplasmic pH and/or Ca changes in cytosol involve the electrical activity of higher plants cells (Trebacz et al., 1997).

In the late 1980s, several models of the electrical activity of single ion channel of plant cells were formulated (Schroeder, 1989; Van Duijn, 1993). Until now, few Studies on the quantitative description of electrical activity of whole cell in plant were reported, especially for simulating the electrical signals in plant induced by environmental variations. In the present study, we will establish an theory model to calculate the rest potential and change of membrane potential of plant cell evoked by environmental factors based on modifying Hodgkin and Huxley model (H-H model). H-H model was based on experimental data that were available at that time from voltage-clamp studies, which those data were subject to limitations in available voltage-clamp techniques and their application to animal. With the development of patch clamp technique (single-cell and single-channel recording techniques) in the 1980s, the limitations of voltage-clamp measurements were overcome and the intracellular and extracellular ionic environments of plant cells could be controlled. The data from patch clamp recordings not only provide the basis for a quantitative description of channel kinetics but also can indicate membrane ionic currents of a cell. The aim of us is to use published experimental data to provide formulation of a modified H-H model to describe electrical signals in plant induced by stimuli. In this paper, the Ca^{2+}, Cl^- and K^+ ions current were included in the equation (Na^+ ion current is a significant phase in H-H model of animal, but not involve APs of plant cells) and change of H^+ was also taken into account (Trebacz et al., 1997; Maathuis et al., 1997; Krole and Trebacz, 2000). This paper focuses on the process for establishment of the equation.

2. TRANSMEMBRANE POTENTIAL – A MODEL OF REST POTENTIAL

In order to calculate the transmembrane potential, a model of steady state was shown in Fig. 1. According to published data of transmembrane potential from the experiment, the measured membrane potential are often more negative than those obtained form the Goldman-Hodgkin-Kazt equation written as formula (1) to give the diffusion potential (Taiz and Zeiger, 1998). Cells in stem and roots of young seedlings generally have membrane potentials of –130 mV to –110 mV, whereas the calculated diffusion potential are always only –80 mV to –50 mV. Thus, in addition to diffusion potential, the membrane potential has other component. The evidence of many researches indicates that the excess voltage is provided by the plasma H^+ pump in steady state. Furthermore, the transmembrane potential should be given as (2).

$$\Delta V = \frac{RT}{F}\left(\ln\frac{P_{K^+}C_{K^+}{}^{o} + P_{Ca^{2+}}C_{Ca^{2+}}{}^{o} + P_{Cl^-}C_{Cl^-}{}^{o}}{P_{K^+}C_{K^+}{}^{i} + P_{Ca^{2+}}C_{Ca^{2+}}{}^{i} + P_{Cl^-}C_{Cl^-}{}^{i}}\right) \quad (1)$$

ΔV being diffusion potential, R, T and F having their usual meaning; Pk^+ PCa^{2+}, PCl^- represent the membrane permeabilities for K^+, Ca^{2+}, Cl^-, respectively.

$$V_m = \Delta V + \frac{RT}{F}\ln\frac{H_i}{H_o} \quad (2)$$

Hi and Ho being inside and outside H^+ concentration.

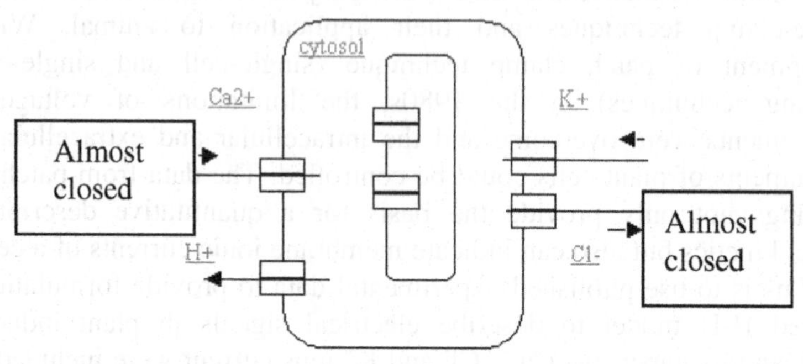

Fig. 1. A model of steady state

3. A MODIFIED H-H MODEL OF HIGHER PLANT CELL

Based on previous studies, a scheme illustrating the ion mechanism for membrane potential was shown in Fig. 2. A formula of transient depolarization induced by stimulation was given by (3). In the depolarization phases of Aps, K^+, Ca^{2+}, Cl^- ion current take place, thus the H-H model should be modified by using Cl^- ion current instead of Na^+ ion current of animal. K^+ efflux and the activity of H^+ pump cause the repolarization of the membrane potential and complete return to the steady state. Here, we suggest that the change of H^+ ion efflux is a constant during whole procedure. In this model, the temperature induced membrane potential changes in higher plant cells is related to Ca^{2+} channel according to the physiological experimental results (Plieth, 1999). The general approach is based on a numerical reconstruction of the ventricular action potential by using Hodgkin-Huxley-type formalism. The rate of change of membrane potential (V) is given by

$$Istim = Cm\frac{dV}{dt} + \sum I_j \tag{3}$$

where *Cm* is the membrane capacitance, *Istim* is a stimulus current, and Σ Ij is the sum of ionic currents: IK, a potassium current; ICa, a calcium current; ICl-, an anion current. The ionic currents are determined by ionic gates, whose gating variables are obtained as a solution to a coupled system of nonlinear ordinary differential equations. The ionic currents, in turn, change V, which subsequently affects the ionic gates and currents. These differential equations about the ionic gates are of the form as formula (4) given by H-H model. y represents any gating variable, T y is its time constant, and y∞ is the steady-state value of y. α y and β y are voltage-dependent rate constants given as (5) and fitted from the published experimental data, Ej is the reversal potential of ion. Ion current can be written as formula (6); Especially, the phase of temperature influence on Ca^{2+} current described by (7), $[Ca^{2+}]_i$ represent concentration of Ca^{2+} in cytosol, T is temperature. The free parameters are: K_1, K_m, *const1* to *const6*. The integration algorithm used to solve the differential equations is based on the Runge-Kutta methods.

$$\frac{dy}{dt}=\frac{(y_\infty - y)}{\tau_y}, \tau_y = \frac{1}{(\alpha_y+\beta_y)}, y_\infty = \frac{\alpha_y}{(\alpha_y+\beta_y)} \tag{4}$$

$$\begin{aligned}\alpha_y &= const(\alpha)\times f(\exp(V-E_j))\\ \beta_y &= const(\beta)\times f(\exp(V-E_j))\end{aligned} \tag{5}$$

$$\begin{aligned}I_j &= cons_j \times y_i^h \times y_n^m \times (V-E_j)\\ j &= K^+, Cl^-\end{aligned} \tag{6}$$

$$\begin{aligned}I^1_{Ca^{2+}} &= cons\times y_i^r \times y_n^r \times (V-E_{Ca^{2+}})\\ E_{Ca^{2+}} &= const1 - const2\ln([Ca^{2+}]_i)\end{aligned} \tag{7}$$

$$\frac{d([Ca^{2+}]_i)}{dt} = (const3 + const4\times I^1_{Ca^{2+}} + \frac{const5\times(-\frac{1}{2}(\frac{dT}{dt}-|\frac{dT}{dt}|))^2}{K_1^2+(-\frac{1}{2}(\frac{dT}{dt}-|\frac{dT}{dt}|))^2} - const6\times\exp(K_Q(T-T_0))\times\frac{[Ca^{2+}]_i^2}{[Ca^{2+}]_i^2+K_m^2})$$

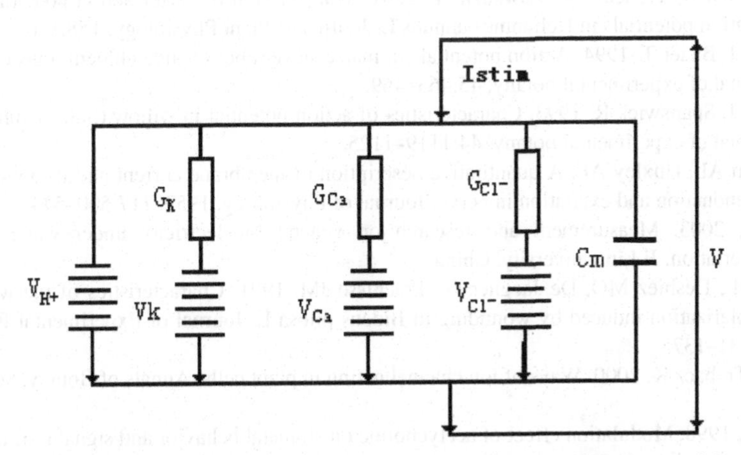

Fig. 2. A scheme illustrating the ion mechanism for membrane potential

4. SUMMARY

A theory model has been developed by means of modifying Hodgkin-Huxley equation. Three predominant ionic current, potassium current, calcium current, and anion current involve in the equation to describe bioelectrical activity of higher plants. Our model allows us to study the influence of temperature parameter to reach threshold. Our results are in accordance with experiments. Most result and analysis of simulation by using the model will be reported in a forthcoming paper (Part-II).

ACKNOWLEDGEMENTS

This research was supported by the National Natural Science Foundation of China (Grant 60571027) and National High Technology Research and Development Program of China (863 Program, Grant 2006AA10Z202).

REFERENCES

Burdon-Sanderson J, 1873. Note on the electrical phenomena which accompany irritation of the leaf of Dionaea muscipula. Proceedings of the Royal Society of London, 21:495–496

Davies E, 1987. Action potentials as multi-functional signals in plants: a unifying hypothesis to explain apparently disparate wound response. Plant Cell and Environment, 10:623–631.

Dziubinska H, Fielk M, Koscielniak J, Zawadzki T, 2003. Variation and action potentials evoked by thermal stimuli accompany enhancement of ethylene emission in distant non-stimulated leaves of Vicia faba minor seedlings. Journal of Plant Physiology, 160:1203–1210.

Dziubinska H, Tredeusz K, Zawadzki T, 2001. Transmission route for action potentials and variation potentials in Helianthus annuus L. Journal of Plant Physiology, 158:1167–1172.

Fromm J, Bauer T, 1994. Action potentials in maize sieve tubes change phloem translocation. Journal of experimental botany, 45:463–469.

Fromm J, Spanswick R, 1993. Characteristics of action potential in willow (Salix vimnlis. L). Journal of experimental botany, 44:1119–1125.

Hodgkin AL, Huxley AF. A quantitative description of membrane current and its application to conduction and excitation in nerve. Journal of Physiology, 1952, 117:500–544.

Hu JH, 2003. Measurement and research on vegetal bioelectricity under water stress, Dissertation. Ji Lin University, China.

Julien JL, Desbiez MO, De Jaegher G., Frachisse JM, 1991. Characteristics of the wave of depolarization induced by wounding in Bidens pilosa L. Journal of Experimental Botany, 42:131–137.

Krole, Trebacz K, 2000. Ways of ion channel gating in plant cells. Annals of Botany, 86:449–469.

Leng Q, 1998. Modulation effect of acetycholine on stomatal behavior and signal transduction in higher plant. Ph.D. Dissertation, China Agriculture University, China.

Lou CH, 1996. The messenger transmission of chemical wave in higher plant. Acta. Biophysics. Sinica, 12(4):739–745.

Maathuis FJM, Ichida AM, Sanders D, Schroeder JI, 1997. Roles of higher plant K+ channels. Plant Physiology, 114:1141–1149.

Plieth C, 1999. Temperature sensing by plants: calcium-permeable channels as primary sensors - a model. Journal Membrane Biology, 172:121–127.

Ren HY, Wang XC, Lou CH, 1993. The universal existence of electrical Signals and its physiological effects in higher plants. Acta Phyto-physiological. Sinica, 19(1):97–101.

Schroeder JI, 1989. Quantitative analysis of outward rectifying K channel currents in guard cell protoplasts from Vicia faba. Journal of Membrane Biology, 107:229–235.

Stankovic B, Davies E, 1996. Both action potentials and variation potentials induce proteinase inhibitor gene expression in tomato. FEBS. Letters, 390:275–279.

Stankovic B, Witters DL, Zawadzki T, Davies E, 1998. Action potentials and variation potentials in sunflower: An analysis of their relationships and distinguishing characteristics. Physiologia plantarum, 103:51–58.

Taiz L, Zeiger E, 1998. Plant physiology. Sinauer Associates INC., USA, 130–131.

Trebacz K, Simonis W, Schonknecht G, 1997. Effects of anion channel inhibitors on light-induced potential changes in the liverwort Conocephalum conicum. Plant and cell physiology, 38:550–557.

Van Duijn B, 1993. Hodgkin-Huxley analysis of whole-cell outward rectifying K+ currents in protoplasts from tobacco cell suspension cultures. Journal of Membrane Biology, 132: 77–85.

Volkov AG, Dunkley TC, Labady AJ, Brown CL, 2005. Phototropism and electrified interfaces in green plants. Electrochimica Acta, 50:4241–4247.

DEVELOPMENT OF A DATA MINING APPLICATION FOR AGRICULTURE BASED ON BAYESIAN NETWORKS

Jiejun Huang[1,*], Yanbin Yuan[1], Wei Cui[1], Yunjun Zhan[1]

[1] *School of Resource and Environmental Engineering, Wuhan University of Technology, Wuhan, China, 430070*

* *Corresponding author, Address: School of Resource and Environmental Engineering, Wuhan University of Technology, 122 Luoshi Road, Wuhan, hubei, 430070, P. R. China, Tel: +86-27-62421206, Email: Hjjtk@21cn.com*

Abstract: Data mining is a process by which the data can be analyzed so as to generate useful knowledge. It aims to use existing data to invent new facts and to uncover new relationships previously unknown even to experts. Bayesian network is a powerful tool for dealing with uncertainties, and has a widespread use in the area of data mining. In this paper, we focus on development of a data mining application for agriculture based on Bayesian networks. Let features (or objects) as variables or the nodes in Bayesian network, let directed edges present the relationships between features, and the relevancy intensity can be regarded as confidence between the variables. Accordingly, it can find the relationships in the agricultural data by learning a Bayesian network. After defining the domain variables and data preparation, we construct a model for agricultural application based on Bayesian network learning method. The experimental results indicate that the proposed method is feasible and efficient, and it is a promising approach for data mining in agricultural data.

Keywords: data mining, Bayesian network, model, agriculture

1. INTRODUCTION

Data mining provides an information technology to develop and utilize the data; it is very helpful for making decision by extracting regulations, patterns and models from large databases. Data mining has proved to have surprisingly broad application. Many scholars pay attention to application of

Huang, J., Yuan, Y., Cui, W. and Zhan, Y., 2008, in IFIP International Federation for Information Processing, Volume 258; Computer and Computing Technologies in Agriculture, Vol. 1; Daoliang Li; (Boston: Springer), pp. 645–652.

data mining for agriculture (Lee et al., 1998; Bajwa et al., 2004; Abdullah et al., 2006; Andujar et al., 2006). Expert systems and geographical information systems have been used to help with an implementation of a land suitability evaluation model (Kalogirou, 2002). It shows that applying Dempster-Shafer Theory in image classification can yield thematic maps with accuracies that can support their operational use, and the potential for applying soft-classification procedures based on the Dempster-Shafer Theory of Evidence was demonstrated (Lein, 2003). Fuzzy set and interpolation techniques are applied for land suitability evaluation for maize in Northern Ghana (Braimoh et al., 2004). A linear mixture model (LMM) approach is applied to classify land covers in the eastern Nile delta of Egypt. It indicates that the LMM is a promising approach for distinguishing the different land cover types and to classify the different vegetation types using Landsat ETM+ data (Ghar et al., 2005). Furthermore, computational intelligence is used to agriculture. A novel model of land suitability evaluation is built based on computational intelligence (Liu et al., 2005).

Bayesian networks are the method for uncertainty reasoning and knowledge representation that was advanced at the end of the 20th Century. It is a probabilistic graphical model, which has been used for probabilistic reasoning in expert systems. Bayesian networks proved to have surprisingly broad applications, such as medical diagnoses, image interpretation, pattern recognition, in particular, knowledge discovery and data mining (Heckerman, 1997; Helman et al., 2004). Bayesian network as an application to agriculture is studied by some authors. An application of belief networks to assess the impact of climate change on potato production is used as an illustration, and used simulated data from a mathematical model which forecasts the impact of climate change on potato production (Gu et al., 1994). The potential of Bayesian networks to assess the yield response of winter wheat to fungicide programmes has been shown, and a Bayesian network that fits the experimental data has been produced (Tari, 1996). An example of application of Bayesian networks for modeling landuse changes is proposed (Benferhat et al., 2004). In addition, Bayesian networks have been used in the sustainable planning of the Eastern Mancha aquifer. The results can be obtained through Bayesian networks are the partial substitution of groundwater with surface water, the improvement of irrigation efficiency and the adequate control of water use (Martin, 2007).

In this paper, we focus on development of a data mining application for agriculture based on Bayesian networks. We believe that it will provide a potentially useful tool in the domain. In next section, we discuss Bayesian network for data mining. In section 3 we discuss an information-theoretical approach to learning Bayesian network. In section 4, we propose an example of application of Bayesian networks for agricultural land gradation. Finally, we draw conclusions and present the future works.

2. BAYESIAN NETWORKS FOR DATA MINING

A Bayesian network is a directed acyclic graph representing the causal relationships between variables that associate conditional probability distributions to variables given their parent variables. It is represented at two levels, qualitative and quantitative. At the qualitative level, we have a directed acyclic graph in which nodes represent variables and directed arcs describe the conditional independence relations embedded in the model. At the quantitative level, the dependence relations are expressed in terms of conditional probability distributions for each variable in the network. Suppose a data set D is given, which is defined by n variables V={ V_1, V_2, ..., V_n}, each variable respond to a node, let G represents a DAG, L is a set of directed links, P is a set of conditional probability distributions associated with every node. Using Bayes chain rule, and let Pa_i is the set of parents of the variable V_i, so we can get the joint probability distribution:

$$P(\mathbf{V}) = \prod_{i=1}^{n} P(V_i \mid Pa_i) \tag{1}$$

A Bayesian network represents a joint probability distribution of a set of random variables of interest. Information we can obtain comes in the form of evidence about a subset of variables. The basic task of inference is to update the joint probability distribution of the variable set conditioned on the given set of evidence. Bayesian networks as data mining have several aspects:

(1) Causality discovery. Bayesian networks give a graphical representation of the domain problems and results. As for a given problem, we assume that it is fully described by a finite set of random variables. Each variable is fully defined in a finite frame, i.e. set of all possible states. The set of relations among variables is called the structure of the Bayesian network, which represents the qualitative knowledge about the problem domain. If all the variables are identified, and each variable is defined with a frame, we say that we have a Bayesian network structure which is a representation of the causality of the variables. Here the structure *S* refers to a set of directed edges.

$$S = \{U \rightarrow V \mid U, V \in \mathbf{V}\} \tag{2}$$

Where for each directed link $U \rightarrow V$, *U* is a parent of *V*. On the ground that, we can discover the causality between variables by finding the directed links which we called learning Bayesian network structure.

(2) Uncertainty reasoning. Inference in data mining is rigorous based on Bayesian probability theory. This rigor will not decrease along with inference passages of any length. Inference always remains rigorous irrespective of the size of the network and how far the information variables are from the target variables in the network. There is no need to distinguish between forward reasoning and backward reasoning. Bayesian network are

capable of learning, the structure can be constructed by training the data sets. And the resulting model has a clear interpretation. The conditional probabilities associated with relations correspond to the quantitative aspects of the expert knowledge. Furthermore, information in the form of evidence may come into the network from any location (variables or nodes) in the network, and such incoming information will then be propagated throughout the rest of the network.

(3) Adaptive learning. Adaptivity is closely related to learning. When adaptivity is concerned, learning becomes a decision problem. Bayesian networks can incorporate expert knowledge and historical data for decision-making. In addition, data mining based on Bayesian network can be controlled by artificial conditioning. For example, in predictive mining, the data model usually consists of a large sample set of cases, with each case containing a certain number of features. Formulating a predictive problem trains the system to "learn" which patterns match predefined criteria within existing cases and which don't, and to accept or reject new cases based on these criteria.

3. LEARNING BAYESIAN NETWORK STRUCTURE FROM DATA

The main obstacle for using Bayesian networks is to construction the domain model. Creation of Bayesian models is a complex task involving participation of a knowledge engineer and domain experts. A Bayesian method was developed for the induction of Bayesian networks from data (Cooper et al., 1992). An information-theory based approach to learning Bayesian networks from data was provided (Cheng et al., 2002). And many authors have studied on learning Bayesian networks and proposed some relative algorithms (Huang et al., 2005; Tsamardinos et al., 2006). Learning a Bayesian network from data involves two tasks: Estimating the probabilities for the conditional probability tables (learning parameters) and deriving the structure of the network. Although ideally the structure and parameters should be learned simultaneously, finding the optimal structure of the network is the most difficult part of the whole problem. It comprises a heuristic search through the space of possible structures. Candidate structures are evaluated by calculating how well the network fits the data.

Fundamental to various approaches to learning Bayesian networks are statistical learning theory, Bayesian learning theory, and computational learning theory. In this section, we introduce an information-theoretical approach: Minimum Description Length (MDL) criterion. Learning Bayesian network structure from a data set can be regarded as a problem of explaining the given set of data using a learned Bayesian network as a model. Given a data set **D** out of the data space $\mathcal{D}$, the MDL principle selects the best model M_{best} out of the model space $\mathcal{M}$ with

$$M_{best} = \underset{M \in \mathcal{M}}{\arg\min}\, L(\mathbf{D}, \mathrm{M}) \tag{3}$$

The model space $\mathcal{M}$ is a Cartesian product of the structure space $\mathcal{S}$ and the parameter space Θ_S given structure S

$$M = (\mathcal{S}, \Theta_S) = \mathcal{S} \times \Theta_S, \qquad S \in \mathbf{S} \tag{4}$$

The joint description length of a given data set and a model – a Bayesian network- can be defined as

$$\begin{aligned} L(\mathbf{D}, M) &= L(\mathbf{D}, \Theta, S) \\ &= L(\mathbf{D} \mid \Theta, S) + L(\Theta \mid S) + L(S) \\ &= L(\mathbf{D} \mid \Theta) + L(\Theta \mid S) + L(S) \end{aligned} \tag{5}$$

For structure learning purpose, we may only need to evaluate

$$L(\mathbf{D}, S) = L(\mathbf{D} \mid S) + L(S) \tag{6}$$

According to the MDL criterion, a data set can be represented by the Bayesian network structure, whose description length as

$$L(g, x^N) = H(g, x^N) + \frac{k(g)}{2} \log N \tag{7}$$

Where $H(g, x^N)$ is the empirical entropy for a network structure g

$$H(g, x^N) = \sum_{j \in J} H(j, g, x^N) \tag{8}$$

$k(g)$ is the number of independent conditional probabilities embedded in the network structure g

$$k(g) = \sum_{j \in J} k(j, g) \tag{9}$$

In this sense, learning Bayesian network from data is the problem same as optimal problem. The best model of all alternative models will be the one with the shortest total description length. This information-theoretical approach can avoid explicitly defining the structure prior and easy to be comprehended. Nevertheless, it is difficult for tackling the incomplete or soft data problem and the computation would be too complex if the network has many nodes.

4. DATA MINING BASED ON BAYESIAN NETWORKS FOR AGRICULTURE

In this section, we propose an example of application of Bayesian networks for agricultural land gradation. The data set contains 2 000 cases, part of the data and the variables with theirs status showed in Table 1. The domain problem has 6 variables; each of them has several attributes. And one variable responds to one node in the model respectively. The variables and their implications describe as follows. 1) Soil texture: the relative proportions of sand, silt, and clay particles in a mass of soil. 2) Organic matter: consists of plant and animal material that is in the process of decomposing. 3) Gradient: A measure of slope (soil-surface), i.e. the rate of inclination for land topography changes. 4) Drainage: The capability of draining off the water when the field doesn't need the water. The means of draining collectively, as a system of conduits, trenches, etc. 5) Soil pH: indicates the acidity of the soil, it can be determined by having a soil analysis carried out, and has a range approximately from 0 to 14; 6) Land grades: the quality of the agricultural land measured by the natural and economic characteristics.

Table 1. Part of data and the variables with theirs status

Land Code	Soil texture	Organic matter	Gradient	Drainage	Soil pH	Land grades
0616	sand	2.06	3	2	6.45	III
0896	silt	1.38	3	2	6.55	IV
1025	sand	2.06	2	2	6.42	III
1380	sand	1.38	2	3	6.42	IV
1620	silt	1.38	2	2	6.42	IV
1698	clay	1.95	2	3	6.42	I
1806	clay	1.38	3	3	6.50	II
1912	clay	1.30	3	3	6.85	II

After defining the domain variables and data preparation, we can get a Bayesian network model for agricultural land gradation by using the approach described in previously section. Based on the model, and learning the parameters of each node with the dataset by using Bayes criterion, the complete network including conditional probability distributions was got.

The experimental study is done on 200 cases to test its validity. The gradation accuracy is 87.5%. Then we compare results given by Bayesian networks with the ones of naive bayes which are simple Bayesian networks, and decision tree method, the results by Bayesian networks is the best one (Table 2). The experimental results validate the practical viability of the proposed approach for data mining in agricultural data.

Table 2. The accuracy of result by using different methods

Methods	Test data set	Correct	Accuracy
Naïve bayes	200	156	78.0%
Decision tree	200	166	83.0%
Bayesian network	200	175	87.5%

5. CONCLUSIONS AND FUTURE WORKS

In this paper, we use Bayesian networks as a data mining method for agriculture. Firstly, we presented an overview of Bayesian network for data mining. Secondly, we discussed an information-theoretical approach to learning Bayesian network structure. Then we propose an example of application of Bayesian networks for agricultural land gradation. Furthermore, we compare results given by Bayesian networks with Naive bayes and decision tree. From the practice of applying Bayesian network, it can deal with all kinds of data timely, and has other functions such as agricultural land evaluation and agricultural machine diagnosis. Bayesian networks as data mining for agriculture have several characteristics. Its representation and reasoning can be carried out simultaneously, and combined with prior knowledge and observed data. Moreover, it can overcome the noise of data set, and provide the scientific evidences in decision making for exploiting agriculture resource. It is undoubted that Bayesian networks will be a promising approach for data mining and get the surprisingly success in the application domains. At the same time, Bayesian network is not almighty. It can not obtain satisfying results from small data or sparse data. Therefore, the future work should be focused on how to deal with sparse data and missing values. Furthermore, we will apply Bayesian network to other agricultural domains with Geological information system and remote sensing.

ACKNOWLEDGEMENTS

The work presented in the paper was supported by the National Natural Science Foundation of P. R. China (No. 40601076, 40572166 and 40571128), the Natural Science Foundation of Hubei Province (2007ABA322) and Open Research Fund of State Key Laboratory of Information Engineering in Surveying, Mapping and Remote Sensing (WKL(06)0303).

REFERENCES

Abdullah A, Hussain A. Data mining a new pilot agriculture extension data warehouse. Journal of Research and Practice in Information Technology, 2006, 38(3): 229–248

Andujar J M, Aroba J, et al. Contrast of evolution models for agricultural contaminants in ground waters by means of fuzzy logic and data mining. Environmental Geology, 2006, 49(3): 458–466

Bajwa S G, Bajcsy P, Groves P, Tian L F. Hyperspectral image data mining for band selection in agricultural applications. Transactions of the American Society of Agricultural Engineers, 2004, 47(3): 895–907

Benferhat S, Cavarroc M, Jeansoulin R. Modeling landuse changes using bayesian networks. Proceedings of the IASTED International Conference on Artificial Intelligence and Applications, 2004, 615–620

Braimoh A k, Vlek Paul L, Stein Alfred. Land Evaluation for Maize Based on Fuzzy Set and Interpolation, Environmental Management, 2004, 33(2): 226–238

Cheng J, Greiner R, Kelly J, et al. Learning Bayesian networks from data: An information-theory based approach. Artificial Intelligence, 2002, 137(1–2): 43–90

Cooper G F, Herskovits E A. Bayesian method for the induction of Bayesian networks from data. Machine Learning, 1992, 9: 309–347

Ghar M A, Renchin T, Tateishi R, Javzandulam T. Agricultural land monitoring using a linear mixture model. International Journal of Environmental Studies, 2005, 62(2): 227–234

Gu Y, Peiris D R, Crawford J, et al. An application of belief networks to future crop production. Proceedings of the 10th Conference on Artificial Intelligence for applications, San Antonio, Texas, 1994, 305–309

Heckerman D. Bayesian Network for data mining, Data mining and knowledge discovery, 1997, 1: 79–119

Helman P, Veroff R, Atlas S, et al. A Bayesian network classification methodology for gene expression data. Journal of Computational Biology, 2004, 11(4): 581–615

Huang J J, Pan H P, Wan Y C. An algorithm for cooperative learning of Bayesian network structure from data. Lecture Notes in Computer Science, 2005, 3168: 86–94

Kalogirou S. Expert systems and GIS: An application of land suitability evaluation. Computers, Environment and Urban Systems, 2002, 26(2–3): 89–112

Lee S W, Kerschberg L. Methodology and life cycle model for data mining and knowledge discovery in precision agriculture. Proceedings of the IEEE International Conference on Systems, Man and Cybernetics, 1998, 3: 2882–2887

Lein J K. Applying evidential reasoning methods to agricultural land cover classification International Journal of Remote Sensing, 2003, 24 (21): 4161–4180

Liu Y, Jiao L. Model of land suitability evaluation based on computational intelligence. Wuhan Daxue Xuebao (Xinxi Kexue Ban), 2005, 30(4): 283–287 (in Chinese)

Martin de Santa Olalla F, Dominguez A, Ortega F, et al. Bayesian networks in planning a large aquifer in Eastern Mancha, Spain. Environmental Modelling and Software, 2007, 22(8): 1089–1100

Tari F. A Bayesian Network for predicting yield response of winter wheat to fungicide programmes. Computer and electronics in agriculture, 1996, 15: 111–121

Tsamardinos I, Brown L, Aliferis C. The max-min hill-climbing Bayesian network structure learning algorithm. Machine Learning, 2006, 65(1): 31–78

FORECAST RESEARCH OF DROPLET SIZE BASED ON GREY THEORY

Ping Liang[1,2,*], Hongping Zhou[1], Jiaqiang Zheng[1]
[1] *College of Mechanical and Electrical Engineering, Nanjing Forestry University, Nanjing, China, 210037*
[2] *Department of Mechanical Engineering, Huaiyin Institute Technology, Huai'an, China, 223003*
* *Corresponding author, Address: College of Mechanical and Electrical Engineering, Nanjing Forestry University, 159 longpan Road, Nanjing, 210037, P. R. China, Tel: +86-25-85442189, Fax: +86-25-85424189, Email: hyitlp@sina.com*

Abstract: The droplet size can influence the application of a spray system. We apply grey theory in establishing a model of forecasting droplet size. The example used in this study shows that this method is precise in forecasting the droplet size, and is helpful to choose the other parameters of a spray system within a certain droplet size.

Keywords: grey theory, droplet size, forecast

1. INTRODUCTION

The effect of spray is determined by the droplet size which is an important index of the function of a sprayer performance. Rapidly and accurately measuring the size and distribution of droplets is required in the study, production, and application of a sprayer. As people deeply realize the influence of droplet size to sprayer performances, the study in droplet size is attached more and more importance. In a spray system, droplet size is affected by the air pressure of the system, the flow, the consistence of the liquid, the angle of spraying, and types of nozzle. Each biological target demands a specific droplet size. The more droplets can be captured on the target, the better the preventing and controlling harmful insect's effect will be. Therefore, accurately measuring droplet size and controlling the

Liang, P., Zhou, H. and Zheng, J., 2008, in IFIP International Federation for Information Processing, Volume 258; Computer and Computing Technologies in Agriculture, Vol. 1; Daoliang Li; (Boston: Springer), pp. 653–658.

distribution of droplet size are very important for the understanding and studying the procedure of spraying.

The analysis of droplet size of a spray system is usually based on a great deal of data collected from a large quantity of samples, which is time-consuming, laborious and is not accurate. Applying grey theory in forecasting droplet size greatly simplifies the routine method and can effectively increase the validity of the results. This method makes best of the internal relevancy of similar samples and gets corresponding analysis data by less samples. At the same time, the other parameters of a spray system corresponding to certain droplet sizes can be forecasted by establishing a grey model.

2. GREY THEORY

Grey theory is brought forward and developed by Dr. Julong Deng in the eighties in the twentieth century (Deng, 1997). Its research targets are those objects with incomplete information, uncertain concepts, and mechanisms with uncertain relationship. Its tasks are: (1) Invent a new method of establishing a grey model to overcome the weakness of probability and to find the rule among the confused, limited and discrete data; (2) Use the method to carry out respective analysis, forecast, decision and programming. Grey forecast can be processed with small quantity of data in a short time.

2.1 Definition and classification of grey forecast

In grey theory, white refers to the information that is complete awareness, black refers to the information that is scanty and grey refers to the information that is incomplete. The system with incomplete information is a grey system. Grey forecast can find out and master the developing rule of a system, and make out measurable forecast of the target system conditions by processing original data and establishing a grey mathematics model. Usually we build differential equation based on original data so that random disturbance among original data weaken and the contained complete information is strengthened by accumulated generation. The general form of a grey model is GM (n, h), in which n is progression of grey differential equation and h is the number of the variable.

Grey forecasts can be classified into different types of forecasts according to the respective function. For example: sequence forecast, abnormal forecast, topological forecast and system forecast and so on. This work mainly focuses on grey sequence forecast.

2.2 The building of grey number arrange forecast model

Grey sequence forecast model GM (1, 1), the model used for forecasting time data, is set up in the following steps:

1 Supposes: there is an initial data as following,

$$X^{(0)} = (x^{(0)}(1), x^{(0)}(2), \cdots, x^{(0)}(n)) \tag{1}$$

2 $X^{(0)}$ is processed by 1-AGO (accumulating generation operator), we can get

$$X^{(1)} = (x^{(1)}(1), x^{(1)}(2), \cdots, x^{(1)}(n)) \tag{2}$$

Therein to, $X^{(1)}(k) = \sum_{i=1}^{k} x^{(0)}(i),\ k = 1,2,\cdots,n$

3 $X^{(1)}$ can establish whitenization function show as

$$\frac{dx^{(1)}}{dt} + ax^{(1)} = b \tag{3}$$

Then we calls formula (3) is a one-step and single-variable grey differential equation model GM (1, 1)

4 a, b are elements of parameters vector $\hat{a}$, that is

$$\hat{a} = [a, b]^T \tag{4}$$

Building repeated additive matrix B and constant vector Y_n, that is

$$B = \begin{bmatrix} -0.5(x^{(1)}(1) + x^{(1)}(2)) & 1 \\ -0.5(x^{(1)}(2) + x^{(1)}(3)) & 1 \\ \cdots & \cdots \\ -0.5(x^{(1)}(n-1) + x^{(1)}(n)) & 1 \end{bmatrix},$$

$$Y_n = \begin{bmatrix} x^{(0)}(2) \\ x^{(0)}(3) \\ \vdots \\ x^{(0)}(n) \end{bmatrix}$$

5 Make out grey parameter $\hat{a}$ using least square, then

$$\hat{a} = [a \quad b]^T = [B^T B]^{-1} B^T Y_n \tag{5}$$

6 The value of differential equation in whitenization format is

$$\hat{X}^{(1)}(k+1) = (x^{(0)}(1) - \frac{b}{a})e^{-ak} + \frac{b}{a} \tag{6}$$

$k = 1,2,\cdots,n$

7 Calculate the simulation value of $X^{(1)}$ by the following formula

$$\hat{X}^{(1)} = (x^{(1)}(1), x^{(1)}(2), \cdots, x^{(1)}(n)) \tag{7}$$

8 Get out the simulation value of $X^{(0)}$ by IAGO (Inverse accumulating generation operator)

$$\widehat{X}^{(0)} = (\widehat{x}^{(0)}(1), \widehat{x}^{(0)}(2), \cdots, \widehat{x}^{(0)}(n))$$
$$\widehat{x}^{(0)} = (\widehat{x}^{(1)}(k) - \widehat{x}^{(1)}(k-1) \quad (8)$$

Parameter a is developing coefficient which reflects development trend of $\widehat{X}^{(0)}$ and $\widehat{X}^{(1)}$; Parameter b is grey action factor which come from background data and reflects variety relation of data.

2.3 The test of grey sequence forecast model

The test system forecasts the quality of a grey sequence forecast model. Whether a model can fully satisfy the requirements of practice is not determined until it is been carefully tested. Usually we test a grey forecast model by methods of residual error (α) test, probability of minor error (p) test and degree of grey incidence (ξ) test. All of these methods evaluate the precision of a model by reviewing residual error, a model with high precision demands small average relative error and simulation error, and large relevancy (ξ). The model is considered to be reliable when ξ>0.60, the quotient of square ratio C is minor, and probability of minor error P is large. For the Given values of α, ξ, C and P, we can classify forecast precision by good, general, eligibility and bad (shown as Table 1).

Table 1. Class table of testing model precision

Accuracy class	Residual error (α)	Quotient of square ratio (C)	Probability of error (P)	Degree of grey incidence (ξ)
Good	1%	0.35	0.95	0.90
General	5%	0.5	0.80	0.80
Eligibility	10%	0.65	0.70	0.70
Bad	20%	0.80	0.60	0.60

3. FORECAST OF DROPLET SIZE BASED ON GREY THEORY

Due to the complexity of forming droplet and variety of spray specification, distribution of droplet size shows a great difference. The distribution of droplet of hydraulic sprayer is determined by detailed analysis. The methods of measuring droplet size include mechanical method, electric method and optical method, among which optical method with a laser droplet analyzer is most accurate. In the process of measuring droplet size, the collection of data is carried out by the laser droplet analyzer of model Winner313. Table 2 is the data collected by the laser droplet analyzer of model Winner313.

Table 2. Droplet size data by laser droplet analyzer of model Winner313

Water pressure (MPa)	0.10	0.12	0.14	0.16	0.18	0.20
Droplet size (μm)	189	183	176	171	168	162

Table 2 data processing: $X^{(0)}$ carries out one step AGO, then we get the data matrixes B and Y_n.

$$B = \begin{bmatrix} -280.5 & 1 \\ -460 & 1 \\ -633.5 & 1 \\ -803 & 1 \\ -968 & 1 \end{bmatrix}, Y_n = \begin{bmatrix} 183 \\ 176 \\ 171 \\ 168 \\ 162 \end{bmatrix}$$

Here, $\hat{a} = \begin{bmatrix} a & b \end{bmatrix}^T = [B^T B]^{-1} B^T Y_n$= (0.02913, 190.3226) T,
According to formula (6), we can get

$$\hat{X}^{(1)}(k+1) = (189 - \frac{190.3226}{0.02913})e^{-0.02913k} + \frac{190.3226}{0.02913}$$
$$=-6344.6099\, e^{-0.02913k} + 6533.6099$$

Therefore, we can get the simulation value of $X^{(0)}$ by IAGO

$$\hat{X}^{(0)} = (\hat{x}^{(0)}(1), \hat{x}^{(0)}(2), \cdots, \hat{x}^{(0)}(6))$$
$$= (189, 182.15, 176.92, 171.84, 166.91, 162.12)$$

Meanwhile, we can obtain droplet size under other water pressure value. According to formula (6), when k is other value, such as k=7, we can also get

$\hat{x}^{(0)}(7)$=157.46

In the practical measure, when water pressure value is 0.22MPa, droplet size value is 156μm.

In order to evaluate the precision of the model forecast, the model value and the actual value error may be examined by using the residual error. The residual error and the relative error may be expressed as:

Residual error $\varepsilon(k) = X_k^{(0)} - \hat{X}_k^{(0)}, k = 1,2,\cdots,n$

Relative error $\Delta k = \left|\varepsilon(k) / X_k^{(0)}\right| \times 100\%, k = 1,2,\cdots,n,$

Table 3 shows the application of the forecast model in forecasting the droplet size of a hydraulic spray system. The forecast data and error have been listed. What needs to be explained is that GM (1, 1) forecast model is

Table 3. Droplet size forecasted data and error

k	$X^{(0)}(k)$	$\hat{X}^{(0)}(k)$	$\varepsilon(k)$	Δk
1	189	189	0	0
2	183	182.15	0.85	0.46%
3	176	176.92	-0.92	-0.52%
4	171	171.84	-0.84	-0.49%
5	168	166.91	1.09	0.65%
6	162	162.12	-0.12	-0.74%
7	156	157.46	-1.46	0.94%

usually only used to forecast the monotone increasing function and monotone decreasing function. In the hydraulic spray system, the droplet size distribution (volume distribution) follows the normal distribution (Ma et al. 1999) that is droplet size peak value exists in a spray process. Therefore, in the droplet size forecast, the model may be established according to the process and the monotone interval. And the forecast can be done by the partition forecast and the generalized analysis so that it can satisfy the droplet size forecast demands.

4. CONCLUSION

A droplet size forecast model has been established based gray system theory in this paper. The results show that: (1) The predicted values almost matched with the measured values under the water pressure from 0.10MPa to 0.22MPa and their relative errors range were between 0.49%–0.94%. (2) The method of grey forecast is simple, is high precise in forecast, and is easy to be simulated by a computer. Meanwhile, it can be used for different nozzle types of droplet size gray forecast model systems. Besides the droplet size, it is also helpful for determining other parameters of a spray system.

REFERENCES

Deng Ju-Long 1997, Grey Control System (second edition). Huazhong University of Science and Technology, (in Chinese).

Liu Si-Feng, Guo Tian-Bang, Dang Yue-Guo 2004, Grey Systems Theory and Applications (third edition). Science Press, (in Chinese).

Ma ChengWei, Yan HeRong, Yuan DongShun, Cui Ying-An 1999, Droplet Size Distribution of Hydraulic Nozzle. Transactions of the Chinese Society of Agricultural Machinery 30: 33-39 (in Chinese).

RESEARCH ON MACHINE VISION BASED AGRICULTURAL AUTOMATIC GUIDANCE SYSTEMS

Bin Liu[1], Gang Liu[1,*], Xue Wu[2]
[1] *Key Laboratory of Modern Precision Agriculture System Integration Research, China Agricultural University, Beijing, China, 100083*
[2] *College of Mechanical Engineering and Automation, Beijing Technology and Business University, Beijing, China, 100037*
* *Corresponding author, Address: P.O. Box 125, Key Laboratory of Modern Precision Agriculture System Integration Research, China Agricultural University, 17 Tsinghua East Road, Beijing, 100083, P. R. China, Tel: +86-10-62736741, Fax: +86-10-62736746, Email: pac@cau.edu.cn*

Abstract: With the concept of precision agricultural was proposed, the research of agricultural automatic vehicle was paid more attention to in the world. The fundamental element of machine vision based agricultural automatic navigation system was presented in this paper. It includes path finding, location & path tracing and motion controlling. The image processing, automatic control and sensor fusion techniques are the key for autonomous vehicle systems.

Keywords: Agricultural vehicle, Machine vision, Guidance, Automatic control

1. INTRODUCTION

With the research and practice of precision agriculture nowadays, new mode of agricultural production and the application of new technology promote the development of agricultural vehicle. In the 1990s, automation, information and intelligence which have been promoted rapidly by the computer and information collecting technology play the key role in vehicle automation of modern agriculture. With the rapid advancement in

Liu, B., Liu, G. and Wu, X., 2008, in IFIP International Federation for Information Processing, Volume 258; Computer and Computing Technologies in Agriculture, Vol. 1; Daoliang Li; (Boston: Springer), pp. 659–666.

electronics, computers and computing technologies, the intelligent agricultural vehicle with autonomous navigation system has been used for many field operations such as planting, fertilizing, tillage, and harvesting.

According to the navigation methods of agricultural vehicle guidance systems, various navigation technologies include: (1) mechanical navigation (2) GPS (Global Positioning System) navigation (3) machine vision navigation (Han et al., 1990).

GPS-based guidance system is a kind of absolute navigation way which is based on some known position points from GPS receiver; in contrast, machine vision guidance is a relative position and heading method based on tracking forward directrix. Compared with GPS-based guidance, machine vision based guidance has great potential for implementation of agricultural vehicle navigation system. Real-time characteristic for detecting obstacle and high cost performance are the superiority of machine vision guidance. As a result, machine vision guidance has been an interest for agriculture researchers.

2. KEY TECHNIQUES OF MACHINE VISION BASED AGRICULTURAL VEHICLE GUIDANCE SYSTEM

The intelligent navigation system based on machine vision has three key elements: path finding, location & path tracing and motion controlling, as shown in Fig. 1. Firstly, path finding is the key in achieving accurate control of the vehicle. Image information collected by image sensor from crop row structure is used for detecting navigation features and finding out the effective navigation course (directrix). Agricultural vehicle is mainly be used in nature environment, therefore image information process technology must have the features of real-timing, robustness and adaptability for nature light. Secondly, according to the result of path finding and position & heading information from other sensors, location & path tracing is to locate the relative position and heading of agricultural vehicle in crop rows. Path planning is then providing an offset that is used with a steering gain to directly control the wheel position. Finally, motion controller is to adjust oriented angle of front wheel according to the navigation planning information. Then agricultural vehicle can be guided in crop rows autonomously by a machine vision-based navigation system.

In a vision-based vehicle guidance system, finding guidance information (path finding) from crop row structure is the key in achieving accurate control of the vehicle operation. A number of image information processing techniques have been used in path finding. In nature environment, the fact that farmland sense is complicated and non-structural make image processing more difficult. So control parameters which are offered by the machine vision navigation system cannot keep up with the vehicle's status shift. The final result is that validity of navigation control will be affected.

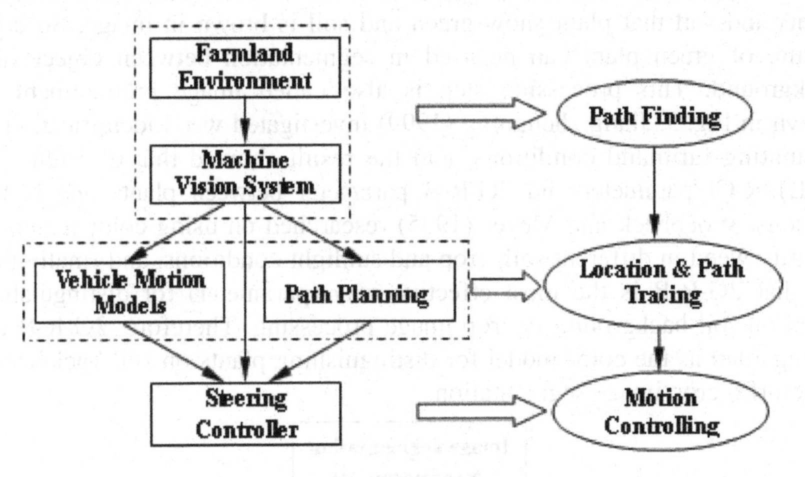

Fig. 1. Key elements of intelligent navigation system

3. VISION-BASED IMAGE PROCESSING TECHNOLOGY

The main target of image processing in the machine vision navigation is to perceive guidance course and to detect obstacle, further, to locate relatively. Path finding and surroundings Perceiving of the vision-based guidance system can be defined as four processing, as shown in Fig. 2.

The key of navigation system research based on machine vision is to isolate target from background, yet traditional edge-detection method is unsatisfactory in crop images processing. One of the problems encountered in outdoor image acquisition is the unevenness of the lighting conditions, in time and space. The variations of the sun elevation and the nebulosity contribute to change the global illumination of the scene observed by the camera. Another problem is that in farmland images the most outline feature of aqueduct, ridge of field and crop is lost. The edge of objects cannot be detected by using traditional processing method. For this reason, a new edge-detection method based on color features has been developed by researchers.

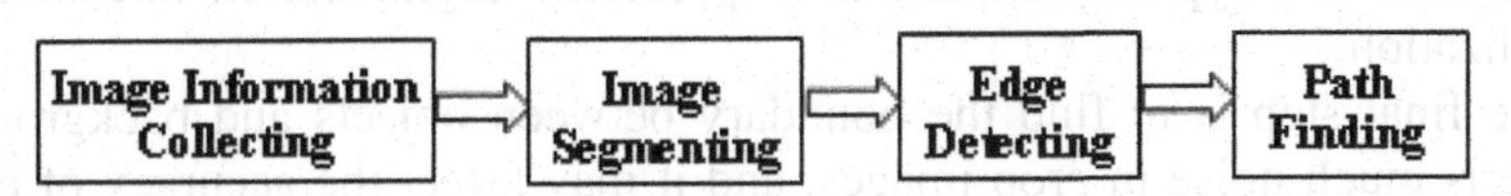

Fig. 2. Four steps of image processing in vision-based guidance

3.1 Color features based segmentation

Crop images from a standard RGB sensor include the most direct information of color features. There are the obvious difference between plants and soil that plant show green and soil is brown in image. So color feature of green plant can be used in segmentation between objects and background. This processing step is also called image pretreatment, as shown in Fig. 3. Jiang zhengrong (1999) investigated weed identification by simulating farmland conditions, and the result showed that difference of (G-B)/|R-G| parameter and 2G-R-B parameter between plants and soil is obvious. Woebbeck and Meyer (1995) researched on using color feature to identify weed in different soil, crop and sunlight conditions, and finally they led that 2G-R-B is the most effective way (parameter) for distinguishing weed on soil background in crop image processing. Therefore, 2G-R-B can be regarded as the color model for distinguishing plants on soil background in colorful crop image segmentation.

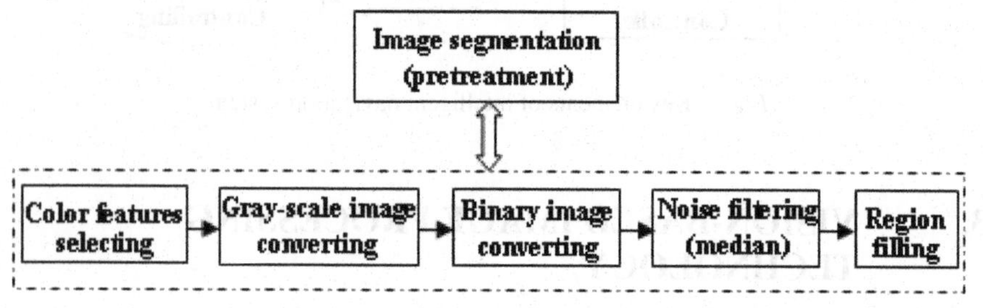

Fig. 3. Image segmentation

The first step is to manipulate the image in such a way that the living plant tissue is emphasized on Green color feature in comparison with all other objects in the image. This is done by first dividing the color image into its red, green and blue channels. Subsequently, an indicator value for living plant tissue is computed as: Indicator = 2×Green – Red – Blue.

To visualize the indicator values as a grey-scale image, they may be mapped to the range of 0-255 by simple linear mapping. This image is well suited for subsequent convert gray-scale image to binary image.

The second step is to convert gray scale image into a binary image. It is also called image thresholding. Selecting appropriate threshold is very important for separating crop pixels from the background. However, it is very difficult that conforming threshold for binary image in real-time navigation. The typical solution is to predefine segmentation threshold in initialization.

The final step is to find the boundary between objects and background. There is much noise in crop images, and it may affect the accuracy of edge detection algorithm. The median filter is well suited to remove spikes and thin lines from the pre-treated image. So the median filter can be applied on the final step. Moreover, there are some holes in bi-level images. Filling holes is an effective method to remove the holes.

Objective region is separated from the background after pre-treated image.

3.2 Edge detection

Followed by pretreatment, Morphological algorithm can be introduced to detect the edge of objective region. Edge detection based on morphology can be regard as:

$$\beta (A) = A-(A\Theta B)$$

Where

β (A) is the edge of an objective region A.

B is the suitable structural element. The structure 3×3 element can be used.

3.3 Guidance directrix extraction

After the edge of objective region detected, guidance directrix can be gained through image processing. Usually, Hough transform is adapted. The Hough transform is widely used for localization of linear objects in images (Sonka et al., 1993).

Hough transform is a kind of linear description way in which point (x, y) in Descartes coordination system is converted into a sine curve (ρ=xcosθ+ysinθ) in polar coordinate system. Subsequently, a number of points in Descartes coordination system will be converted into the corresponding curves. The intersection of these curves in a certain range can be used to find out the best fitting line. According to the transform equation, every point in image can be represented by the parameters (ρ,θ) of polar coordination system, and the maximum value of the intersections in polar coordination system indicates that a line has been found.

4. LOCATION AND TRAJECTORY TRACKING

Location and Trajectory tracking are important for agricultural vehicle control. Most guidance operations require that the vehicle follow some nominal trajectory or directrix. Local information can be the source of directrix and sensed directly. These include crop rows, swath edges, tilled/untilled boundaries. In parallel swathing, the directrixis parallel lanes to some prior path. The directrix might also be defined by a desired course from a map or script. Many research efforts in guidance have dealt with guidance in straight lanes and thus the role of the navigation planner is much simplified. For curved row guidance, the navigation planner must consider the sensor information and vehicle motion to navigate in the desired course. The navigation planner also plays an important role in autonomous operation, providing different machine 'behaviors' based on field conditions (Reid et al., 2000).

A finite state approach was developed by CMU in their autonomous windrower (Hoffman, 1996). Primary states of the machine triggered different behaviors or operations. Some basic states defined which side of the harvester was uncut crop, identified the proximity of a turn, and the implemented of turning functions. In each state of operation the machine performed one or more behaviors. For example, harvesting a rectangular field of alfalfa in a simple serpentine pattern required combinations of the following states with associated behaviors (Reid J.F. et al., 2000):

- Cut on right
- Near right turn
- Turn right
- Cut on left

5. AGRICULTURAL VEHICLE CONTROL TECHNOLOGY

The research on agricultural vehicle control technique is focus on designing the steering controller systems which is the main purpose. The steering controller is the actuator that converts a control signal from a feedback controller to an appropriate mechanical adjustment in steering angle. Agricultural equipment often operates on unprepared, changing and unpredictable terrain. In the case of automatic or autonomous operation, steering controllers should be able to provide appropriate steering actions in response to the variations in equipment operation states, traveling speed, tire cornering stiffness, ground conditions, and many other parameters influencing steering dynamics.

Since most modern agricultural vehicles employ some form of hydraulic steering system, various factors such as electrohydraulic (E/H) steering unit, dynamic feature of vehicle, road conditions, and vehicle speed and so on should be taken in account adequately in designing steering controller (Reid et al., 2000).

A classical model based steering controller was developed by Stombaugh et al. (1998) for high speed agricultural tractor. They found that the steering controller must compensate for the dynamics of both the vehicle and the steering system when these two systems were in the same frequency range.

Benson et al. (1998) designed a PID steering controller in frequency domain for an agricultural tractor guided by a GDS. Its closed loop transfer function was obtained experimentally. Test results showed that steering controller played an important rule in achieving satisfactory automated guidance of an agricultural tractor.

Wu et al. (1998) developed a methodology for designing E/H steering PID controllers for agricultural vehicles. In their approach, a linearized dynamic model of the steering system was used to design the controller.

In the research on controlling system, many problems need to be solved. Those problems include how to modify electrohydraulic controller, how to keep up the stability of electrohydraulic control system, how to deal with sideslip of vehicle's turning in high speed, how to realize real-time control, how to solve deadband and nonlinear question of operation system, how to control automatically agriculture vehicle in upland.

6. CONCLUSIONS

In study of automatic navigation system of agriculture based on machine-vision, there still have several questions to be solved as followed.

Firstly, the real-time characteristic of image processing method should be fulfilled. When sense image collected in nature environment become more and more complicate, the time for processing image will increase greatly. If the speed of image process is even low and nonstructural environment is unknown or changeable, not only the traditional detecting-modeling-planning way but also detecting-action control way will not realize effective control.

Secondly, multi-sensors-based information fusing technique will play an important role in intelligent guidance system. Although machine vision navigation has high flexibility, its navigating accuracy is affected by environment and surroundings features. Through fusing variety of information from GPS and other sensors, accuracy and dependability of navigation will be advanced. Furthermore, multi-sensors-based information fusing technique can solve special problems such as without effective surroundings feature for turning around in the end of farmland.

Finally, effective control algorithm is crucial. In the complicate environment, there has nonlinear and random disturbance in moving vehicle. Because intelligent controlling method doesn't rely on mathematical models, it has good controlling effect and robustness. In the environment which can be known or predicted, investigators can establish effective fuzzy ruler by experiment, simulation and expert experience. Yet persons have no way to predict all the conditions in complicated environment, controlling system of guidance system should have the capability of self-learning and self-adapting.

ACKNOWLEDGEMENTS

This paper is supported by the national 863 projects: Control Technique and Product Development of Intelligent Navigation of Farming Machines (2006AA10A304).

REFERENCES

Benson, E. R., Reid, J. F., Zhang, Q., Machine-vision based guidance system for agricultural grain harvester using cutedge detection [J]. Biosystem Engineering, 2003, 86(4): 389–398.

Benson, E., Stombaugh, T., Noguchi, N., Will, J., Reid, J.F., 1998. An evaluation of a geomagnetic direction sensor for vehicle guidance in precision agriculture applications. ASAE Paper 983203. ASAE. St. Joseph, MI.

Han, S., Zhang, Q., Reid, J.F., A guidance directrix approach to vision-based vehicle guidance systems [J]. Computers and Electronics in Agriculture, 2004, 43: 179–195.

Jiang Zhengrong, Implementation and Application of Computer Technique for Identification of Weed [J]. Weed science, 1999, 4: 2–5. (in Chinese)

Reid, J.F., Zhang, Q., Noguchi, N., Agricultural automatic guidance research in North America [J]. Computers and Electronics in Agriculture, 2000, 25: 155–167.

Søgaard, H.T., Olsen, H.J., Determination of crop rows by image analysis without segmentation [J]. Computers and Electronics in Agriculture, 2003, 38: 141–158.

Stombaugh, T., Benson, E., Hummel, J.W., 1998. Automatic guidance of agricultural vehicles at high field speeds. ASAE Paper 983110. St. Joseph, MI.

Sun Yuanyi, Zhang Shaolei, Li Wei, Guidance lane detection for pesticide spraying robot in cotton fields [J]. Journal of Tsinghua University (Science & Technology), 2007, 47(2): 206–209. (in Chinese)

Tian Haiqing, Ying Yibin, Zhang Fangming, Development of Automatic Control Technique for Agricultural Vehicle Guidance [J]. Transactions of The Chinese Society of Agricultural Machinery, 2005, 36(7): 148–152. (in Chinese)

Leemans, V., Destain, M.-F., Application of the Hough Transform for Seed Row Localisation using Machine Vision [J]. Biosystems Engineering, 2006, 94(3): 325–336.

Woebbecke, D.M., Meyer, G.E., Von Bargen, K. et al. Color indices for weed identification under various soil, residue, and lighting conditions [J]. Trans of the ASAE, 1995, 38(1): 259–269.

Wu, D., Zhang, Q., Reid. J.F., Qiu, H., Benson, E.R., 1998. Model Recognition and Simulation of an E/H Steering Controller on Off-Road Equipment. In: Nair, S.S., Mistry, S.I. (Eds). Fluid Power Systems and Technology 1998, ASME, New York, pp. 55–60.

Zhang Fangming, Ying Yibin, Review of Machine Vision Research in Agricultural Vehicle Guidance [J]. Transactions of The Chinese Society of Agricultural Machinery, 2005, 36(5): 133–136. (in Chinese)

A BIOECONOMIC MODEL BY QUANTITATIVE BIOLOGY TO ESTIMATE SWINE PRODUCTION

Hui Yuan [2], Surong Xiao [1], Qiujuan Wang [2], Keliang Wu [2,*]
[1] *Shanghai Xiangxin Livestock, Ltd. 201302*
[2] *College of Animal Science and Technology, China Agricultural University, Beijing, 100094*
[*] *Corresponding author, Tel: +86-10-62733445, Email: liangkwu@cau.edu.cn*

Abstract: A bioeconomic computer model was constructed to simulate biological and economic inputs and outputs for life cycle swine production. Parameters and relationships used in model were developed and verified by comparison with experimental results in the literature. The bioeconomic model was constructed by several modules such as growth and development, pregnancy, lactation, and replacement gilt etc. The result is: (1) the bioeconomic computer model was efficient way to describe pig production system and research factors' effect and their interactions. (2) Traits in the model were: oestrus traits; mature weight and feed requirements of sows; longevity of sows; litter size; growth rate and daily feed intake of young pigs and fatteners; mortality rate of pigs. (3) Sow longevity is 2.12 years and yearly culling rate is 47.20%. Yearly farrowing sow is 2.08 and total numbers of farrowings per gilt is 4.62. (4) Average litter size total born is 11.23, litter size born alive is 10.40, litter size weaned is 8.80.

Keywords: Quantitative Genetics, Economics, Computer Simulation, Swine, System Analysis

1. INTRODUCTION

A deterministic computer model is a theory and research methodology (Emmans and Fisher, 1986) in many sense, which integrated multi-discipline, such as nutrition, genetics, physiology, meats and economics to describe a production system of livestock.

Yuan, H., Xiao, S., Wang, Q. and Wu, K., 2008, in IFIP International Federation for Information Processing, Volume 258; Computer and Computing Technologies in Agriculture, Vol. 1; Daoliang Li; (Boston: Springer), pp. 667–675.

The methodology of simulation pig production by computer model was named quantitative biology of the pig (Kyriazakis, 1999). That is pig production biological processes can be described by model methodology.

Model methods were applied in simulation livestock production system. For example, De Hatog et al. (1995) reviewed the pig growth models. De Vries (1989) estimated economic values of traits in pig breeding by computer simulation model. The swine simulation model were reported by other researchers (Emmans, 1995; Poma et al., 1991; Schinckel et al., 1996; Whittemore, 1995).

The objective of present study is to construct a pig production model that are in agreement with Chinese farm by means of quantitative biology of pig. Dealing with the biological processes which underlie pig production, a computer simulation model was built to describe the production procedures of mating, pregnancy, farrowing, lactation and growth production.

2. MODEL DESCRIPTION

2.1 Condition and strategy

According to current swine production process of Chinese farm, the methodology of swine quantitative biology is applied to build different stage in swine production system that is mating, pregnancy, farrowing, lactation, growth and development. The different biological procedures were described by mathematical formula and computer program written in FORTRAN–90.

When a computer model, which was in agreement with current situation and production level of Chinese swine production, was built, some features should be taken into account:

First, the model is on the basis of the production performance of the world's main breed (such as Large White and Landrace).

Second, the model can reflect production state of special genetic level exactly.

Third, for the sake of the maximum genetic potential of swine and Chinese actual situation in swine industry sufficiently, the ad lib feeding system is implemented at the stage of growth, and controlled feeding system is executed at the stage of replacement gilts, sow.

Final, the value of the model's biological parameters is determined by some composite factors including the production consequences of Chinese swine farm and the results of experiment research in the literature.

2.2 Modules construction

The different modules, which include mating, pregnancy, farrowing, lactation, growth and development, were constructed according to swine life cycle of production systems.

2.2.1 Growth

The growth is one of the most important stages of swine, directly relating with the level and the production benefit of pig production.

The growth module is based on the growth and development of swine's body components, namely this module is respectively constructed according to the growth and development scale of the four body components (water, protein, lipid and ash). The equations (WU Keliang, 2004) are used to predict the four body compartment and live body weight.

Liveweight (LWT), $LWT = 217.498\, e^{-4.4134\, e^{-0.0113\, xday}}$ (1)

Body protein retention (Pr), Pr = 0.304 EBWT0.8288 (2)

Body lipid retention (Lipid), Lipid = 0.00647 EBWT1.8422 (3)

Body water gain (Wa), Wa = 1.984 EBWT0.6921 (4)

Body ash reserves (Ash), Ash = 0.0238 + 0.0651 EBWT (5)

The live weight (LWT) of growing swine is represented by Equation Gompertz, xday represents swine's growth age in days. EBWT is the empty body weight which is calculated from two approaches as follows:

□ The EBWT takes up 95% of the live weight, i.e. EBWT = 0.95LWT.

□ The EBWT is the algebraic sum of the four body components as described in the following equation: EBWT = Pr + Lipid + Wa + Ash.

At the growth stage, growing pigs are fed ad libitum. And the requirement of growth is determined by deposition rate of body protein and lipid, which is calculated from the following equation:

$$MEI = MEm + bp \bullet \Delta P + bF \bullet \Delta F \quad (6)$$

Where MEI is daily metabolizable energy intake (Mcal/d); MEm is daily maintenance requirement (Mcal/d); bp and bF is energy costs of protein and lipid deposition (Mcal/d), respectively.

The energy value of lipid and protein are respectively 9.5 Mcal/kg and 5.7 Mcal/kg (den Hartog and vander Peet-Schwering, 1995). The feed conversion rate from metabolizable energy to lipid and protein are respectively 0.84 and 0.54, namely the energy costs of protein and lipid deposition per kilogram are 10.56 Mcal and 12.84 Mcal, respectively.

Consequently, the value of bp and bF determined by that research are respectively 10.56 Mcal ME/kg and 12.84 Mcal ME/kg.

The maintenance requirement can be concluded from the following equation of maintenance requirement model presented by Tess (1983):

$$ME_M = 0.126 Mcal \bullet LBM^{0.83} \quad (7)$$

Where MEm = The maintenance requirement of metabolizable energy, LBM = lean body mass.

2.2.2 Replacement gilt

The performance measurement for the growing pigs is terminated until their weight reach 90 kilogram. And then, after culling, the eligible gilts enter into replacement gilts group. The replacement gilts are generally restrictedly fed, which feed supply is 2.3 kg/d, and the rate of body deposition of body protein and lipid is 1:1 and the conversion ratio from feed ME to body components are 0.54 and 0.84 respectively.

The replacement gilts' breeding age in days is determined by two components: age at puberty and estrus period. The initial value of mating day in the model is the second estrus.

2.2.3 Gestation

The growth and development of fetus is key important for swine production. But there are not so many research reports on the growth and development of fetus. The fetus weight gain are more slowly at early stage of pregnancy (0–80 day), but more quickly at late stage of pregnancy (80–115 day). It is said that approximately 60% of weight gain of fetus is in the late stage of pregnancy.

Gestation simulation in the model is based on the following equation (Noblet, 1990).

fw = exp(9.095-17.69 * exp(-0.0305*gday) + 0.0878 * tb) (8)

Where fw = the weight of fetus development; gday = pregnancy time (day); tb = litter size.

In this model, the growth and development of fetus in primiparous or multiparous sow is the same. The weight of embryonic membrance and amnionic fluid increase constantly with fetus's growth and development stages. It is studied that the weight of embryonic membrance and amnionic fluid is about 10 kg when the sow is delivering (Tess, 1983; Noblet, 1990).

Feed intake and feeding management of pregnant sow is adopt control feeding, relating with Chinese farm. The detail feeding schemes are shown as follows: 2.3 kg feed was available in early stage of pregnancy (0–84 day), 3.4 kg in the late phase of pregnancy (85–114 day), the energy (Mcal/kg) of feed is 2.65 megacalorie or 11.09 megajoule (MJ/kg), and the contents of protein is 12.0%.

2.2.4 Lactation

As known, more nutrient substance is essential for the lactating sow. King (1994) applied isotope dilution method to study the milking production of sows. The maximum daily milk potential of sows was likely to overrun 10 kg and the nutrient substance passing udder via blood was almost 7 kg. The simulation of milk yield in this model adopt equation of Whittemore and Morgan (1990):

my = a * exp(-0.025*lday) * exp(-exp(0.5-0.1*lday)) (9)

Where my = milking yield; a = a variable parameter which is from 18 to 30 and its value is determined by sow's genetic background; In the model, the value of a is 24; lday = lactating time in days.

The milk yield is affected by parity number, litter size and stage of lactation, especially parities between number one and two. The milking production of primiparous sow is 80% of production of parous sow.

The nutritional requirement of lactating sow is divided into the lactating requirement and the maintenance requirement of sow.

$$ME_l = ME_M + b_{MK} * \Delta MK \quad (10)$$

Where MEI = intake of metabolizable energy of sow; MEm = metabolizable energy for maintenance requirement; bMK = the ratio of feed conversion to milk; □MK = variation of milk yield.

2.2.5 Levels of genetic trait and parameters of swine production

The confirmation of pig production parameters and performance is difficult, because of too many links and complicated factors. The performance of swine herd is related not only with genetic structure of swine, but also with other factors, for instance, nutrition environmental factors and management pattern and decision.

The fertility traits and culling rates listed in table 1. The confirmation of production parameters in model is based on the followings: (1) the production performance form the JiangXi breeding swine farm, YunNan breeding swine farm and ShunXinLong Breeding Company in Beijing. (2) the surveyed data from other areas in China, for example, the price of semen or insemination. (3) the reported results in the literatures.

Table 1. Fertility traits and culling rates of sows per cycle number

	Parity number									
	1	2	3	4	5	6	7	8	9	10
IWO (d)	–	12	9	8	8	7	7	7	7	7
Marginal Culling rate (%)	–	14.7	16.1	17.5	18.9	20.3	22.4	24.5	26.6	30.1
Litter size born alive	9.4	10.1	10.6	10.9	11.0	10.9	10.8	10.7	10.6	10.5

* IWO is interval wean-oestrus (days)

3. RESULTS

3.1 Results for growth performance

The simulation performance of growth and development is listed in table 2. As shown in table 2, average weight of newborn piglets is 1.25 kg, the weight of weaned piglet at 28 days is 7.40 kg. the body weight of feeder pig at 55 days is to reach 25 kg. It takes 143.6 days for pigs to reach 90 kg or 154.9 days to reach 100 kg. The average daily growth (ADG) during the whole process reaching 90 kg and 100 kg is respectively 626.74 g and 645.58 g.

Table 2. Results for production performance in growth stage

Ages in days	LWT (kg)	LBM (kg)	Protein (kg)	Lipid (kg)
0	1.25	1.2	0.1595	0.0145
wean (28 d)	7.40	6.1058	1.0175	1.0945
55 d	25.00	12.01	2.1725	2.6133
143.6 d	90.00	60.86	12.1405	23.46
154.9 d	100.00	66.28	13.3676	28.73

3.2 Sow's reproduction performance

Simulated performance of the sow herd is listed in table 3. the farrowing per sow yearly is 2.08, the total number of born on average is 11.23, and litter size weaned is 8.82. The longevity of sow in breeding herd is 2.12 years, the total number of farrowings for sows was 4.62, the culling rate of sow per year is 47.20%. The outputs of model about sows are in agreement with actual production situation in China.

Table 3. Simulated average performance of sows

Trait	Average value
Average longevity of sow (year)	2.12
Cullings Sow per year (%)	47.20
Total born	11.23
Number born alive	10.40
Litter size weaned	8.82
Litter per sow per year	2.08
Total no. of farrowings for gilts	4.62
Mortality in suckling (%)	15.23

Table 4. Simulated distribution of farrowings

Distribution (%)	Cycle number									
	1	2	3	4	5	6	7	8	9	10
	20.00	17.27	14.55	12.27	10.00	8.18	6.36	5.00	3.64	2.73

The simulated distribution of life cycle number is in table 4. The reproduction performance with sow parity is to reach a peak among parity 3–5, so the reasonable distribution of the parity is the key importance for breeding herd.

3.3 The growth and development of fetus

The effect of stage of pregnancy on weight of fetus is shown in figure 1. as shown in figure 1, the weight gain of fetus is slowly in early stage of pregnancy (0–84d), but almost 2/3 weight gain in late 30 d of pregnancy.

3.4 Milk yield

The milk yield of lactating sow is shown in figure 2. the peak of milk yield is approximately 12 kg at 17 days in lactation. Lactating sows seldom gain weight during lactation and generally lose weight. Sow body weight and chemical component changes during parity 1 and 5 listed in table 5. The body weight and content of body lipid for primiparous sow last permanently, but lose in some scale for sows.

Table 5. Sow bossdy weight and chemical component changes during parity 1 and 5

Days in lactation	Parity 1			Parity 5		
	lipid	LBM	LWT	lipid	LBM	LWT
1	33.68	118.98	160.70	39.80	144.97	194.51
10	34.13	120.94	163.24	39.69	144.49	193.88
20	33.98	120.29	162.39	39.27	142.73	191.59
28	33.86	119.79	161.75	38.94	141.36	189.79

4. DISCUSSION

The model used required a lot of information. The equations in model were not same according to researched works. However, the model can be evaluated in several ways to judge its suitability in fulfilling its designed purpose. First, the model was constructed in terms of biological process of performance at different stages in production system, so the detailed description of individual module of the model provides a verification of the model. Second, model output or the output of subunits of the model can be validated directly, by comparing simulated results with experimental results not used to construct the model.

The model of swine production is foundation works. The further research, for example, the design of optimum breeding programs, can be fulfilled.

REFERENCES

Amer, P. R., G. C. Fox and C. Smith 1994 Economic weights from profit equations: appraising their accuracy in the long run. Anim. Prod. 58:11-18.

Bourdon, B. M. 1998 Shortcoming of current genetic evaluation systems. J. Anim. Sci. 76:2308-2323.

Cameron, N. D. 1997 Selection indices and prediction of genetic merit in animal breeding. CAB INTERNATIONAL Wallingford, UK.

Chen, Y. C. 1998 Pig industry in China. Proc. of intern. conf. of pig production. Beijing 1-5.

Chen, R. S. 1996 Pig production in China. In: Pig Production edited by M. R. Taverner and A. C. Dunkin. Elsevier Science Publishers.

Cleveland, E. R., P. J. Cunningham and E. R. Peo, Jr. 1982 Selection for lean growth in swine. J. Anim. Sci. 54(4):719-727.

Clutter, A. C. and E. W. Brascamp 1998 Genetcs of performance traits. In "The genetics of the pig" edited by M. F. Rothschild and A. Ruvinsky. CAB INTERNATIONAL.

De Hartog, L. A. and C. M. C. van der Peet-Schwering 1995 The use of growth models for piga in practice. Pig News and Information. 16(2):51N-53N.

De Vries, A. G. 1989 A model to estimate economic values of traits in pig breeding. Livest. Prod. Sci. 21:49-66.

De Vries, A. G. 1989 A method to incorporate competitive position in the breeding goal. Anim. Prod. 48:221-227.

Emmans, G. C. 1995 Ways of describing pig growth and food intake using equations. Pig News and Information. 16(4):113N-116N.

Emmans, G. C. and I. Kyriazakis 1999 Growth and body composition. In: Kyriazakis, I. (eds). A Quantitative Biology of the Pig. CAB Publishing. 181-197.

Fredeen, H. T. 1980 Pig breeding: current programs vs. Future production requirements. Can. J. Anim. Sci. 60: 241-251.

Gibson, J. P. and J. W. Wilton 1998 Defining multiple-trait objectives for sustainable genetic improvement. J. Anim. Sci. 76:2303-2307.

Harris, D. L. 1998 Livestock improvement: Art, Science, or Industry? J. Anim. Sci. 76:2294-2302.

Herrero, M. R. H. Fawcett and J. B. Dent 1999 Bio-economic evaluation of farm management scenarios using integrated simulation and multiple-criteria models. Agricultural Systems 62:169-188.

King, R. H., J. LeDividich and F. R. Dunshea 1999 Lactation and neonatal growth. In: Kyriazakis, I. (eds). A Quantitative Biology of the Pig. CAB Publishing. 155-180.

Kyriazakis, I. 1999. A Quantitative Biology of the Pig. CAB Publishing.

Mclaren, D. G., D. S. Buchanan and J. E. Williams 1987 Economic evaluation of alternative cross-breeding systems involving four breeds of swine. I. The simulation model. J. Anim. Sci. 65:910-918.

Mclaren, D. G., D. S. Buchanan and J. E. Williams 1987 Economic evaluation of alternative crossbreeding systems involving four breeds of swine. II. System Efficiency. J. Anim. Sci. 65:919-928.

Pomar, C., D. L. Harris and F. Minvielle 1991 Computer simulation model of swine production systems: I. Modeling the growth of young pigs. J. Anim. Sci. 69:1468-1488.

Pomar, C., D. L. Harris and F. Minvielle 1991 Computer simulation model of swine production systems: II. Modeling body composition and weight of female pigs, fetal development, milk production, and growth of suckling pigs. J. Anim. Sci. 69:1489-1502.

Schinckel, A. P. and C. F. M. de Lange 1996 Characterization of growth parameters needed as inputs for pig growth models. J. Anim. Sci. 74:2021-2036.

Shields, R. G., Jr., D. C. Mahan and F. M. Byers 1983 Efficiacy of deuterium oxide to estimate body composition of growing swine. J. Anim. Sci. 57(1):66-73.

Smith, C., D. E. Dickerson, M. W. Tess and G. L. Bennett 1983 Expected relative responses to selection for alternative measures of life cycle economic efficiency of pork productiion. J. Anim. Sci. 56(6):1306-1314.

Taverner, M. R. and A. C. Dunkin. 1996 Pig production World Animal Science, C10. Elsevier Science Publishers.

Tess, M. W., G. L. Bennett and G. E. Dickersion 1983 Simulation of genetic changes in life cycle efficiency of pork production. I. A bio-economic model. J. anim. Sci. 56(2):336-353.

Tess, M. W., G. L. Bennett and G. E. Dickersion 1983 Simulation of genetic changes in life cycle efficiency of pork production. II. Effects of components on efficiency. J. anim. Sci. 56(2): 54-379.

Van der Peet-schwering, C. M. C., L. A. den Hartog and H. J. P. M. Vos 1999 Application of growth models for pigs in practice – Review. Asian-Aus. J. Anim. Sci. 12(2):282-286.

Weatherup, R. N., V. E. Beattie, B. W. Moss, D. J. Kilpatrick and N. Walker 1998 The effect of increasing slaughter weight on the production performance and meat quality of finishing pigs. Anim. Sci. 67:591-600.

Webb, A.J. 1996 Future challenges in pig genetics. Pig news and Information. 17(1):11-16.

Webb, A. J. 1986 Selection regime by production system interaction in pig improvement: A review of possible csuses and solutions. Livest. Prod. Sci. 14:41-54.

Whittemore, C. T., J. C. Kerr and N. D. Cameron 1995 An approach to prediction of feed intake in growing pigs using simple body measurements. Agri. Syst. 47: 235-244.

Whittemore, C. T. 1986 An approach to pig growth modeling. J. Anim. Sci. 63:615-621.

Whittemore, C. T., J .B. Tullis and G. C. Emmans 1988 Protein growth in pigs. Anim. Prod. 46: 437- 445.

Wu, Ch. X. 1998 Pig breeding in China. Proc. of intern. conf. of pig production. Beijing 6-8

Zhang, H. F., Zhang Z. Y. 1998 Today and yesterday of swine industry in China & sustainable development in strategy. Proc. of intern. conf. of pig production. Beijing 703-705.

AN INTEGRATED APPROACH TO AGRICULTURAL CROP CLASSIFICATION USING SPOT5 HRV IMAGES

Chang Yi[1,2,*], Yaozhong Pan[1,2], Jinshui Zhang[1,2]

[1] *College of Resources Science and Technology, Beijing Normal University, Beijing, China, 100875*

[2] *State key Laboratory of Earth Surface Processes and Resource Ecology System (Beijing Normal University), Beijing, China, 100875*

* *Corresponding author, Address: 05 Shuo, College of Resources Science and Technology, Beijing Normal University, 19 Xinjiekouwai Street, Beijing, 100875, P. R. China, Tel: +86-13581929372, Email: yichang531@ires.cn*

Abstract: An integrated method that incorporates the advantages of per-parcel and per-pixel approaches as well as spectral and spatial characteristics was proposed for crop classification of a typical agricultural area in south-east China using SPOT5 HRV data. The co-occurrence texture was employed to evaluate the heterogeneity of the image data. The average parcel textures determined each parcel defined by the crop boundaries to be classified whether on a per-parcel or per-pixel basis. The optimal threshold in the span of texture ranges was detected by trend analysis, which assigned the proportions of each approach in the integration, thus to produce the best integrated classification. It was suggested that this integrated approach can be effectively implemented to produce crop classification maps with higher accuracy from satellite images of medium and high spatial resolution in a complex agricultural environment, where both homogeneous and heterogeneous crop fields occur side by side.

Keywords: per-parcel, per-pixel, remote sensing, agricultural crop classification, SPOT5

1. INTRODUCTION

In recent years, advances in satellite imaging technology have boosted multiple spatial applications in which land cover information is an essential

Yi, C., Pan, Y. and Zhang, J., 2008, in IFIP International Federation for Information Processing, Volume 258; Computer and Computing Technologies in Agriculture, Vol. 1; Daoliang Li; (Boston: Springer), pp. 677–684.

prerequisite (Lo and Choi, 2004). For the agricultural application, it requires a quantitative processing of digital images with high accuracy and reliability.

One crucial technique of crop mapping from remotely sensed data is the automated image classification, which usually operates on a per-pixel basis to categorize pixels separately into one of the pre-determined classes according to their spectral characteristics. However, as for agricultural applications, groups of pixels that represent the same crop type may not necessarily have the same spectral information due to the variation in soil moisture conditions, nutrient limitations or pests and diseases (De Wit and Clevers, 2004). Also, boundaries that cross pixels are an additional problem, as the spectral information of the pixel is then a combination of the reflectance from two or more land cover types (Smith and Fuller, 2001). Thus, doubts cast on the reliability of per-pixel classification which often resulted in misclassification and then a speckled appearance.

With the recent development of 'integrated' GIS, a per-parcel approach has been more and more introduced in mapping agricultural landscape. The basic idea behind this method is that agricultural field boundaries integrated with remotely sensed data divide the image into homogeneous units of image pixels, which enables pixels contained within a parcel to be processed in coherence. A parcel-based representation is most appropriate for mapping agricultural land cover by the use of crop field boundaries which can eliminate the classification errors due to the within-field spectral variability and mixed-pixels along the boundaries of fields (Dean and Smith, 2003). However, it is based on the assumption that only one crop type dominates one field, which is not always true in reality. Problems will therefore occur when mosaics of crop types distribute within one parcel structure and represent a heterogeneous landscape (Dean and Smith, 2003).

The aim of this study is to develop an integrated classifier which can switch between per-parcel and per-pixel classification to meet the accuracy requirement for both homogeneous and heterogeneous landscapes.

2. STUDY AREA AND DATA

2.1 Study area

The selected area is located at Xuzhou city, which is situated in Jiangsu Province in the southeast of China, measuring approximately 20km^2 (34° 20'46"N–34°22'57"N; 117°37'56"E–117°41'11"E) (Fig. 1). The study area is representative of the agricultural regions of the southeast level plain of China. The main crops grown in the area include rice, corn, soybean and

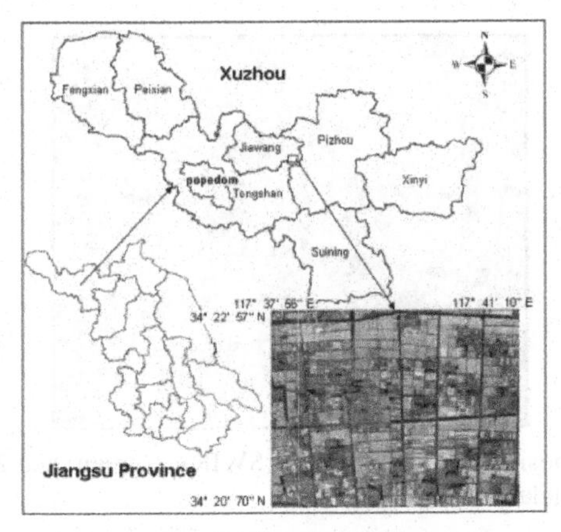

Fig. 1. The study area

mulberry. The structure of landscape is characterized by both homogeneous large rectangle rice fields and fields that are quite heterogeneous due to the irregularly scattered multiple crop types such as corn, soybean and mulberry.

2.2 Data

Optical satellite imagery obtained by SPOT5 HRV has been used. The optimal acquisition periods of optical satellite imagery are determined by the phonological characteristics of the main crops (Fig. 2). The SPOT5 HRV images used for classification were acquired on 17 August, 2006. All images were cloud free and of good quality. A subset (500×400 multi-spectral) and (2000×1600 panchromatic) was extracted from the full scene for the study.

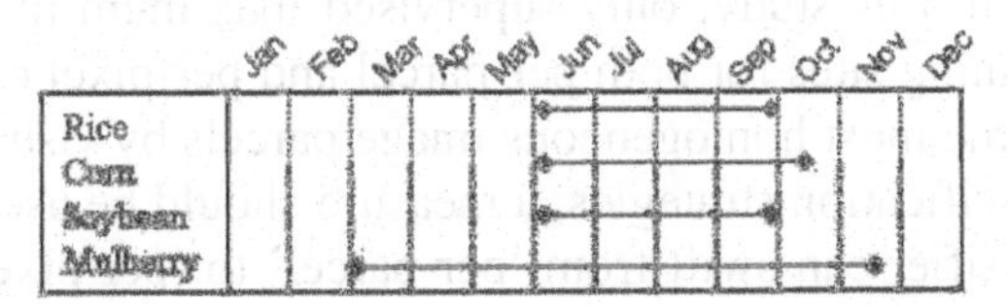

Fig. 2. The phonological characteristics of the main crops in the area

A multi-spectral image of 10-m resolution bands and the 2.5-m panchromatic band were fused using the Brovey algorithm to be used as a base map for vector digitizing afterwards. To ensure the accuracy of classification, the images were geometrically corrected to UTM (Zone-50) projection and datum WGS84 using 20 ground control points (GCPs). The registration errors were controlled no more than 0.5 pixels in localized areas. The vector field boundary data was delineated through manual on-screen digitalization based on the linear feature, such as ditches, roads and tree lines in the fused base map and stored as vector polygons. The SPOT5 multi-spectral image with the digitized vector field boundary data set overlaid is

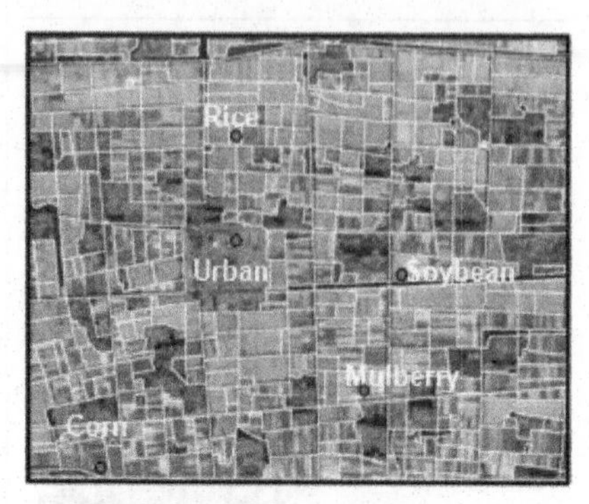

Fig. 3. A color composite of SPOT5 bands 4(SWIR). 1(green) and 2(red) with the digitized agricultural field boundaries overlaid

illustrated in Fig. 3. For crop classification only, the non-crop land cover types such as urban area, woods and water body were clipped out beforehand.

3. METHODS

The use of a mean spectral response for each image object in per-parcel classification against individual pixel spectral responses for each pixel in per-pixel classification was investigated in the integrated classification. In view of the main characteristics of each method, the spatial distribution of land features plays a major role in the success of classification.

A single per-parcel classification and a single per-pixel classification were respectively performed beforehand on the SPOT5 multi-spectral image. The results of each were employed as a basis and contrasts for the integrated classification. In this study, only supervised maximum likelihood classifier was used. Training sites for both per-parcel and per-pixel classification were selected from the most homogeneous image parcels by visual interpretation.

Besides classification strategies, a measure should be used to set a point at which the classifier can swift from 'per-parcel' to 'per-pixel' or versus. The selection of the measure and the determination of the point where the optimal swift happens are crucial to the integrated method. As discussed above, the intended measure should work to represent the spatial variance which is highly related to the method choice and the classification performance. Texture is just such a measure that functions in such a way that texture operators transform input image into texture coded in grey values. In practice, this study use a texture operator based on a co-occurrence matrix that measure the entropy in a 5×5 pixel window on the panchromatic band. The texture information was then incorporated into each parcel by per-parcel calculation of average textures. To determine the best swift point between per-parcel and per-pixel methods, a threshold must be denoted during the range of average parcel textures, above which, the results from per-parcel classification should be replaced by the per-pixel classification results to avoid the weakness of per-parcel method on heterogeneous objects. However, such a threshold can be hardly set unless an analysis operates on a gradual transition from per-parcel to per-pixel classification. The expected optimal point for method swift can then be identified.

4. RESULTS AND DISCUSSION

The final crop classification results from complete per-parcel classification and complete per-pixel classification are shown in Fig. 4 (a) and (b) respectively. The overgeneralization of heterogeneous parcels dominated by irregular crop mixture resulted from per-parcel classification and the speckled appearance at field boundaries and within homogeneous fields resulted from per-pixel classification can be readily recognized.

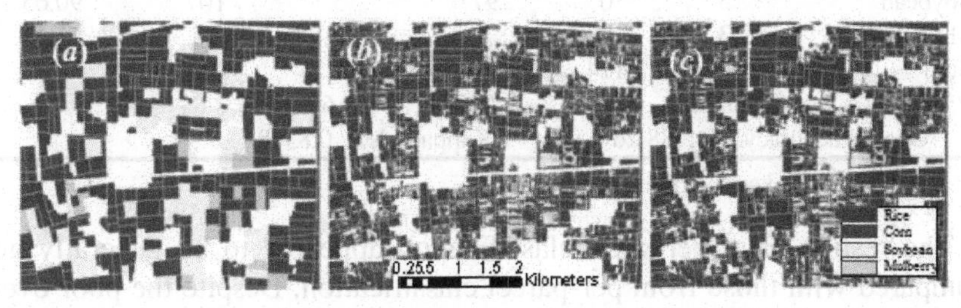

Fig. 4. (a) Crop classification map produced by per-parcel classification; (b) Crop classification map produced by per-pixel classification; (c) Crop classification map produced by an integration classification method with the optimal accuracy

For accuracy evaluation, reference data was established at a pixel level with the combination of ground data and interpretation results from the high resolution SPOT5 images. A total of 443 check points were selected for each classification results at the same locations. Two error matrices for each sole classification results are shown in Table 1 and 3. The overall, producer's and user's accuracies were calculated. A kappa index was also computed.

Table 1. Error matrix of the per-parcel crop classification

	Reference data					
	Rice	Corn	Soybean	Mulberry	Sum	U. Acc. (%)
Rice	85	17	6	7	115	73.91
Corn	2	120	44	10	176	68.18
Soybean	8	37	55	7	107	51.40
Mulberry	2	12	22	9	45	20.00
Sum	97	186	127	33	443	
P. Acc. (%)	87.63	64.51	43.31	27.27		
Overall accuracy (%): 60.72			Kappa index: 0.4385			

U. Acc., User's Accuracy; P. Acc., Producer's Accuracy.

Table 2. Error matrix of the optimal integrated method of crop classification

	Reference data					
	Rice	Corn	Soybean	Mulberry	Sum	U. Acc. (%)
Rice	88	2	0	25	115	76.52
Corn	0	173	0	1	176	98.30
Soybean	5	5	92	5	107	85.98
Mulberry	3	0	1	41	45	91.11
Sum	96	180	95	72	443	
P. Acc. (%)	91.67	96.11	96.84	56.94		
Overall accuracy (%): 88.94			Kappa index: 0.8451			

See Table 1 for key to abbreviations.

Table 3. Error matrix of the per-pixel crop classification

	Reference data					
	Rice	Corn	Soybean	Mulberry	Sum	U. Acc. (%)
Rice	80	1	0	34	115	69.57
Corn	0	173	2	1	176	98.30
Soybean	5	0	97	5	107	90.65
Mulberry	3	0	1	41	45	91.11
Sum	88	174	100	81	443	
P. Acc. (%)	90.91	99.43	97.00	50.62		
Overall accuracy (%): 88.26			Kappa index: 0.8368			

See Table 1 for key to abbreviations.

The results from per-pixel classification appeared to be generally good compared with those from per-parcel classification. Despite the poor overall performance, results from per-parcel classification for rice were fairly good Hence, the major negative affect on the per-parcel classification resorted to those three crop types except rice. The obvious difference between rice fields and parcels of the other three crop types is their spatial variance, low in the large rectangle rice field against much high in a parcel with scattered distribution of other three crop types.

The average texture value ranged from 0 to 3.154. The density split technique was adopted to divide the texture range equivalently into 20 sub-ranges, and then 19 thresholds were generated (Table 4). The per-parcel classification results of those parcels whose average texture values were above the settled threshold would be replaced by correspondent results from per-pixel classification. Thus, 19 integrated classification maps with their error matrices were produced in accordance with different threshold ranges.

Table 4. Texture ranges denoted by different thresholds

Index	1	2	3	4	5	6	7	8	9	10
Texture	>2.97	>2.81	>2.66	>2.50	>2.34	>2.19	>2.03	>1.88	>1.72	>1.56
Index	11	12	13	14	15	16	17	18	19	
Texture	>1.41	>1.25	>1.09	>0.94	>0.78	>0.63	>0.47	>0.31	>0.16	

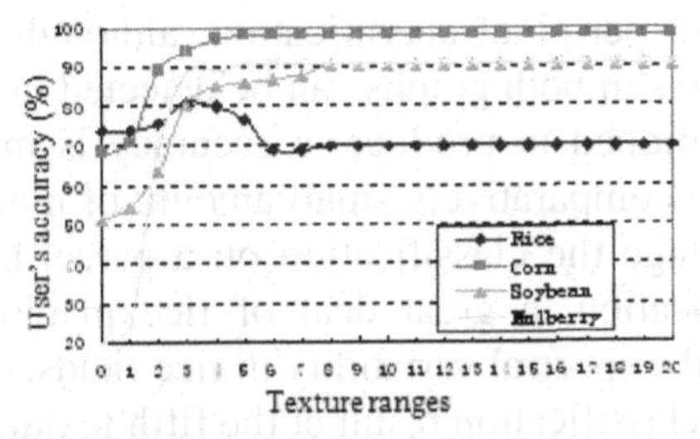

Fig. 5. Changes of overall accuracy and kappa index of classification during the transition from absolute 'parcel-based' to absolute 'pixel-based'
Notes: the numbers in the abscissa correspond to the index in Table 4. The left end '0' represents an absolute parcel-based classification while the right end '20' represents an absolute pixel-based classification

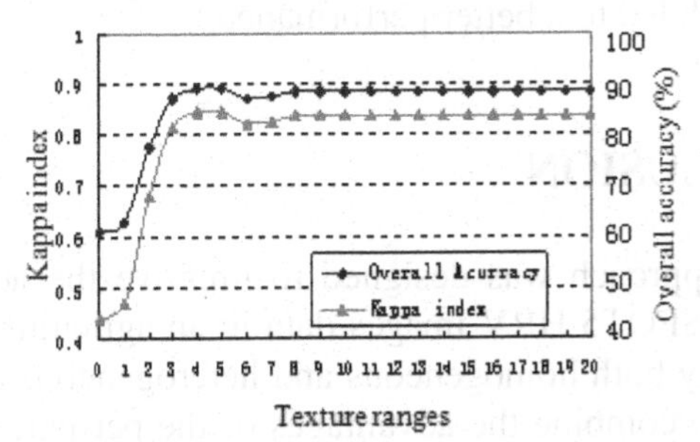

Fig. 6. Changes of user's accuracy of each crop type from absolute 'per-parcel' to absolute 'per-pixel'. See Fig. 5 for notes

The performance of these integrated classification together with that of the absolute per-parcel and per-pixel classification are organized in Fig. 5 to 7.

As shown in Fig. 5, with an increase of per-parcel classification results on heterogeneous parcels replaced by per-pixel classification results, both the overall accuracy and kappa index of the integrated classification are improved dramatically to a top level which is a little bit better than those of per-pixel classification alone. Nevertheless, the increase of accuracy is soon taken place by a decline. Afterwards, the accuracies of integrated classification retain to be almost the same as that of per-pixel classification due to an increasingly large part of per-pixel classification that shelters the advantages of per-parcel classification in processing less variable areas.

Further light was shed by examination the user's and the producer's accuracies. Fig. 6 and 7 present them by giving the explicit classification performance on each crop type, which can deduce the latent causes of the overall classification performance represented by Fig. 5. In the case of user's accuracy, all of the three crop types except rice have an obvious improvement. The only exception for rice is shown clearly by its distinct curve. As for producer's accuracy, the situation for rice was not as good as that in its user's accuracy as far as per-parcel classification was concerned to be compared with per-pixel classification, although the slight similarity of the curve structures in both graphs can be detected in scrutiny. Another thing worth to be mentioned in producer's accuracy is mulberry. For per-parcel classification, the comparatively small amount of mulberry distributed in the study area challenge the classification on a parcel basis at most; while for per-pixel classification, a great deal of rice pixels were misclassified to mulberry due to the spectral variability in rice fields.

An integrated classification result at the fifth texture range, the nearest one to the overall optimal choice, is given in Fig. 4 (c) and its error matrix in Table 2. The crop classification map produced by integrated classification inherited the smoothness of per-parcel classification in homogeneous parcels as well as the detail of per-pixel classification in parcels of high heterogeneity, which led to a better performance.

5. CONCLUSION

The integrated approach was designed to improve the accuracy of crop classification from SPOT5 HRV images data in an agricultural environment in China, typified by both homogeneous and heterogeneous crop fields. The design strategy is to combine the advantages of the per-parcel and per-pixel approaches with both spectral and spatial information. Although the optimal integrated approach represents a better performance than each of the two conventional ones to some degree, the best choice for a classification assignment largely depends on the spatial distribution of the real world. Anyway, an active analysis on the spatial characteristics of the image data is a reasonable prerequisite for any effective information extraction from remotely sensed data.

REFERENCES

De Wit A J W and Clevers J G P W. Efficiency and accuracy of per-field classification for operational crop mapping. International Journal of Remote Sensing, 2004, 25(20): 4091–4112.

Dean A M and Smith G M. An evaluation of per-parcel land cover mapping using maximum likelihood class probabilities. International Journal of Remote Sensing, 2003, 24 (14): 2905–2920.

Lo C P and Choi J. A hybrid approach to urban land use/cover mapping using Landsat 7 Enhanced Thematic Mapper Plus (ETM+) images. International Journal of Remote Sensing, 2004, 25(14): 2687–2700.

Smith G M and Fuller R M. An integrated approach to land cover classification: an example in the Island of Jersey. International Journal of Remote Sensing, 2001, 22(16): 3123–3142.

STUDY ON ESTABLISHMENT SOYBEAN CONTROLLABLE STRUCTURAL MODEL

Zhongbin Su[1,2,*], Ping Zheng[2], Hongmin Sun[2], Jicheng Zhang[2], Xiaoming Li[2]

[1] *Computer Science and Technology Academy Department, Beijing Institute of Technology, Beijing, China, 100081*

[2] *Engineering Academy, Northeast Agricultural University, Harbin, China, 150030*

* *Corresponding author, Address: Computer Science and Technology Academy Department, Beijing Institute of Technology, Beijing, 100081, P. R. China, Tel: +86-0451-55190170, Fax: +86-0451-55190170, Email: suzb001@163.com*

Abstract: By analyzing soybean structural and physiological characteristic sufficiently, the paper proposes a plant structural controllable modeling method for soybean. It selects the key structural factors firstly, and collects the relevant field data. Then it extracts the organ structure and position laws along with their vertical and horizontal change to establish controllable structural model. Combined with structural growth model based on growth function, it could establish soybean dynamic structural model, which has great significance to optimize the field cultivation measure and design ideal plant type.

Keywords: soybean; structural model; controllable model; dynamic model; visualization

1. INTRODUCTION

Virtual plant is one of the hot and difficult issues in digital agriculture research, which need to be solved urgently. It could study complex agricultural ecosystem directly and discover the plant principles that are observed difficultly by traditional research methods. It is helpful to understand thoroughly the crop structural and physiological law, optimize field cultivation measure, and design ideal plant type (Cao, 2005). Virtual plant also has vital significance to increase crops yield. At present, main research theories for virtual plant are L-system, AMAP model, particle system and three dimension reconstructions and ect. (D. Barthelemy, 2003).

Su, Z., Zheng, P., Sun, H., Zhang, J. and Li, X., 2008, in IFIP International Federation for Information Processing, Volume 258; Computer and Computing Technologies in Agriculture, Vol. 1; Daoliang Li; (Boston: Springer), pp. 685–693.

These theories for modeling have their own value. Each one is suitable to simulate specific plant type. They all have certain application scales and some limitation (Guo, 2007).

Soybean is an important food and economic crop. Its growth simulation modeling research is quite mature. Typical example is American soybean growth simulation modeling, named "Soygro", which is used to forecast soybean growth developmental state, yield and simulate the balance of soil's moisture. The US has developed three decision support systems for soybean growth simulation modeling named "Soybean dss86" and "Pcyield" on the basis of "soygro" (Wang, 1982). But at present the soybean structural simulation research is relatively less. S. Chuai-Aree et al. developed "plantVR" based on L-system and growth equation technology. The simulation result of soybean structure is quite vivid, but this software considered less in environmental factors and community growth. So it has little instruction significance to soybean production. By analyzing above questions, this paper proposed one kind of controllable structural modeling method on the basis of understanding soybean structural and physiological characteristic. The structural model established by this method could flexibly display vertical distribution state of leaves and other organs. And it also expresses the trifoliolate leaf change in plant community. It provides a new research technique for establishing structural model interactive with environment, optimizing field cultivation measure, and designing ideal plant type (Su, 2005).

2. MATIERIALS AND METHOD

2.1 Data measurement

Field measurements were conducted at Xiangfang experimental field farm of Northeast Agricultural University from 2006 to 2007. Soybean cultivars Dongnong42 seed was sown 0.2 m apart in east-west-oriented rows that were 0.1 m apart. The result plant population is higher than that commonly used by local farmers and was chosen to maximize competition among around plants, the aim here being to analyze the plant structure with no branching. The plots were irrigated and fertilizer inputs were such as to avoid any mineral and water limitation to plant growth. There was one plant in a hole avoiding competition among one hole's plants. No plant disease, pest symptoms were observed.

The soybean plant has many structural characteristics, like alternate phyllotaxis, trifoliolate leaves, network vein and so on. So it took the relevant plans for the data collection. Plant structural data had been collected since the period of cotyledon. Each week 8 plants were taken randomly back to the laboratory for measuring. The collection data included fresh weight, length and diameter of each internode, fresh weight and area of each leaf; the angles between petioles and main stem; the angle between trifoliolate leaves and petiole.

Fresh weight was taken by electronic balance with accuracy 0.01g. The leaf area was measured by CI-203 Portable Laser Area Meter of CID Corporation with accuracy 0.001 cm2 and error ±1%. The length of internode and petiole was taken by straightedge with accuracy 0.1 cm. The diameter of internode and petiole was taken by vernier caliper with accuracy 0.001 cm. The angles were taken by protractor with accuracy 0.01°

2.2 Controllable modeling method

Plant structural model includes static model and dynamic model. Static model is established by plant structural data measured by three-dimensional digital methods. The model of this method is established directly by these data. It is only used to research the condition and state about plant space structure. Its drawback is transferring a great deal of data collected directly. It is not suitable to reflect plant structural dynamic characteristics. Dynamic model is established based on plant topology and geometry change laws. It is established by extracting plant growth laws. It is the main research direction of virtual plant modeling.

In the process of the establishing soybean structural dynamic model, the paper divides dynamic model into growth model and controllable model, which respectively manifests the growth process with time elapse and growth process with environmental change. In the process of establishing soybean growth model, it generally used growth equation. For example, Logistic equation is often used to simulate internode growth. In the previous simulation study, it has proven that the growth equation can effectively simulate soybean growth with time elapse.

In the process of establishing soybean controllable model, the influent factors considered are more complicated. It should take into account the number and position of plant organs on vertical and horizontal change and organ physiological change law affected by the external environment. The controlled model is established by extracting these factors' laws on the shape of organs, location, biomass and etc. The model could control the plant growth structure change and build plant physiological and ecological development process more really.

3. RESULTS AND ANALYSIS

3.1 Main stem and petiole

3.1.1 Growth model based on L-system

The topological structure of "Dongnong 42" has strongly regularity that trifoliolate leaves petioles with same structure alternately grow on main stem. Three small leaves and petioles respectively compose a petiole region. On comparative analysis of existing model construction theories, the paper selects L-system to establish soybean topological model. First, it gives some symbols with special meanings in L-system, as shown in table 1 (Zheng, 2006).

Table 1. Some symbols' meanings in L-system

Symbol	Meanings	Symbol	Meanings
I	Length of internodes	P	Leafstalk
i	Litter leafstalk	L	Trifoliolate Leaves
P l	Left branch	L	Simple leaf
P r	Right branch	A	Growth point
P	Trifoliolate Leaves Region	Angle	Intersect angles

Through many experimental comparisons and observations, this paper designs topological qualitative model in L-system below:

```
{ Iteration = N
Angle=n
V={I, i, Pl, Pr, P, p, A, L, l, z, [, ], +, -, &, ^, \, /, |}
w : I [+z][-z]A
(N = 1)    P1 : A->I[+iL][-iL]A
(N > 1)    if N%2<>0  P2 : A->I[+Pl]A
else            P3 : A->I[-Pr]A
Pl ->p[\iL][/iL][-iL]
Pr -> p[\iL][/iL][+iL]}
```

The amount of variable "Iteration" is conformed by days of soybean growth and partition of physiological period. For example, if given 126 days, "Iteration" equals 18, and it will create a new character string:

I[+Pr]I[-Pl]I[+Pr]I[-Pl]I[+Pr]I[-Pl]I[+Pr]I[-Pl]I[+Pr]I[-Pl]I[+Pr]I[-Pl]I[+Pr]I[-Pl]I[+Pr]A. (1)

Now the letters in production (1) are expressed by turtle interpret. It can obtain static structure of "Dongnong 42" at about 120 days. Then it builds growth equations by collecting and analyzing the data of internode length and petiole etc. in the production (1). This method could establish growth model with time elapse.

3.1.2 Controllable model simulation

The samples of plants are measured every one-week. It observed and recorded the length of internode, the petiole of trifoliolate leaves, and the angles between main stem and petioles. This paper adopts angle data (as shown in table 2) to analysis and explanation. Matlab analyzes the collect data and finds that the numerical of the angles between main stem and petioles are similar to a quadratic curve by vertical distribution. This research did parameter fitting with Matlab, and obtained regression equation as shown in expression 2.

$$y = 0.0212x^2 - 0.2036x + 0.9678 \quad 1 \le x \le 12$$
$$y = -0.5236x + 7.854 \quad 12 < x \le 14$$
$$y = 0.5236 \quad x = 15 \tag{2}$$

where x is the phyllotaxis between main stem and trifoliolate leave, and y is angle degree.

Table 2. The angle data between main stem and petiole

Phyllotaxis	1	2	3	4	5	6	7	......	13	14
Angle	0.80	0.64	0.55	0.50	0.50	0.55	0.62	......	1.04	0.5

The units for angle are radian.

By this method, the paper extracts every soybean structural data with phyllotaxis to establish controllable model. The controllable model is expressed in delphi 7.0 on OpenGL to obtain the topological structure simulating results of "Dongnong 42". Fig. 1 showed three main stem type of soybean by this method. It can fully reflect relationship of the angles between main stem and petiole along with vertical direction change. It has great significance to further design the ideal plant type.

Soybean is field crop. Its inter plant distances and row distances are narrow. In the growth process it has more competition with surrounding plants. That is to say, the plant structure is affected by plant cultivation

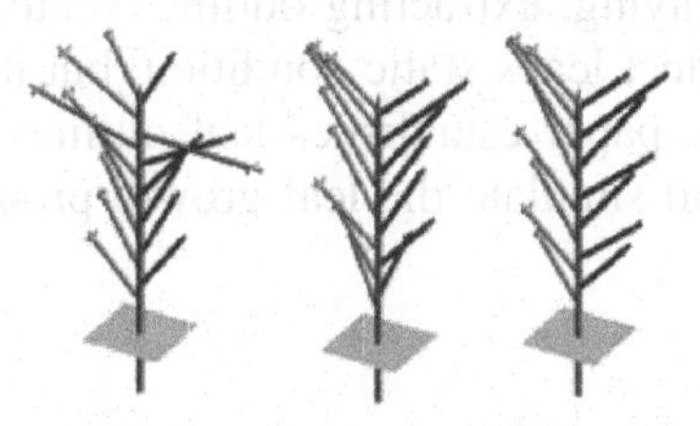

Fig. 1. Petiol's azimuth vertical change simulation

position and plant density. The symbol "+" and "-" in equation (2) are expressed space angles between main stem and petioles. The paper changes them by analyzing the growth state of the surrounding plants state. Then it extracts the laws of horizontal angels' laws and plant's location. The method could simulate the horizontal change law of soybean structure. It is helpful to simulate soybean community growth, as shown in Figure 2.

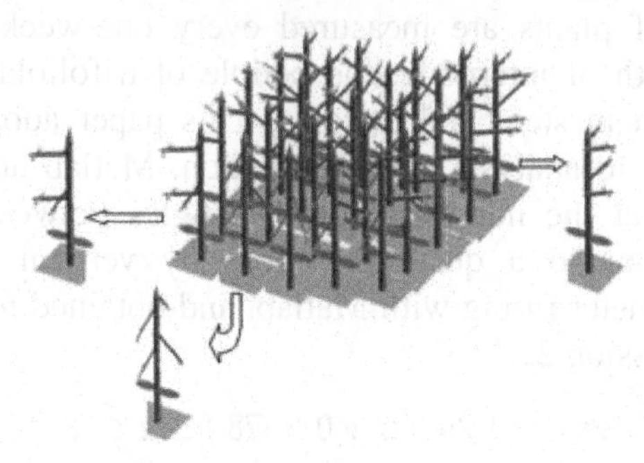

Fig. 2. Petiole's azimuth horizontal change simulation

3.2 Trifoliolate leaf

Crop production is related with the photosynthesis closely. It could guarantee high and stable production by selecting cultivars with high photosynthetic efficiency (HPE) to improve the absorption intensity of light energy and taking reasonable field cultivation measure. Soybean leaf is an important organ to carry on the photosynthesis and make the organic nourishment. The simulating study thoroughly of leaf can provide basis and methods for designing plants type and selecting cultivars. According to soybean leaf' features, such as alternate phyllotaxis, trifoliolate leaf, network veins and so on, the paper selects suitable methods to establish growth models and controllable models to increase 3D effect and realistic feeling. It could provide methods to establish more reasonable plant structure and design ideal plant type (Kang, 2006).

3.2.1 Leaf outline

In the previous study, leaf outline could be obtained by a series of digital processing that is graying, extracting outline, vectorization in VC platform. This method can reflect leaf's static condition, but it lacks the description of the leaf growth. The paper establishes leaf outline by growth function and Bezier curve. It could simulate the leaf growth process with high speed and efficiency.

First, collect the length and width data of each trifolioate leaf, as shown in Table 3.

Table 3. Data of leaf's Length and Width

Time	1	3	6	9	13	15	18
Width	0.30	1.00	1.50	2.70	4.80	5.90	6.10
Length	0.50	1.80	3.50	5.80	10.50	11.70	12.00

The units for Time are day; Width is cm; Length is cm.

Then, the data is regressed by Richards equation, as shown in equation (3),

$$P_1(y)=\frac{6.02299}{(1+10799.224e^{-0.6932751t})^{1/3.59}}$$
$$P_2(y)=\frac{12.1673}{(1+1748.59e^{-0.62778t})^{1/2.765}} \qquad (3)$$

Select the leaf petiole nod "p_0 "as the origin of coordinates, and regard "p_1 "and "p_2" as apexes of the leaf's width and length, as shown in Fig. 3 (a). Line segments "p_2p_4" and "p_4p_5" respectively express tangents, and "p_2" and "p_1" express the tangential point. "p" expresses a random point in the curve "p_0p_1". It introduces a parameter "k". According to De Casteljau algorithm it could get the equation (4).

$$P=(1-k)^2P_0+2k(1-k)P_5+k^2P_1 \qquad (4)$$

When k range from 0 to 1, "P" expresses quadric Bezier curves named curve "P_0P_1" which is defined by three apexes "P_0", "P_1" and "P_5". So as curve "P_1P_2". This method can draw left part of leaf outline through these steps, as shown in Fig. 3. Symmetrical rotating from the left part can draw the right part. This method could create leaf outline based on growth function and Bezier curves. It gets a smooth outline figure that describes leaf growth by limited iteration.

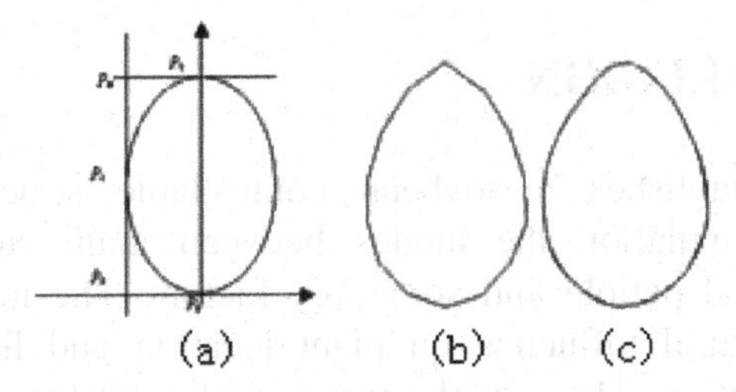

Fig. 3. Outline curve fitting process (a) outline's sketch; (b) outline drawn by 10 iteration; (c) outline drawn by 50 iteration

3.2.2 Controllable model simulation

After chapter 2.1.2 research, it builds controllable model of main stem and petiole. It can simulate leaf petiole azimuth change along with phyllotaxis and plant location. Three leaves think as a whole because of trifoliolate leaf. Through observation and experiments measurement, left and right leaves' main veins and middle leaf identify a plane, named trifoliolate leaf plane as shown in Fig. 4 (a). The angle between the plane and petiole lies on trifoliolate leaf phyllotaxis and light source. The left and right leaves appropriately rotate with their main veins according with light intensity to increase or decreases energy absorption, as shown in Fig. 4 (b), (c). In the leaf controllable model establishment process, it selects parameters to express trifoliolate leaf plane and petiole, left and right leaves and this plane. Then extract the law about two angles along with phyllotaxis and light source to establish leaf structural model. The model considers more about leaf physiology characteristic to give prominence to organ of absorbing light. It could increase soybean structural model's flexibility.

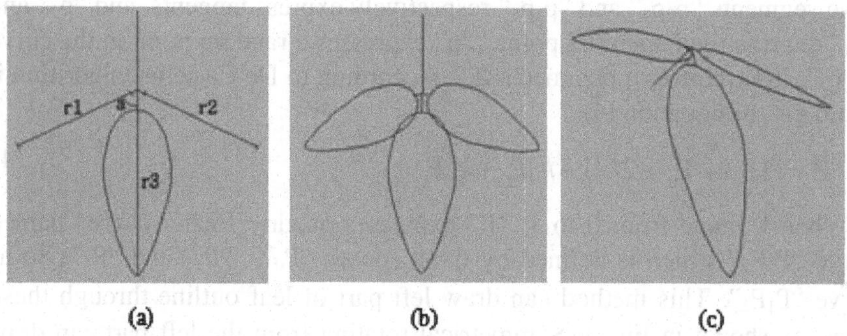

Fig. 4. Controllable leaf structural model (a) trifoliolate leaf plane; (b) left and right leaves in the plane; (c) some angles degree between two leaves and plane

4. CONCLUSION

The paper establishes a soybean controllable structural model by analyzing and simulation the angles between main stem and petiole, trifoliolate leaf and petiole and some key factors. The model implements preliminary structural influence on plant location and light source from environmental factors. The model accords with soybean structural growth laws better. It has positive effect in establishment interactive model with external environmental factors. In further study, the model need combine with soybean physiology to strengthen further analysis of structural collection data. It would be helpful to extract the laws between plant location and structure, environmental factors and structure, yield and structure to establish more realistic soybean structural model.

ACKNOWLEDGEMENTS

This study has been funded by Chinese 863 Plan (#2006AA10Z231), Heilongjiang Natural Science Foundation of China (#C200607), and Program for Innovative Research Team of Northeast Agricultural University, "IRTNEAU".

REFERENCES

Cao Wei-xing. Agricultural Information Science. Bei Jing: China Agriculture Publishing, 2004. (in Chinese)

D. Barthelemy. Botanical Background for Plant Architecture Analysis and Modeling. In: Plant Growth Modeling and Applications Proceedings. Beijing: Tsinghua University Press. 2003:1-20

F. Normand, J. Chadoeuf, R. Habib. Modeling Asynchronous Flowering. In: Plant Growth Modeling and Applications Proceedings. Beijing: Tsinghua University Press. 2003:76-84

Guo Xinyu, Zhao Chunjiang, Xiao Boxiang, et al. Design and implementation of three-dimensional geometric morphological modeling and visualization system formaize. Journal of Agricultural Engineering. 2007, 4, 144-149. (in Chinese)

Kang Li, Su Zhong-bin, Zheng Ping, et al. Research on Modeling Leaf Based on L-system, Agricultural Machinery Research, 2006, 7, 180-182. (in Chinese)

Song Yonghong, Guo Yan, Li Baoguo, et al. Virtual maize model□.plant morphological constructing based on Organ biomass accumulation. Ecological Journal. 2003, 23(12): 2579-2586. (in Chinese)

Su Zhongbin, Meng Fanjiang, Kang Li, et al. Virtual plant modeling based on Agent technology. Agricultural Engineering. 2005, 21 (8):114-116. (in Chinese)

Wang Jingwen, Yi Tianfu. Soybean plant type mathematical models and canopy light of the vertical distribution. Journal of Northeast Agriculture. 1982, 3, 1-9. (in Chinese)

Zheng Ping, Su Zhongbin, Kang Li. Modeling of Virtual Soybean Topology Based on Growth Function, Agricultural Machinery Research, 2006, 7, 193-194. (in Chinese)

STUDY ON MAIZE LEAF MORPHOLOGICAL MODELING AND MESH SIMPLIFICATION OF SURFACE

Xinyu Guo [1,*], Chunjiang Zhao [1], Boxiang Xiao [1,2], Shenglian Lu [1], Changfeng Li [1]

[1] *National Research Center for Information Technology in Agriculture, Beijing, China, 100097*

[2] *Institute of computer technology, Dalian University of Technology, Dalian, China, 116024*

[*] *Corresponding author, Address: P.O. Box 2449-26, Beijing, 100097, P. R. China, Tel: +86-10-51503422, Fax: +86-10-51503750, Email: guoxy@nercita.org.cn*

Abstract: According to the need of canopy visualization calculation in the digital plant research, we introduced a method, using Non-Uniform Rational B-Splines (NURBS) interpolation and multi-line segment splitting algorithm, to reconstruct the 3D morphological structure of maize leaf with a complexity controllable mesh. Using the data cloud obtained by digitizer, construct the surface of maize leaf by calculating the knot vectors and reverse calculating surface control points by difference calculation. The final visualization effect is realistic. According to leaf morphological characteristics, leaf surface mesh can be simplified by using inverse calculation of multi-line segment splitting algorithm, and the surface main characteristics can be maintained simultaneously. This method can be used in canopy visualization calculation and light distribution calculation. Results showed that it can improve the calculation efficiency obviously without increase the calculation error.

Keywords: maize leaf; NURBS; geometric modeling; mesh simplification; light distribution calculation

1. INTRODUCTION

The morphological structure of plant is always the important subject in biology, agronomy and other related area. With the rapid development of computer technologies, now we can model the complex 3D architecture of

Guo, X., Zhao, C., Xiao, B., Lu, S. and Li, C., 2008, in IFIP International Federation for Information Processing, Volume 258; Computer and Computing Technologies in Agriculture, Vol. 1; Daoliang Li; (Boston: Springer), pp. 695–702.

plants (Room et al., 1996; Reffye et al., 1997). For most crops, the aboveground morphological characteristics of them are mainly decided by upward growing stalk and leaves, which grow in different angles and orientations. Therefore the modeling of leaf is always one of the important research focuses, and many methods have been reported. Espana et al. (1999) modeled maize leaf with a rectangular 2D parameterized plane which was similar to the leaf blade. Deng XY et al. (2004) proposed a method to model static leaf based on cardinal spline and triangulation modeling. Liu XD et al. (2002, 2004) used Bezier curve and NURBS surface to model plant leaves. Zhao CJ et al. (2004) introduced a characterized method for simulating maize leaf based on its shape features. All the leaf models mentioned above were made purely from geometric point of view, without the concerns of visualization computing. In fact, a large number of computing are involved in the analysis of crop morphology and architecture, such as computing the light distribution in canopy (Ross, 1981; Chelle et al., 1998, 1999, Wang et al. 2004), this makes the speed of computing become a bottleneck in simulating and visualizing. Because of this, it is very necessary to refine the geometry models of crop leaves. Based on the fact that most of the shapes and surfaces of crop organs are irregular, while Non-Uniform Rational B-Splines (NURBS) is efficient for the representation of crop organs' geometry (Zhu, 2003), and it has the properties of Rational B-Splines and Non-Uniform Splines. In this article, we take the maize leaf as an example, using the point data cloud obtained by 3D digitizer, proposed a method for modeling maize leaf. This model is reconstructed by using NURBS interpolation and multi-line segment splitting algorithm; and the output mesh was utilized in the visualization and calculation of light distribution of canopy.

2. MATIERIALS AND METHOD

2.1 Modeling maize leaf

The shape of maize leaf has some obvious features: it grows on the main stalk, with a long, thin and stiff midrib supports it upward, and the edge is undulate. Using the method described by Zheng et al. (Zheng et al., 2004), obtain a group of characteristic points (data points) through digitizer, representing as $Q_{i,j}$ (i=0,1,…,m; j = 0,1,..,n). In order to use these points to generate NURBS surface, we define the orientation of leaf midrib as u direction, and the vertical orientation to that is v direction, just as shown in Fig. 1 (a). Then from these data points, we can calculate the knot vectors, and the basis function. After that the control points of the NURBS surface can be calculated, as shown in Fig. 1 (b). Equation (1) defines arbitrary points on NURBS surface:

$$P(u,v)=\frac{\sum_{i=0}^{n}\sum_{j=0}^{m}B_{i,k}(u)\cdot B_{j,h}(v)\cdot W_{i,j}\cdot V_{i,j}}{\sum_{i=0}^{n}\sum_{j=0}^{m}B_{i,k}(u)\cdot B_{j,h}(v)\cdot W_{i,j}}, \tag{1}$$

where $V_{i,j}$ is the control points, $W_{i,j}$ is the weight factor, while $B_{i,k}(u)$ and $B_{j,h}(v)$ is k order B-spline basis function of u direction and h order B-spline basis function of v direction respectively.

Using the solved control points and knot vectors, the leaf surface model can be established, and then the NURBS surface can be generated. Fig. 1 (c) shows the wireframe of a maize leaf blade mesh generated by this method, in which the step length of interpolation is equal to one fifth of the length between two knots. That is to say, four points are interpolated between each pair of data points. Fig. 1 (d) demonstrates the rendering model with relatively more realistic effect.

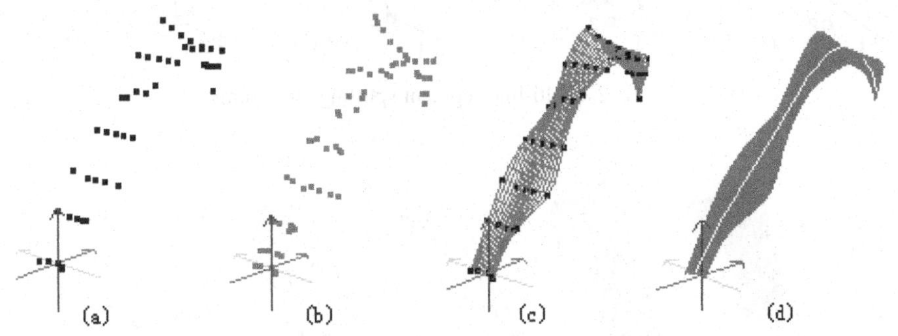

Fig. 1. The maize leaf model using NURBS interpolation

2.2 Refining the mesh

The surface displaying on computer screen consists of a number of mesh grids. In general, the more exquisite the surface is divided, the better rendering effect the surface has. But the smaller the grid is, the more grids are needed to generated, and this may lead to low efficiency in computing. To deal with this problem, this paper proposes a method to refine the mesh of crop leaves. User can refine the mesh of leaves according to their actual needs, and can balance the rendering effect and computing efficiency easily.

We define the mesh generated by adjacent data points as standard mesh, while the mesh generated by interpolation as reference mesh (shown in Fig. 1 (c)). When the users need to refine the reference mesh, they can reduce the number of interpolation points by increasing the step span of interpolation.

The minimum number of interpolation points can be zero, which means the reference mesh is refined to the standard mesh. On the other hand, the standard mesh can be refined with an inverse operation of multi-line segment splitting algorithm, which can be described as follows: to a group of points, such as shown in Fig. 2, (1) select the start point and end point, and connect these two points into a line, then select the first-class characteristic point, which is the point that has the biggest distance to the line, as shown in Fig. 2 (a); (2) replace the initial line with the line connecting the characteristic point to the uplevel characteristic point, then select the second-class characteristic point; (3) continue the process until all the points are selected as characteristic point.

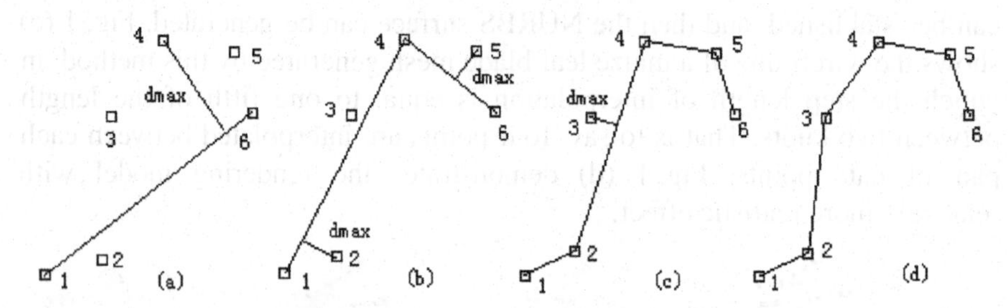

Fig. 2. Multi-line segment splitting algorithms

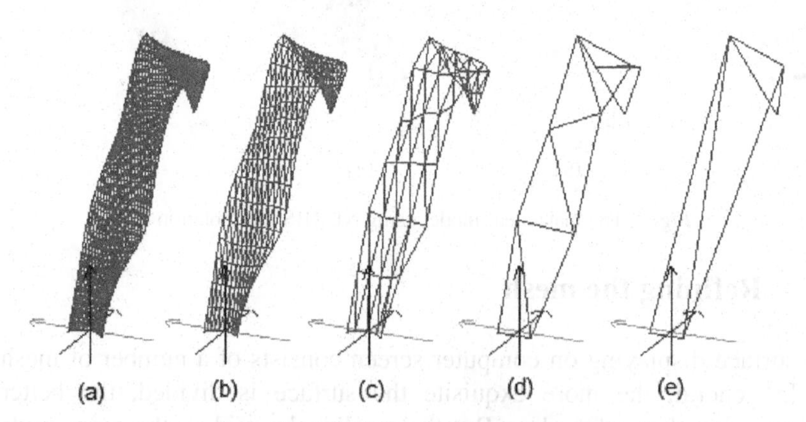

Fig. 3. Refined mesh shape of maize leaf

(a) Mesh with 1800 triangles (the reference mesh)
(b) Mesh with 440 triangles
(c) Mesh with 72 triangles (the standard mesh)
(d) Mesh with 8 triangles
(e) Mesh with 3 triangles

3. RESULTS

3.1 Maize leaf area and projection area

Table 1. Maize leaf area and projection area of different mesh scales

Triangle number	1800	440	72	8	3
Leaf area (m^2)	0.123	0.123	0.122	0.118	0.102
Leaf projection area (m^2)	0.808	0.807	0.801	0.785	0.689

We use the simplified geometric model of maize leaf to calculate the characteristic values, which reflect the main characteristics of leaf morphological structure, and the result was listed in Table 1. It can be found from Fig. 3 and Table 1 that the simplified scheme we proposed can significantly reduce the number of triangles of leaf model. When the number of triangles reduced from 1800 to 8, both the inaccuracy of leaf area and leaf projection area are no more than 5 percent, but the inaccuracy increase evidently when the number of triangles is lower than 8. Therefore our method is suitable for modeling the morphological structure, visualization calculation and analysis of maize leaf; it can dynamically meet the needs of visualization effect and computation efficiency.

3.2 Visualization and calculation of light distribution canopy

Using the above method, we reconstructed the canopy of maize and compared the results of a series of refined meshes.

The data used in our experiment was obtained in 2005 with a 3D digitizer 3SpaceFastrak (0.08cm precision), manufactured by an America-based company Polhemus. The canopy we calculated consists of nine maize plants arranging as 3 rows and 3 columns in the field. The row spacing is 60 cm, and the column width is 30 cm. The height of plant is 245cm. We expanded the dimension by multiplying the measured canopy up to 81 maize plants (Wang et al., 2005), which were arranged in 9 rows and 9 columns. Then the 3D canopy is reconstructed with series types of leaves, whose mesh consists of 440, 72, 36 and 4 triangles respectively, and the shape characteristic parameters in different refined mesh are also calculated. Fig. 4 and Table 2 show that more realistic effect can be achieved with the increase number of triangles in the leaf mesh, and more detailed description to the canopy. Meanwhile, leaf area tends to increase.

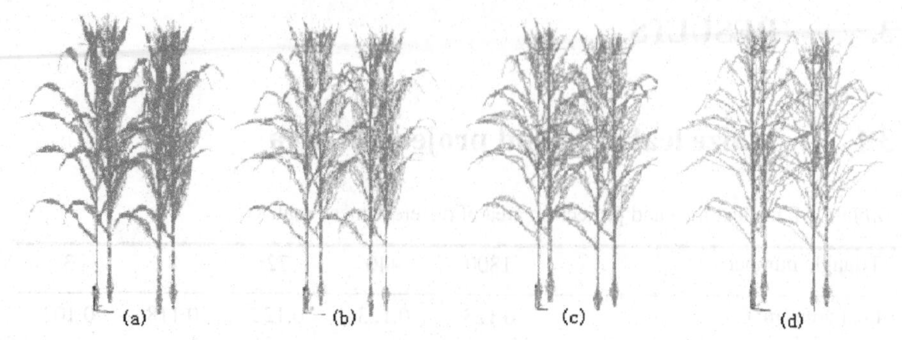

Fig. 4. The maize canopy with different mesh scales

(a) Mesh with 440 triangles
(b) Mesh with 72 triangles
(c) Mesh with 36 triangles (the standard mesh)
(d) Mesh with 4 triangle

Table 2. Maize canopy morphological character parameters of different mesh scales

Character parameter	Mesh number			
	440	72	36	4
Leaf area (m^2)	76.5	75.42	74.79	60.53
Total leaf projection area (m^2)	31.6	30.9	31.76	24.32
Vertical projection leaf area (m^2)	10.89	10.7	10.82	8.94

Table 3. Rate of transmitting direct light within rectangular area at different canopy heights calculated with different refined mesh

Height (cm)	Number of triangles			
	440	72	36	4
0	0.059	0.058	0.064	0.16
50	0.13	0.13	0.12	0.22
100	0.23	0.24	0.24	0.28
150	0.36	0.38	0.39	0.52
200	0.68	0.71	0.74	0.83
225	0.87	0.9	0.91	0.96

The radiation the crop received in nature is mainly direct radiation light from sun and diffuse light in sky. Our work implements the calculating of maize canopy direct radiation based on the mesh generated by the method described in (Mariscal et al., 2004; Wang et al., 2005). The simulating of direct radiation light can be performed as follows: simulate the sun light using a cluster of rays with an angle; the density can be determined by real requirement of light intensity. For each ray, determine whether it collides with the triangles on the mesh; and then sort the triangles collided with the ray according to their depth value. The triangle which has the minimum depth (with maximum z-coordinate) is radiated by the sun directly; triangles with secondary minimum depth receive second-class radiation from the sun, and we assume all others triangles do not receive any light from the sun. Above algorithm used the finite element calculation, the result meets closely to real observation. But there's too much computation in this method, and the calculation time is affected by the number of mesh grids. But using the approach introduced in this article, which uses the controllable mesh generation algorithm, the number of mesh grids can be simplified easily, and with the simplified mesh the calculation efficiency could be upgraded.

Take the light intensity calculation of the canopy at 12 AM at the stage of maize spin silk period for example; table 3 shows different rate of transmitting direct light within rectangular area at different canopy heights of different refined mesh. The selected area is a 180cm×90cm plane residing inner of the canopy. We can see from Table 3 that the variation of calculation inaccuracy is not evident when the number of triangles in each leaf surface decreasing from 440 to 36, but the computation is decreased notably. More exactly, the computing time decreases to 10 percent, this means a lot to the efficiency improvement of calculating light distribution. Users can improve the calculation by using the method of refining leaf model.

4. DISCUSSION

We find that when calculating the characteristic parameters and light distribution of maize canopy, the leaf area, total projection area and vertical projection area do not change evidently if the number of triangles of the leaf surface mesh has been simplified from 440 to 36, and variation of the light transmission within rectangular area at different canopy heights are lower than 10 percent, while the computation time decrease to 10 percent below. When the number of triangles is less than 36, the leaf area, total projection area and vertical projection area decrease, and the variation increase gradually. We can conclude that the refining scheme our proposed can improve the computation evidently when the number of triangles of the leaf surface is no more than 36, and this improvement is obtained without the cost of increasing evidently the calculation variation. Our method provides a balance between the rendering effects and computing efficiency in crop leaves morphological model.

ACKNOWLEDGEMENTS

This study has been funded by National High Tech R&D Program of China (2007AA10Z224), Key Technologies R&D Program of China (2006BAD10A01), Beijing Natural Science Foundation (4062015).

REFERENCES

Chelle M, Andrieu B, Radiative models for architectural modeling, Agronomie, 19, 1999, 225-240

Chelle M, Andrieu B, The nested radiosity model for the distribution of light within plant canopies, Ecological Modelling 111, 1998, 75-91

De Reffye P, Houllier F, Modelling plant growth and architecture:some recent advances and applications to agronomy and forestry, Current Science, 73, 1997, 984-992

Deng XY, Zhou SQ, Guo XY, A Static Leaf Model Based on Cardinal Spline and Triangle Faces, Computer Engineering and Applications 25, 2004, 199-201

Espana ML, Baret F, Aries F et al., Modeling maize canopy 3D architecture application to reflectance simulation, Ecological Modeling, 122, 1999, 25-43

Liu XD, Cao YF, Liu GR, The Modeling of Rice Leaf based on NUBRS, Microelectronics & Compute, 21, 2004, 117-119

Liu XD, Jiang LH, Zhao JJ et al., The Modeling and Display of Plants Based on the Bezier Curve, Computer Engineering and Applications, 13, 2002, 97-98

Mariscal MJ, Martens SN, Light-transmission profiles in an old-growth forest canopy: Simulations of photosynthetically active radiation by using spatially explicit radiative transfer models, Ecosystems, 7, 2004, 454-467

Room PM, Hanan JS, Prusinkiewicz P, Virtual plants: new perspectives for ecologists, pathologists and agricultural scientists, Trends in Plant Science, 1, 1996, 33-38

Ross J, The Radiation regime & architecture of plant stands. Dr. W Junk Publishers. The Hague, The Netherlands. 1981

Wang XP, Li BG, Guo Y et al., Measurement and Analysis of the 3D Spatial Distribution of Photosynthetically Active Radiation in Maize Canopy, ACTA AGRONMICA SINICA, 30, 2004, 568-576

Wang XP, Guo Y, Li BG et al., Modelling the three dimensional distribution of direct solar radiation in maize canopy, ACTA ECOLOGICA SINICA, 25, 2005, 7-12

Zhao CJ, Zheng WG, Guo XY, The Computer Simulation of Maize Leaf, Journal of Biomathemat, 19, 2005, 395-399

Zheng WG, Guo XY, Zhao CJ, Geometry Modeling of Maize Canopy, Transactions of the Chinese Society of Agricultural Engineering, 20, 2004, 152-154

Zhu XX, Modeling Technology on Free-Form Surface, Beijing: Science Press, 2003

RESEARCH AND SYSTEM REALIZATION OF FOOD SECURITY ASSESSMENT IN LIAONING PROVINCE BASED ON GREY MODEL

Bin Xi[1], Yunhao Chen[1,2,*], Hongchun Cai[1], Yang Liu[1]

[1] *College of Resources Science and Technology, Beijing Normal University, Beijing, China, 100875*

[2] *Academy of Disaster Reduction and Emergency Management Ministry of Civil Affairs & Ministry of Education, Beijing Normal University, Beijing, China, 100875*

[*] *Corresponding author, Address: College of Resources Science and Technology, Beijing Normal University, No. 19 Xin Jie Kou Wai Street, Beijing, 100875, P. R. China, Tel: +86-10-58806098, Fax: +86-10-58806098, Email: Cyh@ires.cn*

Abstract: Food security is a key component of state security, and every country, especially some great powers in the world, has been paying more attention to national food security and its security system. This paper aimed at applying grey model to food security assessment in the province scale from statistical data and developing a prototype of food security assessment system in Liaoning Province. Firstly, the use of grey models is a novel concept in food security assessment and the principles of GM (1, 1) are explained systematically. Second, the framework of food security assessment is created based on three-tier architecture and the assessment system adopts the object-oriented development method and component technology. To show the validity and feasibility, the precision inspection is made under strict mathematical calculation. In the end, the realization of system results is presented partially. It is suggested that food security assessment based on grey model is an effective attempt to solve food security problems.

Keywords: food security assessment, grey model, three-tier architecture

Xi, B., Chen, Y., Cai, H. and Liu, Y., 2008, in IFIP International Federation for Information Processing, Volume 258; Computer and Computing Technologies in Agriculture, Vol. 1; Daoliang Li; (Boston: Springer), pp. 703–711.

1. INTRODUCTION

Food security is a key component of state security, and every country, especially some great powers in the world, has been paying more attention to national food security and its security system. According to the food white book released by Chinese State Council in October 1996, the food demand in 2000 is 50 million ton (Wang et al., 2007). However, the annual food yield from 2000 is less than 46.5 million ton (Feng, 2007). The 2005 food yield increased to 48.4 million ton, but it didn't reach the predictive yield. There are some complicated relationship among food security, population and cultivated land. The average cultivated land area decreased to 40% of 1950' area on the account of the inverse trend between population and cultivated land. We can foresee that the decreasing trend of average cultivated land is difficult to reverse and food yield will be confronted with severe challenges during a long time.

The grey system theory is a multidisciplinary theory dealing with those systems for which we lack information (Deng, 2007). Because multiple factors in related to the assessment of food security will be influenced, such as population, cultivated land, food production, some are unknown, some are known, which have distinct dynamic characteristics and uncertainty, thus we belong them to grey system theory. Therefore, with the help of mathematical model resulted from grey system, we could do some research on food security dynamic change and trend so as to acquire the basis of food security system equilibrium.

Food security assessment consists of evaluating annual food total yield of Liaoning Province and its 14 prefecture level cities, market demands and food supply-demand assessment, etc. The following 4 elements, population, cultivated land area, food yield and structure, are the principal factors to measure land security. Besides them, food yield fluctuating coefficient, market supply coefficient, and cultivated land change coefficient should be paid some attention while doing comprehensive analysis.

2. GREY PREDICTION MODEL

2.1 Grey Model Theory

Grey system model or GM (m, n) is a modeling method based on the concept of grey model and take differential fitting method as core. the parameters m and n are the order of model differential equation and the sequences number of modeling, respectively. The more the order is, the more complex calculation and the less precision improved. Therefore, GM (1, 1) is adopted to be the prediction model (Mi et al., 2007). Considering an original time series is expressed as

$$\{x^{(0)}(k)\} \qquad (k = 1, 2, \Lambda, n) \tag{1}$$

such a quantity was made an accumulating generation and it becomes

$$x^{(1)}(k) = [x^{(1)}(1), x^{(1)}(2), \cdots, x^{(1)}(n)] \tag{2}$$

where $x^{(1)}(k)$ is a progressive function relationship, which is modeling GM (1, 1).

2.2 GM (1, 1) Modeling and Its Solution Process

According to the GM (1, 1), the white differential equation is computed from equation (2).

$$\frac{dx^{(1)}(k)}{dk} + ax^{(1)}(k) = u \tag{3}$$

where a is developing coefficient that controls system development. The factor u is grey action to reflect the change of time series.

$$x^{(1)}(k) = \sum_{i=1}^{t} x^{(0)}(i), \quad (k = 1, 2, \Lambda, n) \tag{4}$$

In terms of least-square error method, the parameters in GM (1, 1) are expressed as

$$M = [a, u]^T = (B^T B)^{-1} B^T Y_n \tag{5}$$

where

$$B = \begin{bmatrix} -0.5[x^{(1)}(2) + x^{(1)}(1)] & 1 \\ -0.5[x^{(1)}(3) + x^{(1)}(2)] & 1 \\ \vdots & \vdots \\ -0.5[x^{(1)}(n) + x^{(1)}(n-1)] & 1 \end{bmatrix},$$

$$Y_n = \begin{bmatrix} x^{(0)}(2) \\ x^{(0)}(3) \\ \vdots \\ x^{(0)}(n) \end{bmatrix}.$$

Such a quantity $-0.5[x^{(1)}(2)+x^{(1)}(1)]$ could be defined to be nearest neighbor mean of $x^{(1)}(k)$. The factor M is computed and the value of a and u are put into (3), then solving the differential equation where $x^{(1)}(1)=x^{(0)}(1)$ is the initial value and gives

$$\hat{x}^{(1)}(k)=\left[x^{(1)}(1)-\frac{u}{a}\right]e^{-ak}+\frac{u}{a},(k=1,2,\Lambda,n) \tag{6}$$

Using the predicted value given by (6) will have the predicted value $\hat{x}^{(0)}(k)$, when the time is at k.

$$\hat{x}^{(0)}(k)=\hat{x}^{(1)}(k)-\hat{x}^{(1)}(k-1) \tag{7}$$

2.3 GM (1, 1) Precision Inspection

The precision inspection should be made to determine whether the predicted value is believable or not. This paper adopts post-variance test method to do inspection. The variance proportion c and small error probability p in post-variance test method is expressed as

$$c=S_2/S_1 \tag{8}$$

where S_2 and S_1 are mean square errors of residual and actual sequences, respectively.

$$p=\left\{\left|\varepsilon(k)-\bar{\varepsilon}\right|<0.6745S_1\right\} \tag{9}$$

where $\varepsilon(k)$ and $\bar{\varepsilon}$ are the residuals sequence and its mean, respectively (Ye et al., 2005).

The less value the quantity c is, the better prediction of GM model. The more the value p demonstrates the more probability of little error and the higher precision. Table 1 is the value c and p under different precision levels.

Table 1. Standard of GM (1, 1) precision estimation

Level of precision	Excellent	Good	Pass	No Pass
c	<0.35	0.35-0.50	0.50-0.65	>0.65
p	>0.95	0.95-0.80	0.80-0.70	<0.70

3. APPLICATION OF GM (1, 1) IN FOOD SECURITY PREDICTION

3.1 Introduction of Data Source

According to the available information and the actual situation on Liaoning Province, the selected data consists of annual main corps yield, population, cultivated land area, food demand and main corps per area yield in Liaoning Province and its' 14 regional cities. All data comes from "Liaoning Statistical Yearbook" which was input into "food security.mdb" ACCESS database. The database includes the following data tables: cultivated land area and population of regional cities, food demand per capita, primary products yield per area in the prefecture level cities, and so on.

Table 2. Data Structure of Food Yield Data Table

Field name	Type	Description
year	long	year
cur_sum	double	total food yield of this year (10^4t)
rice	double	total rice yield of this year (10^4t)
wheat	double	total wheat yield of this year (10^4t)
maize	double	total maize yield of this year (10^4t)
sorghum	double	total sorghum yield of this year (10^4t)
millet	double	total millet yield of this year (10^4t)
tubers	double	total tubers yield of this year (10^4t)
soybean	double	total soybean yield of this year (10^4t)
others	double	total others yield of this year (10^4t)

The above Table 2 is the data structure of food yield data table, one example of all the data tables.

3.2 Food Security Assessment Flow Based on GM (1, 1)

Food security assessment based on GM (1, 1) is divided into three procedures:

(1) The description of problem. Given the formalized representation of problem to satisfy the demand of computer expressing and disposing, including the prediction of annual total food yield, market total demands prediction, yield-demands warning, and food integrated assessment warning in Liaoning Province and its' 14 prefecture level cities. Assessing result can be expressed in the patterns of curve, histogram and pie charts, and the food yield is forecasted grounded on diversified demand standard.

(2) The division of problem. Four task presentations and self-precision inspection system of the food security assessment will be solved simply and obviously.

(3) The solution of problem. The brief and clear system interface and self-generated result files are good to the problem solving.

The framework of food security assessment system can be expressed as Figure 1.

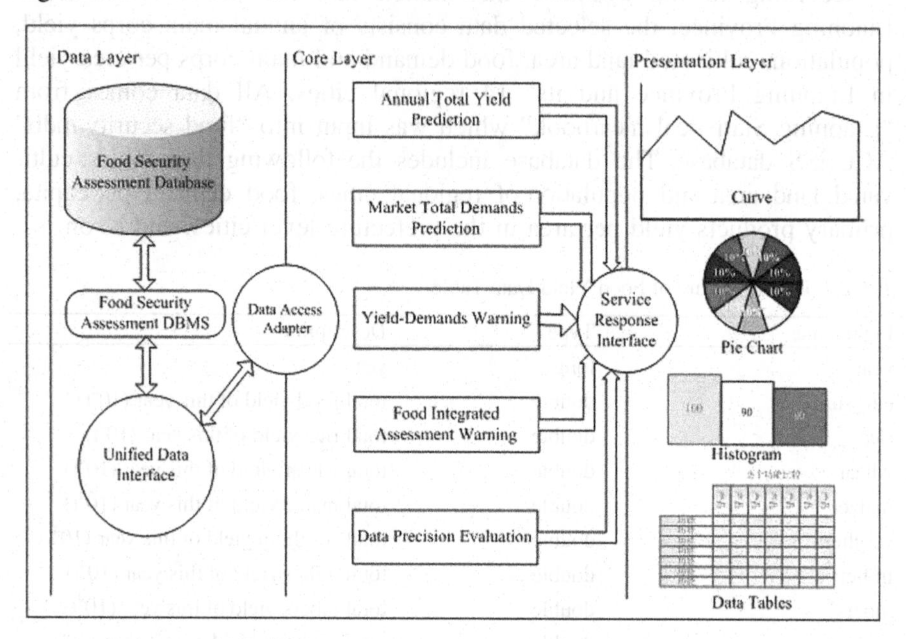

Figure 1. Food Security Assessment System Framework

There is a three-tier architecture used in the food security assessment system, which contains data layer, core layer and presentation layer and they are responsible for achieving data access, function processing and the display of assessment results, respectively (Cai et al., 2006).

(1) Data layer

Food security assessment system contains plenty of statistical data tables, and the function of data layer is to do some operations about reading, writing and deleting the statistical data from the assessment database. It is capable of reducing the dependence of core layer to concrete data and increasing the transparence of database access and system flexibility. Assessment system is managed by specialized DBMS, which supplies unified data interface and data access adaptor to transport information to core layer (Liu et al., 2006).

(2) Core layer

Core layer is the kernel of assessment system, which is responsible for the implementation of system tasks. It responds to the requests of data access adaptor, operations of database, transaction of system tasks, and then reflected on the presentation layer through service response interface.

(3) Presentation layer

Based on the three-tier architecture model, presentation layer is mainly responsible for the results of assessment display. Presentation will respond according to the reception from core layer, and show the assessment results in the form of curves, histogram, pie charts and tables. During the processing of requests responding and results generating, core layer will connect with data layer to store the assessment outcome if necessary.

3.3 Precision Evaluation

GM (1, 1) model is created based on the time series formed by the total food total yield data of Liaoning Province in year of 1980–2005. We define 3 as the system prediction interval to predict yield, so there are 8 values to compare with the 8 actual yields under the precision evaluation (Table 3).

Obviously, it has a high credibility from the comparison of the value c and p in Table 3 and the level of precision in Table 1. The following system running of food security integrated assessment is illustrated in Figure 3.

Table 3. Result of Precision Evaluation

Year	Interval	Actual yield (10^4t)	Predicted (10^4t)	Relative error
1983	3 years	1485.2	1314.3	11.51%
1986	3 years	1222.2	1106.3	9.48%
1989	3 years	968.2	802.3	17.14%
1992	3 years	1588.6	1522.7	4.15%
1995	3 years	1390.5	1246.6	10.35%
1998	3 years	1960.0	1804.1	7.95%
2001	3 years	1419.9	1254.0	11.68%
2004	3 years	1955.7	1659.8	15.13%
c		0.438		
p		0.875		

3.4 The Realization of Food Security Assessment

The food security assessment system is composed of 4 models, Annual Total Yield Prediction, Market Total Demands Prediction, Yield-Demands Warning and Food Integrated Assessment Warning. The two picture Fig. 2 and Fig. 3 display the models of Market Total Demands Prediction and Food Integrated Assessment Warning, respectively.

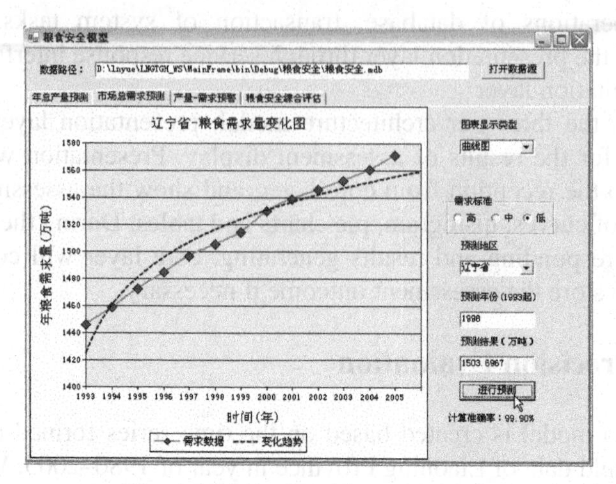

Figure 2. Market Total Demands Prediction

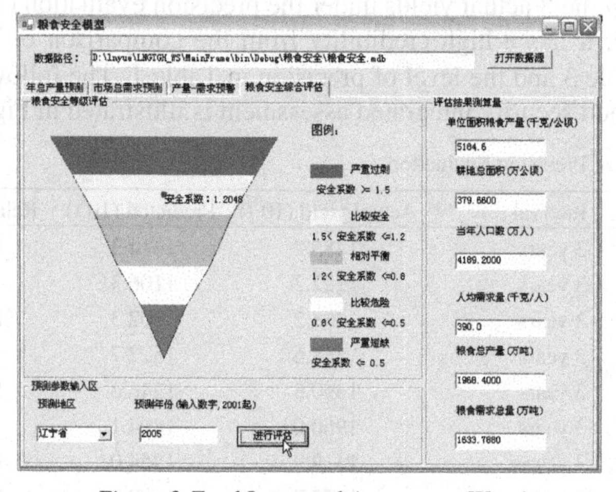

Figure 3. Food Integrated Assessment Warning

We can see from Figure 3 the safe coefficient of food security assessment of Liaoning Province in 2005 is 1.2048, which is in the light green region and indicates the security degree is relative safe.

4. CONCLUSIONS

Grey model, which is widely used in the field of science and technology, has a good performance.

The use of grey models is a novel concept in food security assessment, and it provides a new scientific method or insight for analytical prediction in food security field. In the paper, the result indicates high validity and feasibility of food security assessment based on GM (1, 1) model.

Moreover, the framework of food security assessment is created based on three-tier architecture and the assessment system adopts the object-oriented development method and component technology.

At last, the realization of the assessment system presented successfully. It is suggested that food security assessment based on grey model is an effective attempt to solve food security problems.

ACKNOWLEDGEMENTS

This study has been funded by China National 863 Plans Projects (Contract Number: 2006AA120102).

REFERENCES

Cai Hongwei, et al. Research and appliance of the flexible three-tier architecture. Journal of Zhejing Sci-Tech University, 2006, 23(2): 178–181 (in Chinese).

Deng Julong. The Journal of Grey System. http://www.researchinformation.co.uk/grey.php, 2007.

Feng Zhiming. Future food security and arable land guarantee for population development in China. Population Research, 2007, 31(2): 15–29 (in Chinese).

Liu Hongtao, et al. Management system based on the B/S mode three-layer structure. Seismological Research of Northeast China, 2006, 22(1): 75–80 (in Chinese).

Mi Hongyan, et al. Application of the Grey Model GM (1, 1) to forecast of building settlement. Journal of Southwest Forestry College, 2007, 27(1): 81–84 (in Chinese).

Wang Wenlong, et al. China food security problem: not only the surface but also the bottom. Reformation and Strategy, 2007, 4: 39–41 (in Chinese).

Ye Mingquan, et al. Seasonal artificial neural network forecasting model and its application in the GM (1, 1) residual error correction, Computer Engineering and applications, 2005, 41(1): 194–196 (in Chinese).

CROP DISEASE LEAF IMAGE SEGMENTATION METHOD BASED ON COLOR FEATURES

Lidi Wang*, Tao Yang, Youwen Tian
College of Information & Electrical Engineering, Shenyang Agricultural University
** Corresponding author, Address: College of Information & Electrical Engineering, Shenyang Agricultural University, 120 Dongling Road, Shenyang, 100161, P. R. China, Tel: +86-24-88487129, Email: wanglidi@gmail.com*

Abstract: The color feature has been taken an important role in color image segmentation, especially in the fields of automatic detection of crop disease based on leaf image. In this paper an effective method for image segmentation of cucumber leaf images is proposed. First, the color space model is analyzed. Then a kind of color feature is applied to obtain the feature map, which combines RGB model and HSI model. Finally, the morphological method is used to accomplish the image segmentation. This method has been shown effective through experiments.

Keywords: crop disease, color image segmentation, feature extraction, color space

1. INTRODUCTION

Image segmentation is an important topic in image processing task. It can effect the progress of the whole recognition which is becoming more and more important in the agricultural automation. For example, in the field of crop disease automatic recognition. According to the crop leaf image, the disease type can be confirmed with the image processing method and expert's knowledge. Howerever, the segmentation's accuracy cannot satisfy the current practical needs. There are primarily four types of segmentation techniques: thresholding, boundary-based, region-based, and hybrid techniques. (Frank Y, 2005) Thresholding is the representative method and depends on the feature map and threshold value. It is based on the assumption that clusters in the histogram correspond to either background or

Wang, L., Yang, T. and Tian, Y., 2008, in IFIP International Federation for Information Processing, Volume 258; Computer and Computing Technologies in Agriculture, Vol. 1; Daoliang Li; (Boston: Springer), pp. 713–717.

objects of interest that can be extracted by separating these histogram clusters. In this paper, combining the RGB model and HSI model, a kind of color feature is applied. Along with the threshold selection and morphological method, an image segmentation method is proposed.

The remainder of this paper is organized as follows. The color space is explained in section II. Section III includes the color feature selection and their characters. Section IV explains the image segmentation method. Finally, the conclusions and discussions of this paper are given in section V.

2. COLOR SPACE

Colors are the important feature in color image processing, especially in crop images. Color provides important information for humans to recognize images which can be illuminated under a very wide range of conditions. Commonly used well-known color spaces include (T. Gevers, 1999): (for display and printing processes) RGB, CMY; (for television and video) YIQ, YUV; (standard set of primary colors) XYZ; (uncorrelated features) I1I2I3; (normalized color) rgb, xyz; (perceptual uniform spaces) and (for humans) HSI. Therefore, in this paper, we concentrate on the following standard, color features: intensity I, RGB, hue H and saturation S. Every color space has its advantages and disadvantages.

The HSI color model is cylindrical with the intensity axis coinciding with the achromatic diagonal of the RGB system. Saturation is the radius from the intensity axis and hue is the angle with respect to the red direction. From a perceptual point of view, color can be described in HIS color model by three attributes. The Hue H is a value which represents the main color of the pixel in the RGB triplet; the saturation S describes the pureness of the color, and I represents the amount of light received by the sensor. It depends on the lighting conditions and on the light source emissivity (Marcos, 1998).

Let R, G and B, obtained by a color camera, represent the 3-D sensor space

$$C = \int_{\lambda} p(\lambda) f_c(\lambda) d\lambda \tag{1}$$

for $C \in (R,G,B)$, where $p(\lambda)$ is the radiance spectrum and $f_c(\lambda)$ are the three color filter transmission functions.

To represent the RGB-sensor space, a cube can be defined on the R, G, and B axes. White is produced when all three primary colors are at M, where M is the maximum light intensity, say M=255. The main diagonal-axis connecting the black and white corners defines the intensity

$$I(R,G,B) = R + G + B \tag{2}$$

The transformation from RGB to describe the color impression hue H is given by

$$H(R,G,B) = \arctan\left(\frac{\sqrt{3}(G-B)}{(R-G)+(R-B)}\right) \quad (3)$$

and saturation S measuring the relative white content of a color as having a particular hue by

$$S(R,G,B) = 1 - \frac{\min(R,G,B)}{(R+G+B)} \quad (4)$$

The saturation is a percentage between 0 and 100% (0% for achromatic colors as white, greys or black and 100% for the vivid colors). Hue is measured as an angle between 0 and 360° (0°, 120°, 240° for the basic red R, green G and blue B). (M. Herbin, 1993)

3. FEATURE EXTRACTION METHOD

As analyzed in section 2, varies color features are used in image segmentation. In RGB color space, three features named color mean often be used:

$$\begin{cases} \overline{R} = \frac{1}{n}\sum_{i=1}^{n} R_i \\ \overline{G} = \frac{1}{n}\sum_{i=1}^{n} G_i \\ \overline{B} = \frac{1}{n}\sum_{i=1}^{n} B_i \end{cases} \quad (5)$$

In HSI color space, three parameters H, S and I are commonly used to reflect corresponding characters of the image.

Take cucumber disease leaf image as an example. The feature map of original image is computed as shown in figure 1.

From the figure 1, it is easily to show the pixels in the disease area of the leaf have two distinct characters: one is their R value is much higher than others, the other is their H value is much lower than others. So a feature can be defined:

$$RH = R - H \quad (6)$$

Using this feature, image segmentation could be accomplished more effectively. The details will be shown in the next section.

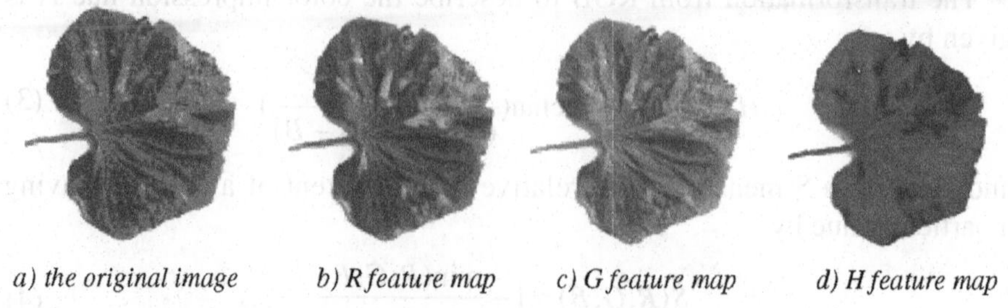

a) the original image *b) R feature map* *c) G feature map* *d) H feature map*

Fig. 1. Cucumber disease leaf image and its different feature maps

4. IMAGE SEGMENTATION METHOD

The details of the color image segmentation method are shown as follows:

Step 1: leaf target extraction. Using the region increasing method to obtain the whole leaf image without the background.

Step 2: computing color feature map. Using the RH feature discussed in the section 3, the feature map can be obtained.

Step 3: threshold selection. Select a suitable threshold value according experience. A binary image can be obtained.

Step 4: morphology processing. Wipe out the little holes and noises by morphological method.

The experiment results are shown as figure 2.

a) the original image

b) leaf target extraction results

c) RH feature map

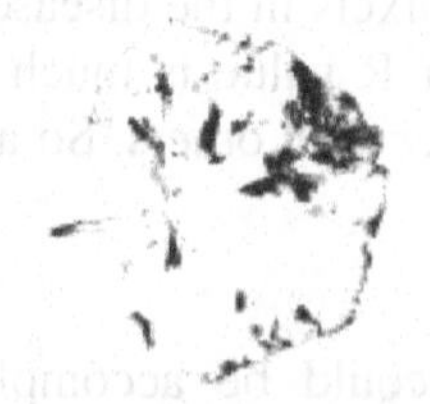

d) thresholding result

e) morphological processing result

Fig. 2. Image segmentation results based on the color feature

5. CONCLUSION

Through experiments results it can be concluded that the proposed method can accomplish the image segmentation effectively. Owing to the complex of the disease leaf image's colors, figures and textures, the shortage of this method is the threshold selection method, and this is the hard work in thresholding method of image segmentation. In our experiment, the Otsu algorithm has been used, but cannot obtain the effective result. More effectively automatic threshold value confirming method is the next work to do.

ACKNOWLEDGEMENTS

This research is supported by Natural Science Foundation of Liaoning Province, China (No. 20052125)

REFERENCES

Frank Y. Shih, Shouxian Cheng. Automatic Seeded Region Growing for Color Image Segmentation. Image and Vision Computing 2005, 23(10): 877–886.

M. Herbin, F.X. Bon, A., Venot etc. Assessment of Healing Kinetics Through True Color Image Processing. IEEE Transactions on Medical Imaging, 1993, 12(1): 39–43.

Marcos Cordeiro D'Ornellas, Rein Van Den Boomgaard, Jan-mark Geusebroek. Morphological Algorithms for Color Images Based on a Generic-programming Approach. Anais do XI SIBGRAPI, 1998: 1–8.

T. Gevers, A.W.M. Smeulders. Color-based Object Recognition. Pattern Recognition, 1999, 32(11): 453–464.

RESEARCH ON THE SPATIAL VARIABILITY OF SOIL MOISTURE BASED ON GIS

Changli Zhang [1] , Shuqiang Liu [2,*] , Junlong Fang [1] , Kezhu Tan [3]
[1] *Engineering College, Northeast Agricultural University, Harbin, Heilongjiang, China, 150030*
[2] *Heilongjiang Institute of Technology, Harbin, Heilongjiang, China, 150050*
[3] *Chengdong College, Northeast Agricultural University, Harbin, Heilongjiang, China, 150030*
* *Corresponding author, Address: Academic Administration, Heilongjiang Institute of Technology, 999 Hongqi Road, Harbin, Heilongjiang, 150050, P. R. China, Tel: +86-0451-88028684, Fax: +86-13946117239, Email: shuqiangliu@hotmail.com*

Abstract: With the help of GPS and measuring instrument of soil moisture, soil moisture was measured and analyzed. As using Geo-statistics to the study of spatial variability of soil moisture and use ArcGIS 9.0, get the spatial distribution map of soil water property with Kriging interpolation. The research result showed that all soil spatial characters are normal distribution and the spatial distribution of soil water property accord with the fact. Geo-statistics Methods is the most appropriate methods in all of Mathematical Methods for Geo-statistics. The spatial distribution map of soil water property what got with Kriging interpolation can make the spatial distribution of the entire plot, more accurate and reliable. Getting a veracious spatial distribution map of soil water speciality was very important and useful for adjusting precision fertilization and precision irrigation in time. It also offered the theoretical foundation of the connection studying between soil water speciality and enhancing the yield.

Keywords: Geographical information system, spatial interpolation, Geo-statistics, spatial variability

1. INTRODUCTION

Soil was inhomogeneous and continuous nature. Actual instance in the field indicated, in the synchronization, soil speciality also had obvious difference on the different spacial situation; this property was named the

Zhang, C., Liu, S., Fang, J. and Tan, K., 2008, in IFIP International Federation for Information Processing, Volume 258; Computer and Computing Technologies in Agriculture, Vol. 1; Daoliang Li; (Boston: Springer), pp. 719–727.

spatial variability of soil moisture (Meng et al., 1992). The most intuitionistic mode of the spatial variability of soil moisture was the spatial distribution map of soil water. Therefore, the all-important work and base of Precision Agriculture is producing the spatial distribution map of soil moisture reliably. This study researched the spatial variability of soil moisture based on Geo-statistics, and applied Kriging interpolation of ArcGIS 9.0 to produce the spatial distribution map of soil moisture, it could offer the scientific evidence for farm management system.

2. BASIC THEORY AND METHOD OF THE SPATIAL VARIABILITY

2.1 Regionalized variables theory

When a variable assumed spacial distribution, it was regionalized variable (Jun et al., 2000; Zhang et al., 1995). The variables reflected distributing character of some spatial property. Regionalized variables had two important characters: the first one was that regionalized variables Z(X) was a random function, its character were local, stochastic, exceptional; the second one was that regionalized variables had ecumenical and average structure quality, the variables Z(X) and Z (X + h) had correlation in some extent at point X and point X + h that its acentric spatial distance was h. At some significance, this was structural character of regionalized variables. Soil moisture and other farmland information were all regionalized variables, therefore it can use regionalized variables theory to study their spatial variable laws.

2.2 Semi-varionram function

Farmland information was fully random variables, analyzed by traditional statistics. The most research only considered mean value and dispersion coefficient of all the observable value, and not considered the difference of the observable point. But actually, most of farmland information was a spatial-temporal continuous variants, not only had randomicity but also had structure. Therefore, it could use semi-varionram function of studying spatial variance of regionalized variables in Geo-statistics to describe spatial variability of farmland information.

Semi-varionram function was a function which described the spatial variability structure of Soil Moisture, scaling spatial correlative extent of known points, calculated by this formula:

$$\gamma(h) = \frac{1}{2n}\sum_{i=1}^{n}(z(x_i) - z(x_i + h))^2 a$$

In which: h was the distance of known points, frequently be lag; n was the number of conjugated swatch points disjoined by h; z was attribute value. Semi-varionram was augmenting along with h augmenting.

Semi-varionram function was generally expressed by variance curve, it was a function map with r(h) to h. Fig. 1 is a representative sketch map of Sphere Model semi-varionram function. C_0 was fundus variance, expressed spatial variance by semi-varionram, it was generally metrical error by observation variance. The a was mutative distance, it expressed that there was distance upper limit in the sampling data. When h≤a, the observation value between random two points had relativity that which was augmenting along with h falling; When h>a, there were no relativity. C_0+C was r(h), it reflected spatial variance intensity of some observational variable in the research region. The eigenvalue of function expressed variable character of observational variable.

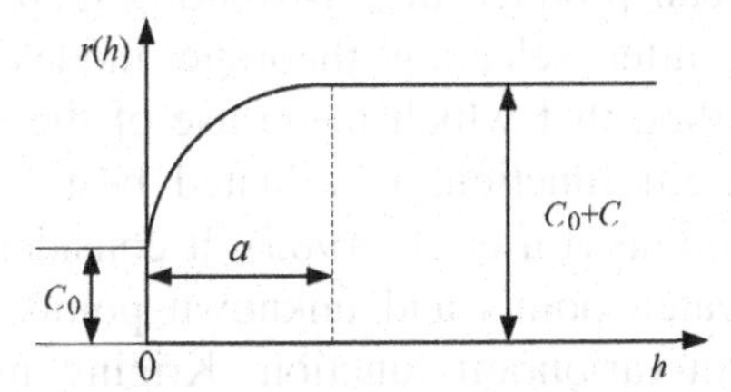

Fig. 1. Semi-varionram of spherical fitted model

2.3 Theoretic model of semi-varionram function

When describing quantificationally variable character of all the research region, it need produce semi-varionram map. It need seek theoretic model of semi-varionram for produce semi-varionram map (Zhang et al., 1995; Feng et al., 2000; Huang et al., 2002; Li et al., 1998), this model attached straightway oneself to Kriging calculation. On account of Kriging formula positive quality, actually there were only some theoretic model of semi-varionram function, generally, Spherical Model, Index Model, Gauss Model, Power Model, Sine Model.

1) Spherical Model

$$\gamma(h) = \begin{cases} 0 & h = 0 \\ c_0 + c_1\left\{\frac{3}{2}\frac{|h|}{a} - \frac{1}{2}\left(\frac{|h|}{a}\right)^3\right\} & 0 < |h| \le a \\ c_0 + c_1 & |h| > a \end{cases}$$

$$h = a,\ r(h) = C_0 + C.$$

2) Index Model

$$\gamma(h)=\begin{cases}0 & h=0\\ c_0+c_1\left(I-\exp\left(-3|h|/s\right)\right) & h\neq 0\end{cases}$$

3) Gauss Model

$$\gamma(h)=\begin{cases}0 & h=0\\ c_0+c_1\left(I-\exp\left(-3\left(\frac{|h|}{a}\right)^2\right)\right) & h\neq 0\end{cases}$$

2.4 Used Kriging interpolation to produce the distributing map of soil moisture

With semi-varionram function, it can reflect spatial variable rule of region variable exactly. We can produce the distributing map of soil moisture by Kriging interpolation after selecting theoretic model of semi-varionram function. It was a method that which made use of the structure of original data and semi-varionram function, it valuated best and truly localizable variable of unknown points (Lu et al., 1985). It considered distance and the relation of known swatch points and unknown points by the structure of original data and semi-varionram function. Kriging interpolation got the value of unknown points by endowing known points with weight, it expressed:

$$Z(x_0)=\sum_{i=1}^{n}\lambda_i Z(x_i)$$

In which: $Z(x_0)$ was the value of unknown points, $Z(x_i)$ was known swatch points around unknown points, λ_i was the power that point i to unknown points. For satisfying non-Biased and optimality, passed establishing Kriging formula to ascertain weight coefficient (Li et al., 2006):

$$\begin{cases}\sum_{j=1}^{n}\lambda_j\gamma(x_i,x_j)+\mu=\gamma(x_i,X)\\ \sum_{i=1}^{n}\lambda_i=1\end{cases}$$

In which: $\gamma(x_i,x_j)$ was covariance function in sampling points, $\gamma(x_i,X)$ was covariance function between sampling points and interpolation points, μ was Lagrange multiplier.

3. DESIGN OF THE EXPERIMENT

Experiment farmland was a soybean field of Heilongjiang Daxijiang farm, its area was about 9.8 hm2. June 27, 2006, soil moisture was measured with the help of GPS and measuring instrument of soil moisture, grid was setting by 15m□15m, about 250 points, sampling deepness was 10 cm.

Sampling mode of soil moisture was point sampling (Joseph K. Berry et al., 1999; Wollenhaupt N.C. et al., 1997; Hao et al., 2002), inerratic grid sampling disjoin the farmland to equal area grid. It was simple and exercisable, was the most effective one of sampling method. This research considered sampling convenience and actual condition, compartmentalized grid along ridge in a field generally.

4. ANALYZED THE DATA

4.1 Calculated the statistical eigenvalue of soil moisture

Based on classical statistics, we calculated statistical eigenvalue of the data of soil moisture in the experiment farmland; experimental results were shown in Table 1. Variance coefficient were respectively 119.2, 117.3, 127.6, all belong to strong variability.

Table 1. Statistical description of soil moisture

Date	Minimum value	Maximal value	Average value	Standard value	Variance	Variance coefficient
June 27, 2006	0.09	0.357	0.224	0.267	0.071	119.2%
July 17, 2006	0.086	0.33	0.208	0.244	0.06	117.3%
August 8, 2006	0.099	0.45	0.275	0.351	0.123	127.6%

4.2 Testing normal distribution of soil moisture data

Testing for normal distribution of soil moisture data was the precondition of using Kriging interpolation of Geo-statistics to analyze soil moisture data. If sampling data submitted to normal distribution, its sampling points should be linear. In ArcGIS, we tested normal distribution to soil moisture data by normal QQ plot (Fig. 2), the result indicated that most points accord with normal distribution. Several points left the beeline overmuch, after testing, it was eligible.

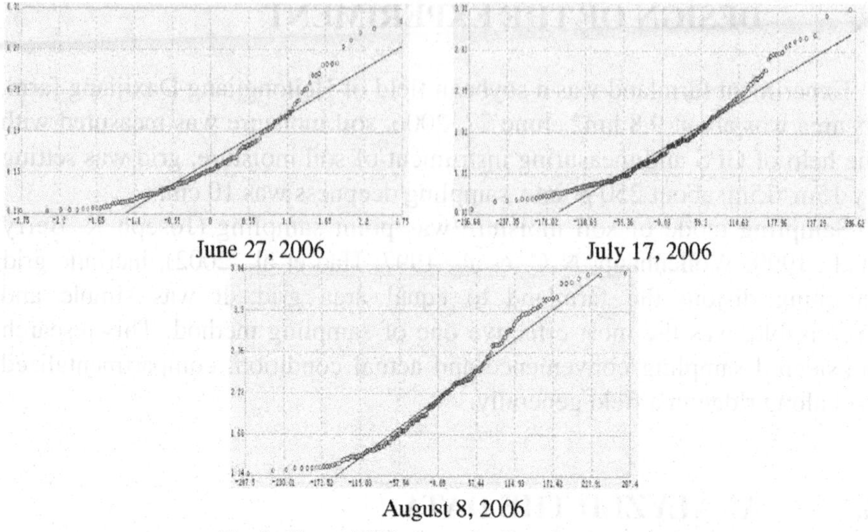

Fig. 2. Normal QQ plots of soil moisture properties

4.3 Spatial interpolation analyzed soil speciality

For describing true and intuitionistic spatial distribution of soil moisture, the research used Spatial Analyst module in ArcGIS 9.0, applied the Kriging to get the distributing map of soil moisture. Experimental results were shown in Fig. 3, Fig. 4, Fig. 5.

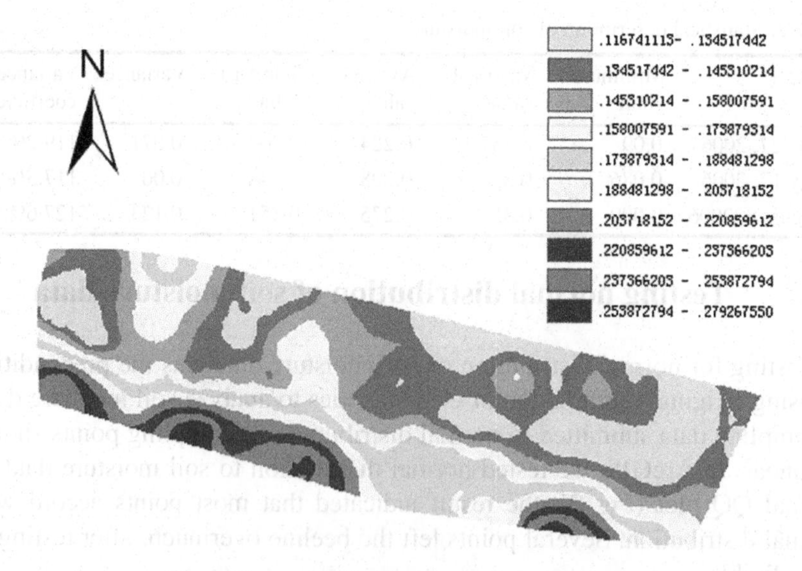

Fig. 3. The spatial distribution map of soil moisture

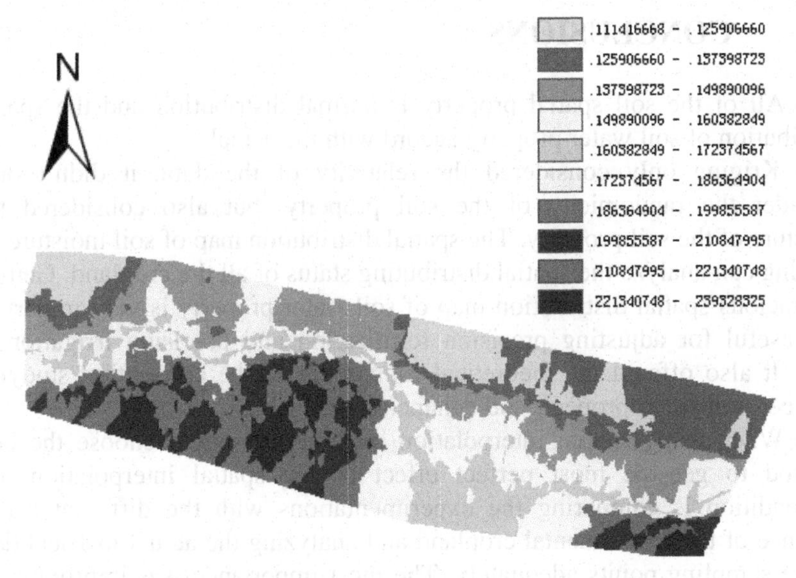

Fig. 4. The spatial distribution map of soil moisture

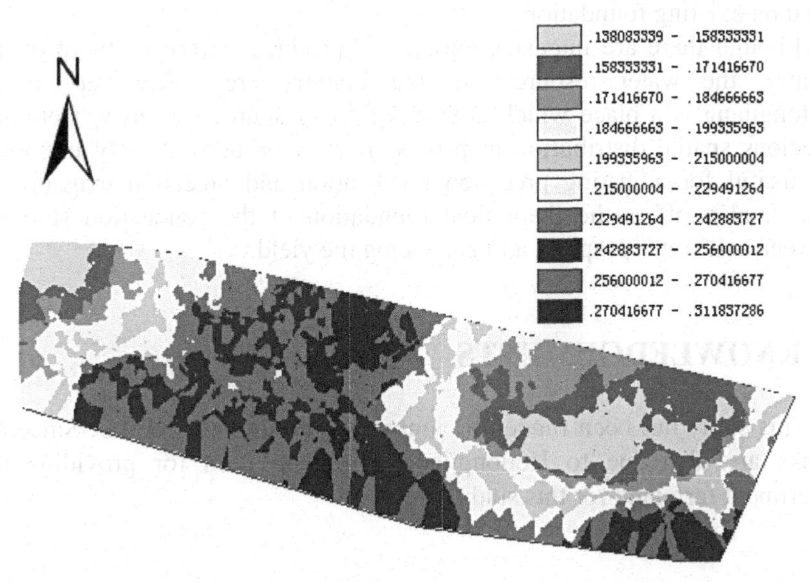

Fig. 5. The spatial distribution map of soil moisture

We saw the spatial distribution of soil moisture. In this farmland, there were low of the north and high of the south. It was consistent with the fact.

5. CONCLUSIONS

1) All of the soil spatial property is normal distribution and the spatial distribution of soil water property accord with the actual.

2) Kriging only considered the relativity of the data, it didn't only consider the randomicity of the soil property, but also considered the structure of the soil property. The spatial distribution map of soil moisture by Kriging can analyze the spatial distributing status of all the cropland. Getting a veracious spatial distribution map of soil water property is very important and useful for adjusting precision fertilization and precision irrigation in time. It also offered the theoretical foundation of the connection studying between soil water property and enhancing the yield.

3) When using spatial interpolation methods, it should choose the best method to get the most perfect effect of the spatial interpolation, the precondition is comparing the experimentations with the different actual instance of the experimental cropland and analyzing the actual metrical data of the sampling points adequately. The most importance was improving on interpolation methods, putting forward a new better scientific interpolation based on existing foundation.

Although there are intensive agriculture produces in the northern of our country, the water resources of our country are scarce very much. Heilongjiang is a place which is short of water source seriously. Getting a veracious spatial distribution map of soil water property is very important and useful for adjusting precision fertilization and precision irrigation in time. It also offers the theoretical foundation of the connection studying between soil water property and enhancing the yield.

ACKNOWLEDGEMENTS

This study has been funded by innovating program of NEAU. Sincerely thanks are also due to Heilongjiang Daxijiang farm for providing the experiment farmland for this study.

REFERENCES

Feng X.Z., Bo Y.C., Shi Z.T. 2000. Snow Depth in North Xinjiang Region Estimated by Kriging Interpolation. Journal of Glaciolgy and Geocryology. 22(4):358-361.

Hao P.F., Liu G., Jiang W.K. 2002. Approaches to processing soil fertility spatial distribution information for precision agriculture. Journal of Agricultural University of Hebei. 25: 277-278.

Hua M., Wang J. 1992. Soil Physics. BeiJing: BeiJing agriculture publishing company.

Huang Sh.W., Jin J.Y., Yang L.P. 2002. Spatial Variability of Soil Nutrients in Grain Crop Region of Yutian County. Chinese Journal of Soil Science. 33(3):188-193.

Joseph K. Berry. The Precision Farming Primer. BASIS, Inc.1999.

Li J., You S.C., Huang J.F. 2006. Spatial interpolation method and spatial distribution characteristics of monthly mean temperature in China during 1961-2000. Ecology and Environment. 15(1):109-114.

Li J.M., Li Sh.X. 1998. Spatial variance of several nutrition element in the soil. Agricultural Research in the Arid Areas. 16(2):58-64.

Lu W.D., Zhu Y.L., Sha J. 1985. The elementary study on spatial variability of soil speciality. Journal of North China Institute of Water Conservancy and Hydroelectric Power. (9): 10-21.

Wang J., Fu B.J., Qiu Y. 2000. Spatiotemporal Variability of Soil Moisture in Small Catchment on Loess Plateau—Semivariograms. ACTA GEOGRAPHICA SINICA. 55(4):428-438.

Wollenhaupt N.C., Mulia D.J., Gotway Crowford C.A. 1997. Soil Sampling and Interpolation Techniques for Mapping Spatial Variability of Soil Properties. The Site-Specific Management for Agricultural Systems. ASA-CSSA-SSSA, 777 S. Segoe Rd., Madison, WI53711, USA. 19-53.

Zhang X.F., Van Eijkeren J.C.H., Heemink A.W. 1995. On the Weighted Least—Squares Method for Fitting a Semivariogram Model. Computers &Geosciences. 21(4):605-608.

DESIGN AND FULL-CAR TESTS OF ELECTRIC POWER STEERING SYSTEM

Jingbo Zhao[*], Long Chen, Haobin Jiang, Limin Niu
School of Automobile and Traffic Engineering, Jiangsu University, Zhenjiang, China, 212013
** Corresponding author, Address: P.O. Box 243, School of Automobile and Traffic Engineering, Jiangsu University, 301 Xuefu Road, Zhenjiang, Jiangsu, 212013, P. R. China, Tel: +86-511-82054123, Fax: +86-511-88791221(ext. 2009), Email:zhaojbl128@yahoo.com.cn*

Abstract: Electric Power Steering (EPS) is a full electric system, which reduces the amount of steering effort by directly applying the output from an electric motor to the steering system. In this paper, the constitutions and its operational mechanism of electric power steering system, and the construction and the equivalent circuit of the DC motor used in EPS were introduced; and the EPS hardware framework based on the ARM was presented and the EPS motor control strategy was designed. The full-car tests were performed and the results confirmed that the system designed was stable and credible, and can meet the requirements of steering performance.

Keywords: automobile, Electric Power Steering, EPS, motor, controller

1. INTRODUCTION

Electric power steering (EPS) system has attracted much attention for their advantages. It uses power only when the steering wheel is turned by the driver, it consumes approximately one-twentieth the energy of conventional hydraulic power steering systems and, as it does not contain any oil, it does not pollute the environment both when it is produced and discarded. Additionally, the software built into the EPS controller results in high performance and easy tuning during the development of prototypes of EPS systems (Jiang Haobin et al., 2006).

Fig. 1 shows a vehicle with a column-type EPS in which the reduction gear is located directly under the steering wheel. The EPS system consists of a

Zhao, J., Chen, L., Jiang, H. and Niu, L., 2008, in IFIP International Federation for Information Processing, Volume 258; Computer and Computing Technologies in Agriculture, Vol. 1; Daoliang Li; (Boston: Springer), pp. 729–736.

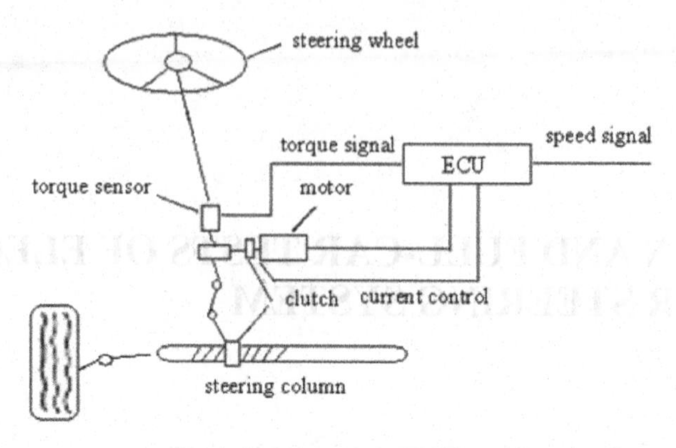

Fig. 1. A columm - type EPS system

torque sensor, which senses the driver's movements of the steering wheel as well as the movement of the vehicle; an ECU, which performs calculations on assisting force based on signals from the torque sensor and vehicle sensor; a motor, which produces turning force according to output from the ECU; and a reduction gear, which increases the turning force from the motor and transfers it to the steering mechanism (Toshinori Tanaka et al., 2003).

The main purpose of electric power steering system is, of course, to provide assist to the driver. This is achieved by the torque sensor, which measures the driver's torque and sends a signal to the controller proportional to this torque. The torque information is processed in the controller and an assist command is generated. This assist command is further modulated by the vehicle speed signal, which is also received by the controller. This command is given to the motor, which provides the torque to the assist mechanism. The gear mechanism amplifies this torque, and ultimately the loop is closed by applying the assist torque to the steering column.

2. DESIGN OF EPS CONTROLLER

The motor for EPS is a permanent magnetic field DC motor. Attached to the power steering gear assembly, it generates steering assisting force. This study introduced the motor control strategy, the design of the EPS Electric Control Unit (ECU), and the full-car test system. The full-car tests were performed and the results were analyzed.

2.1 Motor control strategy

Fig. 2 illustrates the construction of a DC motor, consisting of a stator, a rotor, and a commutation mechanism (Massachusetts Institute of Technology). The stator consists of permanent magnets, creating a magnetic field in the air gap between the rotor and the stator. The rotor has several windings arranged symmetrically around the motor shaft. An electric current applied to the motor is delivered to individual windings through the brush-commutation mechanism, as shown in the figure. As the rotor rotates the polarity of the current flowing to the individual windings is altered. This allows the rotor to rotate continually.

The actual DC motor is not a loss-less transducer, having resistance at the rotor windings and the commutation mechanism. Furthermore, windings may exhibit some inductance, which stores energy. Fig. 3 shows the schematic of the electric circuit, including the windings resistance R and inductance L.

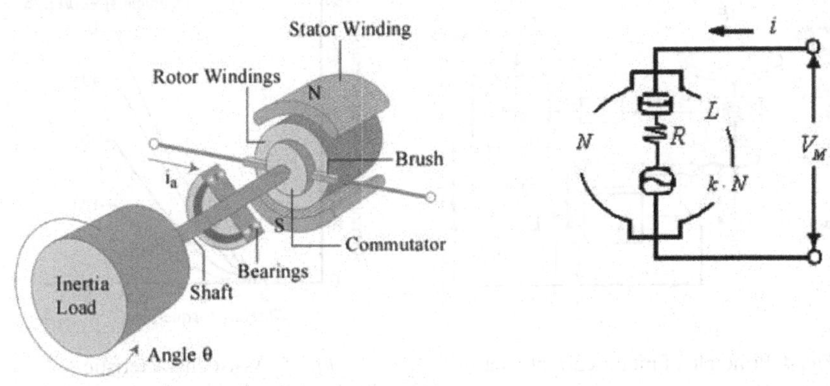

Fig. 2. Construction of DC motor

Fig. 3. Equivalent circuit

In the equivalent circuit of the motor, the relationship between the terminal voltage V_M, the impendance L, the resistanc R, the induced voltage constant k, the revolution speed N, the current i, and the time t, is expressed by the following equation (Ronald K. et al.,1999).

$$V_M = L \cdot (di/dt) + R \cdot i + k \cdot N \tag{1}$$

$$\approx R \cdot i + k \cdot N \tag{2}$$

And it is known that the current i is proportional to the motor torque T_M.

As can be understood from Eq. (2), the motor can be controlled based on the so-called motor current control method which is shown in Fig. 4.

In the motor current control method, the target motor current I_T, which is proportional to the motor assist torque T_M, is determined from the signal output T from the torque sensor, and control is performed so that there is no difference between this target current value I_T and the value detected through feedback from the current sensor I_M.

In this method, the target value, which is determined by the assist characteristic based on the input of torque sensor and vehicle speed sensor, for the motor current is set so that it is equal to the vehicle speed response type derived from the signal of the vehicle speed sensor. The assist characteristic is shown in Fig. 5 (Takayuki Kifuku et al., 1997).

The typical control system of EPS device is shown in the block diagram of Fig. 6 (Masahiko Kurishige et al., 2001; Ji-Hoon Kim, 2002). The target current setting unit determines the reference current i_r to the motor based on the driving conditions, and the controller computes the control signal which minimizes the error between i_r and the actual current i_a.

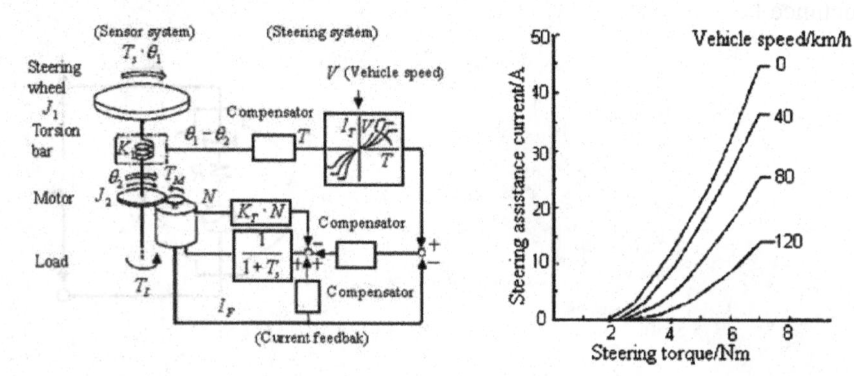

Fig. 4. Principle of motor current control

Fig. 5. Assist characteristic

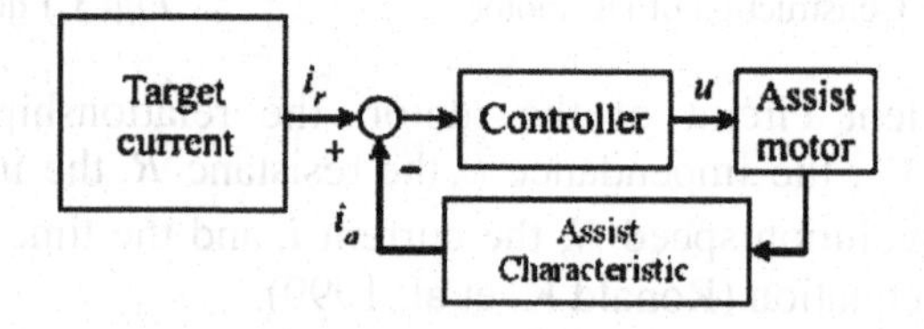

Fig. 6. Block diagram of EPS control system

2.2 Design of EPS controller

The EPS controller is designed based on motor current control method. The EPS controller (refer to Fig. 7) consists of an interface circuit that coordinates the signals from the various sensors, an A/D converter and a PWM unit that are all built into an one-chip microprocessor, a watchdog timer (WDT) circuit that monitors the operation of this microprocessor, the motor-drive circuit that consists of power MOSFETs in an H bridge circuit driven by pulse width modulation (PWM) over a 20kHz carrier (Ronald K. et al., 1999; Takayuki Kifuku et al., 1997; Yuji Kozaki et al., 1999; Masahiko Kurishige et al., 2001).

The ECU conducts a search for data according to a table lookup method based on the signals input from each sensor and carries out a prescribed calculation using this data to obtain the assist force.

In addition, trouble diagnosis for the sensors and the microprocessor is also carried out. When a problem is detected, power to the motor is interrupted, an indicator lamp illuminates, and the problem condition is memorized. Then this problem mode flashes on a display as necessary (International Rectifier Application Notes; Geoffrey Walker, 1998; Jeff Burns et al., 2000; Sergio Fissore et al., 2000).

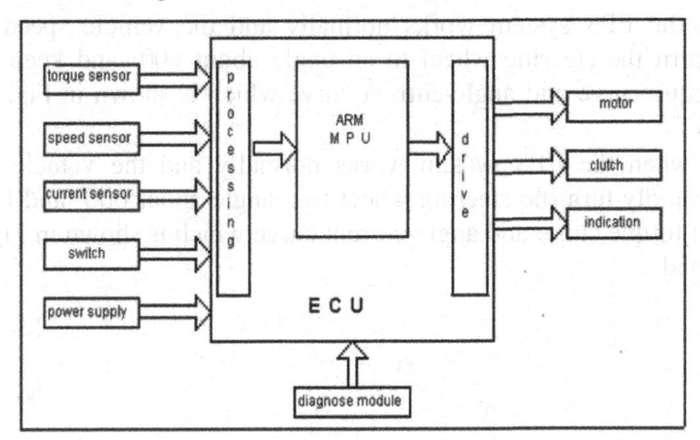

Fig. 7. Framework of EPS controller

3. TESTS AND ANALYSIS

On the basis of the motor control strategy and design of EPS controller, full-car tests are performed. The framework of test system is shown in Fig. 8.

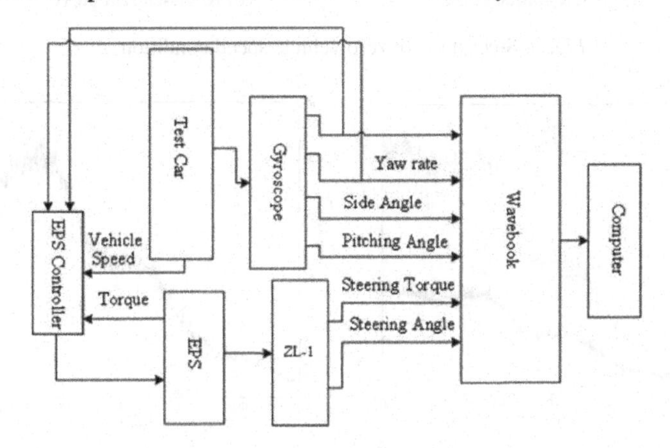

Fig. 8. Test system

The test system is made up of the following parts. (1) sensors: ZL-1 steering parameters test system, vehicle speed sensor, VG400 gyroscope (Grossbow co.), (2) controller: Electric Control Unit with ARM S3C44B0X, the main input signals include torque signal, vehicle speed signal and motor current signal, (3) signal collection system: WaveBook 512H, computer and DASYLab8.0.

3.1 Steering handiness test

When the EPS system works normally and the vehicle speed is zero, rapidly turn the steering wheel to an angle about 600° and keep still, the angle-torque curve and angle-current curve which is shown in Fig. 9 can be obtained.

Also, when the EPS system works normally and the vehicle speed is 5km/h, rapidly turn the steering wheel to a angle about 600° and keep still, the angle-torque curve and angle-current curve which is shown in Fig. 10 can be obtained.

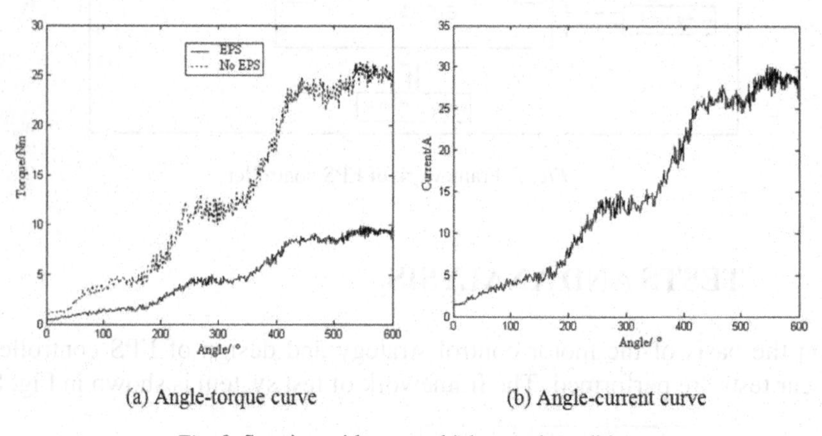

(a) Angle-torque curve (b) Angle-current curve

Fig. 9. Steering with zero vehicle speed condition

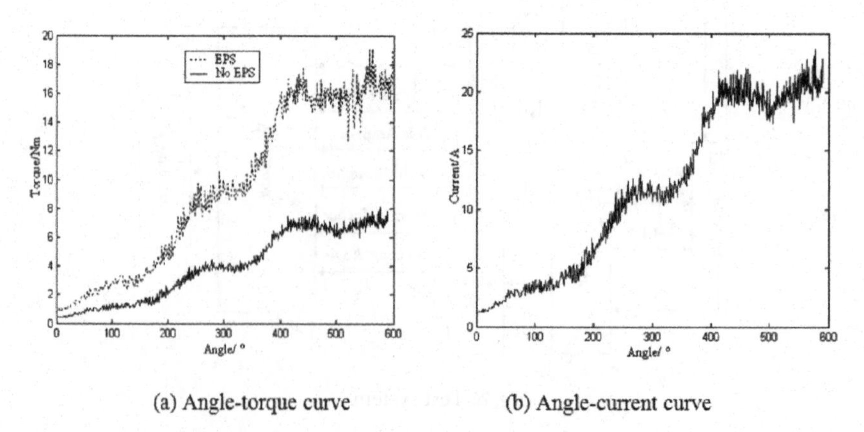

(a) Angle-torque curve (b) Angle-current curve

Fig. 10. Steering with 5km/h vehicle speed condition

And, the detailed comparation is listed as Table 1, when there is Esther handiness is obvious. When the vehicle speed is zero, the handiness is increased 69.5%, and 5km/h 56.3%.

Table 1. Steering handiness test efficiency

Condition	Item	No EPS	EPS
Zero speed	Peak torque/Nm	26.42	9.658
	Average torque/Nm	14.11	4.308
	Handiness target	3.275	
5km/h speed	Peak torque/Nm	19.11	7.8
	Average torque/Nm	8.713	3.806
	Handiness target	2.289	

3.2 Step input test

When the EPS system works normally and the vehicle speed is respectively zero and 5km/h, rapidly turn the steering wheel to a angle and keep still for several seconds, the corresponding motor current response curves are shown in Fig. 11.

It's revealed that the motor current response is rapid and can rapidly achieve stabilization, and can fully meet the demand of real time.

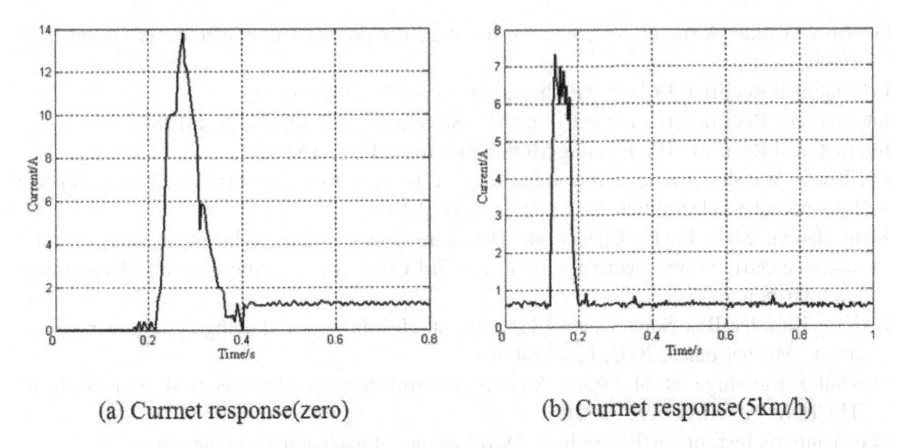

(a) Curmet response(zero) (b) Curmet response(5km/h)

Fig. 11. Motor current response

4. CONCLUSIONS AND DISCUSSIONS

The results show that the EPS controller designed is stable and credible, and can meet the requirements of steering performance.

The demands for faster speed, higher quality, and reduced power requirements in vehicles are continually increasing. In order to respond to these demands, research and development is under way on the application of electronic control with the aim of further improving functions and performance. Features that are being proposed include the introduction of intelligent control strategy and the application of power steering, which responds to the driving environment by varying the assist amount in accordance to fit the sensitivities of human operators.

ACKNOWLEDGEMENTS

The authors acknowledge support of Zhu Yongjun in designing the steering tests and the controller. Chen Dayu provided the matlab/simulink support. This work was supported by National Natural Science Foundation of China under Grant No. 50475121 and Jiangsu Provincial High-tech Project of China under Grant No. BG2004025. The contents of this paper reflect the view of the authors who are responsible for the facts and accuracy of the data presented in herein. The contents do not necessarily reflect the official views or policies of Jiangsu University. This paper do not constitute a standard, specification or regulation.

REFERENCES

Geoffrey Walker. A motor controller for the solar car project.The university of Queensland, 1998

International Rectifier, IR2110 datasheet, Data sheet NO, PD-6.011E

International Rectifier. Bootstrap Component Selection For Control IC's, DT98-2a

International Rectifier. HV Floating MOS-Gate Driver IC's, AN978

Jeff Burns, Suresh Chengalva. Integrated motor drive unit-A mechatronics packaging concept for automotive electronics, SAE paper, 2000-01-0132

Jiang Haobin, Zhao Jingbo, Chen Long. Hardware design and experiment research of automotive electric power steering system, The 3rd China-Japan Conference on Mechatronics 2006 Fuzhou, 68-71

Ji-Hoon Kim, Jae-Bok Song. Control logic for an electric power steering system using assist motor, Mechatronics, 2002, 12:447-459

Masahiko Kurishige et al. Static Steering-Control System for Electric-Power Steering, TECHNICAL REPORTS, 2001

Massachusetts Institute of Technology, Development of machanical engineering

Ronald K. Jurgen. Automotive electronics handbook, Second edition, McGraw-Hill, Inc, 1999

Sergio Fissore, Leigh Cormie. Development of an electronic power assisted steering package, SAE paper, 2000-01-3069

Takayuki Kifuku, Shun'ichi Wada. An Electric Power-Steering System, TECHNICAL REPORTS, 1997

Toshinori Tanaka, Motors for electric power steering, TECHNICAL REPORTS, 2003

Yuji Kozaki, Shozo Sekiya et al. Electric Power Steering (EPS), Motion & Control, No. 6, 1999